T0137010

Lecture Notes in Computer Science   13945

Founding Editors

Gerhard Goos
Juris Hartmanis

Editorial Board Members

Elisa Bertino, *Purdue University, West Lafayette, IN, USA*
Wen Gao, *Peking University, Beijing, China*
Bernhard Steffen ⓘ, *TU Dortmund University, Dortmund, Germany*
Moti Yung ⓘ, *Columbia University, New York, NY, USA*

The series Lecture Notes in Computer Science (LNCS), including its subseries Lecture Notes in Artificial Intelligence (LNAI) and Lecture Notes in Bioinformatics (LNBI), has established itself as a medium for the publication of new developments in computer science and information technology research, teaching, and education.

LNCS enjoys close cooperation with the computer science R & D community, the series counts many renowned academics among its volume editors and paper authors, and collaborates with prestigious societies. Its mission is to serve this international community by providing an invaluable service, mainly focused on the publication of conference and workshop proceedings and postproceedings. LNCS commenced publication in 1973.

Xin Wang · Maria Luisa Sapino ·
Wook-Shin Han · Amr El Abbadi · Gill Dobbie ·
Zhiyong Feng · Yingxiao Shao · Hongzhi Yin
Editors

Database Systems for Advanced Applications

28th International Conference, DASFAA 2023
Tianjin, China, April 17–20, 2023
Proceedings, Part III

 Springer

Editors
Xin Wang 🆔
Tianjin University
Tianjin, China

Maria Luisa Sapino 🆔
University of Torino
Turin, Italy

Wook-Shin Han
POSTECH
Pohang, Korea (Republic of)

Amr El Abbadi
University of California Santa Barbara
Santa Barbara, CA, USA

Gill Dobbie 🆔
University of Auckland
Auckland, New Zealand

Zhiyong Feng
Tianjin University
Tianjin, China

Yingxiao Shao 🆔
Beijing University of Posts
and Telecommunications
Beijing, China

Hongzhi Yin 🆔
The University of Queensland
Brisbane, QLD, Australia

ISSN 0302-9743   ISSN 1611-3349 (electronic)
Lecture Notes in Computer Science
ISBN 978-3-031-30674-7   ISBN 978-3-031-30675-4 (eBook)
https://doi.org/10.1007/978-3-031-30675-4

© The Editor(s) (if applicable) and The Author(s), under exclusive license
to Springer Nature Switzerland AG 2023
This work is subject to copyright. All rights are reserved by the Publisher, whether the whole or part of
the material is concerned, specifically the rights of translation, reprinting, reuse of illustrations, recitation,
broadcasting, reproduction on microfilms or in any other physical way, and transmission or information
storage and retrieval, electronic adaptation, computer software, or by similar or dissimilar methodology now
known or hereafter developed.
The use of general descriptive names, registered names, trademarks, service marks, etc. in this publication
does not imply, even in the absence of a specific statement, that such names are exempt from the relevant
protective laws and regulations and therefore free for general use.
The publisher, the authors, and the editors are safe to assume that the advice and information in this book
are believed to be true and accurate at the date of publication. Neither the publisher nor the authors or the
editors give a warranty, expressed or implied, with respect to the material contained herein or for any errors
or omissions that may have been made. The publisher remains neutral with regard to jurisdictional claims in
published maps and institutional affiliations.

This Springer imprint is published by the registered company Springer Nature Switzerland AG
The registered company address is: Gewerbestrasse 11, 6330 Cham, Switzerland

Preface

It is our great pleasure to present the proceedings of the 28th International Conference on Database Systems for Advanced Applications (DASFAA 2023), organized by Tianjin University and held during April 17–20, 2023, in Tianjin, China. DASFAA is an annual international database conference which showcases state-of-the-art R&D activities in database systems and advanced applications. It provides a premier international forum for technical presentations and discussions among database researchers, developers, and users from both academia and industry.

This year we received a record high number of 652 research paper submissions. We conducted a double-blind review following the tradition of DASFAA, and constructed a large committee consisting of 31 Senior Program Committee (SPC) members and 254 Program Committee (PC) members. Each valid submission was reviewed by at least three PC members and meta-reviewed by one SPC member, who also led the discussion with the PC members. We, the PC co-chairs, considered the recommendations from the SPC members and investigated each submission as well as its reviews to make the final decisions. As a result, 125 full papers (acceptance ratio of 19.2%) and 66 short papers (acceptance ratio of 29.3%) were accepted. The review process was supported by the Microsoft CMT system. During the three main conference days, these 191 papers were presented in 17 research sessions. The dominant keywords for the accepted papers included model, graph, learning, performance, knowledge, time, recommendation, representation, attention, prediction, and network. In addition, we included 15 industry papers, 15 demo papers, 5 PhD consortium papers, and 7 tutorials in the program. Finally, to shed light on the direction in which the database field is headed, the conference program included four invited keynote presentations by Sihem Amer-Yahia (CNRS, France), Kyuseok Shim (Seoul National University, South Korea), Angela Bonifati (Lyon 1 University, France), and Jianliang Xu (Hong Kong Baptist University, China).

Four workshops were selected by the workshop co-chairs to be held in conjunction with DASFAA 2023, which were the 9th International Workshop on Big Data Management and Service (BDMS 2023), the 8th International Workshop on Big Data Quality Management (BDQM 2023), the 7th International Workshop on Graph Data Management and Analysis (GDMA 2023), and the 1st International Workshop on Bundle-based Recommendation Systems (BundleRS 2023). The workshop papers are included in a separate volume of the proceedings also published by Springer in its Lecture Notes in Computer Science series.

We are grateful to the general chairs, Amr El Abbadi, UCSB, USA, Gill Dobbie, University of Auckland, New Zealand, and Zhiyong Feng, Tianjin University, China, all SPC members, PC members, and external reviewers who contributed their time and expertise to the DASFAA 2023 paper-reviewing process. We would like to thank all the members of the Organizing Committee, and the many volunteers, for their great support in the conference organization. Lastly, many thanks to the authors who submitted their papers to the conference.

March 2023   Xin Wang
                                                     Maria Luisa Sapino
                                                     Wook-Shin Han

Organization

Steering Committee Members

Chair

Lei Chen   Hong Kong University of Science and
                                  Technology (Guangzhou), China

Vice Chair

Stéphane Bressan   National University of Singapore, Singapore

Treasurer

Yasushi Sakurai   Osaka University, Japan

Secretary

Kyuseok Shim   Seoul National University, South Korea

Members

Zhiyong Peng   Wuhan University of China, China
Zhanhuai Li   Northwestern Polytechnical University, China
Krishna Reddy   IIIT Hyderabad, India
Yunmook Nah   DKU, South Korea
Wenjia Zhang   University of New South Wales, Australia
Zi Huang   University of Queensland, Australia
Guoliang Li   Tsinghua University, China
Sourav Bhowmick   Nanyang Technological University, Singapore
Atsuyuki Morishima   University of Tsukuba, Japan
Sang-Won Lee   SKKU, South Korea
Yang-Sae Moon   Kangwon National University, South Korea

Organizing Committee

Honorary Chairs

Christian S. Jensen   Aalborg University, Denmark
Keqiu Li   Tianjin University, China

General Chairs

Amr El Abbadi   UCSB, USA
Gill Dobbie   University of Auckland, New Zealand
Zhiyong Feng   Tianjin University, China

Program Committee Chairs

Xin Wang   Tianjin University, China
Maria Luisa Sapino   University of Torino, Italy
Wook-Shin Han   POSTECH, South Korea

Industry Program Chairs

Jiannan Wang   Simon Fraser University, Canada
Jinwei Zhu   Huawei, China

Tutorial Chairs

Jianxin Li   Deakin University, Australia
Herodotos Herodotou   Cyprus University of Technology, Cyprus

Demo Chairs

Zhifeng Bao   RMIT, Australia
Yasushi Sakurai   Osaka University, Japan
Xiaoli Wang   Xiamen University, China

Workshop Chairs

Lu Chen   Zhejiang University, China
Xiaohui Tao   University of Southern Queensland, Australia

Panel Chairs

Lei Chen   Hong Kong University of Science and
                                  Technology (Guangzhou), China
Xiaochun Yang   Northeastern University, China

PhD Consortium Chairs

Leong Hou U.   University of Macau, China
Panagiotis Karras   Aarhus University, Denmark

Publicity Chairs

Yueguo Chen   Renmin University of China, China
Kyuseok Shim   Seoul National University, South Korea
Yoshiharu Ishikawa   Nagoya University, Japan
Arnab Bhattacharya   IIT Kanpur, India

Publication Chairs

Yingxiao Shao   Beijing University of Posts and
                                  Telecommunications, China
Hongzhi Yin   University of Queensland, Australia

DASFAA Steering Committee Liaison

Lei Chen   Hong Kong University of Science and
                                  Technology (Guangzhou), China

Local Arrangement Committee

Xiaowang Zhang   Tianjin University, China
Guozheng Rao   Tianjin University, China
Yajun Yang   Tianjin University, China
Shizhan Chen   Tianjin University, China
Xueli Liu   Tianjin University, China
Xiaofei Wang   Tianjin University, China
Chao Qiu   Tianjin University, China
Dong Han   Tianjin Academy of Fine Arts, China
Ying Guo   Tianjin University, China

Hui Jiang	Tianjin Ren'ai College, China
Kun Liang	Tianjin University of Science and Technology, China

Web Master

Zirui Chen	Tianjin University, China

Program Committee Chairs

Xin Wang	Tianjin University, China
Maria Luisa Sapino	University of Torino, Italy
Wook-Shin Han	POSTECH, South Korea

Senior Program Committee (SPC) Members

Baihua Zheng	Singapore Management University, Singapore
Bin Cui	Peking University, China
Bingsheng He	National University of Singapore, Singapore
Chee-Yong Chan	National University of Singapore, Singapore
Chengfei Liu	Swinburne University of Technology, Australia
Haofen Wang	Tongji University, China
Hong Gao	Harbin Institute of Technology, China
Hongzhi Yin	University of Queensland, Australia
Jiaheng Lu	University of Helsinki, Finland
Jianliang Xu	Hong Kong Baptist University, China
Jianyong Wang	Tsinghua University, China
K. Selçuk Candan	Arizona State University, USA
Kyuseok Shim	Seoul National University, South Korea
Lei Li	Hong Kong University of Science and Technology (Guangzhou), China
Lina Yao	University of New South Wales, Australia
Ling Liu	Georgia Institute of Technology, USA
Nikos Bikakis	Athena Research Center, Greece
Qiang Zhu	University of Michigan-Dearborn, USA
Reynold Cheng	University of Hong Kong, China
Ronghua Li	Beijing Institute of Technology, China
Vana Kalogeraki	Athens University of Economics and Business, Greece
Vincent Tseng	National Yang Ming Chiao Tung University, Taiwan
Wang-Chien Lee	Pennsylvania State University, USA

Xiang Zhao	National University of Defense Technology, China
Xiaoyong Du	Renmin University of China, China
Ye Yuan	Beijing Institute of Technology, China
Yongxin Tong	Beihang University, China
Yoshiharu Ishikawa	Nagoya University, Japan
Yufei Tao	Chinese University of Hong Kong, China
Yunjun Gao	Zhejiang University, China
Zhiyong Peng	Wuhan University, China

Program Committee (PC) Members

Alexander Zhou	Hong Kong University of Science and Technology, China
Alkis Simitsis	Athena Research Center, Greece
Amr Ebaid	Google, USA
An Liu	Soochow University, China
Anne Laurent	University of Montpellier, France
Antonio Corral	University of Almería, Spain
Baoning Niu	Taiyuan University of Technology, China
Barbara Catania	University of Genoa, Italy
Bin Cui	Peking University, China
Bin Wang	Northeastern University, China
Bing Li	Institute of High Performance Computing, Singapore
Bohan Li	Nanjing University of Aeronautics and Astronautics, China
Changdong Wang	SYSU, China
Chao Huang	University of Notre Dame, USA
Chao Zhang	Tsinghua University, China
Chaokun Wang	Tsinghua University, China
Chenyang Wang	Aalto University, Finland
Cheqing Jin	East China Normal University, China
Chih-Ya Shen	National Tsing Hua University, Taiwan
Christos Doulkeridis	University of Pireaus, Greece
Chuan Ma	Zhejiang Lab, China
Chuan Xiao	Osaka University and Nagoya University, Japan
Chuanyu Zong	Shenyang Aerospace University, China
Chunbin Lin	Amazon AWS, USA
Cindy Chen	UMass Lowell, USA
Claudio Schifanella	University of Torino, Italy
Cuiping Li	Renmin University of China, China

Damiani Ernesto	University of Milan, Italy
Dan He	University of Queensland, Australia
De-Nian Yang	Academia Sinica, Taiwan
Derong Shen	Northeastern University, China
Dhaval Patel	IBM Research, USA
Dian Ouyang	Guangzhou University, China
Dieter Pfoser	George Mason University, USA
Dimitris Kotzinos	ETIS, France
Dong Wen	University of New South Wales, Australia
Dongxiang Zhang	Zhejiang University, China
Dongxiao He	Tianjin University, China
Faming Li	Northeastern University, USA
Ge Yu	Northeastern University, China
Goce Trajcevski	Iowa State University, USA
Gong Cheng	Nanjing University, China
Guandong Xu	University of Technology Sydney, Australia
Guanhua Ye	University of Queensland, Australia
Guoliang Li	Tsinghua University, China
Haida Zhang	WorldQuant, USA
Hailong Liu	Northwestern Polytechnical University, China
Haiwei Zhang	Nankai University, China
Hantao Zhao	ETH, Switzerland
Hao Peng	Beihang University, China
Hiroaki Shiokawa	University of Tsukuba, Japan
Hongbin Pei	Xi'an Jiaotong University, China
Hongxu Chen	Commonwealth Bank of Australia, Australia
Hongzhi Wang	Harbin Institute of Technology, China
Hongzhi Yin	University of Queensland, Australia
Huaijie Zhu	Sun Yat-sen University, China
Hui Li	Xidian University, China
Huiqi Hu	East China Normal University, China
Hye-Young Paik	University of New South Wales, Australia
Ioannis Konstantinou	University of Thessaly, Greece
Ismail Hakki Toroslu	METU, Turkey
Jagat Sesh Challa	BITS Pilani, India
Ji Zhang	University of Southern Queensland, Australia
Jia Xu	Guangxi University, China
Jiali Mao	East China Normal University, China
Jianbin Qin	Shenzhen Institute of Computing Sciences, China
Jianmin Wang	Tsinghua University, China
Jianqiu Xu	Nanjing University of Aeronautics and Astronautics, China

Jianxin Li	Deakin University, Australia
Jianye Yang	Guangzhou University, China
Jiawei Jiang	Wuhan University, China
Jie Shao	University of Electronic Science and Technology of China, China
Jilian Zhang	Jinan University, China
Jilin Hu	Aalborg University, Denmark
Jin Wang	Megagon Labs, USA
Jing Tang	Hong Kong University of Science and Technology, China
Jithin Vachery	NUS, Singapore
Jongik Kim	Chungnam National University, South Korea
Ju Fan	Renmin University of China, China
Jun Gao	Peking University, China
Jun Miyazaki	Tokyo Institute of Technology, Japan
Junhu Wang	Griffith University, Australia
Junhua Zhang	University of New South Wales, Australia
Junliang Yu	The University of Queensland, Australia
Kai Wang	Shanghai Jiao Tong University, China
Kai Zheng	University of Electronic Science and Technology of China, China
Kangfei Zhao	The Chinese University of Hong Kong, China
Kesheng Wu	LBNL, USA
Kristian Torp	Aalborg University, Denmark
Kun Yue	School of Information and Engineering, China
Kyoung-Sook Kim	National Institute of Advanced Industrial Science and Technology, Japan
Ladjel Bellatreche	ISAE-ENSMA, France
Latifur Khan	University of Texas at Dallas, USA
Lei Cao	MIT, USA
Lei Duan	Sichuan University, China
Lei Guo	Shandong Normal University, China
Leong Hou U.	University of Macau, China
Liang Hong	Wuhan University, China
Libin Zheng	Sun Yat-sen University, China
Lidan Shou	Zhejiang University, China
Lijun Chang	University of Sydney, Australia
Lin Li	Wuhan University of Technology, China
Lizhen Cui	Shandong University, China
Long Yuan	Nanjing University of Science and Technology, China
Lu Chen	Swinburne University of Technology, Australia

Lu Chen	Zhejiang University, China
Makoto Onizuka	Osaka University, Japan
Manish Kesarwani	IBM Research, India
Manolis Koubarakis	University of Athens, Greece
Markus Schneider	University of Florida, USA
Meihui Zhang	Beijing Institute of Technology, China
Meng Wang	Southeast University, China
Meng-Fen Chiang	University of Auckland, New Zealand
Ming Zhong	Wuhan University, China
Minghe Yu	Northeastern University, China
Mizuho Iwaihara	Waseda University, Japan
Mo Li	Liaoning University, China
Ning Wang	Beijing Jiaotong University, China
Ningning Cui	Anhui University, China
Norio Katayama	National Institute of Informatics, Japan
Noseong Park	George Mason University, USA
Panagiotis Bouros	Johannes Gutenberg University Mainz, Germany
Peiquan Jin	University of Science and Technology of China, China
Peng Cheng	East China Normal University, China
Peng Peng	Hunan University, China
Pengpeng Zhao	Soochow University, China
Ping Lu	Beihang University, China
Pinghui Wang	Xi'an Jiaotong University, China
Qiang Yin	Shanghai Jiao Tong University, China
Qianzhen Zhang	National University of Defense Technology, China
Qing Liao	Harbin Institute of Technology (Shenzhen), China
Qing Liu	CSIRO, Australia
Qingpeng Zhang	City University of Hong Kong, China
Qingqing Ye	Hong Kong Polytechnic University, China
Quanqing Xu	A*STAR, Singapore
Rong Zhu	Alibaba Group, China
Rui Zhou	Swinburne University of Technology, Australia
Rui Zhu	Shenyang Aerospace University, China
Ruihong Qiu	University of Queensland, Australia
Ruixuan Li	Huazhong University of Science and Technology, China
Ruiyuan Li	Chongqing University, China
Sai Wu	Zhejiang University, China
Sanghyun Park	Yonsei University, South Korea

Sanjay Kumar Madria	Missouri University of Science & Technology, USA
Sebastian Link	University of Auckland, New Zealand
Sen Wang	University of Queensland, Australia
Shaoxu Song	Tsinghua University, China
Sheng Wang	Wuhan University, China
Shijie Zhang	Tencent, China
Shiyu Yang	Guangzhou University, China
Shuhao Zhang	Singapore University of Technology and Design, Singapore
Shuiqiao Yang	UNSW, Australia
Shuyuan Li	Beihang University, China
Sibo Wang	Chinese University of Hong Kong, China
Silvestro Roberto Poccia	University of Turin, Italy
Tao Qiu	Shenyang Aerospace University, China
Tao Zhao	National University of Defense Technology, China
Taotao Cai	Macquarie University, Australia
Thanh Tam Nguyen	Griffith University, Australia
Theodoros Chondrogiannis	University of Konstanz, Germany
Tieke He	State Key Laboratory for Novel Software Technology, China
Tieyun Qian	Wuhan University, China
Tiezheng Nie	Northeastern University, China
Tsz Nam (Edison) Chan	Hong Kong Baptist University, China
Uday Kiran Rage	University of Aizu, Japan
Verena Kantere	National Technical University of Athens, Greece
Wei Hu	Nanjing University, China
Wei Li	Harbin Engineering University, China
Wei Lu	RUC, China
Wei Shen	Nankai University, China
Wei Song	Wuhan University, China
Wei Wang	Hong Kong University of Science and Technology (Guangzhou), China
Wei Zhang	ECNU, China
Wei Emma Zhang	The University of Adelaide, Australia
Weiguo Zheng	Fudan University, China
Weijun Wang	University of Göttingen, Germany
Weiren Yu	University of Warwick, UK
Weitong Chen	Adelaide University, Australia
Weiwei Sun	Fudan University, China
Weixiong Rao	Tongji University, China

Wen Hua	Hong Kong Polytechnic University, China
Wenchao Zhou	Georgetown University, USA
Wentao Li	University of Technology Sydney, Australia
Wentao Zhang	Mila, Canada
Werner Nutt	Free University of Bozen-Bolzano, Italy
Wolf-Tilo Balke	TU Braunschweig, Germany
Wookey Lee	Inha University, South Korea
Xi Guo	University of Science and Technology Beijing, China
Xiang Ao	Institute of Computing Technology, CAS, China
Xiang Lian	Kent State University, USA
Xiang Zhao	National University of Defense Technology, China
Xiangguo Sun	Chinese University of Hong Kong, China
Xiangmin Zhou	RMIT University, Australia
Xiangyu Song	Swinburne University of Technology, Australia
Xiao Pan	Shijiazhuang Tiedao University, China
Xiao Fan Liu	City University of Hong Kong, China
Xiaochun Yang	Northeastern University, China
Xiaofeng Gao	Shanghai Jiaotong University, China
Xiaoling Wang	East China Normal University, China
Xiaowang Zhang	Tianjin University, China
Xiaoyang Wang	University of New South Wales, Australia
Ximing Li	Jilin University, China
Xin Cao	University of New South Wales, Australia
Xin Huang	Hong Kong Baptist University, China
Xin Wang	Southwest Petroleum University, China
Xinqiang Xie	Neusoft, China
Xiuhua Li	Chongqing University, China
Xiulong Liu	Tianjin University, China
Xu Zhou	Hunan University, China
Xuequn Shang	Northwestern Polytechnical University, China
Xupeng Miao	Carnegie Mellon University, USA
Xuyun Zhang	Macquarie University, Australia
Yajun Yang	Tianjin University, China
Yan Zhang	Peking University, China
Yanfeng Zhang	Northeastern University, China, and Macquarie University, Australia
Yang Cao	Hokkaido University, Japan
Yang Chen	Fudan University, China
Yang-Sae Moon	Kangwon National University, South Korea
Yanjie Fu	University of Central Florida, USA

Yanlong Wen	Nankai University, China
Ye Yuan	Beijing Institute of Technology, China
Yexuan Shi	Beihang University, China
Yi Cai	South China University of Technology, China
Ying Zhang	Nankai University, China
Yingxia Shao	BUPT, China
Yiru Chen	Columbia University, USA
Yixiang Fang	Chinese University of Hong Kong, Shenzhen, China
Yong Tang	South China Normal University, China
Yong Zhang	Tsinghua University, China
Yongchao Liu	Ant Group, China
Yongpan Sheng	Southwest University, China
Yongxin Tong	Beihang University, China
You Peng	University of New South Wales, Australia
Yu Gu	Northeastern University, China
Yu Yang	Hong Kong Polytechnic University, China
Yu Yang	City University of Hong Kong, China
Yuanyuan Zhu	Wuhan University, China
Yue Kou	Northeastern University, China
Yunpeng Chai	Renmin University of China, China
Yunyan Guo	Tsinghua University, China
Yunzhang Huo	Hong Kong Polytechnic University, China
Yurong Cheng	Beijing Institute of Technology, China
Yuxiang Zeng	Hong Kong University of Science and Technology, China
Zeke Wang	Zhejiang University, China
Zhaojing Luo	National University of Singapore, Singapore
Zhaonian Zou	Harbin Institute of Technology, China
Zheng Liu	Nanjing University of Posts and Telecommunications, China
Zhengyi Yang	University of New South Wales, Australia
Zhenya Huang	University of Science and Technology of China, China
Zhenying He	Fudan University, China
Zhipeng Zhang	Alibaba, China
Zhiwei Zhang	Beijing Institute of Technology, China
Zhixu Li	Fudan University, China
Zhongnan Zhang	Xiamen University, China

Industry Program Chairs

Jiannan Wang   Simon Fraser University, Canada
Jinwei Zhu   Huawei, China

Industry Program Committee Members

Bohan Li   Nanjing University of Aeronautics and
                             Astronautics, China
Changbo Qu   Simon Fraser University, Canada
Chengliang Chai   Tsinghua University, China
Denis Ponomaryov   The Institute of Informatics Systems of the
                             Siberian Division of Russian Academy of
                             Sciences, Russia
Hongzhi Wang   Harbin Institute of Technology, China
Jianhua Yin   Shandong University, China
Jiannan Wang   Simon Fraser University, Canada
Jinglin Peng   Simon Fraser University, Canada
Jinwei Zhu   Huawei Technologies Co. Ltd., China
Ju Fan   Renmin University of China, China
Minghe Yu   Northeastern University, China
Nikos Ntarmos   Huawei Technologies R&D (UK) Ltd., UK
Sheng Wang   Alibaba Group, China
Wei Zhang   East China Normal University, China
Weiyuan Wu   Simon Fraser University, Canada
Xiang Li   East China Normal University, China
Xiaofeng Gao   Shanghai Jiaotong University, China
Xiaoou Ding   Harbin Institute of Technology, China
Yang Ren   Huawei, China
Yinan Mei   Tsinghua University, China
Yongxin Tong   Beihang University, China

Demo Track Program Chairs

Zhifeng Bao   RMIT, Australia
Yasushi Sakurai   Osaka University, Japan
Xiaoli Wang   Xiamen University, China

Demo Track Program Committee Members

Benyou Wang	Chinese University of Hong Kong, Shenzhen, China
Changchang Sun	Illinois Institute of Technology, USA
Chen Lin	Xiamen University, China
Chengliang Chai	Tsinghua University, China
Chenhao Ma	Chinese University of Hong Kong, Shenzhen, China
Dario Garigliotti	Aalborg University, Denmark
Ergute Bao	National University of Singapore, Singapore
Jianzhong Qi	The University of Melbourne, Australia
Jiayuan He	RMIT University, Australia
Kaiping Zheng	National University of Singapore, Singapore
Kajal Kansal	NUS, Singapore
Lei Cao	MIT, USA
Liang Zhang	WPI, USA
Lu Chen	Swinburne University of Technology, Australia
Meihui Zhang	Beijing Institute of Technology, China
Mengfan Tang	University of California, Irvine, USA
Na Zheng	National University of Singapore, Singapore
Pinghui Wang	Xi'an Jiaotong University, China
Qing Xie	Wuhan University of Technology, China
Ruihong Qiu	University of Queensland, Australia
Tong Chen	University of Queensland, Australia
Yile Chen	Nanyang Technological University, Singapore
Yuya Sasaki	Osaka University, Japan
Yuyu Luo	Tsinghua University, China
Zhanhao Zhao	Renmin University of China, China
Zheng Wang	Huawei Singapore Research Center, Singapore
Zhuo Zhang	University of Melbourne, Australia

PhD Consortium Track Program Chairs

Leong Hou U.	University of Macau, China
Panagiotis Karras	Aarhus University, Denmark

PhD Consortium Track Program Committee Members

Anton Tsitsulin	Google, USA
Bo Tang	Southern University of Science and Technology, China

Hao Wang	Wuhan University, China
Jieming Shi	Hong Kong Polytechnic University, China
Tsz Nam (Edison) Chan	Hong Kong Baptist University, China
Xiaowei Wu	State Key Lab of IoT for Smart City, University of Macau, China

Contents – Part III

Retrieval

Text Processing

Graph and Networks

Meta-Path Based Social Relation Reasoning in a Deep and Robust Way

Xuhui Jiang[1,2], Yinghan Shen[1,2], Yuanzhuo Wang[1,3(✉)], Huawei Shen[1,2], Chengjin Xu[4], and Shengjie Ma[5]

[1] Data Intelligent System Research Center, Institute of Computing Technology, Chinese Academy of Sciences, Beijing, China
{jiangxuhui19g,shenyinghan17s,wangyuanzhuo,shenhuawei}@ict.ac.cn
[2] School of Computer Science and Technology, UCAS, Gloucester, UK
[3] Big Data Academy, Zhongke, Zhengzhou, China
[4] International Digital Economy Academy, Shenzhen, China
[5] Gaoling School of Artificial Intelligence, Renmin University of China, Beijing, China

Abstract. Social relation reasoning in heterogeneous social networks (HSNs) should not only infer whether people are connected, but also why they know each other and what their relationship is. This task is challenging because the heterogeneous social network is non-trivial to leverage and the collected data is always noisy and incomplete, which leads that the existing rule-based and representation methods cannot efficiently and accurately reason social relations. To reason social relations in noisy and incomplete HSNs, we shift from simply inferring meta-path based social relations by rule-based methods to reasoning by social relation learning. Then, we present an inductive model named Heterogeneous Graph Variational AutoEncoders (HGVAE) towards robust social relation learning on noisy and incomplete HSN data. In HGVAE, the well-designed heterogeneous graph encoder, multi-signal decoder, and variational inference mechanism bring prominent robustness and significant performance improvement. Extensive experiments are conducted to compare our methods against a set of the most representative baseline methods. The efficiency analysis, robust analysis and ablation studies confirm the value of our proposed approach for modelling meta-path based social relations buried in various HSNs.

Keywords: Heterogeneous Social Network · Relation Learning

1 Introduction

Social relations are derived from human social behaviors and are defined as the associations between individuals. Social relation reasoning is dedicated to completing social relation information, which is a fundamental procedure for

---

X. Jiang and Y. Shen—Indicates equal contributions.

© The Author(s), under exclusive license to Springer Nature Switzerland AG 2023
X. Wang et al. (Eds.): DASFAA 2023, LNCS 13945, pp. 3–20, 2023.
https://doi.org/10.1007/978-3-031-30675-4_1

understanding the semantics within connections between people, and supports a
wide range of person-oriented applications, such as social influence analysis [12].
It not only infers the existence but also explores relations at the semantic level.

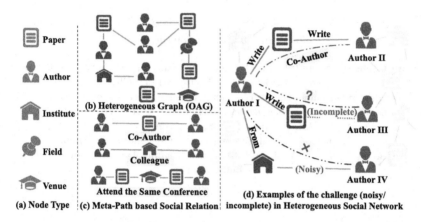

Fig. 1. An example of a HSN. Meta-paths among author nodes reflect various social
relations, and noisy/ incomplete data will cause difficulty recognizing social relations.

Most of the existing social relation reasoning methods [23] are designed for
homogeneous networks that only contain person-type nodes. However, real-world
social networks tend to be heterogeneous in which diverse types of nodes related
to people are widespread. In heterogeneous social networks (HSNs), social relations reflected by indirect connections (meta-path) through other types of nodes
are called meta-path based social relations. For example, as shown in Fig. 1, two
scholars are associated through a co-authored paper. The diverse meta-paths
between person pairs contain abundant semantic information of social relations
(e.g., APA (Co-Author), AIA (Colleague), APFPA (Attend the same conference)). Considering the semantic information on the social relation between
researchers, we can distinguish whether they are intimate partners or strangers
who just attend the same conference.

Leveraging this heterogeneous information for social relation reasoning is
challenging. Earlier approaches date back to rule-based methods [17], which
mainly leverage human-designed rules (e.g., meta-path) to reason social relations. These methods are unstable and inefficient due to graph data's exponential
algorithm complexity and quality. Recently, as representation learning has gained
great success, many heterogeneous graph representation learning methods have
been proposed, which are also adapted to reason social relations. Among them,
random walk-based methods [1,3] utilize the meta-path to guide random walks;
Graph neural network methods adopt a message-passing mechanism combined
with homogenization process [22], attention mechanism [6] or auxiliary tasks [7]
for node representation learning. These methods avoid the complicated search
and matching procedures of rule-based methods. However, they mainly focus

on representation learning of nodes but lack modelling relations. Hence, they perform poorly in reasoning social relations in HSNs.

Another challenge is that the HSN data collected in the real world is noisy and incomplete. The primary reason for this phenomenon is that part of the data is deliberately erased or unsuccessfully gathered by collectors. The absence or mistake of any link in multi-hop paths will lead to several difficulties, such as reasoning meta-path based social relations, the sampling of meta-path based random walk (e.g., metapath2vec [1]), the homogenization process of heterogeneous graph models (e.g., HAN [22]), or auxiliary tasks that leverage meta-path information for node representation (e.g. SELAR [7]).

Given the challenges mentioned above, a proper social relation reasoning method requires an efficient and robust solution that combines the advantages of neural networks and meta-paths. Two key points remain to be considered: (1) How to use heterogeneous information for social relation reasoning. (2) How to improve the model robustness under the noisy/ incomplete data situation.

Therefore, instead of using rule-based methods, we first introduce a representation learning paradigm that automatically learns rich semantic information in a deep way for social relation reasoning. Then, to realize efficient and robust social relation reasoning, we present a novel inductive method named Heterogeneous Graph Variational AutoEncoders (HGVAE). Specifically, we integrate the heterogeneous graph neural network as the encoder to capture both heterogeneous graph structures and person attributes for learning social relation representation. Considering that both the person attributes and semantic correlations of social relations are essential for social relation reasoning, we model the global representations of meta-paths, and propose the multi-signal decoder. It disentangles the semantic features of person pair attributes and meta-paths from the relationship embeddings learned by social relation learning so that the relation embeddings can retain both types of information. HGVAE also adopts a variational inference mechanism towards robustly reasoning social relations with the consideration of incomplete/noisy characteristics.

We evaluate HGVAE on three HSN datasets against representative baselines. Experimental results show that HGVAE achieves superior performance over the baselines, demonstrating its social relation learning competency. Next, we conduct ablation studies and a series of in-depth analyses to verify the contributions of each component of HGVAE. Then, we conduct an extensive experiment to verify the robustness of HGVAE under noisy/ incomplete situations. Finally, we introduce a case study to vividly illustrate the superior of HGVAE compared with rule-based methods and other deep learning methods.

In general, our main contributions are as follows: (1) To reason social relations, we introduce a new representation learning paradigm that leverages rich semantic information of meta-paths in a deeper way. (2) We present an inductive model called HGVAE towards efficient and robust social relation reasoning. (3) We conduct extensive experiments with robust analysis and ablation studies on three real-world datasets to evaluate the performance of HGVAE.

2 Task and Related Works

2.1 Social Relation Reasoning

Social relation reasoning dedicates to infer the relation existing between node pairs, and mines the semantic knowledge of the relation. Real-world social networks consist of various nodes or links containing rich semantic information to model real-world graph data. Efficient and effective leveraging of this information is non-trivial. To study the complex interactions of person nodes and other types of nodes, here we follow the basic definition of *Heterogeneous Graph* and *Meta-path* proposed by [18], and introduce the above concepts to model social networks called *Heterogeneous Social Network* (HSN). It is worth noting that different meta-paths have different semantic information.

Rule-based method [17] mainly designs the meta-path rules, and utilizes multi-hop graph searching and pattern matching for social relation reasoning. They are simple to deploy, while the shortcomings of these methods are also evident. They consume high-computing resources on multi-hop searches and strongly depend on data integrity. Social relation learning [23] methods aim to infer the explicit relation in social networks through deep learning. RELEARN [23] utilizes multi-modal information in social networks for explicit relation learning on social networks. However, RELEARN ignores different node and edge types which is helpful to understanding social relations. Thus, it is essential to propose a method that can fully leverage meta-path semantic information and evade the negative impacts of noisy and incomplete data in HSN.

In this paper, we model social relations by referring to heterogeneous information in multi-hop paths between various types of nodes. Specifically, we extend the link prediction task to a multi-hop and heterogeneous situation, and follow SELAR [7] to formalize social relation reasoning, which aims to predict existing meta-paths between two nodes under the condition that collected data exist the absences or the mistakes of links in multi-hop paths.

Task Formulation. A heterogeneous social network can be formalized as $G = \{\mathcal{V}, \mathcal{E}, \mathcal{A}, \mathcal{R}, \mathcal{T}\}$ where $\mathcal{V}$ denotes the node set and $\mathcal{E}$ denotes the edge set. Each node $v_i \in \mathcal{V}$ is associated with a node type mapping $\phi : \mathcal{V} \longrightarrow \mathcal{A}$, and each edge $e_{ij} \in \mathcal{E}$ is associated with an edge type mapping $\psi : \mathcal{E} \longrightarrow \mathcal{R}$, where $\mathcal{A}$ and $\mathcal{R}$ denote the set of node types and edge types. Besides, meta-path based social relation is a predefined meta-path set $\mathcal{T}$ between persons. Given an HSN G, the social relation reasoning aims to predict existing meta-paths denoted as $S = \{t_1, t_2, \cdots\}, t \in \mathcal{T}$ between a person node pair.

2.2 Related Works

Autoencoder Architectures (AE). AE [5] and VAE [8] are canonical representation learning frameworks that consist of the encoder to compress and the decoder to restore. Previous researches demonstrate the effectiveness of VAE that combines Bayesian inference mechanism for accurate imputations of missing data [14]. Recently, VGAE [9] transfers VAE to the field of graphs. HetHG-VAE [2] transfers the VGAE to the hypergraph situation for direct link prediction. These methods mainly focus on predicting the direct-linked edges via graph generation, but are not suitable for social relation reasoning tasks.

Graph Neural Networks (GNN). GNN aims to learn the representation of graph-structured data utilizing the deep learning method. GCN [10] focus on the message passing of neighbours on the graph and directly define the convolutional operation for aggregation. GraphSAGE [4] improves the aggregation operation for inductive graph representation learning. GAT [19] integrates an attention mechanism to identify neighbour importance. However, these models ignore heterogeneous information ubiquitous in real-world social networks.

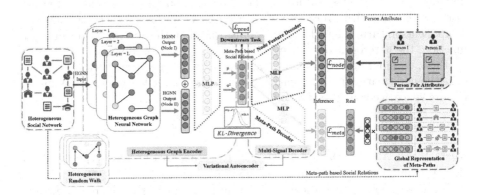

Fig. 2. The overall architecture of HGVAE. Given an HSN as the input, HGVAE learns the representation of meta-path based social relations through the heterogeneous graph encoder and multi-signal decoder in the VAE framework.

Recent studies intend to extend GNN to model HSN. R-GCN [15] uses different convolutional layers to model the relation between heterogeneous nodes. SELAR [7] introduces meta-path prediction as the auxiliary task, and designs a portable component to enhance node representation for primary downstream tasks. HAN [22] utilizes homogenization procedures and transfers GAT into heterogeneous graphs by introducing a hierarchical attention mechanism considering the importance of both different node neighbours and different meta-paths. HGATE [21] introduces a heterogeneous graph attention network into AE architecture. HGT [6] proposes a multi-head attention mechanism considering different types of nodes and edges. However, these methods mainly focus on representation learning of nodes, not social relation learning. Hence, these methods perform poorly on social relation reasoning tasks. To the best of our knowledge, our model is the first to learn the representation of meta-path based social relations in HSN for robust social relation reasoning.

3   Heterogeneous Graph Variational Autoencoders

To reason social relations in a deep and robust way, we propose a method named Heterogeneous Graph Variational AutoEncoders(HGVAE), as shown in Fig. 2.

3.1   Heterogeneous Graph Encoder

How to effectively embed both structural information and person attribute information of HSNs is our primary consideration in designing the encoder part of HGVAE. Inspired by the recent progress of heterogeneous graph neural networks in graph mining, we propose a Heterogeneous Graph Encoder based on the heterogeneous graph neural network. Taking an HSN G with the attributes of each node x as input, the heterogeneous graph encoder aims to learn the representation of meta-path based relations denoted as Z. To model the influence of different types of nodes, we follow the heterogeneous graph attention mechanism [6]. The detailed procedure of each propagation layer l includes four core components in turn: the node feature transformation, the heterogeneous message function, the heterogeneous attention and the aggregation function.

First, considering the different feature spaces of multi-type nodes, we introduce the node feature transformation for projection node feature x to the unified feature space according to the node types ϕ_i, expressed as:

$$h_i^{(0)} = W_{P(\phi_i)} \cdot x_i, \tag{1}$$

where W_P is a linear projection chosen according to the node type ϕ_i. After that, the initial representation $h_i^{(0)}$ of node i is obtained.

Next, we adopt a multi-head mechanism to calculate the different view representation of passed message $m^{(l)}(s, e, t)$ of layer l from a source node v_s to a target node v_t, which is in consideration of the node types of source/ target node ϕ_s, ϕ_t and the edge type ψ_e in heterogeneous graphs, expressed as:

$$m^{(l)}(s, e, t) = \bigoplus_{\tau \in [1, \theta]} \left((W_{M(\phi_s)}^{\tau} h_s^{(l-1)}) W_{\psi_e}^{MSG} \right), \tag{2}$$

where $\oplus$ denotes the concatenate operation, θ is the head numbers, W_M^{τ} and W^{MSG} are linear projections considering the type of nodes and edges.

Then we adopt multi-head attention to calculate attention weights $\alpha^{(l)}(s, e, t)$ between the source node v_s and target node v_t according to the types of nodes ϕ_s, ϕ_t and edges ψ_e in the heterogeneous graph, expressed as:

$$head_{\tau}^{(l)}(s, e, t) = \left((W_{Q(\phi_s)}^{\tau} h_s^{(l-1)}) W_{\psi_e}^{ATT} (W_{K(\phi_t)}^{\tau} h_t^{(l-1)})^T \right) \frac{1}{\sqrt{\xi}}, \tag{3}$$

$$\alpha^{(l)}(s, e, t) = \underset{\forall s \in N(t)}{\text{Softmax}} \left(\bigoplus_{\tau \in [1, \theta]} head_{\tau}^{(l)}(s, e, t) \right), \tag{4}$$

where W_Q^τ, W_K^τ and W^{ATT} denote different linear projection of τ-th head that is chosen according to the type of nodes and edges. ξ is an adaptive scaling to the attention, k is the number of attention heads.

Finally, we aggregate the information from the neighbors of the target node with attention weights to update the target node representation, expressed as:

$$h_t^{(l)} = W_{\phi_t}^{AGG} \left(\sum_{\forall v_s \in N(v_t)} \left(\alpha^{(l)}(s,e,t) \cdot m^{(l)}(s,e,t) \right) \right) + h_t^{(l-1)}, \qquad (5)$$

where W^{AGG} denotes the linear projection considering the target node types.

Through a heterogeneous graph neural network, HGVAE calculates the representation of each node of layer l denoted as $h^{(l)}$, which contains not only the feature of nodes but also the structural information of the heterogeneous graph.

For each node pair v_i and v_j, we concatenate their representations h_i and h_j learned by heterogeneous graph neural network as the initial features of meta-path based social relations. Considering the real-world noisy and incomplete data, we adopt the Gaussian distribution to model the representation $Z \sim \mathcal{N}\left(\mu, \sigma^2\right)$ of meta-path based social relation for training a more robust model. Specifically, we adopt two multi-layer perceptron (MLP) [13] denoted as f_{MLP} with a *ReLU* activation function to output the means μ_{ij} and variances σ_{ij} of meta-path based relation z_{ij}, expressed as:

$$\mu_{ij} = ReLU\left(f_{\mathrm{MLP}_\mu}[h_i \oplus h_j]\right), \qquad (6)$$

$$\sigma_{ij} = ReLU\left(f_{\mathrm{MLP}_\sigma}[h_i \oplus h_j]\right). \qquad (7)$$

Through sampling from learned Gaussian distribution, we can sample the embedding of meta-path based social relation z_{ij}.

3.2 Multi-Signal Decoder

We propose a multi-signal decoder to restore the HSN and disentangle the feature according to the learned representations of meta-path based social relations. The proposed multi-signal decoder ensures the relation embedding preserves both attribute information and meta-path information about person pairs. Especially, on the noisy and incomplete HSN data, two key points are considered when we design the multi-signal decoder: (1) person attributes can reflect social relations, such as the two people with the same research interests are more likely to establish a co-author relation; (2) there is a potential correlation between different meta-path based social relations. For example, people related to the same institute (AIA) are more likely to be co-authors (APA).

Considering the above two key points, for a specific person pair, we design two decoders (i.e., the node feature decoder and meta-path decoder) to describe the node features and the representation of existing meta-paths, respectively. In general, the evidence lower bond (ELBO) of HGVAE is expressed as:

$$\mathcal{L}_{\text{HGVAE}} = \mathbb{E}_{q_{\Theta_e}(Z|G)} \left[(1 - \gamma) \log p_{\Theta_d}(A|Z) + \gamma \log p_{\Theta_d}(D|Z) \right] - KL[q_{\Theta_e}(Z|G)\|p(\hat{Z})]$$
$$= (1 - \gamma) \cdot \mathcal{L}_{node} + \gamma \cdot \mathcal{L}_{meta} - KL[q_{\Theta_e}(Z|G)\|p(\hat{Z})], \tag{8}$$

where q_{Θ_e} and p_{Θ_d} are learned distributions of encoder and decoder, respectively. $\mathcal{L}_{node}$ and $\mathcal{L}_{meta}$ respectively denote the loss function of the node feature decoder and the meta-path decoder. $A, D \in G$ denote the restored node and meta-path features. γ is the hyper-parameter that controls the importance of each decoder part during training.

We follow the VGAE [9] to adopt KL-divergence constraint in the loss function, $q_{\Theta_e}(Z|G)$ and $p(\hat{Z}) \sim \mathcal{N}(0, I)$ respectively denote the learned and the real distribution of latent representation Z and the $\hat{Z}$. The detailed $KL[q(Z|G)\|p(\hat{Z})]$ calculation formula is shown as follows:

$$KL[q_{\Theta_e}(Z|G)\|p(\hat{Z})] = KL[q_{\Theta_e}(Z|G)\|\mathcal{N}(0, I)]$$
$$= \sum_{(i,j),v_i,v_j \in \mathcal{V}} \frac{1}{2}\{\|\mu_{ij}\|^2 + \|\sigma_{ij}\|^2 - dim_z - 2\log\sigma_{ij}\}, \tag{9}$$

where dim_z denotes the dimension of vector z, $\|.\|$ denotes L2 distance.

Node feature decoder aims to reconstruct the attributes $\hat{x}$ of person pairs. We adopt L2 distance as the node feature decoder loss function, expressed as:

$$\mathcal{L}_{node} = \mathbb{E}_{q_{\Theta_e}(Z|A)} \log p_{\Theta_d}(A|Z) = \sum_{(i,j),v_i,v_j \in \mathcal{V}} \mathbb{E}_{z \sim q_{\Theta_e}} \log p_{\Theta_d}(a_{ij}|z_{ij})$$
$$= \sum_{(i,j),v_i,v_j \in \mathcal{V}} \|(x_i \oplus x_j) - ReLU(f_{\text{MLP}_{node}}(z_{ij}))\|, \tag{10}$$

where a_{ij} denotes the restored node feature of v_i and v_j.

Meta-path decoder aims to restore the correlation and semantic information of existing meta-paths between person pairs. For capturing the global structure feature of the HSN, and leveraging it to model the semantic correlation of meta-paths, HGVAE adopts the **Heterogeneous Random Walk** [3] to obtain the global representation of meta-paths denoted as r via training data. Then we average the global representations of existing predefined meta-paths between the person pair as the representation of existing meta-paths. Here we adopt L2 distance as the evaluation, expressed as:

$$\mathcal{L}_{meta} = \mathbb{E}_{q_{\Theta_e}(Z|D)} \log p_{\Theta_d}(D|Z) = \sum_{(i,j),v_i,v_j \in \mathcal{V}} \mathbb{E}_{z \sim q_{\Theta_e}} \log p_{\Theta_d}(d_{ij}|z_{ij})$$
$$= \sum_{(i,j),v_i,v_j \in \mathcal{V}} \|Average(\hat{S}_{ij}^k r_k) - ReLU(f_{\text{MLP}_{meta}}(z_{ij}))\|, \tag{11}$$

where K denotes predefined meta-paths, d_{ij} denotes the restored meta-path feature of v_i and v_j, r_k denotes the global representation of meta-path k, $Average(.)$ denotes the average operation. When the k-th type meta-path in K exists between (v_i, v_j), $S_{ij}^k = 1$; otherwise $S_{ij}^k = 0$, the symbol $\hat{\cdot}$ denotes ground truth.

3.3 HGVAE Joint Training

HGVAE Sampling Algorithm. We develop a sampling algorithm to realize inductive training of HGVAE. First, we sample a subgraph on the original HSN data. Specifically, we randomly choose an initial person node and sample its neighbours with a constraint on the number of each type of node. Then, we choose each person node as the source and expand predefined meta-paths until it reaches a target person node. Finally, we retrieve other existing meta-paths between person pairs of meta-path based social relations represented as multi-hot vectors. HGVAE sampling algorithm can automatically annotate existing meta-paths $\hat{S}_{ij}^k$ of person pairs (v_i, v_j) without manually labeling.

HGVAE Training Procedure. In general, HGVAE follows the training procedure of the counterpart of VAE architecture, which utilizes the loss of reconstruction for training. Especially, HGVAE can jointly utilize the loss of the downstream task for iterative optimization.

For social relation reasoning, we adopt $f_{\mathrm{MLP}}(.)$ with the $Sigmoid(.)$ function to fit the task, and *Binary Cross Entropy loss* to measure the difference between multi-label prediction and ground truth, expressed as:

$$S_{ij}^k = Sigmoid(f_{\mathrm{MLP}_{pred}}(z_{ij})^k), \tag{12}$$

$$\mathcal{L}_{pred} = \sum_{(i,j),v_i,v_j \in \mathcal{V}} -\frac{1}{K} \sum_{k=1}^{K} \left(\hat{S_{ij}^k} log(S_{ij}^k) + (1 - \hat{S_{ij}^k}) log(1 - S_{ij}^k) \right), \tag{13}$$

where S_{ij}^k denotes the prediction result of meta-path k.

To control the importance of $\mathcal{L}_{pred}$ and $\mathcal{L}_{\mathrm{HGVAE}}$, we set a hyper-parameter α. The general loss function $\mathcal{L}$ is expressed as:

$$\mathcal{L} = (1 - \alpha) \cdot \mathcal{L}_{pred} + \alpha \cdot \mathcal{L}_{\mathrm{HGVAE}}. \tag{14}$$

Here we denote the parameters of the Heterogeneous Graph Encoder, Multi-Signal Decoder, and downstream task as Θ_e, Θ_d, and Θ_p, respectively. Through the graph sampling strategy, HGVAE can be trained inductively. The pseudocode of the joint training algorithm is shown as *Algorithm 1*.

Algorithm 1. HGVAE Joint Training

Require: HSN: G; Node features: X; Batch size: B; Number of batches: N
 1: **start training**:
 2: Calculate global meta-path representations r by **Heterogeneous Random Walk**.
 3: **for** $n = 1 : N$ **do**
 4:    Sample B pairs of nodes with node features X and existing meta-paths s.
 5:    Use **Heterogeneous Graph Encoder** to compute μ_n and σ_n, $Z \sim \mathcal{N}(\mu_n, \sigma_n^2)$.
 6:    Draw B random variable vectors $\epsilon_n \sim \mathcal{N}(0, I)$.
 7:    Compute the embedding of meta-path based relation $z_n = \mu_n + \sigma_n \epsilon_n$.
 8:    Use **Multi-Signal Decoder** to compute loss $\mathcal{L}_{node}$, $\mathcal{L}_{meta}$, and KL-divergence.
 9:    Update $\{\Theta_e, \Theta_d\}$ with gradient backpropagation.
10:    Calculate the **Downstream Task** loss $\mathcal{L}_{pred}$.
11:    Update $\{\Theta_e, \Theta_p\}$ with gradient backpropagation.
12: **end for**
13: **end training**

---

4   Experiments

4.1   Dataset

We adopt three real-world HSN datasets for experiments. The detailed statistics of the datasets are shown in Table 1.

$\mathbf{OAG_{ML}}$ is a machine learning domain researcher social network extracted from Open Academic Graph [25], which describes research interactions between authors through heterogeneous nodes. This dataset contains five types of nodes, including Author (A), Paper (P), Field (F), Venue (V), Institute (I).

Douban-Movie dataset is a social network about movies extracted from the Douban site. It describes the social interactions between users with rich heterogeneous information around movies. This dataset contains five types of nodes, including User(U), Group(G), Movie(M), Actor(A), Director(D).

Douban-Book dataset is also a social network about books extracted from the Douban site, which is widely used to describe the social interactions between users with rich heterogeneous information. This dataset contains five types of nodes, including User(U), Group(G), Book(B), Location(L), Author(A).

Table 1. The detailed descriptions of the three real-world experiment datasets.

Dataset	#Nodes	#Edges	Node Types	Predefined Meta-path
OAG_{ML}	227,144	4,167,680	A, P, F, V, I,	APA, APVPA, APPA, APFPA, AIA
Douban-Movie	37,557	1,687,273	U, G, M, A, D	UMU, UMAMU, UMDMU, UGU
Douban-Book	49,150	2,182,982	U, G, B, L, A	UBU, UBABU, ULU, UGU

4.2   Baseline Methods

We compare HGVAE with eight representative baselines. The first three baselines (i.e., **GCN** [10], **GraphSAGE** [4], **GAT** [19]) are designed for homogeneous social networks. The rest baselines (i.e., **HIN2vec** [3], **R-GCN** [15], **HAN** [22], **HGATE** [21], **HGT** [6]) are capable of modelling heterogeneous social networks. All baseline methods are implemented via Deep Graph Library[1] [20]. We treat social relation reasoning as a multi-label classification task for these methods. Therefore, we follow the procedures of SELAR [7] that concatenate the embeddings of person pairs learned by these methods and adopt an MLP to output existing meta-paths between person pairs. Since the definitions and problem settings of social relation reasoning are pretty different from link prediction, GAE, VGAE [9] and HetHG-VAE [2], which mainly follow graph generation processes for directly link prediction, are not competent to transfer to our task. Besides, we conduct ablation studies and give a detailed analysis in Sect. 5.5.

4.3   Implementation Details

Model Settings. For all baselines of our experiments, we kept the same common parameter settings. Specifically, we set the hidden dimension 100, the learning rate 1e-3, the dropout rate 0.3, the batch size 256, the depth of graph layers 2, and the training epochs 500. We followed the 6:2:2 splitting ratio in training/validation/testing data, and adopted the same optimizer *AdamW* [11] for training. For fair comparisons, we modified the depth of MLP to guarantee that the layers of each neural method are the same. For other unique parameters, we tried different parameter settings of these baselines and chose the best of each baseline. All experimental results are performed 10 times to eliminate the influence of randomness, and we take the average as the results.

Feature Initialization. In OAG$_{ML}$, we refer to the implementation [6] for feature Initialization. Specifically, *Paper* nodes feature is the average of title word vectors calculated by pretrained XLNet [24], the *Author* nodes feature is the average of their published paper representations. For other types of nodes, we follow the HGT that uses metapath2vec [1] to train node embeddings. In *Douban-Movie* and *Douban-Book*, we adopt node embeddings proposed by [16].

Evaluation Metrics. To measure the performance of the methods from different perspectives on the social relation reasoning task, we adopt two evaluation metrics: Macro F1-Score (Macro-F1) and Hamming Loss (Ham. Loss) which are widely used in multi-label classification tasks and also fit our task.

---

[1] https://github.com/dmlc/dgl/.

4.4 Experiment Results

The main experiment results are shown in Table 2. HGVAE significantly improved Macro F1-Score and Hamming loss on three real-world datasets compared with other representative baseline methods. Overall, the experiment results convincingly demonstrate the effectiveness of HGVAE.

Table 2. Experiment results of social relation reasoning.

Method	OAG_{ML}		Douban-Movie		Douban-Book	
	Macro-F1	Ham. Loss	Macro-F1	Ham. Loss	Macro-F1	Ham. Loss
GCN	0.599	0.312	0.619	0.313	0.521	0.392
GraphSAGE	0.622	0.306	0.639	0.308	0.560	0.310
GAT	0.618	0.319	0.625	0.300	0.573	0.394
HIN2vec	0.629	0.322	0.512	0.399	0.462	0.356
R-GCN	0.617	0.315	0.659	0.308	0.608	0.374
HAN	0.649	0.310	0.698	0.263	0.645	0.275
HGT	0.655	0.294	0.714	0.247	0.664	0.271
HGATE	0.637	0.314	0.701	0.268	0.636	0.314
HGVAE	**0.689**	**0.228**	**0.750**	**0.209**	**0.716**	**0.239**
HGVAE (w/o *decoder*)	0.656	0.253	0.699	0.265	0.667	0.252
HGVAE (w/o *heterogeneous*)	0.637	0.311	0.693	0.257	0.655	0.260
HGVAE (w/o *variational*)	0.680	0.261	0.725	0.232	0.695	0.255

We can observe that HGVAE and other heterogeneous graph-based methods consistently outperform all the homogeneous graph-based methods (i.e., GCN, GraphSAGE, GAT). The homogeneous graph-based methods ignore the diverse types of nodes in graphs, which suffer seriously from vocabulary gaps, and can hardly be adopted for social relation prediction.

Compared with the heterogeneous graph-based method (i.e., HIN2Vec, R-GCN, HAN and HGT), HGVAE achieves performance gains overall baselines by 3% ~25%. Compared to HAN and HGATE, HGVAE achieves 4% ~7% performance improvement. HAN does not consider the absence of links in multi-hop paths during homogenization. In contrast, HGVAE can be directly trained on the heterogeneous graph, which is more flexible. Compared to HGT, which is the best baseline for most cases, the Macro-F1 improvements of HGVAE on OAG_{ML}, Douban-Movie and Douban-Book datasets are 3.5%, 3.6% and 5.2%, respectively. The performance improvement than HGT demonstrates that the proposed method's multi-signal decoder and variational inference components have the superior ability to learn the relation representation for prediction.

4.5 Ablation Studies

Effect of Heterogeneous Graph Encoder. In HGVAE (w/o *heterogeneous*), to understand the impact of heterogeneous information for reasoning social relations, we replace our heterogeneous graph encoder with GAT [19]. For the comparison with HGVAE (w/o *heterogeneous*) and other baselines, original HGVAE generally achieves significant performance improvement (6% ~12%). In contrast, HGVAE(w/o *heterogeneous*) ignores heterogeneous information. Thus the performance decreases sharply. The phenomena prove the significance of heterogeneous information and HGVAE can leverage it.

Effect of Variational Mechanism. In HGVAE (w/o *variational*), we alter the VAE into an AE, which intends to verify the importance of the variational mechanism. For the comparison with HGVAE (w/o *variational*), the performance of HGVAE with a variational inference mechanism is better (1% ~2%).

Effect of Multi-Signal Decoder. In HGVAE (w/o *decoder*), to verify the importance of the multi-signal decoder, we drop the loss of it, and only preserve the loss of the downstream task for training. Without the Multi-Signal Decoder, the performance of HGVAE decreases sharply (3% ~5%). The significant performance gap between full HGVAE and HGVAE (w/o *decoder*) indicates the power of the multi-signal decoder in capturing complex signals on HSNs for social relation reasoning.

4.6 In-depth Model Analysis

In this section, we conduct in-depth analyses from the following views: **RQ1**. Is HGVAE efficient for social relation reasoning? **RQ2**. How is the robustness of HGVAE on noisy and incomplete social network data? **RQ3**. How does the learned social relation representation benefit from each component of HGVAE? **RQ4**. How do the different training ratios impact the performance of HGVAE?

Efficiency Analysis (RQ1). We analyze the efficiency of HGVAE from two aspects: The parameter quantity and the computational efficiency. The detailed statistics of parameter quantity, training time (per batch) and predicting time (per 100 samples) are shown in Table 3.

Table 3. The statistics of parameter quantity, training and predicting time.

Models	GCN	GraphSAGE	GAT	R-GCN	HAN	HGT	HGVAE	Rule-based
#Param	147,505	138,205	518,905	1,408,405	2,939,445	1,102,103	1,155,287	–
Training Time (ms)	158.6	190.2	396.1	433.6	4168.5	1247.4	1251.3	–
Predicting Time (ms)	1.1	1.2	1.8	2.9	7.6	13.6	13.7	533000.0

For parameter quantity, HGVAE (1155k) is similar to HGT (1102k) and one-third of that of HAN (2939k), but the effect is 4%~7% superior to the baselines, which demonstrates HGVAE does not rely on more parameters but a well-designed architecture for better performance.

The computational efficiency of HGVAE is also competitive. For the training procedure on OAG$_{ML}$ dataset, the average training time per batch of HGVAE

(1247 ms) is better than HAN (4168 ms), which needs time-consuming homogenization processes. HGVAE also has a comparable training cost to other graph neural network baselines. We record the average predicting time for 100 samples to infer social relations. HGVAE directly reasons the social relations between node pairs, which leads to the speed of HGVAE (13.7 ms) being far faster than the rule-based method (533000 ms).

We also analyze the time complexity of HGVAE. Therefore, the time complexity of the training procedure is $O(NB\theta + NB^2)$, where N is the number of batches, B is the batch size, θ is the number of attention heads and NB^2 is the time complexity of sampling procedure. The complexity of HGVAE is theoretically constant. Considering that the number of batches is related to the network scale, the time complexity of HGVAE is $O(|V|)$ where $|V|$ is the number of nodes. It is also far superior to the rule-based method based on graph searching $O(|V|^2)$. As a result, the high efficiency of HGVAE makes it possible to apply to large-scale heterogeneous social network data.

Robust Analysis (RQ2). We conduct robust experiments to horizontally observe the performance of listed models under unideal circumstances of different extents. Specifically, we randomly add/ delete a proportion of edges of training data mimicking noisy/ incomplete conditions for comparing method robustness. As shown in Fig. 3, when facing data with more proportion of disturbance, the F1-Score of HGVAE decreases slowly (-3%) than other methods (-4% ~-8%). These phenomena illustrate that the variational mechanism, which can flexibly model the distribution of social relations, significantly improves robustness, especially on unreliable data. Besides, notably, HGVAE((w/o *variational*) also performs better than baselines, which verifies that other components (e.g., multi-signal decoder) can benefit the model robustness.

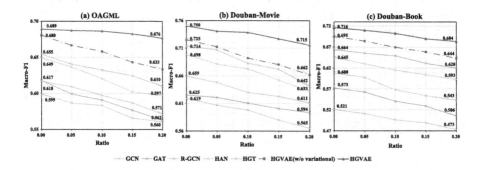

Fig. 3. Comparison of method robustness through randomly adding/deleting edges.

Hyper-parameter Analysis (RQ3, RQ4). To address RQ3 and RQ4, we analyze the influence of γ, α and different training ratios.

Hyper-parameter (γ and α). The hyper-parameter γ controls the weight of loss $\mathcal{L}_{meta}$ and $\mathcal{L}_{node}$ in loss function $\mathcal{L}_{\text{HGVAE}}$. The hyper-parameter α controls the weight of $\mathcal{L}_{HGVAE}$ and $\mathcal{L}_{pred}$ in loss function $\mathcal{L}_{general}$. By adjusting the different combinations of γ and α, we analyze the importance of components in HGVAE in the social relation reasoning task.

As shown in Fig. 4, when $\gamma = 0$ or $\gamma = 1$, the semantic features of person attributes or meta-paths features are entirely erased. The results show that the performance of HGVAE also drops significantly. These phenomena prove that both two features contribute to social relation reasoning. The importance of the two types of information and the corresponding decoders are different on different data with the consideration of the characteristic (i.e., person attributes and meta-paths) of datasets.

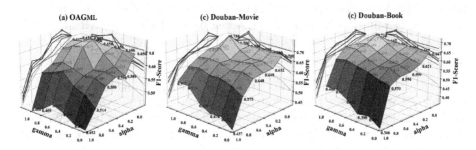

Fig. 4. Comparison of different hyper-parameter γ and α value on the three datasets.

Different values of α also significantly affect the performance of HGVAE. This result demonstrates that the multi-signal decoder, which dedicates to restoring the heterogeneous graph information, can benefit the model performance.

We also notice that HGVAE achieves comparable performance without considering downstream task loss function $\mathcal{L}_{pred}$ ($\alpha = 1$). This phenomenon shows the unsupervised learning ability of HGVAE, which can utilize various information generatively through the VAE framework for reasoning.

Training Ratio. We set the ratios of training data to the total data to 60%, 40%, and 20% to analyze the impact of training ratios as shown in Fig. 5.

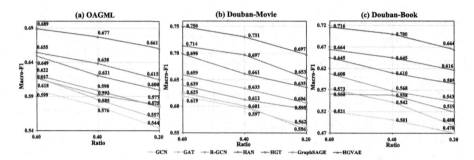

Fig. 5. Comparison of method performance under different training ratios.

As the training ratio decreases, the performances of HGVAE do not decrease (−2% ~−6%) obviously than other baselines. This phenomenon indicates that HGVAE has superior performance with less training data. Notably, the performance of HGVAE when using 20% training data in an inductive way is better than most baselines when using 60% training data, which answers RQ4.

4.7   Case Study

To better illustrate the motivation of our proposed task and the performance of HGVAE, we choose a representative case study from the experiment result as shown in Fig. 6. Given a person pair *(ID 4023, ID 142)* in OAG_{ML}. We randomly deleted the links between them, which resulted in rule-based methods being unable to reason the *colleague* relationship of the person pair, and cost an expensive searching time (2.8 s). The most straightforward idea is that this problem can be solved by link prediction and meta-path reaching. However, this cumbersome enumeration process is inefficient when reasoning all types of predefined meta-paths, and can not effectively handle the noisy links which are also essential for reasoning social relations. Differently, HGVAE is an inductive method that can directly infer the various meta-path based social relations between person pairs. Utilizing HGVAE, the *colleague* type relation of the person pair can be completed in an inductive way with only 10.2 ms time cost.

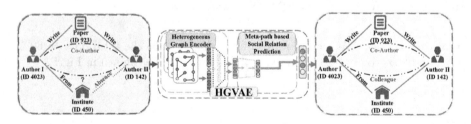

Fig. 6. A case study of social relation reasoning. Given a person pair (Author 4023, Author 142), in the heterogeneous social network (**OAG_{ML}**), HGVAE can reason social relations (Co-Author, Colleague) under the incomplete data situation.

5   Conclusion

This paper presents a study of social relation reasoning on heterogeneous social network data. To address the challenges such as multi-hop and heterogeneous situations, we shift from simply inferring meta-path based social relations by rule-based methods to reasoning by deep learning. Then, we propose an inductive method named HGVAE, which directly predicts existing meta-paths by handling the problem of the absence or the mistakes of multi-hop paths in heterogeneous social networks. The extensive experiments with ablation studies and robust analysis demonstrate that HGVAE performs superior to the baselines.

Acknowledgments. Thanks to reviewers for their helpful comments on this paper. This paper is funded by the National Natural Science Foundation of China (Nos.62172 393, U1836206 and U21B2046), Zhongyuanyingcai program-funded to central plains science and technology innovation leading talent program (No.204200510002) and Major Public Welfare Project of Henan Province (No.201300311200).

References

1. Dong, Y., Chawla, N.V., Swami, A.: metapath2vec: scalable representation learning for heterogeneous networks. In: Proceedings of KDD (2017)
2. Fan, H., et al.: Heterogeneous hypergraph variational autoencoder for link prediction. IEEE Trans. Pattern Anal. Mach. Intell. **44**, 4125-4138 (2021)
3. Fu, T., Lee, W., Lei, Z.: Hin2vec: explore meta-paths in heterogeneous information networks for representation learning. In: Proceedings of the 2017 CIKM (2017)
4. Hamilton, W.L., Ying, Z., Leskovec, J.: Inductive representation learning on large graphs. In: Proceedings of NeurIPS (2017)
5. Hinton, G.E., Osindero, S., Teh, Y.W.: A fast learning algorithm for deep belief nets. Neural computation (2006)
6. Hu, Z., Dong, Y., Wang, K., Sun, Y.: Heterogeneous graph transformer. In: Proceedings of WWW (2020)
7. Hwang, D., Park, J., Kwon, S., Kim, K.-M., Ha, J.-W.: Self-supervised auxiliary learning with meta-paths for heterogeneous graphs. In: Proceedings of NeurIPS (2020)
8. Kingma, D.P., Welling, M.: Auto-encoding variational Bayes. In: Proceedings of ICLR (2014)
9. Kipf, T.N., Welling, M.: Variational graph auto-encoders. arXiv preprint arXiv:1611.07308 (2016)
10. Kipf, T.N., Welling, M.: Semi-supervised classification with graph convolutional networks. In: Proceedings of ICLR (2017)
11. Loshchilov, I., Hutter, F.: Decoupled weight decay regularization. In: Proceedings of ICLR (2019)
12. Nolan, J.M., Schultz, P.W., Cialdini, R.B., Goldstein, N.J.: Normative social influence is underdetected. Personal. Soc. Psychol. Bullet. **34**, 913-923 (2008)
13. Ramchoun, H., Idrissi, M.A.J., Ghanou, Y., Ettaouil, M.: Multilayer perceptron: architecture optimization and training. In: IJIMAI (2016)
14. Rezende, D.J., Mohamed, S., Wierstra, D.: Stochastic backpropagation and approximate inference in deep generative models. In: Proceedings of ICML (2014)

15. Schlichtkrull, M., Kipf, T.N., Bloem, P., van den Berg, R., Titov, I., Welling, M.: Modeling relational data with graph convolutional networks. In: Gangemi, A., et al. (eds.) ESWC 2018. LNCS, vol. 10843, pp. 593–607. Springer, Cham (2018). https://doi.org/10.1007/978-3-319-93417-4_38
16. Shi, C., Hu, B., Zhao, W.X., Yu, P.S.: Heterogeneous information network embedding for recommendation. In: IEEE TKDE (2019)
17. Sun, Y., Barber, R., Gupta, M., Aggarwal, C.C., Han, J.: Co-author relationship prediction in heterogeneous bibliographic networks. In: 2011 International Conference on Advances in Social Networks Analysis and Mining (2011)
18. Sun, Y., Han, J.: Mining heterogeneous information networks: a structural analysis approach. ACM SIGKDD Explor. Newslett. **14**, 20–28 (2013)
19. Velickovic, P., Cucurull, G., Casanova, A., Romero, A., Liò, P., Bengio, Y.: Graph attention networks. In: Proceedings of ICLR (2018)
20. Wang, M., et al.: Deep graph library: A graph-centric, highly-performant package for graph neural networks. arXiv preprint arXiv:1909.01315 (2019)
21. Wang, W., et al.: HGATE: heterogeneous graph attention auto-encoders. In: IEEE TKDE (2021)
22. Wang, X., Ji, H., Shi, C., Wang, B., Ye, Y., Cui, P., Yu, P.S.: Heterogeneous graph attention network. In: Proceedings of WWW (2019)
23. Yang, C., et al.: Relation learning on social networks with multi-modal graph edge variational autoencoders. In: Proceedings of WSDM (2020)
24. Yang, Z., Dai, Z., Yang, Y., Carbonell, J.G., Salakhutdinov, R., Le, Q.V.: XLNet: generalized autoregressive pretraining for language understanding. In: Proceedings of NeurIPS (2019)
25. Zhang, F., Liu, X., Tang, J., Dong, Y., Yao, P., Zhang, J.: OAG: toward linking large-scale heterogeneous entity graphs. In: Proceedings of KDD (2019)

SRACas: A Social Role-Aware Graph Neural Network-Based Model for Popularity Prediction of Information Cascades

Zhenhua Huang[1], Yuhang He[2], Shaojie Wang[1], Zhenyu Wang[3], Ruifeng Xu[2(✉)], and Sharad Merothra[4]

[1] Anhui University, Hefei, China
wsj.ahu@gmail.com
[2] Harbin Institute of Technology (Shenzhen), Shenzhen, China
xuruifeng@hit.edu.cn
[3] South China University of Technology, Guangzhou, China
[4] University of California Irvine, Irvine, USA

Abstract. Popularity prediction of information cascades is a fundamental and challenging task in social network data analysis. Social roles impact users' behaviors and change the structure and popularity of information cascades. Existing deep learning-based methods utilize several independent sub-cascade graphs or paths to learn cascade representations, which lose vital information about social roles and dynamics between sub-cascades at different moments. We propose a social role-aware cascade (SRACas) model that exploits the social influences of nodes on previous and subsequent sub-cascade graphs within an observation window to facilitate the social role learning of nodes. A temporal-aware differential loss is also proposed to discriminate the structures of neighboring sub-cascades and captures the dynamics of sub-cascades. Under the techniques of local graph attention, social role-aware attention, and temporal-aware loss, SRACas learns a better latent representation of cascades at both the node level, sub-cascade level, and cascade level. Moreover, there lacks a platform with standard prepossessing procedures that allow convenient configuration and fair competition between information cascade prediction models. An open platform Open-Cas is built with uniform preprocesses to verify the faithful performance of the compared methods. Extensive experiments show that SRACas achieved significant improvements over existing methods on classic real-world datasets.

Keywords: Information Cascade · Graph Neural Networks · Social Role · Popularity Prediction

---

Z. Huang and Y. He—Equal Contribution.

© The Author(s), under exclusive license to Springer Nature Switzerland AG 2023
X. Wang et al. (Eds.): DASFAA 2023, LNCS 13945, pp. 21–30, 2023.
https://doi.org/10.1007/978-3-031-30675-4_2

1   Introduction

Popularity prediction of information cascades (also called cascade prediction in [3,12]) benefits many practical applications, e.g., fake news and rumor control, viral marketing, advertising, scientific impact quantification [11]. Understanding the nature of cascades and predicting their future popularity has drawn the interest of many scholars. However, cascade prediction remains challenging due to complex social influences, and its accuracy is still unsatisfactory.

The deep learning-based approaches have recently achieved state-of-the-art performances in information cascade prediction [2,3,6,8,9,12]. DeepCas [6], DeepHawkes [2], and TopoLSTM [8] are mainly based on recurrent neural networks. Spatial features are also found helpful in improving the performance of popularity prediction and are learned by the graph neural networks (GNNs) in current works [3,9,12]. Social roles impact retweet behaviors and significantly influence information diffusion [10]. However, current methods, e.g., CasCN [3], VaCas [12] and CasFlow [9] learn node features in isolated sub-cascade graphs, ignoring the social roles of nodes and influences between nodes in different sub-cascades. As described in Fig. 1, the sub-cascade graphs at t_i in two cases are topologically the same. But the node A and B hold different social roles and influences in the cases. A is an opinion leader that leads several following retweets, while B is a common node. Node representations learned independently in a sub-cascade graph can hardly distinguish the situations and reflect the nodes' social roles. However, by reviewing previous and subsequent sub-cascades, we can easily discriminate the social roles of node A and B and know their impacts on information diffusion. As represented in t_i and t_{i+1} in Fig. 1(b), the structure and size of cascades do not always change drastically over time. The adjacent sub-cascades may be slightly different from the previous sub-cascades. The existing models rarely consider the subtle differences between adjacent sub-cascade graphs and lose subtle dynamic information of cascades. Moreover, current models based on dense adjacency matrices are memory-intensive and inefficient or perform much worse in larger cascades.

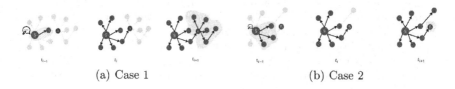

(a) Case 1   (b) Case 2

Fig. 1. Cases describing how social roles impact information diffusion. The purple node A is an opinion leader, and green node B is a common node. (Color figure online)

To address the abovementioned problems of cascade prediction, we propose a novel social role-aware cascade prediction model named SRACas. It utilizes local graph attention and a social role-aware attention mechanism which enables each node to consider both its neighboring nodes' representation and features of sub-cascade graphs from previous and subsequent moments. The social roles and influences are embedded in nodes' representations in this way. A temporal-aware differential loss is also employed to learn the subtle differences between sub-cascade graphs, which benefits the training and convergence of models. To support larger cascades, sparse matrices and efficient computations are also applied.

In addition, although there is a survey on cascade prediction [11], a framework aggregating current cascade models is lacking. It is necessary to unify experimental steps before training models and reorganize the implementations of current models. Therefore, a framework, OpenCas, is built to address the issue.

Our main contributions are as follows:

- **Novel Cascade Model.** We propose a novel cascade model, SRACas, which employs local graph attention and a social role-aware attention mechanism to learn better social role representations of nodes that leverages bidirectional sub-cascades. A temporal-aware differential loss is also applied in SRACas to capture the subtle differences and dynamics of sub-cascade graphs, which is another key to learning better temporal features of cascades.
- **Support Larger Cascades.** Some advanced cascade prediction models (e.g., CasCN [3], VaCas [12], and CasFlow [9]) are unable to work or perform much worse in larger cascades. In contrast, the performance of our approach is most stable in larger cascades leveraging graph sparse encoding.
- **Much Better Performance.** Extensive experiments on two real-world scenarios demonstrate that SRACas significantly outperforms strong baselines, with the MSLE reduced by from 12.0% to 14.6%, from 6.9% to 9.7% in the dataset Sina Weibo and APS Citation, respectively.
- **OpenCas.** We built a cascade prediction framework OpenCas, which enables convenient and fair comparative experiments between different models under the same data preprocessing and partitions. OpenCas improves the performance of classic models and simplifies environment configurations[1]

2 Preliminaries

Cascade Graph. A source of a cascade can be an academic publication, a tweet, a microblog, etc. A cascade graph can be represented as a sub-cascade graph sequence that evolves from an initial source node v_0 at time t_0. Node v_i participates in the cascade at time t_i. A subcascade $\mathcal{G}^{t_j}$ is a graph at moment t_j. We formalize the cascade sequence as $C_i = \{\mathcal{G}^{t_0}, \mathcal{G}^{t_1}, ..., \mathcal{G}^{t_n}\}$, where $\mathcal{G}^{t_j} = (V^{t_j}, E^{t_j}, t_j)$ is a snapshot graph of the cascade C at time t_j. V^{t_j} and E^{t_j} are the sets of nodes and edges of sub-cascade $\mathcal{G}^{t_j}$ until time $t_j \geq 0$, respectively. $V^{t_j} = \{v_0(t_j), v_1(t_j), v_2(t_j), ..., v_i(t_j)\}$. The set of edges E^{t_j} records how information propagates between users in V^{t_j}.

Popularity Prediction. Following previous works [2,3,6,12], the popularity prediction of information cascades is formalized as a regression problem. Given

---

[1] The details of OpenCas are referred to https://github.com/zhenhuascut/OpenCas.

a cascade $C_i = \{\mathcal{G}^{t_0}, \mathcal{G}^{t_1}, ..., \mathcal{G}^{t_n}\}$, the incremental popularity is defined as $\Delta S = |V^{t_p}| - |V^{t_o}|$, where t_o and t_p are the observation time and the prediction time, respectively. The main objective is to train a model f that learns the representation of a cascade within the observation time t_o to predict ΔS.

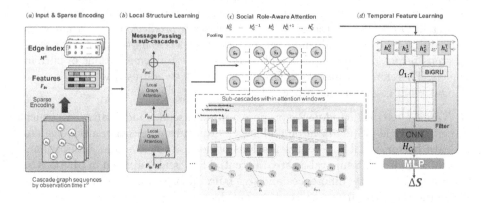

Fig. 2. Overview of SRACas.

3 Proposed Method

The overall training process of SRACas is sketched in Fig. 2. It mainly contains the following modules:

3.1 Local Structure Learning

Given a sub-cascade graph $\mathcal{G}^t$, the first step is to capture the basic structural information and obtain node-level representations. Different from CasCN [3] that neighboring nodes have similar weights. We believe that different nodes play different roles in information diffusion, e.g., opinion leaders, structural holes, and peripheral nodes. Local graph attention similar to GAT [7] is used to calculate the importance of a node and aggregate the features of nodes. The α_{uv} and alignment coefficients e^l_{uv} are calculated by:

$$\alpha_{uv} = Softmax_u(e^l_{uv}) = \frac{exp(e^l_{uv})}{\sum_{k \in N(v) \cup \{v\}} exp(e^l_{kv})} \tag{1}$$

$$e^l_{uv} = attn^l(x^l_u, x^l_v) = f_{prop}(h^l_u, h^l_v) \tag{2}$$

where h^l_u is the node embedding in graph aggregation layer l. The function f_{prop} can be in different forms. One simple form is $f_{prop} = (W^l_a)^T[W^l h^l_u, W^l h^l_v]$, where $W^l_a \in R^{2F^{l+1}}$ is a shared attention weights. x^t_v is calculated by a residual network $x^t_v = f_0(h^0_v) + f_1(h^1_v) + h^L_v$. $x^t_v \in R^{F^L}$, f_0 and f_1 are dense layers that change dimension size of h^l_v to F^L. L is the number of layers set to two in this paper.

3.2 Social Role-Aware Attention

The social role-aware attention considers the influences of nodes from nodes of previous and following sub-cascades within $\mathcal{W}$ temporal intervals (attention window), in which the roles a node plays in information diffusion are learned.

The social influence of a node v by node $u \in V^{t-1}$ at the previous moment is calculated by :

$$\alpha_{uv}^b = \frac{exp(f_{back}(x_u^{t-1}, x_v^t))}{\sum_{k \in V^{t-1}} exp(f_{back}(x_k^{t-1}, x_v^t))} \tag{3}$$

$$x_v^{t-1} = \sigma\left(\sum_{u \in V^{t-1}} x_u^{t-1} \alpha_{uv}^b\right) \tag{4}$$

where α_{uv}^b is the impacts of node u to node v in V^{t-1}. f_{back} is the aggregation method between x_v^{t-1} and x_v^t. Here, we choose the Euclidean distance of nodes as the function f_{back}.

x_v^{t+1} is calculated in the same way as in x_v^{t-1}. The updated representation x_v^{update} of node v is then calculated by combing the representations of v from sub-cascades within attention window size $\mathcal{W}$. When $\mathcal{W} = 1$, $f_{combine}(x_v^{t-1}, x_v^t, x_v^{t+1}) = x_v^{t-1} + x_v^t + x_v^{t+1}$. $f_{combine}$ can also be designed in a more complicated form, which is left for future works. A larger attention window size (e.g., $\mathcal{W} = 2$, $f_{combine}(x_v^{t-2}, x_v^{t-1}, x_v^t, x_v^{t+1}, x_v^{t+2})$) brings a broader view of sub-cascades but increases the calculation complexity.

Graph Pool. After obtaining nodes' representations, the features of nodes at the moment are pooled to produce a presentation of the sub-cascade graph h_G^t.

3.3 Temporal Feature Learning

Social role-aware attention has learned part of cascades' temporal characteristics, as well as the features of social roles. To learn the long-term temporal dependency of cascades, we apply a temporal feature learning model. Given input representations of sub-cascades $H_G = \{h_G^0, h_G^1, h_G^2, ..., h_G^T\}$, the bidirectional gated recurrent units (BiGRU) computes updated hidden states and produces new representations by concatenating outputs of the forward GRU and backward GRU: $O_{1:T} = \{o_0, o_1, o_2, ..., o_T\} = \overrightarrow{GRU}(H_G) || \overleftarrow{GRU}(H_G)$, $O_{1:T} \in R^{T*2M}$, M is the output hidden size of BiGRU.

We can simply use the last output o_T as the embedding features of a cascade. Motivated by the idea [5], convolutional neural networks are used to capture the long-term dependencies of H_G and produce the final representation H_{C_i} of a cascade C_i.

3.4 Prediction and Loss Function

H_{C_i} is fed into a two-layer multi-layer perception (MLP) to produce predicted popularity increment $\Delta \tilde{S}_i = MLP(H_{C_i})$. To capture subtle dynamics of cascades, a temporal-aware differential loss function $\mathcal{L}_{dif}$ is applied to predict the differential size of the sub-cascades by:

$$\mathcal{L}_{dif} = \frac{1}{TN_c} \sum_{i=1}^{N_c} \sum^{T} (\Delta S_{di} - \Delta \tilde{S}_{di})^2 \tag{5}$$

$$\Delta S_{di} = (|\mathcal{G}_f| - |\mathcal{G}_p|), \Delta \tilde{S}_{di} = MLP(h_{\mathcal{G}}^f - h_{\mathcal{G}}^p) \tag{6}$$

where ΔS_{di} is the differential size between two neighboring sub-cascade graphs (a following $\mathcal{G}_f$ and a previous $\mathcal{G}_p$). N_c is the number of cascades. A differential representation of sub-cascades is calculated by subtracting $h_{\mathcal{G}}^f$ and $h_{\mathcal{G}}^p$. The differential representation is then fed into an MLP layer to predict ΔS_{di}.

The final loss is calculated by:

$$\mathcal{L}(\Delta S_i, \Delta \tilde{S}_i) = \frac{1}{N_c} \sum_{i=1}^{N_c} (log\Delta S_i - log\Delta \tilde{S}_i)^2 + \lambda \mathcal{L}_{dif} \tag{7}$$

where λ is a hyperparameter and is set to 1.0 by default.

3.5   Sparse Encoding

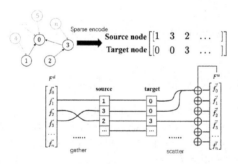

Fig. 3. Sparse graph encoding and computation.

Different from CasCN, VaCas, and CasFlow, which use a full matrix, SRACas applies a sparse matrix to save GPU memories and accelerate computations. The sparse matrix representation records source nodes and target nodes that indicate the edges and directions, as shown in Fig. 3. A matrix is sparsely encoded as:

$$M^s = \begin{bmatrix} s_1 & s_2 & s_3 & s_1 & \ldots & s_i \\ t_1 & t_2 & t_2 & t_3 & \ldots & t_i \end{bmatrix} \tag{8}$$

where s_i and $t_i (i = 1, 2 \ldots, N)$ are a source node and a target node. The gather process aggregates the source nodes with the same target t_i (denoted as the set P_i). The process of the matrix computation is $M^s F^d$, where $F^d \in R^{NF_{in}}$.

The multiplication is improved by accessing the memory and computing in a thread-per-matrix-row pattern [4].

$$F^u_{t_i} = (M^s \cdot F^d)_{t_i} = \sum_{k \in P_i} F^d_k \tag{9}$$

where $F^u_{t_i} \in R^{F_{in}}$ are the representation of node i after a simple sum scatter.

4 Experiments

Two classic tasks are introduced to evaluate the effectiveness of SRACas [2]. One task is to predict the future retweet size of a microblog. The other task predicts the future citation counts of a paper.

4.1 Experimental Setups

Following data preprocessing setups in DeepHawkes [2], the observation time window t_o is set to 1 h, 2 h, and 3 h in the Sina Weibo dataset. The cascades that have less than five retweets within the observation time t_o are removed in experiments. In the dataset APS Citation, we remove papers with fewer than ten citations within the observation time [2]. We set $t_o = 5$ years, 7 years, and 9 years to predict the size in the 20th year.

The MSLE (mean square log-transformed error) and mSLE (median square log-transformed error) [2] are used as evaluation metrics. We consider the classic and advanced models as strong baselines (Table 1).

4.2 Prediction Performance

Table 1. Performance on the entire cascade set.

t_o	Sina Weibo						APS Citation					
	1hour		2hours		3hours		5years		7years		9years	
Metric	MSLE	mSLE	MSLE	mSLE	MSLE	mSLE	MSLE	mSLE	MSLE	mSLE	MSLE	mSLE
DeepCas	3.383	0.810	3.255	0.813	3.186	0.845	1.772	0.730	1.594	0.693	1.579	0.732
TopoLSTM	3.535	1.247	3.386	1.108	3.244	1.101	1.728	0.721	1.615	0.736	1.560	0.644
DeepHawkes	2.307	0.648	2.125	0.597	2.105	0.644	1.347	0.604	1.337	0.598	1.244	0.519
CasCN*	2.862	0.699	2.796	0.774	2.707	0.681	1.397	0.618	1.355	0.563	1.178	0.545
CasFlow*	2.325	0.650	2.256	0.672	2.113	0.668	1.352	0.548	1.345	0.591	1.316	0.571
SRACas	**1.970**	**0.560**	**1.870**	**0.499**	**1.826**	**0.477**	**1.253**	**0.527**	**1.207**	**0.510**	**1.076**	**0.472**

Note that OpenCas optimizes implementations of current models, so performances of some baselines are improved compared to the original papers. Some recent models (e.g., CasCN and CasFlow) are only suitable for predicting cascades with small popularity size [3,9]. Larger cascades form in real-world scenarios, and they are more common in the Sina Weibo dataset than in the APS

Citation dataset. Larger cascades increase the difficulties of cascade prediction and require higher GPU memories, making some models unsuitable. Therefore, a smaller dataset from the Sina Weibo dataset is also used to validate the performance of models by selecting the cascades with no more than over 100 nodes within the observation window following setups [3,9]. The size distributions of the small and large cascades are similar in the APS Citation and Twitter, which is meaningless to be divided.

Prediction on the Entire Cascade Set: The entire cascade set preserves cascades with fewer than 1000 nodes within the observation window as described in DeepHawkes [2]. CasCN and CasFlow are infeasible to run experiments directly on the entire set due to their high GPU demands. We apply CasFlow and CasCN by limiting the first 100 participants in the observation window (denote as CasFlow* and CasCN*). SRACas applies sparse encoding to reduce GPU memory usage and enable experiments on the entire cascades. SRACas follows the same training strategies in experiments on both entire and smaller cascade sets. The baseline models outperform their original implementations by re-implementing part of the codes and optimizing training strategies on OpenCas. For example, the performance of DeepCas and DeepHawkes is better than their original papers or previous reports, with the MSLE reduced from 3.63 to 3.38 and from 2.44 to 2.31 [2] at one hour, respectively. SRACas consistently outperforms the baselines by significant margins.

Prediction on Smaller Cascades Set: All the baseline models work on the smaller dataset, but SRACas performs much better, as shown in Table 2. MSLE of DeepCas and DeepHawkes is reduced from 2.958 [3] to 2.494 and from 2.441 [3] to 2.023, respectively. The performance of DeepCas and DeepHawkes is underestimated in previous works. The second lowest MSLE (2.023 by DeepHawkes) is reduced to 1.810 (SRACas) by around 10.5 % at 1 h on Sina Weibo.

Compared with prediction performance on the smaller set, MSLE of CasFlow and DeepHawkes on the entire set is highly decreased by 14% (substantial performance decrease) as shown in Table 2. The performance of CasCN and TopoLSTM even dropped by more than 30% and 40%, respectively. However, SRACas is the most stable: the MSLE only increased from 1.810 to 1.970, and the mSLE increased from 0.534 to 0.560.

Table 2. Performance on Sina Weibo (smaller set).

	Sina Weibo (Smaller Set)					
t_o	1hour		2hours		3hours	
Metric	MSLE	mSLE	MSLE	mSLE	MSLE	mSLE
DeepCas	2.494	0.758	2.386	0.801	2.316	0.856
TopoLSTM	2.679	0.715	2.595	0.724	2.579	0.703
DeepHawkes	2.023	0.649	1.963	0.536	1.941	0.546
CasCN	2.042	0.553	1.906	0.532	1.867	0.530
CasFlow	2.038	0.547	1.988	0.527	1.920	0.519
SRACas	**1.810**	**0.534**	**1.687**	**0.470**	**1.622**	**0.486**

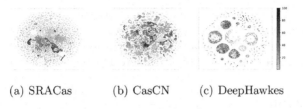

(a) SRACas   (b) CasCN   (c) DeepHawkes

Fig. 4. Subcascade representations in Sina Weibo.

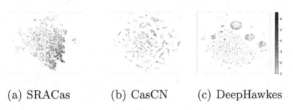

(a) SRACas   (b) CasCN   (c) DeepHawkes

Fig. 5. Differential temporal representations in Sina Weibo.

4.3 Visualization and Explanation

Discussions on Learned Features: To discover the differences in representations learned by different models. We apply t-SNE [1] to project the cascades' or sub-cascade graphs' representations into two-dimensional points. Meaningful representations will gather according to the sizes. Figure 4 represents the sub-cascade graphs' representations with sizes. The aggregation degree of the learned features by SRACas is higher than CasCN and DeepHawkes on relatively large-size sub-cascade graphs.

Furthermore, we explore the differential representation of sub-cascades in different moments by subtracting the representation of $\mathcal{G}_t$ and $\mathcal{G}_{t-1}$ and coloring the data points with incremental sizes between $\mathcal{G}_t$ and $\mathcal{G}_{t-1}$. As shown in Fig. 5, the aggregation effects of CasCN and DeepHawkes are not as obvious as those of SRACas. The nodes in different moments share the same representations in DeepHawkes, which makes it hard to distinguish the structure of sub-cascades. CasCN also fails to distinguish the difference between neighboring sub-cascades, demonstrating the effectiveness of role-aware social attention.

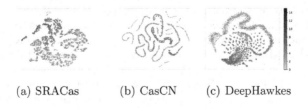

(a) SRACas   (b) CasCN   (c) DeepHawkes

Fig. 6. Cascade representations in Sina Weibo.

Finally, the representations of entire cascades are visualized in Fig. 6. The learned features by all models are related to the increment size to a certain extent. However, compared to CasCN and DeepHawkes, the points cluster more densely and smoothly according to their incremental popularity in SRACas.

5   Conclusion

This paper proposes a novel model SRACas to solve the problem that current models fail to model the social roles of nodes and discriminate the neighboring sub-cascades, utilizing a social-aware attention mechanism and temporal-aware differential loss. The SRACas achieved significant improvements over strong baseline methods and brought cascade prediction performance to a new level in classic real-world datasets. Furthermore, an open platform OpenCas that aggregates cascade prediction models is built, which simplifies the environmental configuration, data preprocessing, and training of models.

Acknowledgements. This work is supported by the NSF of China (No. 71971002) and the NSF of Guangdong Province, China (No. 2019A1515011792).

References

1. Vincent, D.B., Jean-Loup, G., Renaud, L., Etienne, L.: Fast unfolding of communities in large networks. J. Stat. Mech: Theory Exp. **2008**(10), P10008 (2008)
2. Cao, Q., Shen, H., Cen, K., Ouyang, W., Cheng, X.: Deephawkes: bridging the gap between prediction and understanding of information cascades. In: Proceedings of CIKM, pp. 1149–1158 (2017)
3. Chen, X., Zhou, F., Zhang, K., Goce, T., Zhong, T., Zhang, F.: Information diffusion prediction via recurrent cascades convolution. In: Proceedings of ICDE
4. Matthias, F., Jan, E.L.: Fast graph representation learning with PyTorch Geometric. In: Proceedings of ICLR Workshop on RLGM (2019)
5. Kim, Y.: Convolutional neural networks for sentence classification. In: Proceedings of the EMNLP, pp. 1746–1751, Doha, Qatar, October 2014
6. Cheng, L., Jiaqi, M., Xiaoxiao, G., Qiaozhu, M.: Deepcas: an end-to-end predictor of information cascades. In: Proceedings of the 26th WWW, pp. 577–586 (2017)
7. Petar, V., Guillem, C., Arantxa, C., Adriana, R., Pietro, L., Yoshua, B.: Graph attention networks. Proceedings of ICLR (2018)
8. Wang, J., Zheng, V., Liu, Z., Chang, K.: Topological recurrent neural network for diffusion prediction. In: Proceedings of ICDM, pp. 475–484. IEEE (2017)
9. Xu, X., Zhou, F., Zhang, K., Liu, S., Goce, T.: Casflow: exploring hierarchical structures and propagation uncertainty for cascade prediction. TKDE (2021)
10. Yang, Y., et al.: Rain: social role-aware information diffusion. In: Proceedings of AAAI (2015)
11. Fan, Z., Xu, X., Goce, T., Zhang, K.: A survey of information cascade analysis: Models, predictions, and recent advances. ACM Computing Surveys (2021)
12. Zhou, F., Xu, X., Zhang, K., Goce, T., Zhong, T.: Variational information diffusion for probabilistic cascades prediction. In: IEEE INFOCOM, pp. 1618–1627 (2020)

Few-Shot Link Prediction for Event-Based Social Networks via Meta-learning

Xi Zhu[1], Pengfei Luo[1], Ziwei Zhao[1], Tong Xu[1(✉)], Aakas Lizhiyu[2], Yu Yu[2], Xueying Li[2], and Enhong Chen[1]

[1] University of Science and Technology of China, Hefei, China
{xizhu,pfluo,zzw22222}@mail.ustc.edu.cn, {tongxu,cheneh}@ustc.edu.cn
[2] Alibaba Group, Hangzhou, China
{aakas.lzy,greta.yy,xiaoming.lxy}@alibaba-inc.com

Abstract. With the thriving of social network analysis, large efforts have been made on link prediction for event-based social networks (EBSNs). Unfortunately, since society is evolving with constantly emerging social events, it is extremely difficult to accurately capture their semantics and evolution rules at an early stage. Meanwhile, traditional solutions require extensive training from scratch to accommodate new events, leading to lagging predictions and high maintenance costs. To tackle these challenges, we investigate this cross-network few-shot problem and propose a novel meta-learning model for link prediction on new EBSNs. To accurately simulate the few-shot scenarios, we first utilize existing EBSNs to define a task distribution that augments the new event with other observed events. Specifically, we define a unified and generalized target event to be transferred as the few-shot event. Then, we empower a simple but effective event-aware graph attention network to encode existing fine-grained events and the few-shot target events. Furthermore, we follow gradient-based episode learning to obtain transferable knowledge and adapt to unseen EBSNs with sparse connections. Finally, extensive experiments on both public and industrial datasets have demonstrated the performance of fast adaption and even overall performance.

Keywords: few-shot · event-based social networks · meta-learning

1 Introduction

The widespread popularity of social services has brought us a tremendous volume of social interaction data and various spontaneously formed communities. Over the last decade, service providers are motivated to create and promote a series of interest-driven social events to improve information dissemination and further increase the vitality of platforms. Along this line, the concept of **Event-Based**

© The Author(s), under exclusive license to Springer Nature Switzerland AG 2023
X. Wang et al. (Eds.): DASFAA 2023, LNCS 13945, pp. 31–41, 2023.
https://doi.org/10.1007/978-3-031-30675-4_3

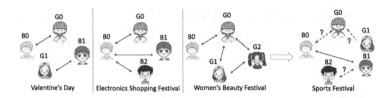

Fig. 1. An example of overlapping EBSNs generated by product share records on an e-commerce platform, where B and G denote boys and girls, respectively.

Social Networks (EBSNs) has emerged which provides explicit fine-grained information to show the diverse social preferences of users.

With the prosperity of social services, society is evolving with successive social events. Taking Fig. 1 as an example, various social-oriented promotions are launched to promote communication within and outside communities. Specifically, couples (e.g., B0&G0, B1&G1) kept in touch on Valentine's Day. The girls (e.g., G0, G1, and G2) would share beauty products, while boys (e.g., B0, B1, and B2) would connect during Electronics Shopping Festival. Notably, the interaction (B0, G0) always holds, exhibiting an event-agnostic relationship, while most users show various social patterns. However, with the advent of the Sports Festival, there is an urgent need to identify and attract users to establish connections in the corresponding EBSN, thus we have to quickly find potential links on the sparsely-estimated network. This motivates us to consider a more challenging setting of graph few-shot learning, which expects to enable fast adaption and high-quality prediction on newly-deployed EBSNs without abundant data. We emphasize that an effective early prediction of the evolution of the novel community will provide indispensable guidance to assist decision-making.

In this paper, we study few-shot link prediction for EBSNs. Intuitively, due to the cold-start problem, new EBSNs would suffer from data deficiency and even distorted topology [1]. A straightforward idea is to apply a multi-relation model to encode all events [6, 8] and retrain it to accommodate new events, which results in lagging predictions and high maintenance costs. Recent years have witnessed the rapid development of few-shot learning (FSL) [1, 2, 4, 10]. As a prevailing paradigm, MAML [2] obtains knowledge from similar tasks and transfers it to unseen tasks with a few instances. We notice that the key is to perfectly simulate the few-shot scenarios, which guarantees the transferability and robustness of the meta-knowledge. Specifically, we have the following observations:

- **The dependencies among EBSNs could be utilized to enhance the semantics and structures of the few-shot event.** Prior arts of FSL on disjoint attributed networks [1,4] are not compatible with densely connected EBSNs from the same domain. Since emerging EBSNs can be noisy, sparse, and unbalanced in scale, previous interactions could be naturally utilized. Correspondingly, we have the challenge to define a task distribution over dense EBSNs, further contributing to knowledge transfer.

– **The shared characteristics of few-shot events could be learned and transferred to guarantee generalizability.** Indeed, the exact semantics of constantly emerging events are difficult to capture [1]. However, the semantics of existing events can be explicitly captured and naturally transferred. Thus, we expect a unified well-initialized representation for the few-shot event as a good starting point for fine-tuning. Correspondingly, we have the challenge to design an effective model as the carrier of meta-knowledge.

We propose a meta-learning framework to address the aforementioned challenges. Briefly, new sparse target events leverage the knowledge from existing dense events, as there are shared semantics and structures worth exploiting. First, to simulate few-shot scenarios, we fully utilize existing EBSNs to define a task distribution, where a cross-network sampling strategy is designed to handle interconnected EBSNs. Afterward, a unified and generalizable target event is proposed to simulate the few-shot event. For each task, fine-grained source events and the special target event are jointly encoded by an event-aware graph attention network (EA-GAT), where both target and auxiliary loss are jointly optimized. Overall, we follow the standard episode learning to learn well-initialized parameters from tasks. In this case, when new events appear, we can individually fine-tune the tasks with a handful of associative instances, and adapt quickly with few resources. Our contribution can be summarized as follows: (1) To the best of our knowledge, we are the first to investigate FSL on interconnected EBSNs, which is universal and significant for early decision-making. (2) To achieve fast adaption, we propose a meta-learning framework for knowledge transfer. To simulate the few-shot scenario, we define a task distribution with network augmentation and learn a unified, generalizable few-shot event. (3) To validate the effectiveness, experiments are conducted on public and industrial datasets to show the superiority of fast adaption and convergence performance.

2 Related Works

Graph Neural Networks (GNNs). Recently, GNNs have been widely adopted to preserve properties and structures on graphs, such as GCN, GAT, and GraphSAGE. Moreover, R-GCN [6] and CompGCN [8] are proposed to model graph heterogeneity. HGT [3] decomposes the interactions with a Transformer-like architecture. However, these approaches learn global parameters with abundant data while we focus on emerging events (relations) with insufficient data.

Graph Few-Shot Learning. Remarkable success has been made on FSL of images and text while the exploration of graphs is still in its infancy, especially in multi-graph settings. Some studies formulate the transferable knowledge as meta-optimizer and metric space, e.g., Prototypical Network [7]. By contrast, Meta-GNN [10] integrates MAML with GNNs and facilitates gradient descent across tasks. However, little literature can be adapted to few-shot link prediction.

Algorithm 1. Meta-Training Task Sampling

Input: $\mathcal{R} = \{R_1, ..., R_N\}, \mathcal{G} = \{G_1, ..., G_N\}$ where $G_i = \{V_i, E_i\}$
Output: The distribution of meta-training tasks $p(\mathcal{T})$

1: **while** not done **do**
2:     Select $R_k \in \mathcal{R}$;
3:     Set R_k to R_{tgt}, Define $\mathcal{R}_{aux} = \mathcal{R} - \{R_k\}$;
4:     Split $E_{tgt} = E_{tgt}^{support} \cup E_{tgt}^{query}$;
5:     Construct $G = (V, E)$, where $E = E_{tgt}^{support} \cup E_{aux}$ and $E_{aux} = \bigcup_{i \neq k} E_i$;
6:     Compose $\mathcal{S} = (\mathcal{S}_{tgt}, \mathcal{S}_{aux})$ from $E_{tgt}^{support}$ and E_{aux}, respectively;
7:     Compose $\mathcal{Q} = (\mathcal{Q}_{tgt}, \mathcal{Q}_{aux})$ from E_{tgt}^{query} and E_{aux}, respectively;
8:     $T = (G, \mathcal{S}, \mathcal{Q})$
9: **end while**

To list a few, G-Meta [4] proposes to reduce parameters by encoding local sub-graphs for large-scale graphs. Meta-Graph [1] recovers graphs with a variational auto-encoder and learns knowledge from disjoint attributed graphs. Despite the progress, these works fail to handle the cross-network dependency for EBSNs.

3   Problem Definition

Definition 1. *Event-Based Social Networks (EBSNs). Suppose the user set is denoted as $\mathcal{V}$ and there have been N events $\mathcal{R} = \{R_1, R_2, ..., R_N\}$. The corresponding EBSNs are denoted as $\mathcal{G} = \{G_1, G_2, ..., G_N\}$, where the i-th EBSN is represented as $G_i = (V_i, E_i)$, where $V_i \subseteq \mathcal{V}$ and $E_i = \{(u, R_i, v) | u, v \in V_i\}$.*

Definition 2. *Few-shot Link Prediction for EBSNs. Assume we have observed the social events $\mathcal{R} = \{R_1, R_2, ..., R_N\}$, and the corresponding EBSNs are $\mathcal{G} = \{G_1, G_2, ..., G_N\}$. For an emerging few-shot event R_{few}, and $G_{few} = (V_{few}, E_{few})$, $V_{few} \subseteq \mathcal{V}$. We follow the few-shot setting and split $E_{few} = E_{few}^{support} \cup E_{few}^{query}$, where $E_{few}^{support} \cap E_{few}^{query} = \emptyset$. As a small fraction of inter-actions are available to support the network inference, i.e. $|E_{few}^{support}| \ll |E_{few}|$, our goal is to predict E_{few}^{query} with limited true edges from $E_{few}^{support}$.*

4   Methodology

In this section, we present the technical details of our model in Fig. 2, including task sampling, event-aware link prediction task, and meta-learning framework.

4.1   Cross-Network Task Sampling

The meta-learning approaches assume there are exploitable, shareable structures across similar tasks. Thus, its success relies heavily on making full use of exist-ing data to create tasks that delicately simulate real-world few-shot scenarios. Specifically, when new events emerge, we enhance the new, sparse, and noisy

EBSN with previous, dense, and complete EBSNs, denoted as auxiliary and target events, respectively. To this end, we leverage existing EBSNs to define a task distribution. Taking Fig. 1 as an example, we possibly choose Women's Beauty Festival as the **target event (few-shot event)**, so the Valentine's Day and Electronics Shopping Festival become **auxiliary events**. Note any of them can be selected as the target event to generate sufficient tasks.

Unfortunately, the support samples are insufficient to express the semantic meaning of the target event. Therefore, we propose a unified, generalizable target event and attempt to make it transferable to unseen few-shot events, i.e. transfer R_{tgt} to R_{few}. In other words, the learnable target event shared across tasks will be adopted to represent the real emerging few-shot event in meta-testing.

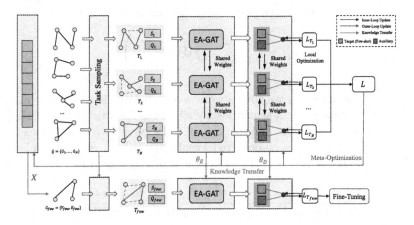

Fig. 2. The proposed meta-learning framework including (1) Few-shot task sampling with network augmentation, (2) EA-GATs, and (3) Joint learning for link prediction. Without loss of generality, we sample B meta-training tasks as a batch.

Meta-Training Tasks. As illustrated in Algorithm 1, we first randomly select R_k from $\mathcal{R} = \{R_1, ..., R_N\}$ as the target event, and set the others are auxiliary events, i.e. $R_{tgt} = R_k, R_{aux} = \mathcal{R} - \{R_k\}$. Then, following the few-shot setting, we allow $G_{tgt} = \{V_{tgt}, E_{tgt}\}$ to be divided into $E_{tgt} = E_{tgt}^{support} \cup E_{tgt}^{query}$, where $E_{tgt} = E_{tgt}^{support} \cap E_{tgt}^{query} = \emptyset$. Next, we utilize E_{aux} to augment the sparse $E_{tgt}^{support}$ and build $G = (V, E)$, where $E = E_{tgt}^{support} \cup E_{aux}$, $E_{aux} = \bigcup_{i \neq k} E_i$. Afterwards, we dynamically sample edges from $E_{tgt}^{support}$ and E_{aux} and compose the support set as $\mathcal{S} = (\mathcal{S}_{tgt}, \mathcal{S}_{aux})$. Similarly, we compose the query set $\mathcal{Q} = (\mathcal{Q}_{tgt}, \mathcal{Q}_{aux})$ from E_{tgt}^{query} and E_{aux}. So far, we have defined a task as $T = (G, \mathcal{S}, \mathcal{Q})$. e can repeat the above steps to sample batches of training tasks. Finally, we target at making prediction on $\mathcal{Q}_{tgt}$ ($\mathcal{Q}_{aux}$ only assists the training) with limited $\mathcal{S}_{tgt}$ to support the inference.

Meta-Testing Tasks. Once event R_{few} comes, we follow the above steps to create a meta-testing task T_{few} in a similar way. Differently, all existing events $\mathcal{R} = \{R_1, ..., R_N\}$ could be auxiliary events as their semantics have been accurately captured. Since R_{tgt} exactly simulates the few-shot event, the prior knowledge of R_{tgt} will be naturally transferred to R_{few} for fine-tuning.

4.2   Event-Aware Link Prediction Task

Event-Aware Graph Attention Networks (EA-GATs). We first denote a generic encoder $\boldsymbol{H}^G = f_E\left(G; \boldsymbol{X}, \theta_E\right)$ parameterized by θ_E. $\boldsymbol{X} \in \mathbb{R}^{|\mathcal{V}| \times d}$ is learnable user embeddings. To incorporate the unified target event, $\tilde{\mathcal{R}} = \mathcal{R} \cup \{R_{tgt}\}$ is involved in the augmented network $G = (V, E)$. Generally, we set $\boldsymbol{X} = \boldsymbol{h}^{(1)}$ and iteratively update the embedding of user u as follows:

$$
\boldsymbol{h}_u^{(l)} = \sigma \left(\sum_{v \in \mathcal{N}_u} \alpha_{uv}^{(l)} \boldsymbol{W}_a^{(l)} \boldsymbol{h}_v^{(l-1)} \right) \tag{1}
$$

where $\alpha_{uv}^{(l)}$ is the attention weight between u and v, $\boldsymbol{W}_a^{(l)}$ is the weight matrix. Following [5], to handle event semantics, we allocate a d_r-dimensional embedding $\boldsymbol{z}_r$ for each event $r \in \tilde{R}$. As multiple events (relations) may exist between any pair of users, suppose the co-occurred events are $\varphi(u, v)$, we accumulate all contributions and compute the raw attention with event-specific representation:

$$
\hat{\alpha_{uv}} = \frac{\sum_{r \in \varphi(u,v)} exp(\boldsymbol{a}^T g(\boldsymbol{W}\boldsymbol{h}_u || \boldsymbol{W}\boldsymbol{h}_v || \boldsymbol{W}_r \boldsymbol{z}_r))}{\sum_{k \in \mathcal{N}_u} \sum_{r \in \varphi(u,k)} exp(\boldsymbol{a}^T g(\boldsymbol{W}\boldsymbol{h}_u || \boldsymbol{W}\boldsymbol{h}_k || \boldsymbol{W}_r \boldsymbol{z}_r))} \tag{2}
$$

where $\boldsymbol{W} \in \mathbb{R}^{d \times d}$ and $\boldsymbol{W}_r \in \mathbb{R}^{d \times d_r}$ are node- and relation-type transformation matrices, and we omit superscript l for simplicity. $\boldsymbol{a}^T \in \mathbb{R}^{3d}$ is a weight vector. $\mathcal{N}_u$ denotes the neighbors of u and $g(\cdot)$ denotes LeakyRELU. Following HGB [5], with residual connections on attention scores, the final attention score is $\alpha_{uv}^{(l)} = \beta \alpha_{uv}^{(l-1)} + (1 - \gamma) \hat{\alpha_{uv}}^{(l)}$, where γ is a scaling factor, i.e., 0.05. Besides, we adopt multi-head attention mechanism to learn from different subspaces. To capture long-range dependency, we stack L layers and use layer normalization to stabilize the training process. The final embeddings are $\boldsymbol{h}_u^g = Pooling\left(\boldsymbol{h}_u^{(1)}, \boldsymbol{h}_u^{(2)}, ..., \boldsymbol{h}_u^{(L)}\right) \in \mathbb{R}^{d_g}$ with concatenation or mean pooling.

Decoder. We define a generic decoder as $\boldsymbol{Y} = f_D\left(\mathcal{A}; \theta_D\right)$ to decode the triples in support or query set, i.e. $\mathcal{A} \in \{S, Q\}$. Inspired by [9], we instantiate the decoder with bilinear score function with $\boldsymbol{W}' \in \mathbb{R}^{(N+1) \times d_g \times d_g}$. Thus for a triple $(u, r, v) \in \mathcal{A}$, the probability y_{uv}^r that u and v holds in the event $r \in \tilde{R}$ is:

$$
y_{uv}^r = \sigma \left(\boldsymbol{h}_u^g \boldsymbol{W}_r' \boldsymbol{h}_v^g \right) \tag{3}
$$

where $\boldsymbol{W}_r' \in \mathbb{R}^{d_g \times d_g}$ is the event-aware matrix of r indexed from $\boldsymbol{W}'$.

Joint Learning with Auxiliary Triples. Given candidate triple set $\mathcal{A} = \{(u, r, v)\}$, the binary cross-entropy loss can be written as:

$$\mathcal{L}(\mathcal{A}) = \frac{1}{|\mathcal{A}|} \sum_{(u,r,v)\in\mathcal{A}} y_{uv}^r \log\left(\hat{y}_{uv}^r\right) + (1 - y_{uv}^r) \log\left(1 - \hat{y}_{uv}^r\right), \mathcal{A} \in \{\mathcal{S}, \mathcal{Q}\} \quad (4)$$

where y_{uv}^r is the ground-truth. Due to the overlapping between observed and few-shot EBSNs, the reconstruction of auxiliary links will be helpful for prediction on new EBSNs. Hence, for $\mathcal{S} = \{\mathcal{S}_{tgt}, \mathcal{S}_{aux}\}$, the task-oriented loss is:

$$\mathcal{L}_{tgt} = \mathcal{L}(\mathcal{S}_{tgt}), \mathcal{L}_{aux} = \mathcal{L}(\mathcal{S}_{aux}), \mathcal{L} = \mathcal{L}_{tgt} + \lambda_0 \mathcal{L}_{aux} + \lambda_1 \|\theta\| \quad (5)$$

where λ_0 is the trade-off parameter and θ denotes all task-level parameters.

4.3 Meta-learning Framework for EBSNs

Inspired by model-agnostic MAML [2], for emerging events, we learn general-purpose parameters from meta-training tasks as the prior knowledge, so that the model could produce good fine-tuning results through a few gradient steps.

Table 1. Statistics of the datasets for the proposed model

Dataset	# events	# nodes	# connections	# pairs	Event Type
DBLP	11	37947	210260	183040	conference
Tmall	8	16961	48782	41186	promotion

Formally, we consider the link prediction model as a function f_θ with $\theta = \{\theta_E, \theta_D, \boldsymbol{X}\}$. When adapting to $\mathcal{T}_i = \{G_i, \mathcal{S}_i, \mathcal{Q}_i\}$ from $p(\mathcal{T})$, we first update the task-level parameters with feeding $\mathcal{S}_i$, which can be expressed as:

$$\theta_i' \leftarrow \theta_i - \alpha \nabla_\theta \mathcal{L}_{\mathcal{T}_i}(f_\theta) \quad (6)$$

We only perform a one-step update here while extending to multiple steps (e.g., $K = 5, 10, 20$) is straightforward. For each meta-training task, we perform gradient descents individually. For the best performance of f_θ with respect to θ across tasks from $p(\mathcal{T})$, we validate each task $\mathcal{T}_i$ with $\mathcal{Q}_i$, so that the meta-objective is to minimize the accumulated loss on queries across sampled tasks:

$$\min_\theta \sum_{\mathcal{T}_i \sim p(\mathcal{T})} \mathcal{L}_{\mathcal{T}_i}(f_{\theta_i'}) = \sum_{\mathcal{T}_i \sim p(\mathcal{T})} \mathcal{L}_{\mathcal{T}_i}\left(f_{\theta - \alpha \nabla_\theta \mathcal{L}_{\mathcal{T}_i}(f_\theta)}\right) \quad (7)$$

Formally, we optimize meta-parameters θ as follows:

$$\theta \leftarrow \theta - \beta \nabla_\theta \sum_{\mathcal{T}_i \sim p(\mathcal{T})} \mathcal{L}_{\mathcal{T}_i}(f_{\theta_i'}) \quad (8)$$

where α, β is the meta-level and task-level learning rate, respectively.

5 Experiments

5.1 Experiment Setup

Data Description. We validate the proposed model on both public and industrial datasets. As summarized in Table 1, (1) **DBLP**[1] is a synthetic dataset, where we select 11 top AI conferences as events, and build EBSNs by connecting the first and other authors of each paper in a pairwise manner as social links. (2) **Tmall**[2] is a real-world dataset collected from 8 category-aware e-commerce promotions. Each social link corresponds to a product-sharing record during the promotion. Users with only one neighbor are eliminated.

Reproducibility Settings. We use PyTorch to implement our model[3]. To avoid accidental deviation caused by different event splitting, we use 5-fold cross-validation and report the mean value of each metric. Following [1,4], we hold a small percentage of true edges as the support set, i.e. $\{10\%, 20\%, 30\%\}$, 10% for validation and the rest for testing. 10 gradient update steps in meta-training and 20 steps in meta-testing are adopted. AdamW and SGD are applied for meta-optimization and task-level optimization, respectively, while their learning rates are turned in $\{1e^{-3}, 1e^{-2}, 1e^{-1}\}$. As for the EA-GAT module, the dimension is fixed to 32. The number of layers is selected in $\{1, 2, 3\}$, and the number of attention heads is tuned in $\{1, 2, 4\}$. As for the task-level loss, λ_0 is tuned in $\{0.2, 0.4, 0.6, 0.8, 1.0\}$, and λ_1 is set to $1e^{-4}$. The batch sizes of auxiliary and target triples are set to 512 and 2048. We tune hyperparameters by grid search and use AUC, Average Precision (AP), and Accuracy as the evaluation metrics.

Table 2. The convergence performance on both DBLP and Tmall datasets. '-' means the metrics are not reported in the original implementation. Best results in **bold**.

Models	DBLP									Tmall								
	30%			20%			10%			30%			20%			10%		
	AUC	AP	Acc	AUC	AP	Acc	AUC	AP	Acc	AUC	AP	Acc	AUC	AP	Acc	AUC	AP	Acc
GCN	71.9	71.9	67.7	71.2	72.9	67.2	69.5	70.9	66.0	72.9	77.5	70.8	71.5	76.6	69.4	70.4	74.9	68.7
GAT	75.0	75.3	68.8	74.6	76.4	68.5	72.5	75.1	67.0	76.6	80.8	70.8	75.6	79.7	70.1	74.8	78.9	69.3
GraphSAGE	76.2	78.8	70.4	74.1	77.0	68.8	72.1	75.6	67.6	76.3	81.1	71.3	75.5	80.4	70.6	74.6	79.5	69.4
R-GCN	70.2	72.4	66.7	69.4	70.4	65.8	67.5	69.2	64.2	60.0	64.8	60.4	57.6	62.5	58.6	56.8	61.3	57.7
CompGCN	77.5	79.8	67.4	76.9	79.4	66.1	74.6	76.9	64.5	76.4	80.9	68.9	74.7	79.2	68.0	74.5	79.0	67.0
HGB	76.7	77.7	69.5	74.9	76.3	67.9	73.6	74.4	66.9	74.8	80.1	69.6	72.4	78.4	66.2	72.0	76.4	65.9
MAML	71.8	73.8	64.2	68.7	71.8	61.6	65.7	69.1	59.6	71.9	75.8	67.0	67.5	70.1	62.5	63.5	66.8	58.6
Meta-Graph	77.2	79.1	-	76.2	78.6	-	73.9	75.7	-	75.8	80.1	-	73.4	78.9	-	73.0	77.0	-
Ours	**81.3**	**84.7**	**72.6**	**78.6**	**82.1**	**70.9**	**75.3**	**79.3**	**69.1**	**80.6**	**84.8**	**74.7**	**79.2**	**83.7**	**73.9**	**78.2**	**82.6**	**72.1**

---

[1] https://dblp.org/.

[2] https://www.tmall.com/.

[3] https://github.com/xizhu1022/FSLP-EBSNs.

Baselines. We compare our method against three categories of baselines: **(1) Single-relation models** operated on homogeneous networks, such as GCN, GAT, and GraphSAGE. **(2) Multi-relation models** that incorporate relation learning, including R-GCN [6], CompGCN [8] and HGB [5]. **(3) Meta-learning models.** Since little literature is applicable to link prediction, we adopt MAML [2] and the closest multi-network FSL work Meta-Graph [1] as baselines. For Meta-Graph, we use GraphSAGE pre-training embeddings as node attributes.

5.2   Experiments Results

Overall Convergence Performance. As shown in Table 2, we evaluate the proposed model against various baselines for final convergence. Here are three findings. First, our model outperforms other models on both datasets with steady improvement in all few-shot settings. For example, as for DBLP in the 30% setting, the absolute gains reach 3.81%/4.97%/2.16% for AUC/AP/Acc, which illustrates the effectiveness of our work. Second, Meta-Graph is slightly inferior to our model probably due to its individual nature and constant attributes. Third, homogeneous models show competitive results, in some cases, even outperform multi-relation models. We argue relations with many samples mistakenly dominate the training process and impair the relation learning with insufficient data.

Table 3. The AUCs of different variants with 20-step finetuning. RD-B indicates the relative decrease w.r.t. the convergence result of its backbone model (denoted in the bracket). RD-O is the relative decrease w.r.t. the best convergence result of our model.

Variants	10%			20%			30%		
	Full	RD-B	RD-O	Full	RD-B	RD-O	Full	RD-B	RD-O
Direct Training (CompGCN)	50.34	-32.52%	-33.12%	50.96	-33.68%	-35.13%	51.11	-34.03%	-37.12%
Train&Finetune (CompGCN)	73.07	-2.05%	-2.92%	73.03	-4.98%	-7.06%	74.73	-3.54%	-8.06%
MAML (GAT)	65.46	-0.39%	-13.3%	67.93	-1.09%	-13.54%	69.17	-3.62%	-14.90%
Ours (EA-GAT)	74.65	-0.83%	-0.83%	76.89	-2.14%	-2.14%	78.87	-2.97%	-2.97%

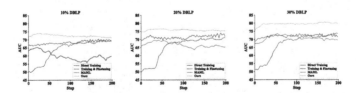

Fig. 3. The AUC curves for convergence of DBLP dataset (200-step finetuning).

Fast Adaption and Convergence Curves. We design the following variants to show the efficiency of rapid learning: **(1) Direct Training** directly trains CompGCN [8], which is among the best traditional multi-relational models (see Table 2). **(2) Training&Finetuning** first extensively trains CompGCN [8] based on existing events, then finetunes with additional few-shot instances. **(3) MAML** is a pure meta-learning method with pure GAT as the meta-model.

Fast Adaption Performance. As shown in Table 3, MAML and our model perform very closely to their convergence results with limited training (see RD-B), which illustrates the remarkable rapid adaptability of meta-learning methods. Notably, we outperform Training&Finetuning which follows the traditional solutions to handle new events. It indicates some common characteristics of few-shot events have been captured to enhance newly-deployed events.

Convergence Curves. As shown in Fig. 3, our model not only outperforms other variants but also achieves fast and stable convergence. Besides, in comparison, our model has a better starting point with more expressive embeddings based on meta-training tasks. However, meta-learning models tend to overfit, especially when the support set is small, which makes early stopping important.

Ablation Study. We conduct experiments on three ablations: **(1) w/o network augmentation** that directly inputs the sparse EBSN to EA-GAT while auxiliary triples are only for optimizing. **(2) w/o auxiliary learning** that removes auxiliary learning in the task-level loss. **(3) MAML** removes both components. According to Fig. 4, all ablations show relatively poor results compared to the full model. Surprisingly, a significant drop is observed without auxiliary learning, which identifies that the reconstruction of source EBSNs boosts user preference learning. Besides, another decrease is found without network augmentation. MAML shows the worst results without them, showing their synergistic effects. Actually, both components are inspired by the interconnection nature of EBSNs. Finally, as support edges decrease, a larger gap is found between the full model and ablations, showing its superiority in few-shot scera.

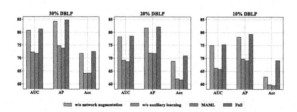

Fig. 4. The ablation results for DBLP dataset.

6   Conclusion

In this work, we studied FSL on new EBSNs. First, we defined a task distribution and considered a unified target event as the few-shot event. Then, for each task, an event-aware link prediction model was proposed with a joint objective. Overall, we followed MAML to achieve knowledge transfer. Finally, the experiments have illustrated the superiority of fast adaption and overall results.

Acknowledgements. This work was supported by the grants from National Natural Science Foundation of China (No. U20A20229, 62072423), the USTC Research Funds of the Double First-Class Initiative (No. YD2150002009), and the Alibaba Group through Alibaba Innovative Research (AIR) Program.

References

1. Bose, A.J., Jain, A., Molino, P., Hamilton, W.L.: Meta-graph: Few shot link prediction via meta learning. CoRR abs/1912.09867 (2019)
2. Finn, C., Abbeel, P., Levine, S.: Model-agnostic meta-learning for fast adaptation of deep networks. In: ICML, vol. 70, pp. 1126–1135 (2017)
3. Hu, Z., Dong, Y., Wang, K., Sun, Y.: Heterogeneous graph transformer. In: WWW, pp. 2704–2710 (2020)
4. Huang, K., Zitnik, M.: Graph meta learning via local subgraphs. In: NeurIPS (2020)
5. Lv, Q., et al.: Are we really making much progress? revisiting, benchmarking and refining heterogeneous graph neural networks. In: KDD. pp. 1150–1160 (2021)
6. Schlichtkrull, M., Kipf, T.N., Bloem, P., van den Berg, R., Titov, I., Welling, M.: Modeling relational data with graph convolutional networks. In: Gangemi, A., Navigli, R., Vidal, M.-E., Hitzler, P., Troncy, R., Hollink, L., Tordai, A., Alam, M. (eds.) ESWC 2018. LNCS, vol. 10843, pp. 593–607. Springer, Cham (2018). https://doi.org/10.1007/978-3-319-93417-4_38
7. Snell, J., Swersky, K., Zemel, R.S.: Prototypical networks for few-shot learning. In: NeurIPS, pp. 4077–4087 (2017)
8. Vashishth, S., Sanyal, S., Nitin, V., Talukdar, P.P.: Composition-based multi-relational graph convolutional networks. In: ICLR (2020)
9. Yang, B., Yih, W., He, X., Gao, J., Deng, L.: Embedding entities and relations for learning and inference in knowledge bases. In: ICLR (2015)
10. Zhou, F., Cao, C., Zhang, K., Trajcevski, G., Zhong, T., Geng, J.: Meta-gnn: on few-shot node classification in graph meta-learning. In: CIKM, pp. 2357–2360 (2019)

Mining Discriminative Sub-network Pairs in Multi-frequency Brain Functional Networks

Jinyi Chen[1], Junchang Xin[1,3]($\boxtimes$), Zhongyang Wang[1,3], Xinlei Wang[1], Sihan Dong[2], and Zhiqiong Wang[2,3]

[1] School of Computer Science and Engineering, Northeastern University, Shenyang 110819, China
{chenjinyi,wangxinlei}@stumail.neu.edu.cn, xinjunchang@mail.neu.edu.cn, wangzhongyang@cse.neu.edu.cn
[2] College of Medicine and Biological Information Engineering, Northeastern University, Shenyang 110819, China
dongsihan@stumail.neu.edu.cn, wangzq@bmie.neu.edu.cn
[3] Key Laboratory of Big Data Management and Analytics (Liaoning Province), Northeastern University, Shenyang 110819, China

Abstract. How to use frequent and discriminative pattern for identifying brain disease is a hot topic in the area of brain functional network topology analysis. Most of the existing researches mine discriminative sub-network from frequent patterns, thus ignoring the underlying comparison relationship of the discriminative patterns within different groups. To solve this problem, we propose a discriminative sub-network pair (DSP) to represent both the intra-group commonality and inter-group specificity of networks. The DSP consists of a paired frequent sub-network mined from the brain networks of different groups within the same or similar node-set and different edge-set. Specifically, the signals are decomposed into multiple frequency bands, then the multi-frequency network is constructed to model the brain activities. We construct the DSP with the most significant distinguishing ability from the frequent patterns that frequently appear in each group. A feature vector is constructed for each subject based on these pairs by drawing on the network motif idea and the classifier is used to detect Alzheimer's disease (AD). Comprehensive experiments on ADNI public datasets demonstrate the effectiveness of DSP in the tasks of AD classification, with an accuracy of 83.33%.

Keywords: Discriminative sub-network pair · Multi-frequency brain functional network · Subgraph mining · Auxiliary diagnosis

1 Introduction

As an irreversible neurodegenerative disease, Alzheimer's disease (AD) is accompanied by clinical symptoms such as cognitive and memory impairment [18]. It can be determined whether the subjects have AD by computer analysis of medical data rather than actual cognitive phenomena. With the development of

© The Author(s), under exclusive license to Springer Nature Switzerland AG 2023
X. Wang et al. (Eds.): DASFAA 2023, LNCS 13945, pp. 42–57, 2023.
https://doi.org/10.1007/978-3-031-30675-4_4

neuroimaging technology, functional Magnetic Resonance Imaging (fMRI) has provided an ideal tool for studying the activity patterns of the whole brain. It is the neuroimaging basis for the auxiliary detection of AD.

Brain activity is frequency-specific, different frequencies can not be considered simply as full-spectrum. In other words, physiological activities at different frequency bands may produce frequency-specific signals and contribute differently to functional connectivity [19, 26]. Researchers found that the correlation of cortical network was concentrated at lower frequency range (0.01–0.06 Hz), while the connections of the edge network were distributed at a wider range (0.01–0.14 Hz) [25]. Ignoring the interaction between different bands may result in the loss of some important information. An effective method to solve this problem is to construct a multi-frequency network.

Once acquired, the network needs to be integrated into a suitable model, researchers try to use graph model to extract clinically relevant information. The traditional methods are analyzing the differences between AD and normal subjects' brain network attributes. They extracted a series of features from the network, such as node degree, centrality [4] and so on. Then they classified the feature vectors with machine learning methods [21]. However, these methods lose some detailed information in the network, such as the network's topology structure and the common topology structure among the networks. They will failed to detect changes in the brain regions and connections that lead to disease. As a typical graph data, the brain network can be analyzed from the perspective of data mining [20]. Subgraph structures (There is no distinction between subnetwork and subgraph in this paper) can represent the common connections, that is, they can represent the "building block" used to convey information in the networks [10].

In this paper, we propose and test the hypothesis that neurodegenerative psychiatric disorders may result from the disruption of certain subgraph in the network in some way. At the same time, we believe that the subgraphs have population stability, that is, the subgraphs are shared within the group, but the two subgraphs obtained from different groups are not completely independent. There is a contrast between their relations. For a certain set of brain regions, the brain networks of two groups will show completely different topological connections, and these different connection combinations can be used as biomarkers to reveal diseases. Therefore, we consider a kind of special subgraph pairs as dictionary elements and construct a feature vector for each network. We propose the discriminative sub-network pair (DSP) to define the comparison relationships within different groups. That is, DSPs are considered as dictionary elements for the brain-related disease identification. The DSP consists of a paired frequent subgraph mined from the networks of different groups. Therefore, the multi-frequency brain functional network is constructed, which fully considers the topological properties of the network at different bands. In addition, a learning framework is developed for the diagnosis of brain-related diseases with DSPs. The contributions of this paper are summarized as below:

- The multi-frequency brain functional network model is constructed by using the frequency-specific functional connectivity, which can reflect deeper inter-action information between brain regions.
- The DSP is elegantly defined to describe the comparison relationships between different groups, which considers both the intra-group commonal-ity and inter-group specificity of networks.
- Extensive experiments are conducted on ADNI public datasets. The results demonstrate the effectiveness of DSP in the task of AD classification.

The rest of this paper is arranged as follows. In Sect. 2, we construct the multi-frequency brain network and present our DSP. In Sect. 3, the proposed method is experimentally verified and discussed. Then, we introduce the related work in Sect. 4. Finally, the paper is summarized in Sect. 5.

2   Methodology

In this work, a new framework for the diagnosis of AD with DSP is developed by multi-frequency functional network. As shown in Fig. 1, the framework can be divided into four main states. First, the resting-state fMRI (rs-fMRI) data is pre-processed, which mainly includes slice timing, head motion correction, normalization, etc. Next, regional mean time series of the signals in each region-of-interest (ROI) are decomposed into multiple frequency bands, and the multi-frequency brain functional networks are constructed by the way of correlation analysis and thresholding. Then, frequent subgraphs are mined from the multi-frequency brain networks at each frequency, and the DSPs are selected within two groups based on the similarity of the edge set, the node set, and the support set. Finally, a feature vector for each subject based on these pairs is constructed by drawing on the network motif idea and the classifier is used to detect AD.

2.1   Multi-frequency Brain Functional Network Construction

The complex system of the brain can be represent by a brain functional network. Nodes usually represent certain regions of the brain, while edges usually represent the functional connectivity among these regions. We decompose regional mean time series of the signal in each ROI into multiple frequency bands and threshold the correlation matrices to analyze the topological properties of the resulting undirected networks.

Suppose that $\mathbf{X} \in \mathbb{R}^{R \times P}$ denotes the matrix of regional mean time series with P time points extracted from a total of R ROIs. The connection strength of any two regions can be measured by the correlation between their time series. For the time series of the i-th ROI $\mathbf{X}_i$ and the j-th ROI $\mathbf{X}_j$, the connection strength $C_{i,j}$ between them can be derived by computing the Pearson's correlation coefficient, as shown as follows:

$$C_{i,j} = \frac{\sum_{m,n=1}^{P} \left(\chi_i^m - \bar{X}_i \right) \left(\chi_j^n - \bar{X}_j \right)}{\sqrt{\sum_{m=1}^{P} \left(\chi_i^m - \bar{X}_i \right)^2} \sqrt{\sum_{n=1}^{P} \left(\chi_j^n - \bar{X}_j \right)^2}} \tag{1}$$

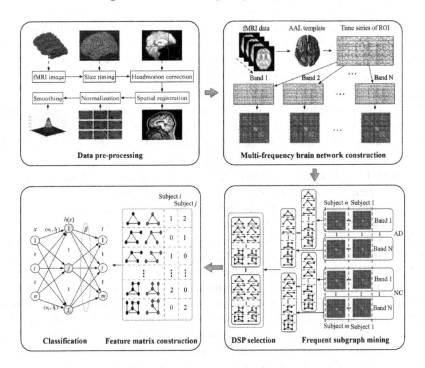

Fig. 1. An overview of the framework

where, $\chi_i^m (m = 1, 2, \ldots, P)$ and $\chi_j^n (n = 1, 2, \ldots, P)$ represent the m-th and n-th time point of the i-th and j-th ROI's time series, respectively. $\bar{X}_i$ and $\bar{X}_j$ are the mean value of all the time points in the corresponding ROI, respectively. P is the number of time points included in the time series.

Assuming that $\mathbf{X}_i$ and $\mathbf{X}_j$ are de-meaned and variance-normalized before calculating $C_{i,j}$, the matrix $\mathbf{C} = [C_{i,j}] \in \mathbb{R}^{R \times R}$ can be further expressed as $\mathbf{C} = \mathbf{X}\mathbf{X}^T$. The network constructed by the above method is a full-spectrum network, which may be incapable to capture subtle differences in disease. Further, multi-frequency brain networks are constructed. Suppose that $g_{k,w}$ and $h_{k,w} (w = 0, \ldots, W_k - 1)$ are the k-th level high-pass filter and low-pass filter, respectively, here, W denotes the width of the initial filter [12]. The corresponding k-th level high-pass filter and low-pass filter are defined by $\widetilde{g}_{k,w} = 2^{-k/2} g_{k,w}$ and $\widetilde{h}_{k,w} = 2^{-k/2} h_{k,w}$, which have the same width W_k. Then, for $\mathbf{X}_i$, its k-th sub-band $\mathbf{X}_i^k$ can be derived as follows:

$$\mathbf{X}_i^k = X_i^{k,H}[P] = \sum_{w=0}^{W_k - 1} \widetilde{g}_{k,w} X_i^{k-1,L}[2P - w] \qquad (2)$$

where, P is length of $\mathbf{X}_i$, and,

$$X_i^{k,L}[P] = \sum_{w=0}^{W_k - 1} \widetilde{h}_{k,w} X_i^{k-1,L}[2P - w] \qquad (3)$$

Within the k-th band, the time series $\mathbf{X}^k \in \mathbb{R}^{R \times P}$ can be obtained, and the corresponding network can be constructed as $\mathbf{C}^k = \mathbf{X}^k(\mathbf{X}^k)^T$. Further, the totally K frequency-specific functional network $\widetilde{\mathbf{C}} = \{\mathbf{C}^1, \mathbf{C}^2, \ldots, \mathbf{C}^K\}$ can be obtained. Obviously, each matrix $\mathbf{C}^k$ in $\widetilde{\mathbf{C}}$ is symmetric, if there is a connection between ROI i and ROI j, a connection between j and i must be there. Generally, the edges with lower connection strength are considered as noisy edges. It is necessary to select a suitable threshold to retain the effective edges with stronger connections and delete the noisy edges with weaker connections. For each $\mathbf{C}^k$, the range of connection strength between ROI i and j $C_{i,j}^k$ is (-1, 1). We suppose that T denotes the threshold, if $|C_{i,j}^k|$ is greater than T, the connection between ROI i and j is considered to be valid, and the value is set to 1. Otherwise, the connection is invalid and the value is 0. The frequency-specific adjacent matrices $\mathbf{A} = \{\mathbf{A}^1, \mathbf{A}^2, \ldots, \mathbf{A}^K\} \in \mathbb{R}^{R \times R \times K}$ are derived by thresholding respective functional connectivity matrices. Further, the multi-frequency brain network $\mathcal{N} = \{\mathcal{N}^1, \mathcal{N}^2, \ldots, \mathcal{N}^K\}$ can be obtained. Pseudocode for constructing multi-frequency brain functional network is provided in Algorithm 1.

According to Eq. 2, each $\mathbf{X}_i$ can be decomposed into K bands layer by layer (Lines 1-2). Within k-th sub-band (Line 3), we get the functional connectivity matrix $\mathbf{C}_k$ by computing the functional connectivity for each possible pair of ROIs (Lines 4-6). The corresponding frequency-specific adjacency matrix $\mathbf{A}^k$ is derived by comparing the elements in the functional connectivity matrix with the preset threshold (Lines 7-11). Further, we construct the brain function network of the corresponding band $\mathcal{N}^k$ (Line 12).

Algorithm 1: Multi-frequency brain functional network construction

---

Input: $\mathbf{X}$: regional mean time series, T: threshold
Output: $\mathcal{N}$: multi-frequency brain functional network

1 **for** *each time series $\mathbf{X}_i$ in $\mathbf{X}$* **do**
2 $\quad \lfloor \; [\mathbf{X}_i^1, \mathbf{X}_i^2, \ldots, \mathbf{X}_i^K] = $ *frequency demultiplication*$(\mathbf{X}_i)$;
3 **for** $k = 1$ *to* K **do**
4 $\quad \mathbf{X}^k = [\mathbf{X}_1^k, \mathbf{X}_2^k, \ldots, \mathbf{X}_R^k]$;
5 $\quad \mathbf{X}^k \leftarrow$ *de-mean and variance-normalize* $(\mathbf{X}^k)$;
6 $\quad \mathbf{C}^k = \mathbf{X}^k(\mathbf{X}^k)^T$;
7 $\quad$ Initialize the adjacency matrix $\mathbf{A}^k$;
8 $\quad$ **for** $i, j = 1$ *to* R **do**
9 $\quad\quad$ **if** $|C_{i,j}^k| \geq T$ **then**
10 $\quad\quad\quad \lfloor \; A_{i,j}^k = 1$;
11 $\quad\quad A_{i,j}^k = 0$;
12 $\quad \mathcal{N}^k = $ *network construction* $(\mathbf{A}^k)$;
13 **return** $\mathcal{N} = \{\mathcal{N}^1, \mathcal{N}^2, \ldots, \mathcal{N}^K\}$;

2.2 Frequent Subgraph Mining

The goal of frequent subgraph mining is finding the subgraph that frequently appear in brain networks. Supposed that $\mathcal{N} = (V, E, L_V, L_E)$ denotes a brain network, where $V(\mathcal{N})$ (resp. $E(\mathcal{N})$) represents the set of nodes (resp.edges) in $\mathcal{N}$, $L_V(\mathcal{N})$(resp. $L_E(\mathcal{N})$) represents the label set of nodes(resp.edges). Given $\mathcal{N}_1 = (V_1, E_1, L_{V_1}, L_{E_1})$ and $\mathcal{N}_2 = (V_2, E_2, L_{V_2}, L_{E_2})$, if $\mathcal{N}_2$ is the subgraph of $\mathcal{N}_1$, it must satisfy that, $V_2 \subseteq V_1$, $E_2 \subseteq E_1$, $L_{V_2} \subseteq L_{V_1}$, $L_{E_2} \subseteq L_{E_1}$, denoted as $\mathcal{N}_2 \in \mathcal{N}_1$. The frequent subgraph is defined by the following definitions.

Definition 1. *frequent degree.* *Given a set of brain functional networks $H = \{\mathcal{N}_1, \mathcal{N}_2, \ldots, \mathcal{N}_m\}$ (m is the number of networks in H), and a subgraph $H_s \in H$, its frequent degree can be described as follows:*

$$FD(H_s|H) = \frac{1}{m} \sum_{i=1}^{m} \delta(i) \qquad (4)$$

where, $\delta(i)$ is a indicative function, $\delta(i) = \begin{cases} 1 & H_s \subseteq \mathcal{N}_i \\ 0 & otherwise \end{cases}$, subgraph H_s appears at most once in each brain network, that is, $FD(H_s|H)$ can also be expressed as the ratio of the number of brain networks in H that contain H_s to the total number m. Obviously, $FD(H_s|H) \leq 1$.

Definition 2. *frequent subgraph.* *Given a set of brain functional networks $H = \{\mathcal{N}_1, \mathcal{N}_2, \ldots, \mathcal{N}_m\}$, and a preset threshold ζ, the subgraph H_s is a frequent subgraph of H, if and only if $FD((H_s|H)) \geq \zeta$.*

Based on the above definition, the problem of mining frequent subgraphs from brain network set can be expressed as finding all frequent subgraphs that satisfy threshold. The idea is to generate candidate subgraph and then determine whether it is frequent. Generating effective candidate subgraphs mainly requires avoiding generating duplicate or redundant subgraphs [16]. To efficiently mine frequent subgraph, we use the method based on depth-first search (DFS) to traverse, code and encode the networks [28]. The method assign a unique minimum DFS code to each brain network based on lexicographical order and establish a DFS tree to enumerate all candidate frequent subgraphs. To determine whether the candidate H_s is frequent, the method counts the frequent degree $FD(H_s, H)$ of the candidate in H and compare it with the preset threshold ζ. The one that meets $FD \geq \zeta$ is a frequent subgraph in the result set. We mine frequent subgraphs in each frequency to obtain the specific-frequency frequent subgraphs with these mining strategies.

2.3 Discriminative Subgraph Pair (DSP) Mining

After mined frequent subgraphs using the aforementioned method, thousands of frequent subgraphs may be obtained. If all these frequent subgraphs are used as markers to detect AD, it is not feasible and meaningless. We design the DSP to describe both the intra-group commonality and inter-group specificity of networks within different groups. We separate the brain network set into two group according to their labels (i.e. AD group as positive group and normal control (NC) as negative group).

Given a brain network set, $H = \{H^+, H^-\} = \{\mathcal{N}_1, \mathcal{N}_2, \cdots, \mathcal{N}_n\}$, where H^+ (resp. H^-) denotes the positive (resp. negative) group. The subgraph set mined from the positive (resp. negative) group is denoted as g^+ (resp. g^-). For a subgraph g, if its frequent degree in the positive (resp. negative) set is much greater than that in the negative (resp. positive) set, it is a discriminative subgraph. The discriminative ability of g^+ is given as shown in Eq. (5). When $d(g^+) < 0$, g^+ has no discriminative ability. The discriminative ability of g^- is similar to g^+.

$$DA(g^+|H) = \log \frac{FD(g^+|H)}{FD(g^-|H)} \tag{5}$$

To make the subgraph more discriminative, the DSP $(g_i, g_j) = \{(g_i, g_j) \mid g_i \in g^+, g_j \in g^-\}$ is constructed. The DSP consists of a paired frequent subgraph mined from different groups within the same or similar node-set and different edge-set. Following, we introduce the structure similarity and support network set similarity. Given a pair (g_i, g_j), its structure similarity $simN(g_i, g_j)$ can be expressed as follows:

$$simN(g_i, g_j) = \frac{\mid E(g_i) \cap E(g_j) \mid}{\mid E(g_i) \mid + \mid E(g_j) \mid - \mid E(g_i) \cap E(g_j) \mid} \tag{6}$$

where $E(g_i)$ and $E(g_j)$ respectively represent the edge set of g_i and g_j.

Definition 3. support network set. *Given a subgraph g, its support network set consists of networks that contain g, denoted as $cov(g, H)$.*

For a pair (g_i, g_j), its support set similarity $simS(g_i, g_j)$ can be expressed as follows:

$$simS(g_i, g_j) = \frac{\mid cov(g_i, H) \cap cov(g_j, H) \mid}{\mid cov(g_i, H) \cup cov(g_j, H) \mid} \tag{7}$$

Based on the above expression of the structural similarity and support set similarity, we introduce the variable μ to define the correlation $corr(g_i, g_j)$ between the subgraph g_i and g_j in the pair, shown in Eq. (8).

$$corr(g_i, g_j) = \mu simN(g_i, g_j) + (1 - \mu)simS(g_i, g_j) \tag{8}$$

where, μ is a weight factor, it can measure the impact of structural similarity and support network set similarity on correlation. Further, we give the definition of the difference between the subgraph g_i and g_j, shown in Eq. (9):

$$diff(g_i, g_j) = 1 - corr(g_i, g_j) \tag{9}$$

Obviously, the larger the value of $diff(g_i, g_j)$, the better the performance of DSP (g_i, g_j), and the more beneficial for subsequent auxiliary diagnosis.

Since the network constructed in this paper uses a fixed brain template, the nodes in the network are unique, and the nature of the network also determines that it must be a connected network. The node sets of subgraph g_i and g_j in pair (g_i, g_j) should be the same or partially correlated, and what we need to find is the edge set with a high difference. The similarity of the node set can be expressed as follows:

$$simNode\left(g_i, g_j\right) = \frac{\mid V\left(g_i\right) \cap V\left(g_j\right) \mid}{\mid V\left(g_i\right) \mid + \mid V\left(g_j\right) \mid - \mid V\left(g_i\right) \cap V\left(g_j\right) \mid} \qquad (10)$$

where $V(g_i)$ and $V(g_j)$ represent the node set of the subgraph g_i and g_j, respectively. The larger the value of $simNode(g_i, g_j)$ is, the higher the similarity of node sets between g_i and g_j.

2.4   Feature Selection and Classification

After obtaining the DSP at different frequency bands, the feature vectors are constructed to represent the comparison relationship of each brain network. But, the feature dimensionality will be rather high, if we directly use the connectivity detail as features from these pairs. An effective alternative method is considering the DSPs as dictionary elements, it is easy to construct an indicative vector based on their occurrence in the brain network to represent the whole network. Based on such vector representation, we further use the extreme learning machine (ELM) [9] for classification.

Given a brain network $\mathcal{N}$ and m DSPs $\{(g_i^k, g_j^k) \mid k = 1, 2, \cdots, m\}$, its feature vector can be describe as a m-dimension vector V_F, with the elements V_F^k denoting the k-th feature based on (g_i^k, g_j^k). The following rule is given to determine the value of V_F^k.

- If $g_i^k \subseteq \mathcal{N}$ and $g_j^k \nsubseteq \mathcal{N}$, then $V_F^k = 1$;
- If $g_i^k \nsubseteq \mathcal{N}$ and $g_j^k \subseteq \mathcal{N}$, then $V_F^k = 2$;
- If $g_i^k \subseteq \mathcal{N}$ and $g_j^k \subseteq \mathcal{N}$, then $V_F^k = 0$;
- If $g_i^k \nsubseteq \mathcal{N}$ and $g_j^k \nsubseteq \mathcal{N}$, then $V_F^k = 0$;

Neuronal oscillations at different frequency bands have different biological and physiological significance, and may contribute differently to functional connectivity, thus affecting network construction. In other words, frequency-specific may lead to different effects of DSPs constructed at different frequency bands on disease classification. Moreover, not all DSPs can show good differences. It is difficult to ensure the efficiency of learning by taking all DSPs as features and as the input of the classifier, which may lead to over-fitting and affect the accuracy of classification. Subsequent experiments also validate our analysis. In this paper, recursive feature elimination (RFE) [14] strategy is used to find optimal feature subset(i.e., DSPs) and reduce the risk of over-fitting. Specifically, the REF is used to obtain the ranking of each feature. Then, based on ranking, some features are selected in turn to form feature subsets for model training and cross-validation, and the feature subset with the highest average score is selected as the final set.

3 Result and Discussion

3.1 Experimental Settings

Data Acquisition. The dataset is selected from the public dataset provided by the Alzheimer's Disease Neuroimaging Initiative (ADNI, http://adni.loni.usc.edu). A total of 256 rs-fMRI data are selected, including two groups of subjects, of which 134 healthy subjects (NC group, 74 male, 60 female, age range: 62–85 years) and 122 AD subjects (AD group, 69 male, 53 female, age range: 58–89 years). To avoid the influence of multiple scanning data of the same subject, only single scanning data of the same subject is selected. Other scanning parameters are as follows: flip angle F_A =80°, time repetition T_R =3.0 s, echo time T_E =30 ms, time points is 140, each time point contains 48 axial time slices, covering the whole brain, and layer thickness is 3.3 mm.

Data Pre-processing. The data are preprocessed by DPABI toolkit (version 4.0) [27] according to the well-accepted pipeline. Specifically, the first 10 unstable time points are discarded. All the time slice are corrected to the $24th$ time slice, which are chosen as the reference slice, to eliminate the time phase difference. By performing spatial head movement correction, we exclude 14 subjects in which the maximum translational distance in three directions exceeded 0.8 mm and the maximum angle of rotation exceeded 0.02°. Then the image of subjects are standardized to the MNI-152 standard space by stretching, compressing, and winding. Finally, the filtering and smoothing process is performed, the selected voxel is 3*3*3 mm^3, and the signal frequency range is 0–0.15 Hz. According to the AAL atlas [22], the rs-fMRI data are divided into 116 ROIs. It is worth nothing that we only use 90 brain regions as ROIs.

Evaluation Metrics. The evaluation of the method is performed on a binary classification problem (AD versus NC). In the experiments, by the limited number of samples, the 10-fold cross-validation strategy is repeated 10 times to evaluate the performance of the methods. Specifically, we randomly divide all subjects into 10 sub-sets with equal size, and select any one of them as testing data, and the rest as training data. We measure the performance of the methods by averaging the results in cross-validation. Also, we evaluate the effectiveness of the brain regions selected by our methods. The classification performance of the methods are evaluated on classification accuracy (ACC), area under ROC curve (AUC), sensitivity (SEN), and specificity (SPE).

Parameter Setting. To verify the influence of the number of frequency bands, the time series are decomposed into 3, 4, 5, 6, and 7 bands. The weight factor μ for measuring the impact of structural similarity and support set similarity on correlation is chosen from {0.4, 0.5, 0.6, 0.7, 0.8}. The threshold ζ for mining frequent subgraph is chosen from {0.6, 0.7, 0.8, 0.9}. The dimension of the feature vector k (i.e., the number of selected DSP) is chosen from {25, 50, 100, 250, 500}.

The network has the best performance, when the connectivity density is [20%, 75%] [29]. Thus, we choose the threshold T for constructing brain network from {0.4, 0.5, 0.6}. It is worth mentioning that these thresholds can ensure all the constructed network connectivity.

3.2   Result Analysis

To verify the effectiveness of our method, we conduct extensive experiments from both multi-frequency and full-frequency with 6 classification methods. (1) network parameters based on full-frequency network (NP-F) [4]: 8 variables including eigenvector centrality are used to exact the features and the ELM classifier is constructed for classification; (2) motifs based on full-frequency network (MO-F): 20 motifs are selected [2], and a 20-dimensional vector for each network is generalized by counting their occurrence on the full-frequency brain network, which are fed into the ELM classifier for classification; (3) discriminative subgraph pair based on full-frequency network (DSP-F): our method without frequency decomposition; (4) network parameters based on multi-frequency network (NP-M): it is similar to (1) except that the network is multi-frequency; (5) motifs based on multi-frequency network (MO-M): it is similar to (2) except that the network is multi-frequency; (6) frequent sub-network based on multi-frequency network(FSN-M): our method without the construction of DSP.

Table 1 shows the classification results using above-mentioned methods, and their respective ROC curves are shown in Fig. 2. It can be seen from Table 1 that our method achieves the best performance among all methods. Specifically, our method achieves 83.33% accuracy, 82.67% sensitivity and 89.53% specificity, all of them are better than other methods. Compared with method NP-F, the three indicators are improved by 17.91%, 21.16% and 20.43%, respectively. Besides, the accuracy of both NP-F and MO-F are less than 70%, they are not good results. The reason may be that these motifs are some existing relatively small subgraphs, such as triangle structure with three nodes and square structure with four nodes, and they may not well characterize the specific substructures in different types of brain networks. The traditional analytical methods based on feature fusion(i.e., network attributes or node attributes) are also not well-capable of capturing relevant structural changes causing the brain-related diseases, our method mining specific subgraphs from real dataset overcomes these weaknesses, as demonstrated by the DSP-F. After adding the frequency-division method, compared with the corresponding full frequency methods(NP-F and MO-F), the performance of NP-M and MO-M has improved to a certain extent, and the accuracy is improved by 5.75% and 4.44%, reaching 71.17% and 73.21%, respectively, as demonstrated in [6,17]. Comparing the FSN-M with our method, it can also be seen that constructing DSPs can capture differential regions and their connections between different groups(AD and NC), which is also helpful to identify brain diseases.

We also select the top 10 DSPs with the highest discriminative ability among 5 frequency bands, and show them in the connection map. As shown in Fig. 3, brain regions are marked by blue dots, and the orange line represents the connection between two brain regions. For each pair of subgraph in Fig. 3, the upper

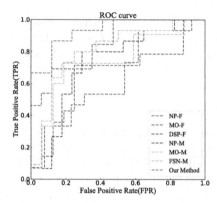

Fig. 2. ROC curves of different methods in the task of AD classification

Table 1. Classification performance in the task of AD classification

Method	Accuracy (%)	Sensitivity (%)	Specificity (%)
NP-F	65.42	61.51	69.10
MO-F	68.77	63.98	73.91
DSP-F	80.27	77.91	82.57
NP-M	71.17	67.23	73.20
MO-M	73.21	69.82	76.92
FSN-M	79.39	76.26	84.83
Our Method	**83.33**	**82.67**	**89.53**

part represents the discriminative subgraph that is intensively mined from AD group, and the lower part represents that mined from NC group. Comparing these pairs, we can find some interesting results. For the subgraph mined from NC subjects, brain regions such as the posterior cingulate gyrus (brain region number are 35, 36), precuneus (brain region number are 67, 68), angular gyrus (brain region number are 65, 66) and hippocampus (brain region number are 37, 38) are mined, and the connection strength between them is significantly higher than that between other brain regions that have not been mined, and these brain regions are also part of the default network [8]. For the subgraph mined from AD group, the connection strength is significantly reduced, while the connections in the medial superior frontal gyrus (brain region numbers are 23, 24) and insula (brain region numbers are 29, 30) are abnormally increased. In the existing references [8], this phenomenon is described as a compensation mechanism. It is worth noting that these conclusions are highly consistent with the existing conclusions obtained from the biological perspective [11,15]. We also count the brain regions that appear frequently in these DSP, such as hippocampus, medial frontal gyrus, posterior cingulate gyrus, amygdala, precuneus and so on. Researchers have demonstrated that these brain regions are associated with

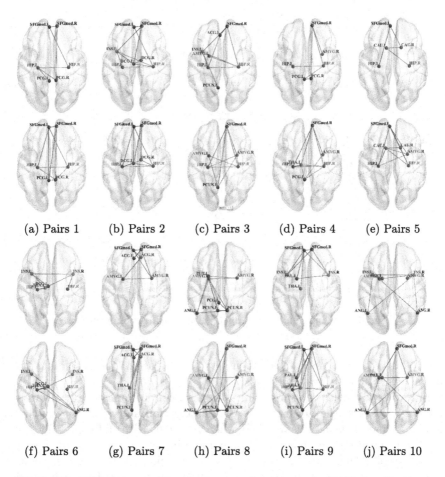

(a) Pairs 1   (b) Pairs 2   (c) Pairs 3   (d) Pairs 4   (e) Pairs 5

(f) Pairs 6   (g) Pairs 7   (h) Pairs 8   (i) Pairs 9   (j) Pairs 10

Fig. 3. The DSP discovered by the framework in the task of AD classification

AD [13]. Comparing the DSP mined with different thresholds ζ, when ζ is 0.8, the number of brain regions included in the subgraph is generally more than the other three values. The reason may be that a lower threshold is more consistent with network discovery. In the analysis of DSP at different frequency bands, we also find that the brain network constructed at different bands have large differences in the connection methods of brain regions. This is an important reason to construct multi-frequency brain network.

3.3 Influence of Parameters

In this section, we present the influence of parameters. There are mainly four variable parameters used in the paper. To evaluate the impact of four parameters on classification performance, we calculate the classification accuracy of each parameter value condition. Firstly, we set the dimension of the feature vector

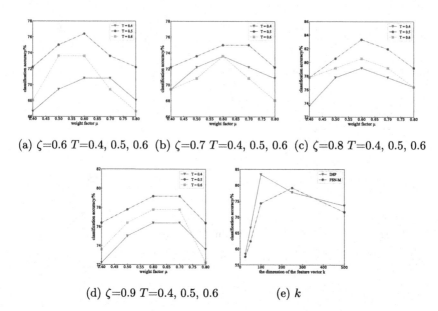

(a) ζ=0.6 T=0.4, 0.5, 0.6 (b) ζ=0.7 T=0.4, 0.5, 0.6 (c) ζ=0.8 T=0.4, 0.5, 0.6

(d) ζ=0.9 T=0.4, 0.5, 0.6   (e) k

Fig. 4. The influence of parameters on classification accuracy

$k = 100$ and keep it unchanged, then test the impact of the remaining three parameters. The results are shown in Fig. 4.

In Fig. 4(a-d), when the threshold ζ and T remain unchanged, and the weight factor μ increases, the accuracy of AD classification shows a trend of firstly increasing and then decreasing. And when $\mu = 0.6$, with different values of ζ and T, the accuracy rate reaches the maximum in most cases. But for the other four values of μ, there is no obvious rule. When keeping ζ and μ unchanged, as T increases, the accuracy of AD classification also shows a trend of firstly increasing and then decreasing. The possible reason is that the high threshold T result in the small connectivity density of the network. And the small-world attribute, which is a natural attribute of the brain functional network, will become less significant. At the same time, a low threshold T will introduce too many weak connection edges (i.e., noisy edges), that is, some edges can't correctly represent the actual connections of the network. Regardless of the values of the parameters ζ and μ, the classification effect achieved by constructing the network with the threshold $T = 0.5$ is the best. In addition, for the parameter ζ, it controls the size of candidate subgraph sets in frequent subgraph mining. The smaller ζ is, the more frequent subgraphs are obtained, that is, more frequent subgraphs will be discovered. However, this is not friendly, because too many frequent subgraphs may interfere with the construction of DSPs, and some of them are not discriminating. To make matters worse, it makes the system run slowly.

Holding T, ζ and μ unchanged and optimal ($T = 0.5$, $\zeta = 0.8$, $\mu = 0.6$), we analyze the influence of the dimension of the feature vector k on the classification accuracy of AD diagnosis, and compare with the method of FSN-M, the results are shown in Fig. 4(e). When k takes 25, 50, 100, the accuracy of AD classification increases rapidly, it increased from 58.33% to 83.33%. When $k = 100$, the accuracy rate reaches the maximum. As k continuously increase, the classification accuracy gradually decreases, but the range of change is relatively gentle. Similar to the changing trend of the accuracy of our method, the method FSN-M also rapidly increases at first and then slowly decreases. When $k = 250$, the accuracy reaches the maximum, which requires more subgraphs than our method. It is worth nothing that the dimension of the feature vector is not as high as possible. The high-dimensional vector may bring some patterns with weak discriminant ability, thereby reducing the performance of classification.

4   Related Work

Due to the complexity of the human brain, the frequency of signals often dominates subtle functional patterns to certain diseases [19, 26]. Multi-frequency brain functional networks provide a unified description of the networks across multiple frequency bands, which can help researchers better understand neurodegenerative diseases [6]. And the network organization such as hubs may differ across different frequency bands, multi-frequency networks are considered in neurophysiological studies. For example, Sasai et al. [17] elucidated that cognitive states distinguished whole-brain functional connections in a frequency-dependent manner and found that diversity of frequency specificity is associated with functional network. Brookes et al. [3] used magnetoencephalography (MEG) recordings to construct multi-frequency brain network. It is shown that there is statistically significant difference between supra-adjacency matrices of the multi-frequency control and AD subjects. Achard et al. [1] applied the discrete wavelet transform method to fMRI signal decomposition and found that the "small world" properties of the networks in different frequency bands were significantly different.

Discriminative subgraphs of functional brain networks can be considered as the "building block" of the entire network, making it easier to capture changes in connection patterns and topological attributes within specific regions of the network [10]. Frequent subgraph-based representation is an efficient way to transfer both global and local information of the network into vector spaces. Analysis of frequently occurring discriminative patterns can reveal some differences in connectivity inter- and intra-groups [24]. Considered the frequent subgraph as a dictionary element and used as an indication vector that can be easily applied to classical machine learning models to analyze brain networks. For example, Guo et al. [7] combined frequent subgraph for feature selection, and used multi-kernel learning methods for depression classification. Van et al. [23] mined frequent subgraph from different cognitive groups, respectively, to reveal differences in connectivity among the groups. However, the discriminative patterns are constructed separately for each group that ignore the underlying comparison relationship of the patterns within different groups. Du et al. [5] used the frequent subgraph and feature selection method to obtain the discriminative subgraph and classification matrix, and used the support vector machine to realize the classification of AD.

5   Conclusions

In this paper, we propose the DSP to describe the comparison relationships within different groups, it can represent both the intra-group commonality and inter-group specificity of networks. We construct a feature matrix with the DSPs and train a classifier for auxiliary diagnosis of AD. To this end, we also construct a multi-frequency brain functional network, which fully considers the topological properties of the network at different frequency bands. The experimental results on ADNI public datasets demonstrate the effectiveness of DSP in the task of AD classification. It can better retain the information of brain activity and be used as specific biological markers to diagnose brain diseases.

Acknowledgements. This work was supported by National Natural Science Foundation of China (62072089); Fundamental Research Funds for the Central Universities of China (N2116016, N2104001, N2019007).

References

1. Achard, S., Salvador, R., Whitcher, B., et al.: A resilient, low-frequency, small-world human brain functional network with highly connected association cortical hubs. J. Neurosci. **26**(1), 63–72 (2006)
2. Battiston, F., Nicosia, V., Chavez, M., et al.: Multi-layer motif analysis of brain networks. Chaos Interdiscip. J. Nonlinear Sci. **27**(4), 047404 (2017). https://doi.org/10.1063/1.4979282
3. Brookes, M.J., Tewarie, P.K., Hunt, B.A., et al.: A multi-layer network approach to MEG connectivity analysis. Neuroimage **132**, 425–438 (2016)
4. De Vos, F., Koini, M., Schouten, T.M., et al.: A comprehensive analysis of resting state fMRI measures to classify individual patients with Alzheimer's disease. Neuroimage **167**, 62–72 (2018). https://doi.org/10.1016/j.neuroimage.2017.11.025
5. Du, J., Wang, L., Jie, B., et al.: Network-based classification of ADHD patients using discriminative sub-network selection and graph kernel PCA. Comput. Med. Imaging Graph. **52**, 82–88 (2016)
6. Gifford, G., Crossley, N., Kempton, M.J., et al.: Resting state fMRI based multilayer network configuration in patients with schizophrenia. NeuroImage Clin. **25**, 102169 (2020). https://doi.org/10.1016/j.nicl.2020.102169
7. Guo, H., Qin, M., Chen, J., et al.: Machine-learning classifier for patients with major depressive disorder: Multi-feature approach based on a high-order minimum spanning tree functional brain network. Comput. Math. Methods Med. **2017**, 4820935 (2017). https://doi.org/10.1155/2017/4820935
8. Hämäläinen, A., Pihlajamäki, M., Tanila, H., et al.: Increased fMRI responses during encoding in mild cognitive impairment. Neurobiol. Aging **28**(12), 1889–1903 (2007). https://doi.org/10.1016/j.neurobiolaging.2006.08.008
9. Huang, G., Zhu, Q., Siew, C.K.: Extreme learning machine: theory and applications. Neurocomputing **70**(1), 489–501 (2006)
10. Kong, X., Yu, P.S.: Brain network analysis: a data mining perspective. ACM SIGKDD Explor. Newsl. **15**(2), 30–38 (2014)
11. Li, W., Antuono, P.G., Xie, C., et al.: Aberrant functional connectivity in papez circuit correlates with memory performance in cognitively intact middle-aged APOE4 carriers. Cortex **57**, 167–176 (2014). https://doi.org/10.1016/j.cortex.2014.04.006

12. Mallat, S.G.: A theory for multi-resolution signal decomposition: the wavelet representation. IEEE Trans. Pattern Anal. Mach. Intell. **11**(7), 674–693 (1989). https://doi.org/10.1109/34.192463
13. Martinez-Murcia, F.J., Ortiz, A., Gorriz, J.M., et al.: Studying the manifold structure of Alzheimer's disease: a deep learning approach using convolutional autoencoders. IEEE J. Biomed. Health Inform. **24**(1), 17–26 (2019)
14. Peng, C., Wu, X., Yuan, W., et al.: MGRFE: Multi-layer recursive feature elimination based on an embedded genetic algorithm for cancer classification. IEEE/ACM Trans. Comput. Biol. Bioinf. **18**(2), 621–632 (2019)
15. Perea, R.D., Rabin, J.S., Fujiyoshi, M.G., et al.: Connectome-derived diffusion characteristics of the fornix in Alzheimer's disease. NeuroImage Clin. **19**, 331–342 (2018). https://doi.org/10.1016/j.nicl.2018.04.029
16. Preti, G., De, G., Riondato, M.: MaNIACS: approximate mining of frequent subgraph patterns through sampling. In: Proceedings of the 27th ACM SIGKDD Conference on Knowledge Discovery & Data Mining, pp. 1348–1358 (2021)
17. Sasai, S., Koike, T., Sugawara, S.K., et al.: Frequency-specific task modulation of human brain functional networks: a fast fMRI study. Neuroimage **224**, 117375 (2021). https://doi.org/10.1016/j.neuroimage.2020.117375
18. Sharma, S., Mandal, P.K.: A comprehensive report on machine learning-based early detection of Alzheimer's disease using multi-modal neuroimaging data. ACM Comput. Surv. **55**(2), 1–44 (2022). https://doi.org/10.1145/3492865
19. Thomas, A.W., Heekeren, H.R., Müller, K.R., et al.: Analyzing neuroimaging data through recurrent deep learning models. Front. Neurosci. **13**, 1321 (2019)
20. Ting, C.M., Samdin, S.B., Tang, M., et al.: Detecting dynamic community structure in functional brain networks across individuals: a multi-layer approach. IEEE Trans. Med. Imaging **40**(2), 468–480 (2020)
21. Tokuda, T., Yamashita, O., Yoshimoto, J.: Multiple clustering for identifying subject clusters and brain sub-networks using functional connectivity matrices without vectorization. Neural Netw. **142**, 269–287 (2021)
22. Tzourio-Mazoyer, N., Landeau, B., Papathanassiou, D., et al.: Automated anatomical labeling of activations in SPM using a macroscopic anatomical parcellation of the MNI MRI single-subject brain. Neuroimage **15**(1), 273–289 (2002)
23. Van Snellenberg, J.X., Slifstein, M., Read, C., et al.: Dynamic shifts in brain network activation during supracapacity working memory task performance. Hum. Brain Mapp. **36**(4), 1245–1264 (2015). https://doi.org/10.1002/hbm.22699
24. Wang, L., Schwedt, T.J., Chong, C.D., et al.: Discriminant subgraph learning from functional brain sensory data. IISE Trans. **54**(11), 1084–1097 (2022)
25. Wu, C.W., Gu, H., Lu, H., et al.: Frequency specificity of functional connectivity in brain networks. Neuroimage **42**(3), 1047–1055 (2008)
26. Wu, D., et al.: Multi-frequency analysis of brain connectivity networks in migraineurs: a magnetoencephalography study. J. Headache Pain **17**(1), 1–10 (2016). https://doi.org/10.1186/s10194-016-0636-7
27. Yan, C., Wang, X., Zuo, X., et al.: DPABI: data processing & analysis for (resting-state) brain imaging. Neuroinformatics **14**(3), 339–351 (2016)
28. Yan, X., Han, J.: gSpan: graph-based substructure pattern mining. In: 2002 IEEE International Conference on Data Mining (ICDM), pp. 721–724. IEEE (2002)
29. Zanin, M., Sousa, P., Papo, D., et al.: Optimizing functional network representation of multivariate time series. Sci. Rep. **2**(1), 1–6 (2012)

Local Spectral for Polarized Communities Search in Attributed Signed Network

Fanyi Yang[1], Huifang Ma[1,2,3(✉)], Wentao Wang[1], Zhixin Li[2], and Liang Chang[3]

[1] College of Computer Science and Engineering, Northwest Normal University, Lanzhou 730070, Gansu, China
mahuifang@yeah.net
[2] Guangxi Key Lab of Multi-source Information Mining and Security, Guangxi Normal University, Guilin 541004, Guangxi, China
[3] Guangxi Key Lab of Trusted Software, Guilin University of Electronic Technology, Guilin 541004, Guangxi, China

Abstract. Signed networks are graphs with edge annotations to indicate whether each interaction is friendly (positive edge) or antagonistic (negative edge). Community search on signed network expects to explore the polarized communities (i.e., two antagonistic subgraphs) containing the set of query nodes. Though previous studies have been proven effective, they generally ignore two insights. First, node attributes provide side information to describe features of nodes. It contributes to optimal results. Secondly, the problem of detecting polarized communities from a global perspective is increasingly limiting and is computationally expensive on large-scale networks. These aspects motivate us to develop a new community search framework searching for Polarized Communities in Attributed Signed network (PCAS). Specifically, we propose a new strategy to combine node attributes with signed topology, which helps to make the most of the different dimensions of information. Furthermore, to search for polarized communities containing the set of query nodes, a sparse indicator-vector is developed based on Rayleigh quotient via solving a linear programming problem. Extensive experimental results on two real-world attributed signed graphs have demonstrated the discovered polarized communities are more accurate and more polarized.

Keywords: Polarized communities search · Local spectral · Rayleigh quotient · Attributed signed network

1 Introduction

Graph (network) is an essential data structure to represent relationships among sets of entities, which is an increasingly common focus of scientific inquiry. As a fundamental problem in network analysis, community search aims to find densely connected subgraphs, i.e., communities containing the query nodes [5,8]. This

© The Author(s), under exclusive license to Springer Nature Switzerland AG 2023
X. Wang et al. (Eds.): DASFAA 2023, LNCS 13945, pp. 58–74, 2023.
https://doi.org/10.1007/978-3-031-30675-4_5

opens up the prospects of user-centered and personalized search, with the potential of producing meaningful answers to users. However, many complex networks in the real world have friendly or antagonistic relationship between nodes. Signed networks act as a simple but powerful model to effectively capture such complex relationships between nodes, whose annotated edge is able to indicate whether each relationship is friendly (positive edges) or hostile (negative edges) [15]. What is more, signed networks can capture novel and interesting structures of real-world phenomena, i.e., polarized communities whose nodes within each community form positive connections, while nodes across different communities are connected by negative links [3,7]. Moreover, owing to the rising rich information in real-world entities, there exist plentiful attributes associated with nodes in signed graph. Attributes provide more descriptive characteristic information about nodes and form the second dimension to represent the underlying messages of networks besides structures [10,22]. Therefore, to provide more accurate and polarized communities, it is compulsory to go beyond modeling topology-based network and take attributes into account.

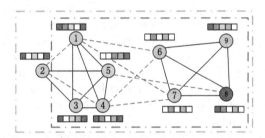

Fig. 1. An example of the polarized communities in an attributed signed network. Solid edges are positive, while dashed edges are negative. The polarized communities where the green box is located utilize topology, while the polarized communities where the red box is located consider both topology and attribute. (Color figure online)

Different from conventional community search, community search in signed networks expects to find polarized communities given query nodes. Figure 1 illustrates an attributed signed network with query nodes v_5 and v_8 and two polarized communities identified by PolarSeeds [18] and our approach. In particular, each node represents a user, each edge represents a friend (solid lines) or foe (dashed lines) relationship between users, and different coloured squares indicate the user's attributes. The green box shows the polarized communities located utilizing topology only while the red box indicates the polarized communities captured with consideration of both topology and node attribute. Intuitively, the latter is more polarized, since the former has two negative edges $((v_1,v_2),(v_2,v_4))$ within one community. This proves that attributes can be utilized as useful complementary information to help find more polarized communities.

Basically, the advantage of performing community search on attributed signed graphs is two-fold. First, complex edge relations and rich node attributes as unique characteristic of attribute symbolic networks can encourage the discovery of new patterns that cannot be found in simple graphs. One such important pattern is polarized communities that is essential to a comprehensive understanding of the entire system. Second, in practice, topology-based network is often noisy, incomplete and sparse due to measurement errors and data access limitations. In contrast, attributes provide complementary information to alleviate this problem which is more robust.

Despite the aforementioned advantages, searching for polarized communities remains a challenging task. First, previous studies mainly relying on the information about direct interactions between nodes are problematic since the networks are noisy due to missing data, repeated measurements, and contradictory observations. Attributes serve as an assistance to provide complementary information. Recent attributed community search studies have also shown that attributes are beneficial for improving the accuracy of community search [11,12]. Whereas, how to integrate the complementary attribute information during searching polarized communities is still an open problem till now. Second, much of the existing research focus on the global view of network to obtain polarized communities [14,16]. Concretely, a global criterion is usually established to determine whether a subgraph is eligible for being polarized communities. This standard is fixed throughout the process of polarized communities detection. Besides, detecting polarized communities in the whole network costs a lot of computing on large-scale networks. Further, in many application scenarios, users are only interested in local polarized communities with specified nodes. Therefore, how to identify personalized polarized communities of interest to users from a local perspective is a challenging task.

To make up for the above challenges, in this paper, we explore a novel problem, called searching for Polarized Communities with local spectral subspace in Attributed Signed network (PCAS). To the best of our knowledge, the presented approach is the first method to search for polarized communities in attributed signed network. Specifically, firstly, the attribute-based signed network is constructed and combined with the topology-based signed network based on signed spectral graph theory. The combined graph takes full advantage of the information in different dimensions of node attributes and topology. Then, the polarized communities are searched by solving a linear programming problem related to a sparse indicator vector based on the Rayleigh quotient. Finally, the polarized communities is obtained by discrete rounding of this sparse indicator vector.

The main contributions are summarized as follows:

1. Providing a unified way to model a new augmented signed graph that takes into account both topology and node attribute.
2. Using the generalized Rayleigh quotient, the sparse indicator vector is solved by Laplacian matrix span eigenspace and discrete rounding is performed to obtain the polarized communities.
3. Conducting experiments on real-world attributed signed social network datasets to assess the effectiveness of our method.

2   Related Work

Attributed Community Search. Community search methods over attributed graph aim to find query-dependent communities such that the community members are densely connected and have homogeneous attribute values. According to the query input, existing studies of community search on attributed graphs can be divided into two categories: taking both nodes and attributes as query inputs and taking only attributes as query inputs. The first category returns the attribute-cohesive community containing the query nodes [2,4]. The second category returns the community related to the query attributes [19,20].

However, the aforementioned algorithms are designed for unsigned networks, which have little utility for signed networks. In this paper, we take into account both negative links and node attributes. It is worth mentioning that our approach takes a different attitude towards attributes in contrast to attributed community search approach. Specifically, existing attributed community search methods regard attributes as a measure of community quality since the homogeneous effect or social influence between pairs of nodes in the attributed network [9,21]. Whereas, it has been shown that similar homogeneous effects are not observed in the attributed signed network [17]. Therefore, attributes should be considered as complementary information to measure the degree of polarization of the communities instead of as a criterion of community quality in attributed signed networks.

Polarized Communities Mining. The signed network has been used to study polarization and other related phenomena due to the presence of both positive and negative links. An interesting and challenging task in this application domain is to detect polarized communities in signed graphs. A number of different methods have been proposed for this task [1,13,18]. Bonchi et al. [1] introduce the 2-Polarized-Communities (2PC) problem, which requires finding two communities. And they prove 2PC problem is NP-hard and devise two efficient solutions with provable approximation guarantees. Xiao et al. [18] propose a spectral method to find polarized communities in signed graphs, and characterize the solutions in close-form using techniques in duality theory.

It is worth noting that our approach is designed from a local perspective. However, different from existing polarized communities search, on one hand, in order to take attributes into account, PCAS propose a new fusion strategy to effectively integrate the network topology and the attribute carried by the nodes. On the other hand, PCAS formalizes polarized communities search problem as a linear programming problem, which has a low complexity while maintaining an accurate covering on the target communities.

3   Preliminaries

An attributed signed graph is represented as an undirected graph $G_1 = (V, E_1^+, E_1^-, \mathbf{F})$, where V is a set of nodes, E_1^+ and E_1^- are the sets of positive and negative edges, respectively. In addition, each node is also associated with a set of attributes. We use $\mathbf{F} \in \mathbb{R}^{n \times k}$ to denote the attribute matrix where k is the number of attributes. $E_1 = E_1^+ \cup E_1^-$ is defined as the set of all edges. The adjacency matrix $\mathbf{A}_1$ of G_1 is denoted as follows: the entry $A_{ij} = 1$ if $(v_i, v_j) \in E_1^+, A_{ij} = -1$ if $(v_i, v_j) \in E_1^-$ and $A_{ij} = 0$ otherwise. Further, the diagonal matrix $\mathbf{D}_1$ denotes the degree matrix of graph G_1 (i.e., $\mathrm{D}_{ii} = \deg(v_i)$), where $\deg(v_i)$ is the degree of node v_i, i.e., the number of all edges adjacent to v_i. The signed Laplacian matrix of G_1 is defined as $\mathbf{L}_1 = \mathbf{D}_1 - \mathbf{A}_1$.

Definition 3.1 (Polarized communities). Given an attributed signed graph $G_1 = \left(V, E_1^+, E_1^-, \mathbf{F}\right)$, polarized communities are defined as two disjoint sets of nodes $C_1, C_2 \subseteq V$, denoted as (C_1, C_2), such that (1) there are relatively few (resp. many) negative (resp. positive) edges within C_1 and within C_2; (2) there are relatively few (resp. many) positive (resp. negative) edges across C_1 and C_2. (3) there are relatively few edges (of either sign) from C_1 and C_2 to the rest of the graph.

Our Objective Function. Given an indicator vector $\mathbf{y}$ defined over C_1 and C_2 (i.e., $y_i = 1$ if $v_i \in C_1$, $y_i = -1$ if $v_i \in C_2$, and $y_i = 0$ otherwise), the generalized Rayleigh quotient $\mathcal{R}_\mathbf{L}(\mathbf{y})$ is defined as:

$$
\begin{aligned}
\mathcal{R}_\mathbf{L}(\mathbf{y}) &= \frac{\mathbf{y}^T \mathbf{L} \mathbf{y}}{\mathbf{y}^T \mathbf{D} \mathbf{y}} \\
&= \frac{4\left|E^+(C_1, C_2)\right| + 4\left|E^-(C_1)\right| + 4\left|E^-(C_2)\right|}{\mathrm{vol}(C_1 \cup C_2)} \\
&\quad + \frac{\left|E(C_1 \cup C_2, V \setminus (C_1 \cup C_2))\right|}{\mathrm{vol}(C_1 \cup C_2)}
\end{aligned}
\tag{1}
$$

where $E^+(C_1, C_2) = \{(v_i, v_j) \in E^+ \mid v_i \in C_1, v_j \in C_2\}, E^-(C_1)$ is the set of negative edges having both endpoints in C_1, $\mathrm{vol}(C_1 \cup C_2) = \sum_{v_i \in C_1 \cup C_2} \deg(v_i)$. In particular, the more polarized the communities (C_1, C_2), the smaller the value of $\mathcal{R}_\mathbf{L}(\mathbf{y})$.

Definition 3.2 (Polarized Communities Search). Given an attributed signed graph $G_1 = \left(V, E_1^+, E_1^-, \mathbf{F}\right)$, a pair of antagonistic node sets (S_1, S_2) as query. We aim to find polarized communities (C_1, C_2) indicated by $\mathbf{y}$ with $S_1 \subseteq C_1, S_2 \subseteq C_2$ such that our objective function is minimal.

4   Method

4.1   Construction of Augmented Signed Graph

In attributed signed networks, from the perspective of topology, the extended structural balance theory implies that a node should sit closer to its friends (with positive links) than non-linked nodes and sit far away from foe (with negative links) [6]. From the perspective of attribute, intuitively, nodes should share more similar attributes with friends than non-linked nodes and few similar attributes with foes. However, existing studies have reached the following two conclusions [17]: (1) *Nodes have more similarity to their positive links than nodes with negative links*; (2) *Nodes have more similarity to their negative links than to unconnected nodes.* Naturally, *a unified augmented signed network should be constructed that satisfies both extended structural balance theory from topology and manifold constraint from attribute.*

Based on the above findings, we construct an attribute-based signed network G_2 via node attributes capturing the manifold structure in terms of attributes in attributed signed network. Specifically, the pairwise similarity $s(i,j)$ between the attribute vectors of nodes v_i and v_j is computed by using the Gaussian similarity kernel:

$$s(i,j) = \exp\left\{-\frac{d^2\left(\mathbf{f}_i, \mathbf{f}_j\right)}{2\sigma^2}\right\} \tag{2}$$

where $d\left(\mathbf{f}_i, \mathbf{f}_j\right)$ denotes the Euclidean distance between the node attribute vector $\mathbf{f}_i$ and $\mathbf{f}_j$ and σ is empirically set to be half of the maximum pairwise distance between any two points.

We sort the above pairwise similarities in descending order and then construct the attribute-based signed network G_2 based on the statistics of the edges in the original attributed signed network G_1. This allows the scale between the constructed network G_2 and the original network G_1 to be consistent. Specifically, the set $\{s_1, s_2, \ldots, s_i, \ldots, s_p\}$ can be obtained, where s_i denotes the i-th pairwise similarity by Eq. (2). Let $\left\{s_{(1)}, \ldots, s_{(l)}, s_{(l+1)}, \ldots, s_{(k)}, s_{(k+1)}, \ldots s_{(p)}\right\}$ be the set that has been sorted by similarity in descending order, where $l = \left|E_1^+\right|$, $k = |E_1|$, $p = \frac{|V|*(|V|-1)}{2}$. We define the following reconstruction rules:

1. Adding positive edges between the nodes corresponding to $\left\{s_{(1)}, \ldots, s_{(l)}\right\}$;
2. Adding negative edges between the nodes corresponding to $\left\{s_{(l+1)}, \ldots s_{(k)}\right\}$;
3. No edges between the remaining nodes in the set of $\left\{s_{(k+1)}, s_{(k+2)}, \ldots, s_{(p)}\right\}$.

Relying on the above construction rules, an attribute-based signed network can be denoted as $G_2 = \left(V, E_2^+, E_2^-\right)$. Like the definition in Sect. 3, the set of all edges is $E_2 = E_2^+ \cup E_2^-$, $\mathbf{A}_2$ and $\mathbf{D}_2$ denote the adjacency matrix and degree matrix, respectively. The signed Laplacian matrix of G_2 is defined as $\mathbf{L}_2 = \mathbf{D}_2 - \mathbf{A}_2$.

Intuitively, we can independently find polarized communities from topology-based and attribute-based signed networks, and fuse them. However, it is not likely to capture the inherent connection between networks and attributes, leading to obtain the non-optimal polarized communities. In addition, individual network may be noisy and incomplete. Therefore, to discover more polarized communities, we need to fuse the topology-based signed network G_1 with the attribute-based signed network G_2 in an adaptive manner. Specifically, we term the hybrid structure of topology-based signed network and attribute-based signed network as augmented signed network denoted as $G = (V, E)$, which encodes topology-based and attribute-based information as a unified signed network. The Laplacian matrix $\mathbf{L}$ of G can be obtained by linearly combining the Laplacian matrices $\mathbf{L}_1$ and $\mathbf{L}_2$ of G_1 and G_2 mentioned above.

$$\mathbf{L} = \alpha_1 \mathbf{L}_1 + \alpha_2 \mathbf{L}_2 \tag{3}$$

where $\mathbf{L}_1$ and $\mathbf{L}_2$ denote the Laplacian matrices of the topology-based signed network G_1 and the attribute-based signed network G_2. The combination factor α_i determines the weight of the influence of each Laplacian matrix. The choice of combination factors will be described in detail in Sect. 4.2. The matrix $\mathbf{L}$ captures the properties of both the original topology and the node attribute, providing rich information for the follow-up search of polarized communities.

The degree matrix of the graph G is denoted as $\mathbf{D}$, and $\mathbf{D}_{ii} = \mathbf{L}_{ii}$. The normalized signed Laplacian matrix of G is defined as $\mathcal{L} = \mathbf{D}^{-1/2}\mathbf{L}\mathbf{D}^{-1/2}$, which is symmetric and positive semi-definite.

4.2   Choice of Combination Factor

The combination factors α_i determine the weights of both Laplacian matrices. In order to assign weights adaptively, we propose a new weighting strategy by utilizing the spectral graph theory of signed networks. In signed spectral graph theory, given a signed network, the smallest eigenvalue γ_1 of the Laplacian matrix $\mathbf{L}$ of the signed network reflects whether the signed network is balanced. A dense and balanced signed network indicates the existence of polarized communities. The eigenvector $\mathbf{q}_1$ corresponding to the eigenvalue γ_1 can be used to partition the nodes of the graph to obtain polarized communities.

Let $\mathbf{q}_1 = (q_{11}, \ldots, q_{1j}, \ldots, q_{1n})$, spectral partitioning aims to find a splitting value x such that the nodes in signed network with $q_{1j} \leq x$ belong to one set, while nodes with $q_{1j} > x$ belong to the other set to form 2-partition. Once a 2-partition is obtained, our generalized Rayleigh quotient $\mathcal{R}_{\mathbf{L}}(\mathbf{y})$ can evaluate the degree of polarization of that partition. As mentioned before, a smaller $\mathcal{R}_{\mathbf{L}}(\mathbf{y})$ of the partition indicates a better polarization of the signed network. A more polarized signed network is expected to have smaller γ_1 and $\mathcal{R}_{\mathbf{L}}(\mathbf{y})$ on the $\mathbf{q}_1$ vector. Thus, a measure of 'relevance' of Laplacian matrix $\mathbf{L}_i$ is defined as:

$$\mathbf{r}_i = e^{-\gamma_1^{(i)} \cdot \mathcal{R}(\mathbf{q}_1^{(i)})}, (i = 1, 2) \tag{4}$$

where $\gamma_1^{(i)}$ is the smallest eigenvalue of the Laplacian matrix $\mathbf{L}_i$ of G_i and $\mathbf{q}_1^{(i)}$ is the corresponding eigenvector. $\mathcal{R}(\mathbf{q}_1^{(i)})$ is the value of $\mathcal{R}_{\mathbf{L}}(\mathbf{y})$ of the partition obtained by $\mathbf{q}_1^{(i)}$. The value of relevance measure r_i lies in [0,1]. Higher value of r_i implies better the degree of polarization. Hence, combination factor α_i corresponding to the weight of Laplacian matrix $\mathbf{L}_i$ is given by:

$$\begin{cases} \alpha_1 = r_1\theta^{-1}, \alpha_2 = r_2\theta^{-2}, \text{if} r_1 \geq r_2 \\ \alpha_1 = r_1\theta^{-2}, \alpha_2 = r_2\theta^{-1}, \text{if} r_1 \leq r_2 \end{cases} \tag{5}$$

where θ is the decay factor. We then normalize the combination factors with a softmax function. This assignment of α_i upweights structure with higher degree of polarization, while dampens the effect of irrelevant ones those having poor structure.

4.3   Local Spectral Subspace

In this subsection, we provide the necessary theoretical base to find a sparse indicator vector in the span of dominant eigenvectors of the Laplacian matrix with larger eigenvalues.

Since the matrix $\mathcal{L}$ is a normalized Laplace matrix of matrix $\mathbf{L}$ and also a real symmetric matrix, $\mathcal{L}$ can always be eigen-decomposed, i.e., $\mathcal{L} = \mathbf{U}\Lambda\mathbf{U}^{\mathrm{T}}$, where $\Lambda = \mathrm{diag}\,(\lambda_1, \lambda_2, \ldots, \lambda_n)$ is arranged in ascending order of eigenvalues, i.e., $\lambda_1 \leq \ldots \leq \lambda_i \leq \ldots \leq \lambda_n$, the matrix formed by the corresponding eigenvectors are denoted as $\mathbf{U} = (\mathbf{u}_1, \mathbf{u}_2, \ldots, \mathbf{u}_n)$. So, the generalized Rayleigh quotient of Eq. (1) can be rewritten as follows.

$$\mathcal{R}_{\mathbf{L}}(\mathbf{y}) = \frac{\left(\mathbf{U}^{\mathrm{T}}\mathbf{D}^{\frac{1}{2}}\mathbf{y}\right)^{\mathrm{T}} \Lambda \left(\mathbf{U}^{\mathrm{T}}\mathbf{D}^{\frac{1}{2}}\mathbf{y}\right)}{\left(\mathbf{U}^{\mathrm{T}}\mathbf{D}^{\frac{1}{2}}\mathbf{y}\right)^{\mathrm{T}} \left(\mathbf{U}^{\mathrm{T}}\mathbf{D}^{\frac{1}{2}}\mathbf{y}\right)} \tag{6}$$

Let $w_i = \dfrac{\left(\mathbf{u}_i^{\mathrm{T}}\mathbf{D}^{\frac{1}{2}}\mathbf{y}\right)^2}{\sum_{i=1}^{n}\left(\mathbf{u}_i^{\mathrm{T}}\mathbf{D}^{\frac{1}{2}}\mathbf{y}\right)^2}$, then Eq. (6) can be rewritten again as

$$\mathcal{R}_{\mathbf{L}}(\mathbf{y}) = \frac{\sum_{i=1}^{n}\lambda_i\left(\mathbf{u}_i^{\mathrm{T}}\mathbf{D}^{\frac{1}{2}}\mathbf{y}\right)^2}{\sum_{i=1}^{n}\left(\mathbf{u}_i^{\mathrm{T}}\mathbf{D}^{\frac{1}{2}}\mathbf{y}\right)^2} = \sum_{i=1}^{n} w_i\lambda_i \tag{7}$$

where $w_i = \dfrac{\left(\mathbf{u}_i^{\mathrm{T}}\mathbf{D}^{\frac{1}{2}}\mathbf{y}\right)^2}{\sum_{i=1}^{n}\left(\mathbf{u}_i^{\mathrm{T}}\mathbf{D}^{\frac{1}{2}}\mathbf{y}\right)^2}$ can be treated as the weight coefficient of the eigenvalues. Equation (7) shows that if a minimum $\mathcal{R}_{\mathbf{L}}(\mathbf{y})$ is desired, then the smaller eigenvalues of the Laplacian matrix $\mathbf{L}$ should have the majority of the weights, i.e., the Laplacian matrix $\mathcal{L}$ should be as large as possible. A larger w_i indicates a smaller angle between $\mathbf{D}^{\frac{1}{2}}\mathbf{y}$ and eigenvector $\mathbf{u}_i$. When we relax $\mathbf{y}$ from $\{-1, 1\}^{n \times 1}$ to $[-1, 1]^{n \times 1}$, the relaxed scaled indicator vector $\mathbf{D}^{\frac{1}{2}}\mathbf{y}$ should be well approximated by a linear combination of the dominant eigenvectors with smaller eigenvalues.

Based on the linear combination of the eigenvectors with larger eigenvalues, the polarized communities search can be formulated as a local spectral-based eigen problem, which can be exploited via an indicator-vector $\mathbf{y}$ in a semi-supervised manner. Mathematically, it is equivalent to solve the following linear programming problem:

$$\min \|\mathbf{y}\|_1 = \mathbf{e}^T\mathbf{y}$$
$$\text{s.t. } (1)\mathbf{y} = \mathbf{Vk},$$
$$(2) \ y_i \in [-1,1] \tag{8}$$
$$(3) \begin{cases} y_i \geq -\frac{1}{|S_1|}, \text{if } i \in S_1 \\ y_i \geq \frac{1}{|S_2|}, \text{if } i \in S_2 \end{cases}$$

where $\mathbf{e}$ denotes the vector of all ones. The column vectors of $\mathbf{V}$ are formed by the dominant eigenvectors of normalized Laplacian matrix $\mathcal{L}$. The vector $\mathbf{k}$ is a coefficient vector, The i-th element of the relaxed indicator-vector $\mathbf{y}$ can be denoted as y_i. Constraint (1) requires $\mathbf{y}$ be in the span of $\mathbf{V}$, and constraint (2) requires the element value range of the relaxed indicator-vector $\mathbf{y}$ should in [-1,1] indicating the likelihood of node v_i belonging to the polarized communities. Constraint (3) enforces the query nodes be in the support of indicator-vector $\mathbf{y}$.

We can solve the above linear programming problem via dual simplex method. After obtaining indicator vector $\mathbf{y}$, we can get the desired target communities $(V_{\mathbf{x}}(t), V_{\mathbf{x}}(-t))$ by using Proposition 1.

Proposition 1. For any non-zero vector $\mathbf{y}$, there exists a $t \in [0, \max_{u \in V} |\mathbf{y}_u|]$ such that [18]: our objective function $\mathcal{R}_{\mathbf{L}}(\mathbf{y})$ is minimum, where, $\mathbf{y}_i = 1$ if $v_i \in V_{\mathbf{y}}(t)$, $\mathbf{y}_i = -1$ if $v_i \in V_{\mathbf{y}}(-t)$, $V_{\mathbf{y}}(t) = \{u \in V \mid \mathbf{y}_u \geq t\}$ and $V_{\mathbf{y}}(-t) = \{u \in V \mid \mathbf{y}_u \leq -t\}$.

Complexity Analysis. The first step is to construct the augmented signed graph, in particular, the attribute-based signed network is constructed in $\mathcal{O}(n^2)$. Then, the eigen-decomposition of $\mathbf{L}_i (i = 1, 2)$ is computed in Eq.(5) which takes $\mathcal{O}(2n^3)$ time for the $(n \times n)$ matrix. Furthermore, the construction of the dominant eigenspace $\mathbf{V}$ in Eq.(8) takes $\mathcal{O}(n^3)$ time. Finally, the polarized communities can be obtained by Proposition 1. A plain way is to try every value in $\mathbf{y}$, which takes $\mathcal{O}(n^2 + n \log_2 n)$ time.

4.4 Discussion on Quality of Polarized Communities

In this subsection, we will discuss how different polarization measures relate to our optimization objective.

(1) HAM is the harmonic mean of Cohesion and Opposition, which is defined as:

$$Cohesion\,(C_1, C_2) = \frac{1}{2} \left[d^+(C_1) + d^+(C_2) \right]$$
$$Opposition\,(C_1, C_2) = d^-(C_1, C_2) \tag{9}$$
$$HAM = \frac{2 \times Cohesion\,(C_1, C_2) \times Opposition\,(C_1, C_2)}{Cohesion\,(C_1, C_2) + Opposition\,(C_1, C_2)}$$

where $d^+(C) = \frac{2|E^+(C)|}{|C|(|C|-1)}$ and $d^-(C_1, C_2) = \frac{|E^-(C_1, C_2)|}{|C_1||C_2|}$.

(2) Polarity counts the number of edges that agree with the polarized structure and penalizes large communities, which is defined as:

$$\text{Polarity}(C_1, C_2) = \frac{|E^+(C_1) \cup E^+(C_2)| + 2|E^-(C_1, C_2)|}{|C_1 \cup C_2|} \tag{10}$$

Jointly analyzing Eqs. (9) and (10), HAM and Polarity are concerned on constraint (1) and constraint (2), i.e., intra-communities edge and inter-communities edge, while the constraint (3) is totally ignored, i.e., the edges between polarized communities and the rest of the network in Definition 3.1 and higher values indicate more polarized communities.

(3) Signed bipartiteness ratio is the classical measure of the quality of the polarized communities.

$$\begin{aligned}
\beta(C_1, C_2) = {} & \frac{2|E^+(C_1, C_2)| + |E^-(C_1)| + |E^-(C_2)|}{\text{vol}(C_1 \cup C_2)} \\
& + \frac{|E(C_1 \cup C_2, V \setminus (C_1 \cup C_2))|}{\text{vol}(C_1 \cup C_2)}
\end{aligned} \tag{11}$$

Signed bipartiteness ratio and our goal penalize inconsistent edges and thus the smaller values imply more polarized communities. Compared to signed bipartiteness ratio, which treats inter-communities edges twice more heavily than intra-communities edges, our objective in Eq. (1) weights intra-communities edges and inter- communities edges as equally important. Besides, our objective regard both intra-communities and inter-communities edges 4 times more heavily than that of edges between polarized communities and the rest of the network. The rationale is as follows: Intuitively, for polarized communities, the significance of constraint (1) and constraint (2) in Definition 3.1 should be equivalent. Constraint (3) serves as a vitamin to measure polarized communities, which prevent the intra/inter-communities edges from dominating the objective. In a word, our optimization objective is consistent with the definition of polarized communities.

5 Experiments

5.1 Datasets Description

For the purpose of this study, we collect two datasets from Epinions and Slashdot [17]. The detailed information for both datasets is shown in Table 1.

Table 1. Dataset statistics.

Dataset	#Users	#Positive edges	#Negative edges	#Attributes	#Community
Epinions	27,215	326,909	58,695	22,367	18
Slashdot	33,407	477,176	158,104	19,875	22

Epinions is a consumer review site where users can view ratings and reviews of products by other users. Epinions provides a trust mechanism where users can determine whether to trust other users. User reviews are used to construct user-attribute matrix $\mathbf{F}$ using bag-of-words.

Slashdot is an information technology website where all news is provided by users. Users can comment on the news published on the site. Users of the site's community can determine whether to add other users to their friends or enemies list. The user attribute matrix $\mathbf{F}$ can be constructed with user comments using bag-of-words.

5.2   Experimental Settings

Baselines Methods. To demonstrate the effectiveness, we compare our proposed PCAS with the following methods:

1. *RE* [1] introduces the 2-Polarized-Communities problem. Two efficient algorithms are devised with provable approximation guarantees.
2. *PolarSeeds* [18]formulates the polarized communities problem in signed graphs as a locally-biased eigen-problem, and uses techniques of linear algebra to approximate the solution.
3. *Timbal* [13] presents an algorithm for finding large balanced subgraphs in signed networks. By relying on signed spectral theory and a bound for perturbations of the graph Laplacian.
4. *PCAS-noAttri* is the variant of PCAS which ignores the external attributes.
5. *PCAS-sameFactor* considers the topology-based signed network G_1 and the attribute-based signed network G_2 in Eq.(3) to have the same weight.

Parameter Settings. Since RE and Timbal is 'query-less', for the purpose of fair comparison, we make the following adaptation to make them comparable. Let the polarization communities obtained by the above two methods be (C_1, C_2), (C_1', C_2') respectively, we use a simple heuristic to select the seed nodes. Specifically, nodes u and v are considered seeds respectively, if $u \in C_1 \cap C_1'$, $v \in C_2 \cap C_2'$ and $(u, v) \in E^-$ and $\deg^+(u) \geq t, \deg^+(v) \geq t$, where t is some pre-defined positive number. We set t to be the average degree of all positive links. In each trial, we randomly select six pairs of antagonistic nodes via the above way from ground-truth as the query nodes.

5.3   Performance Evaluation

Exp-1: The Effectiveness of Finding the Real Communities. We apply the selected methods to search polarized communities and repeat the experiments 100 times and the average results are reported in Table 2. From the result, we summarize several important observations:

(1) *Our proposed PCAS achieves significant improvements over the baselines on all the datasets, which demonstrates the effectiveness of our proposed*

Table 2. Comparisons of average recall, precision, F-Score between our method and baselines.

	Epinions			Slashdot		
	precision	recall	F1	precision	recall	F1
PolarSeeds	0.747	0.711	0.729	0.785	0.769	0.777
RE	0.631	0.618	0.624	0.674	0.658	0.667
Timbal	0.673	0.652	0.662	0.719	0.694	0.706
PCAS-noAttri	0.729	0.706	0.717	0.773	0.756	0.764
PCAS-sameFactor	0.774	0.743	0.758	0.803	0.797	0.801
PCAS	0.793	0.750	0.771	0.821	0.804	0.812

method. Concretely, the superiority of our model arises in the following three aspects: 1) The topology and node attribute in the attributed signed network can be naturally integrated into an augmented signed network; 2) we adaptively assign importance weights to topology and attribute, which can improve the effectiveness of polarized communities search; 3) The polarized communities are identified in local spectral eigenspaces in semi-supervised manner.

(2) *Attributes contribute to the task of polarized communities search in signed networks.* As the results show, on the one hand, compared with PCAS, the performances of PCAS-noAttri becomes worse. Further, PCAS-sameFactor performs better than all baselines. These results prove the ability of PCAS in effectively leveraging attribute information for polarized communities search. On the other hand, PCAS-noAttri outperforms RE and Timbal by a large margin. Meanwhile, PCAS-noAttri can achieve comparable result with PolarSeeds. This indicates the superiority of finding polarized communities in local spectral eigenspace in a semi-supervised manner.

(3) *Topology and node attribute should have different importance.* Although attributes contribute to polarized communities search, they should be distinguished from topology. In Table 2, compared to PCAS , the performance of PCAS-sameFactor has shown various degrees of degeneration. The results demonstrates that via assigning the different importance to topology and node attribute, the proposed PCAS can find more polarized communities.

Exp-2: Polarization Quality Analysis of Polarized Communities. The distributions of all evaluation metrics for four methods are shown in Fig. 2. Overall, our proposed PCAS achieves superior results in all polarity metrics. In particular, in terms of HAM (Fig. 2(a)) and Polarity (Fig. 2(b)), it is surprising that our method improves more significantly than its competitors. Despite HAM and Polarity are not the goals of this paper, the possible reasons for this are as follows: 1) Although the optimization objective of RE is to measure the polarized communities from both the consistent and inconsistent perspectives, it ignores constraint (3) in Definition 3.1 and the node attributes, resulting in poor results.

2) Timbal iteratively removes nodes by relying on the topology to find polarized communities, but it fails to take into account the effect of node attributes during the entire iteration. From Fig. 2(c) and (d), we can observe that PCAS consistently outperforms all baselines, which is expected owing to the fact that our goal is to optimize three constraints in Definition 3.1. Noted that although PolarSeeds also optimizes the three constraints in Definition 3.1, PolarSeeds performs worse than PCAS. The reason is that PolarSeeds only relies on the topology and does not consider the node attributes while our PCAS integrates both topology and node attribute during the search process.

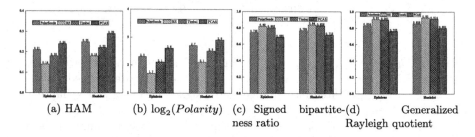

(a) HAM   (b) $\log_2(Polarity)$ (c) Signed   bipartite-(d)   Generalized
                                        ness ratio   Rayleigh quotient

Fig. 2. Distributions of four evaluation metrics on all communities found by PolarSeeds, RE, Timbal and PCAS.

Exp-3:Comparison with the Efficiency of All Baselines. Figure 3 shows the average running time of the baseline for 10 runs on each dataset. We can observe that: (1) Among RE, PolarSeeds and Timbal, RE is the fastest method, which is expected because its running time depends on the computation of the dominant eigenvector of the adjacency matrix. Timbal takes more time due to the numerous iterations to remove nodes. Compared to PCAS, PolarSeeds takes less time as it ignores the attribute information carried by the nodes. Although PCAS requires the calculation of node attribute similarity, PCAS is always among the top three fastest methods. (2) Compared to PCAS, PCAS-noAttri is faster than the PCAS because it ignores the calculation of node attribute information. Compared to PCAS, PCAS-sameFactor takes less time because PCAS-sameFactor computes node attribute information, while PCAS-sameFactor does not need to compute the importance of the topology and attribute.

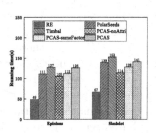

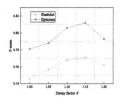

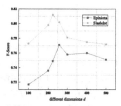

(a) The effect of different values of decay factor θ.

(b) The effect of different dimensions d of the subspace.

Fig. 3. Efficiency evaluation of our model on all datasets.

Fig. 4. Hyperparameter sensitivity analysis on both datasets.

5.4 Parameter Sensitivity Analysis

In the section, we evaluate how different settings of hyper-parameters affect the performance of PCAS. (1)**Impact of decay factor** θ.To analyze the influence of decay factor θ, we vary θ from 1.0 to 1.2 with increment 0.05, and illustrate the performance changing curves on Epinions and Slashdot in Fig. 4. From the results in Fig. 4(a), with the increasing of θ, we find that it brings about a great performance improvement. Compared to $\theta = 1$ (i.e., without decay), the results show that there is a leap in preserving polarization by decaying r_i when $\theta ¿ 1$. However, as θ increases further, the performance decreases sharply. The main reason is that the larger the θ excessively weakens the importance of topology and attribute, which fails to obtain the appropriate weights adaptively. Therefore, we set the parameter θ=1.15. (2)**Impact of the subspace dimension** d. We vary d from 100 to 500 to investigate the influence of the subspace dimension. From the results in Fig. 4(b), we find the performance of PCAS gets better when d increases, which indicates that the subspace spanned by eigenvectors corresponding to larger eigenvalues has better representational ability. However, a larger subspace dimension does not always bring stronger representation ability. One possible reason is that the larger dimension of the subspace, the more noises are introduced. Thus, we set the subspace dimension $d = 224$ on Epinions and $d = 258$ on Slashdot.

5.5 Case Study

To further demonstrate the benefits of PCAS, we give a case study of polarized communities on Epinions. Figure 5 shows the results of PCAS-noAttri and PCAS when we choose antagonistic nodes v_8 and v_{129} as query nodes. The red line indicates positive links (friends) and the green line represents negative links (foes). As is shown in Fig. 5, both PCAS-noAttri and PCAS detect closely connected polarized communities containing query nodes. Here, we have three observations: (1) The polarized communities (C_1, C_2) found by PCAS is shown in Fig. 5(a). It is obvious that there are all dense positive edges (red line) within C_1 and C_2

and all negative edges (green line) across C_1 and C_2 (with $HAM(C_1, C_2)=0.26$, $\log_2(Polarity) = 2.54$, $\beta(C_1, C_2)=0.673$ and our $\mathcal{R_L}(\mathbf{y})=0.753$). (2) As shown in Fig. 5(b), PCAS-noAttri finds the communities (C_1', C_2'). It can be shown that there are all negative edges (green line) across C_1' and C_2', but there are positive edges (red line) and negative edges (red line) within C_1' and C_2' simultaneously (with $HAM(C_1, C_2)=0.23$, $\log_2(Polarity) = 2.31$, $\beta(C_1, C_2)=0.739$ and our $\mathcal{R_L}(\mathbf{y})=0.713$). (3) By comparing Fig. 5(a) with Fig. 5(b), we can notice that PCAS can find more polarized communities, which means that by taking attributes into account, richer information in the attributed signed network can be captured to discover more polarized communities.

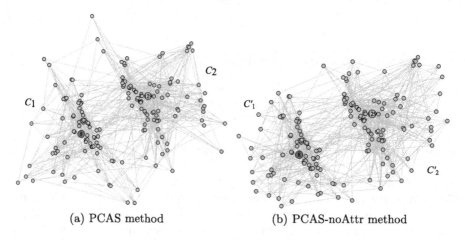

(a) PCAS method   (b) PCAS-noAttr method

Fig. 5. Visualization of polarized communities found via PCAS-noAttri and PCAS in Epinions dataset with v_8 and v_{129} as query nodes.

6   Conclusion and Future Work

In this paper, we focus on the polarized communities search task in attributed signed networks. We propose the community search framework searching for polarized communities in attributed signed network, which searches for two polarized subgraphs on an attributed signed network for given query nodes. We leverage both signed network and node attribute into a unified framework PCAS by incorporating the extended structure balance theory and the relationship between node links and node attributes. Then, a spectral method based on Rayleigh quotient is proposed. Finally, a linear programming problem is designed to detect polarized communities by local eigenspace. Experiments on real-world datasets demonstrate the effectiveness of our method. As for future work, we intend to extend PCAS to find multiple communities. In addition, there are issues worth investigating in terms of developing reasonable polarization metrics.

Acknowledgements. This work is supported by the Industrial Support Project of Gansu Colleges (2022CYZC-11), The National Natural Science Foundation of China (61966009, U22A2099, 62276073, 61762078), Gansu Natural Science Foundation Project (21JR7RA114), NWNU Teachers Research Capacity Promotion Plan (NWNU-LKQN2019-2) and NWNU Graduate Research Project Funding Program (2021KYZZ02101).

References

1. Bonchi, F., Galimberti, E., Gionis, A., Ordozgoiti, B., Ruffo, G.: Discovering polarized communities in signed networks. In: the 28th ACM International Conference on Information and Knowledge Management (CIKM 2019), pp. 961–970 (2019)
2. Fang, Y., et al.: On spatial-aware community search. IEEE Trans. Knowl. Data Eng. **31**(4), 783–798 (2018)
3. Hua, J., Yu, J., Yang, M.: Fast clustering for signed graphs based on random walk gap. Soc. Networks **60**, 113–128 (2020)
4. Huang, X., Lakshmanan, L.V.: Attribute-driven community search. Proc. VLDB Endow. **10**(9), 949–960 (2017)
5. Jiang, Y., Rong, Y., Cheng, H., Huang, X., Zhao, K., Huang, J.: Query driven-graph neural networks for community search: from non-attributed, attributed, to interactive attributed. Proc. VLDB Endow. **15**(6), 1243–1255 (2022)
6. Jung, J., Jin, W., Kang, U.: Random walk-based ranking in signed social networks: model and algorithms. Knowl. Inf. Syst. **62**(2), 571–610 (2020)
7. Kumar, S., Hamilton, W.L., Leskovec, J., Jurafsky, D.: Community interaction and conflict on the web. In: the 2018 World Wide Web Conference on World Wide Web (WWW 2018), pp. 933–943 (2018)
8. Li, Q., Ma, H., Li, J., Li, Z., Jiang, Y.: Searching target communities with outliers in attributed graph. Knowl. Based Syst. **235**, 107622 (2022)
9. Li, Q., Ma, H., Li, Z., Chang, L.: Local spectral for multiresolution community search in attributed graph. In: the International Conference on Multimedia and Expo (ICME 2022), pp. 1–6 (2022)
10. Lin, Z., Kang, Z.: Graph filter-based multi-view attributed graph clustering. In: the 30th International Joint Conference on Artificial Intelligence (IJCAI 2021), pp. 19–26 (2021)
11. Liu, Y., Liu, Z., Feng, X., Li, Z.: Robust attributed network embedding preserving community information. In: 38th IEEE International Conference on Data Engineering (ICDE 2022), pp. 1874–1886 (2022)
12. Luo, X., et al.: ComGA: community-aware attributed graph anomaly detection. In: the Fifteenth ACM International Conference on Web Search and Data Mining (WSDM 2022), pp. 657–665 (2022)
13. Ordozgoiti, B., Matakos, A., Gionis, A.: Finding large balanced subgraphs in signed networks. In: The Web Conference 2020 (WWW 2020), pp. 1378–1388 (2020)
14. Sun, R., Zhu, Q., Chen, C., Wang, X., Zhang, Y., Wang, X.: Discovering Cliques in Signed Networks Based on Balance Theory. In: Nah, Y., Cui, B., Lee, S.-W., Yu, J.X., Moon, Y.-S., Whang, S.E. (eds.) DASFAA 2020. LNCS, vol. 12113, pp. 666–674. Springer, Cham (2020). https://doi.org/10.1007/978-3-030-59416-9_43
15. Tang, J., Chang, Y., Aggarwal, C.C., Liu, H.: A survey of signed network mining in social media. ACM Comput. Surv. **49**(3), 1–37 (2016)

16. Tzeng, R., Ordozgoiti, B., Gionis, A.: Discovering conflicting groups in signed networks. In: advances in Neural Information Processing Systems 33: Annual Conference on Neural Information Processing Systems (NeurIPS 2020) (2020)
17. Wang, S., Aggarwal, C., Tang, J., Liu, H.: Attributed signed network embedding. In: the 2017 ACM on Conference on Information and Knowledge Management (CIKM 2017), pp. 137–146 (2017)
18. Xiao, H., Ordozgoiti, B., Gionis, A.: Searching for polarization in signed graphs: a local spectral approach. In: The Web Conference 2020 (WWW 2020), pp. 362–372 (2020)
19. Xie, X., Zhang, J., Wang, W., Yang, W.: Attributed community search considering community focusing and latent relationship. Knowl. Inf. Syst. **64**(3), 799–829 (2022)
20. Ye, W., Mautz, D., Böhm, C., Singh, A., Plant, C.: Incorporating user's preference into attributed graph clustering. IEEE Trans. Knowl. Data Eng. **33**(12), 3716–3728 (2020)
21. Zhao, Q., Ma, H., Guo, L., Li, Z.: Hierarchical attention network for attributed community detection of joint representation. Neural Comput. Appl. **34**(7), 5587–5601 (2022)
22. Zhao, S., Du, Z., Chen, J., Zhang, Y., Tang, J., Yu, P.S.: Hierarchical representation learning for attributed networks. In: 38th IEEE International Conference on Data Engineering (ICDE 2022), pp. 1497–1498 (2022)

Subgraph Reconstruction via Reversible Subgraph Embedding

Boyu Yang and Weiguo Zheng[✉]

School of Data Science, Fudan University, Shanghai, China
{yangby19,zhengweiguo}@fudan.edu.cn

Abstract. Reconstructing a subgraph through an embedding is very useful for many subgraph-level tasks, e.g., subgraph matching and minimum Steiner tree problem. To support subgraph reconstruction, a naive approach is materializing subgraph embeddings for all possible candidate subgraphs in advance, which is impractical since the subgraphs are exponential to the size of the input graph. Therefore, it is desired to devise a subgraph embedding based on which the subgraph can be reconstructed. To the end, we develop a novel reversible subgraph embedding in this paper. By importing the compressed sensing theory into learning node embeddings, we design a reversible read-out operation such that the aggregation vector can be recovered according to the subgraph embedding, where the aggregation vector acts as a bridge between the adjacency matrix and subgraph embedding. To reconstruct the structure of the subgraph from the decoded aggregation vector, we present a bijective rule by applying a simple transformation between binary number and decimal number with a scale operation. We conduct extensive experiments over real graphs to evaluate the proposed subgraph embedding. Experimental results demonstrate that our proposed method greatly and consistently outperforms the baselines in three tasks.

Keywords: Subgraph Reconstruction · Compressed Sensing · Reversible Subgraph Embedding

1 Introduction

Benefiting from the great representation power on complicated relationships among a huge number of objects, the graph structure is attracting increasing interest and has been widely used in real-world applications. Inspired by the great success achieved by deep learning in several tasks, e.g., speech recognition and image recognition, an extraordinary trend is emerging that equipping graph mining with deep learning models. Generally, a prior step, also called graph embedding, is to transform a graph into a vector or a matrix, enabling tasks on graphs to be winged with off-the-shelf deep learning techniques. Most of the existing algorithms [10,11,16,19,20] proposed for graph representation in the

© The Author(s), under exclusive license to Springer Nature Switzerland AG 2023
X. Wang et al. (Eds.): DASFAA 2023, LNCS 13945, pp. 75–92, 2023.
https://doi.org/10.1007/978-3-031-30675-4_6

past few years focus on solving node-level tasks (such as node classification and link prediction) and graph-level tasks (such as graph classification, graph generation, and so on). In addition to the two kinds of tasks, there is a great demand for subgraph-level analysis and processing. However, designing embeddings for subgraph-level tasks has not received much attention yet despite its importance.

1.1 Motivation

Example 1. Let us consider two classical graph problems.
(1) *Subgraph Matching.* Given a large graph G and a query graph q, the task subgraph matching (SM) is to find the subgraphs in G that are isomorphic to q.
(2) *Minimum Steiner Tree.* Given a graph $G = (V, E)$ with non-negative edge weights and a set of terminal nodes $R \subseteq V$, the minimum Steiner tree problem (MSTP) is defined as finding a tree T in G which covers all nodes in R with the minimum weight, where V and E are the vertices and edges in G, respectively.

Clearly, both SM and MSTP are subgraph-level tasks, and they suffer from intractable computational cost. Thus, a lot of efforts have been made to improve the time efficiency [13,22] We notice that there are some works dealing with subgraph search by importing deep learning components [5,9,26]. However, these works focus on using deep learning models to determine a better searching order. Thus, they still fall into the classical search frameworks.

Considering the superiority of deep learning, a question arises *"can we develop a novel paradigm, leveraging the deep learning techniques thoroughly, to handle the subgraph-level problems?"*. A promising paradigm dealing with subgraph-level tasks consists of two key modules, i.e., *prediction* and *reconstruction*. Specifically, given the queries over graph G, the embeddings of the target subgraphs are inferred through the prediction model, and then the subgraphs are expected to be identified based on the subgraph embeddings. Therefore, an effective subgraph embedding is desired to enable such a paradigm.

Although subgraph plays an important role in graph representation learning [12,25] , there are just a few works [3,6,17] focusing on embedding subgraphs. Sub2Vec [3] learns vector representations such that the next node in the random walk can be predicted by the learned subgraph embedding and node embeddings. SubRank [6] is trained inspired by the random walk proximity measure Personalized PageRank. S2N [17] constructs a new graph G' by coarsely transforming each given subgraph into a "node" in G'. Then it facilitates subgraph-level tasks through node-level tasks in the graph G'. We notice that all the subgraph embedding methods above cannot be used to recover the subgraph directly if the subgraph embeddings are not materialized. In other words, these subgraph embeddings fail to be plugged into the predication-and-reconstruction paradigm above to tackle the subgraph-level tasks.

1.2 Our Approach and Contributions

Motivated by the limits of the existing subgraph embeddings, we elaborate on learning *"reversible"* distributed representations of subgraphs in a low dimensional continuous vector space. The term *"reversible"* means that a subgraph g can be reconstructed precisely just based on the embedding y of g. In this paper, we extend the popular way of graph embedding, which reads out graph by aggregating node embeddings, to a subgraph. Specifically, given the node embedding matrix Φ, any subgraph g can be embedded as the product of Φ and a high dimension vector x (also called as *aggregation vector* which depends on g), *i.e.*,, $y = \Phi x$. The node embeddings can be learned by exploiting popular graph neural networks. During the process, two major challenges have to be addressed.

- How to design a reversible read-out operation such that the aggregation vector x can be recovered according to the subgraph embedding y?
- How to reconstruct the structure of subgraph g based on the aggregation vector x?

To address the first challenge, we resort to compressed sensing theory, which focuses on the exact recovery of sparse linear transformation [7]. We propose an approach to learning node embeddings nearly satisfying Restricted Isometry Property, and consequently, exact decoding is achieved. We handle the second challenge by designing a bijective rule of generating the aggregation vector according to the adjacency matrix of subgraph g, by applying a simple transformation between binary number and decimal number with some scaling operation. The major contributions of this paper are summarized as follows.

- To the best of our knowledge, we are the first to propose a framework for subgraph reconstruction powered by a reversible subgraph embedding;
- We propose a novel loss function for GNN models according to the compressed sensing theory, such that the sparse linear aggregation of node embeddings can be decoded exactly;
- We design a bijective aggregation rule to readout the embedding of the subgraph, which enables that the subgraph can be reconstructed through the aggregation vector;
- Experimental results on real graphs demonstrate the overwhelming superiority of the proposed subgraph embedding in tasks like subgraph reconstruction and graph reconstruction.

2   Problem Definition and Framework

Let $G = \{V(G), E(G)\}$ represents an undirected graph, where $V(G)$ and $E(G)$ denote the set of vertices and edges of G, respectively.

2.1   Problem Formulation

Definition 1. *(Induced Subgraph). A graph g is an induced subgraph of G if (1) $V(g) \subseteq V(G)$, and (2) each two vertices have an edge in $E(g)$ only if the two vertices have an edge in $E(G)$.*

Definition 2. *(Embedded Subgraph). A graph g is an embedded subgraph of G if (1) $V(g) \subseteq V(G)$, and (2) $E(g) \subseteq E(G)$.*

Definition 3. *(Subgraph Reconstruction). Assume that we have obtained a set of learned embeddings for a graph G. Given the embedding for a subgraph g, the subgraph reconstruction task is to reconstruct the topology structure of g according to the available embeddings.*

Notice that the topology structure of neither G nor g can be used in the whole process of the reconstruction task.

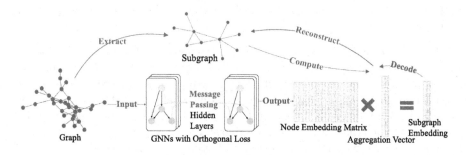

Fig. 1. Framework of subgraph reconstruction via the reversible subgraph embedding.

2.2   Compressed Sensing

Compressed sensing theory, as known as sparse linear regression, indicates that we can solve an underdetermined system of equations with a unique solution under the *restricted isometry hypothesis* [7]. For a linear system $y = \Theta x + \varepsilon$ where $y \in \mathbb{R}^p$, $\Theta \in \mathbb{R}^{p \times n}$ and $x, \varepsilon \in \mathbb{R}^n$ with $p \ll n$, the S-restricted isometry constant (RIC) δ_S of Θ is defined as the smallest quantity such that

$$(1 - \delta_S)\|x\|_{\ell_2}^2 \leq \|\Theta x\|_{\ell_2}^2 \leq (1 + \delta_S)\|x\|_{\ell_2}^2 \tag{1}$$

for all $\|x\|_{\ell_0} \leq S$ [7]. They propose the *restricted isometry principle (R.I.P)* stating that if S obeys $\delta_S + \delta_{2S} + \delta_{3S} < 1$, then we can recover any sparse x in noiseless case with support size no more than S by solving ℓ_1-minimization

$$\min_{\tilde{x} \in \mathbb{R}^n} \|\tilde{x}\|_{\ell_1} \quad \text{subject to} \quad \Theta \tilde{x} = y. \tag{2}$$

In most applications, the sensing matrix Θ is designed as the product of a signal basis matrix Ψ and a measurement matrix Φ. In this paper, we focus on a training based way to design the measurement matrix Φ when Ψ is given.

2.3 Framework

In this paper, we propose a reversible embedding technique for small subgraphs (compared with the entire graph). Figure 1 gives the framework powered by reversible subgraph embedding, consisting of three major steps, where the recovery process is indicated via red lines.

Step 1: *Learning node embeddings.* The reversible property demands that the node embeddings can be recovered when given the subgraph embedding. Thus, it is required to design node embeddings satisfying such a constraint. Our approach is motivated by the compressed sensing theory, that is, if we can design a set of node embeddings that satisfy *Restricted Isometry Property*, any sparse linear aggregation of the node embedding can be recovered exactly.

Step 2: *Aggregation vector generation.* To enable the mutual transformation between the aggregation vector and adjacency matrix of the subgraph, a bijection should be established. In this paper, we propose a simple but effective bijective rule to generate an aggregation vector from the adjacency matrix.

Step 3: *Subgraph embedding generation.* Obtaining node embedding matrix and aggregation vector, the subgraph embedding is straightforward to be produced by taking their product.

3 Reversible Subgraph Embedding

3.1 Node Embedding

We will introduce how to train node embeddings which satisfy *R.I.P.*. The main idea is to combine the popular graph neural network models (GNNs) with a loss function specifically designed for compressed sensing. The loss function, also called as orthogonal loss function, is formulated as

$$\mathcal{L} = \frac{1}{2}\|(\Phi\Psi)^{\mathsf{T}}(\Phi\Psi) - \mathbf{I}\|_F^2, \tag{3}$$

where Φ is the output of GNNs and Ψ is the signal basis matrix. However, as the size of matrix grows, making the gram matrix of $\Phi\Psi$ as close as the identity matrix is a pretty strict constraint. Hence, Abolghase et al. [1] propose a advanced loss function $\mathcal{L} = \frac{1}{2}\|(\Phi\Psi)^{\top}(\Phi\Psi) - H\|_F^2$, where H is updated in each training epoch following the rule

$$H_{ij} = \begin{cases} 1, & i = j \\ \Theta_{ij}, & i \neq j \text{ and } |\Theta_{ij}| < \mu \\ \mu \cdot \mathrm{sgn}(\Theta_{ij}), & \text{otherwise} \end{cases} \tag{4}$$

Here, $\Theta = (\Phi\Psi)^{\top}(\Phi\Psi)$ and μ is a predefined threshold and $\mathrm{sgn}(x) = \begin{cases} 1, & x > 0 \\ -1, & x < 0 \end{cases}$.

In our implementation, we use the discrete cosine transform (DCT) matrix as Ψ and compute Φ through a three-layer graph convolutional networks (GCNs).

In the following parts, we will briefly introduce DCT and GCNs. Discrete cosine transform [4] is one of the most popular transformation techniques in data compressing. The orthogonal version of DCT matrix Ψ is defined as

$$\Psi_{i,j} = a_i \cos(\frac{(2j+1)\pi i}{2N}), \; i,j = 0,1,\ldots,N-1, \tag{5}$$

where

$$a_i = \begin{cases} \sqrt{\frac{1}{N}}, & i = 0 \\ \sqrt{\frac{2}{N}}, & i = 1,2,\ldots,N-1 \end{cases}. \tag{6}$$

Graph convolutional networks are a popular approach for semi-supervised learning on graph-structured data, which is based on an efficient variant of convolutional neural networks. The graph spacial convolution layer is defined as

$$X^{(l+1)} = \sigma(\tilde{D}^{-\frac{1}{2}}\tilde{A}\tilde{D}^{-\frac{1}{2}}X^{(l)}W^{(l)}). \tag{7}$$

Here, $X^{(l)}$ and $W^{(l)}$ denote the output and trainable weight matrix of l-th layer respectively, $\tilde{A} = A + I_N$ is the adjusted adjacency matrix where A and I_N are adjacency matrix and identity matrix, and $\tilde{D} = \mathrm{diag}(\sum_j A_{1j}, \ldots, \sum_j A_{Nj})$ is the diagonal matrix taking degree of each node as elements.

3.2   Subgraph Embedding

If we establish a bijection from the adjacency matrix of a subgraph g to a sparse vector x, we can also construct a bijection from the adjacency matrix to subgraph embedding $y = \Theta x$. Let us consider a subgraph g of G. We define g's adjacency matrix $A^g \in \mathbb{R}^{n \times n}$ in the view of the whole graph rather than restricting it to the scale (i.e., the number of vertices contained in g). We set all elements in corresponding rows and columns of A to 0 for those nodes not contained in g. Then, we can not only understand the structure of g but also locate it on G.

Definition 4. *(Adjacency Matrix). Given a graph G with size n and one of its subgraph g, the adjacency matrix of g, denoted as A^g, is defined as*

$$A_{ij}^g = \begin{cases} 1, & \text{if } i \text{ connect to } j \text{ in } g \\ 0, & \text{otherwise} \end{cases}, \; \forall \, i,j \in [n] = \{1,2,\ldots,n\}.$$

Since every element of the adjacency matrix is 0 or 1, we can view each row or column of A as a binary number. Hence, we can encode A as a vector by transforming each row of A into a decimal number. However, this approach is often affordable because a binary number with length n refers to $O(2^n)$. Thus, we just record the nodes in g by the minor adjacency matrix as defined next.

Definition 5. *(Minor Adjacency Matrix). Given a subgraph g with size d, the minor adjacency matrix of g, denoted as $\tilde{A}^g$, is defined as*

$$\tilde{A}_{\phi(i),\phi(j)}^g = \begin{cases} 1, & \text{if } i \text{ connect to } j \text{ in } g \\ 0, & \text{otherwise} \end{cases}, \; \forall \, \phi(i), \phi(j) > 0.$$

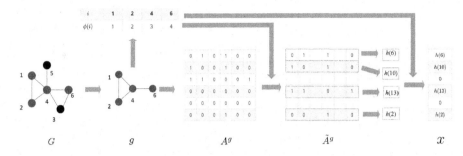

Fig. 2. Illustration of computing adjacency vector for subgraph g.

where $\phi(\cdot)$ is the mapping of node index as

$$\phi : [n] \to [d] \quad \phi(k) = \begin{cases} \sum_{i=1}^{k} \mathbf{1}_{\{i \in g\}} & \text{if } k \in g \\ -1 & \text{if } k \notin g \end{cases}, \forall k \in G.$$

Finally, we formulate our implementation mathematically. Let $f(\cdot)$ and $f^{-1}(\cdot)$ denote the transformation from binary to decimal and from decimal to binary, respectively. Similarly, let $h(\cdot)$ denote the scaling function and $h^{-1}(\cdot)$ as its inverse function. The adjacency vector x can be computed following

$$x_i = \begin{cases} h(f(\tilde{A}^g_{\phi(i)})) & \text{if } \phi(i) > 0 \\ 0 & \text{if } \phi(i) = -1 \end{cases}, \forall i \in [n], \tag{8}$$

where x_i and $\tilde{A}_i$ refer to the i-th element of x and i-th row of $\tilde{A}$, respectively. The ultimate embedding y of subgraph g is computed as $y = \Theta x$.

Example 2. Given the graph G and its subgraph g as shown in Fig. 2, we can directly get the node index mapping $\phi(\cdot)$ and g's adjacency matrix A^g. Then, we extract the minor adjacency matrix $\tilde{A}^g$ from A^g following $\phi(\cdot)$. Each row in $\tilde{A}^g$ can be viewed as a binary number and compressed into a demical number through $h(f(\cdot))$. For instance, $(0, 1, 1, 0)$, the first row of $\tilde{A}^g$, is compressed as $h(6)$, where $h(\cdot)$ is the scaling function. Finally, we map each element in the adjacency vector as the corresponding row of $\tilde{A}^g$ after compressed according to $\phi(\cdot)$, where those elements not in g are set to 0.

4   Subgraph Reconstruction

In this section, we present a reconstruction strategy. Suppose that we are given y, the embedding of an unknown subgraph, we can reconstruct the subgraph by following two steps.

- Decoding the embedding into the corresponding adjacency vector;
- Reconstruct the subgraph according to the adjacency vector.

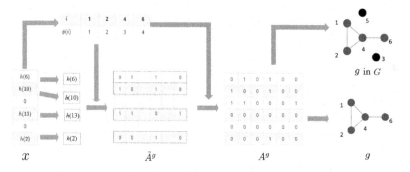

Fig. 3. Reconstructing subgraph g based on its adjacency vector and locating it in G.

The first step actually conducts the classic decoding task in compressed sensing. However, due to the loss of precision in computing, we cannot directly apply the exact recovery algorithm depending on the noiseless case assumption. To address the problem, we propose a decoding algorithm particularly according to the encoding method. The key point is that in our encoding approach, each element in the support set of adjacency matrix should be larger than a designed threshold μ. For instance, if we set $h(x) = \ln(1+x)$, we have $\mu = \ln 2$ (we suppose that there is no isolated node in the subgraph). According to this property, we can correct the solutions based on the classical decoding algorithm. First, we get a basic solution $\hat{x}$ by fitting a LASSO regression model, that is

$$\hat{x} = \arg\min_{x} \frac{1}{2}\|y - \Theta x\|_{\ell_2}^2 + \alpha\|x\|_{\ell_1} \tag{9}$$

Then we examine the support set of $\hat{x}$, denoted as $\text{supp}(\hat{x})$. If every element in $\text{supp}(\hat{x})$ is larger than the threshold μ, we just return $\hat{x}$ as the solution. If there is any element in $\text{supp}(\hat{x})$ smaller than the threshold μ, we remove it from $\text{supp}(\hat{x})$ and resolve the equation system

$$y = \Theta_{\text{supp}(\hat{x})}\hat{x}_{\text{supp}(\hat{x})}, \tag{10}$$

where the $\Theta_{\text{supp}(\hat{x})}$ and $\hat{x}_{\text{supp}(\hat{x})}$ means the sub matrix or sub vector only contains the rows or elements in $\text{supp}(\hat{x})$. Intuitively, this correction step can be repeated until all elements in $\text{supp}(\hat{x})$ are larger than the threshold μ, while one step is enough in most cases as shown in our experiments.

After getting the adjacency vector $\hat{x}$, we also set a map $\phi' : [n] \to [d]$ from the index of $\hat{x}$ to $\text{supp}(\hat{x})$ as

$$\phi'(k) = \begin{cases} \sum_{i=1}^{k} \mathbf{1}_{\{i \in \text{supp}(\hat{x})\}} & \text{if } k \in \text{supp}(\hat{x}) \\ -1 & \text{if } k \notin \text{supp}(\hat{x}) \end{cases}, \forall k \in [n]. \tag{11}$$

Then, we can reconstruct $\tilde{A}$ following the way

$$\tilde{A}_i = h^{-1}\big(f^{-1}(x_{\arg_j\{\phi'(j)=i\}})\big), \forall i \in [d]. \tag{12}$$

Finally, we can reconstruct A according to $\tilde{A}$ as

$$
A_{ij} = \begin{cases} \tilde{A}_{\phi'(i)\phi'(j)} & \text{if } \phi'(i), \phi'(j) > 0 \\ 0 & \text{otherwise} \end{cases}. \tag{13}
$$

In the reconstruction process, we only need to input the embedding y of subgraph g and the node embedding matrix Θ, totally independent of the structure of g and G.

Example 3. Figure 3 presents an example of how to reconstruct the subgraph g according to its adjacency vector x. Once x is obtained, we can establish the mapping of index from x to its support set $\text{supp}(x)$. Then, we use $h^{-1}(f^{-1}(\cdot))$ to decode each nonzero element of x into a row in g's minor adjacency matrix $\tilde{A}^g$ following the mapping $\phi(\cdot)$. Finally, following the mapping $\phi(\cdot)$ again, we augment $\tilde{A}^g$ into g's adjacency matrix A^g. According to A^g, it is straightforward to infer the structure of g. Moreover, g can be located in G at the same time.

5 Experiments

5.1 Experiment Setup

To demonstrate the effectiveness, four tasks are conducted. (1) Subgraph reconstruction: predicting the adjacency matrix of a subgraph g based on the learned embeddings; (2) Graph reconstruction: predicting the adjacency matrix of graph G just depends on the learned embeddings; (3) Node classification: predicting the labels of unlabelled nodes by looking at a set of labeled nodes; (4) Graph classification: predicting the label of each unlabelled graph.

Dataset Details: As shown in Tables 1 and 2, the widely used benchmarks for the tasks above are collected [18,21,27]. Note that although graph reconstruction seems like a graph-level task, the existing methods adopt node-level metrics for evaluation currently. So, we place the datasets used in the graph reconstruction task into the former table.

Reproducibility: The source codes are available at: https://github.com/ RevSubEmb/Reversible_Subgraph_Embedding. The dimension of each hidden convolution layer is set as 256 while the dimension of the output layer and the regularization coefficient in the decoding algorithm differs in different tasks, which will be introduced in corresponding subsections. All the experiments are conducted on a machine with an NVIDIA RTX 2080 GPU (12 GB memory), Intel Xeon(R) CPU (2.50 GHz), and 256 GB of RAM.

5.2   Subgraph Reconstruction

Given a graph G, we will randomly generate one subgraph g from G to test whether we can recover g through learned embeddings, e.g., g's embedding y and node embedding matrix Θ in our approach. Generally, there are two kinds of graphs, i.e., *induced subgraph* (an edge is added between two nodes in g only if the edge exists in G) and *embedded subgraph* (two nodes are not necessary to have an edge in g even if they have an edge in G). Thus, we investigate the reconstruction over the two kinds of subgraphs.

Evaluation on Induced Subgraph Reconstruction. The existing node embeddings, e.g., Node2Vec [10] and Attention Walk [2], can be extended to support subgraph reconstruction. Specifically, we compute the inner product for each pair of nodes in the subgraph as an indicating score and then choose the highest m pairs of nodes to add edges between each selected pair of nodes. Here m denotes the number of edges in the given subgraph. We investigate induced subgraph with $|g|$ equals to 5, 10, 15, and 20 on three datasets respectively, for each $|g|$ (i.e., the number of nodes in g) 100 induced subgraphs are generated randomly. Table 3 lists the averaged precision of the methods. RSE-Θ outperforms Node2Vec and Attention Walk greatly as it recovers the subgraphs exactly.

Evaluation on Embedded Subgraph Reconstruction. The previous node embedding-based methods cannot be used to predict the subgraph when it is not an induced one. In contrast, our approach has no such limit. The experiments are conducted on three datasets: Cora, Citeseer, and ARXIV-GRQC. The dimensions of the node embeddings are set as 128, 256, and 256, respectively. For each dataset, we start with subgraphs with fixed size $|g| = 2$, then we iteratively increase $|g|$ from 2 to 30. For each size $|g|$, we separately generate 100 subgraphs through random walk. The implementation details and the results of the test are shown in Fig. 4. We can see that for pretty small $|g|$, our algorithm accurately reconstructs almost all subgraphs. For each dataset, there exists a threshold d, we call it as an invalid point, such that the success rate of exact reconstruction reduces sharply once the subgraph size $|g|$ reaches d. The invalid points of Cora, Citeseer, and ARXIV-GRQC in our test are 33, 35, and 31, respectively.

Table 1. Datasets for subgraph (graph) reconstruction and node classification.

Graphs	# Nodes	# Edges	# Classes
Cora	2708	5429	7
Citeseer	3327	4732	6
Pubmed	19717	44338	3
ARXIV-GRQC	5242	14484	–

Table 2. Datasets for graph classification.

Datasets	# Graphs	# Average nodes	# Classes
MUTAG	188	17.9	2
PTC	344	25.5	2
IMDb-BINARY	1000	19.8	2
IMDb-MULTI	1500	13.0	3

The reversible property of our subgraph embedding depends on compressed sensing theory. There is an upper bound of subgraph size for exact recovery. That is, as the subgraph grows larger, the corresponding adjacency vector becomes denser. Then we cannot decode it exactly when the adjacency vector is not sparse enough. The training method gives a bound $p > \mathcal{O}(|g| \log \frac{n}{|g|})$ in [1], where p denotes the dimension of embedding and $|g|$ denotes the number of vertices in a subgraph. However, since we combine the loss function with GCNs, extra constraints are introduced, and we need to re-explore this bound.

5.3 Graph Reconstruction

The basic idea of the existing methods, e.g., Node2Vec [10] and STRAP [28], is to predict whether there is an edge between nodes u and v based on the inner product of corresponding node embeddings. However, we can address this task in a novel way benefiting from the proposed reversible subgraph embedding. Due to the reversible property of proposed embeddings for small subgraphs, we can

Table 3. Averaged precision of induced subgraph reconstruction, where $|g|$ denotes the number of nodes of subgraph g and numerics in brackets represent standard deviation.

Datasets	$|g|$	Node2Vec	Attention Walk	RSE-Θ
Cora	5	69.88 ($\pm$ 10.18)	53.31 ($\pm$ 17.9)	**100**
	10	68.84 ($\pm$ 11.15)	30.93 ($\pm$ 16.95)	**100**
	15	63.94 ($\pm$ 7.92)	24.92 ($\pm$ 14.38)	**100**
	20	65.58 ($\pm$ 9.46)	22.26 ($\pm$ 10.95)	**100**
Citeseer	5	66.36 ($\pm$ 12.94)	44.05 ($\pm$ 20.90)	**100**
	10	59.60 ($\pm$ 10.11)	30.04 ($\pm$ 21.84)	**100**
	15	55.65 ($\pm$ 9.64)	37.69 ($\pm$ 24.24)	**100**
	20	56.28 ($\pm$ 11.77)	33.38 ($\pm$ 24.05)	**100**
GRQC	5	70.47 ($\pm$ 11.51)	69.50 ($\pm$ 21.07)	**100**
	10	70.20 ($\pm$ 11.39)	51.54 ($\pm$ 22.71)	**100**
	15	70.98 ($\pm$ 12.08)	47.60 ($\pm$ 24.83)	**100**
	20	73.74 ($\pm$ 11.77)	39.38 ($\pm$ 22.91)	**100**

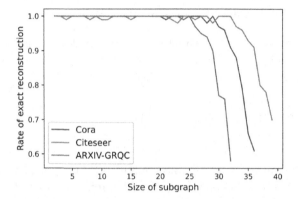

Fig. 4. Results of embedded subgraph reconstruction on Cora, Citeseer and ARXIV-GRQC. For parameters in compressed sensing decoding algorithm, we set the regularization coefficient α to 1×10^{-7}, 1×10^{-6} and 1×10^{-5} on Cora, Citeseer and ARXIV-GRQC, respectively, while we set the threshold μ to 0.25 in all three graphs.

partition G into small subgraphs and reconstruct these subgraphs separately. Finally, the reconstructed subgraphs are combined to form the final solution. To reduce the influence brought by the partition algorithm, we simply reconstruct the ego network of each node, that is, the induced subgraph of G that contains the node and its neighbors only. To handle high-degree nodes, we set a threshold D. If the degree of any node is larger than D, we just randomly sample D nodes from their neighbors and reconstruct the subgraph induced by the selected nodes.

We test the graph reconstruction task on three datasets: Cora, Citeseer, and ARXIV-GRQC. The dimensions of embedding are set as 128, 128, and 256 respectively. To evaluate the performance of our method, we report precision, recall, and the number of incorrectly predicted edges. The results are shown in the top of Table 4. We can see that our method almost exactly recovers these three graphs. Furthermore, the proposed method always keeps 100% precision in three datasets. It is because we choose the threshold D a bit smaller than the invalid point d, and our subgraph reconstruction algorithm works perfectly on subgraphs with these sizes (as shown in Sect. 5.2). Since we randomly sample neighbors when a node's degree is larger than D, some edges may not be covered by the selected subgraphs, which incurs a subtle decline in performance.

As shown in Table 4, we report the metric precision@k of Node2Vec and STRAP, which denotes the precision computed by selecting the top-k pairs of nodes in the ranked candidate edges. Here, we set k to the number of edges of each graph. Clearly, those works are all far from reconstructing the graph exactly, which confirms our effectiveness of the proposed reversible subgraph embedding.

Table 4. Results of graph reconstruction task, where precision (P) and recall (R) are written in the percentage form. The threshold of degree is set to 20 for all three graphs.

Datasets	Cora	Citeseer	GRQC
RSE-P (%)	**100.00**	**100.00**	**100.00**
RSE-R (%)	**99.98**	**100.00**	**99.97**
# False edges	6	0	4
Node2Vec (precision, %)	48.77	34.47	44.07
STRAP (precision, %)	72.68	57.91	71.81

Table 5. Results of node classification task with 20 instances per class as training set. The column Φ and Θ means using Φ and Θ as node embeddings, respectively. The rows means how many hops of neighbors are used to compute subgraph embedding as node representation, where 0-hop means just using the node embedding itself.

Datasets	Cora		Citeseer		PubMed	
	Φ	Θ	Φ	Θ	Φ	Θ
RSE (0-hop)	32.8	13.8	43.7	16.6	49.4	37.4
RSE (1-hop)	55.2	30.8	55.1	25.9	62.0	37.5
RSE (2-hop)	67.3	45.2	56.0	32.7	72.9	35.6
Sub2Vec	29.8		31.2		36.8	
SubRank	61.9		42.3		65.4	
Node2vec	63.1		45.6		51.1	
Attention Walk	67.9		51.5		70.2	
SSP	82.8		74.3		80.1	

5.4 Node Classification

We apply our method to the node classification task on datasets Cora, Citeseer, and Pubmed. The dimensions of node embedding are set to 128, 128, and 256 on the three graphs. We use two different splits of these datasets in the node classification task. Both two splits select 500 instances for validation and 1000 for test. One split selects 20 instances each class for training, while the other split uses all nodes except the validation set and test set as the training set.

Let RSE denote our proposed subgraph embedding in the paper. As mentioned before, either Θ or Φ in the proposed model can be used as node embeddings in different tasks. Intuitively, Θ is more suitable for reconstruction-related tasks, while Φ is more suitable for data mining tasks. Hence, we will evaluate both cases in this task. Moreover, to evaluate the proposed subgraph embedding, we can also represent a node using the embedding of the induced subgraph generated by its 1-hop or 2-hop neighbors. The classifier is implemented by employing a three-layer MLP that takes $\tanh(\cdot)$ and $\text{sigmoid}(\cdot)$ as the activate function in the former two layers and the output layer, respectively.

Baselines can be divided into two categories. One is trained independently from the downstream task, e.g., Node2Vec [10] and Attention Walk [2]. The other one is the end-to-end model such as SplineCNN [8] and SSP [14], which usually performs much better than the models in the former category. That is because the extra knowledge from the downstream task can be learned and integrated into the model. The results are presented in the Tables 5 and 6. As observed, exploiting Φ as node embedding performs much better than using Θ, which accords with intuition. More detailed analysis will be given in Sect. 5.6.

Table 6. Results of node classification (all nodes except validation/test set as the training set).

Datasets	Cora		Citeseer		PubMed	
	Φ	Θ	Φ	Θ	Φ	Θ
RSE (0-hop)	50.9	31.9	59.5	18.3	56.5	40.7
RSE (1-hop)	77.0	43.7	73.0	37.6	73.7	50.1
RSE (2-hop)	80.5	66.0	71.9	50.3	80.2	71.3
Sub2Vec	56.2		47.4		71.2	
SubRank	75.6		61.1		81.8	
Node2Vec	74.0		56.7		77.8	
Attention Walk	57.3		56.4		76.6	
SplineCNN	89.48		79.20		88.88	
SSP	87.60		79.50		88.46	

Table 7. Results of graph classification task (the average accuracy on ten fold validation sets, with standard deviation in brackets)

Datasets	MUTAG	PTC	IMDb-BINARY	IMDb-MULTI
RSE-Θ	58.04 ($\pm$7.28)	43.63 ($\pm$6.69)	45.90 ($\pm$ 4.41)	32.67 ($\pm$ 2.78)
RSE-Φ	68.57 ($\pm$6.67)	58.11 ($\pm$3.83)	79.50 ($\pm$3.38)	52.60 ($\pm$2.67)
Sub2Vec	66.49 ($\pm$ 2.28)	53.49 ($\pm$ 7.53)	53.30 ($\pm$ 2.01)	40.53 ($\pm$ 4.46)
GIN	89.40 ($\pm$5.60)	64.60 ($\pm$7.00)	75.10 ($\pm$5.10)	52.30 ($\pm$2.80)
SEG-BERT	89.80($\pm$ 6.71)	64.84($\pm$ 6.77)	74.70 ($\pm$ 3.74)	50.60($\pm$ 3.73)

5.5   Graph Classification

Different from the tasks above, each dataset in the graph classification contains hundreds of graphs, but each of which is very small, which makes it difficult for our method to generate subgraph embeddings. To relieve the problem, we combine all the graphs in each dataset as a "super" graph, on which the node embeddings are trained. Note that the input graph of our method is not necessary to be a connected graph. So, we can view each graph in the dataset as a subgraph of the super graph, and we can compute its embedding following the subgraph embedding method proposed in this paper. The classifier is implemented in the same way as that in the node classification task. Since the amount of data is too small for universal data splitting in transductive learning, we use 10-fold cross validation as what [27] does. We compare our method RSE with two state-of-the-art algorithms GIN [27] and SEG-BERT [29]. The results are reported in Table 7. Our model performs roughly as well as the GIN on most datasets except MUTAG. Because GIN is an end-to-end model while our embedding method learns the representation of graphs without using any information of the classification task, the gap is acceptable. Furthermore, we find that $RSE\text{-}\Phi$ outperforms GIN on datasets IMDb-BINARY and IMDb-MULTI.

5.6   Discussion

Next, we discuss two questions based on the empirical results above.

– Why does the proposed subgraph embedding show significant superiority in both subgraph and graph reconstructions?
– Why does use Φ as node embedding usually performs better than using Θ in graph classification?

Overwhelming Superiority in (Sub-)Graph Reconstruction. Our proposed method exhibits overwhelming advantages in both subgraph and graph reconstructions, almost recovering the subgraphs and graphs precisely. The reason is that our node embeddings are trained under the guidance of compressed sensing theory such that the readout option can be decoded exactly in most cases. Moreover, the transformation between adjacency vector and adjacency matrix is bijection which ensures precise reconstruction of the (sub-)graph.

Φ **as Node Embedding vs. Θ as Node Embedding.** Θ is trained under the guidance of $R.I.P.$ to realize exactly sparse decoding. Furthermore, there is another equivalent explanation of the training purpose, that is, trying to make any k columns of Θ linear independent. Therefore, using Θ as node embedding heavily peels the commonness between nodes and that is why it performs poorly on classification-related tasks. However, it is also the reason why using Φ as node embedding performs better. Since $\Theta = \Phi\Psi$, where Ψ is a signal basis matrix, we can view Φ as the coordinates in the space spanned by Ψ. So, Φ reflects how to strip the correlation of nodes, delivering more information in classification tasks.

6   Related Work

Early graph embedding methods view each node in the graph as a word, then generate node sequences randomly based on some specifically designed rules, such as random walk and deep walk [20], and finally maximize the co-occurrence probability to create the representation. DeepWalk [20] generates training data by truncated random walk, others like LINE [23] by at most one-step random walk, Node2Vec [10] by truncated high order random walk, Attention Walk [2] by attention guided random walk, and so on. Some other graph embedding techniques based on matrix factorization have been proposed [19,31]. They can be applied to reconstruct a graph by predicting edges based on proximities, however, fail to recover the graph exactly. Most graph neural network models use the loss of downstream classification tasks as the objective function to train the model. Since they are specifically designed in end-to-end form for node classification or graph classification task, models like GAT [24], GResNet [30], and SSP [15], usually perform much better than probabilistic and factorization based methods on the node classification task.

As a probabilistic embedding method, Sub2Vec [3] learns a distributed representation for each subgraph such that the likelihood of preserving neighborhood and structural properties in the feature space is maximized. SubRank [6] introduces a new measure of proximity for subgraphs, based on which subgraph embeddings are computed. S2N [17] introduces a translation from subgraphs to nodes. It constructs a new graph by taking each subgraph as a "node" and adding an edge between two "nodes" if the two corresponding subgraphs share at least one node in the original graph. Then popular graph neural networks can be applied on the newly constructed graph.

As for the combination of deep learning methods and subgraph-level tasks, most existing works learn how to explore the search space under the classical search frameworks. GLSearch [5] proposes a novel GNN-based deep Q-network to iteratively search node pairs for the maximum common subgraph detection. Ge et al. present a machine-based active learning component in subgraph matching, suggesting the search order to reduce the search space [9]. RL-QVO [26] deals with the similar task, but adopts the reinforcement learning models.

7   Conclusion

In this paper, we develop a framework to generate reversible distributed representations for subgraphs such that the subgraph can be exactly reconstructed once given the subgraph embedding. To the end, a reversible read-out operation is proposed based on the compressed sensing theory. Moreover, we build a simple but effective bijection between the aggregation vector and adjacency matrix of the subgraph. Extensive experiments on real graphs have been conducted to evaluate the proposed subgraph embedding, and the experimental results confirm the effectiveness of our approach.

Acknowledgement. This work was supported by National Natural Science Foundation of China (Grant No. 61902074).

References

1. Abolghasemi, V., Ferdowsi, S., Sanei, S.: A gradient-based alternating minimization approach for optimization of the measurement matrix in compressive sensing. Signal Process. **92**(4), 999–1009 (2012)
2. Abu-El-Haija, S., Perozzi, B., Al-Rfou, R., Alemi, A.A.: Watch your step: learning node embeddings via graph attention. In: NeurIPS, pp. 9198–9208 (2018)
3. Adhikari, B., Zhang, Y., Ramakrishnan, N., Prakash, B.A.: Sub2vec: feature learning for subgraphs. In: PAKDD, pp. 170–182 (2018)
4. Ahmed, N., Natarajan, T., Rao, K.R.: Discrete cosine transform. IEEE Trans. Comput. **100**(1), 90–93 (1974)
5. Bai, Y., Xu, D., Sun, Y., Wang, W.: GLSearch: maximum common subgraph detection via learning to search. In: ICML, vol. 139, pp. 588–598 (2021)
6. Balalau, O., Goyal, S.: SubRank: subgraph embeddings via a subgraph proximity measure. In: PAKDD, pp. 487–498 (2020)
7. Candes, E.J., Tao, T.: Decoding by linear programming. IEEE Trans. Inf. Theory **51**(12), 4203–4215 (2005)
8. Fey, M., Lenssen, J.E., Weichert, F., Müller, H.: SplineCNN: fast geometric deep learning with continuous b-spline kernels. In: CVPR, pp. 869–877 (2018)
9. Ge, Y., Bertozzi, A.L.: Active learning for the subgraph matching problem. In: Big Data, pp. 2641–2649 (2021)
10. Grover, A., Leskovec, J.: node2vec: scalable feature learning for networks. In: SIGKDD, pp. 855–864 (2016)
11. Hao, Z., et al.: ASGN: an active semi-supervised graph neural network for molecular property prediction. In: SIGKDD, pp. 731–752. ACM (2020)
12. Huang, K., Zitnik, M.: Graph meta learning via local subgraphs. In: NeurIPS (2020)
13. Iwata, Y., Shigemura, T.: Separator-based pruned dynamic programming for steiner tree. In: AAAI, pp. 1520–1527 (2019)
14. Izadi, M.R., Fang, Y., Stevenson, R., Lin, L.: Optimization of graph neural networks with natural gradient descent. arXiv preprint arXiv:2008.09624 (2020)
15. Izadi, M.R., Fang, Y., Stevenson, R., Lin, L.: Optimization of graph neural networks with natural gradient descent. CoRR abs/2008.09624 (2020)
16. Jin, W., Ma, Y., Liu, X., Tang, X., Wang, S., Tang, J.: Graph structure learning for robust graph neural networks. In: SIGKDD, pp. 66–74. ACM (2020)
17. Kim, D., Oh, A.: Efficient representation learning of subgraphs by subgraph-to-node translation. CoRR abs/2204.04510 (2022)
18. Leskovec, J., Kleinberg, J., Faloutsos, C.: Graph evolution: densification and shrinking diameters. TKDD **1**(1), 2-es (2007)
19. Ou, M., Cui, P., Pei, J., Zhang, Z., Zhu, W.: Asymmetric transitivity preserving graph embedding. In: SIGKDD, pp. 1105–1114 (2016)
20. Perozzi, B., Al-Rfou, R., Skiena, S.: Deepwalk: online learning of social representations. In: SIGKDD, pp. 701–710 (2014)
21. Sen, P., Namata, G., Bilgic, M., Getoor, L., Galligher, B., Eliassi-Rad, T.: Collective classification in network data. AI Mag. **29**(3), 93–93 (2008)
22. Sun, S., Luo, Q.: In-memory subgraph matching: an in-depth study. In: SIGMOD, pp. 1083–1098. ACM (2020)
23. Tang, J., Qu, M., Wang, M., Zhang, M., Yan, J., Mei, Q.: Line: large-scale information network embedding. In: WWW, pp. 1067–1077 (2015)

24. Velickovic, P., Cucurull, G., Casanova, A., Romero, A., Liò, P., Bengio, Y.: Graph attention networks. In: ICLR, OpenReview.net (2018)
25. Wang, C., Liu, Z.: Graph representation learning by ensemble aggregating subgraphs via mutual information maximization. CoRR abs/2103.13125 (2021)
26. Wang, H., Zhang, Y., Qin, L., Wang, W., Zhang, W., Lin, X.: Reinforcement learning based query vertex ordering model for subgraph matching, pp. 245–258 (2022)
27. Xu, K., Hu, W., Leskovec, J., Jegelka, S.: How powerful are graph neural networks? In: ICLR (2018)
28. Yin, Y., Wei, Z.: Scalable graph embeddings via sparse transpose proximities. In: SIGKDD, pp. 1429–1437 (2019)
29. Zhang, J.: Segmented graph-BERT for graph instance modeling. arXiv preprint arXiv:2002.03283 (2020)
30. Zhang, J., Meng, L.: Gresnet: graph residual network for reviving deep gnns from suspended animation. CoRR abs/1909.05729 (2019)
31. Zhang, Z., Cui, P., Wang, X., Pei, J., Yao, X., Zhu, W.: Arbitrary-order proximity preserved network embedding. In: SIGKDD, pp. 2778–2786 (2018)

Efficiently Answering Why-Not Questions on Radius-Bounded k-Core Searches

Chuanyu Zong[1]([✉]), Zefang Dong[1], Xiaochun Yang[2], Bin Wang[2], Tao Qiu[1], and Huaijie Zhu[3]

[1] School of Computer Science, Shenyang Aerospace University, Liaoning 110136, China
{zongcy,qiutao}@sau.edu.cn, dongzefang@stu.sau.edu.cn
[2] School of Computer Science and Engineering, Northeastern University, Liaoning 110819, China
{yangxc,binwang}@mail.neu.edu.cn
[3] Laboratory of Big Data Analysis and Processing, Sun Yat-sen University, Guangdong 510275, China
zhuhuaijie@mail.sysu.edu.cn

Abstract. The problem of searching radius-bounded k-cores (RB-k-cores) for a given query vertex is to find cohesive subgraphs satisfying both social and spatial constraints on geo-social networks. However, the search results are effected by two parameters: the cohesive constraint k and the query radius r. Furthermore, as the users lack of enough professional knowledge, it is very hard to provide reasonable parameters for RB-k-core searches. In this case, some users will confuse that why some expected vertices are not included in the search results, which called why-not questions. Therefore, we will investigate the problem of answering why-not questions on radius-bound k-core searches in this paper, which is to find why some expected vertices are missing in the search results and how to make the expected vertices appear in the same RB-k-cores with query vertices. To tackle this problem, we firstly analyze the effect of the two parameters on the search results of RB-k-core searches. Then, we explore two effective algorithms by refining the initial search parameters k and r of RB-k-core searches. Finally, we conduct comprehensive experimental studies on four real geo-social networks to evaluate both the effectiveness and efficiency of our proposed explanation algorithms.

Keywords: Radius-bounded k-core searches · Why-not questions · Query refinement · Explanation

1 Introduction

In recent years, many geo-social networks have emerged, such as Foursquare and Twitter [5]. Geo-social network combines the social network information and geo-spatial information of users. Under this situation, finding subgraphs with high

© The Author(s), under exclusive license to Springer Nature Switzerland AG 2023
X. Wang et al. (Eds.): DASFAA 2023, LNCS 13945, pp. 93–109, 2023.
https://doi.org/10.1007/978-3-031-30675-4_7

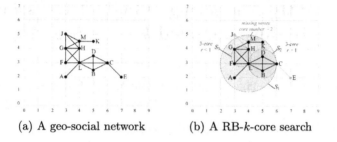

(a) A geo-social network   (b) A RB-k-core search

Fig. 1. An example of RB-k-core search

cohesiveness in geo-social networks has become a popular research topic. Given a geo-social network $G(V, E)$, a query vertex q in V, a positive integer k, and a query radius r, the problem of finding radius-bound cohesive subgraphs in G is to find all k-cores in G such that each k-core H contains query vertex q and all vertices in H fall into a circle with the query radius r. This problem is to search *radius-bounded* k-cores, named as RB-k-cores [5]. This search problem has many real-life applications. Especially, in Facebook, there is a valuable part Event-For-You, which can recommend events to users based on their social relationships and personal locations. For example, a user in Facebook wants to hold a party to play board games by gathering a group of people who are not living far away which are bounded by a circle with a radius r and each of whom has at least k friends in the group. The RB-k-core search can address this problem for the user. Figure 1(a) shows a geo-social network where vertices represent users, edges represent friendships, and locations represent the locations of users. When we set the query vertex L, the cohesive constraint $k = 3$, and the query radius $r = 1$, two RB-k-cores containing the query vertex L can be obtained as detailed in Fig. 1(b), i.e., $S_2 = \{L, D, C, B\}$ and $S_3 = \{G, H, F, L\}$.

However, the results of RB-k-core searches are sensitive to the two input parameters k and r. Additionally, it is very hard that providing appropriate parameters for the RB-k-core searches due to the users lack of enough professional acknowledge. If the two parameters k and r are not set good enough, the search results cannot meet the users' requirements. In this situation, users often have to refine the queries multiple times to find desirable search results. The explanation capability for users supported by *query refining* is thus desirable to help them tune their original queries to obtain desirable query results. Specifically, one often want to ask a why-not question that WHY an expected vertex is NOT in the search results. Why-not question was proposed by Jagadish et al. [1], which asks why an expected tuple is missing in the query result. Answering why-not questions is to explore how to make the expected tuple appear in the query results. So far, many efforts have been made to answer why-not questions based on the query refining model [2], such as refining original queries for SQL queries [2], spatial keyword top-k queries [8], similar graph matching [9], structural graph clustering [10], and so on. However, the existing explanation solutions for why-not questions are far from meeting users' requirements.

In Fig. 1(b), when we set $k = 3$ and $r = 1$, for the query vertex L, two RB-k-cores S_2 and S_3 are obtained. However, L is not satisfied with the search results, the reason is that L confuses that why his friend M is not included in the search results since M is his/her best friend and they are not far away. In this situation, the user L may want to figure out an explanation for the why-not question that why his friend M is missing in the search results and how to make the expected vertex M appear in the search results. For the expected vertex M, we can answer the why-not question by relaxing at least one search parameter, e.g., decreasing the cohesive constraint $k=3$ to $k' = 2$ or increasing the query radius $r = 1$ to $r' = 2$, which can make the missing vertex M be included in the search results. It would be very useful for users that geo-social network system can explore explanations for why-not questions. Therefore, we investigate the problem of answering why-not questions on RB-k-core searches in this paper.

Moreover, to answer why-not questions by refining the original query parameters, one common principle is that the original query results should be retained as much as possible. That is, to guarantee the quality of answers/explanations for why-not questions on RB-k-core searches, the original search results should not be destroyed by the refined queries for RB-k-cores as much as possible. However, such an approach of manually seeking explanations by tuning parameters of RB-k-core is rather tedious, which involves possibly many rounds of RB-k-core refinement. Moreover, In the process of making the expected vertices appear in the search results by refining the search parameters, the search parameters could be over relaxed since many irrelevant vertices besides the expected vertices are included in the search results under the new refined parameters. Therefore, the main challenge of answering why-not questions on RB-k-core searches is how to quickly explore optimal refined search parameters to make expected vertices appear in the search results with minimum irrelevant vertices.

To sum up, the main contributes of this paper are summarized as follows:

(1) We analyze the influences of changing search parameters on the search results of RB-k-core searches from two aspects that decreasing the cohesive constraint k and increasing the query radius r.

(2) We explore two effective explanation algorithms to answer why-not questions on RB-k-core searches by refining the parameters k and r, respectively.

(3) We conduct comprehensive experiments on four real geo-social networks to evaluate our explanation algorithms, which show that they can efficiently answer why-not questions on RB-k-core searches.

2   Related Work and Problem Statement

In this section, we first review related work. Then, we present problem statement.

2.1   Related Work

Group Query over Geo-Social Networks. Yang et al. [3] propose a geo-social group query to find a group of nearby attendees with tight social relation. [4] proposes solutions for finding a spatial-aware community (SAC) that contains the query vertex. Wang et al. [5] propose a radius-bounded k-cores search that aims to find cohesive subgraphs satisfying both social and spatial constraints in large geo-social networks. [6] propose a geo-social community search problem (GCS) to find a social community and a cluster of spatial locations. Zhu et al. [7] explore a new group query problem of continuous geo-social groups monitoring (CGSGM) over moving users.

Why-Not Questions. [1] answers why-not questions by identifying the "culprit" operations which excludes missing tuples. [2] answers why-not questions by generating a refined query whose query results include both original query answers and missing answers. Chen et al. [8] answer why-not questions on spatial keyword top-k queries [9] answers why-not questions in similar graph matching. [10] answers why-not questions on structural graph clustering. Zheng et al. [11] investigate the problem of answering why-not group spatial keyword queries. [13] answers why-not questions on event pattern queries. Zhang et al. [12] propose solutions to answer why-not questions on top-k spatial keyword queries over moving objects. [14] proposes a novel approach to answer why-not questions over nested data. Song et al. [15] answer why questions for subgraph queries.

2.2   Preliminary

Given a geo-social network graph $G(V, E)$, where V denotes the set of vertices in G, and E represents the set of edges in G. The set of neighbors of a vertex u in $G(V, E)$ is formalized as $N_G(u) = \{v \in V | (u, v) \in E\}$. And the degree of the vertex u in G is denoted as $deg_G(u) = |N_G(u)|$. The Euclidean distance between two vertices u and v in G is denoted as $d(u, v)$. We use $C(u, r)$ to denote as the circle centered at vertex u with a radius r. Given a set of vertices H, an induced subgraph of G formed by H is denoted as $G(H)$, a set of binary-vertex-bounded circles with radius r is denoted as $W_r(v, u)$.

Definition 1. k-Core. *Given a geo-social network graph G and an integer k, the k-core of graph G is the maximal subgraph of G, which is denoted as H_k, such that $\forall v \in H_k, deg_{H_k}(v) \geq k$.*

Definition 2. Core Number. *Given a geo-social network graph G and an vertex v, the core number of v is the highest order of a k-core that contains v in G, denoted by $core_G[v]$.*

Definition 3. (Minimum Covering Circle (MCC for short)). *Given a set of vertices H, the minimum covering circle of H is the circle which encloses all the vertices in H with the smallest radius. The vertices which lie on the boundary of a MCC are called as boundary vertices.*

Definition 4. (Radius-Bounded k-Core Search Problem). *Given a geo-social graph $G(V, E)$, a query vertex q, a positive integer k, a query radius r, the radius-bounded k-core (RB-k-core) search problem is to find all RB-k-cores in G, and each RB-k-core G_k^r should satisfies the following three constraints:*

(1) Connectivity constraint. G_k^r contains q and is connected.
(2) Social constraint. $\forall v \in G_k^r$, $deg_{G_k^r}(v) \geq k$.
(3) Spatial constraint. The MCC of G_k^r has a radius $r' \leq r$.
(4) Maximality constraint. There exists no another RB-k-core $G_{k^r}' \supseteq G_k^r$ satisfying (1), (2), and (3).

2.3 Problem Definition

To answer why-not questions on RB-k-core searches based on query refinement model, we formalize the RB-k-core search as a RB-k-core query $Q(k, r)$. There are many possible refinements for the original RB-k-core query can be generated for a given why-not question, which can make missing vertex be included in the new search results of the refined RB-k-core queries. Thus, it is necessary to define a penalty function to evaluate the quality of refined RB-k-core queries so that only the *"good"* refined RB-k-core queries are returned as possible answers for why-not question on RB-k-core search. To ensure that the missing vertex appear in the result, ideally the searching results of any possible refined query Q' will grow over the original query result R. This part of the growing vertices is considered as irrelevant vertices and should be minimized. Therefore, the penalty function for a refined RB-k-core query Q' is defined as the quotient between the number of vertices that appear in the result R' of Q' and the number of vertices that appear in the result set R of Q, which is formalized:

$$Penalty(Q') = \left| \bigcup_{i=1}^{n} C_i' \right| \div \left| \bigcup_{i=1}^{n} C_i \right|, where\ C_i'\ in\ R', C_i\ in\ R \qquad (1)$$

Based on the above penalty function, we formally define our problem.

Problem Definition. Given a geo-social graph $G(V, E)$, a query vertex q, a RB-k-core query $Q(k, r)$, the query result R of Q. A why-not question on RB-k-core search contains an expected vertex ω; Answering this why-not question is to seek a refined RB-k-core query $Q'(k', r')$ by refining the parameters k or r to obtain the new result R', such that the following two conditions should be satisfied: (1) There exists a RB-k-core in R' which contains ω and q; (2) The penalty $Penalty(Q')$ for the refined RB-k-core query Q' is minimal.

3   Influence Factors Analysis

In this section, we analyze the effect of r and k on RB-k-core search results.

3.1   The Analysis of k

The parameter k constrains the results of RB-k-core searches in terms of structural cohesiveness. We help missing vertex ω that do not satisfy the cohesiveness constraint to enter the results by refined k to k'. And k' must be less than k.

In Fig. 1(b), when r is fixed and $k=3$ is modified to $k'=2$, M can satisfy the structural cohesiveness requirement. M can exist in a connected 2-core with L, as seen in Fig. 2(a). The sets $\{G, H, F, L, M\}$ in circle S_2 together form a connected 2-core set satisfying the constraint under the refined query $Q'(2,1)$.

Property 1. For the result sets R'_k of any refined query $Q'(k', r)$, and the result sets R''_k of refined query $Q''(k'', r)$, if $k' > k''$, there must exist:

$$\bigcup_{i=1}^{n} C'_i \subseteq \bigcup_{i=1}^{n} C''_i, \text{where } C'_i \text{ in } R'_k, C''_i \text{ in } R''_k. \tag{2}$$

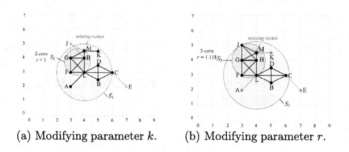

(a) Modifying parameter k.   (b) Modifying parameter r.

Fig. 2. Influence factors for RB-k-core

From Property 1, we know that the penalty value of the refinement $Q'(k', r)$ is negatively related to k', i.e., the maximum possible refined k' is $core_{G(C(q,r))}[\omega]$. Also, we can simply prove that the refined k' is necessarily smaller than k.

3.2   The Analysis of r

The RB-k-core result sets are also affected by the query radius r, which means that if r is not set properly, the users will not get the expected results either.

When k is fixed and r grows to $d(L, J)/2 = 1.118$ as shown in Fig. 2(b). Since there exists a connected 3-core set $\{J, M, G, H, F, L\}$ containing L and M, and the radius of the MCC formed by this set is 1.118, by refining $r=1$ to $r'=1.118$, we obtain a refined query $Q'(3, 1.118)$ that can get the missing vertex M.

Property 2. For the result sets R'_r of any refined query $Q'(k, r')$, and the result sets R''_r of any refined $Q''(k, r'')$, if $r'' > r'$, there must exist:

$$\bigcup_{i=1}^{n} C'_i \subseteq \bigcup_{i=1}^{n} C''_i, where\ C'_i\ in\ R'_r, C''_i\ in\ R''_r. \tag{3}$$

It follows from Property 2 that the penalty value of the refined query $Q'(k, r')$ is positively related to r', i.e., the optimal refined r' is the minimum radius that makes the missing vertex ω satisfy the k constraint. Also, we can simply prove that the refined r' is necessarily larger than the original query condition r.

4 Explanation Algorithms

In this section, we explore two effective explanation algorithms to answer why-not questions on RB-k-core searches by refining the original query parameters.

4.1 Modifying the Cohesiveness Constraint k

RotC and *RotC*$^{+}$ are the best RB-k-core search algorithms in [5], whose essential idea is based on the concept of binary-vertex-bounded circles, which reduces the search cost by sharing a large amount of reusable information. *RotC* and *RotC*$^{+}$ consider that there is only one vertex difference between adjacent binary-vertex-bounded circles. As shown in Fig. 3, there are only three binary vertex bounded circles constructed with L as pole, but there are six vertices in the set. Obviously, there are many vertex differences among adjacent bounded circles. By comparing two adjacent circles and splitting the different vertices between the two circles into several independent enter vertices and leave vertices, we improve this process. *RotC*$^{+}$ builds a social network for each enter circle, and validates the RB-k-core. We optimize this process as follows: only the current circle is an enter circle and the next circle in the sequence is a leave circle. Only then do we construct the social network as well as verify the RB-k-core.

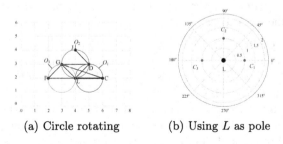

(a) Circle rotating   (b) Using L as pole

Fig. 3. Multi-vertex intersection

A basic idea for finding the best answer is to iteratively subtract k by one and then execute the query algorithm until the expected vertex appears in the result set. This produces a very large number of useless calculations, especially when the difference between k' and k is large. The basic idea may require many rounds of iterations. The efficiency of the basic idea is very low. In order to get the best answer quickly, we propose our refined k' algorithm. The algorithm combines with optimized and modified $RotC^+$ to obtain the best refined k' incrementally. Combining the properties of missing vertex, we propose three pruning strategies and an early termination condition.

Lemma 1. *Given a social network G, a query radius r, a positive integer k, a query vertex q, and a missing vertex ω, a vertex p may be a vertex in the RB-k-core result satisfying the constraints if and only if $p \in (C(\omega, 2*r) \cap C(q, 2*r))$.*

Pruning Strategy 1(ψ_{k_1}). According to Lemma 1, we can safely prune the vertices outside the range $C(\omega, 2*r) \cap C(q, 2*r)$ from G. We can also prune the vertices that are not connected to q and ω in the range.

Pruning Strategy 2(ψ_{k_2}). Based on Lemma 2, we can prune any vertex whose core number or degree is smaller than or equal to the current best refined k'.

Lemma 2. *Given a social network G, a current best refined k', for any vertex n, the vertex can be safely pruned if there exist $(deg_G(n) \le k' \vee core_G[n] \le k')$.*

Proof. Given the current optimal refined k', by Property 1, we know that if there exists a refined k'' with a lower penalty. There must be $k'' > k'$. Any MCC satisfying radius r in which there exists a k''-core result R_u must have any vertex $n_i \in R_u$ with $(deg_G(n_i) > k' \wedge core_G[n_i] > k')$.

Lemma 3. *The upper limit of the best possible modification of k' is $core_{G_k}[\omega]$, G_k is a graph after pruning according to Lemma 1.*

Lemma 4. *A binary vertex bounded circle whose center lies outside the range $C(\omega, r) \cap C(q, r)$ can be safely pruned.*

According to Lemma 3, the algorithm can be terminated early when the current best refined k' is $core_{G_k}[\omega]$.

Pruning Strategy 3(ψ_{k_3}). Using ω and $N_{G_k}(\omega)$ with core number value greater than or equal to $core_{G_k}[\omega]$ as the pole pre-execution algorithm can help get a larger k' faster. And obviously, the faster we find a relatively large k' value, the more effective our ψ_{k_2} will be.

Algorithm 1 introduces the refined k' algorithm, and we first prune the graph G (Lines 1-2) based on the ψ_{k_1} of the refined k' algorithm, and then compute the core number of G_k and obtain the early termination condition k_{max}. After pre-acquiring the current best k' using ψ_{k_3}, we use the grouping-based pre-processing in the $RotC^+$ algorithm strategy to further prune the vertices that

---

Algorithm 1: REFK(q, k, r, ω)

Input: A graph $G(V, E)$, a R-Tree built with $G(V, E)$, a query vertex q, a query $Q(k, r)$, a missing vertex ω

Output: Optimal refinement $Q'(k', r)$

1   $k' \leftarrow 1$, $S \leftarrow$ Null;

2   $G_k \leftarrow$ pruning G according to Lemma 1;

3   $k_{max} = core_{G_k}[\omega]$;

4   $S = \omega + n_i$; \ $* n_i \in N_{G_k}(\omega)$ and $core_{G_k}[n_i] \geq k_{max} *$ \

5   $k' \leftarrow$ pruning according to ψ_{k_3};

6   **for** vertex $v \in (G_k - S)$ and $core_{G_k}[v] > k'$ **do**

7      $C =$ Null, $D = \{v\}$;

8   **for** vertex $u \in V(G_k - S)$ and $core_{G_k}[u] > k'$ **do**

9   **if** $v \neq u$ and $d(u, v) \leq 2 * r$ **then**

10   add $W_r(u, v)$ to C; \ $* W_r(u, v)$ is a set of binary-vertex-bounded circles with radius $r *$ \

11   add u to D;

12   **if** $core_{G(D)}[\omega] \leq k'$ or $core_{G(D)}[q] \leq k'$ **then**

13   continue;

14   sort C;

15   **for** $C(c, r) \in C$ **do**

16          $X \leftarrow$ a set of vertices contained in $C(c, r)$;

17   maintain the degree of the vertices in X;

18   **if** There exists an RB-k''-core , $k'' > k'$ **then**

19             $k' = k''$;

20   **if** $k' == k_{max}$ **then**

21   break;\ $*$terminated early$*$\

22 **return** $Q'(k', r)$;

---

do not satisfy the condition from G_k (Lines 3-5). Then, we construct binary-vertex-bounded circles using the vertices that are contained in G_k-S and whose core number satisfy the requirement, classify the enter and leave circles according to the modified classification method, prune them according to Lemma 4, deposit them in C, and ascending them by polar angle. Then, we prune the unsatisfied pole according to Lemma 2 (Lines 6-14). Finally, we obtain the set X of vertices in each MCC range in C and maintain its vertex degree. The adjacent binary vertex bounded circle after correction have only one vertex difference, so this step can be constructed quickly by a vertex change. When the current circle is a enter circle and the next circle is a leave circle, we construct the graph $G(X)$, if there exists a RB-k''-core($k'' > k'$) containing ω in $G(X)$, we update the current best refined k'. By improving the algorithm in [16], the cohesiveness judgement can be obtained in $O(n + m)$ (Lines 15-21). If the current k' reaches the early termination condition, the algorithm ends and the best refinement $Q'(k', r)$ is returned for the missing vertex ω (Line 22).

For the expected vertex M, as shown in Fig. 4(a). First, according to Lemma 1, $\{A, B, C, E, J\}$ is pruned to obtain the set $\{G, F, L, H, M, K, D\}$ and $k_{max}=2$. Then, in the process of constructing binary-vertex-bounded circles using H, G and M as poles, respectively. According to ψ_{k_3} and determining RB-k-core, the MCC composed of the set $\{M, G, H, L, F\}$ satisfies r less than or equal to 1, and the graph composed of this set is a 2-core. And then k' is updated to 2. Following that, according to Lemma 2, $\{D, K\}$ will be pruned, and then F and L will be used as poles to see if there is a better solution, respectively. Then, early termination is trigged, and the best refined k' is finally obtained as 2.

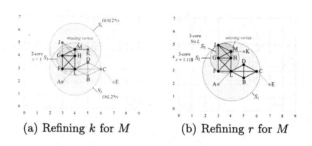

(a) Refining k for M (b) Refining r for M

Fig. 4. Refining parameters for the missing vertex M

4.2   Modifying the Spatial Constraints r

When $d(\omega, q) > 2 * r$, REFK cannot be used to make ω appear in the query results. This motivates us to develop a refined r' algorithm to answer why-not questions on RB-k-core searches.

Our algorithm consists of two parts, an expansion algorithm is used to quickly find a subgraph of G that may contain the best refined r'. An refined r' algorithm is used to find the best refined r'. Given a graph $G(C)$ in the range $C(q, 2 * r)$, expansion algorithm is to iteratively add vertices to $G(C)$ until $G(C)$ satisfies Property 3. We design several expansion strategies to quickly expand $G(C)$. The goal of the refined r' algorithm is to find a circle with minimum radius in $G(C)$ that contains the missing and query vertices. Combining two prunes and an early termination, we design the refined r' algorithm based on a three-vertices circle.

Property 3. Given a positive integer k, there exists a social network $G_r \in G$. We can obtain an RB-k-core containing ω by modifying r if and only if there exists a connected k-core containing q and ω in G_r.

Property 4. Suppose there exists a social network $G_r \in G$ containing a connected k-core of ω and q. G_r must contain ω and at least k neighboring vertices of ω in G with a core number greater than or equal to k.

Algorithm 2: $\text{EXR}(q, r, k, \omega)$

Input: A graph $G(V, E)$, a R-Tree built with $G(V, E)$, a index of core number
of G, a query vertex q, a query $Q(k, r)$, a missing vertex ω

Output: The modified lower bound r_{min} for r, a graph $G(S_c)$

1 $DisSet = $ Null;

2 **for** $v_i \in N_G(\omega)$ and $core_G[v_i] \geq k$ **do**

3 $\quad$ calculate $d(v_i, q)$;

4 $\quad$ add $< v_i, d(v_i, q) >$ to $DisSet$;

5 sort $DisSet$;

6 $r_{min} = max(d(\omega, q), d(u_k, q))/2, (u_k \in DisSet)$;

7 $c{=}q$, $r_c{=}2 * r_{min}$, $S_c \leftarrow$ vertex sets $V(C(c, r_c))$;

8 **for** $u_i \in DisSet, (i \geq k)$ **do**

9 $\quad$ construct the graph $G(S_c)$ using S_c;

10 $\quad$ **if** There exists a RB-k-core in $G(S_c)$ containing q and ω **then**

11 $\quad\quad$ $r_{min} = r_c$;

12 $\quad\quad$ break;

13 $\quad$ $r_c = (d(u_i, c) + r_c)/2$, update c;

14 $\quad$ **if** $v_j \in (C(c, r_c) - S_c)$ and $core_G[v_j] \geq k$ **then**

15 $\quad\quad$ add v_j to S_c;

16 **return** $G(S_c)$ and r_{min};

Property 5. $\forall G_n \in G, core_{G_n}[\omega] < k$. Adding n_i to G_n may make $core_{G_n}[\omega] \geq k$ if and only if vertex $n_i \in G, core_G[n_i] \geq k$.

In order to answer the why-not questions on RB-k-core searches by modifying r, based on Property 3, we need to develop a method for obtaining the social network for which the best refined r' exists. To improve the speed of acquisition and minimize the size of the social network that satisfies the conditions after expansion. We designed the expansion algorithm.

As shown in Algorithm 2, We first calculate the distance between the vertices with a core number greater than k-1 in $N_G(\omega)$ and q. Then, add these distances and the corresponding vertices to $DisSet$, and ascending sort them (Lines 1-5). According to Property 4, we first get the lower bound r_{min} of refined r' is $max(d(\omega, q), d(u_k, q))/2$, u_k is the kth vertex in the $DisSet$. Define the circle center c as the query vertex q, expand radius r_c as $2 * r_{min}$ (Lines 6-7). If a set of vertices S_c of $C(c, r_c)$ does not satisfy Property 3. According to Property 5, we need to iteratively add vertices with a core number greater than or equal to k into $C(q, r_{min})$. In order to expand the size of the social network that satisfies the condition as small as possible and to obtain it as fast as possible. We take out the next vertex u_i in the $DisSet$ sequence and use $d(u_i, c) + r_c$ as the diameter r_c of the new circle, then, update the circle center c. Add the vertices in $C(c, r_c)$ that satisfying Property 5 to Sc, and loop the above process until there exists a connected k-core containing q and ω. Then, we update the modified lower bound to r_c. Finally, $G(S_c)$ and r_{min} are returned (Lines 8-16).

Algorithm 3: $\text{REFR}(q, r, k, \omega)$

Input: A graph $G(V, E)$, a R-Tree built with $G(V, E)$, a index of core number of G, a query vertex q, a query $Q(k, r)$, a missing vertex ω

Output: The best refinement $Q'(k, r')$

1 $(G(S_c)$ and $r_{min}) \leftarrow \text{EXR}(q, r, k, \omega)$;

2 $r' = \infty$;

3 **for** $v_i \in S_c$, $i = 3$ to $|G(S_c)|$ **do**

4   **for** $v_j \in S_c$, $j = 0$ to i-1 **do**

5   **for** $v_x \in S_c$, $x = j$+1 to i **do**

6   **if** *the triangle constructed by $\{v_i, v_j, v_x\}$ is a non-obtuse triangle* **then**

7   construct $C(c, r)$ of the vertex set $\{v_i, v_j, v_x\}$;

8   **if** $d(c, q) > r$ *or* $d(c, \omega) > r$ **then**

9   continue;*In-process pruning 1 of r * \

10   **if** $r \geq r'$ **then**

11   continue; *In-process pruning 2 of r * \

12   constructing the graph $G(C(c, r))$ of $C(c, r)$;

13   **if** *There exists a RB-k-core in $G(C(c, r))$ containing ω and q* **then**

14   **if** $r < r_{min}$ **then**

15   return r_{min};*terminated early*\

16                  $r' = r$;

17 **return** $Q'(k, r')$;

Lemma 5. *The optimal refined r' must be greater than r_{min}.*

Proof. According to EXR, a connected k-core containing q and ω may exist in $G(S_c)$ only if v_j (the last neighbor vertex of ω that joins S_c in the EXR) exists. Because $RotC^+$ prunes vertices outside the range of $2 * r$, if the current best refined r' is less than r_{min}, v_j will be pruned.

Early Termination Strategy. Based on Lemma 5, when the current best refined $r' \leq r_{min}$, the algorithm can terminate and return r_{min} as the best refinement.

Pruning Strategy 1(ψ_{r_1}) for Refining r. Based on Property 6, we can prune the MCC that do not contain both query and missing vertex.

Pruning Strategy 2(ψ_{r_2}) for Refining r. Based on Property 7, we can safely prune MCC with radius greater than or equal to the current best refined r'.

Property 6. Any three vertices constructed minimum covering circle $C(c, r)$, if $d(c, q) > r$ or $d(c, m) > r$, the MCC cannot contain both q and ω.

Property 7. According to Property 2, any minimum covering circle $C(c, r'')$, if r'' is greater than or equal to the current best refined r', the refinement result penalty found must be greater than the current best refinement.

After the execution of the expansion algorithm, in order to find the best refined r' in $G(S_c)$, based on the above properties, we design a algorithm for refining r. Algorithm 3 first obtains the social network $G(S_c)$ and the refined lower bound r_{min} by calling the expansion algorithm (Line 1). Then, constructs the circle $C(c, r)$ by combining the vertices in S_c three-by-three when the triangle formed by the three vertices is a non-obtuse triangle (Lines 2-5). Then, we prune the generated circle using ψ_{r_1} and ψ_{r_2} (Lines 6-11). We update the current best refinement if the circle is not pruned and there is a RB-k-core containing ω in the network composed of this circle. Also, the early termination is determined (Lines 12-16). Finally, the best refined r' is returned (Line 17).

For the missing vertex M, as shown in Fig. 4(b). First, by executing EXR, the candidate graph $G(S_1)$ consisting of S_1 and $rmin = 1.118$ are obtained. Then, three vertices from $G(S_1)$ are taken for combination. Assuming that the first combination is $\{J, F, L\}$, r' is updated to 1.118 due to the existence of a 3-core containing both M and L in the MCC composed by $\{J, F, L\}$. And then some $MCCs$ satisfying ψ_{r_1} are pruned, such as the MCC composed by $\{G, H, J\}$, and some $MCCs$ are also pruned because the radius is larger than the current best refined r', for example, the MCC consisting of $\{J, F, C\}$ is pruned because its radius of 1.802 is larger than 1.118, and REFR iterates the above process. Finally, we obtain the best refined $r' = 1.118$.

5   Experiments

In this section, we evaluate the performance of our proposed algorithms.

5.1   Experimental Setup

We conducted experiments on four real datasets.

- **Weeplaces(WP)**[1]: This dataset was collected from Weeplaces. The friendship network consisted of 15,799 vertices and 119,931 edges.
- **Brightkite(BK)**[2]: Brightkite used to be a location-based social networking service provider where users shared their location by checking in. The friendship network consisted of 58,228 vertices and 214,078 edges.
- **Gowalla(GW)**[5]: This dataset was collected from Gowalla. The friendship network consisted of 107,092 vertices and 456,830 edges.
- **Foursquare(FS)**[5]: Foursquare contains 2,153,471 users and 27,098,490 social connections. We randomly selected 100,000 users and 404,484 edges.

To the best of our knowledge, there is no work can be used to answer why-not questions on RB-k-core searches in geo-social networks. In our experiments, first, different why-not questions w_1-w_4 were designed for WP, w_5-w_8 for BK, w_9-w_{12} for GW, and w_{13}-w_{16} for FS. Then, we investigate the performance of the REFK algorithm and the REFR algorithm under different parameters.

All algorithms are implemented in GNU C++. The experiments were run on a PC running Ubuntu 20.04, 3.9GHz CPU, 256GB RAM, and 1TB disk.

---

[1] https://www.yongliu.org/.
[2] https://www.comp.hkbu.edu.hk/.

5.2   Performance Evaluation

Exp-1: In Fig. 5, we compare the execution time of the algorithms under different data sizes. It can be observed that the execution efficiency of the algorithms basically increases with the increase in the dataset size. This is because the increase in the size of the dataset leads to an increase in the number of vertices to be processed with the same radius constraint. The algorithms needs to validate more combinations. Figure 5(a) shows a special case. The REFK algorithm gains an increase in execution speed as the dataset size grows from 80% to 100%. By studying the execution process. We found that this is because the core number of the missing vertices and some of their neighbors increased in the last 20% of the size growth. This leads to a larger candidate refined k' obtained after ψ_{k_3}. Therefore, ψ_{k_2} becomes more efficient. So the execution efficiency of REFK increases instead. Then, we found that there is no refinement strategy that presents an overwhelming advantage over another refinement strategy. So next, we investigated the effect of some parameters on the efficiency of algorithms execution.

Exp-2: We investigate the effect of $core_{G_n}[\omega]$ of the two algorithms. As shown in Fig. 6, we fixed the query radius r to be 5km, k to be 10, the dataset size is 100%. In Fig. 6(a), we fixed $d(q,\omega)$ to be $7.5\,\mathrm{km} \pm 0.1\,\mathrm{km}$. And selected missing vertices with $core_{G_n}[\omega]$ values of 5-9 for the experiments. It is found that the efficiency of REFK execution increases as $core_{G_n}[\omega]$ increases. This is because the larger the $core_{G_n}[\omega]$ can help prune more vertices during REFK execution. At the same time, we find that the increases in efficiency are not large. The experiments reveal that REFK is more likely to early termination at lower $core_{G_n}[\omega]$. This is because the termination condition is easier to reach if the missing vertex with a low core number. The balance of prunes and early termination also makes the efficiency of REFK execution largely stable on the dataset WP with fewer vertices. In Fig. 6(b), we constrain $d(q,\omega)$ to be 12.5km±0.1km. and stretch $core_{G_n}[\omega]$ to be 10-14 to investigate the effect of $core_{G_n}[\omega]$ on REFR. Observing the experimental fig, we find that the execution time of REFR decreases as the value of $core_{G_n}[\omega]$ increases. This is because, indeed, the larger the $core_{G_n}[\omega]$ than k, the faster it is possible to find the refined r' that satisfies the constraint during the REFR execution. Then the pruning efficiency is improved.

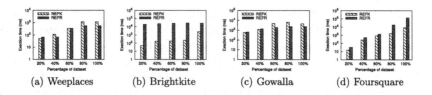

(a) Weeplaces   (b) Brightkite   (c) Gowalla   (d) Foursquare

Fig. 5. Execution time of algorithms under different datasizes

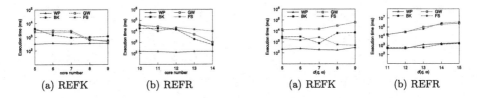

(a) REFK   (b) REFR   (a) REFK   (b) REFR

Fig. 6. Effect of $core_{G_n}[\omega]$ **Fig. 7.** Effect of $d(q, \omega)$

Exp-3: We investigate the effect of the distance between the missing and query vertex on the efficiency of the two algorithms. We fixed r is 5km and k is 10. The dataset size is 100%. In Fig. 7(a), we fixed $core_{G_n}[\omega]$ to be 7 ± 1 and studied the effect of $d(q, \omega)$ on REFK by scaling the distance to be 5-9. The execution time of REFK increases with distance on the three datasets, and the execution time of REFK decreases on the FS dataset. The execution time increases because the farther the missing vertex is from the query vertex, the weaker the cohesiveness constraint of the two vertices, and the more difficult it is to find a higher value of k' quickly. However, the increase in distance means that more vertices can be pruned by ψ_{k_1}. This constrains the execution time from growing by a great magnitude. ψ_{k_1} works very well in the FS dataset, with a much higher percentage of vertices pruned than the other three datasets. And in FS, REFK's early termination is achieved quickly, so the execution efficiency of REFK increases instead. In Fig. 7(b), we fixed $core_{G_n}[\omega]$ as 12 ± 1, and set the distance as 11-15 to study the effect of $d(q, \omega)$ on REFR, the algorithm's execution time grows rapidly as $d(q, \omega)$ increases. It is the growth of $d(q, \omega)$ that causes EXR to return a larger r_{min} value and a larger candidate social network.

Exp-4: Then we investigate the execution efficiency of the algorithms on different why-not questions on different datasets. In Fig. 8(a) we observe that there is not much difference in the execution speed of algorithms on w_1-w_4. This is due to the relatively balanced distribution and the small size of the WP. Figure 8(b) investigates the performance of algorithms on the BK dataset. w_6 and w_7 have the same number of vertices to be processed. However, the structural difference causes the REFR algorithm to trigger early termination quickly on w_7. But the REFE algorithm can obtain a faster execution speed on w_6. Since the missing vertex does not satisfy the space requirement, w_1 and w_8 cannot return refinement result by the REFK algorithm. In Fig. 8(c), the candidate processing

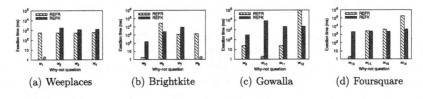

(a) Weeplaces   (b) Brightkite   (c) Gowalla   (d) Foursquare

Fig. 8. Running time of algorithms under different why-not questions

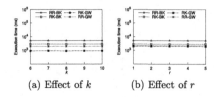

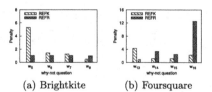

| (a) Effect of k | (b) Effect of r | (a) Brightkite | (b) Foursquare |

Fig. 9. Effect of query Q **Fig. 10.** Penalty for different w_i

vertices size of w_{12} is the largest, but the efficient of ψ_{k_1} and ψ_{k_3} prune a very large number of vertices. It ensures that REFK is executed efficiently. However, the REFR algorithm cannot guarantee the structural cohesiveness of the missing vertex within a 100km radius circle on w_{12}. The huge vertices size leads to the REFR timeouts. Also, we observe that REFK is slower than REFR on the GW dataset. This may be due to the fact that the vertices in the GW dataset are more discrete under the same cohesiveness constraint. During the execution of REFR, w_{16} in Fig. 8(d) handles the maximum number of vertices in the 16 why-not questions. So it shows the worst execution efficiency on REFR. For w_{10} and w_{13}, since the missing vertex does not satisfy the cohesiveness requirement, we cannot return to refinement result using the REFR algorithm.

Exp-5: We investigate the effect of query attributes r and k on the efficiency of the algorithms on BK and GW. By observing the experimental results in Fig. 9, we can find that the initial r and k have almost no effect on the algorithms execution efficiency. The algorithms' execution efficiency is only related to the structure of the vertices in the dataset, the difference between the query and the missing vertex, and the properties of the missing vertex.

Exp-6: We investigate the penalties under different why-not questions on the two datasets BK and FS. As shown in Fig. 10(a), the penalties of REFK on BK is generally higher than that of REFR. This is because on BK, REFK usually leads to more irrelevant vertices joins due to a smaller refined k', while REFR brings a smaller refined r'. On FS, as shown in Fig. 10(b), refined k' is usually larger, and r' usually be expanded to larger values. So the REFR algorithm yields a higher penalty on FS. This is also matched by our Property 1 and Property 2. That is, larger r or k value changes bring larger penalties.

6   Conclusion

To answer why-not questions on radius-bounded k-core searches, we first analyze the influences of changing parameters on the results of RB-k-core searches. Then, we propose two explanation algorithms by refining the parameters k and r, respectively. Finally, we conduct extensive experiments on four real geo-social networks, which show that our algorithms can return high-quality refined parameters which can make the missing vertices appear in the results. In future work, we will explore effective explanation algorithms for answering why-not questions on RB-k-core searches by refining the two parameters k and r simultaneously.

Acknowledgements. The work is partially supported by the National Key Research and Development Program of China (2020YFB1707901), National Natural Science Foundation of China (Nos. U22A2025, 62072088, 62232007, 62002245, 62102271, 61802268), Ten Thousand Talent Program (No. ZX20200035), Liaoning Distinguished Professor (No. XLYC1902057), Natural Science Foundation of Liaoning Province (Nos. 2022-MS-303, 2022-BS-218, 2022-MS-302).

References

1. Chapman, A., Jagadish, H.V.: Why not '?'. In: SIGMOD, pp. 523–534 (2009)
2. Tran, Q., Chan, C.: How to ConQueR Why-not Questions. In: SIGMOD, pp. 15–26 (2010)
3. Yang, D., Shen, C., Lee, W., Chen, M.: On socio-spatial group query for location-based social networks. In: KDD, pp. 949–957 (2012)
4. Fang, Y., Cheng, R., Li, X., Luo, S., Hu, J.: Effective community search over large spatial graphs. In: PVLDB, pp. 709–720 (2017)
5. Wang, K., Cao, X., Lin, X., Zhang, W., Qin, L.: Efficient computing of radius-bounded k-cores. In: ICDE, pp. 233–244 (2018)
6. Kim, J., Guo, T., Feng, K., Cong, G., Khan, A., Choudhury, F.: Densely Connected User Community and Location Cluster Search in Location-Based Social Networks. In: SIGMOD, pp. 2199–2209 (2020)
7. Zhu, H., et al.: Continuous geo-social group monitoring over moving users. In: ICDE, pp. 312–324 (2022)
8. Chen, L., Lin, X., Hu, H., Jensen, C.S., Xu, J.L.: Answering why-not spatial keyword top-k queries via keyword adaption. In: ICDE, pp. 697–708 (2016)
9. Islam, M.S., Liu, C., Li, J.: Efficient answering of why-not questions in similar graph matching. In: TKDE, pp. 2672–2686 (2015)
10. Zong, C., et al.: Answering why-not questions on structural graph clustering. In: DASFAA, pp. 255–271 (2018)
11. Zong, B., et al.: Answering why-not group spatial keyword queries. In: TKDE, pp. 26–39 (2020)
12. Zhang, W., Li, Y., Shu, L., Luo, C., Li, J.: Shadow: answering why-not questions on top-k spatial keyword queries over moving objects. In: DASFAA, pp. 738–760 (2021)
13. Song, S., Huang, R., Gao, Y., Wang, J.: why not match: on explanations of event pattern queries. In: SIGMOD, pp. 1705–1717(2021)
14. Diestelkamper, R., Lee, S., Herschel, M., Glavic, B.: To not miss the forest for the trees - a holistic approach for explaining missing answers over nested data. In: SIGMOD, pp. 405–417 (2021)
15. Song, Q., Namaki, M.H., Lin, P., Wu, Y.: Answering why-questions for subgraph queries. In: TKDE, pp. 4636–4649 (2022)
16. Batagelj, V., Zaversnik, M.: An O(m) Algorithm for Cores Decomposition of Networks. In: cs.DS, pp. 129–145 (2003)

Multi-scale Community Detection in Subspace of Attribute

Cairui Yan[1], Huifang Ma[1,2(✉)], Yuechen Tang[1], Xiaohong Li[1], and Zhixin Li[2]

[1] College of Computer Science and Engineering, Northwest Normal University,
Lanzhou 730070, Gansu, China
mahuifang@yeah.net
[2] Guangxi Key Lab of Multi-source Information Mining and Security,
Guangxi Normal University, Guilin 541004, Guangxi, China

Abstract. The goal of attributed community detection is to search a partition of the network such that there is high cohesion within each group and low coupling between two groups. We argue that multiple partitions of the attributed network should be captured with different semantics and community detection should be approached from the perspective of attribute subspace. In this paper, we integrate spectral wavelets with attribute subspace, and develop a framework of Multi-scale Community Detection in Subspace of Attribute (MCDSA). Our idea is to implement graph partitioning via scale-dependent modularity and independent attribute subspaces, thus making our model more flexible and effective. In MCDSA, communities at each scale have independent attribute subspace, which is helpful to analyze the importance of each attribute under different network partition, better revealing the relationship between nodes. Extensive experiments on multiple benchmark datasets show that, the quality of community detection can be remarkably enhanced under the regime of attribute subspaces, achieving the state-of-the-art performance.

Keywords: Community detection · Multi-scale · Spectral wavelets · Attribute subspace

1 Introduction

Community detection (CD), one of the most vital and fundamental tasks in network analysis, has a broad range of applications in various domains, such as functional prediction and sub-market identification. With a given network, a community is defined as a cohesive set of nodes with more connections inside than outside. Due to the fact that the network in general can be modelled as a graph over vertices and edges, CD is regarded as a graph clustering problem, where each community is corresponding to one of the clusters in the graph. In addition to network topology, the entities, i.e., nodes are usually associated with attribute that is important for making sense of communities. E.g., papers in citation networks have areas of keywords. Such networks with node attributes are

© The Author(s), under exclusive license to Springer Nature Switzerland AG 2023
X. Wang et al. (Eds.): DASFAA 2023, LNCS 13945, pp. 110–119, 2023.
https://doi.org/10.1007/978-3-031-30675-4_8

named as attributed graphs [1]. Attributed graphs are broadly used to represent social networks, gene and protein interactions.

Despite the significant progress that has been achieved, current CD methods still face challenges when applied to real-world networks. First, CD is flexible in nature, and it is almost impossible to generate high-quality communities directly using a predefined community size. Fortunato [3] proves that an excellent method should be able to locate communities of different sizes to ensure that one is closest to the real community. However, most of the existing CD can only find a single partition of the network, which often does not meet our requirements. Secondly, with the increase of attribute information, researchers start to explore the relationship between topology and attribute information. Existing studies have revealed that the performance of CD could be substantially boosted with the help of additional information (i.e., attributes) compared to topology-only approaches. Most of the existing studies simply regard attributes are of equal importance to each community on the network, which is not necessarily the case however.

In light of the aforementioned shortcomings of the existing CD methods on attributed graph, we propose a novel **M**ulti-scale **C**ommunity **D**etection in **S**ubspace of **A**ttribute (MCDSA), that is, to find all communities with various semantics such that nodes are densely-connected and share homogeneous properties within the same community. In order to deal with nodes with different semantics, we define the scale range and introduce the idea of spectral wavelet to get scale-dependent network partition. In order to tackle difference of semantic information between communities under certain partition, we provide interpretable attribute subspaces for each community partition. In summary, our contributions are three folds: (1) To our knowledge, we introduce spectral wavelet into attributed network partitioning for the first time and define a scale-dependent modularity on the node-attributed graph inspired by the node attribute, modularity and graph wavelets; (2) We propose an adaptive spectral clustering solution for the attribute subspaces on attributed graph, where each partition is equipped with a unique set of attribute subspace weights assigned; (3) We demonstrate the effectiveness of MCDSA by conducting extensive experiments on synthetic and real-world datasets with vividly designed cases.

2   Preliminaries

Given an attributed graph via a triple $G = (V, E, \mathbf{F})$, where $V = \{v_1, v_2, ..., v_N\}$ and E are set of nodes and edge respectively, N is the number of nodes. $\mathbf{F} \in \mathbb{R}^{N \times F}$ is node attributes matrix with i-th row f_i represents the attribute value of node v_i using an F-dimensional vector. We normalize each dimension of the feature between $[0, 1]$. The topology of attributed graph G can be represented by its adjacency matrix $\mathbf{A} = [a_{ij}] \in \{0, 1\}^{N \times N}$, where $a_{ij} = 1$ denotes the existence of an edge between nodes v_i and v_j. The degree matrix $\mathbf{D} \in \mathbb{R}^{N \times N}$ is a diagonal matrix with diagonal element $d_{ii} = \sum_{i=1}^{N} a_{ij}$ as the degree of node v_i. In our algorithm, we also define the range of scales $S = [s_{min}, s_{max}]$. Let us define the normalized Laplacian matrix of the graph as $\mathbf{L} = \mathbf{I}_N - \mathbf{D}^{-1/2}\mathbf{A}\mathbf{D}^{-1/2}$. $\mathbf{L} \in \mathbb{R}^{N \times N}$ is real symmetric, therefore diagonalizable: its spectrum is composed of its sorted eigenvalues $(\lambda_l)_{l=1...N}$, so that $0 = \lambda_1 \leq \lambda_2 \leq \lambda_3 \leq \cdots \leq \lambda_N \leq 2$; and of the matrix $\mathbf{X} \in \mathbb{R}^{N \times N}$ of its normalized eigenvectors: $\mathbf{X} = (\mathbf{X}_1 | \mathbf{X}_2 | \dots | \mathbf{X}_N)$. The framework of MCDSA is depicted in Fig. 1.

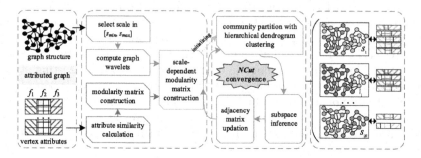

Fig. 1. Overview of the MCDSA framework. Input: topology of the attributed graph and attribute information (suppose there exist three attributes, i.e., f_1, f_2, f_3); Output: network partitioning at multiple scales and the corresponding attribute subspaces.

3   Method

3.1   Scale-dependent Node Representation

Spectral Graph Wavelets. Graph wavelets were defined in [4] using the graph Fourier modes. Its construction is based on band-pass filters defined in the graph Fourier domain, generated by stretching a band-pass filter kernel $g(\cdot)$ by a scale parameter $s > 0$. The stretched filter has matrix representation $\hat{G}_s = \text{diag}(g(s\lambda_1), \ldots, g(s\lambda_N))$ that is diagonal on the Fourier modes. Hence, the wavelet basis $\mathbf{\Psi}_s \in \mathbb{R}^{N \times N}$ at scale s reads as: $\mathbf{\Psi}_s = (\psi_{s,1} | \psi_{s,2} | \dots | \psi_{s,N}) = \mathbf{X}\hat{G}_s\mathbf{X}^\top$. We use the band-pass filter kernel $g(\cdot)$:

$$g(x; \alpha, \beta, x_1, x_2) = \begin{cases} x_1^{-\alpha}x^\alpha & \text{for } x < x_1 \\ p(x) & \text{for } x_1 \leq x \leq x_2 \\ x_2^\beta x^{-\beta} & \text{for } x > x_2 \end{cases} \qquad (1)$$

where $p(\text{x})$ is taken as the unique cubic polynomial interpolation that respects the continuity of g and its derivative g'.

Scale-sensitive Modularity. Since the attribute information of the attributed graph facilitates the accurate identification of clusters, we add the attribute information of the nodes to $\mathbf{A}$ for the initial community partitioning. Specifically, we transform $\mathbf{A}$ into a weighted adjacency matrix $\mathbf{A}^{(0)}$, where the strength of connected edges between nodes is the attribute similarity between nodes. Gauss kernel function is used to compute the similarity between nodes.

The local modularity $\mathbf{B}$ evaluates the target communities using only local information in the network and is defined as:$b_{ij} = a_{ij} - \frac{d_i d_j}{2M}$, where M is the total number of edges of the graph. Since the task of this paper is attributed graph oriented, we integrate the attribute information into the modularity matrix $\mathbf{B}^{(t)}$:$b_{ij}^{(t)} = a_{ij}^{(t)} - \frac{d_i^{(t)} d_i^{(t)}}{2M'}$, where $d_i^{(t)} = \sum_{j=1}^{N} a_{ij}^{(t)}$, M' is the weighted sum of all edges of attributed graph. Then, we incorporate $\mathbf{B}^{(t)}$ with the wavelet basis $\mathbf{\Psi}_s$ to obtain the new $\mathbf{B}_s^{(t)}$ at each scale s:$\mathbf{B}_s^{(t)} = \mathbf{\Psi}_s \mathbf{B}^{(t)} = \mathbf{X} \hat{G}_s \mathbf{X}^{\top} \mathbf{B}^{(t)}$.

It is worth noting that the current $\mathbf{B}_s^{(t)}$ already contains information about the structure and attributes of each node in the network at each scale s, where each row in $\mathbf{B}_s^{(t)}$ is the current view of the node. The relevant distances between nodes are calculated: $d_s(v_a, v_b) = 1 - \left(\mathbf{B}_{s,a}^{(t)} \right)^{T} \mathbf{B}_{s,a}^{(t)}$.

Fast Wavelet Transform. When the size of the nodes in the graph exceeds 1000, the Laplacian's diagonalization becomes computationally prohibitive and the exact computation of the Fourier matrix $\mathbf{X}$ is no longer possible. Hammond et al. [4] devised an efficient way to bypass the Laplacian's diagonalization and obtained an approximation of the wavelets by using Chebyshev polynomials to approximate the filters [8]. We will write $\mathcal{FWT}_s$ the operator corresponding to this fast wavelet transform at scale s. Then the wavelet basis $\mathbf{\Psi}_s$ at scale s can be efficiently approximated by: $\mathbf{\Psi}_s \sim \mathcal{FWT}_s(\mathbf{I}_N)$, where $\mathbf{I}_N$ is the identity matrix of size N.

3.2 Community Detection with Attribute Subspaces

Community Partitioning Based on Hierarchical Dendrogram Clustering. To group the nodes, the connectivity constraint is then imposed to a hierarchical complete linkage clustering algorithm to output a dendrogram, making sure that each node is clustered into a group of nodes to which it has path in the original network. For the multiple possible divisions that exist in the dendrogram, we choose the classical k-means as the judgment benchmark, thus forming an initial network division at scale s.

In the light of the above discussion, we now have a set of partitions $\mathcal{P} = \{\mathbf{P}_s\}_{s \in S}$, one for each scale. Let $\mathbf{P}_s \in \mathbb{R}^{N \times J_s}$ be a matrix to indicate a community clustering membership at scale s, where $\mathbf{P}_s = (\mathbf{1}_{s,C_1} | \mathbf{1}_{s,C_2} | ... | \mathbf{1}_{s,C_{J_s}})$, $\mathbf{1}_{s,C_j}$ is the binary indicator function of community C_j at scale s (i.e., $\mathbf{1}_{s,C_{ij}} = 1$ if node v_i is in C_j, else 0), J_s is the total number of community divisions.

Construction of Attribute Subspaces. Inspired by subspace clustering, our approach will assign a relevant semantic information to each clustering result, i.e., probe for relevant node features. For symbolic clarity, we will focus on subspace inference tasks for network partition $\mathbf{P}_s$ at a specific scale s in this section. For any community at scale s, the attribute weight that makes the nodes in the community similar to each other is calculated such that the internal nodes in the community are similar to each other based on the attributes and not similar to the external nodes. In other words, the weight should be such that the nodes have a small distance from each other based on the attributes. We devise an operator named Entropy Weight Constraint (EWC) on the subspace inference to infer the truthful community division. EWC defines an attribute subspace weight method without iterative optimization and uses the objective function to compute the attribute subspace weight vector. The objective function is:

$$H_{s,C}(\mathbf{w}) = \frac{1}{2}\sum_{i\in C}\sum_{j\in C,i\neq j}\sum_{t=1}^{F} w_t\left(f_{v_it} - f_{v_jt}\right)^2 + \gamma\sum_{t=1}^{F} w_t\log_2 w_t \qquad (2)$$

where C is the target community at scale s, w_t represents the weight of the t-dimension attribute. The former term of Eq.(2) is the distance between the nodes within the current community, and the latter represents the negative entropy value, the positive parameter γ is the incentive power that controls the multidimensional weights. To minimize $H(\mathbf{w})$, each target community one can obtain the subspace weight vector $\mathbf{w} = [w_1, ..., w_t, ..., w_F]$, and $\mathbf{w}$ only has higher values in the few dimensions corresponding to the attributes in the subspace, while the remaining dimensions are smaller. The attributes subspace weight satisfies a constraint, namely $\sum_{t=1}^{F} w_t = 1, w_t > 0$. By minimizing $H(\mathbf{w})$, we obtain:

$$w_t = \frac{\exp\left(-D_t/\gamma\right)}{\sum_{t=1}^{F}\exp\left(-D_t/\gamma\right)}, D_t = \frac{1}{2}\sum_{i\in C}\sum_{j\in C,i\neq j}\left(f_{v_it} - f_{v_jt}\right)^2 \qquad (3)$$

The derivation process is detailed in the Appendix.

Construction Normalized Cut in Subspace Projections. We use the normalized cut (*NCut*) because it has excellent advantages in avoiding unbalanced cuts.

$$NCut_{\mathbf{W}}^{(t)}(\mathbf{P}_s) = \sum_{k=1}^{J_s}\frac{\mathbf{p}_k^T \cdot \mathbf{A}^{(t)} \cdot (\mathbf{1}_N - \mathbf{p}_k)}{\mathbf{p}_k^T \cdot \mathbf{W}_s \cdot \mathbf{1}_N} \qquad (4)$$

The nominator calculates the sum of weights over outgoing edges of cluster k, while the denominator calculates the sum of weights over internal and outgoing edges of cluster k. Thus, the normalized cut trades off low inter-cluster connectivity and high intra-clustering connectivity. Due to the fact that the convergence of *NCut* depends on the matrix $\mathbf{A}^{(t)}$, the $\mathbf{A}^{(t)}$ has a dramatic impact on the optimization of *NCut*. Here, we enrich $\mathbf{A}^{(t)}$ by acting the kernel transformation $k(\mathbf{x}, \mathbf{y})$ on the node features:$a_{u,v}^{(t)} = k(\mathbf{x}, \mathbf{y}) \cdot \mathbb{I}((u,v) \in E)$, where $\mathbf{x}=f(u)$ and $\mathbf{y}=f(v)$ are the feature vectors of vertices u and v, and $\mathbb{I}$ is the indicator function. In this paper, we consider limiting the range to kernels with non-negative derivatives, so we use Gaussian kernels as kernel functions and use the weighted Euclidean norm:$\|\mathbf{x} - \mathbf{y}\| := \sqrt{(\mathbf{x} - \mathbf{y})^T \operatorname{diag}(\mathbf{W})(\mathbf{x} - \mathbf{y})}$, where $\sum_{i=1}^{D} w_i = 1, w_i \geq 0$.

Straightforwardly, we project the entire graph into the subspace $\mathbf{P}_k$ and gauge the quality of cluster k via analysing the level of separation of cluster k. It is irrelevant how well the k-th cluster is divided from the other clusters in the network. Consequently, the weight matrix is formalized as:

$$\mathbf{W}_{\mathbf{P},\mathbf{W}}^{(t)} = \sum_{k=1}^{K}\mathbf{W}_{\mathbf{s}_k}\circ\left(\mathbf{p}_k \cdot \mathbf{1}_N^T\right) \qquad (5)$$

Note that the weighted adjacency matrix $\mathbf{A}^{(t)}$ is unsymmetrical, making it not interpretable and unsuitable for subsequent spectral clustering. Thus, it is reasonable to average the sum of matrices $\mathbf{A}$ and $\mathbf{A}^T$ to achieve symmetry. What is more, if two nodes belong to the same community, their similarity is evaluated in the corresponding subspace. In terms for two nodes belonging to different communities, their similarity is measured as the average of the two similarities which are calculated in each individual subspace.

4 Experiments

4.1 Experimental Setup

Datasets. Our proposed MCDSA is evaluated on six single-scale datasets and eight SP datasets [7] with different parameters. The statistics of all datasets are summarized in Table 1 and 2. Moreover, we provide all the data websites in the supplement for reproducibility.

Baselines. To verify the effectiveness of MCDSA, we compare our MCDSA with two categories representative methods: (1) Classical community detection methods on attributed graph: NAGC [6], AGGR [11] and NotMle [10]; (2) GCNs-based attributed graph community detection method: HCD [5], GDCL [9] and SOA [2]. F_1 and conductance(con) are used to measure the accuracy of detected local communities. Also, we use the quality metric of community attribute similarity(CAS):$CAS(C) = \frac{1}{|C|^2} \sum_{t \in F} \sum_{u \in C} \sum_{v \in C} w_t (f_{ut} - f_{vt})^2$. It should be noted that all of the above-mentioned evaluation metrics are specific to a particular community at the scale s, so the performance of the model needs to be evaluated by averaging the results of the experiments of all communities at the scale s.

Table 1. Statistics of single-scale datasets used in the experiments.

Dataset		$|V|$	$|E|$	$|F|$	k
categorized	Texas	187	283	1703	5
	Washington	230	366	1703	5
	Facebook	1,046	27,783	576	9
numerical	Disney	124	333	28	9
	ArXiv	856	2,660	30	19
	Enron	13,533	176,967	18	40

4.2 Performance Comparison

Performance Evaluation. We first make a comparison of MCDSA with the baselines for the CD task. We perform 10 times experiments on each dataset. In particular, since all baseline methods are single-scale, we record the best results at all scales. Table 3. reports the main results of all methods.

Table 2. Statistical of the eight SP datasets with different parameters.

Dataset	SP1	SP2	SP3	SP4	SP5	SP6	SP7	SP8
ρ	$\rho=1$				$\rho=2$			
$\overline{k}$	$\overline{k}=11$	$\overline{k}=13$	$\overline{k}=15$	$\overline{k}=17$	$\overline{k}=10$	$\overline{k}=15$	$\overline{k}=20$	$\overline{k}=25$
$|E|$	3,593	4,226	4,767	5,504	3,175	4,759	6,331	7,955

Table 3. Attributed community search performance compared with other approaches.

Dataset	Metric	NAGC	AGGR	NotMle	HCD	GDCL	SOA	MCDSA
Texas	F_1	0.84	0.83	0.84	0.85	**0.94**	0.92	0.93
	CAS	0.25	0.25	0.26	0.21	0.17	0.17	**0.15**
Washington	F_1	0.79	0.73	0.82	0.79	0.87	0.87	**0.89**
	CAS	0.35	0.27	0.25	0.27	0.25	0.23	**0.21**
Facebook	F_1	0.82	0.79	0.84	0.81	0.89	0.89	**0.91**
	CAS	0.26	0.27	0.24	0.24	0.23	0.22	**0.18**
ArXiv	con	0.34	0.23	0.21	0.21	0.16	**0.14**	0.15
	CAS	0.26	0.25	0.22	0.25	0.18	0.14	**0.13**
Enron	con	0.28	0.24	0.23	0.24	0.22	0.21	**0.17**
	CAS	0.34	0.26	0.28	0.27	0.23	0.19	**0.16**
4Area	con	0.37	0.29	0.31	0.3	0.24	0.26	**0.21**
	CAS	0.31	0.3	0.29	0.31	0.27	0.24	**0.19**

Observations: (1) The performance of the traditional methods are worse than that of the GCNs-based methods generally. The reason might be that the traditional methods pay too much attention to the topology of the network, which results in the underfitting of the performance. GCNs-based methods consider the impact of node neighbors on the target node so as to enhance the accuracy of CD. (2) MCDSA generally achieves better performance than all baselines in most cases, which shows that the spectral approach combines the advantages of independent attribute subspaces. GCNs-based approaches typically embed attributes as vectors and propagate them to neighborhood representations as the structure progresses, which will not be sufficient to exploit the impact of attributes on community formation, even the GCNs-based approach degrades performance in networks with sparse topologies. (3) Although multiple matrix factorizations are introduced in this paper, due to the use of Chebyshev polynomials approximation, it will greatly reduce the complexity of the algorithm, and benefit the CD of large networks.

Module Analysis. In this section, we evaluate experiments on different SP networks, validate the validity of multi-scale community detection. At each scale, we record the total number of partitions of the current network and give the convergence of *NCut* in Table 4.

With the increase of scale s, the number of network partition decreases sharply. This is because as s grows, the constraints of the network topology gradually relax, resulting in a smaller detection space. In particular, when the number of scales approaches the maximum, the number of network partitions will also increase, owing to the fact that the weight of the attribute information of nodes increase when the effect of topology constraints is reduced, to get a more meaningful division of the community. The change of the number of network partitions further proves that the formation of community is controlled by topology and attribute information.

Table 4. The *NCut* of SP at each scale with different parameter settings.

Scales		SP1	SP2	SP3	SP4	SP5	SP6	SP7	SP8
s=12.20	#partitions	242	64	64	65	640	640	640	4
	NCut	208.771	31.769	32.046	33.412	639.284	639.544	639.662	2.1
s=15.80	#partitions	66	53	65	3	230	640	640	72
	NCut	34.076	24.876	33.049	0.616	191.466	639.544	639.662	50.462
s=20.45	#partitions	46	19	3	2	153	65	5	5
	NCut	22.028	5.493	0.587	0.301	118.609	44.245	1.991	1.963
s=26.47	#partitions	16	3	17	22	6	4	3	17
	NCut	4.019	0.462	5.112	10.085	2.295	1.27	0.968	8.046
s=34.26	#partitions	15	3	3	10	3	3	16	25
	NCut	3.692	0.356	1.502	2.209	0.746	0.76	7.121	15.949
s=44.36	#partitions	3	4	8	3	16	15	11	5
	NCut	0.341	0.511	4.511	0.328	7.921	6.549	4.564	1.618
s=57.42	#partitions	3	3	5	3	4	9	5	4
	NCut	0.333	0.323	1.502	0.325	1.184	6.068	2.201	1.164
s=74.33	#partitions	3	3	9	3	10	3	3	4
	NCut	0.347	0.392	3.719	0.328	2.644	1.347	0.795	1.174

Sensitivity Evaluation. As verified in Sect. 3.2, γ hold a pivotal function in EWC. Under the MCDSA, the performance of different γ on various datasets is shown in Fig. 2. It can be observed that when $\gamma=0.5$, the experimental results of MCDSA on all datasets are optimal. γ represents the contribution of node structure information and attribute information during community detection. It is concluded that although the topology of nodes is more intuitive, the presence of attribute information cannot be ignored either.

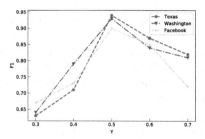

Fig. 2. The effect of parameter γ in real-world datasets.

5 Conclusion

In this paper, we propose MCDSA on attributed graph, a multi-scale attributed community detection method with multi-semantic community partitions from the perspective of attribute subspace deduction. MCDSA is able to find all communities with various semantics such that nodes are densely-connected and share homogeneous properties within same community. Extensive experimental results show that the performance of our method is better than that of comparison method, and it can be applied to categorical and numerical attributed networks.

Acknowledgment. This work is supported by the Industrial Support Project of Gansu Colleges (2022CYZC-11), National Natural Science Foundation of China (62276073, 61762078), Gansu Natural Science Foundation Project (21JR7RA114) and NWNU Teachers Research Capacity Promotion Plan (NWNU-LKQN2019-2).

Appendix

In Eq. (5), the minimization of the objective function $H(\mathbf{w})$ is a constrained nonlinear optimization problem for which the solution is uncertain. For this we introduce the Lagrange multiplication to generate an unconstrained minimization problem as shown in Eq. (6).

$$\min H'(\mathbf{w},\delta) = \sum_{i=1}^{C}\sum_{j=i+1}^{C}\sum_{t=1}^{F} w_t\left(f_{v_it}-f_{v_jt}\right)^2 + \gamma\sum_{t=1}^{F} w_t\log_2 w_t - \delta\left(\sum_{t=1}^{F} w_t - 1\right)$$
(6)

where δ is containing the Lagrange multiplier corresponding to the constraint. Set the gradient of $H'(\mathbf{w},\delta)$ with regard to w_t and δ to 0, so we have:

$$\frac{\partial H'}{\partial \delta} = (\sum_{t=1}^{F} w_t - 1) = 0$$

$$\frac{\partial H'}{\partial w_t} = \sum_{i=1}^{C}\sum_{j=i+1}^{C}\left(f_{v_it}-f_{v_jt}\right)^2 + \gamma\sum_{t=1}^{F} w_t\left(1+\log_2 w_t\right) - \delta = 0$$
(7)

Let $D_t = \sum_{i=1}^{C} \sum_{j=i+1}^{C} \left(f_{v_i t} - f_{v_j t}\right)^2$, Eq. (8) can be obtained:

$$w_t = \exp\left(\frac{-D_t - \gamma + \delta}{\gamma}\right) = \exp\left(\frac{\delta - \gamma}{\gamma}\right) \times \exp\left(\frac{-D_t}{\gamma}\right) \tag{8}$$

According to Eq. (7), $\sum_{t=1}^{F} w_t = 1$, thus we have:

$$\sum_{t=1}^{F} w_t = \sum_{t=1}^{F} \exp\left(\frac{\delta - \gamma}{\gamma}\right) \times \exp\left(\frac{-D_t}{\gamma}\right) = \exp\left(\frac{\delta - \gamma}{\gamma}\right) \times \sum_{t=1}^{F} \exp\left(\frac{-D_t}{\gamma}\right) = 1 \tag{9}$$

By Eq. (9), we have $\exp\left(\frac{\delta-\gamma}{\gamma}\right) = 1/\sum_{t=1}^{F} \exp\left(\frac{-D_t}{\gamma}\right)$. Substituting the above results back into Eq. (8), Eq. (10) can be obtained:

$$w_t = \exp\left(\frac{-D_t}{\gamma}\right)/\sum_{t=1}^{F} \exp\left(\frac{-D_t}{\gamma}\right) \tag{10}$$

References

1. Chang, Y., Ma, H., Chang, L., Li, Z.: Community detection with attributed random walk via seed replacement. Front. Comput. Sci. **16**(5), 1–12 (2022). https://doi.org/10.1007/s11704-021-0482-x
2. Chen, H., Yu, Z., Yang, Q., Shao, J.: Community detection in subspace of attribute. Inf. Sci. **602**, 220–235 (2022)
3. Fortunato, S.: Community detection in graphs. Phys. Rep. **486**(3–5), 75–174 (2010)
4. Hammond, D.K., Vandergheynst, P., Gribonval, R.: Wavelets on graphs via spectral graph theory. Appl. Comput. Harmonic Anal. **30**(2), 129–150 (2011)
5. Li, T., Lei, L., Bhattacharyya, S., Van den Berge, K., Sarkar, P., Bickel, P.J., Levina, E.: Hierarchical community detection by recursive partitioning. J. Am. Stat. Assoc. **117**(538), 951–968 (2022)
6. Maekawa, S., Takeuch, K., Onizuka, M.: Non-linear attributed graph clustering by symmetric NMF with PU learning. arXiv preprint arXiv:1810.00946 (2018)
7. Sales-Pardo, M., Guimera, R., Moreira, A.A., Amaral, L.A.N.: Extracting the hierarchical organization of complex systems. Proc. Natl. Acad. Sci. **104**(39), 15224–15229 (2007)
8. Shuman, D.I., Vandergheynst, P., Frossard, P.: Chebyshev polynomial approximation for distributed signal processing. In: 2011 International Conference on Distributed Computing in Sensor Systems and Workshops (DCOSS), pp. 1–8. IEEE (2011)
9. Zhao, H., Yang, X., Wang, Z., Yang, E., Deng, C.: Graph debiased contrastive learning with joint representation clustering. In: IJCAI, pp. 3434–3440 (2021)
10. Zhao, Q., Ma, H., Li, X., Li, Z.: Is the simple assignment enough? exploring the interpretability for community detection. Int. J. Mach. Learn. Cybern. **12**(12), 3463–3474 (2021)
11. Zhe, C., Sun, A., Xiao, X.: Community detection on large complex attribute network. In: Proceedings of the 25th ACM SIGKDD International Conference on knowledge Discovery Data Mining, pp. 2041–2049 (2019)

Efficient Anomaly Detection in Property Graphs

Jiamin Hou, Yuhong Lei, Zhe Peng, Wei Lu$^{(\boxtimes)}$, Feng Zhang,
and Xiaoyong Du$^{(\boxtimes)}$

School of Information and DEKE, MOE, Renmin University of China, Beijing, China
{jiaminhou,leiyuhong,pengada,lu-wei,fengzhang,duyong}@ruc.edu.cn

Abstract. Property graphs are becoming increasingly popular for modeling entities, their relationships, and properties. Due to the computational complexity, users are seldom to build complex user-defined integrity constraints; worse, the systems often do not have the capabilities of defining complex integrity constraints. For these reasons, violation of the implicit integrity constraints widely exists and leads to various data quality issues in property graphs. In this paper, we aim to automatically extract abnormal graph patterns and efficiently mine all matches in large property graphs to the abnormal patterns that are taken as anomalies. For this purpose, we first propose a new concept namely *CGPs(Conditional Graph Patterns)*. CGPs have the capability of modeling anomalies in the property graph by capturing both abnormal graph patterns and the attribute (i.e., property) constraints. All matches to any abnormal CGP are taken as anomalies. To mine abnormal CGPs and their matches automatically and efficiently, we then propose an efficient parallel approach called *ACGPMiner (Abnormal Conditional Graph Pattern Miner)*. ACGPMiner follows the generation-and-validation paradigm and does the anomaly detection level by level. At each level i, we generate CGPs with i edges, validate whether CGPs are abnormal, and mine all matches to any abnormal CGPs. Further, we propose two optimizations, pre-search pruning to reduce the search space of match enumerations and a two-stage strategy for balancing the workload in distributed computing settings. Using real-life graphs, we experimentally show that our approach is feasible for anomaly detection in large property graphs.

Keywords: graph · abnormal data · parallel

1 Introduction

The graph model has been showing its effectiveness in modeling entities and their relationships in a wide spectrum of applications scenarios, like knowledge bases, transportation graphs, social networks, etc. Due to the computation complexity of graph models, it is seldom to build complex user-defined integrity constraints

© The Author(s), under exclusive license to Springer Nature Switzerland AG 2023
X. Wang et al. (Eds.): DASFAA 2023, LNCS 13945, pp. 120–136, 2023.
https://doi.org/10.1007/978-3-031-30675-4_9

in large graphs [1]. Worse still, some integrity constraints are not trivial to be expressed, e.g., a person is not allowed to have two nationalities with one as Chinese, but is allowed to have two nationalities with one as "the United States" and the other as "England". Due to the above two reasons, violation of the implicit integrity constraints widely exists and leads to various data quality issues in the graph.

Example 1 (**Motivation Example**). Consider two real subgraphs G_1, and G_2 in Fig. 1(a), which are extracted from a classic knowledge graph YAGO [2]. G_1 shows that two persons Bardas and Nikephoros are each other's children, and this modeling is obviously contradictory. G_2 demonstrates that a person named Preus holds both German and Norwegian nationality at the same time. This modeling is also not correct because Norway does not allow dual citizenship. Interestingly, although Norway does not support dual citizenship, some other countries support dual citizenship, such as the United States and England.   □

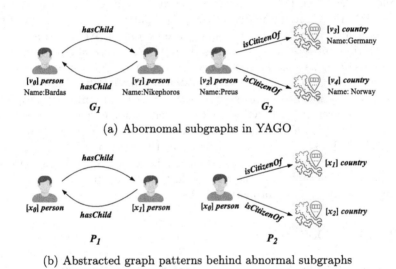

(a) Abornomal subgraphs in YAGO

(b) Abstracted graph patterns behind abnormal subgraphs

Fig. 1. Abnormal subgraphs in YAGO and abstracted graph patterns

Thus far, most of the existing works [3–6] follow the defining-and-identifying paradigm to do graph anomaly detection. Specifically, they first define abnormal graph patterns and then perform graph pattern matching to identify subgraphs that are graph isomorphic to any of the patterns. These subgraphs are considered anomalies. For example, we first define two graph patterns that are P_1 and P_2 in Fig. 1(b), and then identify subgraphs that are graph isomorphic to either P_1 or P_2 in a large graph, e.g., G_1 and G_2, which are considered as anomalies. However, the defining-and-identifying paradigm has two drawbacks. On the one hand, modeling anomalies simply using graph patterns is not adequate.

For example, it is not correct that any matches to P_2 are modeled as anomalies. This is because, as repeatedly discussed, a subgraph G is considered as an anomaly only when G is graph isomorphic to P_2, and the attribute values (resp. country) of vertices of G are confined to a pre-defined set of values, e.g., Norway. On the other hand, the graph patterns need to be defined a priori. In reality, defining the abnormal graph patterns is not trivial, and often, they are modeled after collecting quite a few "manually-reported" anomalies. For example, patterns P_1 and P_2 are modeled by the abstraction of manually-reported anomalies G_1, G_2, and other isomorphic subgraphs can be identified by performing graph pattern matching. Unfortunately, this manually-reported paradigm is incomplete, leaving many more anomalies to be unrevealed.

To mine anomalies from a large property graph automatically and efficiently, in this paper, we first propose a concept namely the conditional graph pattern (a.b.a. CGP) that is able to model the anomalies properly. CGP captures the graph topology as well as the attribute value constraints and hence compared with graph patterns, it takes a more expressive capability to model anomalies. Take G_2 in Fig. 1(a) for example. If we add an attribute constraint $x_1.name =$'Norway' on any match to the graph pattern P_2, then no false positives of anomalies over Norway nationality are produced. We introduce a quantifiable metric namely *abnormality*, based on which we are able to mine CGPs that are considered as anomalies. Furthermore, to make CGPs useful in practice, we propose *ACGPMiner (Abnormal Conditional Graph Pattern Miner)*, an efficient parallel approach to identifying all subgraphs in the property graph that satisfy the requirements of CGPs. ACGPMiner combines pattern mining and attribute discovery in a single process, designs various effective pruning strategies to reduce the search space, and balances the workload in the distributed compute settings using a two-stage strategy. Extensive experiments are conducted on real-life graphs and the results show that our approach is feasible for anomaly detection in large property graphs.

2   Problem Definition

Definition 1 (Property graph). *A property graph G is modeled as a quadtuple (V, E, L, F_A). (1) V is a set of vertices; (2) E is a set of edges, and $E \subseteq V \times V$; (3)each $v \in V$ is labeled $L(v) \in \Theta$ and each $e \in E$ is labeled $L(e) \in \Theta$, where Θ is an alphabet of the node and edge labels in graphs; (4) each vertex $v \in V$ is associated with a set $A = \{A_1, \ldots, A_n\}$ of attributes (i.e., properties). Attribute values of v are denoted as $F_A(v) = (A_1, c_1), (A_2, c_2), ..., (A_n, c_n)$, where c_i $(1 \le i \le n)$ is the attribute value of v over A_i.*

Definition 2 (Subgraph, $\in$). *Given two graphs $G_1 = (V_1, E_1, L_1, F_{A_1})$ and $G_2 = (V_2, E_2, L_2, F_{A_2})$, G_1 is said to be a subgraph of G_2, written as $G_1 \in G_2$, iff (1) $V_1 \subseteq V_2$, $E_1 \subseteq E_2$; (2) for each vertex $v \in V_1$, $L_1(v) = L_2(v)$ and $F_{A_1}(v) = F_{A_2}(v)$; (3) for each edge $(u, v) \in E_1$, $L_1(u, v) = L_2(u, v)$.*

Definition 3 (Isomorphism, $\simeq$). *Two graphs G_1 and G_2 are said to be isomorphic, written as $G_1 \simeq G_2$, iff there is a bijective function $f : V_1 \rightarrow V_2$ satisfying: (1) for each vertex $v \in V_1$, $L_1(v) = L_2(f(v))$; (2) for each edge $(u, v) \in E_1$, $(f(u), f(v)) \in E_2$ and $L_1(u, v) = L_2(f(u), f(v))$.*

Definition 4 (Graph pattern). *A graph pattern is a graph $P[\bar{x}] = (V_P, E_P, L_P, u)$, where (1) V_P (resp. E_P) is a set of vertices (resp. edges); (2) L_P is a function that assigns labels to each vertex $v \in V_P$ (resp. edge $e \in E_P$); (3) $\bar{x}$ is a list of variables, and (4) u is a bijective mapping from $\bar{x}$ to V_P that assigns a distinct variable to each vertex $v \in V_P$.*

Definition 5 (Graph pattern matching). *A match of a graph pattern P in the graph G is a subgraph G_1 of G that is isomorphic to P, i.e., $G_1 \in G$ and $G_1 \simeq P$.*

Example 2 Figure 1 shows two graph patterns: P_1 and P_2. (a) $P_1[x_0, x_1]$ describes two persons x_0 and x_1 who are each other's children. In P_1, 1) V_P are two vertices and E_P are two edges; 2) L_P assigns the label "person" to both two vertices and the label "hasChild" to both two edges; 3) $\bar{x}$ contains two variables x_0 and x_0; and 4) u maps x_0 to the left vertex and x_1 to the right vertex in P_1. A match of the pattern P_1 in G_1 is $x_0 \rightarrow v_0$ and $x_1 \rightarrow v_1$. (b) Similarly, $P_2[x_0, x_1, x_2]$ indicates that a person x_0 is a citizen of both country x_1 and country x_2. A match of pattern P_2 in G_2 is $x_0 \rightarrow v_2$, $x_1 \rightarrow v_3$ and $x_2 \rightarrow v_4$.   □

Graph isomorphism verifies whether two (sub)graphs have an identical structure (i.e. topology). Given a set of (sub)graphs $\mathcal{G} = \{G_i, G_2, ..., G_N\}$, the isomorphism relation divides $\mathcal{G}$ into equivalence classes. Each class is abstracted as a *graph pattern* and subgraphs belonging to the same class are graph isomorphic to each other. In this way, a graph pattern can be considered as a template of all isomorphic subgraphs and a subgraph is considered as an instance (match) of its pattern. However, graph patterns do not contain any attribute information which is required in property graphs. To address this issue, we introduce the conditional graph pattern, given in Definition 6, to impose the attribute constraint on the graph pattern.

Definition 6 (Conditional graph pattern). *A conditional graph pattern (a.b.a. CGP) is defined as $P[\bar{x}](X)$, where $P[\bar{x}]$ is a graph pattern and X is sets of conditions of $\bar{x}$. A condition has the form of $(x.A, c)$, where x is a variable in $\bar{x}$, A denotes an attribute, and c is the attribute value attached to x over A.*

Definition 7 (Conditional graph pattern matching). *A match of a CGP $P[\bar{x}](X)$ in the graph G is a subgraph G_1 of G that 1) is isomorphic to P and 2) satisfies all conditions in X.*

In a CGP $P[\bar{x}](X)$, X can be $\emptyset$, which can be seen as a particular conditional graph pattern without additional conditions, i.e., a simple graph pattern. Moreover, to reduce excessive conditional literals, we select a set of active attributes from G that are of users' interest or are attributes contained in most entities.

Example 3 (1) Consider a CGP $P_1[x_0, x_1](\emptyset)$. The condition of this CGP is empty thus it can be targeted as a graph pattern. (2) Consider a CGP $P_2[x_0, x_1, x_2]$ ($x_1.name = $ '$Norway'$). The condition of this CGP is attached to the variable x_1, which limits the attribute "name" to the value 'Norway'. Compared to P_2 depicting the structure that one person has two nationalities, This CGP additionally requires attribute constraint that the person's one nationality is Norway. With the above CGPs, we can capture the abnormal subgraphs G_1 and G_2 in Fig. 1.   □

In order to measure the degree of anomaly of CGPs, we put forward the concept of *abnormality*. Before introducing the concept of abnormality, we first introduce the concept of support to better explain the abnormality.

Definition 8 *(Support). Consider a graph G, and a CGP $\varphi = P[\bar{x}](X)$, where P has a pivot [7] $z \in \bar{x}$. We define the support of φ as $supp(\varphi, G) = |P(G, X, z)|$, where $P(G, X, z)$ is the set of unique vertices corresponding to the variable z for all matches of φ.*

Given a graph G, a CGP φ and a support threshold λ, we say φ is frequent in G if $supp(\varphi, G) \geq \lambda$. It is intuitive to arise a simple solution that infrequent CGPs can capture abnormal data. But if we consider the entire graph as an instance of a CGP, it cannot occur more than once. It is not enough to simply look for infrequent CGPs. In this paper, we propose the *abnormality* to mine abnormal CGPs as follows.

Definition 9 *(Abnormality). We define the abnormality of a CGP φ as:*

$$abn(\varphi, G) = \frac{supp(\varphi, G)}{supp(\varphi', G)} \qquad (1)$$

Here, φ is generated by adding an edge or a condition to a CGP φ'.

If the support of φ' is very large, while the extended newly CGP φ has little even no support, then φ most likely extends some unreasonable/abnormal information. The abnormality describes the ratio of these two support degrees. What's more, abnormality is an additional condition that is applied in infrequent CGPs. So formula (1) has an implicit condition that $supp(\varphi, G) \leq \lambda$.

The Problem Statement. Given a property graph G, a support threshold $\lambda \geq 0$, and an abnormality threshold $\varepsilon \geq 0$, the anomaly detection problem is to extract abnormal CGPs φ over G with $supp(\varphi, G) \leq \lambda$ and $abn(\varphi, G) \leq \varepsilon$ and then find all matches to any abnormal CGPs.

3   ACGPMiner Approach

3.1   Overview

To mine anomalies from a large property graph automatically and efficiently, we propose *ACGPMiner (Abnormal Conditional Graph Pattern Miner)*, a parallel approach to identifying abnormal subgraphs in the property graphs that satisfy the requirements of CGPs. As shown in Fig. 2, ACGPMiner employs the master-worker paradigm in a multithreaded shared-nothing environment. Communication occurs between the master and workers. The master manages the underlying cluster resources and coordinates the execution of tasks. Workers perform actual specific data processing tasks and report the status of tasks to the master. Each worker in $\{w_0, ..., w_{n-1}\}$ is a process running on multiple cores in $\{c_0, ..., c_{m-1}\}$.

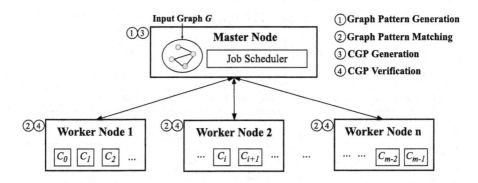

Fig. 2. An overview of ACGPMiner

The core technique of ACGPMiner is to discover abnormal CGPs, based on which we can detect all abnormal subgraphs that match them. To mine abnormal CGPs, ACGPMiner first finds graph patterns P in G, then generates CGPs with P by adding conditions, and finally verifies whether CGPs are abnormal. Specifically, ACGPMiner discovers abnormal CGPs level by level, from smaller CGPs to larger ones. At each level i, it digs out abnormal CGPs with i edges through four significant steps: (1) graph pattern generation to obtain a set of graph patterns; (2) graph pattern matching to find matches for all graph patterns that contribute to CGP candidates; (3) CGP generation to attach conditions to graph patterns for producing candidate CGPs based on matches, and (4) CGP verification for validating whether a CGP is abnormal. In our design, we model graph pattern generation and CGP generation as generating tasks, and model graph pattern matching and CGP verification as computing tasks. Considering that the generating tasks are typically lightweight while computing tasks are prohibitively expensive, ACGPMiner performs generating tasks in the master node and conducts computing tasks in worker nodes in parallel, i.e., assigning multiple workers to compute and execute together computing tasks in parallel.

3.2 Detail Design

We now elaborate on four major steps of ACGPMiner, as shown in Fig. 3.

Graph Pattern Generation. First, we perform graph pattern generation to generate various graph patterns that are candidates for future CGP discovery.

As stated previously, ACGPMiner runs level by level. At each level i, it first generates new graph patterns with i edges. Each graph pattern P' expands a level $i-1$ graph pattern P by adding a new edge (possibly with new nodes). The main techniques of this step have been extensively discussed in numerous works [8–11]. In this paper, we employ the CAM code [8] to guarantee the uniqueness of graph patterns and the FFSM-Join and FFSM-Extend search strategies [11] to generate candidate graph patterns quickly.

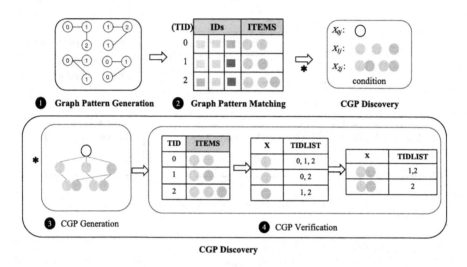

Fig. 3. Four major steps of ACGPMiner

Graph Pattern Matching. Next, we perform graph pattern matching to identify matches for all graph patterns contributing to CGP discovery. ACGPMiner depicts matches in a materialized table view consisting of three sections: *TID*, *IDs*, and *ITEMS*. *TID* is the order of matches; *IDs* characterize the match as a list of unique IDs of vertices; *ITEMS* represent attribute information of each match's vertices as a list of attribute-value pairs. For instance, G_1 is a match of P_1 in Fig. 1. It can be converted to a table view containing *TID* with [0] (assuming it is the first match of P_1), *IDs* with $[v_0, v_1]$ and *ITEMS* with $[(x_0 - name, Bardas), (x_1 - name, Nikephoros)]$. Furthermore, ACGP-Miner provides an incremental method, which extends the stored matches to obtain matches of a larger graph pattern. To obtain matches of P, ACGPMiner performs a join operation $Matches(P') \bowtie Matches(e)$, where P is generated by adding a frequent edge e to P'. When doing the join operator, we also perform an isomorphism check and an automorphism check to reduce the exploration space. Since matches may involve large graph data and are therefore computationally expensive, we perform them in parallel. Specifically, we decompose matches across multiple workers, with each worker computing a portion of graph pattern matches.

CGP Generation. Based on matches of the graph pattern, ACGPMiner performs abnormal CGP discovery. CGP discovery is comprised of two subtasks: CGP generation and verification. We first perform CGP generation to obtain a set of candidate CGPs. Recall that a CGP $P[\bar{x}](X)$ is a graph pattern P coupled with conditions X. To introduce conditions to graph patterns, we utilize the semantic attribute information of matches, i.e., the *ITEMS*. We build conditions by starting the search from singleton X collected from the *ITEMS* part and progressing to a larger X through the set combination level by level. For instance, we can add a singleton $(x_0 - name, Bardas)$ on P_1 to generate a candidate CGP $P_1(x_0.name = \text{'}Bardas\text{'})$.

CGP Verification. After gathering a set of candidate CGPs, ACGPMiner performs CGP verification to check whether a CGP is abnormal. ACGPMiner applies a vertical data format for fast computation. We convert the original data to the vertical-layout data with the format *{X: TIDLIST}*. Here, X is a singleton item, and *TIDLIST* is a list of *TIDs* containing X. For example, the original data of G_1's matches can be converted to $\{\{X : [(x_0 - name, Bardas)], TIDList: [0]\}; \{X : [(x_1 - name, Nikephoros)], TIDList: [0]\}\}$. Based on this design, for getting matches of larger conditions X, we only need to collect the intersection of *TIDLIST* of X's subsets. The support of a CGP with conditions X is the size of *TIDLIST*. Once we get the support of each CGP, we can check whether it is abnormal based on the formula (1).

Pruning. To reduce the search space, we apply the below prunings: (1) Pruning graph patterns: If $supp(P, G) \leq \lambda$, we cease expanding a graph pattern P to produce a larger graph pattern. It is based on the fact that the support of graph patterns is anti-monotonic, meaning that as the graph pattern is expanded, the support decreases. Similarly, no CGP discovery is made if $supp(P, G) \leq \lambda$. (2) Pruning candidate CGPs: ACGPMiner prunes candidate condition sets of length k containing infrequent subsets of length $k - 1$. If X is a frequent condition set, then all of its subsets must also be frequent.

3.3   Discussion

Two computing tasks dominate the cost of ACGPMiner. Both tasks require efficiently generating matches, and there is potential for improvement.

First, there may exist a huge storage overhead in graph pattern matching since we materialize matches of all graph patterns into table views. However, we do not need to materialize matches for infrequent graph patterns since we prune them without making CGP discovery. Inspired by this consideration, we use a pre-search pruning strategy to only materialize matches of frequent graph patterns. A detailed discussion is given in Sect. 4.1.

Second, it is necessary to balance the workload for computing matches in parallel with multiple workers. As previously mentioned, we decompose matches onto multiple workers, where each worker computes a portion of the matches. A basic parallelization strategy allocates all workers to compute matches for a graph pattern. However, not all workers are required to compute each graph pattern's matches. The matches may be small and solvable by a few workers. For this case, assigning all workers could incur significant communication and synchronization costs and waste valuable resources. To solve it, we design a two-stage strategy for load balancing. The discussion is elaborated in Sect. 4.2.

4   Optimizations

4.1   Pre-search Pruning

We use pre-search pruning to reduce the search space of ACGPMiner. The key insight is that we only get all matches when the support is greater than the threshold λ. To do so, we design a *domain structure* and employ a heuristic search strategy for graph pattern matching on this structure.

Domain Structure. For a graph pattern $P[\bar{x} = x_0, x_1, ..., x_k]$, its domain structure of matches can be formalized as $D = (D_{x_0}, D_{x_1}, ..., D_{x_k})$, where D_{x_i} is the set of vertices that match x_i induced by $m(x_i)$ for all matches m of P in G. For each vertex in D_i, we store its unique *ID* and attribute information *ITEM*. In addition, we store its adjacent vertex for searching matches of larger graph patterns. Semantically, D_{x_i} represents the set of domains of the variable x_i. With the domain structure, we can directly get the support of a graph pattern, i.e., the size of D_z where z is the variable representing the pivot pattern node.

Example 4 To better explain domain structure, we show the representation of a graph pattern $P[\bar{x} = x_0, x_1, ..., x_k]$ in Fig. 4(a). Here, (1) $\{x_0, x_1..x_k\}$ are a set of variables of $\bar{x}$; (2) v_x is a vertex of D_{x_1}, and (3) $\{D_{x_k} : v_y\}$ is v_x's adjacent vertex, indicating that v_x is connected to v_y, which is in the domain D_k. As shown in Fig. 4(b), the domain structure can represent P_2's match G_2.   □

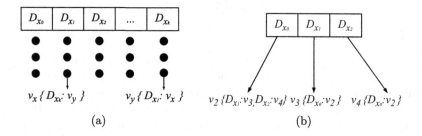

Fig. 4. Domain structure

Graph Pattern Matching. Based on the domain structure, we employ a heuristic search strategy in graph pattern matching to reduce the search space. Specifically, for each vertex $v \in D_z$, we only search for one match containing v. This is due to the fact that our support only considers the size of the vertex specified in the pivot pattern variable. Only if the support meets the threshold

---

Algorithm 1: Graph pattern matching using domain structure

```
// get candidate domain structure
```
1 $D(P)=D(P') \bowtie D(e)$;
```
// calculate the support
```
2 **for** *each vertex u of D_z where z is the pivot* **do**
3   find a match m that assigns u to x ;
4   **if** *not find* **then**
5   Remove u from the domain D

6 *support*=size of D_z ;
```
// get all matches
```
7 **if** *support $\geq \lambda$* **then**
8   **for** *each vertex u of D_z* **do**
9   get all matches consisting u;

---

during the pre-search are all matches containing the v searched. Otherwise, we stop the discovery.

Algorithm 1 details how we use domain structure for incremental graph pattern matching. For a candidate graph pattern P, we first get its candidate domain storage structure by joining matches of P's parent graph pattern P' and its frequent extended edge e (Line 1). We stitch together domain structures D_i corresponding to the connected pattern vertex. Then we apply heuristics to get support on the domain structure (Line 2–6). We iterate over each vertex $u \in D_z$ and search for a match that assigns u to z (Line 3). Note that we perform isomorphism and automorphism checks through searching matches. If the search is unsuccessful, u is removed from D_z (Line 5). Hence, if these vertices are considered in the later pattern matching for getting all matches (Line 7–9), it precludes any further search and reduces the search space. After traversing all vertices of D_z, it is easy to get P's support, which is equal to the size of D_z (Line 6). Only when the support is greater than the threshold λ, do we materialize all matches (Line 7–9).

4.2 Load Balancing

We now discuss the two-stage load balancing strategy, as shown in Fig. 5. The key insight is that candidates with high computational costs can be assigned as many workers as possible, while candidates with low computational costs can be assigned only a few workers. To do so, the first stage builds an approximate computational cost model for deciding the number of workers to compute graph pattern matches. The second stage generates efficient execution plans and assigns workers to perform actual parallel computing. The detailed introduction is as follows.

The First Stage. ACGPMiner generates a pool of jobs, where each job computes matches of one specific graph pattern. In this stage, our goal is to decide how many workers should be assigned to process a job. To do so, we build an approximate computational cost model which predicts the size of the matches of the newly generated graph pattern. Recall that a graph pattern P's matches are generated by a join $Matches(P') \bowtie Matches(e)$. Drawing on the cost estimation of natural joins in relational databases, we build the approximate model in the following.

$$C(P) = \frac{T(P')T(e)}{Max(V(P',c), V(e,c))} \qquad (2)$$

Here, $T(P')$ (resp. $T(e)$) means the count of matches of the candidate graph pattern P' (resp. e). c is the connected pattern node between P' and e. $V(P',c)$ (resp. $V(e,c)$) means the count of distinct values in P' (resp. e) for the variable c. On the basis of the predicted statistics, we can divide one job among n workers for parallel processing. If the predicted cost $C(P)$ is more than a given maximum cost θ, we need $n = max(1, C(P)/\theta)$ workers. Intuitively, if a worker bears data that exceeds the threshold in the future, then the current data will be distributed to other workers. More expensive jobs are assigned to more workers.

The Second Stage. In this stage, we utilize the statistics from the first stage to generate fast execution plans with good load balance. Our goal is to utilize workers as much as possible and not keep them idle for too long. To do so, we handle similar jobs that require almost a similar processing time at the same time. The master continually dispatches jobs to available workers until it becomes empty. Dispatched jobs are prioritized by predicted size; smaller and similar jobs are processed first. Once workers are assigned to jobs, they perform actual parallel computations, i.e., parallel graph pattern matching or parallel CGP verification.

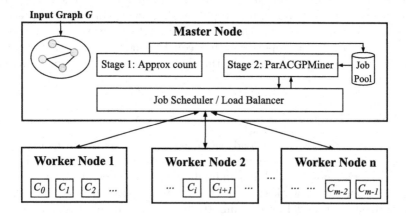

Fig. 5. Two-stage load balancing strategy

5   Evaluation

5.1   Experimental Setup

Implementation. ACGPMiner is implemented in Scala and is built on top of Spark [12]. Spark deals with iterative algorithms in an efficient way and performs the in-memory process to faster the execution. All experiments are conducted on Apache Spark (version 3.1.2). Each executor is set with 5 GB memory and 2 cores. The memory of the driver program is 2 GB. The total cores of each worker are maxed to 8, and the executors of each worker are maxed to 4.

Algorithms. We implement different configurations of ACGPMiner: (1) ParAM: ACGPMiner with the optimizations of pre-search pruning and the two-stage load balancing strategy; (2) DomainAM: ParAM with pre-search pruning but without the two-stage load balancing strategy; (3) TableParAM: ParAM without pre-search pruning but with the two-stage load balancing strategy; and (4) BaseAM: ParAM without pre-search pruning and the two-stage load balancing strategy.

The Compared Method. The CGP is a new proposed definition to mine abnormal data. Existing works were not aimed at mining abnormal CGPs. To compare with other methods, we implement a *Baseline* following the idea proposed in [7], which is one of the most current state-of-the-art methods under the automatic paradigm. We make appropriate changes to fit our topic. Specifically, we run in iterations and discover abnormal CGPs with i edges at each iteration i, which works similarly to our BaseAM algorithm discussed above. The difference is that it uses a brute-force algorithm in attribute discovery. It first lists all candidate CGPs and then iteratively validates each CGP. Our algorithm, however, employs a vertical-data layout for fast validation and does not require scanning through the dataset for each CGP validation. We also implement the Baseline in Scala with Spark for a fair comparison.

Dataset. We use the following two real-life graphs. (a) YAGO: YAGO [2] is a knowledge graph that augments WordNet with common knowledge facts extracted from Wikipedia. We use YAGO with 18 entity types and 36 edge labels. We pick up YAGO with different scales, controlled by the numbers $|V|$ of vertices varying in $\{0.5M, 1M, 1.5M, 2M\}$ and numbers $|E|$ of edges varying in $\{1M, 2M, 3M, 4M\}$. Each entity has an average of 3 attributes in this dataset. (b) DBpedia [13] is another well-known knowledge graph that aims to extract structured content from the information created in Wikipedia. We use DBpedia with 401 entity types, 5M vertices, 268 edge labels, and 14M edges. Each entity in this dataset has an average of 4 attributes.

Hardware Setup. We conduct experiments in an in-house cluster with virtual nodes running CentOS 7.4. Each node is equipped with two Intel(R) Xeon(R) Platinum 8276 CPUs (28 cores × 2 HT), 8 × 128GB DRAM, and 3TB NVMe SSDs.

5.2 Experimental Results

We compare the result of our method ACGPMiner (with all optimizations, i.e., ParAM) to the Baseline using a variety of configurations. We next report our findings.

Exp-1: Effect of Support Threshold λ. We first study the performance by varying support threshold λ on YAGO and DBpedia datasets. We set $k = 3$ and $n = 4$ and report the results in Fig. 6(a) and 6(b). First, the running time of both algorithms grows when the support threshold decreases. It is expected that more graph data will be involved with a smaller support threshold. Second, ParAM outperforms Baseline all the time. For YAGO, ParAM is 1.72x faster on average and up to 1.93x than Baseline. For DBpedia, ParAM is 1.69x faster on average and up to 1.84x than Baseline. It confirms that our proposed algorithm is reliable. Third, the support threshold has smaller impacts on ParAM than Baseline. For YAGO, along with the reduction in the support threshold, the running time of Baseline increases by a factor of 1.46 compared to ParAM's 1.1. For DBpedia, Baseline suffers from 3.07x performance loss while ParAM needs 2.4x more running time. Decreasing the support threshold results in an exponential increase in the number of possible candidates and, thus, the exponential decrease in the performance of the mining algorithm. A feasible algorithm should be able to handle a small support threshold. ParAM is more suitable with a low support threshold since ParAM conducts the pre-search pruning which significantly reduces the overhead of materialized pattern matching.

Exp-2: Effect of Pattern Size k. In this experiment, we evaluate the impact of pattern size k. We study the performance on YAGO and DBpedia datasets by varying k from 2 to 5. We set $n = 4$, $\lambda = 2000$ for YAGO, and $\lambda = 9000$ for DBpeida. The results are shown in Fig. 6(c) and 6(d). First, both ParAM and Baseline algorithms need more time to discover abnormal CGPs with larger patterns. Since more CGPs are discovered with larger patterns. Second, matches may be exponentially large since the graph structure is more complex as the number of pattern edges increases. It is challenging to solve such cases. ParAM outperforms Baseline by varying k on both two datasets. ParAM outperforms Baseline by 10 times on average for YAGO and by 3.8 times on average for DBpedia. It again affirms that our method is feasible for property graphs. Furthermore, pattern size k has more negligible impacts on ParAM than Baseline. For YAGO, the running time of ParAM has increased by 1.7x by varying k from 2 to 5. In contrast, the running time of the Baseline has increased by 17.2x. The result is consistent with the DBpedia dataset.

Exp-3: Scalability with $|G|$. We evaluate the scalability by varying the size of graph $|G| = (|V|, |E|)$ from (0.5M, 1M) to (2.0M, 4M). We fix $k = 3$, $n = 3$ and $\lambda = 8000$. As shown in Fig. 6(e), it takes longer to discover abnormal CGPs for larger graphs, as expected. The execution time of ParAM is 1.72x faster than Baseline on average. When the scale of graphs grows to $(2.0M, 4M)$, it takes up to 1.97x less time to discover abnormal CGPs. Moreover, ParAM is

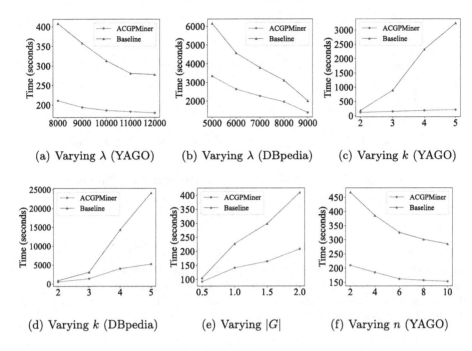

Fig. 6. Performance evaluation of ACGPMiner

less sensitive to the scale of the graph. As the graph scale increases, the running time of ParAM increases by 2.26x while the time of Baseline increases by 3.9x.

Exp-4: Parallel Scalability. In this experiment, we study the parallel scalability by varying the number n of workers from 2 to 10 on YAGO dataset. We fix $k = 3$ and $\lambda = 8000$. As shown in Fig. 6(f), the running time decreases with the increment of workers. Parallel graph pattern mining and CGP verification dominate the cost. Nonetheless, the parallel costs are reduced when more workers are used. ParAM outperforms Baseline by 2.0x on average and up to 2.3x.

5.3   Optimization Analysis

In this experiment, we study the effect of various optimizations. We compare the performance of various optimizations, i.e., the pre-search pruning and the two-stage load balancing strategy, on YAGO and DBpedia datasets by varying support thresholds. We set $k = 3$ and $n = 3$, and the result is shown in Fig. 7.

We next report our findings. First, for both YAGO and DBpedia datasets, ACGPMiner performs best with all optimizations (denoted by ParAM) and performs worst when no optimization is involved (denoted by BaseAM), as expected. ParAM outperforms BaseAM 1.66x on average for YAGO and 1.68x on average for DBpedia. Second, for both datasets, the optimization of the pre-search pruning is more effective than the two-stage load balancing strategy. Compared to BaseAM, DomainAM outperforms 1.62x while TableParAM outperforms 1.12x.

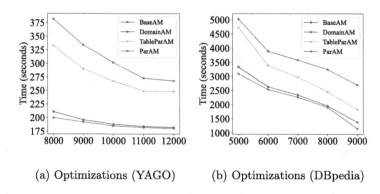

(a) Optimizations (YAGO)   (b) Optimizations (DBpedia)

Fig. 7. The effect of optimizations

It is because we reduce much more search space by applying the pre-search pruning and thus do not need to require computing. Third, we can also observe that the running time takes longer on DBpedia dataset compared to YAGO dataset. It again confirms that the impact of $|G|$ is consistent: the larger G is, the longer ACGPMiner takes.

6 Related Work

Graph Pattern Matching. Given a pre-defined graph pattern, graph pattern matching finds subgraphs in the data graph that are similar (graph isomorphism) to the pattern. Graph pattern matching has been extensively studied in the past decades. A straightforward way is to find matches based on subgraph isomorphism [3]. However, it suffers from huge computational overhead when the graph is big. To reduce time complexity, other methods of a family of graph simulations have been proposed, such as an incremental simulation method [4], a bounded simulation method [5], and a distributed simulation method [6]. However, those methods suffer from either poor expression or in a manual way. First, graph patterns focus on graph structures and ignore the rich attribute information of property graphs. Furthermore, they are required to define graph patterns in advance, which is not trivial and often is developed with a lag after multiple occurrences of abnormal data.

Automatic Discovery of Abnormal Data. Under the automatic paradigm, there exist a few works applying data dependencies to mine abnormal data. Data dependencies are traditionally used to enforce data quality in relations [14–16], and more recently in graphs [1,7,17,18]. As opposed to data dependencies implied in relations, graph dependencies impose the functional dependencies on graph typologies. These works aim to discover laws behind the *normal* graph data. Abnormal data is considered data that does not satisfy these dependencies, which is not intuitive. Also, those works suffer from poor performance since they either use brute force algorithms or do not provide parallel approaches for supporting large-scale graphs.

7   Conclusion

In this paper, we define CGPs to formalize abnormal graph data. CGPs specify the graph structure and attribute conditions in a uniform manner, which can provide a fine-grained paradigm as opposed to graph patterns. We also define the abnormality to measure the degree of exception of abnormal CGPs. Based on the above two notions, we formalize the discovery problem for mining abnormal graph data. To make CGPs useful in practice, we propose a parallel approach, ACGPMiner, for efficiently and automatically mining abnormal CGPs in large-scale graphs. Moreover, we present various optimizations: (1) we design a domain structure and employ a heuristic search strategy for pre-search pruning to reduce search space; (2) we provide a two-stage strategy for load balance. We implement our approach in Scala and build it on top of Spark. Using real-life graphs, we experimentally confirm the effectiveness of our approach.

Acknowledgement. This work was partially supported by National Natural Science Foundation of China under Grant 61972403 and 62072459.

References

1. Fan, W., Lu, P.: Dependencies for graphs. ACM Trans. Database Syst. (TODS) **44**(2), 1–40 (2019)
2. Mahdisoltani, F., Biega, J., Suchanek, F.: Yago3: a knowledge base from multilingual wikipedias. In: 7th Biennial Conference on Innovative Data Systems Research, CIDR Conference (2014)
3. Gou, G., Chirkova, R.: Efficient algorithms for exact ranked twig-pattern matching over graphs. In: Proceedings of the 2008 ACM SIGMOD International Conference on Management of Data, pp. 581–594 (2008)
4. Fan, W., Wang, X., Wu, Y.: Incremental graph pattern matching. ACM Trans. Database Syst. (TODS) **38**(3), 1–47 (2013)
5. Cao, Y., Fan, W., Huai, J., Huang, R.: Making pattern queries bounded in big graphs. In: 2015 IEEE 31st International Conference on Data Engineering, pp. 161–172, IEEE (2015)
6. Ma, S., Cao, Y., Huai, J., Wo, T.: Distributed graph pattern matching. In: Proceedings of the 21st International Conference on World Wide Web, pp. 949–958 (2012)
7. Fan, W., Wu, Y., Xu, J.: Functional dependencies for graphs. In: Proceedings of the 2016 International Conference on Management of Data, pp. 1843–1857 (2016)
8. Huan, J., Wang, W., Prins, J.: Efficient mining of frequent subgraphs in the presence of isomorphism. In: Third IEEE International Conference on Data Mining, pp. 549–552. IEEE (2003)
9. Miyoshi, Y., Ozaki, T., Ohkawa, T.: Frequent pattern discovery from a single graph with quantitative itemsets. In: 2009 IEEE International Conference on Data Mining Workshops, pp. 527–532. IEEE (2009)
10. Jiang, X., Xiong, H., Wang, C., Tan, A.-H.: Mining globally distributed frequent subgraphs in a single labeled graph. Data Knowl. Eng. **68**(10), 1034–1058 (2009)
11. Huan, J., Wang, W., Prins, J.: Efficient mining of frequent subgraphs in the presence of isomorphism. In: Third IEEE International Conference on Data Mining, pp. 549–552, IEEE (2003)

12. Karau, H., Konwinski, A., Wendell, P., Zaharia, M.: Learning spark: lightning-fast big data analysis. O'Reilly Media, Inc. (2015)
13. DBpedia. http://wiki.dbpedia.org/Datasets
14. Kolahi, S., Lakshmanan, L.V.: On approximating optimum repairs for functional dependency violations. In: Proceedings of the 12th International Conference on Database Theory, pp. 53–62 (2009)
15. Chiang, F., Miller, R.J.: Discovering data quality rules. Proc. VLDB Endowment **1**(1), 1166–1177 (2008)
16. Fan, W., Geerts, F., Li, J., Xiong, M.: Discovering conditional functional dependencies. IEEE Trans. Knowl. Data Eng. **23**(5), 683–698 (2010)
17. He, B., Zou, L., Zhao, D.: Using conditional functional dependency to discover abnormal data in rdf graphs. In: Proceedings of Semantic Web Information Management on Semantic Web Information Management, pp. 1–7 (2014)
18. Alipourlangouri, M., Chiang, F.: Keyminer: discovering keys for graphs. In: VLDB workshop TD-LSG (2018)

Edge Coloring on Dynamic Graphs

Zhepeng Huang[1], Long Yuan[1(✉)], Haofei Sui[1], Zi Chen[2], Shiyu Yang[3],
and Jianye Yang[3]

[1] Nanjing University of Science and Technology, Nanjing, China
{hzp,longyuan,pinkypie}@njust.edu.cn
[2] East China Normal University, Shanghai, China
zchen@sei.ecnu.edu.cn
[3] Guangzhou University, Guangzhou, China
{syyang,jyyang}@gzhu.edu.cn

Abstract. Graph edge coloring is a fundamental problem in graph theory and
has been widely used in a variety of applications. Existing solutions for edge
coloring mainly focus on static graphs. However, many graphs in real world are
highly dynamic. Motivated by this, we study the dynamic edge coloring problem
in this paper. Since edge coloring is NP-Complete, to obtain an effective dynamic
edge coloring, we aim to incrementally maintain the edge coloring in a way such
that the coloring result is consistent with one of the best approximate static edge
coloring algorithms when the graph is dynamically updated. Unfortunately, our
theoretical result shows that the problem of finding such dynamic graph edge coloring is unbounded. Despite this, we propose an efficient dynamic edge coloring
algorithm that only explores the edges with color change and their 2-hop incident
edges to maintain the coloring. Moreover, we propose some early pruning rules to
further reduce the unnecessary computation. Experimental results on real graphs
demonstrate the efficiency of our approach.

1 Introduction

Graph edge coloring is a fundamental problem in graph analysis [6]. Given a graph G,
an edge coloring of G is an assignment of colors to the edges such that no two adjacent
edges receive the same color. The goal of edge coloring is to find a coloring with the
smallest number of colors to properly color edges of G.

Applications. Edge coloring can be used in many applications. For example:

- *Privacy-preserving clustering on private network.* In a private network, all vertices
  are independent and private, and each of them knows nothing about vertices other
  than itself and its neighbors [1,8]. Edge coloring is used as a key routine in the
  privacy-preserving EM (Expectation-Maximization) clustering algorithm [32].
- *Transmission scheduling in sensor networks.* In wireless sensor network, transmission scheduling aims to assign time slots to nodes or edge links. Edge coloring has
  long been used for transmission scheduling in wireless networks in which a proper
  edge coloring is first computed, and then each time slot is mapped to a unique link
  with a direction of transmission [7,14].

ⓒ The Author(s), under exclusive license to Springer Nature Switzerland AG 2023
X. Wang et al. (Eds.): DASFAA 2023, LNCS 13945, pp. 137–153, 2023.
https://doi.org/10.1007/978-3-031-30675-4_10

- *Graph visulization.* In social network, graph edge coloring can be used to visualize the network structure for graph exploration [19,29].

Motivation. Due to the wide application scenarios of edge coloring, lots of graph edge coloring algorithms have been proposed in the literature [5,18,24,25]. However, these algorithms mainly focus on the static graphs while many real world graphs are large and frequently updated. For example, Facebook has more than 1.3 billion users and approximately 5 new users join Facebook every second; Twitter has more than 300 million users and 3 new users join Twitter every second [26]. Obviously, the direct solution that uses static algorithms to compute the edge coloring from scratch to handle the graph update is impractical. Motivated by this, we study the dynamic edge coloring problem and aim to propose a new incremental algorithm to maintain the edge coloring when the graph is dynamically updated.

Our Approach. To make the dynamic edge coloring practically applicable in real applications, we should guarantee the effectiveness and efficiency of the designed algorithm simultaneously. For the effectiveness, computing an edge coloring with the smallest number of colors is NP-Complete unfortunately [17], which means it is very unlikely that a polynomial time algorithm exists to compute the optimal edge coloring. On the other hand, [16] shows that using heuristics to perform the edge coloring can not only achieve orders of magnitude speedup in efficiency, but often find an optimal or a nearly optimal coloring for large graphs. Following this observation, we propose an ordering heuristic based static edge coloring algorithm OHEColoring. For a given graph G, OHEColoring can finish the edge coloring in $O(m \cdot d_{max})$ time, where m is the number of edges and d_{max} is the maximum degree in the graph. Obviously, OHEColoring is very efficient to compute the edge coloring for large graphs. Moreover, as verified in our experiments, OHEColoring can also find an optimal or a nearly optimal coloring (refer to Table 1). Therefore, we can address the dynamic edge coloring problem as follows: after each update of the graph, we maintain the edge coloring incrementally and ensure that the coloring is the same as the coloring result obtained by OHEColoring. In this way, the effectiveness of the dynamic coloring can be well guaranteed. The remaining problem is how to maintain such coloring efficiently.

No matter how desirable, we theoretically prove that incrementally maintaining the coloring based on OHEColoring is unbounded [11,12,27] (refer to Theorem 1 in Sect. 4.1), which means we cannot design an algorithm whose running time is polynomial to the minimum amount of work that any incremental algorithm needs to do. Despite this, we reveal the smallest edge color property of OHEColoring and devise an new edge coloring propagation mechanism. Based on the propagation mechanism, we propose a dynamic edge coloring algorithm by iteratively recoloring the edges whose color may be affected due to the graph update. Although this algorithm only needs to recolor a small number of edges in the graph, it may recolor the same edge multiple times, which leads to unnecessary computation. Therefore, we further propose an edge priority based recoloring algorithm in which each edge is recolored only once. Remarkably, our theoretically analysis show that this algorithm only needs to explore the edges with color change and their 2-hop incident edges for the maintenance, and it is practically very efficient as verified in the experiments.

Contributions. We make the following contributions in this paper:

(A) Unboundedness of the Dynamic Edge Coloring. We theoretically prove that incrementally maintaining the coloring based on OHEColoring is unbounded.

(B) An Efficient Dynamic Edge Coloring Algorithm. We propose an efficient dynamic edge coloring algorithm that only explores the edges with color changes and their 2-hop incident edges to maintain the coloring. We also propose several pruning rules to further improve the dynamic edge coloring performance.

(C) Extensive Performance Studies on Real Datasets. We conduct extensive performance studies on six real graphs. The experimental results demonstrate the effectiveness and efficiency of our proposed algorithms.

2 Related Work

Graph analysis has been extensively studied in the literature [1,8,15,19,21–23,29,33–37]. Graph edge coloring is a classic problem in graph analysis. Computing the optimal graph edge coloring is NP-Complete [17]. As graph edge coloring is NP-complete, it is unlikely to be fixed parameter tractable when parametrized by the number of colors. [38] proves that for graphs with treewidth w, an optimal edge coloring can be computed in time $O(nw(6w)^{w(w+1)/2})$, where n is the number of vertices in the graph. [20] shows that it is possible to test whether a graph has a 3-edge-coloring in time $O(1.344^n)$. [4] proves that it is possible to optimally edge-color any graph in time $2^m m^{O(1)}$ and exponential space, or in time $O(2.2461^m)$ and only polynomial space, where m is the number of edges in the graph. [31] shows that edge coloring requires at least d_{max} colors for general graphs, and this is tight for bipartite graphs. Due to the NP-completeness, lots of approximate algorithms are proposed for edge coloring [5,18,24,25]. [16] shows that heuristic algorithm can not only achieve orders of magnitude speedup, but often find a solution close to the optimal coloring for large graphs. Besides, edge coloring is also considered in the context of multigraphs [30], bipartite graphs [13], and planar graphs [9]. There are also extensive work on distributed edge coloring algorithms [2] and online edge coloring problems [28]. Regarding dynamic graph edge coloring, [3,10] study the problem from the theoretical perspective. Nevertheless, these methods involve complex data structures and practically expensive operations, which make them unable to handle large graphs in real applications. Therefore, we have to design new practically effective and efficient dynamic edge coloring algorithms.

3 Preliminaries

Consider an undirected an unweighted graph $G = (V, E)$, where $V(G)$ represents the set of vertices and $E(G)$ represents the set of edges in G. We denote the number of vertices as n and the number of edges as m. Every vertex has a unique ID and we use $id(u, G)$ to denote the id of vertex u. We use $nbr(u, G)$ to denote the neighbors of u in G. The degree of a vertex u, denoted by $deg(u, G)$, is the number of neighbors of u in G. For simplicity, we omit G in the notations if the context is self-evident. Given a graph G, we use d_{max} to denote the largest degree of vertex in G and $\mathbb{N}$ to denote the set of non-negative integers.

Algorithm 1: OHEColoring(G)

---

1 initialize each edge as uncolored;
2 **foreach** $e \in E(G)$ *in decreasing order of* $\prec$ **do**
3 $\quad\mid\quad$ e.color $\leftarrow$ greedyColor(G, e);
4 **Procedure** greedyColor(G, e)
5 $\quad C \leftarrow \{0, 1, \ldots, \deg(u) + \deg(v)\}; \mathbb{C} \leftarrow \emptyset;$
6 **foreach** e' *adjacent to e in G* **do**
7 $\quad\mid\quad$ $\mathbb{C} \leftarrow \mathbb{C} \cup \{e'.\text{color}\};$
8 **return** $\min\{c|c \in C \wedge \notin \mathbb{C}\};$

Definition 1 *(Edge Coloring). Given a graph $G = (V, E)$, an edge coloring of G is a mapping $\mathcal{C} : E(G) \to \mathbb{C}$ from the set of edges $E(G)$ to a set $\mathbb{C}$ of colors such that no two adjacent edges are assigned the same color, where $\mathbb{C} \subset \mathbb{N}$.*

For a graph G and an edge coloring $\mathcal{C}$, we use $|\mathcal{C}(G)|$ to denote number of colors used in $\mathcal{C}$. For an edge e, we use $e.\text{color}(\mathcal{C})$ to denote the color of e assigned by $\mathcal{C}$.

Definition 2 *(Chromatic Index). Given a graph G, the chromatic index of G, denoted by $\chi(G)$, is the smallest number of colors needed for an edge coloring of G.*

Definition 3 *(Optimal Edge Coloring). Given a graph G, the optimal edge coloring of G, denoted by $\varrho(G)$, is an edge coloring of G such that $|\varrho(G)| = \chi(G)$.*

Problem Statement: In this paper, we focus on the problem of dynamic edge coloring, which is defined in the following: Given a graph, we compute the optimal edge coloring $\varrho(G)$ when graph is dynamically updated by inserting/deleting edges. Since computing the optimal edge coloring is an NP-complete problem [17], we resort to approximate solution in this paper.

In this paper, we mainly focus on edge insertion/deletion, because vertex insertion/deletion can be regarded as a sequence of edge insertions/deletions preceded/followed by the insertion/deletion of an isolated vertex.

4   Our Approach

4.1   General Idea and Problem Analysis

As introduced in Sect. 1, to achieve the goal of effectiveness and efficiency simultaneously, we aim to incrementally maintain the coloring and ensure that the dynamic coloring result is consistent with the ordering heuristic based static edge coloring algorithm OHEColoring. Therefore, we present OHEColoring first.

Ordering Heuristic Based Static Edge Coloring Algorithm. Intuitively, the edge with the largest number of adjacent edges potentially produces the highest color. Therefore, OHEColoring processes the edges in the decreasing order characterized by their adjacent edges and each edge is given the color with the smallest number that is not already used by one of its adjacent edges. Specifically, we define the edge order as follows:

Definition 4 *(Edge Dominance $\prec$). Given a graph G and two edges $e = (u, v), e' = (u', v') \in E(G)$ (w.l.o.g, assume $\deg(u) \geq \deg(v)$ and $\deg(u') \geq \deg(v')$), we say e dominates e', denoted by $e \prec e'$, if (1) $\deg(u) > \deg(u')$, or (2) $\deg(u) = \deg(u')$ but $\deg(v) > \deg(v')$.*

For the case that $\deg(u) = \deg(u')$ and $\deg(v) = \deg(v')$, we can further break the tie based on the ids of vertices, and we prefer the smaller id without loss of generality in this paper. Obviously, $\prec$ defines a total order of all edges in G. Following $\prec$, our ordering heuristic based edge coloring is shown in Algorithm 1.

Algorithm 1 first initializes the edges in G as uncolored (line 1). Then, it iterates over the edges in decreasing order of $\prec$ (line 2) and assigns each edge the color returned by procedure greedyColor (line 3). For a given edge e, procedure greedyColor is used to compute the smallest color not assigned to an edge adjacent to e (line 5–8).

Example 1. Consider graph G illustrated in Fig. 1 (a). According to Definition 4, we have $(v_5, v_6) \prec (v_5, v_1) \prec (v_5, v_2) \prec \cdots (v_{11}, v_{12})$. Following Algorithm 1, we color (v_5, v_6) first. As no adjacent edge of (v_5, v_6) is colored, then color 0 is assigned to (v_5, v_6). Then, we color (v_5, v_1), since (v_5, v_6) has been colored with color 0, then (v_5, v_1) is assigned with color 1. We continue the above procedure and the finial coloring is shown in Fig. 1.

For a graph G, the time complexity of OHEColoring to color G is $O(m \cdot d_{max})$. Considering that we have to iterate each edge in a graph and explore the adjacent edges even for verifying the correctness of an edge coloring, OHEColoring is efficient regarding edge coloring. In addition, as verified in our experiments, the number of colors used by Algorithm 1 is nearly optimal. Therefore, we adopt OHEColoring as our underlying static algorithm and aim to incrementally maintain the coloring consistent with OHEColoring when the graph is dynamically updated.

For dynamic algorithms, a criterion for measuring the effectiveness of dynamic algorithms is boundedness [11,12,27]. A dynamic edge coloring algorithm $\mathcal{A}$ regarding OHEColoring is bounded if it computes ΔO such that $\text{OHEColoring}(G) \oplus \Delta O = \text{OHEColoring}(G \oplus \Delta G)$, and its cost can be expressed as a polynomial function of the size |CHANGED| of changes, where ΔG represents the updates to G, ΔO represents

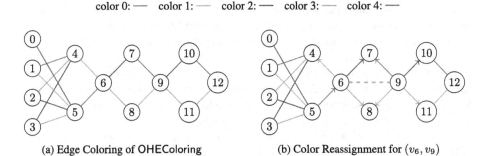

(a) Edge Coloring of OHEColoring   (b) Color Reassignment for (v_6, v_9)

Fig. 1. OHEColoring and Dynamic Edge Coloring

the edges with color change after the updates, $|\text{CHANGED}| = |\Delta G| + |\Delta O|$. Following [11,12,27], $\mathcal{A}$ should be locally persistent, i.e., each edge only stores its auxiliary status information and pointers to its adjacent edges; no global information is maintained between successive calls to the algorithm. No matter how desirable, the boundedness is beyond reach for OHEColoring, which is shown in the following theorem:

Theorem 1. *Given a graph G, the dynamic edge coloring regarding* OHEColoring *is unbounded.*

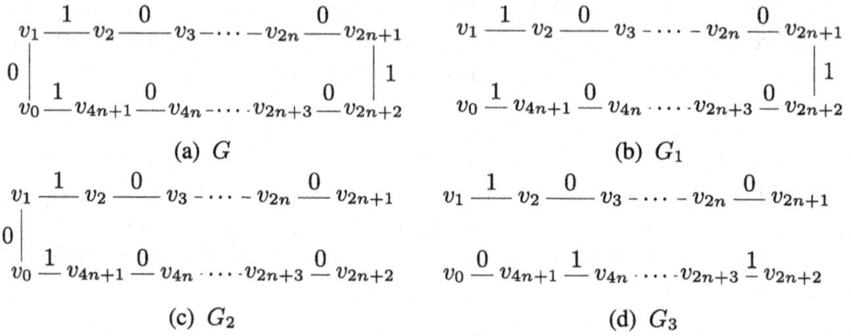

Fig. 2. Unboundedness of Dynamic Edge Coloring

Proof. We prove it by contradiction. Consider the graph G shown in Fig. 2 (a), which is an even length cycle. The assigned color (0 or 1) by OHEColoring is shown near each edge. Let Δ_1 denote the deletion of edge (v_0, v_1) and let Δ_2 denote the deletion of edge (v_{2n+1}, v_{2n+2}). Let $G_1 = G \oplus \Delta_1$, $G_2 = G \oplus \Delta_2$ and $G_3 = G_1 \oplus \Delta_2$. Assume that there exists a bounded dynamic algorithm $\mathcal{A}$ for OHEColoring under a unit edge deletion. Let $\text{trace}(G, \Delta G)$ be the sequence of edge $\mathcal{A}$ visits when ΔG is applied to G. As no color of edges in G_1 and G_2 is affected compared with G, then $|\text{trace}(G, \Delta_1)| = O(1)$ and $|\text{trace}(G, \Delta_2)| = O(1)$ based on the assumption.

Next, we prove that $|\text{trace}(G, \Delta_1)| + |\text{trace}(G, \Delta_2)|$ can not be $O(1)$, which leads to a contradiction against the assumption. We first obtain G_1 by applying Δ_1 to G using $\mathcal{A}$. After that, as $\mathcal{A}$ decides the next operation deterministically based on the status of visited edges, it behaves exactly the same when applying Δ_2 to G and applying Δ_2 to G_1 until an edge e with different status information in two processes is visited. Such e must exist as the results of these two processes are different. As G_1 is obtained by applying Δ_1 to G, $\text{trace}(G, \Delta_1)$ also contains e with information updated. As $\mathcal{A}$ is locally persistent, there exists a path from (v_0, v_1) to e in $\text{trace}(G, \Delta_1)$ and a path from (v_{2n+1}, v_{2n+2}) to e in $\text{trace}(G, \Delta_2)$, implying $|\text{trace}(G, \Delta_1)| + |\text{trace}(G, \Delta)| = \Omega(2n)$ as the length of path between (v_0, v_1) and (v_{2n+1}, v_{2n+2}) is $\Omega(2n)$, which contradicts with the conclusion that $|\text{trace}(G, \Delta_1)|$ and $|\text{trace}(G, \Delta_2)|$ are of constant size. Thus, the theorem holds. □

4.2   A Basic Algorithm

Despite the unboundedness, the situation is not so hopeless. In this section, we present an efficient dynamic edge coloring algorithm that only explores the edges with color change and their 2-hop incident edges to maintain the coloring. For the ease of presentation, we use $adj(e, G)$ to denote the set of adjacent edges of e in G for a given edge e. Moreover, we use $adj^{\prec}(e, G)$ to denote the set of edges in $adj(e, G)$ that dominate e and $adj^{\succ}(e, G)$ to denote the set of edges in $adj(e, G)$ that are dominated by e respectively. For simplicity, we omit G in the notations if the context is self-evident. When an edge (u, v) is inserted into/deleted from G, we use $G + (u, v)/G - (u, v)$ to represent the new graph after the update. According to Algorithm 1, the essence of OHEColoring is to find a coloring C such that the color of every edge satisfies the following property:

Definition 5 *(Smallest Edge Color Property). Given a graph G and an edge coloring C, the color of edge e satisfies the smallest edge color property if $e.color(C(G)) = \min\{c|c \in \mathbb{N} \wedge c \notin \cup_{e' \in adj^{\prec}(e,G)} e'.color(C(G))\}$.*

For simplicity, we call an edge coloring C is smallest edge coloring if all the edge color in C satisfy the smallest edge color property and denote it as $\Psi(G)$. Based on Definition 5, our dynamic edge coloring problem is equal to maintain $\Psi(G)$ when the graph is updated. Clearly, we have the following lemma:

Lemma 1. *Given an graph G and $\Psi(G)$, when an edge e is inserted/deleted, for an edge $e' \in E(G)$, $e'.color(\Psi(G)) = e'.color(\Psi(G \pm e))$ if $e' \prec e$ in G and $G \pm e$.*

Proof. This lemma can be proved directly based on Definition 4, Definition 5 and the procedure of Algorithm 1. □

According to Lemma 1, the colors of the edges that always dominate e before and after the update keep the same in $\Psi(G)$ and $\Psi(G \pm e)$. Thus, these edge do not need to be considered. Apart from these edges, the colors of other edges may violate the smallest edge color property, and thus need to be recolored. To make the colors of these remaining edges satisfy the smallest edge color property, we reassign their colors according to the following recursive 1:

$$e.color(C_{new}) \leftarrow \min\{c|c \in \mathbb{N} \wedge c \notin \cup_{e' \in adj^{\prec}(e)} e'.color(C_{old})\}, \qquad (1)$$

where C_{old} and C_{new} denote the edge coloring before and after color reassignment of edge e. And we have the following lemma:

Lemma 2. *Given a graph G and $\Psi(G)$, when an edge e is inserted/deleted, the coloring C_{new} when Eq. 1 converges for all edges $e \in G$ is $\Psi(G \pm e)$ if we start the reassignment procedure with $\Psi(G)$.*

Proof. We use contradiction to prove it. If the coloring C_{new} is not $\Psi(G \pm e)$, there must exist an edge whose color violates the smallest edge color property. This contradicts with the condition that Eq. 1 converges. Thus, the lemma holds. □

Edge Color Propagation Mechanism. According to Lemma 2, we can obtain $\Psi(G\pm e)$ by recursively reassigning the color of the edges whose colors in $\Psi(G)$ may violate the smallest edge color property regarding $G\pm e$. The remaining problem is how to conduct the recursive color reassignment procedure effectively. A straightforward implement is that we scan the edges whose color may violate the smallest edge color property and reassign their colors based on Eq. 1 round by round and terminate when there is no edge whose color is changed during that round. The drawback of this approach is that in each round, we have to scan all the edges and reassign colors for them even there is only one edge changes its color in that round. On the hand, following the Eq. 1, an edge e' needs to be reassigned its color only if the color of any edge in $\mathrm{adj}^{\prec}(e', G\pm e)$ changes its color. Meanwhile, the color change of e' may further affect the color of edges in $\mathrm{adj}^{\succ}(e', G\pm e)$. Therefore, instead of scanning all the edges, we can start by identifying the direct edges whose color may violate the smallest edge color property due to the insertion/deletion of e and reassign the colors for these edges. For these edges e' whose colors are changed, we take the edges in $\mathrm{adj}^{\succ}(e', G\pm e)$ as candidate edges whose color may violate the smallest edge color property and continue the above process until no edge changes its color. Following this edge color propagation mechanism, we have the following lemma:

Lemma 3. *Given a graph G and $\Psi(G)$, when an edge $e = (u, v)$ is inserted, the different between $e''.\mathrm{color}(\Psi(G+e))$ and $e''.\mathrm{color}(\Psi(G))$ leads to the other edges' color change, where $e'' \in \{(u, v), (u, u'), (v, v')\} \cup \{\mathrm{adj}^{\prec}((u, u'), G) \cap \mathrm{adj}^{\succ}((u, u'), G+e)\} \cup \{\mathrm{adj}^{\prec}((v, v'), G) \cap \mathrm{adj}^{\succ}((v, v'), G+e)\}$, $u' \in \mathrm{nbr}(u', G)$, $v' \in \mathrm{nbr}(v', G)$.*

Proof. We can prove this by contradiction. According to Definition 5, for an edge e, the color of e is determined by the color of $\mathrm{adj}^{\prec}(e, G)$, which means if $\mathrm{adj}^{\prec}(e, G) = \mathrm{adj}^{\prec}(e, G+e)$ and the corresponding colors are same in G and $G+e$, then the color of e is also same in G and $G+e$. When an edge (u, v) is inserted, if the color difference between $\Psi(G)$ and $\Psi(G+e)$ is not caused by the color change of e'' in the lemma, as $\mathrm{adj}^{\prec}(e''', G) = \mathrm{adj}^{\prec}(e''', G+e)$ for other edges e''' and their colors satisfy the smallest color property, then $\Psi(G)$ is always the same as $\Psi(G+e)$, which contradicts with the fact that $\Psi(G)$ may be different from $\Psi(G+e)$. Thus, the lemma holds. □

Lemma 4. *Given a graph G and $\Psi(G)$, when an edge $e = (u, v)$ is deleted, the different between $e''.\mathrm{color}(\Psi(G-e))$ and $e''.\mathrm{color}(\Psi(G))$ leads to the other edges' color change, where $e'' \in \{(u, u'), (v, v')\} \cup \{\mathrm{adj}^{\succ}((u, u'), G) \cap \mathrm{adj}^{\prec}((u, u'), G-e)\} \cup \{\mathrm{adj}^{\succ}((v, v'), G) \cap \mathrm{adj}^{\prec}((v, v'), G-e)\}$, $u' \in \mathrm{nbr}(u', G)$, $v' \in \mathrm{nbr}(v', G)$.*

Proof. It can be proved similarly as Lemma 3. □

Algorithm. According to Lemma 3 and Lemma 4, we can conduct the dynamic edge coloring as follows: we first select such edges e'' as seed edges and continuously propagate the edge color changes until no new edge changes it color. Following this idea, our basic dynamic edge coloring algorithm is shown in Algorithm 2. Given a graph G, DECBasic first inserts/deletes (u, v) into/from G (line 1). Then, it computes the set of edges S whose color changes lead to the difference between $\Psi(G)$ and $\Psi(G \pm (u, v))$

Algorithm 2: DECBasic$(G, (u, v))$

---

1 insert / delete (u, v) from G;
2 $S \leftarrow$ the set of edge e'' following Lemma 3 (insertion case) / Lemma 4 (deletion case);
3 Queue $q \leftarrow \emptyset$; push edges $e'' \in S$ into q;
4 **while** $q \neq \emptyset$ **do**
5 $\quad (u', v') \leftarrow q.\text{pop}()$;
6 $\quad C \leftarrow \{0, 1, \ldots, \deg(u') + \deg(v')\}; \mathbb{C} \leftarrow \emptyset$;
7 $\quad$ **foreach** $e'' \in \text{adj}^{\prec}((u', v'))$ **do**
8 $\quad\quad \mathbb{C} \leftarrow \mathbb{C} \cup \{e''.\text{color}\}$;
9 $\quad c_{\text{new}} \leftarrow \min\{c | c \in C \land \notin \mathbb{C}\}$;
10 $\quad$ **if** $c_{\text{new}} \neq (u', v').\text{color}$ **then**
11 $\quad\quad (u', v').\text{color} \leftarrow c_{\text{new}}$;
12 $\quad\quad$ **foreach** $e'' \in \text{adj}^{\succ}((u', v'))$ **do**
13 $\quad\quad\quad$ **if** $e'' \notin q$ **then** $q.\text{push}(e'')$;

---

step	(v_4,v_1) e_0	(v_4,v_2) e_1	(v_4,v_3) e_2	(v_4,v_6) e_3	(v_6,v_5) e_4	(v_6,v_7) e_5	(v_6,v_9) e_6	(v_9,v_7) e_7	(v_9,v_8) e_8	(v_9,v_{11}) e_9	(v_9,v_{10}) e_{10}	(v_{10},v_{12}) e_{11}	(v_{11},v_{12}) e_{12}	e_{13}	q
Init	0	3	2	1	0	2	-	3	0	1	3	2	0	1	$e_6, e_4, e_3, e_5, e_7, e_8, e_9, e_{10}, e_{11}$
1. e_6	0	3	2	1	0	2	1	3	0	1	3	2	0	1	$e_4, e_3, e_5, e_7, e_8, e_9, e_{10}, e_{11}$
2. e_4	0	3	2	1	0	2	1	3	0	1	3	2	0	1	$e_3, e_5, e_7, e_8, e_9, e_{10}, e_{11}$
3. e_3	0	3	2	2	0	2	1	3	0	1	3	2	0	1	$e_5, e_7, e_8, e_9, e_{10}, e_{11}, e_0, e_1, e_2$
4. e_5	0	3	2	2	0	3	1	3	0	1	3	2	0	1	$e_7, e_8, e_9, e_{10}, e_{11}, e_0, e_1, e_2$
5. e_7	0	3	2	2	0	3	1	4	0	1	3	2	0	1	$e_8, e_9, e_{10}, e_{11}, e_0, e_1, e_2$
6. e_8	0	3	2	2	0	3	1	4	0	1	3	2	0	1	$e_9, e_{10}, e_{11}, e_0, e_1, e_2$
7. e_9	0	3	2	2	0	3	1	4	0	2	3	2	0	1	$e_{10}, e_{11}, e_0, e_1, e_2$
8. e_{10}	0	3	2	2	0	3	1	4	0	2	3	2	0	1	e_{11}, e_0, e_1, e_2
9. e_{11}	0	3	2	2	0	3	1	4	0	2	3	3	0	1	$e_0, e_1, e_2, e_{10}, e_{12}$
10. e_0	0	3	2	2	0	3	1	4	0	2	3	3	0	1	e_1, e_2, e_{10}, e_{12}
11. e_1	0	1	2	2	0	3	1	4	0	2	3	3	0	1	e_2, e_{10}, e_{12}
12. e_2	0	1	4	2	0	3	1	4	0	2	3	3	0	1	e_{10}, e_{12}
13. e_{10}	0	1	4	2	0	3	1	4	0	2	4	3	0	1	e_{12}, e_{13}
14. e_{12}	0	1	4	2	0	3	1	4	0	2	4	3	0	1	e_{13}
15. e_{13}	0	1	4	2	0	3	1	4	0	2	4	3	0	1	$\emptyset$

Fig. 3. Steps of DECBasic for inserting edge (v_6, v_9)

following Lemma 3/Lemma 4 for edge insertion/deletion case (line 2). After that, it initializes an empty query q and pushes the edges in S into q (line 3). After that, it conducts the edge color propagation procedure. Specifically, it first pops an edge (u', v') from q (line 5). Then, it computes a new color c_{new} by iterating the edges in $\text{adj}^{\prec}((u', v'))$ following Definition 5 (line 6–9). If c_{new} is different from the existing color of (u', v'), then, it assigns c_{new} to (u', v') (line 11) and pushes the edges in $eadjout((u', v'))$ for further recoloring (line 12–13). DECBasic finishes when q becomes $\emptyset$ (line 4).

Example 2. Recall the graph G in Fig. 1 (a) and assume a new edge (v_6, v_9) is inserted into G. Figure 3 shows the procedure of DECBasic to handle the insertion. After inserting of edge (v_6, v_9), $e_6, e_4, e_3 \ldots e_{11}$ are computed following Lemma 3 and are pushed into q. After that, DECBasic processes the edge in q continuously. It first processes edge (v_6, v_9) whose details are shown in Fig. 1 (b). For (v_6, v_9), $\text{adj}^{\prec}((v_6, v_9)) = (v_5, v_6)$, and the color of (v_5, v_6) is 0, then (v_6, v_9) is assigned with color 1. After that, the edges in $\text{adj}^{\succ}((v_6, v_9))$ is pushed into q if they are not in q, and the procedure continues. As shown in Fig. 3, the whole coloring maintenance finishes in 15 steps.

Algorithm 3: DECOpt(G, (u, v))

---

1 insert / delete (u, v) from G;
2 $S \leftarrow$ the set of edge e'' following Lemma 3 (insertion case) / Lemma 4 (deletion case);
3 Priority Queue $q \leftarrow \emptyset$; push edges $e'' \in S$ into q;
4 line 4-13 of Algorithm 2;

Theorem 2. *Given a graph G and $\Psi(G)$, when an edge(u, v) is inserted/deleted, Algorithm 2 computes $\Psi(G \pm (u, v))$ correctly.*

Proof. Following Lemma 3 and Lemma 4, the edges caused the color change is pushed in q is pushed in line 3. Moreover, for a specific edge, its color is correctly reassigned in line 11, and the affected edges are pushed in q in line 13. According to Lemma 2, $\Psi(G \pm (u, v))$ correctly computed when Algorithm 2 terminates.   □

Theorem 3. *Given an graph G, the time complexity of Algorithm 2 to handle the insertion/deletion of an edge is $O(m_{\mathsf{basic}} \cdot d_{\mathsf{max}})$, where m_{basic} denote the number of edges pushed in q in Algorithm 2.*

Proof. The time complexity of line 1–3 can be bounded by $O(|S|)$, and these edges are inserted into q, thus $|S| < m_{\mathsf{basic}}$. For each edge $(u', v') \in q$, line 6–10 can be finished in $O(d_{\mathsf{max}})$, line 12 can be finished in $O(d_{\mathsf{max}})$ as well. Thus, the time complexity of Algorithm 2 is $O(m_{\mathsf{basic}} \cdot d_{\mathsf{max}})$.

4.3 An Optimized Approach

Theorem 3 shows that the size of m_{basic} determines the whole complexity of Algorithm 2. However, m_{basic} cannot be well bounded as some edges are pushed into q multiple time in Algorithm 2. Consider the example shown in Fig. 3, e_{10} is pushed into q due to e_6 at step 1 and pop out from q at step 8. However, at step 9, e_{10} is pushed into q again due to the color reassignment of e_{11}. Obviously, processing the same edge multiple times leads to not only the loose bound of Algorithm 2 but also the its inefficiency as verified in our experiments.

Prioritized Dynamic Edge Coloring. Reconsider the example discussed above, the reason leading to the same edge e processes multiple times is that e is processed before the edges in $\mathsf{adj}^{\prec}(e)$. In Fig. 3, $e_{11} \in \mathsf{adj}^{\prec}(e_{10})$, but e_{10} is processed at step 8 while e_{11} is still in q. Therefore, when e_{11} is processed at step 9, e_{10} is pushed into q again. This inspires us that we need to postpone the recoloring of an edge e until all candidate edges in $\mathsf{adj}^{\prec}(e)$ have been recolored. Following this idea, we can define the edge priority based on the edge dominance $\prec$ as follows:

Definition 6 *(Edge Priority).* *Given two edge e and e', if $e \prec e'$, then e has a higher priority than e'.*

According to Definition 6, when conducting the recoloring, we can process the edges based on their edge priority. In this way, the problem that one edge may be processed multiple times can be avoided. Following this idea, the optimized algorithm is shown in Algorithm 3. It shares the same framework of Algorithm 2. The only difference is that we use a priority queue to replace the queue in Algorithm 2.

Step	color														q
	(v_4,v_1)	(v_4,v_2)	(v_4,v_3)	(v_4,v_6)	(v_6,v_5)	(v_6,v_7)	(v_6,v_9)	(v_6,v_8)	(v_9,v_7)	(v_9,v_8)	(v_9,v_{11})	(v_9,v_{10})	(v_{10},v_{12})	(v_{11},v_{12})	
	e_0	e_1	e_2	e_3	e_4	e_5	e_6	e_7	e_8	e_9	e_{10}	e_{11}	e_{12}	e_{13}	
Init	0	3	2	1	0	2	-	3	0	1	3	2	0	1	$e_4,e_6,e_3,e_5,e_7,e_8,e_9,e_{11},e_{10}$
1. e_4	0	3	2	1	0	2	-	3	0	1	3	2	0	1	$e_6,e_3,e_5,e_7,e_8,e_9,e_{11},e_{10}$
2. e_6	0	3	2	1	0	2	1	3	0	1	3	2	0	1	$e_3,e_5,e_7,e_8,e_9,e_{11},e_{10}$
3. e_3	0	3	2	2	0	2	1	3	0	1	3	2	0	1	$e_5,e_7,e_8,e_9,e_{11},e_{10},e_0,e_1,e_2$
4. e_5	0	3	2	2	0	3	1	3	0	1	3	2	0	1	$e_7,e_8,e_9,e_{11},e_{10},e_0,e_1,e_2$
5. e_7	0	3	2	2	0	3	1	4	0	1	3	2	0	1	$e_8,e_9,e_{11},e_{10},e_0,e_1,e_2$
6. e_8	0	3	2	2	0	3	1	4	0	1	3	2	0	1	$e_9,e_{11},e_{10},e_0,e_1,e_2$
7. e_9	0	3	2	2	0	3	1	4	0	2	3	2	0	1	$e_{11},e_{10},e_0,e_1,e_2$
8. e_{11}	0	3	2	2	0	3	1	4	0	2	3	3	0	1	$e_{10},e_0,e_1,e_2,e_{12}$
9. e_{10}	0	3	2	2	0	3	1	4	0	2	4	3	0	1	$e_0,e_1,e_2,e_{12},e_{13}$
10. e_0	0	3	2	2	0	3	1	4	0	2	4	3	0	1	e_1,e_2,e_{12},e_{13}
11. e_1	0	1	2	2	0	3	1	4	0	2	4	3	0	1	e_2,e_{12},e_{13}
12. e_2	0	1	4	2	0	3	1	4	0	2	4	3	0	1	e_{12},e_{13}
13. e_{12}	0	1	4	2	0	3	1	4	0	2	4	3	0	1	e_{13}
14. e_{13}	0	1	4	2	0	3	1	4	0	2	4	3	0	1	$\emptyset$

Fig. 4. Steps of DECOpt for inserting (v_6, v_9)

Example 3. Reconsider the graph shown in Fig. 1 (a), Fig. 4 shows the steps of DECOpt to handle the insertion of edge (v_6, v_9). In Fig. 4, q is a priority queue and the edges are processed based on their priority. Compared with DECBasic, it is clear that DECOpt processes each edge once. As a result, DECOpt only needs 14 steps to finish the coloring maintenance.

Theorem 4. *Given a graph G, when an edge (u, v) is inserted/deleted, let Δ_Ψ denote the edges whose color in $\Psi(G)$ and $\Psi(G \pm (u, v))$ are different, then the number of edges pushed in q by Algorithm 3 can be bounded by m_Δ where $m_\Delta = |\{(u, v)\} \cup \{\mathsf{adj}(u, v)\} \cup \{\cup_{e \in \mathsf{adj}(u,v)}\mathsf{adj}(e)\} \cup \Delta_\Psi \cup \{\cup_{e \in \Delta_\Psi}\mathsf{adj}^\prec(e)\}|$.*

Proof. In line 3 of Algorithm 3, the edges in S is inserted into q, and $S \subseteq \{\{(u, v)\} \cup \{\mathsf{adj}(u, v)\} \cup \{\cup_{e \in \mathsf{adj}(u,v)}\mathsf{adj}(e)\}\}$. Moreover, for each edge e whose color changes, DECOpt pushes e and $\mathsf{adj}^\prec(e)$ into q. As DECOpt pushes each edge into q only once, the total number of edges pushed in q is bounded by m_Δ. □

Theorem 5. *Given a graph G, when an edge (u, v) is inserted/deleted, Algorithm 3 processes the update in $O(m_\Delta(\mathsf{d_{max}} + \log(m_\Delta)))$.*

Proof. In line 3 of Algorithm 3, as $S \subseteq \{\{(u, v)\}\cup\{\mathsf{adj}(u, v)\}\cup\{\cup_{e \in \mathsf{adj}(u,v)}\mathsf{adj}(e)\}\}$, the time complexity of this part can be bounded by $O(m_\Delta)$. For each edge e in q, we have to iterate the edges in $\mathsf{adj}^\prec(e)$ to compute the new color, which can be finished in $O(\mathsf{d_{max}})$. For the priority q, the push/pop operation can be finished $O(1)/O(\log(m_\Delta))$ if we use Fibonacci heap. Therefore, the time complexity of Algorithm 3 can be bounded by $O(m_\Delta(\mathsf{d_{max}} + \log(m_\Delta)))$. □

4.4 Early Pruning

In Algorithm 3, for an edge e with color change, all the edges in $\text{adj}^{\succ}(e)$ are pushed into q for the further color reassignment. However, in same cases, the color change of e will not affect the colors of edges in $\text{adj}^{\succ}(e)$. In this section, we explore different coloring cases between (u,v) and $(v,w) \in \text{adj}^{\succ}((u,v))$, and aim to find some rules that can guarantee that the color of (v,w) is not affected by color change of (u,v), and thus we do not need to push (v,w) into the priority queue to further improve the performance.

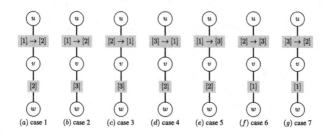

Fig. 5. Different Coloring Cases for (u,v) and $(v,w) \in \text{adj}^{\succ}((u,v))$

Figure 5 shows the different cases when the color of (u,v) changes and we will analyze how the change affects the color of $(v,w) \in \text{adj}^{\succ}((u,v))$. In Fig. 5, the color change of (u,v) is shown near it. For example, in Fig. 5(a), $[1] \rightarrow [2]$ means the color of (u,v) changes from 1 to 2. For ease of presentation, we use $(u,v).\text{color}_{old}$ and $(u,v).\text{color}$ to represent the colors of (u,v) before and after the change and we use $(v,w).\text{color}$ to represent the color of (v,w). We catalog different cases based on the relationships between $(u,v).\text{color}$ and $(v,w).\text{color}$, and we have:

- **case 1**: $(u,v).\text{color} = (v,w).\text{color}$. (u,v) should be recolored obviously in this case. When $(u,v).\text{color} < (v,w).\text{color}$, we have:
- **case 2**: $(u,v).\text{color}_{old} < (v,w).\text{color}$. The color of (u,v) changes from 1 to 2, (v,w) can be possibly recolored with color 1.
- **case 3**: $(u,v).\text{color} < (u,v).\text{color}_{old} < (v,w).\text{color}$. The color of (u,v) changes from 2 to 1, (v,w) can be possibly recolored with color 2.
- **case 4**: $(u,v).\text{color}_{old} > (v,w).\text{color}$. The color of (u,v) changes from 3 to 1. The color of (v,w) is 2, which implies color 0 and 1 have been assigned to the edges in $\text{adj}^{\prec}((v,w))$. Thus, we cannot find a color smaller than 2 to recolor (v,w). Consequently, the color change of (u,v) does not affect the color of (v,w). When $(u,v).\text{color} < (v,w).\text{color}$, we have:
- **case 5**: $(u,v).\text{color}_{old} < (v,w).\text{color}$. The color of (u,v) changes from 1 to 3. In this case, (v,w) can be possibly recolored with color 1.
- **case 6**: $(u,v).\text{color}_{old} > (v,w).\text{color}$. The color of (u,v) changes from 2 to 3, but the color change of (u,v) does not affect the color of (v,w). The reason is similar to case 4.

- **case 7:** $(u,v).\text{color}_{\text{old}} > (v,w).\text{color}$. The color of (u,v) changes from 3 to 2, but the color change of (u,v) does not affect the color of (v,w). The reason is similar to case 4.

Based on the above analysis, when the color of (u,v) changes, we have the following rules on whether $(v,w) \in \text{adj}^\succ(u,v)$ should be pushed into q:

- **Rule 1:** If $(u,v).\text{color} = (v,w).\text{color}$, (v,w) should be pushed into q;
- **Rule 2:** If $(u,v).\text{color} \neq (v,w).\text{color}$ and $(u,v).\text{color}_{\text{old}} < (v,w).\text{color}$, (v,w) should be pushed into q;
- **Rule 3:** If $(u,v).\text{color} \neq (v,w).\text{color}$ and $(u,v).\text{color}_{\text{old}} > (v,w).\text{color}$, (v,w) does not need to be pushed into q;

By applying the above rules, we can further reduce the number of edges that need to explored during the coloring maintenance.

5 Performance Studies

In this section, we evaluate the performance of our proposed algorithms. All experiments are conducted on a machine with Intel Xeon CPU 2.6GHz (32 core) and 128 GB main memory running Linux (Ubuntu Server 22.04.1, 64bit).

Datasets. We evaluate the algorithms on six real-world graphs. Enwiki is downloaded from LAW (https://law.di.unimi.it/datasets.php/) and the remaining datasets are downloaded from SNAP (http://snap.stanford.edu/data/index.html/). The details of the datasets are shown in Table 1.

Table 1. Datasets used in experiments

ID	Datasets G	Type	$|V(G)|$	$|E(G)|$	d_{avg}	d_{max}	#color
G0	Twitter	Social	81,306	1,768,149	43.49	2,799	2,799
G1	Skitter	web	1,696,415	11,095,298	13.08	35,455	35,455
G2	Pokec	Social	1,632,803	30,622,564	16.90	8,753	8,753
G3	LiveJournal	Social	3,997,962	34,681,189	17.18	14,815	14,815
G4	Orkut	Social	3,072,441	117,185,083	76.30	33,313	33,313
G5	Enwiki	Web	5,839,060	135,700,782	23.40	224,964	224,964

Table 2. Efficiency of the dynamic algorithms (in Seconds)

Graph	OHEColoring	Insertion			Deletion		
		DECBasic	DECOpt	DECOpt*	DECBasic	DECOpt	DECOpt*
G0	7.48	0.28311	0.00257	0.00147	0.26103	0.00191	0.00112
G1	45.99	5.72610	0.08744	0.08326	4.14323	0.08543	0.08235
G2	30.10	0.04389	0.00456	0.00379	0.42940	0.00446	0.00361
G3	101.91	0.89463	0.02789	0.02720	0.76926	0.02721	0.02680
G4	723.92	1.25628	0.44875	0.09203	1.05082	0.42707	0.08281
G5	1118.75	8.03554	0.34250	0.34086	7.87554	0.32077	0.31795

Algorithms. We implement and compare four algorithms. All algorithms are implemented in C++ 11, using g++ complier with −O3 optimization.

- OHEColoring: Ordering heuristic edge coloring algorithm (Algorithm 1 in Sect. 4.1).
- DECBasic: Basic dynamic edge coloring algorithm (Algorithm 2 in Sect. 4.2).
- DECOpt: Optimized dynamic edge coloring algorithm (Algorithm 3 in Sect. 4.3).
- DECOpt*: Optimized dynamic edge coloring algorithm with pruning rules (Algorithm 3 in Sect. 4.3 + Rules in Sect. 4.4).

Exp-1: Effectiveness of Static Graph Coloring Algorithm. In this experiment, we evaluate the effectiveness of OHEColoring. Since edge coloring is a NP-complete, the exact solution cannot find the solution in reasonable time. On the other hand, the maximum degree of the graph d_{max} is a lower bound the optimal edge coloring [31]. Thus, to evaluate the effectiveness of OHEColoring we compare the number of colors used by d_{max} and OHEColoring. The results are shown in Table 1.

As shown in Table 1, OHEColoring can always obtain the optimal results on the datasets used in our experiment. For example, on dataset G_4 (Orkut), the maximum degree d_{max} is 33, 313 while the number of colors used by OHEColoring is also 33, 313. Moreover, we also show the running time of OHEColoring on all datasets in Table 2. It can be see that OHEColoring can finish the coloring from scratch very soon as its time complexity is $O(m \cdot d_{max})$. Therefore, OHEColoring is a practically efficient and effective static algorithm for graph edge coloring, which is consistent with the conclusion shown in [16] that heuristic algorithm can not only achieve orders of magnitude speedup, but often find a solution close to the optimal edge coloring for large graphs.

Exp-2: Efficiency of the Algorithms. In this experiments, we evaluate the efficiency of our proposed dynamic algorithms. To test the efficiency, we color the graph using OHEColoring initially, randomly insert/delete 1,000 edges in/from the graph, and record the average processing time for each update for insertion/deletion, respectively. Moreover, we also record the time to color the graph by OHEColoring from scratch for comparison. The results are shown in Table 2.

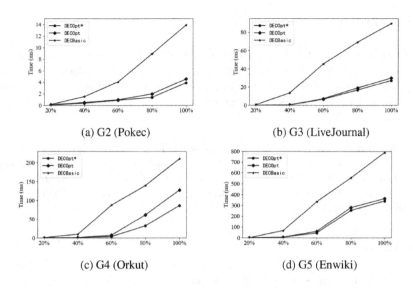

Fig. 6. Scalability when varying $|G|$

As shown in Table 2, DECBasic takes the most time among three dynamic graph algorithms for all cases. For example, on dataset G_3 (LiveJournal), it uses 0.89 s/0.76 s to insert/delete an edge on average. This is because DECBasic does not consider the recoloring orders of the edges in $\mathrm{adj}^{\succ}(e)$ and it may recolor the same edge multiple times, which leads to lots of unnecessary computations. On the other hand, the average processing time of DECOpt for each edge is significantly smaller than that of DECBasic. This is because DECOpt adopts the priority queue to maintain the recoloring orders of edges and it can guarantee each edge is recolored only once during the whole maintaining procedure. Compared with DECOpt, DECOpt* further improve the maintenance performance. This is because lots of the unnecessary edge recoloring are pruned following the rules introduced in Sect. 4.4. Moreover, compared with recoloring the graph using the OHEColoring from scratch, DECOpt* finishes the maintenance in. The experimental results demonstrate the efficiency of our proposed algorithms in maintaining the edge coloring when the graph is dynamically updated.

Exp-3: Scalability When Varying $|G|$. In this experiments, we evaluate the scalability of our proposed algorithms when varying $|G|$. We vary $|E|$ from 20% to 100% of four large graphs, randomly sample 1,000 edges for insertion or deletion, and report the average processing time for each update. The results are shown in Fig. 6.

Figure 6 shows that the average processing time for each update increasing when $|E|$ regarding all three algorithms. This is because as $|E|$ increases, more edges need to be recolored generally. Among our proposed three algorithms, DECBasic consumes the most time for each update while DECOpt is much faster than DECBasic. DECOpt* is the most efficient one among these three algorithms and increases stably when $|E|$ increases. The reasons are the same as discussed in Exp-2. The experimental results show that DECOpt* has a good scalability.

6   Conclusion

In this paper, we study the dynamic edge coloring problem which aims to incrementally maintain the edge coloring when the graph is dynamically updated. We propose an efficient dynamic edge coloring algorithm and three early pruning rules to further reduce the unnecessary computation. Experimental results demonstrate the efficiency of proposed algorithms.

Acknowledgments. Long Yuan is supported by National Key RD Program of China 2022YFF0712100, NSFC61902184, and Science and Technology on Information Systems Engineering Laboratory WDZC20205250411. Zi Chen is supported by CPSF 2021M701214. Shiyu Yang is supported by NSFC61802127, Guangzhou Research Foundation 202201020131 and CCF-Ant Research Fund.

References

1. Archer, A., Lattanzi, S., Likarish, P., Vassilvitskii, S.: Indexing public-private graphs. In: Proceedings of the WWW, pp. 1461–1470 (2017)
2. Barenboim, L., Elkin, M.: Distributed graph coloring: fundamentals and recent developments. In: Synthesis Lectures on Distributed Computing Theory (2013)
3. Barenboim, L., Maimon, T.: Fully-dynamic graph algorithms with sublinear time inspired by distributed computing. In: Proceedings of ICCS, pp. 89–98 (2017)
4. Björklund, A., Husfeldt, T., Koivisto, M.: Set partitioning via inclusion-exclusion. SIAM J. Comput. **39**(2), 546–563 (2009)
5. Borghini, F., Méndez-Díaz, I., Zabala, P.: An exact algorithm for the edge coloring by total labeling problem. Ann. Oper. Res. **286**(1), 11–31 (2020)
6. Cao, Y., Chen, G., Jing, G., Stiebitz, M., Toft, B.: Graph edge coloring: a survey. Graphs Comb. **35**(1), 33–66 (2019)
7. Cheng, M., Yin, L.: Transmission scheduling in sensor networks via directed edge coloring. In: IEEE International Conference on Communications, pp. 3710–3715 (2007)
8. Chierichetti, F., Epasto, A., Kumar, R., Lattanzi, S.: Efficient algorithms for public-private social networks. In: Proceedings of KDD, pp. 139–148 (2015)
9. Cole, R., Kowalik, L.: New linear-time algorithms for edge-coloring planar graphs. Algorithmica **50**(3), 351–368 (2008)
10. Duan, R., He, H., Zhang, T.: Dynamic edge coloring with improved approximation. In: Chan, T.M. (ed.) Proceedings of SODA, pp. 1937–1945 (2019)
11. Fan, W., Liu, M., Tian, C., Ruiqi, X., Zhou, J.: Incrementalization of graph partitioning algorithms. Proc. VLDB Endow. **13**(8), 1261–1274 (2020)
12. Fan, W., Tian, C.: Incremental graph computations: doable and undoable. ACM Trans. Database Syst. **47**(2), 6:1-6:44 (2022)
13. Gabow, H.N., Kariv, O.: Algorithms for edge coloring bipartite graphs and multigraphs. SIAM J. Comput. **11**(1), 117–129 (1982)
14. Gandham, S., Dawande, M., Prakash, R.: Link scheduling in wireless sensor networks: Distributed edge-coloring revisited. J. Parallel Distrib. Comput. **68**(8), 1122–1134 (2008)
15. Hao, K., Yuan, L., Zhang, W.: Distributed hop-constrained s-t simple path enumeration at billion scale. Proc. VLDB Endow. **15**(2), 169–182 (2021)
16. Hilgemeier, M., Drechsler, N., Drechsler, R.: Fast heuristics for the edge coloring of large graphs. In: Proceedings of DSD, pp. 230–239 (2003)

17. Holyer, I.: The np-completeness of edge-coloring. SIAM J. Comput. **10**(4), 718–720 (1981)
18. Karloff, H.J., Shmoys, D.B.: Efficient parallel algorithms for edge coloring problems. J. Algorithms **8**(1), 39–52 (1987)
19. Khurana, U., Nguyen, V.-A., Cheng, H.-C., (Stephen) Chen, X., Shneiderman, B.: Visual analysis of temporal trends in social networks using edge color coding and metric timelines. In: Proceedings of IEEE PASSAT/SocialCom, pp. 549–554 (2011)
20. Kowalik, L.: Improved edge-coloring with three colors. Theor. Comput. Sci. **410**(38–40), 3733–3742 (2009)
21. Liu, B., Yuan, L., Lin, X., Qin, L., Zhang, W.: Efficient (α, β)-core computation: an index-based approach. In: Proceedings of WWW, pp. 1130–1141 (2019)
22. Liu, B., Yuan, L., Lin, X., Qin, L., Zhang, W., Zhou, J.: Efficient (α, β)-core computation in bipartite graphs. VLDB J. **29**(5), 1075–1099 (2020). https://doi.org/10.1007/s00778-020-00606-9
23. Meng, L., Yuan, L., Chen,Z., Lin, X., Yang, S.: Index-based structural clustering on directed graphs. In: Proceedings of ICDE, pp. 2831–2844 (2022)
24. Misra, J., Gries, D.: A constructive proof of vizing's theorem. In: Information Processing Letters (1992)
25. Nemhauser, G.L., Park, S.: A polyhedral approach to edge coloring. Oper. Res. Lett. **10**(6), 315–322 (1991)
26. Ohsaka, N., Maehara, T., Kawarabayashi, K.: Efficient pagerank tracking in evolving networks. In: Proceedings of SIGKDD, pp. 875–884 (2015)
27. Ramalingam, G., Reps, T.W.: On the computational complexity of dynamic graph problems. Theor. Comput. Sci. **158**(1&2), 233–277 (1996)
28. Saberi, A., Wajc, D.: The greedy algorithm is not optimal for on-line edge coloring. In: Proceedings of ICALP, pp. 109:1–109:18 (2021)
29. Sameh, A.: A twitter analytic tool to measure opinion, influence and trust. J. Ind. Intell. Inf. **1**(1), 37–45 (2013)
30. Sanders, P., Steurer, D.: An asymptotic approximation scheme for multigraph edge coloring. In: Proceedings of SODA, pp. 897–906 (2005)
31. Vizing, V.G.: On an estimate of the chromatic class of a p-graph. Discret. Analiz **3**, 25–30 (1964)
32. Yang, B., Sato, I., Nakagawa, H.: Privacy-preserving EM algorithm for clustering on social network. In: Proceedings of PAKDD, pp. 542–553 (2012)
33. Long Yuan, L., Qin, X.L., Chang, L., Zhang, W.: Diversified top-k clique search. VLDB J. **25**(2), 171–196 (2016)
34. Long Yuan, L., Qin, X.L., Chang, L., Zhang, W.: Effective and efficient dynamic graph coloring. Proc. VLDB Endow. **11**(3), 338–351 (2017)
35. Long Yuan, L., Qin, W.Z., Chang, L., Yang, J.: Index-based densest clique percolation community search in networks. IEEE Trans. Knowl. Data Eng. **30**(5), 922–935 (2018)
36. Zhang, J., Li, W., Yuan, L., Qin, L., Zhang, Y., Chang, L.: Shortest-path queries on complex networks: experiments, analyses, and improvement. Proc. VLDB Endow. **15**(11), 2640–2652 (2022)
37. Zhang, J., Yuan, L., Li, W., Qin, L., Zhang, Y.: Efficient label-constrained shortest path queries on road networks: a tree decomposition approach. Proc. VLDB Endow. **15**(3), 686–698 (2021)
38. Zhou, X., Nakano, S.-I., Nishizeki, T.: Edge-coloring partial k-trees. J. Algorithms **21**(3), 598–617 (1996)

Discovering Persistent Subgraph Patterns over Streaming Graphs

Chu Huang[1], Qianzhen Zhang[1(✉)], Deke Guo[1(✉)], and Xiang Zhao[2]

[1] Science and Technology on Information Systems Engineering Laboratory,
National University of Defense Technology, Changsha, China
{zhangqianzhen18,dekeguo}@nudt.edu.cn
[2] Laboratory for Big Data and Decision, National University of Defense Technology,
Changsha, China

Abstract. Streaming graph analysis is gaining importance in various fields due to the natural dynamicity in many real graph applications. Prior subgraph discovery problem over streaming graphs mostly focuses on characteristics like frequency and burstiness. Persistence, as a new characteristic, is getting increasing attention. Persistent subgraph discovery highlights behaviors where a subgraph appears recurrently in many time windows, which is vital for many real-world applications (e.g., anomaly detection). While persistent subgraph discovery enjoys many interesting real-life applications, there is no off-the-shelf solution to compute the persistent pattern efficiently. In this work, we are the first to study the persistent subgraph pattern discovering problem over the streaming graph. We devise an auxiliary data structure called TFD to detect the persistent subgraph patterns in real-time with limited memory usage. TFD maps each subgraph into the corresponding bucket based on hash functions to compute the persistence of each pattern. Then we introduce optimizations to separate persistent and non-persistent patterns, further improving the effectiveness and throughput in space-scarce scenarios. Extensive experiments confirm the superiority of our proposed method.

1 Introduction

A recent development is the proliferation of high throughput, dynamic graph-structured data organized as streaming graphs. For example, consider the knowledge graph DBpedia, which gets updated daily according to a stream of change logs from Wikipedia [4,7,10]. Streaming graph analysis is gaining importance in various fields such as subgraph match [5,12], frequent pattern mining [2,13], and bursting pattern mining [18]. Apart from the above characteristics, another important characteristic - ***persistence***, has received growing attention. Given a pattern P and a streaming graph with T tumbling windows, the persistence of P is defined as the number of windows where P appears. We say P is a persistent pattern if its persistence is larger than a user-defined threshold. Persistent

ⓒ The Author(s), under exclusive license to Springer Nature Switzerland AG 2023
X. Wang et al. (Eds.): DASFAA 2023, LNCS 13945, pp. 154–171, 2023.
https://doi.org/10.1007/978-3-031-30675-4_11

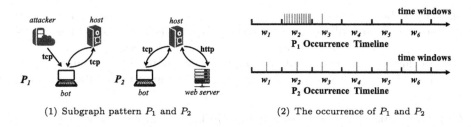

(1) Subgraph pattern P_1 and P_2 (2) The occurrence of P_1 and P_2

Fig. 1. Concealed Cyber-attack detection

pattern often indicates the happening of abnormal or notable events. We next use an example of detecting malicious behaviors to illustrate its basic idea.

Application. Malicious behaviors have patterns[1]. Security analysts can identify malicious behaviors by monitoring the appearances of the patterns (based on the semantics of subgraph isomorphism) in network traffics. As shown in Fig. 1(1), some malicious behaviors try to hide by spreading their communications over many time windows. As a result, these patterns cannot be detected by finding frequent subgraph patterns. To detect such threats, we should use persistence instead of frequency as an indicator. Figure 1 shows two communication patterns and their matching results during corresponding time windows. P_1 is a pattern detected by finding frequent subgraph pattern, which is only a general broadcasting mechanism and cannot give us valuable information. P_2 is a pattern detected by using persistence, representing an attack pattern. P_2 describes information exfiltration, where the victim host takes commands from the bot and exchanges data with compromised websites that lead to a data breach.

Formally, given a streaming graph $\mathcal{G}$, a persistence threshold δ, and an integer k, the continuous persistent pattern discovering problem is to find the k-edge subgraph patterns that appear in at least δ tumbling windows.

Challenges. Albeit important, the problem of persistent patterns discovery lacks a dedicated technique. A straightforward way is to enumerate all possible k-edge subgraphs in each window and then calculate the corresponding patterns of these subgraphs to verify the existence of each pattern at current window. The algorithm needs to calculate and store all k-edge subgraphs at each time window, which consumes a lot of time and memory. What's more, the algorithm needs to re-execute subgraph isomorphism calculation to verify the existence of each pattern in each window, which can be detrimental. In this light, advanced techniques are desiderated to discover persistent patterns efficiently.

Our Solution. In fact, the large scale and high-dynamic of streaming graphs make it memory and time-consuming to discover persistent patterns accurately. It is a natural choice to compute approximations with limited memory efficiently.

Our main idea is as follows: to avoid mapping each newly produced subgraph to all existing candidate subgraph patterns, we propose to design an

---

[1] https://www.verizon.com/business/resources/reports/dbir/.

auxiliary data structure called TFD, which consists of d arrays. Each newly produced subgraph will be mapped into one bucket of the arrays by hash functions $h_1(\cdot), \cdots, h_d(\cdot)$ to calculate the existence of the patterns at each time window. In this way, two isomorphic subgraphs will be mapped to the same bucket. Based on TFD, we can avoid storing any k-edge subgraph and repeated subgraph matching.

Contributions. The major contributions include: 1) We are the first to study the problem of persistent patterns discovering over streaming graphs. 2) We propose an auxiliary data structure called TFD to reduce the redundant subgraph matching calculations in the persistent patterns discovering process. 3) We exploit one optimization to improve further the effectiveness of persistent subgraph patterns discovering, which can achieve time- and memory-efficiency. 4) Extensive experiments confirm that our method outperforms the baseline solution.

2   Problem Formulation

Definition 1 (Streaming Graph). *A streaming graph $\mathcal{G}$ is a constantly growing sequence of directed edges $\{\sigma_1, \sigma_2, \cdots, \sigma_n\}$ where each $\sigma_i = (v_{i_1}^{id_1}, v_{i_2}^{id_2}, t(\sigma_i))$ indicates a directed edge from vertices $v_{i_1}^{id_1}$ to $v_{i_2}^{id_2}$ arriving at time $t(\sigma_i)$ and the superscripts of the vertices are vertex IDs.*

It's worth noting that the throughput of the streaming graph keeps varying. For simplicity of presentation, we only consider vertex-labeled graphs.

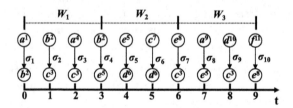

Fig. 2. Streaming Graph

Definition 2 (Tumbling Window). *Tumbling windows are a series of fixed-sized, non-overlapping and contiguous time intervals. A tumbling window, denoted as W_i, has a timespan with fixed-sized duration τ in $\mathcal{G}$.*

Example 1. A streaming graph $\mathcal{G}$ is shown in Fig. 2. Specifically, for the edge σ_2 in $\mathcal{G}$, it shows that σ_2 has two vertices b^2 and c^3, where "b" and "c" are vertex labels and the superscripts are vertex IDs. Besides, the timestamp of σ_2 is shown below it. The streaming graph is divided into three time windows, from start timestamp $t_0 = 0$, each of which has size $\tau = 3$ and is non-overlapping.

Definition 3 (Snapshot graph). *A snapshot graph at timestamp t, denoted as G_t, is a graph induced by all the edges in W_i that have been observed up to and including time t where $t \in W_i$.*

For any $t \in W_i$, at time $t + 1$ we receive an edge insertion e and add it into G_t to obtain G_{t+1}. For each newly inserted edge e in G_{t+1}, we use the notation $E_k(e)$ to denote the set of k-edge subgraphs that contain e in G_{t+1}. Besides, we use G^i to denote the snapshot graph of tumbling window W_i where G^i is a graph induced by all the edges within W_i.

A subgraph $g_k = (V_g, E_g)$ is referred to as a k-edge subgraph if k edges in G_t induce it. We define $\mathcal{P}^k$ as the set of all induced subgraphs with k edges in G^i.

Definition 4 (Subgraph isomorphism). *Two k-edge subgraphs $g_{k_1} = (V_{g_1}, E_{g_1})$ and $g_{k_2} = (V_{g_2}, E_{g_2})$ are isomorphic if there exists a bijection f from V_{g_1} to V_{g_2} such that the following cases hold: 1) $\forall v_i \in V_{g_1}$, $L(v_i) = L(f(v_i) \in V_{g_2})$, and 2) $\forall(v_i, v_j) \in g_{k_1}$, $(f(v_i), f(v_j)) \in g_{k_2}$. The function L preserves the vertex labels.*

The isomorphism relation partitions the set of subgraphs $\mathcal{P}^k$ into m equivalence classes, denoted by $\mathcal{P}_1^k, \cdots, \mathcal{P}_m^k$. Each equivalence class $\mathcal{P}_i^k$ is called a subgraph pattern. Note that $\mathcal{P}_i^k$ can be obtained by deleting the IDs (resp. timestamps) of the vertices (resp. edges) of corresponding k-edge subgraph that is in $\mathcal{P}_i^k$. For simplicity, we use shorthand P_i to denote the generic pattern $\mathcal{P}_i^k$. We define the frequency $fre(P_i, G^i)$ of P_i in each time window as the number of k-edge subgraphs in $\mathcal{P}_i^k$ and PS as the set of different k-edge patterns in $\mathcal{G}$.

Pattern Persistence Measures. Persistence, is a particular pattern of the occurrence behavior in terms of the number of windows that a k-edge subgraph pattern exists in a streaming graph, denoted as $per(P)$. The persistence measure of a pattern P is to count the number of time windows in which P exists. The persistence of a persistent pattern should exceed a user-defined threshold δ. The formal definition of a persistent pattern is as follows.

Definition 5 (Persistent Pattern). *Given a streaming graph $\mathcal{G}$, a k-edge pattern P and a persistence threshold δ. P is a persistent pattern if $per(P) \geq \delta$.*

Problem Statement. Given a streaming graph $\mathcal{G}$, and parameters k, τ, and δ, persistent patterns discovery computes the set of k-edge subgraph patterns that the persistence is greater than persistence threshold δ.

3   The Baseline Solution

A straightforward way (Algorithm 1) to discover persistent patterns over a streaming graph is to enumerate all possible k-edge subgraphs when a new time window comes and then partition the set of k-edge subgraphs into different equivalence classes to verify the existence of each k-edge pattern. If the persistence measure of a k-edge pattern exceeds a user-defined threshold, it will be returned as a persistent pattern. More details are described below.

We use a set PS to store the different k-edge patterns in $\mathcal{G}$. Each item in PS is a tuple $(P, per(P))$, where P is a k-edge pattern, $per(P)$ is the persistence value of pattern P. Whenever a new window W_i appears, findPP updates PS by calling computePer (Line 2–3). Then, for each pattern P in the PS, findPP verifies whether the persistence value of P satisfies the persistence threshold δ (Lines 4–5). Finally, it returns all persistent patterns (Line 6).

Function computePer. computePer first calls findSubgraph (omitted) to calculate the k-edge subgraphs set $\mathcal{P}^k$ in G^i (Line 1). In detail, whenever an edge insertion e occurs at timestamp t ($t \in W_i$), findSubgraph explores a candidate subgraph space in a tree shape in G_t to calculate $E_k(e)$, each node representing a candidate subgraph, where a child node is grown with one-edge extension from its parent node. To avoid duplicate enumeration of a subgraph, findSubgraph checks whether two subgraphs are composed of the same edges at each level in the tree space. After dealing with all edge insertions in W_i, we can obtain the k-edge subgraphs set $\mathcal{P}^k$ in G^i. To compute the corresponding k-edge pattern, computePer calls evaluateFre(omitted) to partition the subgraphs in $\mathcal{P}^k$ into equivalence classes based on subgraph isomorphism calculations, each of which can represent a pattern P (Line 2). If $fre(P, G^i) \geq 1$, computePer further check whether $P \in PS$ through subgraph isomorphism calculation; if so, it sets $per(P) \leftarrow per(P) + 1$. Else it adds $(P, per(P) = 1)$ into the set PS (Line 3–6).

Algorithm 1: findPP

    Input : $\mathcal{G}$ is a streaming graph; δ and k are the parameters.
    Output : the set of persistent patterns $PPSet$.
1 $PatternSet \leftarrow \emptyset$, $PPSet \leftarrow \emptyset$;
2   **foreach** the snapshot graph of tumbling window G^i **do**
3 | $PS \leftarrow$ computePer(G^i, PS);
4   **foreach** pattern $P_i \in PS$ **do**
5 | **if** $per(P) \geq \delta$ **then** $PPSet \leftarrow PPSet \cup \{(P, per(P))\}$;
6   **return** $PPSet$;
    Function *computePer(G^i, PS)*
1 | $\mathcal{P}^k \leftarrow$ findSubgraph(G^i);
2 | evaluateFre($\mathcal{P}^k$);
3 | **foreach** k-edge patten P **do**
4 | | **if** $fre(P, G^i) \geq 1$ **then**
5 | | | **if** $P \in PS$ **then** $per(P) \leftarrow per(P) + 1$;
6 | | | **if** $P \notin PS$ **then** $PS \leftarrow PS \cup (P, per(P) = 1)$;

Algorithm Analysis. There are three main steps in findPP. (1) In the k-edge subgraph enumeration process, given an edge insertion e in G_t, let n be the average number of vertices of the subgraph extended from e with radius k. findSubgraph takes $O(2^{n^2})$ to explore all k-edge subgraphs that contain e. (2) In the PS update process, let σ be the average unit time to verify whether two k-edge subgraphs are isomorphic. evaluateFre takes $O(N \cdot (N^2 - 1) \cdot \sigma)$ time to partition the set of k-edge subgraphs into m equivalence classes. Let M be the number of patterns in PS. computePer takes $O(m \cdot M \cdot \sigma)$ to update PS. (3) findPP takes $O(1)$ to return the persistent patterns.

4   The fastPP Framework

In this section, we first analyze the drawback of findPP and then devise a auxiliary data structure TFD to efficiently mine persistent patterns over a streaming graph, which can significantly reduce memory cost and computational costs.

Why Costly? The algorithm findPP is not scalable enough to handle large streaming graphs. Firstly, to find the k-edge patterns in each time window, findPP needs to calculate and store all k-edge subgraphs at each time window, which consumes a large amount of time and memory. Secondly, in the PS updated process, findPP needs to re-execute subgraph isomorphism calculation for each pattern at the current window to check whether it exists in PS.

One can do it more efficiently with large space savings. Our idea is as follows: we propose a new algorithm for persistent pattern discovery, which can overcome the drawbacks introduced above. In the new algorithm, we design an auxiliary data structure called TFD and add a state field for the counter in each bucket of TFD to calculate the persistence of a pattern in each time window. Specifically, for each newly produced k-edge subgraph, we use a hash function to map it into a fixed position in the TFD. Once the state of the counter indicates that the corresponding bucket has been counted in the current time window, the counter no longer counts. In this way, we can calculate the persistence of the pattern directly, and thus we can avoid storing all subgraphs of the current time window and repeated subgraph isomorphism calculation.

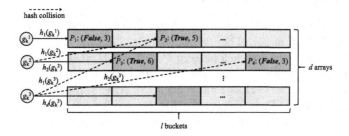

Fig. 3. Data Structure of TFD

TFD (Fig. 3). TFD consists of d arrays, each of which consists of l buckets. Let $B_i[j]$ be the jth bucket in the ith array. Each bucket consists of a key-value pair, where the key is a k-edge pattern, and the value is a counter that aims to count the persistence of the k-edge pattern. The d arrays are associated with d pairwise independent hash functions $h_1(\cdot), \cdots, h_d(\cdot)$ respectively, each of which maps a k-edge subgraph into one bucket of the corresponding array. If two subgraphs are isomorphic, they will be mapped into the same bucket. It is worth noting: Although two subgraphs are not isomorphic, they are mapped into the same bucket due to hashish conflict. In this situation, we will use other hash functions to map one subgraph into one bucket in the corresponding array.

Regardless of the number of subgraphs mapped into one bucket in a time window, we should only increment this counter in the bucket by one due to the characteristic of persistence. Therefore, for persistence calculation, we add a state field for each counter to indicate whether this counter has been incremented in the current time window. The state field has two states: $True$ and $False$. When a new time window comes, the state of every counter is set to $True$. Whenever a k-edge subgraph g_k is mapped into a bucket, only if the state of the counter is $True$, we can increment the counter by one and then turn the state to $False$. Thus we can directly count the persistence of the pattern in each time window.

To map a k-edge subgraph g_k to auxiliary data structure during hashing process, we encode the subgraph g_k as a string representation g_k^{str} by exploiting the graph invariant [18]. The graph invariant encodes subgraphs such that the isomorphic subgraphs are mapped into the same location in the TFD. Specifically, first, for each vertex in the k-edge subgraph, we push the degree and label of a vertex together as its new label denoted $l(v)$. And, for each edge $e = (v_i, v_j, t(e))$, we label $l(e) = (l(v_i), l(v_j))$. Second, we assigned a weight to each edge. The weight is equal to the order in which the edge's corresponding single-edge pattern occurs in the streaming graph. Specifically, we use $w(e)$ to denote the weight of edge e. If $w(e_i) < w(e_j)$, then $e_i < e_j$. Else, if $w(e_i) = w(e_j) \cup l(e_i) < l(e_j)$, then $e_i < e_j$, where $l(e_i) < l(e_j)$ means the vertex degrees of e_i is lexicographically smaller. Third, the coding string encodeSub(g_k) of a subgraph g_k including edges $e_1, \cdots, e_n$ in the g_k where $e_i < e_{i+1}$ is $l(e_1) \cdots l(e_n)$.

4.1   The Fast Algorithm fastPP

The new framework fastPP is shown in Algorithm 2. It first calls initializePer to initialize the TFD (Line 1). Then it updates the TFD by calling updateTFD to calculate the persistence for each pattern P in the TFD when a new time window comes (Line 2–3). After dealing with the current time window, it will set the state of counters in the TFD as $True$ (Line 4–5). Afterwards, for each nonempty bucket $B_i[j]$ in the TFD, it checks whether $B_i[j].value$ satisfies the persistence threshold δ (Line 6–7). Finally, it returns all persistent patterns (Line 8).

Function updateTFD. updateTFD processes the snapshot graphs in W_i in ascending (Line 1). Whenever an edge insertion e occurs at timestamp t ($t \in W_i$), updateTFD calls findSubgraph to calculate $E_k(e)$ (Line 2–3). For each subgraph $g_k \in E_k(e)$, TFD first chooses which hash function $h_i(\cdot)$ to map g_k into one bucket $B_i[h_i(g_k)]$ in array B_i, then checks whether the $B_i[h_i(g_k)].state$ is $True$

Algorithm 2: fastPP

Input   : $\mathcal{G}$ is the streaming graph; δ,k are the parameters.

Output : the set of persistent patterns $PPSet$.

1 TFD $\leftarrow$ initializePer(TFD), $PPSet \leftarrow \emptyset$;

2 **foreach** the time window W_i of $\mathcal{G}$ **do**

3   TFD $\leftarrow$ updateTFD(W_i, TFD);

4   **foreach** bucket $B_i[j]$ in the TFD **do**

5 $B_i[j].state \leftarrow True$;

6 **foreach** nonempty bucket $B_i[j]$ in the TFD **do**

7   **if** $B_i[j].value \geq \delta$ **then** $PPSet \leftarrow PPSet \cup \{(B_i[j].key, B_i[j].value)\}$;

8 **return** $PPSet$;

    Function *updateTFD(W_i, TFD)*

1   **foreach** the snapshot graph G_t at time t where $t \in W_i$ **do**

2   **foreach** edge insertion e at time t **do**

3 $E_k(e) \leftarrow$ findSubgraph(e, G_t);

4   **foreach** subgraph $g_k \in E_k(e)$ **do**

5   **foreach** $i \in [1,d]$ **do**

6   **if** $B_i[h_i(g_k)].state == True$ **then**

7   **if** g_k is isomorphic to $B_i[h_i(g_k)]$ **then**

8 $B_i[h_i(g_k)].value \leftarrow B_i[h_i(g_k)].value + 1$;

9 $B_i[h_i(g_k)].state \leftarrow False$, break;

10   **if** g_k is not isomorphic to $B_i[h_i(g_k)]$ and $B_i[h_i(g_k)] == \emptyset$ **then**

11   calculate the k-edge pattern P from g_k;

12 $B_i[h_i(g_k)] \leftarrow (P, (False, 1))$, break;

13   **return** TFD;

---

to avoid overestimation (Line 4–6). If so, there are two cases: (1) g_k is isomorphic to the pattern P in bucket $B_i[h_i(g_k)]$. updateTFD increases the persistence of P by 1 and sets the $B_i[h_i(g_k)].state$ to $False$ (Line 7–9). (2) g_k is not isomorphic to the pattern P and the bucket is empty. updateTFD first calculates the pattern of g_k, then inserts $(P, (state = False, per(P) = 1))$ into the bucket (Line 10–12). Finally, updateTFD returns the updated TFD (Line 13).

Example 2. Figure 3 demonstrates an example of the hash process. In current time window, subgraph g_k^1 is hashed into bucket $B_1[h_1(g_k^1)]$. However, the state of counter of $B_1[h_1(g_k^1)]$ is $False$, so the TFD does nothing. When considering subgraph g_k^2, we first hash it into $B_1[h_1(g_k^2)]$. Although the state of counter of $B_1[h_1(g_k^2)]$ is $True$, g_k^2 is not isomorphic to P_2. Thus we then hash it into $B_2[h_2(g_k^2)]$. Since the state of the counter of $B_2[h_2(g_k^2)]$ is $True$, the TFD will increment the counter by one and turns the state to $False$. Note that we need to compute the pattern P_5 of the subgraph g_k^3 since g_k^3 is not isomorphic to the pattern in any bucket. Because the bucket $B_d[h_d(g_k^3)]$ is empty, we will insert the key-value pair $(P_5, (False, 1))$ into $B_d[h_d(g_k^3)]$. Note that we will set the state of counters in the TFD as $True$ after the current time window is processed.

Algorithm Analysis. Compared to the baseline solution, fastPP directly maps each newly produced k-edge subgraph into a fixed bucket in the TFD to count the persistence of the pattern, rather than calculates and stores all k-edge subgraphs at each time window, which significantly reduces the memory cost and time cost. What's more, once the state of a counter is $False$, fastPP will not need to verify the existence of the pattern in the corresponding bucket in the current window, which avoids repeated subgraph matching calculations.

5   TFD$^+$: An Optimized Auxiliary Data Structure

In this section, we first analyze the drawback of the algorithm fastPP, and then propose an optimized version of TFD, TFD$^+$ to achieve effectiveness and reduce the computational cost.

Problem Analysis. In the worst case, fastPP needs to check d buckets for each newly produced k-edge subgraph due to hash collisions, which is time-consuming. What's more, if non-persistent patterns take up too many buckets, no sufficient locations exist for persistent patterns, resulting in a lower recall rate when memory is limited.

To achieve higher efficiency and accuracy, we propose fastPP$^+$, which replaces TFD with an optimized version called TFD$^+$.

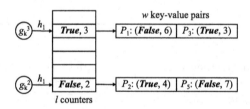

Fig. 4. Auxiliary Data Structure of TFD$^+$

TFD$^+$ (Fig. 4). To achieve time efficiency, the novel auxiliary data structure TFD$^+$ guarantees that every newly produced k-edge subgraph only needs to access one array and one cell. To detect more persistent k-edge patterns, our key idea is to separate persistent and non-persistent k-edge patterns. Specifically, TFD$^+$ consists of two parts. The first part is an array. The array is associated with a hash function $h_1(\cdot)$, mapping each subgraph into one cell in the array, which records the estimated persistence of a non-persistent subgraph. Each cell is a counter. It has l cells, where i^{th} cell in the array is denoted as $C_1[i]$. The second part consists of an array of l buckets, where i^{th} bucket is denoted as $B[i]$. Each bucket corresponds to a cell in the array, i.e., $B[i]$ corresponds to $C_1[i]$. There are w key-value pairs in each bucket, and each key-value pair is similar to the key-value pair of TFD. Let $B[h_1(P)][P]$ represent the corresponding key-value pair of k-edge pattern P if P is stored in $B[i]$, where $B[h_1(P)][P].key$ is the P and the value is the corresponding counter. Initially, all state fields are $True$, and the value of each counter is 0.

Algorithm 3: updateTFD$^+(\mathcal{G}, \text{TFD})$

---

1 line 1-3 of Function updateTFD;
2 foreach subgraph $g_k \in E_k(e)$ **do**
3   **foreach** $B[h_1(g_k)][j]$ in the bucket $B[h_1(g_k)]$ **do**
4   **if** $B[h_1(g_k)][j].state == True$ **then**
5   **if** g_k is isomorphic to $B[h_1(g_k)][j]$ **then**
6 $B[h_1(g_k)][j].value \leftarrow B[h_1(g_k)][j].value + 1$;
7 $B[h_1(g_k)][j].state \leftarrow False$, break;

8   **if** g_k is not isomorphic to any key-value pair in $B[h_1(g_k)]$ and
          $C_1[h_1(g_k)].state == True$ **then**
9 $C_1[h_1(g_k)].value \leftarrow C_1[h_1(g_k)].value + 1$, $C_1[h_1(g_k)].state \leftarrow False$;
10 $B[h_1(g_k)]^{min} \leftarrow$ findMinValue($B[h_1(g_k)]$);
11   **if** $C_1[h_1(g_k)].value > B[h_1(g_k)]^{min}.value$ **then**
12   swapCB($C_1[h_1(g_k)], B[h_1(g_k)], B[h_1(g_k)]^{min}$);

13 return TFD$^+$;

Compared with Algorithm fastPP, Algorithm fastPP$^+$ is only different from the process of updating data structure. Due to the limited space, the specific algorithm is omitted. Algorithm updateTFD$^+$ shows the update process of TFD$^+$.

Algorithm updateTFD$^+$. For each subgraph g_k in $E_k(e)$, there are two cases as follows: (1) g_k is isomorphic to one of key-value pair of the $B[h_1(g_k)]$. If the state of counter in $B[h_1(g_k)][j]$ is $True$, updateTFD$^+$ will check whether g_k is isomorphic to the $B[h_1(g_k)][j]$. If so, updateTFD$^+$ will increment the counter by 1 in $B[h_1(g_k)][j]$ and set the state of counter to $False$ (Line 4–7). (2) g_k is not isomorphic to any key-value pair of the $B[h_1(g_k)]$ and $C_1[h_1(g_k)].state == True$. updateTFD$^+$ increments the counter and sets the state of the counter to $False$. Then, we invoke findMinValue to find the smallest counter in $B[h_1(g_k)]$ to compare the counter in $C_1[h_1(g_k)]$ in order to determine whether the corresponding pattern P is persistent enough to store in the bucket $B[h_1(g_k)]$. We use $B[h_1(g_k)]^{min}$ to store the key-value in $B[h_1(g_k)]$ with minimum persistence. If $C_1[h_1(g_k)].value > B[h_1(g_k)]^{min}.value$, indicating that the estimated persistence of P is larger, P should be stored in $B[h_1(g_k)]$. As a consequence, updateTFD$^+$ swaps $C_1[h_1(g_k)]$ and $B[h_1(g_k)][B[h_1(g_k)]^{min}.key].value$, and swaps P and $B[h_1(g_k)][B[h_1(g_k)]^{min}.key].key$ (Lines 8–12). Finally, updateTFD$^+$ returns the updated TFD$^+$ (Line 13).

Example 3. Let $w = 2$, and Fig. 5 shows an example of the hash process in TFD$^+$. In current time window, we find that subgraph g_k^1 is hashed into $B[h_1(g_k)][1]$. Since g_k^1 is isomorphic to P_5 and the state of the counter of $B[h_1(g_k^1)]$ is *True*, the TFD$^+$ will increment the counter by one and turns the state to *False*. In other example, g_k^3 is not isomorphic to any key-value pair in $B[h_1(g_k^3)]$, so the TFD$^+$ maps the g_k^3 into the counter $C_1[h_1(g_k^3)]$. The counter is updated to $(False, 4)$. Because 4 is larger than the smallest counter (3) in the bucket, the TFD$^+$ sets the key to P_6 that is the pattern of the subgraph g_k^3, and swaps $(False, 4)$ and $(True, 3)$. After that, to map subgraph g_k^4, because g_k^4 is not isomorphic to any key-value pair in $B[h_1(g_k^4)]$, the TFD$^+$ maps g_k^4 into the counter, and the counter $C_1[h_1(g_k^4)]$ is updated to $(False, 4)$.

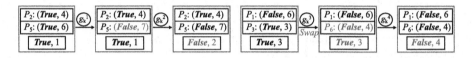

Fig. 5. Data Structure of TFD$^+$

Theorem 1. *Let T be the number of time windows in the streaming graph, $per(P)$ be the persistence of pattern P, $\widehat{per(P)}$ be the estimated persistence reported by our fastPP$^+$ and $l = \frac{2}{\epsilon}$, the approximate ratio returned by the Algorithm fastPP$^+$ is*

$$\mathbb{P}(\widehat{per(P)} \leq per(P) + \frac{\epsilon \sum_{i=1}^{N} P}{w+1}) \geq \frac{1}{2} \tag{1}$$

Theorem 2. *Let $\mathbb{P}$ be the probability that P is stored in buckets, the recall rate returned by the Algorithm fastPP$^+$ is*

$$\mathbb{P} \geq 1 - \frac{\sum_{i=1}^{N} P - per(P)}{w \cdot l \cdot per(P)} \tag{2}$$

Algorithm Analysis. Compared with fastPP, fastPP$^+$ is more efficient since it only needs to access one bucket for each new produced k-edge subgraph. Besides, fastPP$^+$ can separate persistent and non-persistent patterns and only record the persistent patterns by using an additional bucket for each counter, resulting in a higher recall rate.

6   Experiments

In this section, we perform extensive experiments to show the performance of fastPP$^+$ for discovering persistent patterns over streaming graphs. All the algorithms were implemented in C++, and run on a PC with an Intel i7 3.50GHz CPU and 32GB memory. Every quantitative test was repeated 5 times.

Datasets. For each dataset, we divide it into 50 time windows, i.e., $T = 50$.

- *Enron*[2] is an email communication network of 86K entities (e.g., employees), 297K edges (e.g., email), with timestamps corresponding to communication data.
- *Offshore*[3] contains in total 839K offshore entities (e.g., companies), 3.6M relationships (e.g., establish) and 433 labels covering offshore entities and financial activities, with timestamps corresponding to active days.
- *Facebook*[4] is a social network of 415K entities (e.g., users), 2.1M edges (e.g., message), with timestamps corresponding to posts date.

Solutions for Comparison. We compare the following 3 solutions for the persistent subgraph pattern discovery problem.

- findPP: Our baseline method for mining persistent patterns;
- fastPP: Our advanced algorithm framework that uses the TFD;
- fastPP$^+$: fastPP$^+$ that improves fastPP by incorporating the optimized TFD.

Metrics. We use the following four metrics:

- Recall Rate (RR): Ratio of the number of persistent patterns that are reported to the number of persistent patterns.
- Precision Rate (PR): Ratio of the number of persistent patterns that are reported to the number of reported patterns.
- F1 Score: $\frac{2 \times RR \times PR}{RR+PR}$. It is calculated from the precision and recall of the test, which is also a measure of a test's accuracy.
- Throughput: Kilo insertions handled per second (KIPS).

Parameter Settings. There are four parameters: the number of hash functions d (in TFD) and the number of key-value pairs in a bucket w (in TFD$^+$), the number of buckets l (in TFD) and the number of cells in a bucket l (in TFD$^+$), and the persistence threshold δ.

In specific, we vary d from 4 to 16 with a default 12, vary w from 4 to 16 with a default 12, and vary l from 4 to 32 with a default 16. δ could be set by domain scientists based on domain knowledge and is selected from 10 to 30 with a default 20. In addition, we fix the subgraph size $k = 4$. Without otherwise specified, when varying a certain parameter, the values of the other parameters are set to their default values.

---

[2] http://konect.uni-koblenz.de/networks/.
[3] https://offshoreleaks.icij.org/pages/database.
[4] http://socialnetworks.mpi-sws.org.

6.1   Performance Evaluation

EXP-1: Effect of Dataset. We first evaluate the performance of findPP, fastPP and fastPP$^+$ on *Enron, Offshore,* and *Facebook*. Note that, in each dataset, the temporal edges are continuously loaded into memory according to the time sequence of their occurrence to obtain a streaming graph. Therefor, we need reserve space for storing all edges in each dataset. As the edges are organized as a linked list, each edge needs for 48 bytes. To this end, we fix the memory size of *Enron, Offshore* and *Facebook* to 50MB, 250MB, and 150MB, respectively.

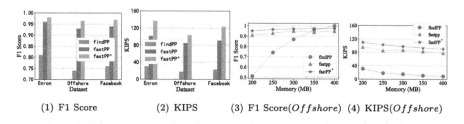

(1) F1 Score   (2) KIPS   (3) F1 Score(*Offshore*) (4) KIPS(*Offshore*)

Fig. 6. Experimental Results on Different Datasets and Varying Memory

F1 score (Fig. 6(1)). Our results show the F1 score of fastPP$^+$ and its competitors on three datasets with default parameters. Similar results can also be observed under the other parameter settings. As shown in Fig. 6(1), we can observe that the F1 score of fastPP$^+$ is higher than all other competitors, and fastPP is also higher than findPP. For example, on *Facebook*, the F1 score of fastPP$^+$ is higher than 95%, fastPP is higher than 90%, however findPP is smaller than 80%. The main reason is that findPP needs enough memory to store all k-edge subgraphs to guarantee accuracy, which will cause low performance when the memory is limited. Compared with findPP, our data structures need not store any subgraph, which is less affected by memory size. The reason why fastPP$^+$ outperform fastPP is that some non-persistent patterns may be mapped into the TFD first, resulting in insufficient locations for subsequent persistent patterns when the number of locations of the TFD is limited. As a result, some persistent patterns are misjudged as non-persistent patterns. Note that the F1 score of findPP can achieve 100% under enough memory because findPP exactly counts each pattern's persistence under enough memory.

Throughput (Fig. 6(2)). From Fig. 6(2), we also observe that the throughput of fastPP$^+$ is always higher than other algorithms, and fastPP is also higher than findPP. In specific, fastPP$^+$ outperforms fastPP by up to 1.4 times on *Enron*, and fastPP outperforms findPP by up to 4.7 times on *Offshore*. This is because findPP first needs to partition $\mathcal{P}^k$ and calculate the k-edge subgraph patterns based on subgraph isomorphism. Then, findPP needs to re-execute subgraph isomorphism calculation for each pattern to check whether it exists in the PS, which also causes high computational costs. In contrast, fastPP directly maps it into a fixed bucket in the TFD to form a corresponding k-edge pattern, which can avoid expensive calculations. What's more, fastPP$^+$ can further improve the efficiency since it guarantees that every subgraph only needs to access one counter and one bucket. The performance of fastPP$^+$ in three datasets is slightly different, and the trends are very similar. The results show the robustness of fastPP$^+$, so in the following experiments, we only use *Offshore* dataset.

EXP-2: Effect of Memory. We evaluate the accuracy and speed of fastPP$^+$ and its competitors with varying memory size on *Offshore*. In the experiment, the memory size ranges from 200MB to 400MB.

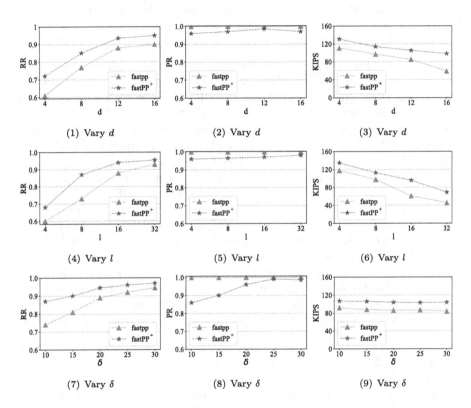

Fig. 7. Experimental Results on Varying Parameters

F1 score (Fig. 6(3)). In this figure, we can see that the increase of memory size has a large effect on findPP, due to insufficient space to store all subgraphs for exactly computing the persistence of each pattern. We also find that the increase of memory size has little effect on that of fastPP and fastPP$^+$, and the impact on fastPP$^+$ is minimal. This is because they use auxiliary data structures without storing all k-edge subgraphs. Since the TFD and TFD$^+$ only store the k-edge patterns and their persistence. And, fastPP$^+$ guarantees non-persistent patterns will be replaced quickly and will not take up much space. Therefore, our algorithms work well with limited memory.

Throughput (Fig. 6(4)). As expected, fastPP$^+$ is much faster than other algorithms. We can see the increase of memory size can decrease the throughput of findPP and has little effect on that of fastPP and fastPP$^+$. The reason is that findPP needs less time to partition $\mathcal{P}^k$ into m equivalence classes. The throughput of findPP is much lower because findPP causes redundant subgraph matching. Due to fewer memory accesses, the throughput of fastPP$^+$ is the best.

EXP-3: Effect of Parameters. We evaluate the RR, the PR, and throughput of fastPP and fastPP$^+$ with varying parameters on *Offshore* using fixed memory size, i.e., 250MB. Note that when varying a parameter, we keep other parameters as default. The results on the other datasets are consistent.

Effect of d/w (Fig. 7(1)–(3)). We vary d (resp. w) in this experiment from 4 to 16. Especially we observe that the increase of d can increase the RR and decrease the throughput of fastPP, and the PR of fastPP is always 1. When w increases, the throughput often decreases, and the RR and the PR often increase. The reason is that for a larger d or w, potential persistent patterns have more opportunities to be stored into the TFD and TFD$^+$, and the RR of fastPP and fastPP$^+$ will be increased. And, for a larger w, there are more positions to store persistent patterns so that more potential persistent patterns will be protected from replacements and hash collisions, and the PR of fastPP$^+$ will increase. Because the TFD counts the persistence of patterns exactly, the PR of fastPP is always 1. However, the throughput of fastPP will decrease since it has to check $d - 1$ more buckets for each edge insertion. And the throughput of fastPP$^+$ will decrease because the number of memory accesses is small when the w is large. Therefore, choosing an appropriate d or w is a trade-off between accuracy and throughput. The larger d or w is, the higher the accuracy is, while the lower the throughput is. If the application demands high throughput, we should decrease the d or w. If the application requires high accuracy, we should increase d or w.

Effect of l (Fig. 7(4)–(6)). The experimental results show that the increase of l can increase the RR and decrease the throughput of fastPP and fastPP$^+$. This is because when l increases, there are more tracks in the TFD and TFD$^+$, and we can detect more patterns simultaneously. However, resulting in more subgraph matching calculations since we need to count the persistence of the pattern in each track of the TFD and TFD$^+$ in each time window. l does not affect the PR of fastPP and fastPP$^+$ because l does not affect the persistence of the k-edge patterns in the TFD and TFD$^+$.

Effect of δ (Fig. 7(7)–(9)). The experimental results show that the increase of δ can increase the RR of fastPP and fastPP$^+$. This is because, for a smaller δ, the ground truth could be huge, and we can only detect the fixed number of patterns in the TFD and TFD$^+$. Therefore, resulting in a lower RR. We also find that the increase of δ can increase the PR of fastPP$^+$ since more false positives can be filtered safely due to persistence constraints. The throughput of fastPP and fastPP$^+$ is insensitive to δ since δ does not affect the number of the k-edge patterns in the TFD and TFD$^+$.

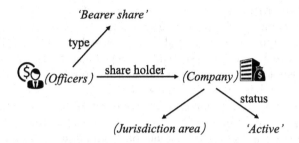

Fig. 8. Real-life persistent pattern (*Offshore*)

6.2 Case Study

We demonstrate the utilities of the persistent pattern discovery problem by conducting a case study on *Offshore*. Figure 8 shows a persistent pattern obtained by our solution. It reveals a significant move of active bearer shares companies from tax heaven "British Virgin Islands" to "Panama". This pattern can help us discover the abnormal behavior of these active bearer shares companies and analyze the causes of the abnormal behavior. Through analysis, we know that the reason for the abnormal behavior is that the British Virgin Islands has cracked down on bearer shares. For the British Virgin Islands, the risk of capital outflows needs to be guarded so offshore financial markets can be better built.

7 Related Work

Streaming Graph Analysis. Many streaming graph analytics tasks have been proposed to mine valuable information, e.g., query processing [14], time constrained continuous subgraph search [9], triangle counting [16], streaming graph summarization [11] and graph sampling method [17]. In this work, we study the persistent pattern discovery problem, which aims to find persistent patterns over a streaming graph.

Frequent Subgraph Pattern Mining in Dynamic Graphs. The studies of frequent subgraph pattern mining in dynamic graphs are related to our work. Nasir et al. [13] presented TipTap, a collection of sampling-based approximation algorithms for mining frequent k-vertex patterns in fully-dynamic graphs. Aslay et al. [2] proposed a sampling-based method to find the latest frequent pattern when edge updates occur on the graph. Ray et al. [15] proposed a heuristic approach to mining a single graph that continuous updates. Abdelhamid et al. [1] proposed a system IncGM+ for continuous frequent subgraph mining that prunes the search space using a set of infrequent subgraphs which are adjacent to frequent subgraphs. While these methods seek to find frequent subgraphs, we seek to find persistent subgraphs.

Persistent Item Mining. There is many studies for mining persistent items. Belth et al. [3] explored the persistence of activity snippets, i.e., sequences of reoccurring edge updates. Zhang et al. [19] proposed a fast and accurate sketch for finding persistent items. The sketch introduced a state flag into each counter of the sketch instead of maintaining a bloom filter to record the existence of the element. Our research is inspired by the work [19] aforementioned, and especially the preliminaries set a good foundation for our work. Dai et al. [6] concentrated on finding the persistent items in data streams. Li et al. [8] proposed the notion of persistent k-core to capture the persistence of a community in temporal networks. Compared to them, our work adopts a different research object pattern and considers a subgraph pattern that appears recurrently in many time windows of the streaming graph. Therefor these methods cannot handle the persistent pattern mining problem.

8   Conclusion

In this paper, we tackle the novel problem of discovering persistent patterns continuously in streaming graphs. We propose an auxiliary data structure called TFD for counting the persistence of a pattern without storing any subgraph and avoiding repeated subgraph matching calculation, which is fast, memory efficient, and accurate. We explore an optimized auxiliary data structure TFD^+ that guarantees that every newly produced k-edge subgraph only needs to access one counter and one bucket. Experimental results have verified that our algorithms can achieve high accuracy and efficiency with limited memory usage in real-time persistent pattern detection. We also demonstrated the utility of the discovered persistent patterns by a case study on *Offshore*.

Acknowledgements. This work is partially supported by National Natural Science Foundation of China under Grant No. U19B2024, National Natural Science Foundation of China under Grant No.6227246 and Postgraduate Scientific Research Innovation Project of Hunan Province under Grant No. CX20210038.

References

1. Abdelhamid, E., Canim, M., Sadoghi, M., Bhattacharjee, B., Chang, Y., Kalnis, P.: Incremental frequent subgraph mining on large evolving graphs. In: ICDE 2018, pp. 1767–1768 (2018)
2. Aslay, Ç., Nasir, M.A.U., Morales, G.D.F., Gionis, A.: Mining frequent patterns in evolving graphs. In: CIKM 2018. pp. 923–932 (2018)
3. Belth, C., Zheng, X., Koutra, D.: Mining persistent activity in continually evolving networks. In: KDD 2020, pp. 934–944 (2020)
4. Chen, Z., Wang, X., Wang, C., Li, J.: Explainable link prediction in knowledge hypergraphs. In: CIKM 2022, pp. 262–271 (2022)

5. Choudhury, S., Holder, L.B., Jr., G.C., Agarwal, K., Feo, J.: A selectivity based approach to continuous pattern detection in streaming graphs. In: EDBT 2015, pp. 157–168 (2015)
6. Dai, H., Shahzad, M., Liu, A.X., Zhong, Y.: Finding persistent items in data streams. Proc. VLDB Endow. **10**(4), 289–300 (2016)
7. Hellmann, S., Stadler, C., Lehmann, J., Auer, S.: DBpedia live extraction. In: Meersman, R., Dillon, T., Herrero, P. (eds.) OTM 2009. LNCS, vol. 5871, pp. 1209–1223. Springer, Heidelberg (2009). https://doi.org/10.1007/978-3-642-05151-7_33
8. Li, R., Su, J., Qin, L., Yu, J.X., Dai, Q.: Persistent community search in temporal networks. In: ICDE 2018, pp. 797–808 (2018)
9. Li, Y., Zou, L., Özsu, M.T., Zhao, D.: Time constrained continuous subgraph search over streaming graphs. In: ICDE 2019, pp. 1082–1093 (2019)
10. Li, Z., Liu, X., Wang, X., Liu, P., Shen, Y.: Transo: a knowledge-driven representation learning method with ontology information constraints. World Wide Web **26**, 297–319 (2023)
11. Ma, Z., Yang, J., Li, K., Liu, Y., Zhou, X., Hu, Y.: A parameter-free approach for lossless streaming graph summarization. In: Jensen, C.S., et al. (eds.) DASFAA 2021. LNCS, vol. 12681, pp. 385–393. Springer, Cham (2021). https://doi.org/10.1007/978-3-030-73194-6_26
12. Min, S., Park, S.G., Park, K., Giammarresi, D., Italiano, G.F., Han, W.: Symmetric continuous subgraph matching with bidirectional dynamic programming. Proc. VLDB Endow. **14**(8), 1298–1310 (2021)
13. Nasir, M.A.U., Aslay, Ç., Morales, G.D.F., Riondato, M.: Tiptap: approximate mining of frequent k-subgraph patterns in evolving graphs. ACM Trans. Knowl. Discov. Data **15**(3), 48:1-48:35 (2021)
14. Pacaci, A., Bonifati, A., Özsu, M.T.: Regular path query evaluation on streaming graphs. In: SIGMOD Conference 2020, pp. 1415–1430 (2020)
15. Ray, A., Holder, L., Choudhury, S.: Frequent subgraph discovery in large attributed streaming graphs. In: Proceedings of the 3rd International Workshop on Big Data, vol. 36, pp. 166–181 (2014)
16. Yang, X., Song, C., Yu, M., Gu, J., Liu, M.: Distributed triangle approximately counting algorithms in simple graph stream. ACM Trans. Knowl. Discov. Data **16**(4), 79:1-79:43 (2022)
17. Zhang, L., Jiang, H., Wang, F., Feng, D., Xie, Y.: T-sample: a dual reservoir-based sampling method for characterizing large graph streams. In: ICDE 2019, pp. 1674–1677 (2019)
18. Zhang, Q., Guo, D., Zhao, X.: Discovering bursting patterns over streaming graphs. In: DASFAA 2022, pp. 441–458 (2022)
19. Zhang, Y., et al.: On-off sketch: a fast and accurate sketch on persistence. Proc. VLDB Endow. **14**(2), 128–140 (2020)

MRSCN: A GNN-based Model for Mining Relationship Strength Changes Between Nodes in Dynamic Networks

Tianbao Wang[1,2], Yajun Yang[1,2(✉)], Hong Gao[3], and Qinghua Hu[1]

[1] College of Intelligence and Computing, Tianjin University, Tianjin, China
{tbwang,yjyang,huqinghua}@tju.edu.cn
[2] State Key Laboratory of Communication Content Cognition, Beijing, China
[3] College of Mathematics and Computer Science, Zhejiang Normal University,
Jinhua, China
honggao@zjnu.edu.cn

Abstract. The relationship strength between individuals in the network is an essential task in network analysis. However, existing measures of relationship strength are mostly artificially predefined, which can only reflect the relationship strength from a single perspective. To compensate for this, we propose a novel GNN-based model for Mining Relationship Strength Changes between Nodes in dynamic networks, named MRSCN, which learns the reasonable relationship strength from networks. To verify the effectiveness of our measure on the relationship strength change, we further propose a novel ϵ-drastic group model. We develop two group mining algorithms. We conduct extensive experiments on real-life dynamic networks to evaluate our models. The results demonstrate the effectiveness of the proposed MRSCN model and the drastic group mining method.

Keywords: Relationship strength change · Drastic group · Dynamic networks

1 Introduction

In the real world, the relationships between various entities are constantly changing and can be described by dynamic networks, such as social networks. The graph sequence has been introduced to model the dynamic network. At each time point, the graph snapshot is taken to capture the status of the network. For example, in a mobile communication network, the calls between participants each day are modeled and captured as a graph snapshot.

Investigating relationship strength among individuals in the network is essential in network analysis. In social networks, the relationship strength indicates

© The Author(s), under exclusive license to Springer Nature Switzerland AG 2023
X. Wang et al. (Eds.): DASFAA 2023, LNCS 13945, pp. 172–182, 2023.
https://doi.org/10.1007/978-3-031-30675-4_12

the possibility that two people become friends even though they may not know each other at the current time. However, the relationship strength also changes over time. Then it is crucial to study the relationship strength change for dynamic network analysis. Based on the relationship strength change, various drastically changing groups can be mined for human beings to track the significant or abnormal changes in networks. For example, in financial trade networks, gang frauds always occur much more frequently than individual frauds. Money laundering syndicates often evade regulation by making small but multiple transactions, resulting in frequent changes in relationship strength between individuals. Mining groups with relationship strengths changing frequently help regulators detect potential risks in time. Due to the importance of the relationship strength between entities, it is necessary to find a reasonable measure of relationship strength.

Several works [1,4] proposed various metrics to measure the relationship strength between nodes, such as connectivity and common neighbors. However, all these artificially predefined measurements can only reflect relationship strength from a single perspective. For the example in Fig. 1, relationship strengths between nodes of the same color change significantly when different metrics are used. It is necessary to design a reasonable relationship strength model to study the relationship strength change for dynamic networks.

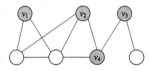

Fig. 1. An illustrative example of existing relationship strength measurements.

In this paper, we propose a novel GNN-based model for Mining Relationship Strength Changes between Nodes in dynamic networks, named MRSCN, which adaptively learns the relationship strength between nodes from networks. Then we further propose the concept of cumulative relational strength change. We also define a drastic group model and investigate the ϵ-drastic group mining problem to verify the usefulness of our measure on relationship strength change. Finally, we conduct extensive experiments to validate the effectiveness of our method.

2   Related Work

In this section, we review related studies on dynamic network analysis, graph embedding, and relationship strength.

As a hot area for researchers in recent years, researchers analyzed dynamic networks from different aspects. Yang et al. [18] developed an algorithm to capture frequently changing components in dynamic networks. Qin et al. [15] proposed a model to mine periodic cliques in dynamic networks. Jia et al. [10]

proposed a method to measure the community consistency. Li et al. [14] developed a new algorithm to find stable communities based on the density-based graph clustering framework. To the best of our knowledge, previous studies have paid little attention to the relationship strength change in dynamic networks.

Graph embedding is used to map nodes to low-dimensional vectors based on network topology. Existing work focuses on preserving structural and attribute information in the embedding, such as DeepWalk [13], Node2Vec [8], DANE [7] and so on. Graph Neural Networks (GNNs) [16], which use a deep learning framework on graph data, have attracted lots of attention. GNNs have been widely used in graph-based problems, such as GCN [12], GAT [17] and GraphSage [9].

Existing research on relationship strength can be divided into two main categories: the first one is based on the network topology to calculate relationship strength, such as common neighbors [14] and connectivity [4]; the second one is a combination of user interactions and user profiles. Guo et al. [11] obtained the fused similarity matrix from different views of user interactions and user profiles. However, there are shortcomings such as artificially predefined or considering structural information from one aspect.

3   Preliminaries

We represent the dynamic network consisting of a sequence of undirected graphs as $\mathcal{G} = (G_1, G_2, \cdots, G_{\|\mathcal{G}\|})$, where $G_t = (V_t, E_t)$ is a graph snapshot at time t with a set of vertices V_t and a set of edges E_t, and $\|\mathcal{G}\|$ is the number of G_t in $\mathcal{G}$. $N(u)$ denotes the neighbors of node u. Given nodes u and v, we denote $d(u,v)$ as the distance between u and v, namely the shortest number of hops from u to v. The i-hop neighbors of node u, denoted as $N_i(u)$, contain all the nodes whose distance to u are i, i.e., $N_i(u) = \{v \in V | d(u,v) = i\}$. The i-hop reachable neighbors of node u, denoted as $N_{\leq i}(u)$, contain all the nodes whose distance to u is no more than i $(i \geq 1)$. Clearly, $N_{\leq i}(u) = \cup_{j=1}^{i} N_j(u)$.

Relationship strength has no uniform definition and refers to the closeness between individuals. To address the problem of existing predefined models capturing a single relationship, we learn the relationship strength adaptively based on GNN. The relationship strength between u and v in G_t is denoted as $rs_t(u, v)$. Furthermore, the relationship strength change between u and v from G_t to G_{t+1} is defined as $\delta_t(u, v) = |rs_{t+1}(u, v) - rs_t(u, v)|$. Our aim is to learn the reasonable relationship strength measurement and propose the drastic group model to verify the effectiveness of the relationship strength measurement.

4   Mining Relationship Strength Changes Between Nodes

In this section, we introduce our novel MRSCN model in dynamic networks. The overview architecture of MRSCN is shown in Fig. 2. The key idea behind our model is to use the GNN model to mine the relationship strength by comprehensively considering global and local information.

4.1   Global Structure Information Capture

In this section, we use random walk and pointwise mutual information to encode the global structure information.

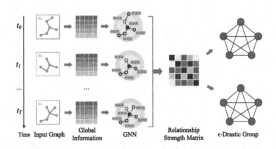

Fig. 2. The overview of the proposed MRSCN.

Calculating Node Co-occurrence Matrix C. We use matrix $\mathbf{C}$ to represent the co-occurrence frequencies between nodes. We use the random walk algorithm here. For each vertex v_i in the graph, we first select it as the starting vertex, which is also the current vertex. We define the current state as $s(t) = v_i$. Then we randomly select the next vertex v_j from the neighbors of v_i. The transition probability of jumping from the current node v_i to v_j is calculated as:

$$p(s(t+1) = v_j | s(t) = v_i) = A_{i,j} / \sum_j A_{i,j}, \qquad (1)$$

where $\mathbf{A}$ is the adjacency matrix. Now, we mark this newly selected vertex v_j as the current vertex and repeat such a vertex sampling process. The algorithm terminates when the length of the vertex sequence reaches a pre-set number called walk length η. We repeat the above procedure γ times for each node and record the starting node v_i and node v_j in the sequence for each walk. For each pair (v_i, v_j), We add one to the values of $\mathbf{C}_{i,j}$ and $\mathbf{C}_{j,i}$ respectively, and finally obtain the node co-occurrence matrix $\mathbf{C}$.

Calculating PPMI Matrix M. Pointwise mutual information [5] is often used to measure the correlation of variables. In this paper, we use it to measure the global relationship between nodes. Based on $\mathbf{C}$, we calculate the PPMI matrix $\mathbf{M} \in \mathbb{R}^{n \times n}$ as:

$$m_{i,j} = max\{\log(\frac{p_{i,j}}{p_{i,*}p_{*,j}}), 0\}. \qquad (2)$$

Algorithm 1. GraphSage Embedding Generation Algorithm

Input: Graph $\mathcal{G}(\mathcal{V}, \mathcal{E})$; input features $\{\mathbf{f}_v, \forall v \in \mathcal{V}\}$; depth K; weight matrices $\mathbf{W}^k$; aggregator functions AGGREGATE_k; neighborhood function $\mathcal{N} : v \to 2^{\mathcal{V}}$
Output: Vector representations $\mathbf{z}_v$ for all $v \in \mathcal{V}$

1: $\mathbf{h}_v^0 \leftarrow \mathbf{f}_v, \forall\, v \in \mathcal{V}$
2: **for** $k = 1, \dots, K$ **do**
3:     **for** $\forall v \in \mathcal{V}$ **do**
4:         $\mathbf{h}_v^k \leftarrow \sigma(\mathbf{W}^k \cdot \text{MEAN}(\{\mathbf{h}_v^{k-1}\} \cup \{\mathbf{h}_u^{k-1}, \forall u \in \mathcal{N}(v)\}))$
5:     **end for**
6:     $\mathbf{h}_v^k \leftarrow \mathbf{h}_v^k / \|\mathbf{h}_v^k\|_2, \forall v \in \mathcal{V}$
7: **end for**
8: $\mathbf{z}_v \leftarrow \mathbf{h}_v^K, \forall v \in \mathcal{V}$

The global structure information is encoded by applying Eq. (2). $p_{i,j}$ is the estimated probability that nodes v_i and v_j occur during the random walk at the same time, i.e., $p_{i,j} = \frac{C_{i,j}}{\sum_{i,j} C_{i,j}}$. $p_{i,*}$ and $p_{*,j}$ are the estimated probability that nodes v_i and v_j occur during the walk respectively, i.e., $p_{i,*} = \frac{\sum_j C_{i,j}}{\sum_{i,j} C_{i,j}}$ and $p_{*,j} = \frac{\sum_i C_{i,j}}{\sum_{i,j} C_{i,j}}$. $m_{i,j}$ is the measure of the global correlation of nodes v_i and v_j. As we are focusing on the semantic relation, our method uses a nonnegative pmi. It is worth noting that matrix $\mathbf{C}$ and matrix $\mathbf{M}$ need to be recomputed for different graph snapshots.

4.2 Graph Neural Network Model

To jointly consider the global and local structure information, we integrate global information into the GNN model. GraphSage is used to generate embeddings by aggregating global features from the node's local neighborhood [9].

For each node, the algorithm iteratively aggregates global information from the node's neighbors. The process of aggregating information from neighbors is capturing local information. Algorithm 1 describes the embedding generation process. We use the row vector $\mathbf{M}_{v,:}$ in $\mathbf{M}$ as the node feature $\mathbf{f}_v$ and the inductive variant of the GCN approach as the aggregator function. The final representation of node v is expressed as $\mathbf{z}_v$, as shown in line 8 in Algorithm 1.

In order to learn representations in a fully unsupervised setting, we apply a graph-based loss function:

$$J_{\mathcal{G}}(\mathbf{z}_u) = -\log\left(\sigma\left(\mathbf{z}_u^{\top} \mathbf{z}_v\right)\right) - Q \cdot \mathbb{E}_{v_n \sim P_n(v)} \log\left(\sigma\left(-\mathbf{z}_u^{\top} \mathbf{z}_{v_n}\right)\right), \qquad (3)$$

where v is a node that co-occurs near u on fixed-length random walk, σ is the sigmoid function, P_n is a negative sampling distribution, and Q defines the number of negative samples.

4.3   Relationship Strength Change Computation

After learning the embeddings of nodes in graphs, we aim to obtain the relationship strength. Here, we use the embedding vector obtained by the GNN model as the key feature of the relationship strength change computation.

We select *cosine* similarity to calculate the relationship strength because we are more interested in the directional similarity of different embeddings than in the absolute values. The relationship strength between u and v in G_t, denoted as $rs_t(u, v)$, can be presented as:

$$rs_t(u, v) = \frac{{\mathbf{z}_u^t}^\top \cdot \mathbf{z}_v^t}{||\mathbf{z}_u^t|| \times ||\mathbf{z}_v^t||}, \tag{4}$$

The relationship strength change between u and v from G_t to G_{t+1}, denoted as $\delta_t(u, v)$, can be calculated as follows:

$$\delta_t(u, v) = |rs_{t+1}(u, v) - rs_t(u, v)| . \tag{5}$$

Thus, the **cumulative relationship strength change** between u and v in the dynamic network is given as follows:

$$\Delta(u, v) = \sum_{t=1}^{||\mathcal{G}||-1} \delta_t(u, v). \tag{6}$$

5   Drastic Group Mining

Based on the relationship strength change between nodes by the MRSCN model, we can further mine drastic groups. In this section, we introduce our *top-k ϵ-drastic group mining* method.

Definition 1. *ϵ-drastic group. A set of nodes $C \subset \mathcal{G}$ is called ϵ-drastic group if the following conditions holds: (1) $\forall v_1, v_2 \in C$, cumulative relationship strength change $\Delta(v_1, v_2) \geq \epsilon$; (2) $|C|$ is maximized; (3) $\sum_{v_1, v_2 \in C} \Delta(v_1, v_2)$ is maximized.*

ϵ-drastic group mining problem has two optimization objectives, namely large coverage and a large sum of relationship strength changes. In reality, they can't be satisfied simultaneously. We consider the problem in the following two cases: (1) priority to make the coverage as large as possible. (2) priority to make the sum of relationship strength changes as large as possible. Based on these two cases, the Coverage-First algorithm (CF) and Strength Change-First algorithm (SCF) are proposed.

First, we need to generate all drastic groups that satisfy $\Delta(v_1, v_2) \geq \epsilon$ for any two nodes v_1, v_2 in the drastic group, which consists of two steps. Firstly, we need to generate a graph containing all the nodes that satisfy $\Delta(v_1, v_2) \geq \epsilon$ for any two nodes v_1, v_2 in the graph. Secondly, we enumerate maximal cliques on the graph to return all drastic groups. Many widely used maximal clique enumeration algorithms can be adopted, such as BasicMCE [6].

Table 1. Datasets statistics.

Dataset	$n = |V|$	$m = |E|$	$T = \|\mathcal{G}\|$	Avg. Degree
Chess	7301	65053	25	17.82
Lkml	30665	197356	24	12.87
Enron	66903	189353	14	5.66
P2P-Gnutella	23089	72131	9	6.25

Coverage-First Algorithm. First, all drastic groups are enumerated and sorted in non-increasing order of their coverages. Drastic groups with the same coverage are sorted in non-increasing order of their sums of cumulative relationship strength changes. Second, all sorted drastic groups are scanned sequentially based on the greedy strategy, i.e., the drastic group with the largest coverage is always given priority. Suppose that $\mathcal{K}$ is the current result set and C is the current scanned drastic group that is being decided whether to be added into $\mathcal{K}$ or not. If C overlaps too much with a drastic group C' in $\mathcal{K}$, C might be discarded. In order to evaluate the degree of overlap between C and C', we propose an indicator named the overlap ratio, denoted by $\frac{|C \cap C'|}{|C|}$. Given a threshold α, for C, if there exists a C' such that $\frac{|C \cap C'|}{|C|} > \alpha$, C will be discarded. Finally, top-k ϵ-drastic groups can be gained from the result set $\mathcal{K}$. The time complexity of CF is $O(n \log n)$, where n is the number of drastic groups.

Strength Change-First Algorithm. First, all drastic groups are enumerated and then sorted in non-increasing order of their sums of cumulative relationship strength changes, and drastic groups with the same cumulative relationship strength change are sorted in non-increasing order of their sizes, which indicate the coverages. Second, all sorted drastic groups are scanned sequentially based on the greedy strategy. As with CF, the overlap ratio is used to evaluate whether the drastic group is added to the result set $\mathcal{K}$. We can obtain top-k ϵ-drastic groups from the result set $\mathcal{K}$. The time complexity of SCF is the same as CF.

6   Experiments

6.1   Experimental Setup

We conduct experiments on four real-world datasets, including Chess, Lkml, Enron and P2P-Gnutella, whose statistics are summarized in Table 1.

In our experiments, we compare MRSCN with three different categories of methods, including DeepWalk [13], Node2Vec [8], GraRep [2] and DNGR [3].

Since most existing metrics are tailored for traditional graphs, we introduce four goodness metrics evaluating drastic groups for dynamic networks, which are motivated by *separability, density, common neighbors* and *clustering coefficient*. Let $\mathcal{C}$ be the mining group. The descriptions of evaluation metrics are as follows.

- *Variability of Separability* (VS) captures the intuition that the variability of the separation between groups and the rest of the network is drastic for the group with violent relationship strength changes between nodes: $\text{VS} = \sum_{t=1}^{||\mathcal{G}||-1} |S(G_{t+1}) - S(G_t)|$, $S(G_t)$ is given by: $S(G_t) = \frac{|\{(u,v_1)\in E_t:u\in C, v_1\in C\}|}{|\{(u,v_2)\in E_t:u\in C, v_2\notin C\}|}$.
- *Variability of Density* (VD) catches the intuition that the group with violent relationship strength changes has drastic connection changes: $\text{VD} = \sum_{t=1}^{||\mathcal{G}||-1} |DS(G_{t+1}) - DS(G_t)|$, $DS(G_t)$ is given by: $DS(G_t) = 2\frac{\sum_{v_j\in C} d_C^t(v_j)}{|C|(|C|-1)}$, where $d_C^t(v_j)$ denotes the degree of v_j in the group C at t timestamp.
- *Variability of Common Neighbors* (VCN) builds on the intuition that the nodes in the group C with violent relationship strength changes have drastic changes in common neighbors: $\text{VCN} = \sum_{t=1}^{||\mathcal{G}||-1} |CN(G_{t+1}) - CN(G_t)|$, $CN(G_t)$ is calculated as: $CN(G_t) = \frac{2}{|C|(|C|-1)} \sum_{v_i,v_j\in C} \frac{|N_t^{\leq 2}(v_i,v_j)|}{\sqrt{|N_t^{\leq 2}(v_i)|\times|N_t^{\leq 2}(v_j)|}}$.
- *Variability of Clustering Coefficient* (VCC) is based on the premise that the connection in pair of nodes with common neighbors in C changes drastically: $\text{VCC} = \sum_{t=1}^{||\mathcal{G}||-1} |CC(G_{t+1}) - CC(G_t)|$, $CC(G_t)$ is given by: $CC(G_t) = \frac{1}{|C|} \sum_{v_j\in C} |\frac{2\#edge(N_t(v_j,C))}{d_C^t(v_j)\cdot(d_C^t(v_j)-1)}|$, where $\#edge(N_t(v_j,C))$ is the number of edges in C whose two end nodes are v_j's neighbors in C.

Intuitively, the group with drastic relationship strength changes between nodes should have high VS, VD, VCN and VCC values.

6.2 Experimental Results

Exp 1. Effectiveness of the ϵ-Drastic Group Mining. In this experiment, we study the effectiveness of the ϵ-drastic group mining. We use the SCF algorithm to evaluate the effectiveness. Table 2 shows the performance of our method compared with other methods.

From the results, we can see that MRSCN performs better than other methods in general. For example, compared to the most powerful compared method DNGR, our MRSCN model reaches nearly 11.1%, 14.8% and 8.1% gain at VD, VCN and VCC, respectively. The experimental results demonstrate that MRSCN has a solid ability to mine relationship strength changes. This is due to the effectiveness of capturing both global structure information and local structure information. In terms of VS, our method has few obvious advantages.

Exp 2. Impact of Parameters. In this part, we analyze the impact of two key parameters in our method, i.e., the embedding dimension d and ϵ. Figure 3 describes the results of our method with varying parameters on VS, VD and VCC in Chess and Lkml, respectively. Similar results can also be observed in the other datasets. VCN follows the same trend as VD and VCC. We first illustrate the performance under various settings of embedding dimension while keeping other parameters fixed, as shown in Fig. 3(a) and Fig. 3(c). We can see that the performance of our method on VD and VCC improves as the embedding size increases and gradually becomes stable when the embedding size increases.

Table 2. Results of the ϵ-drastic group mining.

Methods	Chess				Lkml			
	VS	VD	VCN	VCC	VS	VD	VCN	VCC
DeepWalk	0.255	0.255	0.347	0.512	0.324	0.315	0.341	0.367
Node2Vec	0.281	0.405	0.424	0.508	0.310	0.356	0.385	0.387
GraRep	0.311	0.466	0.403	0.554	0.280	0.349	0.412	0.449
DNGR	0.325	0.468	0.429	0.543	0.314	0.377	0.403	0.436
MRSCN	**0.387**	**0.520**	**0.494**	**0.587**	**0.356**	**0.423**	**0.465**	**0.532**
Methods	Enron				P2P-Gnutella			
	VS	VD	VCN	VCC	VS	VD	VCN	VCC
DeepWalk	0.276	0.208	0.207	0.412	0.257	0.213	0.216	0.336
Node2Vec	0.355	0.261	0.231	0.398	0.290	0.209	0.230	0.342
GraRep	**0.376**	0.254	0.323	0.404	**0.315**	0.218	**0.243**	**0.393**
DNGR	0.314	0.266	0.284	0.436	0.287	0.225	0.207	0.341
MRSCN	0.358	**0.318**	**0.326**	**0.547**	0.266	**0.246**	0.228	0.354

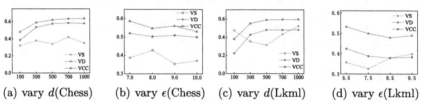

(a) vary d(Chess)   (b) vary ϵ(Chess)   (c) vary d(Lkml)   (d) vary ϵ(Lkml)

Fig. 3. Effectiveness of our method with varying parameters on datasets.

However, we can see from Fig. 3(b) and Fig. 3(d) that both VD and VCC values in different settings of parameter ϵ are irregular, because it cannot guarantee that the total relationship strength change of the group increases with the increase of ϵ.

Exp 3. Ablation Study. To get a better understanding of how different components affect the performance of MRSCN, we conduct ablation tests on three datasets with one variant: MRSCN-G, which removes global features when generating the node embedding. In addition, we choose the best-performing DNGR model as a comparison. The results w.r.t. VD and VCC are shown in Fig. 4.

We can find that the method performs better by combining global features, which shows that capturing global structure information is able to mine relationship strength changes more effectively. Moreover, we find that the basic GNN model works better than DNGR. The experiment demonstrates the effectiveness of mining the relationship strength changes between nodes by jointly considering global and local structure information.

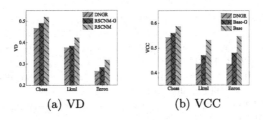

(a) VD   (b) VCC

Fig. 4. Performance comparison with variants of our method on three datasets.

7   Conclusion

In this paper, we propose a novel GNN-based model for Mining Relationship Strength Changes between Nodes in dynamic networks, named MRSCN. We use random walk and pointwise mutual information to capture the global structure information. After that, we learn the reasonable relationship strength change by GNN. Based on MRSCN, we propose the ϵ-drastic group model and develop mining algorithms. We conduct experiments on real-world datasets. The results demonstrate the effectiveness of MRSCN and the drastic group mining method.

Acknowledgements. This work is supported by the National Key Research and Development Program of China No. 2019YFB2101903, the State Key Laboratory of Communication Content Cognition Funded Project No. A32003.

References

1. Adamic, L.A., Adar, E.: Friends and neighbors on the web. Soc. Networks **25**(3), 211–230 (2003)
2. Cao, S., Lu, W., Xu, Q.: Grarep: learning graph representations with global structural information. In: CIKM, pp. 891–900. ACM (2015)
3. Cao, S., Lu, W., Xu, Q.: Deep neural networks for learning graph representations. In: AAAI, pp. 1145–1152. AAAI Press (2016)
4. Cho, J.J., Chen, Y., Ding, Y.: On the (co)girth of a connected matroid. Discret. Appl. Math. **155**(18), 2456–2470 (2007)
5. Church, K.W., Hanks, P.: Word association norms, mutual information and lexicography. In: ACL, pp. 76–83. ACL (1989)
6. Eppstein, D., Löffler, M., Strash, D.: Listing all maximal cliques in sparse graphs in near-optimal time. In: Cheong, O., Chwa, K.-Y., Park, K. (eds.) ISAAC 2010. LNCS, vol. 6506, pp. 403–414. Springer, Heidelberg (2010). https://doi.org/10.1007/978-3-642-17517-6_36
7. Gao, H., Huang, H.: Deep attributed network embedding. In: IJCAI, pp. 3364–3370. ijcai.org (2018)
8. Grover, A., Leskovec, J.: node2vec: Scalable feature learning for networks. In: KDD, pp. 855–864. ACM (2016)
9. Hamilton, W.L., Ying, Z., Leskovec, J.: Inductive representation learning on large graphs. In: NIPS, pp. 1024–1034 (2017)

10. Jia, X., et al.: Tracking community consistency in dynamic networks: an influence-based approach. IEEE Trans. Knowl. Data Eng. **33**(2), 782–795 (2021)
11. Ju, C., Tao, W.: Relationship strength estimation based on wechat friends circle. Neurocomputing **253**, 15–23 (2017)
12. Kipf, T.N., Welling, M.: Semi-supervised classification with graph convolutional networks. CoRR abs/1609.02907 (2016)
13. Perozzi, B., Al-Rfou, R., Skiena, S.: Deepwalk: online learning of social representations. In: KDD, pp. 701–710. ACM (2014)
14. Qin, H., Li, R., Wang, G., Huang, X., Yuan, Y., Yu, J.X.: Mining stable communities in temporal networks by density-based clustering. IEEE Trans. Big Data **8**(3), 671–684 (2022)
15. Qin, H., Li, R., Wang, G., Qin, L., Cheng, Y., Yuan, Y.: Mining periodic cliques in temporal networks. In: ICDE, pp. 1130–1141. IEEE (2019)
16. Scarselli, F., Gori, M., Tsoi, A.C., Hagenbuchner, M., Monfardini, G.: The graph neural network model. IEEE Trans. Neural Networks **20**(1), 61–80 (2009)
17. Velickovic, P., Cucurull, G., Casanova, A., Romero, A., Liò, P., Bengio, Y.: Graph attention networks. CoRR abs/1710.10903 (2017)
18. Yang, Y., Yu, J.X., Gao, H., Pei, J., Li, J.: Mining most frequently changing component in evolving graphs. World Wide Web **17**(3), 351–376 (2014)

TE-DyGE: Temporal Evolution-Enhanced Dynamic Graph Embedding Network

Liping Wang[1], Yanyan Shen[2(✉)], and Lei Chen[1,3]

[1] DSA, The Hong Kong University of Science and Technology (Guangzhou)
(HKUST-GZ), Guangzhou, China
lwang347@connect.hkust-gz.edu.cn

[2] Shanghai Jiao Tong University (SJTU), Shanghai 200240, China
shenyy@sjtu.edu.cn

[3] CSE, The Hong Kong University of Science and Technology (HKUST),
Hong Kong, China
leichen@cse.ust.hk

Abstract. Dynamic graph representation has gradually attached more attention in the research area since many real-world graphs evolve over time. Recent works typically represent the dynamic graph as a sequence of discrete static subgraphs and apply a recurrent neural network (RNN) to capture graph evolution over multiple discrete-time subgraphs. However, these approaches fail to capture the fine-grained temporal evolution leading to introducing temporal invalid information into the structure representation. In this paper, we propose a novel temporal evolution-enhanced dynamic graph embedding method (**TE-DyGE**) to address this limitation. To capture the fine-grained temporal evolution, TE-DyGE first applies a temporal random walk to filter out temporally valid information. Additionally, we introduce temporal-dependent weight to enhance the structure attention network. We evaluate the TE-DyGE in link prediction tasks. The experimental results demonstrate a generally higher performance of TE-DyGE compared with several state-of-the-art related baselines.

Keywords: Dynamic Graphs · Representation Learning

1 Introduction

Graph representation is a fundamental problem for many applications due to its ability to encode the graph structure into a linear space. Low-dimensional vectors, also known as embeddings, capture properties of nodes enabling the graph to be used in various downstream tasks such as node categorization [13,19], link prediction [2], and recommendation [16,24]. Existing graph representation learning works focus primarily on static graphs containing fixed sequences of nodes and edges. However, dynamic graphs arise naturally in the real world because graph structures such as company relation graphs, credit card fraud

© The Author(s), under exclusive license to Springer Nature Switzerland AG 2023
X. Wang et al. (Eds.): DASFAA 2023, LNCS 13945, pp. 183–198, 2023.
https://doi.org/10.1007/978-3-031-30675-4_13

graphs [9], co-authorship networks [11], and user-item interaction graphs [5] constantly evolve. The representation of dynamic graphs contains not only the graph structure information but also the temporal evolution.

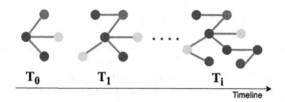

Fig. 1. Discrete-time dynamic graph (DTDG).

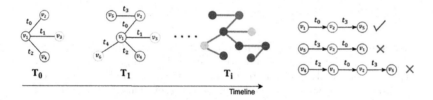

Fig. 2. Visual illustration of temporal evolution-enhanced DTDG (Edges are labeled by time).

For instance, dynamic company relation graphs represent not only the networks between companies but also the volatility of the financial market. The volatility of the financial market is usually a significant feature for analyzing the movements of the financial markets.

The above dynamic graph examples urge graph representation methods to model the temporal evolution of dynamic graphs. The dynamic graphs are evolving with the insertion and deletion of nodes and links at any time [18], leading to the challenge of capturing the temporal evolution of the node based on the changed actions in it and its neighborhoods. In order to represent the complicated time-varying graph structures, existing works divide a dynamic graph by the same time window into a sequence of snapshots [6–8,17] that can be referred to as a *discrete-time dynamic graph* (shown in Fig. 1 (a)).

Discrete-time dynamic networks cover changing topology and feature over time by representing the dynamic graph as multiple static snapshots at different time intervals. Each subgraph contains the final statement of entities during a certain time period. DySAT [14] applies a structure self-attention network to capture the structure information for each snapshot and a temporal self-attention network to join the structure information of discrete snapshots in all time intervals. DyHATR [25] proposes a hierarchical attention model to capture the heterogeneity of static snapshots by node-level and edge-level attention. Then they model the evolution representation among snapshots by an attentive

GRU/LSTM module. EvolveGCN [12] also proposes a combination of GNNs (structure attention network) and RNNs (temporal attention network). Unlike the DySAT and DyHATR, which use the temporal attention network to update the node embedding, they utilize the RNN to regulate the parameters in the GNN model.

The existing works primarily consider the structure information of discrete snapshots resulting in no temporal information inside the snapshots being collected. As shown in Fig. 1 (b), the fine-grained temporal evolution inside the subgraph $(t_0 \rightarrow t_1 \rightarrow t_2)$ cannot be represented. This limitation of existing works leads to encoding meaningless and obsolete information into the graph embedding.

Specifically, when time is respected, some neighboring nodes actually offer no information to the embedding of the target node, e.g., in a social network graph (Fig. 2), we have two messages $m_1 = (v_1, v_2, t_0)$ and $m_2 = (v_2, v_3, t_3)$. The user v_1 sends a message to the user v_2 at the time t_0, and v_2 sent a message to the user v_5 at the time t_3. If $t_0 < t_3$, we can assume that the message m_2 is affected by the message m_1 (temporal valid). In contrast, m_1 does not contribute to the representation of m_2 (temporal invalid). Thus, ignoring the fine-grained temporal evolution may lead to capturing vast quantities of temporally invalid information.

In addition, in many practical settings, the strength of the edges gradually decays over time. In an online fraud network, if two devices share the same IP address and behave similarly, such as visiting the same websites within a short time frame (less than an hour), they likely belong to the same person. Instead, even if two devices have the same IP and exhibit similar behavior over a long period of time, such as a week or a month, it is improbable that they belong to the same user. Due to the lack of temporally weighted edges, the vast majority of existing methods embed obsolete information, which is expected to result in less accurate predictions of future links.

Considering the preceding instances, we present a temporal evolution-enhanced dynamic graph embedding method (TE-DyGE) to address the limitation of ignoring the fine-grained temporal evolution inside the discrete subgraphs. Firstly, we introduce **temporal random walks** to preclude the temporally invalid sequences of edges. Instead of sampling simply the structure neighbors, the temporal random walk samples discrete subgraphs based on the ascending order of the time sequence. A temporal random walk is a sequence of nodes linked by edges with non-decreasing timestamps, thereby capturing the temporally valid information. In subgraph $\mathcal{G}_{T_1}$, for example, the non-sequential walk $\{(v_5, v_2, t_3), (v_2, v_1, t_0), (v_1, v_6, t_4)\}$ will not be sampled because v_6 exists in the past with regard to v_5.

Secondly, we propose a temporally weighted structure attention network, which captures temporal evolution representations of discreet subgraphs by incorporating **time distance** as edge weight. Time distance can be defined as a time difference of two edges, $\mathcal{D}\{(v_2, v_1, t_0), (v_3, v_1, t_1)\} = t_1 - t_0$. The proposed network avoids embedding obsolete information by attaching importance to temporally relevant edges and trying to subdue the old interactions.

We summarize the key contributions of this paper below:

1. We propose a novel discrete-time dynamic graph embedding method named TE-DyGE, which can model fine-grained temporal evolution. To the best of our knowledge, the proposed TE-DyGE is the first attempt to leverage temporal evolution inside the discrete-time subgraph.
2. We apply a temporal random walk to deliver temporal evolution information into the graph embedding network.
3. We propose a temporally weighted structure attention network that introduces the time distances as weights of edges in node embedding.
4. TE-DyGE outperforms several state-of-the-art baselines on the task of link prediction using four real-world datasets.

2 Related Works

This work is related to graph representation learning, especially learning the temporal evolution of dynamic graphs.

2.1 Static Graph Representation Learning

Static graph representation learning can be separated into two groups: unsupervised learning and supervised or semi-supervised learning. Early works focus on unsupervised representation learning, which prefers dimensionality reduction based on spectral graph topology and properties [20]. Inspired by the skip-gram model in the natural language domain, several graph representation works [2,13] use skip-gram to maximize the likelihood of co-occurrence in random walks to learn the node embedding. Specifically, they use random walks to sample a fixed-length sequence of nodes from a graph, then use the skip-gram model to learn the node embedding. Relatively recent static graph representation works focus on supervised and semi-supervised learning. A number of convolution graph neural networks (GCN [3]) extend the convolution neural network by operating the convolution layers to aggregate information from neighbors. Some methods [22] additionally apply an attention mechanism [21] to learn the importance of neighbor nodes. However, the above works fail to consider the dynamic graphs, which evolve over time.

2.2 Dynamic Graph Representation Learning

Existing dynamic graph representation works tend to split a dynamic graph by the same time window into a sequence of snapshots [6]. Several works capture the evolution features by incrementally updating the node embedding. They use the embedding from the previous time step as the initial embedding, However, the long-range variation and the complex time-varying features can not be preserved by the initial embedding. Recent works focus on separating the graph structure representation task and the evolution representation task into two independent modules. DySAT [14] captures the graph structure embedding for each discrete-time subgraph and then applies temporal attention to learning the temporal importance of discrete-time subgraphs. The temporal attention module benefits from capturing the most relevant historical context of the subgraph. Xue et al. [25] applies a similar architecture to dynamic heterogeneous graph representation learning, and they additionally introduce RNN to capture more stable evolution features among discrete subgraphs. Wang et al. [23] proposed CoEvoGNN for modeling the dynamic sequence. They implement GCN, GAT, and GraphSAGE as subgraph structure aggregators and propose a stack temporal attention module for preserving and fusing influences of previous subgraphs' embedding. EvolveGCN [12] uses RNN to learn the temporal evolution and use learned temporal features to regulate the parameters in GCN (subgraph structure aggregator).

3 Problem Definition

Existing discrete-time dynamic networks [14, 25] represent the dynamic graph as a set of static discrete subgraphs. Each subgraph is defined as $\mathcal{G}_t = (\mathcal{V}, \mathcal{E}_t)$, which primarily considers the vertices and edges. In this paper, we model the discrete-time dynamic graph with the timestamps of edges, defined as follows.

Definition 1 (Discrete-time Dynamic Graph in TE-DyGE). *A dynamic graph is described as an order list of graph snapshots,* $\mathbb{G} = \{\mathcal{G}_1, \mathcal{G}_2, ..., \mathcal{G}_T\}$, *where* T *denotes as the number of time steps. Each snapshot is defined as* $\mathcal{G}_t = (\mathcal{V}, \mathcal{E}_t, \mathcal{T})$, *where* $\mathcal{V}$ *is a shared vertices set and* $\mathcal{E}_t$ *is the set of temporal edges between vertices in* $\mathcal{V}$. *And* $\mathcal{T} : \mathcal{E} \rightarrow \mathbb{R}^+$ *maps the edges to their corresponding timestamp.*

Dynamic graph representation aims to learn the node representation $\mathbf{e}_u^t \in \mathbb{R}^d$, where d is the expected output embedding dimension. Each node $u \in \mathcal{V}$ at time step $t = \{1, 2, 3..., T\}$. $\mathbf{e}_u^t$ retains both the structure centered at node u and the temporal evolutionary behaviors such as link/node insertion and deletion up to time step t.

4 Framework

TE-DyGE is developed to describe the dynamic representation of each entity (node) by incorporating two dimensions of temporal evolution: temporal structure neighborhoods and discrete-time subgraphs (listed from fine to coarse granularity). We proposea **fine-grained temporal weighted attention module** by introducing the temporal evolution of each node based on the temporal random walk and time distance. For the coarse level of dimensions, we aggregate the output representations of all discrete-time subgraphs by **a coarse-grain temporal attention module**. Figure 3 shows a brief architecture of the TE-DyGE. We describe the details methodology in this section.

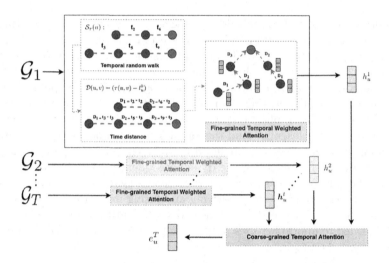

Fig. 3. Framework of TE-DyGE

4.1 Fine-grained Temporal Weighted Attention

To obtain the temporal structure embedding of each node, we design a fine-grained temporal weighted attention module. In this module, we seek to incorporate fine-grained temporal evolution via - (i) temporal neighborhood sets generated by the temporal random walk, and (ii) a graph structure self-attention layer based on temporally weighted edges.

Temporal Neighbors. Instead of relying only on the structural neighbors to aggregate node representation (e_u^t), TE-DyGE samples the time-complex interaction sequences using temporal random walks. The random walk method is used by network embedding methods [2,13] to build the corpus of node context. A temporal random walk incorporates temporal dependence into the random walk method, so providing the graph embedding network with information regarding temporal evolution. Following the work of Nguyen et al. [10] and Shekhar et al. [15], we define a temporal random walk as a sequence of nodes linked by edges with non-decreasing timestamps. The set of possible temporal structure neighborhoods of a node u in subgraph $\mathcal{G}_t = (\mathcal{V}, \mathcal{E}_t, \mathcal{T})$ at time τ is described as $\mathcal{N}_\tau(u) = \{(v, \tau') \mid e = (u, v, \tau') \in \mathcal{E}_t \wedge \mathcal{T}(e) > \tau\}$. We can select the next node in a temporal random walk from $\mathcal{N}_\tau(u)$.

Definition 2 (Temporal Random Walk). *A temporal random walk of length k from node v_1 to v_k in graph $\mathcal{G}_t = (\mathcal{V}, \mathcal{E}_t, \mathcal{T})$ is a sequence of vertices $\mathcal{S}_\tau = \langle v_1, v_2, \ldots, v_k \rangle$ such that $(v_i, v_{i+1}, \tau) \in \mathcal{E}_t$ for $1 \le i < k$. The timestamps are in valid temporal order: $\tau(v_i, v_{i+1}) \le \tau(v_{i+1}, v_{i+2})$ for $1 \le i < (k-1)$.*

We define $\mathbb{S}_\tau$ as the space of all temporal random walks in a subgraph $\mathbf{G}$. $\mathcal{S}_\tau(u)$ is a set of temporal random walks from $\mathbb{S}_\tau$, which represents the temporal neighborhoods of node u. The temporal neighborhoods of a node u can be generated based on the temporal random walk, over which we apply a self-attention network to generate embeddings for nodes. We next introduce a temporally weighted graph embedding network for more intuitively oriented temporal-based considerations.

Temporally Weighted Structure Self-attention. The input of this layer is temporal random walks sampled from a subgraph $\mathcal{G}_t$, and a set of initializing input node embeddings $\{\mathbf{x}_u \in \mathbb{R}^D, \forall u \in \mathcal{V}\}$, where D is the input embedding dimension.

Temporally Weighted Edges. TE-DyGE extends a structure attention network [14,25] to model fine-grained temporal graph structure embedding by incorporating temporally weighted edges. Intuitively, temporally recent edges indicate more important linkages between entities and more direct interactions. In order to measure the temporal weight of edges, we introduce the time distance, which is the "in-between" time of consecutive edges in a temporal random walk. Time distance is defined to assess the different weight between a current interaction and an interaction from the distant past. It is formally defined as follows:

Definition 3 (Time Distance). *The time distance of node v to the target node u is described as $\mathcal{D}(u,v) = (\tau(u,v) - t_u^0)$, where $t_u^0 = min\{\tau(u,v_i) \mid v_i \in \mathcal{S}_\tau(u)\}$ is the creation time of a temporal random walk from the node u.*

The time distance depends on the initial edge $e_i = (u,v,t^0)$ of the temporal random walk. And the creation time varies across distinct temporal random walks. Thus, we define the temporal weight for a particular edge $(u,v,\tau(u,v))$ with respect to the node u based on the time distance as follows:

$$w_{uv} = \frac{\exp\{\mathcal{D}(u,v)\}}{\sum_{j \in \mathcal{S}_\tau(u)} \exp\{\mathcal{D}(u,j)\}} \tag{1}$$

where $\mathcal{S}_\tau(u)$ represents the set of temporal neighbors of node u. We apply the *softmax* function to normalize the temporal weight. The temporal weight w_{uv} is then employed in the structure self-attention layer.

Temporal Structure Self-attention. The temporal structure self-attention layer aims to aggregate features of a target node u by learning the importance of each temporal neighborhood. The function of the structure weight coefficient can be defined as follow:

$$\alpha_{uv} = softmax(e_{uv}) = \frac{\exp(e_{uv})}{\sum_{j \in \mathcal{S}_\tau(u)} \exp(e_{uj})} \tag{2}$$

$$e_{uv} = \sigma\left(\boldsymbol{a}^T\left[\boldsymbol{W}^s\boldsymbol{x}_u \| \boldsymbol{W}^s\boldsymbol{x}_v\right]\right) \forall(u,v,\tau) \in \mathcal{E}_t \tag{3}$$

where $\boldsymbol{W}^s$ is a shared weight transformation matrix for each node in the subgraph; $\boldsymbol{a}$ is a weighted vector used to parameterize the feed-forward layer α; $\sigma(\cdot)$

denotes the activate function and ∥ represents the operation of concatenation. The learned coefficient α_{uv} is normalized across all temporal neighbors of node u by the *softmax* function. We apply the LeakyReLU non-linearity to compute the structure attention weights.

Existing works solely aggregate the latent embedding of neighbors using the learned structure attention weight coefficients. Unlike previous efforts, we include the temporal weight in the node representation function. The final representation of a node u in subgraph $\mathcal{G}_t$ can be computed as follows:

$$h_u^t = \sigma \left(\sum_{u \in \mathcal{N}_v} w_{uv} \times \alpha_{uv} \boldsymbol{W}^s \boldsymbol{x}_u \right) \tag{4}$$

In our implementation, we use an exponential linear (ELU) activation for the output embedding of node u. With the support of temporal weight and structural attention coefficient, representations of nodes are aggregation over neighbors that are close in time.

4.2   Coarse-grained Temporal Attention Module

Decoupling nodes' temporal evolution and subgraphs' temporal history into independent layers contributes to the efficiency [14]. Thus, we further propose a coarse-grained temporal attention module to capture the temporal history of discrete-time subgraphs. This module offers the capacity to model sequential information and learn evolutionary patterns from discrete-time subgraphs. The input of the coarse-grained temporal attention module is node embeddings for different subgraphs obtained from the temporally weighted structure attention layer. As an example, the input for a node u is defined as $\{\boldsymbol{x}_u^1, \boldsymbol{x}_u^2, ..., \boldsymbol{x}_u^T \mid \boldsymbol{x}_u^t \in \mathbb{R}^{D'}\}$, where D' is the dimension of the input embedding. We assume that the output embedding for node u is $\{\boldsymbol{h}_u^1, \boldsymbol{h}_u^2, ..., \boldsymbol{h}_u^T \mid \boldsymbol{h}_u^t \in \mathbb{R}^{F'}\}$ with dimensionality F'. The input and output representations are packed across time to $\boldsymbol{X}_u \in \mathbb{R}^{T \times D'}$ and $\boldsymbol{H}_u \in \mathbb{R}^{T \times F'}$ respectively.

In contrast to temporally weighted structure attention, which focuses on the representation of temporal neighbors, coarse-grained temporal attention relies solely on the temporal evolution of each discrete-time subgraph. We apply the scaled dot-product attention [21] to learning the node embedding at different subgraphs. We denote queries, keys, and values as $\boldsymbol{Q}_u = \boldsymbol{X}_u \boldsymbol{W}_q$, $\boldsymbol{K}_u = \boldsymbol{X}_u \boldsymbol{W}_k$ and $\boldsymbol{V}_u = \boldsymbol{X}_u \boldsymbol{W}_v$ respectively by mapping input into different feature spaces, where $\boldsymbol{W}_q, \boldsymbol{W}_k$ and $\boldsymbol{W}_v$ are trainable parameters. The coarse-grained temporal self-attention function is defined as:

$$\mathbf{H}_u = \boldsymbol{\beta}_u \cdot \boldsymbol{V}_u = \text{softmax} \left(\frac{(\mathbf{X}_u \mathbf{W}_q)(\mathbf{X}_u \mathbf{W}_k)^T}{\sqrt{F'}} + \mathbf{M} \right) \cdot (\mathbf{X}_u \mathbf{W}_v) \tag{5}$$

where $\boldsymbol{\beta}_u \in \mathbb{R}^{T \times T}$ is the attention weight matrix and $\mathbf{M} \in \mathbb{R}^{T \times T}$ denotes the mask matrix, which is used to enforce the auto-regressive property. To preserve the auto-regressive property and encode the temporal order, we permit each time step t to attend over all preceding time steps. And we define the mask matrix as follows:

$$M_{ij} = \begin{cases} 0, & i \leq j \\ -\infty, & \text{otherwise} \end{cases} \tag{6}$$

where i, j denotes the different time steps. The attention weight becomes zero when $M_{ij} = -\infty$. Furthermore, we utilize multi-head attention to enhance the capability of TE-DyGE to capture multiple facets or types of graph evolution. Finally, the node u can be represented as:

$$\boldsymbol{H}_u = \text{Concat}\left(H_u^1, H_u^2, \ldots, H_u^k\right) \quad \forall u \in \mathcal{V} \tag{7}$$

where k is the number of attention heads, and $\boldsymbol{H}_u \in \mathbb{R}^{T \times F'}$.

4.3   Objective Function

The aim of TE-DyGE is to capture both the fine-grained temporal evolution of each node and the structural evolution of discrete-time subgraphs. The objective function is defined to preserve the temporal structure neighbors of a node across multiple time steps. We apply a binary cross-entropy loss function in each time step to encourage node u to have similar embedding features with its temporal neighbors.

$$L = \sum_{t=1}^{T} \sum_{u \in \mathcal{V}} \Bigg(\sum_{v \in \mathcal{S}_\tau^t(u)} -\log\left(\sigma\left(< e_u^t, e_v^t >\right)\right)$$
$$-n_p \cdot \sum_{u' \in P_n^t(u)} \log\left(1 - \sigma\left(< e_{u'}^t, e_v^t >\right)\right) \Bigg) \tag{8}$$

where σ is the sigmoid function; $< \cdot >$ is the inner product operation; $P_n^t(u)$ is a negative sampling distribution for subgraph $\mathcal{G}_t$ and n_p is the negative sampling ratio.

5   Experiments

The experiments are designed to investigate the quantity of the proposed TE-DyGE framework. Experiments are conducted using a wide array of temporal graphs with distinct structural and temporal properties derived from a variety of application sectors. We propose three types of experiments to evaluate and analyze the proposed in different aspects:

1. **Performance of TE-DyGE.** We compare the TE-DyGE with several state-of-the-art static and dynamic graph representation learning methods on the link prediction task.

2. **Effectiveness of Proposed Contributions.** We evaluate the temporal random walk and temporal weighted edge separately and verify the effectiveness of fine-grained temporal evolution.
3. **Parameters Sensitivity.** We analyze the TE-DyGE in varying parameters.

5.1 Datasets

To analyze the quantity of the generated embedding, we conduct experiments on communication real-world networks. We consider two publicly available dynamic network datasets: Enron [4], and UCI [11]. Links in Enron represent the message interactions between core employees in the Enron corporation. UCI describes social interactions in an online community. All these datasets can indicate a complex-evolved real-world communication network.

Table 1. Statistics of datasets.

Attribute	Communication	
	Enron	UCI
# of Nodes	143	1,809
# of Links	2,347	16,822
# of Time steps	12	13

5.2 Baselines

To evaluate the benefits of using temporal information for link prediction, we first provide comparisons with static graph embedding techniques. We integrate all discrete-time subgraphs into a static graph to provide the entire history of the dataset.

- **node2vec** [2]: Utilizes second-order random walk sampling to capture node representations.
- **GAT** [22]: A static graph embedding method that employs the attention mechanism.

We develop off-the-shelf implementations for node2vec[1] and GAT[2] by referencing their original implementations. Additionally, we evaluate the TE-DyGE against the state-of-the-art **discrete-time** dynamic graph embedding baselines.

- **DyGEM** [1]: Capture evolution feature by incrementally learning graph autoencoders with differing layer sizes.
- **DySAT** [14]: Obtain the dynamic representation by decoupling the structure and temporal information into independent modules. It uses the attention mechanism in both models.
- **EvolveGCN** [12]: Use an RNN to evolve the GCN parameters to capture the dynamic evolution of the graph.

---

[1] https://github.com/eliorc/node2vec.
[2] https://github.com/PetarV-/GAT.

5.3   Experimental Setup

The task of link prediction aim to predict the links at final subgraph $\mathcal{G}_T$ based on the learned node representation over the previous subgraph $\{\mathcal{G}_1, \mathcal{G}_2, ..., \mathcal{G}_{T-1}\}$.

Implementation. We randomly sample an equal number of negative samples (pairs of nodes without links) for training and testing. We set the length of the temporal random walk as 8 on all datasets. And the number of heads for multi-head attention is set to 8. The dimension of the output embedding is 32. We use Adam and stochastic gradient descent to optimize and update parameters in the proposed model.

Metric. We train a Logistic Regression classifier and use the area under the ROC curve (AUC) metric to evaluate the performance. In order to ensure the fairness of the experimental results, we repeatedly run all baselines and our models ten times and report the best values.

Environment. All the experiments are run on a machine with GeForce RTX 2080 Ti GPU.

5.4   Effectiveness of TE-DyGE

The results of link prediction are shown in Table 2, which indicates that the proposed method, TE-GyGE, achieves the best performance over all datasets among static graph representation baselines and dynamic graph representation baselines. Specifically, dynamic graph representation methods have an approximately better AUC score when compared to static graph representation methods. This proves that aggregating the temporal evolution of nodes is responsible for the superior performance of dynamic graph representation methods on link prediction tasks. We can also find that DySAT often achieves comparable performance to our model. One possible reason is that they use an attention mechanism to aggregate the graph structure embedding and joint the structure and temporal modeling with multi-attention aggregators.

We further compare the performance at each time step among the baselines and the proposed method (Fig. 4). We observed that TE-DyGE often has a higher AUC score at the first time step than other methods. It's probable that the time distance initializes the temporal importance of edges, resulting in a such excellent performance.

5.5   Effectiveness of Proposed Contributions

We first conduct an ablation study on two contributions: temporal random walk and temporally weighted structure attention layers. We conduct the ablation experiment over all datasets. Figure 5 represents the effectiveness of the TE-DyGE without the temporal random walk (the blue bar) and the TE-DyGE

Table 2. Link prediction results (* and † denote the TE-DyGE without the temporal random walk and the TE-DyGE without the temporally weighted edges respectively).

	Enron	UCI
Method	AUC	AUC
node2vec	69.35 ± 1.2	62.35 ± 0.8
GAT	72.58 ± 0.8	71.32 ± 0.5
DyGEM	67.57 ± 0.3	75.89 ± 0.5
DySAT	85.27 ± 0.3	80.56 ± 0.3
EvolveGCN	77.58 ± 1.45	81.43 ± 0.5
TE-DyGE	87.64 ± 0.2	83.26 ± 0.5

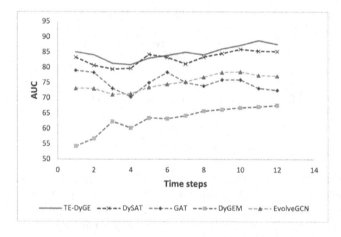

Fig. 4. Performance at each time step (solid line denotes TE-DyGE)

without the temporal weighted edge (the brown bar) separately. The results indicate that DE-DyGE outperforms the ablation variants by a margin of $2\%AUC$ on average.

On the other hand, to evaluate the effectiveness of fine-grained temporal evolution, we conduct an ablation study by independently removing the fine-grained temporal evolution and the coarse-grained temporal evolution from TE-DyGE.

– **No Fine-grained**: We simply remove the temporal random walk and the temporal weight in the fine-grained temporal attention module. Thus the TE-DyGE can jointly capture the structure and temporal history of discrete-time subgraphs.
– **No Coarse-grained**: We remove the coarse-grained temporal attention module for this experiment, Because the node embeddings are jointly optimized based on Eq. 8, the removal of the coarse-grained temporal attention module primarily results in no explicit subgraphs' temporal evolution modeling.

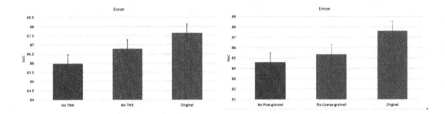

Fig. 5. Ablation study on temporal random walk and temporal weighted edge (Left); Ablation study on fine-grained temporal evolution and coarse-grained temporal (Right)

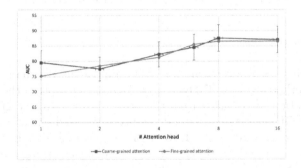

Fig. 6. Effects of the number of heads in multi-head attention

From Fig. 5, we can find a reasonable result that TE-DyGE without fine-grained evolution has an acceptable performance since it can capture both graph structure and evolution features. The fine-grained temporal evolution enhances the dynamic graph representation model by about 3% increase in AUC.

5.6 Parameters Sensitivity

We analyze the critical parameters: the number of multi-heads and the dimension of output embedding for TE-DyGE. The effects of a number of multi-heads in both the fine-grained temporal attention module and the coarse-grained temporal attention module are examined. The rest parameters are kept fixed when we vary the analyzed parameter. Figure 6 demonstrates that multi-head attention on both the fine-grained attention and coarse-grained attention layers enhances the TE-DyGE. The performance stabilizes at 8 attention heads, which is adequate for capturing graph evolution from the most number of latent facets. Similarly, the TE-DyGE can benefit from the final output embedding with 32 dimensions (Fig. 7).

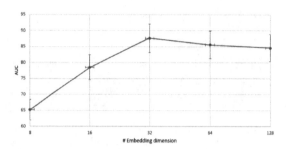

Fig. 7. Effects of the number of output dimensions

6 Conclusion

This paper proposes a novel temporal evolution-enhanced dynamic graph representation method named TE-DyGE. TE-DyGE can capture the fine-grained temporal evolution information from the discrete-time dynamic graph. In particular, TE-DyGE aggregates the node embedding based on sampled temporal valid neighbors with introducing the temporally weighted edges in each discrete-time subgraph. TE-DyGE then joints the representation of subgraphs at different time steps by attention aggregator. This allows the proposed method to represent not only the temporal history across all subgraphs but also the temporal evolution of each node. Our experimental results demonstrate considerable performance improvements for TE-DyGE over static and dynamic graph embedding baselines.

Acknowledgments. The authors would like to thank the anonymous reviewers for their insightful reviews. This work is supported by the National Key Research and Development Program of China (2022YFE0200500), Shanghai Municipal Science and Technology Major Project (2021SHZDZX0102) and SJTU Global Strategic Partnership Fund (2021 SJTU-HKUST). Lei Chen's work is partially supported by National Science Foundation of China (NSFC) under Grant No. U22B2060, the Hong Kong RGC GRF Project 16213620, RIF Project R6020-19, AOE Project AoE/E-603/18, Theme-based project TRS T41-603/20R, China NSFC No. 61729201, Guangdong Basic and Applied Basic Research Foundation 2019B151530001, Hong Kong ITC ITF grants MHX/078/21 and PRP/004/22FX, Microsoft Research Asia Collaborative Research Grant, HKUST-Webank joint research lab grants. Yanyan Shen is the corresponding author.

References

1. Goyal, P., Kamra, N., He, X., Liu, Y.: DynGEM: deep embedding method for dynamic graphs. arXiv preprint arXiv:1805.11273 (2018)
2. Grover, A., Leskovec, J.: node2vec: Scalable feature learning for networks. In: Proceedings of the 22nd ACM SIGKDD International Conference on Knowledge Discovery and Data Mining, pp. 855–864 (2016)
3. Kipf, T.N., Welling, M.: Semi-supervised classification with graph convolutional networks. arXiv preprint arXiv:1609.02907 (2016)

4. Klimt, B., Yang, Y.: Introducing the Enron corpus. In: CEAS 2004 - First Conference on Email and Anti-Spam, 30–31 July 2004, Mountain View, California, USA (2004) https://www.ceas.cc/papers-2004/168.pdf
5. Kumar, S., Zhang, X., Leskovec, J.: Predicting dynamic embedding trajectory in temporal interaction networks. In: Proceedings of the 25th ACM SIGKDD International Conference on Knowledge Discovery & Data Mining, pp. 1269–1278. KDD 2019 (2019)
6. Leskovec, J., Kleinberg, J., Faloutsos, C.: Graph Evolution: Densification and shrinking diameters. ACM Trans. Knowl. Disco. Data **1**, 2 (2007)
7. Li, T., Zhang, J., Yu, P.S., Zhang, Y., Yan, Y.: Deep dynamic network embedding for link prediction. IEEE Access, pp. 29219–29230 (2018). https://doi.org/10.1109/ACCESS.2018.2839770
8. Lu, Y., Wang, X., Shi, C., Yu, P.S., Ye, Y.: Temporal network embedding with micro-and macro-dynamics. In: Proceedings of the 28th ACM International Conference on Information and Knowledge Management, pp. 469–478 (2019)
9. Neville, J., Şimşek, Z., Jensen, D., Komoroske, J., Palmer, K., Goldberg, H.: Using relational knowledge discovery to prevent securities fraud. In: Proceeding of the Eleventh ACM SIGKDD International Conference on Knowledge Discovery in Data Mining - KDD 2005, p. 449. Chicago, Illinois, USA (2005)
10. Nguyen, G.H., Lee, J.B., Rossi, R.A., Ahmed, N.K., Koh, E., Kim, S.: Continuous-time dynamic network embeddings. In: Companion of the The Web Conference 2018 on The Web Conference 2018 - WWW 2018, pp. 969–976. ACM Press. https://doi.org/10.1145/3184558.3191526. http://dl.acm.org/citation.cfm?doid=3184558.3191526
11. Panzarasa, P., Opsahl, T., Carley, K.M.: Patterns and dynamics of users' behavior and interaction: network analysis of an online community. J. Am. Soc. Inf. Sci. Technol. **60**(5), 911–932 (2009). https://doi.org/10.1002/asi.21015
12. Pareja, A., et al.: EvolveGCN: evolving graph convolutional networks for dynamic graphs. In: Proceedings of the Thirty-Fourth AAAI Conference on Artificial Intelligence (2020)
13. Perozzi, B., Al-Rfou, R., Skiena, S.: Deepwalk: Online learning of social representations. In: Proceedings of the 20th ACM SIGKDD International Conference on Knowledge Discovery and Data Mining, pp. 701–710 (2014)
14. Sankar, A., Wu, Y., Gou, L., Zhang, W., Yang, H.: DySAT: deep neural representation learning on dynamic graphs via self-attention networks. In: Proceedings of the 13th International Conference on Web Search and Data Mining, pp. 519–527. ACM, Houston TX USA (2020)
15. Shekhar, S., Pai, D., Ravindran, S.: Entity resolution in dynamic heterogeneous networks. In: Companion Proceedings of the Web Conference 2020, pp. 662–668. ACM. https://doi.org/10.1145/3366424.3391264. http://dl.acm.org/doi/10.1145/3366424.3391264
16. Shi, C., Hu, B., Zhao, W.X., Philip, S.Y.: Heterogeneous information network embedding for recommendation. IEEE Trans. Knowl. Data Eng. **31**(2), 357–370 (2018)
17. Singer, U., Guy, I., Radinsky, K.: Node embedding over temporal graphs. arXiv preprint arXiv:1903.08889 (2019)
18. Skarding, J., Gabrys, B., Musial, K.: Foundations and modelling of dynamic networks using Dynamic Graph Neural Networks: a survey. IEEE Access, pp. 79143–79168 (2021)

19. Tang, J., Qu, M., Wang, M., Zhang, M., Yan, J., Mei, Q.: Line: large-scale information network embedding. In: Proceedings of the 24th International Conference on World Wide Web, pp. 1067–1077 (2015)
20. Tenenbaum, J.B., de Silva, V., Langford, J.C.: A global geometric framework for nonlinear dimensionality reduction. Science **290**(5500), 2319–2323 (2000)
21. Vaswani, A., et al.: Attention is all you need. In: Proceedings of the 31st International Conference on Neural Information Processing Systems, pp. 6000–6010. NIPS2017, Curran Associates Inc., Red Hook, NY, USA (2017)
22. Veličković, P., Cucurull, G., Casanova, A., Romero, A., Liò, P., Bengio, Y.: Graph attention networks. In: International Conference on Learning Representations (2018). https://openreview.net/forum?id=rJXMpikCZ
23. Wang, D., et al.: Modeling co-evolution of attributed and structural information in graph sequence. IEEE Trans. Knowl. Data Eng. **35**, 1817–1830 (2021)
24. Wen, Y., Guo, L., Chen, Z., Ma, J.: Network embedding based recommendation method in social networks. In: Companion Proceedings of the The Web Conference 2018, pp. 11–12. WWW 2018 (2018)
25. Xue, H., Yang, L., Jiang, W., Wei, Y., Hu, Y., Lin, Yu.: Modeling dynamic heterogeneous network for link prediction using hierarchical attention with temporal RNN. In: Hutter, F., Kersting, K., Lijffijt, J., Valera, I. (eds.) ECML PKDD 2020. LNCS (LNAI), vol. 12457, pp. 282–298. Springer, Cham (2021). https://doi.org/10.1007/978-3-030-67658-2_17

Mining Top-k Frequent Patterns
over Streaming Graphs

Xi Wang[1], Qianzhen Zhang[2(✉)], Deke Guo[2(✉)], and Xiang Zhao[3]

[1] Institute for Quantum Information and State Key Laboratory of High Performance
Computing, National University of Defense Technology, Changsha, China
[2] Science and Technology on Information Systems Engineering Laboratory,
National University of Defense Technology, Changsha, China
{zhangqianzhen18,dekeguo}@nudt.edu.cn
[3] Laboratory for Big Data and Decision, National University of Defense Technology,
Changsha, China

Abstract. Mining top-k frequent patterns is an important operation
on graphs, which is defined as finding k interesting subgraphs with the
highest frequency. Most existing work assumes a static graph. However,
graphs are dynamic in nature, which is described as streaming graphs.
Mining top-k frequent patterns in streaming graphs is challenging due
to the streaming nature of the input and the exponential time complexity of the problem. A naive solution is to calculate approximations of
the frequent patterns in the streaming graph and then find the top-k
answers, which is a memory- and time-consuming method. In this paper,
we design a novel auxiliary data structure, FPC, to detect valid subgraph
patterns and their frequency in real-time. We first convert each newly
produced subgraph into a sequence and then map it into corresponding
tracks in FPC based on hash functions. We theoretically prove that FPC
can provide unbiased estimation and then give an error bound of our algorithm. In addition, we propose a vertical hashing and candidate buckets
sampling technique to further improve FPC with higher space utilization
and higher accuracy. Extensive experiments confirm that our approach
generates high-quality results compared to the baseline method.

1 Introduction

With the proliferation of graph data, mining frequency-based subgraph patterns
have been extensively investigated due to its wide applications [6,15], including bioinformatics, security, and social sciences. The goal of frequent subgraph
mining is to find subgraphs whose appearances exceed a user-defined threshold.
Such subgraphs might be indicative of important protein interactions, common
social norms, or frequent activity between users.

Existing studies on frequency-based subgraph patterns mining [5,8,11,13]
mainly focus on static graphs. Nevertheless, many graphs in the real world are
dynamic in nature, which continuously evolve over time. For example, consider
the knowledge graph DBpedia, which gets updated every day according to a
stream of change logs from Wikipedia [9]. These graphs are treated as streaming

© The Author(s), under exclusive license to Springer Nature Switzerland AG 2023
X. Wang et al. (Eds.): DASFAA 2023, LNCS 13945, pp. 199–216, 2023.
https://doi.org/10.1007/978-3-031-30675-4_14

graphs. In specific, a streaming graph is defined as an unbounded sequence of directed edges $\{e_1, e_2, \cdots, e_x\}$ where each item e_i indicates an edge between two labeled vertices arriving at a particular time t_i.

In this research, we address the problem of mining frequent patterns over a streaming graph. As an example of the importance of frequent patterns and their practical uses in a streaming graph, consider a music streaming service, and a graph of its users, where users (nodes) are temporarily connected by an edge when they listen to the same song within a certain interval of time. Analyzing the behavior of frequent patterns over time allows us to study the evolving taste of users in music and use this information for song recommendations.

An important limitation of discovering all frequent patterns in a streaming graph is that the user-defined threshold is hard to set. If the threshold is set too low, few patterns are found, and the user may miss valuable information. On the contrary, if the threshold is set too high, millions of subgraphs may be found, and algorithms may have very long execution times, or even run out of memory or storage space. To this end, we focus on the top-k version of the frequent pattern mining problem, which is finding k interesting patterns with the highest frequency over a streaming graph.

Challenges. The large scale and high dynamicity make it both memory- and time-consuming to keep track of the exact changes in the frequencies of all patterns at all times. It is a natural choice to resort to efficiently compute approximations. Recent advances have resulted in the development of frequent patterns mining in evolving graphs [3,16], which can be revised to provide top-k frequent patterns as new data arrives. They propose to maintain a uniform sample of subgraphs via reservoir sampling [19], which in turn allows ensuring the uniformity of the sample when an edge insertion occurs. Based on the sample of subgraphs, we can estimate the frequency of different patterns, after which we rank all patterns for top-k results.

Since the estimation accuracy depends on the sample size, the algorithm needs to maintain a large number of subgraphs of the streaming graph for mining top-k frequent patterns accurately, which is memory-consuming. Moreover, the algorithm needs to conduct expensive subgraph matching calculations for these sampled subgraphs to estimate the frequency of each pattern after all updates have occurred at the current timestamp, which is time-consuming. In this light, advanced techniques are desiderated to mine top-k frequent patterns efficiently.

Our Solution. Based on the above discussion, existing methods cannot yield high efficiency for continuous top-k frequent pattern mining over streaming graphs in both spaces and mining time. Our paper aims for a new way to solve the problem. Our main idea is as follows: instead of using the sampling techniques to maintain the subgraphs, we propose to design an auxiliary data structure called FPC to detect valid patterns and their frequency in real-time. We use k buckets, each new produced subgraph will be mapped into one bucket by hash functions $h_1(\cdot), \cdots, h_k(\cdot)$ to estimate the frequency directly. In this way, we can avoid storing any subgraph of the streaming graph in the mining process.

Contributions. In short, the major contributions we have made are summarized below:

- We are the first to study the top-k version of frequent patterns mining problem over streaming graphs. We propose an auxiliary data structure called FPC to store the potential top-k results with accuracy and efficiency guarantee under limited memory.
- We provide an unbiased estimation of the frequencies for the patterns stored in FPC with a theoretical guarantee. In addition, we design a new graph invariant that maps each subgraph to its sequence space representation in the FPC for deriving high efficiency.
- We propose an optimal strategy called vertical hashing to map a subgraph into more than one bucket in FPC and replace the least frequent pattern with the newly produced subgraph pattern. In this way, we can achieve higher space utilization and higher accuracy.

Extensive experiments using three real-life streaming graphs show that our proposed method outperforms the baseline solution in terms of efficiency, memory size and estimation accuracy.

2 Preliminaries

Definition 1 (Streaming graph). *A streaming graph is an unbounded time evolving sequence of items $\{e_1, e_2, e_3, \cdots, e_n\}$, where each item $e_i = (v_1, v_2, t(e_i))$ indicates a directed edge from vertex v_1 to v_2 arriving at time $t(e_i)$. This sequence continuously arrives from data sources like routers or monitors at high speed. It should be noted that the throughput of the streaming graph keeps varying. There may be multiple (or none) edges arriving at each time point.*

Figure 1 shows a constantly time evolving sequence of items and a streaming graph G formed by the data items. We use $G_t = (V_t, E_t)$ to denote the graph observed up to time t.

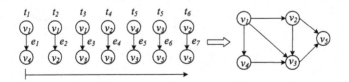

Fig. 1. Streaming Graph

A subgraph $S^m = (V_S, E_S)$ is referred to as a m-edge subgraph if it is induced by m edges in G_t. For any $t \geq 0$, at time $t + 1$, we receive an edge insertion e and add it into G_t to obtain G_{t+1}. For each newly inserted edge e in G_{t+1}, we use $S^m(e)$ to denote the set of m-edge subgraphs that contain e in G_{t+1}.

Definition 2 (Subgraph isomorphism). *Two subgraphs $S_1^m = (V_1, E_1)$ and $S_2^m = (V_2, E_2)$ are isomorphic if there is a bijection function $f: V_1 \rightarrow V_2$ such that 1) $\forall v \in V_1$, $L(v) = L(f(v))$, and 2) $\forall (v_i, v_j) \in E_1$, $(f(v_i), f(v_j)) \in E_2$. The function L preserves the vertex labels.*

Let $\mathcal{C}$ be a set of m-edge subgraphs that have an isomorphism relation. We call the generic graph $P = (V_P, E_P, L)$, which is isomorphic to all the members of $\mathcal{C}$, the m-edge pattern of $\mathcal{C}$, where V_P is a set of vertices in P, E_P is a set of directed edges with size m and L is a function that assigns a label for each vertex in V_P. Note that, P can be obtained by deleting the IDs (resp. timestamps) of the vertices (resp. edges) of any k-edge subgraph in $\mathcal{C}$. We define the frequency of any m-edge pattern as the number of subgraphs in $\mathcal{C}$.

Problem Statement. Given a streaming graph G, and the user-defined parameters k, m. The problem of top-k frequent pattern mining consists of finding k interesting m-edge patterns with the highest frequency at each timestamp.

3   The Baseline Solution

In literature [3], the algorithm enumerated all frequent patterns over a streaming graph. It used a reservoir sampling [19] technique to maintain a uniform sample of subgraphs when an edge update occurs and then estimated the frequency of each pattern via subgraph matching. To obtain a reasonable baseline, we extend the algorithm proposed in [3] to mine top-k frequent m-edge patterns over streaming graphs. Figure 2 shows the framework of our baseline solution FPM.

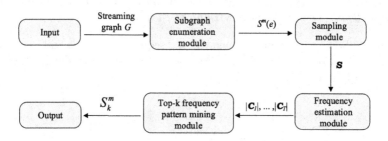

Fig. 2. Solution framework of FPM

Whenever an insertion edge e occurs at timestamp t, FPM first needs to calculate $S^m(e)$. Specifically, FPM explores a candidate subgraph space in a tree shape in G_t, each node representing a candidate subgraph, where a child node is grown with a one-edge extension from its parent node. The intention is to find all possible subgraphs with size m grown from e. To avoid duplicate enumeration of a subgraph, FPM checks whether two subgraphs are composed of the same edges at each level in the tree space.

After obtaining all the m-edge subgraphs in $S^m(e)$, FPM needs to decide whether to add the subgraphs s_i in $S^m(e)$ to a sample set $\mathcal{S}$ with a fixed size M. If $|\mathcal{S}| < M$, FPM directly adds s_i into the sample set $\mathcal{S}$. Otherwise, if $|\mathcal{S}| = M$, FPM removes a randomly selected subgraph in $\mathcal{S}$ and inserts s_i with probability M/N. Here, N is the total number of m-edge subgraphs encountered from time t to time 0. After that, a uniform sample set $\mathcal{S}$ of subgraphs can be maintained at any time t, and all the subgraphs are selected with equal probability M/N.

Then FPM partitions the set of subgraphs in $\mathcal{S}$ into T equivalence classes based on subgraph isomorphism, denoted by $\mathcal{C}_1$, ..., $\mathcal{C}_T$, and calculates the number of m-edge subgraph pattern P of each equivalence class $\mathcal{C}_i$ ($i \in [1,T]$). As proofed in [3], for any isomorphism class $i \in [1,T]$, $|\tilde{f}_i\text{-}f_i| <= \epsilon/2$ holds with probability at least $1 - \delta/T$, when we set $M = log(1/\delta) \cdot \frac{4+\epsilon}{\epsilon^2}$, where $\tilde{f}_i$ is the estimated frequency, f_i is the real frequency and $0 < \epsilon$.

Finally, FPM calculates the k interesting m-edge patterns set S_k^m with the highest frequency by arranging all $|\mathcal{C}_i|$ ($i \in [1,T]$) in a descending order.

Complexity Analysis. In the m-edge subgraphs enumeration process, given an insertion edge e in G_t, let n be the number of vertices of the subgraph extended from e with radius m. FPM takes $O(2^{n^2})$ to find the m-edge subgraphs that contain e. In the subgraph sampling process, for each newly produced subgraph m-edge subgraph, FPM takes $O(1)$ to add it into the reservoir. In the frequency estimation process, FPM takes $O((M^3 - M) \cdot \eta)$ to partition the m-edge subgraphs in $\mathcal{S}$ into T equivalence classes, where η is the average unit time to verify whether two m-edge subgraphs are isomorphic. And then for each timestamp, FPM takes $O(T \cdot \eta)$ to update the frequency of each pattern in $\mathcal{S}$. Finally, it takes $O(k \cdot log_2 T)$ to obtain the top-k frequency pattern results.

4 TopKF: A Progressive Solution

In this section, we first analyze the drawbacks of the baseline solution, and then introduce our progressive solution to significantly reduce the memory and computational cost in quest of top-k frequent pattern mining.

4.1 Problem Analysis

Why Costly? The baseline algorithm is not scalable enough to handle large streaming graphs due to the following two limitations:

1) **Large memory consumption:** As mentioned above, to ensure the accuracy of the sampling results, it is necessary to guarantee the sampling subgraph set with a certain size M, i.e., $M >= log(1/\delta) \cdot \frac{4+\epsilon}{\epsilon^2}$. Thus, there are a lot number of m-edge subgraphs need to be stored at each timestamp, which consumes a large amount of memory.

2) **Large computational cost:** In the frequency estimation process, whether it is to classify the sample set or update the frequency of the pattern at each time, subgraph matching is required. Since it is an NP-complete problem, the

large number of subgraph matching computations can lead to high latency when dealing with streaming graphs.

Our Idea. Based on the above analysis and to significantly improve the algorithm, we design an auxiliary data structure called FPC to efficiently estimate the frequencies of each m-edge pattern. Specially, for each newly produced m-edge subgraph, we use a hash function to map it into a fixed position in FPC. In this way, we can count the frequency of the pattern directly without storing any subgraphs, thus avoiding the repeated subgraph isomorphism calculation.

FPC Structure(Fig 3(a)). FPC consists of l buckets. Let $B[i]$ be the i-th bucket in FPC. Each m-edge subgraph s_i in $S^m(e)$ is mapped into one bucket $B[h(s_i)]$ through a hash function $h(\cdot)$. Each bucket consists of two parts: a counter $B[i].count$ and a subgraph set $B[i].sub$. The counter can provide an unbiased estimation frequency for each newly produced m-edge pattern. The subgraph set consists of d cells. Each cell is used to store a key-value pair and a flag $< ID, fre, flag >$. The key is the subgraph pattern ID, which uniquely identifies a set of isomorphic subgraphs, the value is its estimated frequency, and the flag is used to check whether the frequency is exact or has an error. In specific, we use another hash function $s(\cdot)$ to map s_i to $\{+1, -1\}$. For each subgraph s_i mapped into the bucket $B[h(s_i)]$, it will be recorded in one or both of the two parts.

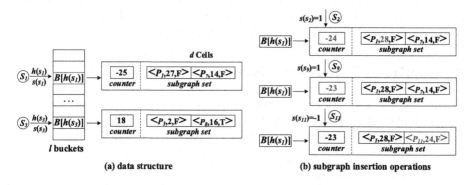

(a) data structure   (b) subgraph insertion operations

Fig. 3. Data structure and insertion examples of FPC

4.2   The Progressive Algorithm Framework

Based on FPC, we propose a new algorithm TopKF to calculate top-k frequent patterns with higher effectiveness and efficiency. It first calls InitialFPC to initialize the auxiliary data structure FPC. Then it calls UpdateFPC to handle subgraph insertion and estimate the frequency of each pattern P in FPC, after which it reports the k interesting m-edge patterns set S_k^m with the highest frequency.

Function UpdateFPC. For each $s_i \in S^m(e)$, UpdateFPC first maps s_i to sequence and then chooses a hash function $h(\cdot)$ with a value range $[0, l)$ to map s_i into bucket $B[h(s_i)]$, where l is the number of the buckets in FPC. Specifically, there are three cases demonstrated how to insert s_i into $B[h(s_i)]$:

Case 1: If s_i is isomorphic to the pattern p_j in $B[h(s_i)].sub$, according to the flag of p_j, there are two situations. 1) the flag of p_j is true, UpdateFPC increments its frequency by 1; 2) the flag of p_j is false, UpdateFPC not only increases its frequency, but also adds $B[h(s_i)].count$ by $s(s_i)(s(s_i) \in \{-1, +1\})$.

Case 2: If s_i is not isomorphic to any pattern stored in $B[h(s_i)].sub$ and $B[h(s_i)].sub$ is not full, UpdateFPC first calculates the pattern p_i of s_i by deleting its vertex IDs and edge timestamps, and then insert $< p_i, 1, true >$ into $B[h(s_i)].sub$.

Case 3: If $B[h(s_i)].sub$ is full and there is no pattern isomorphic to s_i, UpdateFPC uses a replacement strategy to guarantee $B[h(s_i)].sub$ to maintain the patterns with higher frequency. It uses the counter $B[h(s_i)].count$ to unbiasedly estimate the frequency of each pattern denoted as $\tilde{fre}_i = B[h(s_i)].count *$ $s(s_i)$. If $\tilde{fre}_i$ is smaller than the minimum fre in $B[h(s_i)].sub$, UpdateFPC inserts the pattern p_i of s_i into $B[h(s_i)].count$, i.e., adding $B[h(s_i)].count$ by $s(s_i)$; Otherwise, after inserting the p_i into $B[h(s_i)].count$, UpdateFPC replaces the pattern p_n who has the minimum fre with p_i. Specifically, UpdateFPC modifies the key-value pair $< ID, fre, flag >$ of p_n by setting the ID field with p_i, the frequency field to $\tilde{fre}_i + 1$ and the flag field to false. If the flag of the replaced pattern p_n is true, then p_n is inserted into $B[h(s_i)].count$, i.e., adding $B[h(s_i)].count$ by $p_n.fre * s(p_n)$.

Example 1. In Fig. 3(b), we take the bucket $B[h(s_1)]$ as an example to show the subgraph insertion process of UpdateFPC. When s_2 arrives, it is mapped into the bucket $B[h(s_1)]$, and $s(s_2) = 1$. The pattern p_1 is isomorphic to s_2, so UpdateFPC increases the frequency of p_1 to 28. Since the flag of p_1 is false, UpdateFPC also needs to insert s_2 to $B[h(s_1)].count$ and set it to $-25 + s(s_2) = -24$ (in case 1). When s_9 is mapped into the bucket $B[h(s_1)]$, and $s(s_9) = 1$. There is no pattern isomorphic to s_9 in $B[h(s_1)].sub$ and $B[h(s_1)].sub$ is full, thus we have $\tilde{fre}_9 = B[h(s_1)].count \cdot s(s_9) = -23$. Because $\tilde{fre}_9$ is smaller than the minimum frequency 14 in $B[h(s_1)].sub$, UpdateFPC only needs to add s_9 to $B[h(s_1)].count$ and set it to $-24 + s(s_9) = -23$ (in situation 1 of case 3). When s_{11} is mapped into the bucket $B[h(s_1)]$, and $s(s_{11}) = -1$, the $\tilde{fre}_{11} = 23 > 14$. Thus, UpdateFPC inserts the pattern p_{11} to $B[h(s_1)].count$, and replaces the ID field of that cell with p_{11}, sets the frequency field to $\tilde{fre}_{11} + 1 = 24$, and further sets the flag field to false (in situation 2 of case 3).

Subgraph Sequence Representation. We define a function $m: s_i \rightarrow Seq_i$ to map graph s_i to its sequence space representation Seq_i. The goal of this conversion procedure is to map the subgraph into a string representation making each subgraph have a unique code, which is a well-known technique called **graph**

invariants. If m-edge subgraph s_i is isomorphic to subgraph s_j, then $m(s_i)$ $= m(s_j)$. There are several possible graph invariants [12,20], but most of them impose a lexicographic order among the subgraphs, which is clearly as complex as graph isomorphism. In this paper, we generate a degree sequence via both degree and label as our graph invariant that can achieve higher efficiency. Specially, we first compress the degree and label of a vertex v together as its new label $l(v)$. Then, for each edge $e= (v_i, v_j, t(e))$, we label $l(e)=(l(v_i), l(v_j))$. After that, we use the weight of each edge to determine the edge mapping order $O(\cdot)$ of a subgraph. We assign each single-edge pattern a weight $w(\cdot)$ in the streaming graph, which is equal to the order of the occurrence of the pattern. If $w(e_i) <$ $w(e_j)$, then $O(e_i) < O(e_j)$. Else, if $w(e_i) = w(e_j)$, the order of the encoding sequence is determined according to the vertices degree. If $l(e_i) < l(e_j)$, then $O(e_i) < O(e_j)$. Here, ties are broken arbitrarily. Finally, for a subgraph s_j, we obtain the mapping sequence $m(s_j) = \{l(e_1), \ldots, l(e_n)\}$, where $O(e_i) < O(e_{i+1})$.

Algorithm Analysis. Compared to the baseline solution FPM, TopKF does not need to store the sampled m-edge subgraphs since it uses hash functions to map each m-edge subgraph into a fixed bucket in the FPC. This significantly reduces memory consumption. And the proposed unbiased estimation avoids a large number of repeated subgraph matching calculations. Note that, the memory size of TopKF mainly depends on parameters l and d. Users can tune the parameter to make a trade-off between accuracy and speed depending on the application requirements. As shown in our experiments, the precision rate increases as d becomes larger while throughput decreases. Because in a bucket, more cells can reduce the replace operation when the bucket is full and increase the checking cost when there are two subgraphs mapped into the same bucket.

4.3   Mathematical Analysis

In this section, we provide a performance analysis for TopKF. The storage of each pattern in FPC is accurate, which means for a pattern p and a subgraph s, its frequency will be added up if and only if they are isomorphic subgraphs. Thus, the error is mainly caused by unbiased estimation. Firstly, we prove that our algorithm can provide an unbiased estimated frequency. Then, we show the variance and the error bound of unbiased estimation.

Proof of Unbiasedness. For each pattern p_i, FPC can provide an unbiased estimated frequency $\tilde{fre}_i$. If p_i is in the subgraph set and its flag is true, $\tilde{fre}_i$ is equal to the corresponding frequency field in the subgraph set. Otherwise, $\tilde{fre}_i = B[h(s_i)].count * s(p_i)$.

Theorem 1. *The estimation of $\tilde{fre}_i$ is unbiased, i.e., $E(\tilde{fre}_i) = fre_i$, here fre_i is the real frequency.*

Proof. For an insertion subgraph s_i, the expected increment to $\tilde{fre}_i$ is 1 if p_i (p_i is pattern of s_i) is the next pattern insert into $B[h(s_i)].sub$ and 0 otherwise. Let $\tilde{fre}_i{}'$ be the estimated frequency after the next subgraph comes. According to whether the flag of p_i is true and whether it is the next pattern, there are four cases should be considered and analyzed.

Case 1: The flag of p_i is true and p_i is the next pattern.

The corresponding frequency recorded in the subgraph set is increased by 1, i.e., $p_i.fre' = p_i.fre + 1$, where $p_i.fre$ denotes the recorded value of p_i in $B[h(s_i)].sub$. p_i is still in the subgraph set and its flag is true. Thus, we have $\tilde{fre}_i{}' = p_i.fre' = \tilde{fre}_i + 1$.

Case 2: The flag of p_i is true and p_i is not the next pattern.

The corresponding frequency recorded in the subgraph set stays the same. If p_i is still in the subgraph set, we have $\tilde{fre}_i{}' = p_i.fre' = \tilde{fre}_i$. Otherwise, p_i is eliminated from the subgraph set, then p_i is inserted into the counter and its flag becomes false. Then $B[h(s_i)].count' = B[h(s_i)].count + \tilde{fre}_i$. Thus, $\tilde{fre}_i{}' = B[h(s_i)].count' \cdot s(p_i) = B[h(s_i)].\ count \cdot s(p_i) + \tilde{fre}_i \cdot s(p_i)^2$. Since $s(p_i)$ is a hash function with a random value of $\{1, -1\}$, so $E(s(p_i)) = 0$, $E(s(p_i)^2) = 1$. We have $E(B[h(s_i)].count \cdot s(p_i)) = B[h(s_i)].count \cdot E(s(p_i)) = 0$. Thus, $E(\tilde{fre}_i{}') = E(B[h(s_i)].count \cdot s(p_i)) + \tilde{fre}_i \cdot E(s(p_i)^2) = \tilde{fre}_i$.

Case 3: The flag of p_i is false and p_i is the next pattern.

If there is no pattern whose flag is false replaced from the subgraph set, we have $\tilde{fre}_i{}' = (B[h(s_i)].count + s(p_i)) \cdot s(p_i) = \tilde{fre}_i + 1$. Otherwise, p_i replaces the pattern p_n which has the lowest frequency in the subgraph set and its flag is true. We have $B[h(s_i)].count' = B[h(s_i)].count + s(p_i) + \tilde{fre}_n \cdot s(p_n)$. Thus, our estimation satisfies $\tilde{fre}_i{}' = B[h(s_i)].count' \cdot s(p_i) = \tilde{fre}_i{}' + 1 + \tilde{fre}_n \cdot s(p_n) \cdot s(p_i)$. Finally, we have $E(\tilde{fre}_i{}') = \tilde{fre}_i{}' + 1 + E(\tilde{fre}_n \cdot s(p_n)) \cdot E(s(p_i)) = \tilde{fre}_i + 1$.

Case 4: The flag of p_i is false and p_i is not the next pattern.

Let p_l be the next pattern. We have $\tilde{fre}_i = B[h(s_i)].count \cdot s(p_i)$. If the flag of p_j is true, it does not influence the counter. Thus, $\tilde{fre}_i{}' = \tilde{fre}_i$. Otherwise, $B[h(s_i)].count$ is added by $s(p_l)$. If there is no pattern whose flag is true removed from the subgraph set, we have $\tilde{fre}_i{}' = (B[h(s_i)].count + s(p_l)) \cdot s(p_i) = \tilde{fre}_i + s(p_i) \cdot s(p_l)$. Since $s(p_i)$ and $s(p_l)$ are independent, then we have $E(s(p_i) \cdot s(p_l)) = E(s(p_i)) \cdot E(s(p_l)) = 0$. Thus, $E(\tilde{fre}_i{}') = \tilde{fre}_i + E(s(p_i) \cdot s(p_l)) = \tilde{fre}_i$. Otherwise, p_l replaces the pattern p_n which has the lowest frequency in the subgraph set and its flag is true. We have $B[h(s_i)].count' = B[h(s_i)].count + s(p_l) + \tilde{fre}_n \cdot s(p_n)$. Thus, our estimation satisfies $\tilde{fre}_i{}' = B[h(s_i)].count' \cdot s(p_i) = \tilde{fre}_i + s(p_l) \cdot s(p_i) + \tilde{fre}_n \cdot s(p_n) \cdot s(p_i)$. As proved before, we have $E(s(p_l) \cdot s(p_i)) = 0$ and $E(\tilde{fre}_n \cdot s(p_n) \cdot s(p_i)) = 0$. Thus, $E(\tilde{fre}_i{}') = \tilde{fre}_i + E(s(p_l) \cdot s(p_i)) + E(\tilde{fre}_n \cdot s(p_n) \cdot s(p_i)) = \tilde{fre}_i$.

In summary, we have proved that the expected increment to $\tilde{fre}_i$ is 1 if p_i is the next pattern and 0 otherwise, which indicates that we always have $E(\tilde{fre}_i) = fre_i$. In other words, our estimation is unbiased.

Variance and Error Bound. Here, we show the variance and the error bound of our estimation for each pattern in Theorems 2 and 3.

Theorem 2. *Let $p_1, p_2, \cdots, p_n$ be the pattern inserted into $B[h(s_i)]$. We can get the bound of the variance of our estimation*

$$Var(\tilde{fre}_i) \leq \sum_{p_j \neq p_i} (fre_j)^2 \tag{1}$$

Proof. If the flag of p_i is true, we have $\tilde{fre}_i = fre_i$. Otherwise, $\tilde{fre}_i = (\Sigma_{p_j \in \varphi} fre_i \cdot s(p_j)) \cdot s(p_i)$, where φ is the set of patterns whose flag is false. As proofed in Theorem 1, we have $E(\tilde{fre}_i) = fre_i$, thus the $Var(\tilde{fre}_i) = E_{s(p_j) \in \{1,-1\}}((\Sigma_{p_j \in \varphi, j \neq i} fre_i \cdot s(p_j)) \cdot s(p_j))^2 = E_{s(p_j) \in \{1,-1\}}(\Sigma_{p_j \in \varphi, j \neq i} fre_j \cdot s(p_j))^2$. Since $s(p_j)$ and p_i are independent, the cross terms have the same chance to be 1 and -1. Thus the expectation of their sum is 0. Therefore, we have $Var(\tilde{fre}_i) = E_{s(p_j) \in \{1,-1\}}(\Sigma_{p_j \in \varphi, j \neq i} fre_j \cdot s(p_j))^2 \leq \Sigma_{p_j \neq p_i}(fre_i)^2$.

Theorem 3. *Let $l = \frac{e}{\epsilon^2}$, then $P(|\tilde{fre}_i - fre_i| \geq \epsilon ||fre||_2) \leq \frac{1}{e}$*

Proof. Based on Chebyshev's theorem, we can obtain that $P(|\tilde{fre}_i - fre_i| \geq \sqrt{e\Sigma p_j \neq p_i (fre_i)^2}) \leq \frac{Var(\tilde{fre}_i)}{\sqrt{e\Sigma p_j \neq p_i (fre_i)^2}} \leq \frac{1}{e}$. For patterns in $B[h(s_i)]$, we have $\Sigma_{p_j}(fre_j)^2 = \frac{1}{l}(||fre||_2)^2$. Then $P(|\tilde{fre}_i - fre_i| \geq \epsilon ||fre||_2) \leq P(|\tilde{fre}_i - fre_i| \geq \epsilon \sqrt{l \cdot \Sigma_{h(p_j)=h(p_i)} fre_j^2}) \leq P(|\tilde{fre}_i - fre_i| \geq \sqrt{e\Sigma p_j \neq p_i (fre_i)^2}) \leq \frac{1}{e}$.

5   FPC$_S$: Augmented Auxiliary Data Structure

In this section, we propose FPC$_S$, which is an augmented version of FPC by adding vertical hashing and candidate buckets sampling techniques.

Vertical Hashing. In FPC, if frequent replacement operations occur in a bucket, the expected patterns with high frequency may also be replaced, causing a lower accuracy. To this end, we assign a fingerprint $f(s_i)(0 \leq f(s_i) \leq F)$ (occupies less than 8 bits) to each subgraph s_i in FPC$_S$ where $f(s_i) = h(s_i)\%F$ and compute a sequence of bucket addresses $\{B_x[h(s_i)] | 1 \leq x \leq r\}$ for s_i $(0 \leq h(s_i) < l)$. We store its pattern p_i into the first bucket with an empty cell and add the key-value $< ID, fre, f(s_i), flag >$ into the corresponding position. In this way, FPC$_S$ can provide more candidate buckets for a newly produced subgraph, reducing the errors caused by constant replacement.

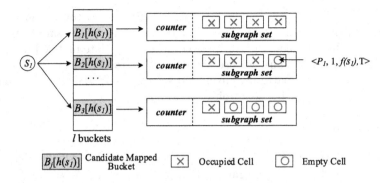

Fig. 4. Subgraph insertion example of FPC_S

To compute the sequence of bucket addresses, we first use the linear congruence method to generate a sequence of r random values $\{q_x(\cdot)|1 \leq x \leq r\}$ with $f(\cdot)$ as seeds. The linear congruence method is as follows: select a timer a, small prime b and a module p, then

$$
\begin{cases}
q_1(s_i) = (a \times f(s_i) + b)\%p \\
q_x(s_i) = (a \times q_{x-1} + b)\%p, (2 \leq x \leq r)
\end{cases}
\tag{2}
$$

Using the linear congruence method can make sure enough sequences are generated and there will be no repetitive numbers in the sequence. Then we generate the sequence of bucket addresses as following

$$
\{B_x[h(s_i)]|B_x[h(s_i)] = (B[h(s_i)] + q_x(s_i))\%m, 1 \leq x \leq r\}
\tag{3}
$$

Example 2. Figure 4 shows an example of subgraph insertion in FPC_S. The inserted subgraph s_1 is mapped to 3 buckets with hash address sequence $B_1[h(s_1)]$, $B_2[h(s_1)]$ and $B_3[h(s_1)]$ via vertical hashing. And then FPC_S searches all the cells in candidate mapped buckets for patterns that are isomorphic to s_1. If no, s_1 is stored in the first empty cell with key-value pair $< p_1, 1, f(s_1), T >$.

Candidate Buckets Sampling. With the vertical hashing technique, we need to check $r \times d$ candidate mapped cells whenever inserting a subgraph, which will be time-consuming. To improve the updating speed while guaranteeing a reasonable collision rate, we propose a bucket sampling technique. That is, instead of considering all the $r \times d$ cells, we only select w candidate mapped buckets as a sample from the bucket address sequence $\{B_x[h(\cdot)]\}$. We only check these w buckets in updating and queries. The method to select these w buckets for a subgraph s_i is also a linear congruence method. We first compute a w length sequence based on Eq. 4, and then choose w buckets with the address.

$$
\{B_{\lfloor \frac{q_x(s_i)}{r} \rfloor \%r}[h(s_i)], 1 \leq x \leq w\}
\tag{4}
$$

Note that, with the hash address sequence, FPC_S may store subgraph patterns with different hash values in the same bucket. It is unnecessary to perform subgraph matching for every pattern stored in the candidate mapped buckets in updating and queries. Thus, we use the fingerprint to efficiently filter a larger amount of undesirable patterns. Given two subgraphs s_1 and s_2, if and only if $f(s_1) = f(s_2)$ they may be isomorphic subgraphs.

6   Experiments

In this section, we report and analyze experimental results. All the algorithms were implemented in C++ and ran on a PC with an Intel i7 3.50GHz CPU and 32GB memory. In all experiments, we use BOB Hash[1] with different initial seeds to implement the hash functions. Every quantitative test was repeated 5 times, and the average was reported.

Datasets. We use three real-life datasets:

- *Twitter* [2] is a social network containing 4.9M entities (e.g., users) and 32M edges (e.g., mentions), which is the largest one of the three datasets.
- *Facebook* [18] contains 415K entities (e.g., users) and 2.1M edges (e.g., message). If user X posts a message on the wall of another user Y, then a directed edge from node X is created to node Y with timestamps to the post's date.
- *Enron* [1] is an email communication network of 86K entities (e.g., ranks of employees), 297K edges (e.g., email), with timestamps to communication data.

Algorithms. We implement and compare three algorithms:

- FPM: Our baseline method via reservoir sampling;
- TopKF: Our advanced algorithm with the auxiliary data structure FPC;
- TopKF_S: Our augmented algorithm uses the auxiliary data structure FPC_S with some optimizations.

Metrics. We use the following five metrics:

- Average Relative Error (ARE): $\frac{1}{\Psi}\sum_{s_i \in \Psi} |fre_i - \tilde{fre}_i|/fre_i$, where fre_i is the real frequency of subgraph s_i, $\tilde{fre}_i$ is its estimated frequency, and Ψ is the query set. In the experiments, we query each actually frequent subgraph once in the auxiliary data structure to search top-k frequent subgraph patterns.
- Recall Rate (RR): The ratio of the number of correctly reported patterns to the number of correct patterns.
- Precision Rate (PR): The ratio of the number of correctly reported patterns to the number of reported patterns.
- F1 Score: $\frac{2 \times RR \times CR}{RR + PR}$, the harmonic average of precision and recall, and it is also a measure of a model's accuracy.
- Throughput: Kilo insertions per second (Kips).

---

[1] http://burtleburtle.net/bob/hash/evahash.html.

Parameter Settings. To evaluate the performance of TopKF_S and its competitors, we first fix $d = 24$, $k = 400$ and vary the value of memory size ranges from 100 MB to 300 MB in 50 MB increments. And then, to evaluate the influence of some key parameters, we fix the memory usage of FPC_S and FPC and vary d from 12 to 48 in 12 increments and vary k from 200 to 800 in 200 increments to see how parameters influence the performance of the auxiliary data structure. For other parameters, we fix the subgraph size $m = 3$, fix the vertical hashing range $r = 6$ and address sample probability $w = 3$. Without otherwise specified, when varying a certain parameter, the values of the other parameters are set to their default values.

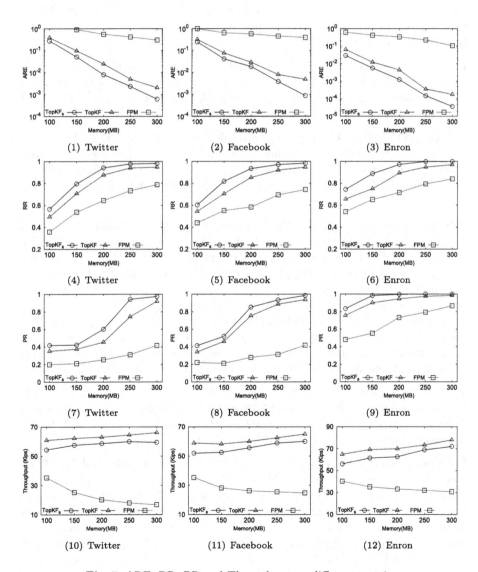

Fig. 5. ARE, RR, PR and Throughput on different metrics

6.1   Experiments on Different Metrics

In this section, we evaluate the performance of $TopKF_S$, TopKF, FPM with varying memory sizes on three real-life datasets. We mainly show the result of ARE, RR, PR and Throughput of the three algorithms with default parameters. Similar results can also be observed under the other parameter settings.

ARE (Fig. 5(1)-(3)): We find that, the ARE of $TopKF_S$ is much lower than all other competitors. Specifically, $TopKF_S$ outperforms TopKF by up to 3.29 times and FPM by up to 492.24 times on *Twitter*. On the other two datasets, the ARE of $TopKF_S$ is around 5.51, and 442.18 times lower than TopKF and FPM, respectively. Moreover, the increase of memory can decrease the ARE of the three algorithms. The ARE of $TopKF_S$ can achieve around 3.7×10^{-5} under the memory of 300 MB on *Enron*.

RR (Fig. 5(4)-(6)): We find that, the RR of $TopKF_S$ is around 1.10, and 1.46 times higher than TopKF and FPM on *Twitter*, respectively. And on the other two datasets, we can find similar results. Specifically, the CR of $TopKF_S$ is around 1.19 and 1.60 times higher than FPC and FPM, respectively.

PR (Fig. 5(7)-(9)): The experiment results show that the PR of $TopKF_S$ is around 1.27, and 3.04 times higher than TopKF and FPM, respectively. And $TopKF_S$ can achieve around 1 under the memory of 150 MB on *Enron*, 250 MB on *Facebook* and 300 MB on *Twitter*, which is smaller than TopKF and FPM achieve.

Effectiveness Analysis. All the experiment results show that TopKF and $TopKF_S$ have higher accuracy than the baseline solution on each dataset. The reason is that FPM needs to sample a large number of subgraphs to guarantee accuracy and store all the sampled subgraphs for estimating the frequency of each pattern exactly. Thus, when memory is not sufficient, FPM achieves a low accuracy. While $TopKF_S$ and TopKF use an auxiliary data structure without storing any sampled m-edge subgraph. It only stores the m-edge patterns and their frequency, we can store it into memory directly. In addition, $TopKF_S$ can further improve the accuracy since FPC_S can provide more candidate mapped buckets with vertical hashing technique to reduce errors caused by replacement.

Throughput (Fig. 5(10)-(12)): The experiment results show that the insertion throughput of TopKF is around 1.13, and 3.57 times higher than $TopKF_S$ and FPM, respectively. Moreover, the insertion throughput of $TopKF_S$ and TopKF increases as the memory increases, while FPM does the opposite.

Efficiency Analysis. The experiment results show that $TopKF_S$ and TopKF are more efficient than FPM, because FPM carries out expensive subgraph matching calculations to estimate the frequency of each subgraph pattern after all updates have occurred at the current timestamp. In contrast, $TopKF_S$ and TopKF use an auxiliary data structure to unbiasedly estimate the frequency of each pattern in real-time which can avoid redundant calculations. In addition, the throughput of TopKF is also higher than $TopKF_S$, because FPC only needs one memory access for each subgraph insertion. While FPC_S keeps a hash address set, leading to an extra check time. In addition, under a small memory, the throughput of FPM is higher because it needs less time to partition the set of subgraphs in $\mathcal{S}$ into T equivalence classes. As memory increases, FPM needs to process more sampled subgraphs. While TopKF_S and TopKF can achieve higher throughput with memory increasing, since they have more buckets to handle the subgraph insertion, greatly reducing the hash collisions.

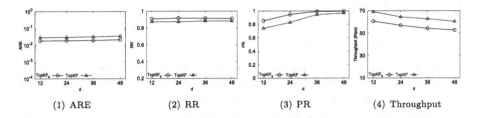

(1) ARE   (2) RR   (3) PR   (4) Throughput

Fig. 6. Performance varying d on *Twitter*

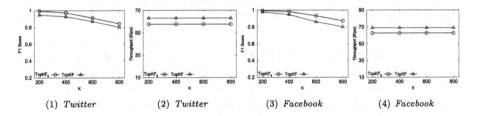

(1) *Twitter*   (2) *Twitter*   (3) *Facebook*   (4) *Facebook*

Fig. 7. Performance varying k on two datasets

6.2 Experiments on Varying Parameters

In this section, we evaluate the performance of TopKF_S, TopKF with varying parameter d on *Facebook* using bounded-size memory, i.e., 200 MB, and then evaluate F1 Score and throughput of TopKF_S, TopKF, FPM with varying parameter k on *Twitter* and *Facebook* with the same memory. Note that, when varying a parameter, we keep other parameters as default. The results on the other datasets are consistent.

Effect of d (Fig. 6(1)-(4)): The experimental results show that the increase of d can increase the recall rate and decrease the throughput of TopKF_S and TopKF. While the ARE and RR are not influenced explicitly. The reason could be that for a larger d, it is more likely to store those patterns that can be top-d patterns in a bucket. Moreover, with the increase of d, the throughput of TopKF_S and TopKF decrease since they need to check more cells in each bucket and execute more subgraph matching calculations for subgraph insertion. Therefore, users can adjust d to strike a good trade-off between precision and speed.

Effect of k (Fig. 7(1)-(4)): Our experimental results show that the F1 Score of TopKF_S and TopKF decrease as the k increases. And there is a minor change in the throughput of TopKF_S and TopKF with the varying k. This is because TopKF_S and TopKF store all the patterns with high frequency in a fixed-size auxiliary data structure, regardless of the value of k. After all the subgraphs at the current moment are inserted, TopKF_S and TopKF can directly search the top-k frequent patterns from the data structure.

7   Related Work

We categorize the related work as follows.

Frequent subgraph Mining. Inokuchi et al. [10] first introduced the problem of Frequent Subgraph Mining (FSM). It uses a generate-and-select method based on anti-monotonicity. However, it generates numerous candidates that are infrequent or do not exist in the database, causing a lot of time overhead [12]. To solve this problem, [20] proposed gSpan algorithm which adopts a pattern-growth approach to recursively grow patterns by scanning the graph database. Though FSM has also been widely studied on a single static graph. None of the proposed approaches [4,5,8,11,13], either exact or approximate, are applicable to streaming graphs.

Top-k Frequent Patterns Mining. While top-k frequent patterns mining adopts a different approach, combining the mining and ranking phases into one. Since they do not need to mining all the patterns, they are more effective than traditional frequent subgraphs mining methods and not require user to set a precise minimum threshold. TGP [14] algorithm was designed to find the top-k closed subgraphs. It uses a structure called Lexicographical pattern net to quickly check if a subgraph is closed. The FS^3 [17] algorithm was proposed to find an approximate set of top-k frequent subgraphs. It invented a novel Markov Chain Monte Carlo sampling which performs a random walk over the p-size subgraphs of the graph. However, this sampling is non-uniform. Later, another algorithm kFSIM [7] was proposed to mine top-k fixed size frequent subgraphs. But it still cannot guarantee the completeness and accuracy of the results.

8   Conclusion

Top-k frequent pattern mining has been wildly used in numerous fields. In this work, we tackle the novel problem of mining top-k frequent patterns continuously in a streaming graph. We propose an auxiliary data structure called FPC to unbiasedly estimate the frequency of the pattern and use a replacement strategy to maintain a much smaller pattern set, which is fast, memory efficient, and accurate. We prove its unbiasedness mathematically and design a new graph invariant that map each subgraph to its sequence space. In addition, we explore an augmented version FPC_S by some other optimization strategies to speed up the frequency estimation process and improve its accuracy. Experimental results show that our algorithms can achieve high accuracy and efficiency with limited memory usage in real-time top-k frequent pattern mining.

Acknowledgement. This work is partially supported by National Natural Science Foundation of China under Grant No. U19B2024,62272469.

References

1. Enron. http://www.cs.cmu.edu/enron/
2. Snap. http://snap.stanford.edu/
3. Aslay, Ç., Nasir, M.A.U., Morales, G.D.F., Gionis, A.: Mining frequent patterns in evolving graphs. In: Proceedings of the 27th ACM International Conference on Information and Knowledge Management, Torino, Italy, pp. 923–932. ACM (2018)
4. Bringmann, B., Nijssen, S.: What is frequent in a single graph? In: Washio, T., Suzuki, E., Ting, K.M., Inokuchi, A. (eds.) PAKDD 2008. LNCS (LNAI), vol. 5012, pp. 858–863. Springer, Heidelberg (2008). https://doi.org/10.1007/978-3-540-68125-0_84
5. Chen, C., Yan, X., Zhu, F., Han, J.: gApprox: mining frequent approximate patterns from a massive network. In: Proceedings of the 7th IEEE International Conference on Data Mining, Omaha, Nebraska, USA. pp. 445–450. IEEE (2007)
6. Chen, Z., Wang, X., Wang, C., Li, J.: Explainable link prediction in knowledge hypergraphs. In: Proceedings of the 31st ACM International Conference on Information & Knowledge Management, Atlanta, GA, USA, pp. 262–271 (2022)
7. Duong, V.T.T., Khan, K., Jeong, B., Lee, Y.: Top-k frequent induced subgraph mining using sampling. In: Proceedings of the Sixth International Conference on Emerging Databases: Technologies, Applications, and Theory, Jeju Island, Republic of Korea, pp. 110–113 (2016)
8. Elseidy, M., Abdelhamid, E., Skiadopoulos, S., Kalnis, P.: GRAMI: frequent subgraph and pattern mining in a single large graph. Proc. VLDB Endow. **7**(7), 517–528 (2014)
9. Hellmann, S., Stadler, C., Lehmann, J., Auer, S.: DBpedia live extraction. In: Meersman, R., Dillon, T., Herrero, P. (eds.) OTM 2009. LNCS, vol. 5871, pp. 1209–1223. Springer, Heidelberg (2009). https://doi.org/10.1007/978-3-642-05151-7_33
10. Inokuchi, A., Washio, T., Motoda, H.: An apriori-based algorithm for mining frequent substructures from graph data. In: Zighed, D.A., Komorowski, J., Żytkow, J. (eds.) PKDD 2000. LNCS (LNAI), vol. 1910, pp. 13–23. Springer, Heidelberg (2000). https://doi.org/10.1007/3-540-45372-5_2
11. Khan, A., Yan, X., Wu, K.: Towards proximity pattern mining in large graphs. In: Proceedings of the ACM SIGMOD International Conference on Management of Data, Indianapolis, Indiana, USA, pp. 867–878. ACM (2010)
12. Kuramochi, M., Karypis, G.: Frequent subgraph discovery. In: Proceedings of the 2001 IEEE International Conference on Data Mining, San Jose, California, USA, pp. 313–320 (2001)
13. Kuramochi, M., Karypis, G.: Finding frequent patterns in a large sparse graph. In: Proceedings of the Fourth SIAM International Conference on Data Mining, Lake Buena Vista, Florida, USA, pp. 345–356. SIAM (2004)

14. Li, Y., Lin, Q., Li, R., Duan, D.: TGP: mining top-k frequent closed graph pattern without minimum support. In: Cao, L., Feng, Y., Zhong, J. (eds.) ADMA 2010. LNCS (LNAI), vol. 6440, pp. 537–548. Springer, Heidelberg (2010). https://doi.org/10.1007/978-3-642-17316-5_51

15. Li, Z., Liu, X., Wang, X., Liu, P., Shen, Y.: TransO: a knowledge-driven representation learning method with ontology information constraints. World Wide Web (WWW) **26**(1), 297–319 (2023). https://doi.org/10.1007/s11280-022-01016-3

16. Nasir, M.A.U., Aslay, Ç., Morales, G.D.F., Riondato, M.: TipTap: approximate mining of frequent k-subgraph patterns in evolving graphs. ACM Trans. Knowl. Discov. Data **15**(3), 1–35 (2021)

17. Saha, T.K., Hasan, M.A.: Fs^3: a sampling based method for top-k frequent subgraph mining. In: 2014 IEEE International Conference on Big Data (IEEE BigData 2014), Washington, DC, USA, pp. 72–79 (2014)

18. Viswanath, B., Mislove, A., Cha, M., Gummadi, P.K.: On the evolution of user interaction in Facebook. In: Proceedings of the 2nd ACM Workshop on Online Social Networks, Barcelona, Spain, pp. 37–42. ACM (2009)

19. Vitter, J.S.: Random sampling with a reservoir. ACM Trans. Math. Softw. **11**(1), 37–57 (1985)

20. Yan, X., Han, J.: gSpan: graph-based substructure pattern mining. In: Proceedings of the 2002 IEEE International Conference on Data Mining, Maebashi City, Japan, pp. 721–724 (2002)

An Efficient Index-Based Method for Skyline Path Query over Temporal Graphs with Labels

Linlin Ding, Gang Zhang, Ji Ma, and Mo Li[✉]

School of Information, Liaoning University, Shenyang, China
limo@lnu.edu.cn

Abstract. Recently, with the proliferation of path query-based applications, extensive research has concentrated on the path query problem. Existing works either focus on the path query over temporal graphs or skyline path query over static graphs, ignoring the temporal information. Meanwhile, edge-labeled temporal graphs are widespread in our daily life, such as traffic networks labeled with "expressway" or "provincial road". In this paper, we define a novel skyline path query over temporal graphs with labels. To handle this problem, we first devise an index named MP-index, divided into two phrases of MP nodes search and Mout set construction. Based on this index, we propose an efficient TMP algorithm to provide the skyline path query over temporal graphs with labels. Finally, extensive experiments show the effectiveness and efficiency of our proposed algorithm.

Keywords: Temporal Graph · Skyline Paths · Query Algorithms

1 Introduction

With the proliferation of path query-based applications, such as searching the shortest path in traffic networks [13,16], discovering temporally connected components [9], research efforts [1,8,21] have been devoted to some key techniques in path queries over temporal graphs. Formally, edges in a temporal graph are always associated with temporal information, that is, a node can connect with another at specific time instances. Consequently, path query over temporal graphs is more complicated than that of static graphs. Yang et al. [20] proposed an algorithm to calculate the optimal cost path under the time-dependent graph. Wu et al. [17] first defined the earliest arrival path, the latest departure path, the fastest path and the shortest path over temporal graphs. Chen et al. [2] used the idea of two-hop labeling to query the earliest arrival path over temporal bipartite graph. However, most existing research focused on path queries with a single factor (time costs, distance, etc.), which cannot solve the problem of multiple factors paths over temporal graphs.

Skyline path query extends skyline query to solve the problem of path query with multi-factors. Tian et al. [14] suggested two pruning strategies for finding

© The Author(s), under exclusive license to Springer Nature Switzerland AG 2023
X. Wang et al. (Eds.): DASFAA 2023, LNCS 13945, pp. 217–233, 2023.
https://doi.org/10.1007/978-3-031-30675-4_15

skyline paths on road networks. Gong et al. [5] proposed a hierarchical index and skyline path query in a multi-constrained road network. Shekelyan et al. [12] introduced the multiple standard linear skyline path query, and designed the algorithm to calculate the linear skyline path.

However, the existing skyline path queries ignore two main factors: (i) temporal information and (ii) label elements, which lead to a huge gap between academic research and real application and cannot satisfy users' diversified query intentions.

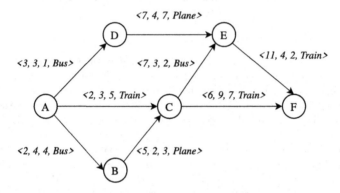

Fig. 1. The traffic network.

Example 1. Figure 1 shows a traffic network, and the 4-tuple $<start\ time,\ time\ costs,\ distance,\ type>$ over an edge represents the start time, time costs, physical distance, and the type this edge. Assuming a user wants to travel from A to E, the user not only wants to arrive smoothly at the least cost of time and distance, but wants to take a plane during the trip. There are three routes from $A \rightarrow E$. If we consider the time conflict in temporal graph, the route $A \rightarrow B \rightarrow C \rightarrow E$ are excluded, as users from A to B cannot catch up with flights to C. In the remaining two routes, although the route $A \rightarrow C \rightarrow E$ dominates the route $A \rightarrow D \rightarrow E$, it does not meet users' needs to take a plane. So the final result should start from A to E via D.

The above example requires a new type of path query, skyline path query over edge-labeled temporal graphs, which provides a set of non-dominance skyline paths from two given nodes with a specific label. The most intuitive way is to adopt the skyline path query method at every time instance to find the skyline paths in the entire time interval. In that way, we need to recompute the skyline paths at every time instant, leading to substantial re-computational costs. Similarly, existing path query methods over temporal graphs can only meet single-factor query intention, but cannot solve the skyline path query.

Therefore, in this paper, we first present a novel skyline path query over edge-labeled temporal graphs. To solve this problem, we propose an index called Main-Point(MP), which includes MP discovering, and Mout set construction. Based on this index and pruning strategy, we propose a path query algorithm called TMP. Moreover, this index also supports the dynamic update of the temporal graphs. Finally, extensive experiments over real datasets have demonstrated the efficiency and effectiveness of our proposed algorithm.

In general, the contributions of this paper are as follows:

- According to many practical applications in real life, we propose a novel problem of skyline path query over temporal graph with labels.
- We propose a MainPoint(MP) index, and design a TMP query algorithm based on MP and pruning strategies to solve the skyline path query over temporal graph with labels.
- Finally, the experimental results show that our method has better query performance and shorter query time.

The rest of the paper is organized as follows. We introduce related works in Sect. 2. In Sect. 3, we formally define the skyline path query. Then we propose MP index and TMP algorithm in Sects. 4 and 5, respectively. We show the experimental results in Sect. 6, and conclude the paper in Sect. 7.

2 Related Work

Temporal Path. Wu et al. [17] proposed the reachability query algorithm to judge whether two nodes on the temporal graph are reachable. Wang et al. [16] proposed TD-G-tree index to support shortest path queries under time dependence. To find the optimal departure time with the least travel time, Ding et al. [4] proposed a path query algorithm based on classical Dijkstra. Yuan et al. [21] studied the constrained shortest path query on the time dependent graph, but its weight on the edge of the time period does not change.

Although time factor is considered in these path queries, most researches only focus on path queries with single constraint (e.g. fastest path, earliest arrival path), which still cannot solve the path query problem with multiple constraints.

Skyline Path. Tian et al. [14] firstly extended skyline query to skyline path query on road networks. Kriegel et al. [7] focused on the skyline on paths in a road network with evolving preferences, such as distance, diving time, and gas consumption. Yang et al. [18] focused on constructing a multi-cost, time-dependent, uncertain graph model of a road network based on GPS data from vehicles that traversed the road network.

Skyline path queries solve the problem of multi-constraint path queries, but they mainly concentrate on static graphs without considering the temporal information. Additionally, they ignore the impact of edge labels.

Label Path. Rice et al. [11] proposed a hierarchical graph index technology with reduced hierarchy to solve shortest path queries restricting the type of roads or modes of travel. Hassan et al. [6] studied the dynamic graph with labeled edges and introduced an Edge-Disjoint Partitioning method to solve the shortest path query with edge label constraints. Valstar et al. [15] defined the reachability query problem with label constraints, and solves the reachability problem with label constraints by searching some landmarks to establish an index.

Like temporal paths, label paths are also with single constraints. Most of the studies ignore the path problem with multiple constraints. Therefore, we define the problem of skyline path query over temporal graphs with labels. As far as we know, we are the first to propose this problem.

3   Problem Definition

Let $G = (V, E)$ be a labeled temporal graph, where V represents the set of nodes, and E represents the set of edges. An edge $e \in E$ is a 6-tuple (u, v, s, λ, d, l), where $u, v \in V$, s is the starting time, λ is the traversal time go from u to v starting at time s and ending at $s + \lambda$, d is the cost from u to v, and l is the label of each edge.

A temporal path P_{v_s, v_t} from v_s to v_t can be represented as a sequence of edges $P_{v_s, v_t} = < e_1, e_2, \cdots, e_n >$, where $e_i = (v_i, v_{i+1}, s_i, \lambda_i, d_i, l)$ for each $i \in [1, n]$. Notice that there is a constraint between adjacent edges, that is $s_i + \lambda_i \leq s_{i+1}$, because we cannot go from b to c before the previous path $a \rightarrow b$ arrived.

The starting and ending time of P_{v_s, v_t} is s_1 and $s_n + \lambda_n$, respectively. So the time cost is $\lambda_{P_{v_s, v_t}} = s_n + \lambda_n - s_1$, the distance is $d_{P_{v_s, v_t}} = \sum_{i=1}^{n} d_i$, and the label set is the union of sub-path labels, i.e. $l_{p_{u, v}} = \bigcup_{i=1}^{n} l_i$. Notice that, we focus on directed temporal graphs, since undirected graphs can be easily transformed through two bi-directed edges.

Then we can define the path domination relation between temporal paths.

Definition 1 (Path domination). *Given two paths P_{v_s, v_t} and P'_{v_s, v_t} with different time costs, distances and labels, P_{v_s, v_t} dominate another path P'_{v_s, v_t} iff $\lambda < \lambda'$, $d \leq d'$ (or $\lambda \leq \lambda'$, $d < d'$), and $l \supseteq l'$, denoted as $P'_{v_s, v_t} \ll P_{v_s, v_t}$.*

Definition 2 (Skyline Path Query). *Given two nodes u and v in temporal graph G and label set L, the skyline path query aims to find a set of paths $P_{set}(u, v)$, such that $\forall (u, v, t_s, C) \in P_{set}(u, v)$ is a skyline path that cannot dominated by any other path from u at t_s to v with label L, where $C = (\lambda, d, l)$ represents that the total time cost by node u arriving at v is λ, the sum of distances is d, and the label constraint set is $l \supseteq L$.*

Example 2. As shown in Fig. 1. The path $P_{A \rightarrow B \rightarrow C}$ is not a temporal path because it violate the temporal order. In contrast, $P_{A \rightarrow C \rightarrow E}$ is a temporal path where the time cost is $7 + 3 - 2 = 8$, distance cost is $5 + 2 = 7$ and label set is $\{Train\} \cup \{Bus\} = \{Train, Bus\}$. In the set of temporal paths from A to E, $P_{A \rightarrow D \rightarrow E}$ and $P_{A \rightarrow C \rightarrow E}$ don't dominate each other, they are both skyline paths.

4   MP-Index Construction

4.1   Main Points Discovering

As we know, the degree of a node measures the total number of edge connections, representing its importance in the entire graph [3, 15, 19]. For example, suppose we travel from one city to another in the railway transportation network. We will pass through some relatively large central cities with many railway connections. Inspired by these observations, we design an index based on the nodes with a large degree and record the connections with it.

In the beginning, we will find the nodes with a relatively large degree, and define them as the main points. The main idea in finding these main points is to greedy remove the boundary nodes until the entire graph cannot be connected through the main points, where boundary nodes are the nodes with a degree of 1. Notice that, we conduct the main points discovering over the sum of all temporal edges.

Specifically, we first sort the nodes in descending order of their degrees and initialize them as a candidate set. We also define the node with the current largest degree as a judge node and put it in the Main Point Set (MPSet). Then in each iteration, we delete the current judge node and update the degree of its neighbors. The neighbors who become boundary nodes after the update will also be deleted from the candidate set during this period. When the candidate set is empty, we will obtain the main point set, called MPSet.

Example 3. As shown in Fig. 1, we get the candidate set $S = \{C, A, E, B, D, F\}$ after sorting by degree. In the 1-iteration, we will find the node with the current largest degree, i.e., C, and regard it as part of MPSet. And the degree of its neighbors will be minus 1. As a consequence, B and F will be deleted due to their degrees becoming 1. In the 2-iteration, the judge node becomes A and will be added to MPSet. After that, we remove its neighbors D will from S. In the 3-iteration, the remaining candidate node E will be regarded as part of MPSet. Finally, we get the MPset=$\{C, A, E\}$.

Algorithm 1: findMP Algorithm

Input: A temporal graph G
Output: MPset
1 MPset= $\emptyset$, sortorder= $\emptyset$
2 sortorder $\leftarrow$ sort nodes $v \in G$ in descending order according to $degree(v)$
3 **while** $sortorder \neq \emptyset$ **do**
4   MPset $\leftarrow$ MPset $\cup\, v$ // v is the first node in sortorder
5   Remove node v from sortorder
6   **for** $w \in pre(v) \cup suc(v)$ **do**
7           $degree(w) \leftarrow degree(w) - 1$
8   **while** $degree(w) = 1$ **do**
9   Remove node w from sortorder
10 **return** MPset;

The pseudo-code of findMP algorithm is shown as Algorithm 1. Given the temporal graph G, the findMP algorithm aims to find the node set of main points, i.e. MPSet. At first, we initialize the MPSet and sort the nodes in descending order according to their degree (Lines 1-2). Then add the node with large degree to MPset each time, and delete its adjacent edges and boundary nodes until the sortorder set is empty (Lines 3-9). Finally, the MPset is returned (Line 10).

Time Complexity. The time complexity of Algorithm 1 is $O(N * d_{max})$, where N is the number of nodes and d_{max} is the maximum degree of the node.

4.2 Mout Set Recording

After finding the main point nodes, we want to construct an efficient index based on them. In this subsection, we first introduce the concept of $Mout$ set to record the consequential nodes of main point nodes.

Definition 3 (*Mout*). *Mout* (v) *represents the set of other MP nodes that v can reach. For each 4-tuple $(w, t_s, t_e, C) \in Mout(v)$, it indicates node v departs at t_s and arrives at MP w at t_e, $C = (\lambda_{v,w}, d_{v,w}, l_{v,w})$ represents that the total time cost by node v arriving at w is $\lambda_{v,w}$, the sum of distances is $d_{v,w}$, and the label constraint set is $l_{v,w}$.*

$MPset$ and $Mout$ construct are the core of the MP-index, where $MPset$ shows the main point nodes and $Mout$ shows the eligible connections between main point nodes. Since $MPset$ has been found in the previous subsection, our task is to find the $Mout$. The main idea of finding $Mout(v)$ is to traverse MP nodes in reverse topological order, and record eligible connections with 4-tuple between v and w into $Mout(v)$ in each iteration.

Specifically, we first topologically sort each node in temporal graphs according to Formula (1), where $d_{in}(v)$ denotes the in-degree of node v, and $pre(v)$ represents incoming neighbor nodes of node v. Next, we traverse each MP node in reverse topological order to record the eligible connections over the sum of all temporal edges. In each iteration, we conduct an improved breadth-first-search (BFS) algorithm to traverse the out-edges of a particular MP node and record connections with eligible 4-tuple between MP nodes. The reason to utilize reverse topological order is to reduce colossal re-computational costs. In detail, the nodes with small topology numbers no longer need to repeatedly traverse those with a larger one, because these nodes have been traversed in previous visits. Moreover, the eligible connections recorded by each MP node will be transmitted to their upper MP node.

$$T(v) \begin{cases} 0 & d_{in}(v) = 0 \\ max_{u \in pre(v)}(T(u) + 1) \ otherwise \end{cases} \tag{1}$$

In addition, there will still be some redundant paths during the construction process. According to Definition 1, we have the following pruning strategy when building the index.

When we traverse from a particular MP node v, we suppose that there are two paths $P_{v,w}$ and $P'_{v,w}$ between them, and $P_{v,w} \ll P'_{v,w}$. Only the 4-tuple of $P'_{v,w}$ will be transmit to $Mout(u)$, and it is unnecessary to transmitted the 4-tuple of $P_{v,w}$, because their domination relationship.

Example 4 As shown in Fig. 1, we assume that node F and C are MP nodes. When we visted node C, since $P_{C \to F} \ll P_{C \to E \to F}$, we pass $\forall (V, t_s, t_e, C') \in Mout(F)$ to $Mout(C)$ through path $P_{C \to E \to F}$ where $C'.\lambda \geq 15$.

Algorithm 2: Mout Construction

Input: A temporal graph G, MPset
Output: $Mout$

1 MPset $\leftarrow$ sort node $u \in$ MPset in descending order of topology number
2 **for** $u \in MPset$ **do**
3 $\quad$ $Mout(u) \leftarrow \emptyset$
4 $\quad$ $Q \leftarrow \emptyset$, $Q.push(v)$ // $v \in suc(u)$
5 $\quad$ **while** Q *is not empty* **do**
6 $\quad\quad$ $w \leftarrow Q.pop()$
7 $\quad\quad$ **if** $w \in MPset$ *and* $dis(u,w) \leq k$ **then**
8 $\quad\quad\quad$ $C_{u,w} = (\lambda, d, l)$;
9 $\quad\quad\quad$ $Mout(u) \leftarrow (w, t_s.t_e, C_{u,w})$
10 $\quad\quad\quad$ **if** $(w, t'_s, t'_e, C'_{u,w}) \ll (w, t_s, t_e, C_{u,w})$ **then**
11 $\quad\quad\quad\quad$ Remove $(w, t'_s, t'_e, C'_{u,w})$ from $Mout(u)$;
12 $\quad\quad\quad\quad$ **for** $(z, t''_s, t''_e, C_{w,z}) \in Mout(w)$ **do**
13 $\quad\quad\quad\quad\quad$ **if** $t''_s \geq max\{t_e, t'_e\}$ *and* $dis(u,z) \leq k$ **then**
14 $\quad\quad\quad\quad\quad\quad$ $Mout(u) \leftarrow (z, t_s, t''_e, C_{u,z})$
15 $\quad\quad$ **if** $w \notin MPset$, $z \in suc(w)$, $t_e(u,w) \leq t_s(w,z)$, $dis(u,z) \leq k$ **then**
16 $\quad\quad\quad$ $Q.push(z)$
17 $\quad$ **return** $Mout(u)$
18 **return** $Mout$;

Meanwhile, in order to make our index suitable for large graphs, we introduce a constant k to limit the size of index construction, where k means that MP nodes can reach within k hops. The value of k is related to the size of the temporal graph and the number of topological layers. We limit the size of $Mout(v)$ to all MP node elements obtained by traversing k steps starting from v.

Algorithm 2 describes the process of building $Mout$ sets. We first sort the MP nodes according to their topology number (Line 1). For each MP node u, we construct a $Mout(u)$ for it (Line 3). Then we initialize an empty queue Q, and add out-neighbors of u to the queue (Line 4). Subsequently, we judge whether the queue Q is empty, if not, we will conduct the iteration (Lines 5-16). In the t-th trial, for nodes that are MP nodes and can be reached in k steps, it will be added into $Mout(u)$ (Lines 7-9). If two 4-tuple in the set are completely dominated, then the dominated elements are removed from $Mout(u)$, and the eligible elements are added to $Mout(u)$ according to pruning strategy (Lines 10-14). If the node is not MP, then the subsequent neighbor node that conforms to the temporal relationship will be added to the queue (Lines 15-16).

In addition, considering the temporal graph are evolve over time, our index also supports for the graph updates. There are two cases for a newly added node, i.e., v, according to the node v connected. Case (i): if the node that v connects to is an MP node, the index will not be changed, because MPset and $Mout$ set remain unchanged. Case (ii): if the node that v connects to is a non-MP node, i.e., u, where v is also a boundary node. As a consequence, u becomes an MP node, and the corresponding MPset and $Mout(z)$ will also be modified, where z is the MP node that can reach u within k steps.

Time Complexity. The time complexity of Algorithm 2 is $O(N_{mp} * k * d_{max})$ in the worst case, where N_{mp} represents the number of MP and $k * d_{max}$ means the maximum number of elements in the Mout set, which is actually a constant.

5   Skyline Path Query Based on MP Index

5.1   Bidirectional Topology

Before querying, we first consider the problem that the two nodes given by the user are unreachable. For MP index, this is the worst case of query. We need to return results after traversing all nodes between nodes, which will reduce the efficiency index. Therefore, before querying, we introduce a bidirectional topology to judge some obviously unreachable nodes. If the two nodes are unreachable, the query results will be returned directly.

Assuming the given two nodes are u and v, we utilize two variables to judge whether they are reachable quickly. The first variable is the forward topological sorting value in Formula (1). Furthermore, the second variable is the reverse topological sorting value given in Formula (2), where D_{max} represents the maximum path length in the temporal graph. Based on these two variables, we have the Theorem 1 to judge whether two nodes are reachable.

$$RT(v) \begin{cases} D_{max} & d_{out}(v) = 0 \\ min_{u \in suc(v)}(RT(u) - 1) & otherwise \end{cases} \tag{2}$$

Theorem 1. *In the temporal graph, for the given two nodes u and v, if $T(u) \geq T(v)$ or $RT(u) \geq RT(v)$, then u to v must not be reachable in the temporal graph. That is, there is no path between u and v.*

Proof. Suppose that u can reach v when $T(u) \geq T(v)$. Then there is at least one path from u to v. Therefore, in topological sorting, the access order of v must be after u, that is, $T(u) < T(v)$. This contradicts the assumption. The proof of $RT(u)$ is similar.

5.2   Query Algorithm Based on MP-Index

Since the MP index only stores some main point nodes and their connection information, we can easily tackle the skyline path query between these main point nodes in linear time. However, the query node may not be the main point node in practical application scenarios. So we have the following theorem to guarantee the availability of the skyline path query between non-MP nodes.

Theorem 2. *The skyline path query of any non-MP node can be converted to skyline path query of MP node.*

Proof. According to Theorem 1, at least one of the two nodes on the edge is an MP node. So if a node is a non-MP node, its adjacent node must be an MP node. Therefore, the skyline path query of a non-MP node can be converted to the skyline path query of the adjacent MP node.

For a given starting node u and target node v, the given nodes of query are divided into the following three cases. (1) Both u and v are MP. (2) One of u and v is MP. (3) u and v are not MP. We will give the corresponding solutions in each case.

Due to the limited size of the set, we can get the results directly from $Mout$ that can be reached in k steps. When the path is exceed k steps, we need to do further analysis. For query nodes u and v, we first record the $td = T(v) - T(u)$ to represent the topological distance between two nodes, and judge when the two nodes meet. Then we can tackle the case (1) according to Theorem 2.

Algorithm 3: TMP skyline path query

Input: TMP(G, u, v, td, L)
Output: $P_{set}(u, v)$

1 $P_{set}(u, v) \leftarrow \emptyset$;
2 **if** $T(u) \geq T(v)$ *or* $RT(u) \geq RT(v)$ **then**
3 $\quad$ **return** $P_{set}(u, v)$

4 **if** $u, v \in MPset$ **then**
5 $\quad$ **return** based-query(TMP(G, u, v, td, L))

6 **if** $u \in MPset$ **then**
7 $\quad$ **for** $w \in pre(v)$ **do**
8 $\quad\quad$ $td = td - 1$, $P_{set}(u, w) \leftarrow$ based-query(TMP(G, u, w, td, L))
9 $\quad\quad$ **if** $(w, t_s, t_e, C) \in P_{set}(u, w)$ *and* $t_e \leq t_s(w, v)$ **then**
10 $\quad\quad\quad$ $P_{set}(u, v) = P_{set}(u, v) \cup (v, t_s, t_e(w, v), C_{u,v})$, **return** $P_{set}(u, v)$

11 **if** $v \in MPset$ **then**
12 $\quad$ **for** $w \in suc(u)$ **do**
13 $\quad\quad$ $td = td - 1$, $P_{set}(w, v) \leftarrow$ based-query(TMP(G, w, v, td, L))
14 $\quad\quad$ **if** $(v, t_s, t_e, C) \in P_{set}(w, v)$ *and* $t_s \geq t_e(u, w)$ **then**
15 $\quad\quad\quad$ $P_{set}(u, v) = P_{set}(u, v) \cup (v, t_s(u, w), t_e, C_{u,v})$, **return** $P_{set}(u, v)$

For case (2) and case (3), it is more complicated. For case (2), we convert u or v to adjacent outgoing or incoming nodes, and then the topological distance will be updated as $td = td - 1$. For case (3), we convert u and v at the same time, convert u to an adjacent outgoing node, and convert v to an adjacent incoming node, and then $td = td - 2$.

We then compare td with k. If $td \leq k$, we search the 4-tuple in $Mout(v)$ and add the qualified temporal path to the final result set, and terminate the process. Else if $td > k$, it indicates that the information between two nodes is not completely saved in the existing $Mout$ set, we need to further analysis this situation. First, we traverse the $Mout(w)$ where w is the k-th hop nodes in the $Mout(u)$. If there are corresponding query nodes and conform to the temporal relationship, we will accumulate the distance and time, add the union set of the label set to the final result set. Note that, we do not need to traverse all the elements in $Mout(w)$, according to the path dominance relationship and the k-hop range. We also design the following two pruning strategies for unqualified elements to speed up the query.

Algorithm 4: TMP skyline path query (Continued)

---

1 **for** $w \in suc(u)$ and $z \in pre(v))$ **do**
2    $td = td - 2$, $P_{set}(w, z) \leftarrow$ based-query(TMP(G, w, z, td, L))
3   **if** $(z, t_s, t_e, C) \in P_{set}(w, z)$ and $t_s \geq t_e(u, w)$,$t_e \leq t_s(z, v)$ **then**
4       $P_{set}(u, v) = P_{set}(u, v) \cup (v, t_s(u, w), t_e(z, v), C_{u,v})$, **return** $P_{set}(u, v)$

5 **Function** *based-query(TMP(G, u, v, td, L))*:
6   **if** $\exists (v, ts, te, C) \in Mout(u)$ and $C.l \subseteq L$ **then**
7       $P_{set}(u, v) \leftarrow (v, t_s, t_e, C)$;

8   **if** $td \leq k$ **then**
9   **return** $P_{set}(u, v)$;

10   **for** $(w, t_s, t_e, C) \in Mout(u)$ and $dis(u, w) = k$ **do**
11   **if** $(w, t_s, t_e, C) \ll (v, t_s^{'}, t_e^{'}, C^{'}) \in P_{set}(u, v)$ **then**
12   **continue**
13   based-query(TMP$(G, w, v, td - k, L)$)

14   **return** $P_{set}(u, v)$

The Element Pruning Strategy within the td Hop. When we visit the k-th hop element w in $Mout(v)$, the limit range changes from td hop to $td - k$ hop. If the hop number in $Mout(w)$ is larger than $td - k$, it is no longer traversed.

Candidate Path Pruning Strategy. If a path can be dominated by another in the candidate set, the subsequent traversal will not continue from it, because the path starting from this element must not be the optimal path.

Algorithm 3 describes the TMP algorithm. Give two nodes u and v, the TMP algorithm returns the skyline temporal path from them. We first initialize $P_{set}(u, v)$ to record the result (Line 1). Then, we judge whether the two nodes are reachable according to the bidirectional topology (Lines 2-3). If u and v are both MP nodes, function based-query is called (Lines 4-5). If only one of the two nodes is an MP node, we convert it to the outgoing neighbor MP node or incoming neighbor MP node for calculation, and remove the ones that do not meet the temporal relationship to obtain the result set (Lines 6-15). If both nodes are not MP nodes, we convert the nodes simultaneously to the outgoing neighbor MP node and the incoming neighbor MP node, and then call function based-query to calculate, and finally get the result set (Lines 16-19). The function of based-query gives the first case of query, where both query nodes are MP. If there are eligible elements and the number of hops is within k steps, the result is directly obtained (Lines 21-24). When the number of hops is greater than step k, first prune the k-th hop element according to the pruning strategy, and then call based-query repeatedly to get the result (Lines 25-28).

Time Complexity. The worst case is when the query nodes u and v are not MP nodes. In that case, we need to visit all outgoing neighbor nodes of u and all incoming data nodes of v. The time complexity is $O(d_{in} * d_{out})$, where d_{in} and d_{out} represent the number of incoming neighbor nodes and outgoing neighbor nodes, respectively.

Correctness. The TMP algorithm is based on the MP-index, which records the main point nodes and the crucial connections with their eligible 4-tuple information. The TMP algorithm considers all the situations of the query nodes, that is, both MP nodes or non-MP nodes, or one is an MP node and another not. We thoroughly consider how to obtain the corresponding skyline path between the two given nodes in different situations. Through the above steps, the integrity of the query algorithm and the correctness of the query results are ensured.

6   Experiment

6.1   Experimental Settings

Datasets. We select 10 datasets to test our proposed algorithm, which are Bitcoin, DBLP, epinions, mathoverflow, wiki, Digg, collegeMsg, Elec, Askubuntu, and Facebook. Among them, mathoverflow, collegeMsg and Askubuntu datasets are from SNAP[1], remaining datasets from KONECT[2]. The edges of these datasets contain time elements, but no label elements. So we modified datasets on this basis, and randomly added label elements at the edge to meet our algorithm.

The summary of datasets is shown in Table 1. $|V|$ and $|E|$ denote the number of nodes and edges of temporal graph, respectively. d_{avg} and d_{max} denote the average and maximum degrees of nodes in temporal graph, respectively.

---

[1] http://snap.stanford.edu/data/.

[2] http://konect.cc/.

Algorithms. Our experiments are conducted against the following designs:

- TD algorithm: Because there is no algorithm exactly the same as the problem studied in this paper, we improved the Dijkstra [10] algorithm and designed the TD algorithm based on the temporal label graph.

Table 1. Datasets

Dataset	$|V|$	$|E|$	d_{avg}	d_{max}
collegeMsg	1,899	6,452	6.79	564
wiki	2,356	6,432	5.46	320
Bitcoin	5,881	35,592	12.1	1,298
Elec	7,118	103,675	29.1	1,167
DBLP	12,590	49,759	7.90	714
Digg	30,398	87,627	5.76	310
Facebook	47,985	257,576	10.73	11,250
mathoverflow	88,580	506,550	11.43	14,320
epinions	131,828	841,372	12.76	3,622
Askubuntu	560,180	4193,430	14.97	91,751

- TL-NODE: NODE [5] is an index-based algorithm to solve the multiple constrained skyline path in static graphs, ignoring the temporal information and label factors. We adjust it and design a TL-NODE algorithm to solve our defined problem.

Table 2. Index construction time (ms) and index size (MB)

Dataset	MP		Topo		TMP	
	Time	Size	Time	Size	Time	Size
collegeMsg	377.11	5.0	5.21	0.5	382.32	5.5
wiki	524.75	6.2	10.76	1.2	535.51	7.4
Bitcoin	753.21	17.4	50.14	2.1	803.35	19.5
Elec	2,472.57	54.1	123.40	3.4	2,595.97	57.5
DBLP	1,730.73	36.7	60.24	6.8	1,790.97	43.5
Digg	2,247.28	67.2	156.10	12.5	2,403.38	79.7
Facebook	4,874.47	102.3	322.45	24.6	5,196.92	126.9
mathoverflow	3,861.25	348.5	495.12	38.5	4,356.37	387
epinions	5,562.66	587.2	562.32	54.6	6,124.98	641.8
Askubuntu	105,888.85	1,003.4	1,234.56	87.6	107,123.41	1,091

6.2   Index Evaluation

Index Construction Time. We first evaluate the index construction time of the two phrases of our proposed algorithm by setting the parameter k to 20. For simplicity, MP is short for the Main Points Index, Topo denotes the bidirectional topology, and TMP includes MP index and bidirectional topology.

In Table 2, we can observe that the construction time of MP index is far more than that of Topo comparing the second column and the fourth column. And the total index construction time (i.e. TMP) on the collegeMsg is less than one second, and on the Askubuntu is less than 110 s.

We also compare the index construction time of our proposed TMP algorithm with the modified TL-NODE algorithm. As shown in Fig. 2 (a), the construction time of MP and TMP is significantly better than TL-NODE. The reason is that the TL-NODE algorithm will increase the construction costs after adding temporal factors, and the index needs to be incremental recomputed. Besides, TL-NODE does not have a pruning strategy to reduce redundant computations. Even in some smaller graphs, the construction time of TMP is nearly 10 times faster than that of TL-NODE. In larger graphs, the construction time of TMP is also significantly better than that of TL-NODE.

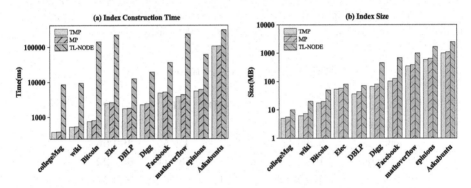

Fig. 2. Index construction time and size

Index Size. We first compare the index size of the two phrases of our proposed algorithm. Table 2 shows the sizes of MP and TMP in different datasets under the default parameter $k = 20$. We can observe that the sizes of MP and TMP are between 5MB and 1.1GB through all the selected datasets. Then, we compare our proposed algorithm with the TL-NODE algorithm. As shown in Fig. 2 (b), the TL-NODE index size is between 10 MB and 2.5 GB among all selected datasets. We can also observe that the index size of our proposed algorithm is smaller than that of TL-NODE. Moreover, in some datasets, such as Digg and Facebook, TMP is more than 5 times smaller than TL-NODE.

6.3 Query Time Evaluation and k Value Influence

Query Time Evaluation. We randomly set 100 query nodes from datasets. The ratio of reachable nodes to unreachable nodes is 1:1. Table 3 shows the average query time of the four algorithms on datasets. It can be seen that TD algorithm performs the worst. Especially for large graphs, its query time will increase exponentially. TL-NODE is a hierarchical index structure algorithm. We can see in collegeMsg and Bitcoin datasets, the query time of TL-NODE is less than that of MP algorithm, but with the increase of data volume, the advantages are no longer obvious. TMP adds a bidirectional topology before establishing the MP index. Queries between unreachable nodes are filtered out, so the perform is better than MP.

k **Value Influence.** In building the TMP index, we limit the number of traversal steps to handle large-scale graphs. As a result, the different steps may affect the index construction time and query time. Therefore, we conduct experiments to evaluate the influence of different k values. We set different values for k as 10, 20, 30, and 40, respectively. Table 4 shows the TMP index's construction time and average query time when k takes different values. From the table, we can see that the time of TMP index construction increases with the increase of k value, because the number of layers determines the scale of traversal. At the same time, as the value of k increases, the query time decreases. This is because more path information is stored in the index, which can cover more queries.

Table 3. Average query time (ms)

Dataset	TD	TL-NODE	MP	TMP
collegeMsg	4,023.76	264.15	351.61	**280.15**
wiki	6,206.45	785.23	456.55	**300.66**
Bitcoin	10,235.20	621.36	786.11	**651.22**
Elec	120,034.45	2,340.54	1,236.16	**560.37**
DBLP	835,610.36	3,011.54	865.45	**756.47**
Digg	–	6,874.21	1,357.52	**566.68**
Facebook	–	10,463.26	6,147.56	**1,235.17**
mathoverflow	–	21,354.33	6,248.66	**1,635.25**
epinions	–	50,214.21	10,521.14	**6,215.33**
Askubuntu	–	82,457.66	23,556.35	**12,548.64**

Table 4. The influence of k value

Dataset	TMP Construct Time (ms)				Average Query Time (ms)			
	k = 10	k = 20	k = 30	k = 40	k = 10	k = 20	k = 30	k = 40
collegeMsg	210.34	382.32	476.55	523.21	60.54	25.15	20.00	14.33
wiki	323.15	535.51	756.60	875.35	82.16	30.66	22.30	18.98
Bitcoin	652.10	803.35	924.00	1,025.30	267.54	100.35	89.64	58.75
Elec	987.20	2,595.97	3,412.33	4,687.00	945.55	560.37	422.14	400.66
DBLP	795.31	1,790.97	2,234.14	4,221.98	758.66	356.47	287.45	200.00
Digg	1,244.00	2,403.38	3,287.00	5,978.45	943.78	566.68	420.04	350.78
Facebook	2,446.37	5,196.92	6,678.75	8,796.00	2,653.00	1,235.17	852.77	504.00
mathoverflow	1,958.00	4,356.37	5,673.11	7,446.71	3,574.24	1,635.25	1,021.45	884.22
epinions	3,566.36	6,124.98	9,874.24	11,234.22	15,243.00	6,215.33	5,567.58	4,324.54
Askubuntu	69,860.21	107,123.41	184,655.21	243,654.67	35,448.30	12,548.64	10,004.50	8,798.66

6.4 Influence of the Number of Labels

The number of query labels may impact the query results, so we conduct experiments to evaluate the influence of the different numbers of labels. We use the datasets of collegeMsg, Elec, Facebook, and Askubuntu to verify the impact of the number of labels on the query time. We set the label quantity values to 2, 4, 6, 8, and 10, respectively, and the ratio of reachable to unreachable is 1:1. As shown in Fig. 3, as the number of labels increases, the query time of all algorithms

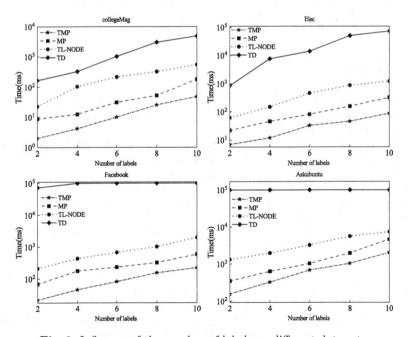

Fig. 3. Influence of the number of labels on different data sets

will increase accordingly. Because when the number of labels increases, the paths that meet the query conditions will increase. Similarly, we can see that the TD algorithm consumed much time (in Facebook and Askubuntu datasets, if the query time exceeds 100s, it is expressed as 100s). Moreover, the TL-NODE algorithm uses a hierarchical structure based on backbone and clustering, increasing efficiency in judging skyline paths with labels. However, redundant labels are not pruned when building the index, and the overall query time is higher than MP and TMP. The query time of MP is slightly longer than that of TMP because TMP judges some unreachable nodes and reduces the query time.

7 Conclusion

In this paper, we define a novel skyline path query over edge-labeled temporal graphs. To Solve this problem, we first design an index called MP, consisting of MP discovering and Mout set construction. Moreover, we propose a TMP algorithm based on MP-index and bidirectional topology strategy to solve the skyline path query over temporal graphs with labels. Finally, through experiments on real datasets and comparison with other algorithms, the experimental results show that our algorithm performs well.

Acknowledgements. This study was funded by the National Natural Science Foundation of China (Nos.62072220, 61502215); National Key Research and Development Program of China (No.2022YFC3004603); Natural Science Foundation of Liaoning Province (Nos.2022-KF-13-06, 2022-BS-111, LJKFZ20220174); Central Government Guides Local Science and Technology Development Foundation Project of Liaoning Province (No.2022JH6/100100032).

References

1. Cai, T., et al.: Incremental graph computation: anchored vertex tracking in dynamic social networks. In: TKDE, pp. 1–14 (2022). https://doi.org/10.1109/TKDE.2022.3199494
2. Chen, X., Wang, K., Lin, X., Zhang, W., Qin, L., Zhang, Y.: Efficiently answering reachability and path queries on temporal bipartite graphs. In: PVLDB (2021)
3. Cheng, J., Shang, Z., Cheng, H., Wang, H., Yu, J.X.: Efficient processing of k-hop reachability queries. VLDB J. **23**(2), 227–252 (2014)
4. Ding, B., Yu, J.X., Qin, L.: Finding time-dependent shortest paths over large graphs. In: EDBT, pp. 205–216 (2008)
5. Gong, Q., Cao, H.: Backbone index to support skyline path queries over multi-cost road networks. In: EDBT, pp. 2–325 (2022)
6. Hassan, M.S., Aref, W.G., Aly, A.M.: Graph indexing for shortest-path finding over dynamic sub-graphs. In: SIGMOD, pp. 1183–1197 (2016)
7. Kriegel, H.P., Renz, M., Schubert, M.: Route skyline queries: a multi-preference path planning approach. In: ICDE, pp. 261–272 (2010)
8. Li, M., Xin, J., Wang, Z., Liu, H.: Accelerating minimum temporal paths query based on dynamic programming. In: ADMA, pp. 48–62 (2019)

9. Nicosia, V., Tang, J., Musolesi, M., Russo, G., Mascolo, C., Latora, V.: Components in time-varying graphs. Chaos Interdiscip. J. Nonlinear Sci. **22**(2), 023101 (2012)
10. Noto, M., Sato, H.: A method for the shortest path search by extended Dijkstra algorithm. In: SMC, vol. 3, pp. 2316–2320 (2000)
11. Rice, M., Tsotras, V.J.: Graph indexing of road networks for shortest path queries with label restrictions. In: PVLDB, vol. 4, pp. 69–80 (2010)
12. Shekelyan, M., Jossé, G., Schubert, M.: Linear path skylines in multicriteria networks. In: ICDE, pp. 459–470 (2015)
13. Song, X., Li, J., Cai, T., Yang, S., Yang, T., Liu, C.: A survey on deep learning based knowledge tracing. Knowl.-Based Syst. **258**, 110036 (2022)
14. Tian, Y., Lee, K.C., Lee, W.C.: Finding skyline paths in road networks. In: SIGSPATIAL, pp. 444–447 (2009)
15. Valstar, L.D., Fletcher, G.H., Yoshida, Y.: Landmark indexing for evaluation of label-constrained reachability queries. In: SIGMOD, pp. 345–358 (2017)
16. Wang, Y., Li, G., Tang, N.: Querying shortest paths on time dependent road networks. In: PVLDB, vol. 12, pp. 1249–1261 (2019)
17. Wu, H., Cheng, J., Ke, Y., Huang, S., Huang, Y., Wu, H.: Efficient algorithms for temporal path computation. TKDE **28**(11), 2927–2942 (2016)
18. Yang, B., Guo, C., Jensen, C.S., Kaul, M., Shang, S.: Multi-cost optimal route planning under time-varying uncertainty. In: ICDE (2014)
19. Yang, S., et al.: Robust cross-network node classification via constrained graph mutual information. Knowl.-Based Syst. **257**, 109852 (2022)
20. Yang, Y., Gao, H., Yu, J.X., Li, J.: Finding the cost-optimal path with time constraint over time-dependent graphs. In: PVLDB, vol. 7, pp. 673–684 (2014)
21. Yuan, Y., Lian, X., Wang, G., Ma, Y., Wang, Y.: Constrained shortest path query in a large time-dependent graph. In: PVLDB, vol. 12, pp. 1058–1070 (2019)

Efficient and Scalable Distributed Graph Structural Clustering at Billion Scale

Kongzhang Hao[1]([✉]), Long Yuan[2], Zhengyi Yang[1], Wenjie Zhang[1], and Xuemin Lin[3]

[1] The University of New South Wales, Sydney, Australia
{k.hao,zhengyi.yang,wenjie.zhang}@unsw.edu.au
[2] Nanjing University of Science and Technology, Nanjing, China
longyuan@njust.edu.cn
[3] Shanghai Jiao Tong University, Shanghai, China

Abstract. Structural Graph Clustering (SCAN) is a fundamental problem in graph analysis and has received considerable attention recently. Existing distributed solutions either lack efficiency or suffer from high memory consumption when addressing this problem in billion-scale graphs. Motivated by these, in this paper, we aim to devise a distributed algorithm for SCAN that is both efficient and scalable. We first propose a fine-grained clustering framework tailored for SCAN. Based on the new framework, we devise a distributed SCAN algorithm, which not only keeps a low communication overhead during execution, but also effectively reduces the memory consumption at all time. We also devise an effective workload balance mechanism that is automatically triggered by the idle machines to handle skewed workloads. The experiment results demonstrate the efficiency and scalability of our proposed algorithm.

Keywords: graph · clustering · distributed processing

1 Introduction

With the proliferation of graph applications, research efforts have been devoted to many problems in analyzing graphs [4,14,15,18,22,23,27,28,30,31]. Among them, graph clustering (e.g. [17,19,25]) is a fundamental problem and many different clustering algorithms have been proposed, while most of them partition the entire set of vertices into disjoint clusters. For real graphs, however, it is usual that not all vertices are members of clusters: some vertices are hubs [10] that bridge many clusters, and some vertices are just outliers [25]. Following this idea, *structural graph clustering* (SCAN) [25] is proposed to distinguish the different roles of the vertices and to uncover overlapping clusters. In the real world, SCAN is essential for many applications (e.g. [7,10,12,24]). For example, the identified hubs are believed to play an important role in viral marketing [7] and epidemiology [24]. Additionally, the identification of hubs in the WWW improves the search for relevant authoritative web pages [12].

© The Author(s), under exclusive license to Springer Nature Switzerland AG 2023
X. Wang et al. (Eds.): DASFAA 2023, LNCS 13945, pp. 234–251, 2023.
https://doi.org/10.1007/978-3-031-30675-4_16

Motivation. In real-world applications, the scale of graphs is large and grows exponentially. The size of the graphs can usually exceed the memory size of a single machine, which gives rise to the need for distributed algorithms [8,9,13,26]. In the literature, a number of distributed algorithms for SCAN have been proposed [11,19,32,33]. Specifically, ParallelSCAN [32], SparkSCAN [33] and CASS [11] are parallelized and implemented on big-data engines (e.g. Hadoop MapReduce [20] and Apache Spark [29]). However, it was verified in [19] that they all suffer from high communication overhead, which leads to their inefficiency in processing billion-scale graphs. Recently, DSCAN [19] was proposed to reduce the communication cost and hence improve clustering efficiency. DSCAN first removes edges of adjacent vertices that are locally found to be dissimilar, and then stores the adjacency lists of all remote vertices in each machine such that the clustering is run locally. Nevertheless, because DSCAN requires storing the remote adjacency lists in advance, the per-machine memory consumption can be exponentially larger than the original graph partition. In the worst case, DSCAN even needs to store the whole graph in each machine. This leads DSCAN to easily run out of memory as verified in our experiments.

Motivated by these, in this paper, we aim to develop an efficient and scalable distributed algorithm for the problem of SCAN. The developed algorithm should not only have a low communication overhead, but also maintain a low memory consumption.

Our Idea. Our general idea to overcome these challenges is simple: instead of processing the clustering for all vertices together in each machine, we propose a fine-grained framework for distributed clustering that divides the local vertices in each machine into batches and processes each of them separately. Based on this new clustering framework, we are able to not only reduce the communication overhead, but also control the memory consumption. However, to make our idea practically applicable, the following challenges still need to be addressed: (1) how to efficiently perform the fine-grained clustering without storing all remote adjacency lists in each machine? (2) how to keep the correctness of clustering results across different batches? (3) how to handle the skewed workload which commonly occurs in distributed systems?

Contributions. In this paper, we address the efficiency and scalability issues in existing works and make the following contributions: (1) we propose a new distributed algorithm for SCAN based on our fine-grained framework for distributed clustering. Besides keeping a low communication overhead during execution, our new algorithm can also effectively control the memory consumption during the clustering, which significantly improves the scalability and efficiency of existing algorithms; (2) we design an effective work-stealing mechanism to handle unbalanced workloads; (3) we conduct extensive performance studies using large real-world graphs. The experiment results demonstrate that our proposed algorithm is efficient and scalable for SCAN in billion-scale graphs.

2 Preliminary

Let $G = (V, E)$ denote an unweighted undirected graph, where $V(G)$ is the set of vertices and $E(G)$ is a set of directed edges. We denote the number of vertices as $|V(G)|$ and the number of edges as $|E(G)|$. For a vertex $u \in V(G)$, we use $\mathcal{N}(u)$ to denote the neighbors of u. The structural neighborhood of a vertex u, denoted by $N(u)$, refers to the set of neighbors of u plus u itself (i.e. $N(u) = \mathcal{N}(u) \cup \{u\}$).

Definition 1. (Structural Similarity) *The similarity between two vertices u and v, denoted by $\sigma(u, v)$, is defined as the number of common vertices in $N(u)$ and $N(v)$ normalized by the geometric mean of their cardinalities: $\sigma(u, v) = \frac{|N(u) \cap N(v)|}{\sqrt{|N(u)||N(v)|}}$.*

Given a similarity threshold $0 < \varepsilon \leq 1$, we call two vertices u and v are *similar* if $\sigma(u, v) \geq \varepsilon$, or *dissimilar* otherwise. The ε-neighborhood of u, denoted by $N_\varepsilon(u)$, is defined as the subset of vertices in $N(u)$ that are similar to u.

Definition 2. (Core Vertex) *Given an integer $\mu \geq 1$, a vertex u is a core vertex if $|N_\varepsilon(u)| \geq \mu$, or a non-core vertex otherwise.*

A cluster C is a subset of $V(G)$ with $|C| \geq 2$. For a core vertex u and a vertex v, if v is similar to u, then u and v belong to the same cluster. A vertex u that is not in any cluster is a *hub* vertex if its neighbors belong to two or more clusters, and it is an *outlier* vertex otherwise.

Problem Statement. Given a graph $G = (V, E)$, parameters $0 < \varepsilon \leq 1$ and $\mu \geq 1$, the problem of distributed SCAN aims to compute the roles of all the vertices and the clusters in G in the distributed context.

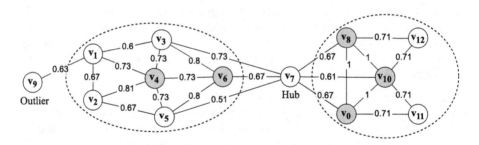

Fig. 1. SCAN on Graph G with $\varepsilon = 0.7$, $\mu = 4$

Example 1. Consider the graph G shown in Fig. 1. When running SCAN with $\varepsilon = 0.7$ and $\mu = 4$ on G, the similarity value $\sigma(u, v)$ is shown on each edge (u, v). The core vertices are marked in grey. SCAN identifies two clusters, namely $\{v_1, v_2, v_3, v_4, v_5, v_6\}$ and $\{v_0, v_8, v_{10}, v_{11}, v_{12}\}$, and isolates vertex v_7 as a hub and vertex v_9 as an outlier, which are shown in the figure.

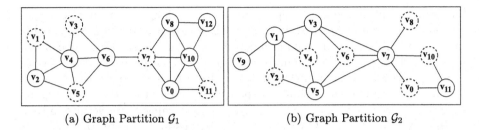

(a) Graph Partition $\mathcal{G}_1$ (b) Graph Partition $\mathcal{G}_2$

Fig. 2. Example Partitions of G

Graph Storage. Given a data graph G, we use the widely used *hash partitioning* by default, which is also used in the existing works on distributed SCAN algorithms [11,19,32,33]. Specifically, we assume the data graph is partitioned by vertices, that is, for each vertex $v \in V$, we store it with its adjacency list $(v; \mathcal{N}(v))$ in one of the partitions. Each partition is referred to as $\mathcal{G}$. We call a vertex that resides in the local partition a *local vertex*, or a *remote vertex* otherwise. Figure 2 demonstrates an example hash partitioning of G from Fig. 1, which is equally partitioned into two machines by its vertices. In each partition, the local vertices and remote vertices are marked in solid and dashed vertices respectively. As can be seen, every *local vertex* has its adjacency list stored in the same machine. Although we use the *hash partitioning* in this paper, our approach is orthogonal to partitioning, i.e., it flexibly supports all graph partitioning methods.

3   Related Works

In this section, we extend the description of the related works on SCAN introduced in Sect. 1. We roughly divide the existing works into two categories: distributed solutions and centralized solutions. We focus on the distributed solutions for this problem. We also briefly introduce the centralized solutions.

The Distributed Solutions. In the literature, the distributed algorithms for SCAN have been well studied [11,19,32,33]. In [32], a distributed SCAN algorithm PSCAN was proposed based on MapReduce [6]. The map function refines the adjacency list of the vertices, while the reduce function calculates the structural similarity between vertices. However, as the Hadoop MapReduce model requires storing intermediate results into a distributed file system for each iteration, it is thus unsuitable for SCAN that includes iterative tasks as shown in [33]. To improve the inefficiency of PSCAN, SparkSCAN [33] and CASS [11] were further proposed based on Apache Spark [29] with the advantages of its iterative computation. Nevertheless, it was recently verified in [19] that PSCAN, SparkSCAN and CASS all suffer from high communication overhead introduced by their based big-data engines, which leads to the inefficiency and long processing time for billion-scale graphs. Following the observation, DSCAN was proposed in [19], which reduces the communication cost by first locally removing unpromising edges (u, v) with $min\{\frac{|N(u)|}{|N(v)|}, \frac{|N(v)|}{|N(u)|}\} < \varepsilon^2$ in each machine, and then supplementing the local graph partition by fetching and storing the neighbor sets of all remote vertices. By doing this in advance, the SCAN algorithm can be run individually in each machine without requiring any separate communication, which consequently improves efficiency. However, it is verified in our experiments that when storing the adjacency lists of remote neighbors, the memory consumption in each machine is exponentially larger than the original graph partition, which easily exceeds per-machine memory capacity (as verified in Exp-1). Moreover, although DSCAN reduces the size of stored remote adjacency lists by first removing unpromising edges locally, it is experimentally verified in our paper that DSCAN still runs out of memory on billion-scale graphs unless ε is relatively large (e.g. $\varepsilon = 0.9$), when most edges have been removed.

The Centralized Solutions. A number of centralized algorithms have been proposed to accelerate SCAN [2,3,16,21]. Recently, pSCAN [2] introduces dynamic pruning techniques to reduce the amount of similarity computation. Specifically, it introduces similar and effective degrees, which are the lower and upper bounds of $(|\mathcal{N}_\varepsilon| - 1)$ for each vertex. While conducting the clustering, it records the similar and effective degrees for the vertices and their neighbors, and uses them for early termination in the core checking. In addition to the sequential algorithms, a number of parallel algorithms for SCAN are also proposed [3,16,21]. However, they all assume that each thread has direct access to the complete graph, which cannot be trivially applied to our problem. As the pruning techniques in centralized solutions can effectively reduce computation cost, our algorithm adopts similar pruning methods to improve the overall performance.

4   Our Approach

4.1   A Fine-grained Framework for Clustering

Based on the above analysis, we have to make a fresh start and design a new efficient and scalable distributed algorithm tailored for distributed SCAN. The design goals of our new approach are: (1) *Low memory consumption.* Our approach should have a low memory consumption in each machine in the distributed system; (2) *Low communication overhead.* Our approach should effectively reduce the communication overhead caused by the shuffling of intermediate results; (3) *Parallelism.* Our approach should fully utilize the computation resources of the distributed system; (4) *Load balance.* As a distributed solution, our approach should be able to handle the situation with imbalanced workload.

As analyzed in the above section, although DSCAN can reduce the communication overhead of big-data engine-based solutions, it leads to a high memory consumption. Revisiting DSCAN's algorithm procedure, it can be observed that DSCAN's high memory consumption results from the prior storing of a large number of adjacency lists of remote vertices. Inspired by this, we process clustering in a finer granularity to reduce the size of stored remote adjacency lists at each time, which is based on the following lemma:

Lemma 1. *Given three clusters found by SCAN, namely C_0, C_1 and C_2, if $C_1 \bigcup C_2 = C_0$ and $C_1 \bigcap C_2 = \emptyset$, then there exists $v_1 \in C_1$ and $v_2 \in C_2$ such that v_1 and v_2 are structural similar.*

Proof. The lemma can be proven directly based on the definition of *structural connectivity* in SCAN.

According to Lemma 1, the clustering results for local vertices $V(\mathcal{G})$ in each machine can be obtained by first dividing $V(\mathcal{G})$ into disjoint batches $S = \{S_0, S_1, \cdots, S_n\}$, then clustering separately for each $S_v \in S$ and the clusters are updated once a new batch is computed. By doing so, instead of storing all remote adjacency lists in every machine for all time, the remote adjacency lists for only each batch of vertices are fetched when needed and stored in a fixed-size cache. The benefits of adopting this approach are threefold: (1) the *memory consumption* is well-managed as the number of vertices in each batch and the cache size are both controlled; (2) the *parallelism* is not sacrificed since the multiple vertices are processed in each step; (3) the *communication overhead* is low because there is no exchange of intermediate results and only the remote adjacency lists that do not exist locally are fetched and cached. In order to make the idea practical, we first introduce the overview of our algorithm FgSCAN in Algorithm 1. The detailed implementation of our algorithm and the load-balancing mechanism are delayed to the following sections.

Algorithm 1: FgSCAN($G = \{\mathcal{G}_1, \cdots, \mathcal{G}_n\}, B$)

---
1 **foreach** *machine* **do**
2   **foreach** $S_v \in V(\mathcal{G}_i)$ *divided into* $\lceil \frac{|V(\mathcal{G}_i)|}{B} \rceil$ *batches* **do**
3   CoreCheck(S_v);
4   CoreCluster(S_v);
5 All machines send their clustering results to the first machine;

Algorithm. As shown in Algorithm 1, given a list of partitions $\{\mathcal{G}_1, \cdots, \mathcal{G}_n\}$ of G and a batch size B defining the maximum number of vertices in each batch, all machines start the clustering together. Each machine first divides the vertices in its local partition $\mathcal{G}_i$ into batches, each with size B (line 2). For each batch S_v, FgSCAN runs CoreCheck and CoreCluster on S_v. Specifically, CoreCheck determines the role for each v in S_v by checking whether v is core or non-core, while CoreCluster determines the belonging cluster for each core vertex v in S_v by unioning v with its similar neighbors (lines 3-4). Finally, the cluster ids of all vertices are sent to the first machine, which are combined there (line 5).

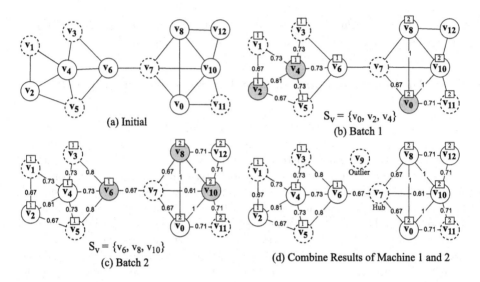

Fig. 3. Example of FgSCAN with $\varepsilon = 0.7, \mu = 4$

Example 2. Reconsider G and its partitions shown in Fig. 1 and Fig. 2, and assume the parameters for SCAN are set as $\varepsilon = 0.7$ and $\mu = 4$. We set batch size $B = 3$. The execution of FgSCAN is shown in Fig. 3. Note that Fig. 3 only shows the detailed execution of FgSCAN on machine 1 as the execution on machine 2 is similar. The local and remote vertices are marked as solid and dashed vertices. For clearness of presentation, in Fig. 3, the vertices processed in each batch are marked in grey and the similarity $\sigma(u, v)$ between each vertex pair (u, v) that is computed is shown on each edge (u, v). Initially, the similarities between all vertex pairs are unknown, as demonstrated in Fig. 3 (a).

Initially, the batch of vertices $\{v_0, v_2, v_4\}$ is processed, as shown in Fig. 3 (b). After running CoreCheck, v_0 and v_4 are found to be core vertices as the numbers of their structural similar neighbors are both greater than or equal to μ. In CoreCluster, the similar neighbors of v_4, namely $\{v_1, v_2, v_3, v_5, v_6\}$, are unioned with v_4 and assigned a cluster id of 1, which are shown by the labels above vertices in Fig. 3 (b). Similarly, v_0 is unioned with its similar neighbors $\{v_8, v_{10}, v_{11}\}$ with a cluster id of 2. Note that v_7 is not assigned to the same cluster as v_0 because it is dissimilar to v_0. Subsequently, the batch of vertices $\{v_6, v_8, v_{10}\}$ is processed, which is shown in Fig. 3 (c). After running CoreCheck, v_6, v_8 and v_{10} are found as core vertices. However, because v_6 and v_8 are both dissimilar to v_7, vertex v_7 cannot be unioned into any of the clusters. The processing of the last batch $\{v_{12}\}$ in the next iteration is not demonstrated in Fig. 3 because no change is made on clustering. Finally, the clustering results of machine 1 and machine 2 are combined, shown in Fig. 3 (d).

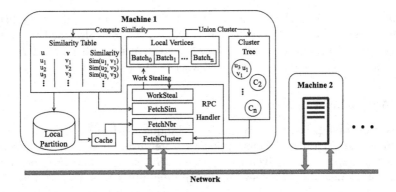

Fig. 4. Architecture

4.2 Architecture

Our framework adopts a share-nothing architecture in a n-machine cluster as shown in Fig. 4. Each machine consists of the following components:

RPC Handler. RPC handler is used to communicate between machines, which is supported with a RPC server and a RPC client. The server is responsible for answering incoming requests from other machines, while the requests are sent through the client. There are four RPC functions, namely FetchSim(), FetchNbr(), QueryCluster() and WorkSteal():

- FetchSim(): FetchSim() takes a list of vertex pairs $\{(u_0, v_0), \cdots, (u_n, v_n)\}$ as its arguments and returns a list of results $\{r_0, \cdots, r_n\}$, where each result r is either *true*, *false* or *unknown*, based on whether each vertex pair is similar, not similar or unknown yet.
- FetchNbr(): FetchNbr() takes a list of vertices $\{v_0, \cdots, v_n\}$ as its arguments and returns their neighbor sets $\{(v_0, \mathcal{N}(v_0)), \cdots, (v_n, \mathcal{N}(v_n))\}$.
- FetchCluster(): FetchNbr() takes a list of vertices $\{v_0, \cdots, v_n\}$ as its arguments and returns their cluster ids $\{C_0, \cdots, C_n\}$.
- WorkSteal(): WorkSteal() dynamically steals others' workload when the current machine is idle.

Similarity Table T_{sim}**:** There is a local similarity table in each machine that records the pairwise similarity between the local vertices and their neighbors. Given a vertex pair (u, v) in T_{sim}, $T_{sim}(u, v)$ returns *true* if u and v are similar, or *false* otherwise.

Cluster Tree T_c**:** There is a union-find tree [5] in each machine that efficiently maintains a set of nodes partitioned into disjoint clusters. Given a vertex u, T_c.cluster(u) returns the cluster id of u. Moreover, given two vertices u and v, T_c.union(u, v) merges the two clusters of u and v into the same cluster.

Cache C: There is a shared cache that stores the fetched adjacency lists of remote vertices. The shared cache has a fixed-size and is designed to be lock-free. Given a list P of $(v, \mathcal{N}(v))$ pairs, C.add(P) inserts all pairs in P to the cache and removes random existing pairs when the cache is full.

4.3   Algorithm Implementation

This section first presents the pruning technique used in CoreCheck, then the implementation details of CoreCheck and CoreCluster based on the discussed architecture. Specifically, given a vertex $u \in G$, let sim_u and $diff_u$ denote the number of u's neighbors that are similar and dissimilar to u, respectively. Given a vertex u, if $sim_u \geq \mu - 1$, then u is immediately a core vertex. Additionally, if $|\mathcal{N}(u)| - diff_u < \mu - 1$, then u is immediately a non-core vertex. The correctness of these two pruning rules can be proved directly based on Definition 2. Following the idea, by recording sim_u and $diff_u$ for each vertex u during core checking, CoreCheck stops computing the similarity between u and its neighbors as soon as u is found to be a core or non-core vertex. The detailed implementation of CoreCheck and CoreCluster are as follows:

Algorithm 2: CoreCheck$(S_v, \mathcal{G})$

---

1 $S_f \leftarrow \forall (u, v)$ *where* $u \in S_v, v \in \mathcal{N}(u)$ *and* $(u, v) \notin T_{sim}$;
2 $R \leftarrow$ FetchSim(S_f); Add the similarity results R to T_{sim};
3 **parallel foreach** $u \in S_v$ **do**
4 $\quad$ $sim_u \leftarrow 0$; $diff_u \leftarrow 0$;
5 $\quad$ **foreach** $v \in \mathcal{N}(u)$ **do**
6 $\quad\quad$ **if** $T_{sim}(u, v) = true$ **then**
7 $\quad\quad\quad$ $sim_u \leftarrow sim_u + 1$;
8 $\quad\quad\quad$ **if** $sim_u \geq \mu - 1$ **then**
9 $\quad\quad\quad\quad$ u.role $\leftarrow Core$; **break**;
10 $\quad\quad$ **else if** $T_{sim}(u, v) = false$ **then**
11 $\quad\quad\quad$ $diff_u \leftarrow diff_u + 1$;
12 $\quad\quad\quad$ **if** $|\mathcal{N}(u)| - diff_u < \mu - 1$ **then**
13 $\quad\quad\quad\quad$ u.role $\leftarrow NonCore$; **break**;
14 FetchNbr$(\bigcup_{u \in S_v} v \in \mathcal{N}(u)$ *s.t.* $(u, v) \notin T_{sim}$ *and* v *is remote*$)$;
15 **parallel foreach** $u \in S_v$ **do**
16 $\quad$ **foreach** $v \in \mathcal{N}(u)$ *s.t.* $(u, v) \notin T_{sim}$ **do**
17 $\quad\quad$ $T_{sim}(u, v) \leftarrow$ CompSim(u, v);
18 $\quad\quad$ Repeat lines 6-13;

CoreCheck. CoreCheck checks if a batch of local vertices are core vertices. As shown in Algorithm 2, given a batch of local vertices S_v and a graph partition $\mathcal{G}$, the vertex pairs of every vertex u in S_v and u's neighbor v whose similarity to u is unknown locally are added to S_f (line 1). Then their similarities are fetched remotely by calling FetchSim() (line 2). For each vertex u in S_v, variables sim_u

---

Algorithm 3: CoreCluster(S_v, $\mathcal{G}$)

1 FetchCluster($\bigcup_{u \in S_v} \mathcal{N}(u)$);
2 **parallel foreach** $u \in S_v$ *s.t.* u.role $= Core$ **do**
3   **foreach** $v \in \mathcal{N}(u)$ *s.t.* $T_{sim}(u, v) = true$
           and T_c.cluster(u) $\neq T_c$.cluster(v) **do**
4 | T_c.union(u, v);
5 $S_f \leftarrow \emptyset$;
6 **foreach** $u \in S_v$ *s.t.* u.role $= Core$ **do**
7   **foreach** $v \in \mathcal{N}(u)$ *s.t.* $(u, v) \notin T_{sim}$, v *is remote*
           and T_c.cluster(u) $\neq T_c$.cluster(v) **do**
8 | S_f.add(v);
9 FetchNbr(S_f);
10 **parallel foreach** $u \in S_v$ *s.t.* u.role $= Core$ **do**
11   **foreach** $v \in \mathcal{N}(u)$ *s.t.* $(u, v) \notin T_{sim}$ *and* T_c.cluster(u) $\neq T_c$.cluster(v) **do**
12          $T_{sim}(u, v) \leftarrow$ CompSim(u, v);
13   **if** $T_{sim}(u, v) = true$ **then** T_c.union(u, v);

---

and $diff_u$ are initialized to record the number of similar and dissimilar neighbors of u (lines 3-4). For each neighbor v of u, if u and v are similar, sim_u is increased by one and u is a core vertex if the number of u's similar neighbors is larger than or equal to $\mu - 1$ (lines 6-9); otherwise, if u and v are dissimilar, $diff_u$ is increased by one and u is not a core if the maximum possible number of similar neighbors left is smaller than $\mu - 1$ (lines 10-13). After this, for each vertex u in S_v, CoreCheck fetches the neighbor sets of all u's neighbors vs if the similarity between u and v is unknown and v is remote (line 14). Then, for every u in S_v and for each of u's neighbor v, their similarity is computed locally and CoreCheck repeats lines 6-13, which decides the role for each vertex u (lines 15-18).

CoreCluster. CoreCluster clusters a list of local vertices and their neighbors. As shown in Algorithm 3, given a batch of local vertices S_v and a graph partition $\mathcal{G}$, the cluster ids of the neighbors of every u in S_v are first fetched remotely (line 1). Then for every core vertex u in S_v, for any of u's neighbor v whose cluster id is different from u's and u is similar to v, the clusters of u and v are merged into the same cluster (lines 2-4). Then, for each core vertex u in S_v, for each of u's remote neighbor v whose cluster id is different from u's and their similarity are unknown, the neighbor set of v is fetched remotely (lines 6-9). After this, CoreCluster computes the similarity of core vertex u in S_v and u's neighbors with unknown similarity and different belonging clusters (lines 10-12). If u and v are similar, their belonging clusters are merged (line 13).

Example 3. Reconsider the partition $\mathcal{G}$ of G shown in Fig. 2 (a) and assume the parameters for SCAN are set as $\varepsilon = 0.7$ and $\mu = 4$. Figure 5 demonstrates the execution of CoreCheck and CoreCluster when processing a batch of vertices.

Given the initial batch $S_v = \{v_0, v_2, v_4\}$ in Fig. 5 (a), CoreCheck checks whether any of them is a core vertex in Fig. 5 (b). Firstly, as the similarities between all vertex pairs are unknown, the adjacency lists of remote neighbors of

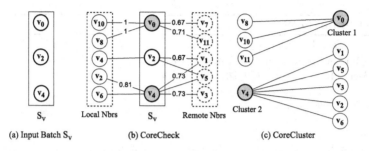

Fig. 5. Running Example of FgSCAN with $\varepsilon = 0.7, \mu = 4$

v_0, v_2 and v_4 are fetched together by FetchNbr(), as shown by the right column of vertices in Fig. 5 (b). Then CoreCheck starts core checking for vertices in S_v by computing the similarity between them and their neighbors. In Fig. 5 (b), only the similarity $\sigma(u, v)$ between each vertex pair (u, v) that is computed is shown on the edges. For v_0, it can be seen that v_7 is dissimilar to v_0 and $\{v_{11}, v_8, v_{10}\}$ are all similar to v_0. Thus, $sim_{v_0} \geq \mu - 1$ and v_0 is a core. For v_2, after v_1 is found to be dissimilar to v_2, CoreCheck notices that v_2 has at most two neighbors whose similarity to it can be potentially true, which is smaller than $\mu - 1$ (i.e. $|\mathcal{N}(v_2)| - diff_{v_2} < \mu - 1$). Hence, v_2 is a non-core and the similarity computation for (v_2, v_4) and (v_2, v_5) is skipped. For v_4, after $\{v_1, v_2, v_3\}$ are all found similar to it, CoreCheck finds that $sim_{v_4} \geq \mu - 1$. Then, v_4 is immediately a core and the similarity computation for (v_4, v_5) and (v_4, v_6) is skipped. The core vertices v_0 and v_4 are labelled in grey in Fig. 5 (b). In CoreCluster shown in Fig. 5 (c), the similar neighbors of each core vertex are clustered together with it. As can be seen, v_0 is clustered with $\{v_8, v_{10}, v_{11}\}$ and v_4 is clustered with $\{v_1, v_2, v_3, v_5, v_6\}$.

4.4 Load Balance

In the real world, there are some skewed graphs that cause the split workload for the machines unbalanced. Existing solutions [11,19,32,33] all distribute workload based on the pre-partitioned local vertices in each machine, which may still suffer from load skew during clustering. As a result, we address the straggler problem by devising a dynamic work stealing based mechanism. Its main idea is that the idle workers automatically "steal" the unfinished workload from the busy workers to accelerate the whole process. Following the idea, we implement dynamic work stealing to accommodate FgSCAN's architecture and execution.

In FgSCAN, the computation in each machine is divided into multiple iterations and a batch of vertices is processed in each round. Dynamic work stealing is triggered when any machine M completes computing all batches of vertices assigned to it. In this case, work stealing happens as follows: (1) the machine will send a StealWork() RPC to a busy machine M'; (2) when M' receives the request, it removes half of the unprocessed batches from its task pool and sends the removed batches to machine M; (3) M receives the vertex batches and continues the clustering. It can be easily verified that no machine will become idle unless each machine has at most one unprocessed batch of vertices. In Algorithm 1, StealWork() is added after line 4 to trigger dynamic work-stealing once a machine finishes processing all its batches.

4.5   Algorithm Analysis

Theorem 1. *Clusters are correct and complete after running* FgSCAN.

Proof. After running CoreCheck on all batches of vertices, the roles of all vertices are known. Besides, CoreCluster strictly follows Lemma 1, which means the clustering of u and v only happens when u is a core and u and v are similar. Thus, the clustering is correct. Moreover, when running CoreCluster on all batches of vertices, all similar edges where it is possible to force cluster union are explored. As a result, the clusters are complete.   □

Theorem 2. *The memory consumption of* FgSCAN *in each machine is bounded by* $O(|\mathcal{G}| + min\{\sum_{v \in \cup_{u \in S_v} \mathcal{N}(u)} |\mathcal{N}(v)|, |C|\})$, *where* $|\mathcal{G}|$ *is the size of the graph partition,* S_v *is any of the vertex batches and* $|C|$ *is size of the fixed-size cache.*

Proof. According to the procedure of FgSCAN, each machine by default maintains the graph partition $\mathcal{G}$ in its main memory, occupying a size of $O(|\mathcal{G}|)$. In each iteration, the adjacency lists of neighbors of a maximum of $|S_v|$ vertices are fetched, leading to a maximum size of the stored remote adjacency lists as $O(\sum_{v \in \cup_{u \in S_v} \mathcal{N}(u)} |\mathcal{N}(v)|)$, while the maximum size of the cache is $O(|C|)$. Additionally, the role and cluster for each $v \in V(G)$ are maintained, whose total sizes are both bounded by $O(|V(\mathcal{G})|)$. As a result, the total memory in each machine is bounded by $O(|\mathcal{G}| + min\{\sum_{v \in \cup_{u \in S_v} \mathcal{N}(u)} |\mathcal{N}(v)|, |C|\})$.   □

5   Evaluation

In this section, we evaluate the efficiency of the proposed algorithms. All the experiments except Exp-2 are performed on a local cluster of 10 machines, each with one 4-core Intel Xeon CPU E3-1220, 16GB memory, 1T disk, connected via a 10Gbps network, running Red Hat Linux 7.3, 64 bit. Each machine runs 4 workers. For Exp-2, we use a machine with one 20-core Intel Xeon CPU E5-2698 and 128 GB main memory running Red Hat Linux 7.3, 64 bit.

Datasets. We evaluate our algorithms on eight real-world graphs. The size of the graphs is shown in Table 1. All datasets are downloaded from SNAP (http://snap.stanford.edu/data), KONECT (http://konect.cc/networks) and LAW (http://law.di.unimi.it/datasets.php).

Algorithms. We compare FgSCAN with the existing distributed SCAN algorithms, namely ParallelSCAN, SparkSCAN, CASS and DSCAN, and the state-of-the-art sequential algorithm pSCAN. All the algorithms except the big-data engine-based solutions are implemented in Rust 1.43. For big-data engine-based solutions (i.e. ParallelSCAN, SparkSCAN, CASS), we implement them on their

Table 1. Statistic of the datasets

Dataset	Name	$|V|$	$|E|$	Max degree	Average degree
BerkStan	BK	685 K	7 M	84,230	22.18
LiveJournal	LJ	4 M	68 M	20,333	17.9
Orkut	OR	3 M	117 M	33,313	38.1
DBpedia	DB	18 M	172 M	472,799	18.85
Twitter-WWW	TW	42 M	1.46 B	2,997,487	70.5
Friendster	FS	65 M	1.81 B	5,214	27.5
Twitter-MPI	TM	52 M	1.96 B	3,691,240	74.7
UK-2007	UK	134 M	5.51 B	6,366,528	41.2

original platforms. We implement DSCAN and pSCAN following their original implementations. FgSCAN is implemented with RPC [1], while adopting similar optimization as DSCAN that removes unpromising local edges before execution.

In the experiments, the workload balance mechanism is enabled for FgSCAN by default. We set the size of the fixed-size cache to 5 GiB and the batch size B to 1000. The time cost is measured as the amount of wall-clock time elapsed during the program's execution. If an algorithm cannot finish in $10{,}000$ s or runs out of memory, we denote the processing time as INF. Moreover, if an algorithm runs out of memory, we also mark the case with a × on the top of the figure. In the experiments, we set the default parameters for SCAN as $\varepsilon = 0.5$ and $\mu = 5$.

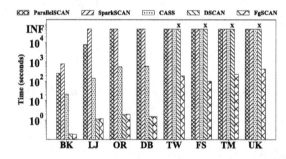

Fig. 6. Processing Time on Each Dataset   **Fig. 7.** Centralized Comparison

Exp-1: Efficiency on Different Datasets. In this experiment, we evaluate the efficiency of the algorithms on all datasets. We use the default parameters and report the average processing time of the algorithms in Fig. 6. As can be seen, our proposed algorithm FgSCAN always significantly outperforms the big-data engine-based algorithms, namely ParallelSCAN, SparkSCAN and CASS. This is because FgSCAN effectively reduces the communication overhead. Moreover, although DSCAN demonstrates a similar performance as FgSCAN, it runs out of memory on all of the billion-scale graphs TW, FS, TM and UK. Comparatively, FgSCAN does not have the problem and can finish clustering efficiently in all the billion-scale graphs, which verifies the effectiveness of the fine-grained framework in controlling memory consumption.

Exp-2: Comparison with Single-Threaded Algorithm. In this experiment, we evaluate the performance of FgSCAN on a single machine, compared to the existing state-of-the-art single-threaded solution pSCAN [2]. We evaluate the performance of FgSCAN on TW by increasing the number of cores used, while pSCAN is always executed with one core. As shown in Fig. 7, FgSCAN requires 2 cores to outperform pSCAN in the worst case, which clearly demonstrates that FgSCAN only introduces little overhead.

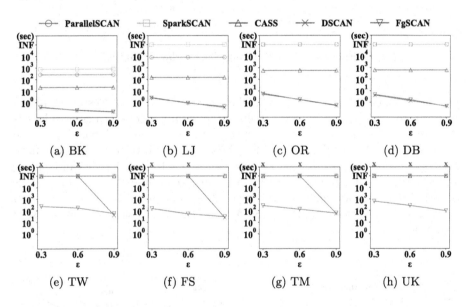

Fig. 8. Efficiency When Varying Similarity Threshold ε

Exp-3: Efficiency when varying ε and μ. In this experiment, we evaluate the efficiency when varying similarity threshold ε and neighborhood threshold μ. In Fig. 8, we run the experiments with $\varepsilon = 0.3, 0.6$, and 0.9, respectively. As shown, the running time of big-data engine-based algorithms is steady for different ε values due to exhaustively computing all structural similarities and thus irrelevant to ε. When ε grows larger, FgSCAN and DSCAN both run faster because there is more unnecessary computation that can be pruned. Moreover, when $\varepsilon = 0.3$ and 0.6, DSCAN fails to cluster the billion-scale graphs due to

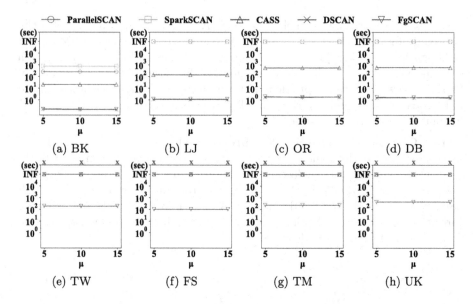

Fig. 9. Efficiency When Varying Neighborhood Threshold μ

the out-of-memory issue; in contrast, when $\varepsilon = 0.9$, DSCAN successfully finishes clustering on all the billion-scale graphs without running out of memory. This is because when ε is large, a huge number of edges can be removed in each machine in advance, which considerably reduces the size of stored remote adjacency lists. Moreover, in Fig. 9, we run experiments with $\mu = 5, 10$, and 15. As can be seen, all algorithms perform quite steadily regarding the different values of μ. This is because all big-data engine-based algorithms need to compute all structural similarities regardless of μ, while DSCAN and FgSCAN can prune unnecessary computation to achieve a similar performance when μ varies.

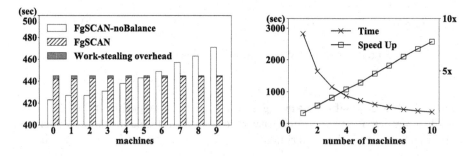

Fig. 10. Load Balancing

Fig. 11. Scalability

Exp-4: Effectiveness of Load Balance. In this experiment, we evaluate the effectiveness of our load balance mechanism. We report the processing time of each machine when the load balance mechanism is enabled or not on UK. As shown in Fig. 10, without load balance mechanism, the processing time in each machine is quite unbalanced. In contrast, after enabling work-stealing, all machines have a similar running time and the average time taken for work-stealing overhead occupies only 1.2% of total processing time, shown by the grey fillings.

Exp-5: Scalability. In this experiment, we evaluate the scalability of FgSCAN. We report the average processing time and speedup on UK when varying the number of machines in Fig. 11. As can be seen, FgSCAN demonstrates almost linear scalability as the number of machines increases in the cluster. The scalability of FgSCAN comes from its high parallelism as analyzed in Sect. 4.1.

6 Conclusion

In this paper, we study the problem of distributed SCAN. We first propose a novel fine-grained clustering framework. Based on it, we devise a new distributed SCAN algorithm, FgSCAN, which not only keeps a low communication overhead, but also effectively reduces the memory consumption and prunes unnecessary computation during the clustering. In addition, we design an effective work-stealing mechanism to handle unbalanced workload. The experiment results demonstrate the efficiency and scalability of our proposed algorithm.

Acknowledgements. Long Yuan is supported by National Key RD Program of China 2022YFF0712100, NSFC61902184, and Science and Technology on Information Systems Engineering Laboratory WDZC20205250411.

References

1. Birrell, A., Nelson, B.J.: Implementing remote procedure calls. ACM Trans. Comput. Syst. **2**(1), 39–59 (1984)
2. Chang, L., Li, W., Lin, X., Qin, L., Zhang, W.: pSCAN: fast and exact structural graph clustering. In: ICDE, pp. 253–264 (2016)
3. Che, Y., Sun, S., Luo, Q.: Parallelizing pruning-based graph structural clustering. In: Proceedings of ICPP, pp. 1–10 (2018)
4. Chen, X., Peng, Y., Wang, S., Yu, J.X.: DLCR: efficient indexing for label-constrained reachability queries on large dynamic graphs. Proc. VLDB Endow. **15**(8), 1645–1657 (2022)
5. Cormen, T.H., Leiserson, C.E., Rivest, R.L., Stein, C.: Introduction to algorithms. MIT press (2022)
6. Dean, J., Ghemawat, S.: Mapreduce: Simplified data processing on large clusters. In: OSDI, pp. 137–150 (2004)
7. Domingos, P., Richardson, M.: Mining the network value of customers. In: Proceedings of SIGKDD, pp. 57–66 (2001)

8. Hao, K., Yang, Z., Lai, L., Lai, Z., Jin, X., Lin, X.: PatMat: a distributed pattern matching engine with cypher. In: Proceedings of CIKM, pp. 2921–2924 (2019)
9. Hao, K., Yuan, L., Zhang, W.: Distributed hop-constrained s-t simple path enumeration at billion scale. Proc. VLDB Endow. **15**(2), 169–182 (2021)
10. Kang, U., Faloutsos, C.: Beyond 'Caveman communities': hubs and spokes for graph compression and mining. In: ICDM, pp. 300–309 (2011)
11. Kim, J., et al.: CASS: a distributed network clustering algorithm based on structure similarity for large-scale network. PLoS ONE **13**(10), e0203670 (2018)
12. Kleinberg, J.M.: Authoritative sources in a hyperlinked environment. J. ACM (JACM) **46**(5), 604–632 (1999)
13. Lai, L., et al.: Distributed subgraph matching on timely dataflow. Proceed. VLDB Endow. **12**(10), 1099–1112 (2019)
14. Liu, B., Yuan, L., Lin, X., Qin, L., Zhang, W., Zhou, J.: Efficient (α,β)-core computation: an index-based approach. In: WWW, pp. 1130–1141 (2019)
15. Liu, B., Yuan, L., Lin, X., Qin, L., Zhang, W., Zhou, J.: Efficient (α, β)-core computation in bipartite graphs. VLDB J. **29**(5), 1075–1099 (2020). https://doi.org/10.1007/s00778-020-00606-9
16. Mazumder, S., Liu, B.: Context-aware path ranking for knowledge base completion. In: Sierra, C. (ed.) IJCAI, pp. 1195–1201 (2017)
17. Meng, L., Yuan, L., Chen, Z., Lin, X., Yang, S.: Index-based structural clustering on directed graphs. In: ICDE, pp. 2831–2844 (2022)
18. Peng, Y., Bian, S., Li, R., Wang, S., Yu, J.: Finding top-r influential communities under aggregation functions. In: ICDE, pp. 1941–1954 (2022)
19. Shiokawa, H., Takahashi, T.: DSCAN: distributed structural graph clustering for billion-edge graphs. In: DEXA, pp. 38–54 (2020)
20. Shvachko, K., Kuang, H., Radia, S., Chansler, R.: The Hadoop distributed file system. In: MSST, pp. 1–10 (2010)
21. Takahashi, T., Shiokawa, H., Kitagawa, H.: SCAN-XP: parallel structural graph clustering algorithm on intel xeon phi coprocessors. In: NDA, pp. 1–7 (2017)
22. Wang, K., Lin, X., Qin, L., Zhang, W., Zhang, Y.: Accelerated butterfly counting with vertex priority on bipartite graphs. VLDB J. **32**, 1–25 (2022)
23. Wang, K., Zhang, W., Lin, X., Qin, L., Zhou, A.: Efficient personalized maximum biclique search. In: ICDE, pp. 498–511 (2022)
24. Wang, Y., Chakrabarti, D., Wang, C., Faloutsos, C.: Epidemic spreading in real networks: an eigenvalue viewpoint. In: SRDS, pp. 25–34 (2003)
25. Xu, X., Yuruk, N., Feng, Z., Schweiger, T.A.: SCAN: a structural clustering algorithm for networks. In: Proceedings of SIGKDD, pp. 824–833 (2007)
26. Yang, Z., Lai, L., Lin, X., Hao, K., Zhang, W.: HUGE: an efficient and scalable subgraph enumeration system. In: SIGMOD, pp. 2049–2062 (2021)
27. Yuan, L., Qin, L., Lin, X., Chang, L., Zhang, W.: Diversified top-k clique search. VLDB J. **25**(2), 171–196 (2016)
28. Yuan, L., Qin, L., Zhang, W., Chang, L., Yang, J.: Index-based densest clique percolation community search in networks. TKDE **30**(5), 922–935 (2018)
29. Zaharia, M., et al.: Apache spark: a unified engine for big data processing. Commun. ACM **59**(11), 56–65 (2016)
30. Zhang, J., Li, W., Yuan, L., Qin, L., Zhang, Y., Chang, L.: Shortest-path queries on complex networks: experiments, analyses, and improvement. Proc. VLDB Endow. **15**(11), 2640–2652 (2022)
31. Zhang, J., Yuan, L., Li, W., Qin, L., Zhang, Y.: Efficient label-constrained shortest path queries on road networks: a tree decomposition approach. Proc. VLDB Endow. **15**(3), 686–698 (2021)

32. Zhao, W., Martha, V., Xu, X.: PSCAN: a parallel structural clustering algorithm for big networks in mapreduce. In: AINA, pp. 862–869 (2013)
33. Zhou, Q., Wang, J.: SparkSCAN: a structure similarity clustering algorithm on spark. In: BDTA, pp. 163–177 (2015)

Hierarchical All-Pairs SimRank Calculation

Liangfu Zhang, Cuiping Li$^{(\boxtimes)}$, Xue Zhang, and Hong Chen

Key Laboratory of Data Engineering and Knowledge Engineering of Ministry
of Education, School of Information, Renmin University of China, Beijing, China
{liangfu_zhang,licuiping,xue.zhang,chong}@ruc.edu.cn

Abstract. *All-pairs* SimRank calculation is a classic SimRank problem.
However, *all-pairs* algorithms suffer from efficiency issues and accuracy
issues. In this paper, we convert the non-linear simrank calculation into
a new simple closed formulation of linear system. And we come up with
a sequence of novel algorithms to efficiently solve the linear system with
accuracy guarantees. To reduce the memory consumption and improve
the computational efficiency, we build a hierarchical framework to calculate the *all-pairs* SimRank scores, which includes locally coarse calculation and globally refine calculation. We first solve the local linear systems
generated from the subgraphs, then we refine the SimRank scores on the
full graph from the residuals of the local structures. We also show that
our algorithms outperform the state-of-the-art *all-pairs* SimRank computation algorithms on real graphs.

Keywords: SimRank Calculation · All Pairs · Graph Algorithms

1 Introduction

SimRank [6] is a widely used link-based similarity measurement between two
vertices in graphs, and SimRank has been widely used in many applications,
such as recommendation systems [1], web mining [8], spam detection [17] and
social network analysis [10]. The link-based similarity measurement is based on
two intuitive statement: (i) two objects in a graph are similar if they are linked
by similar objects, and (ii) two identical objects have the greatest similarity of
1.0. The recursive definition allows SimRank model to capture the similarity
scores of two vertices based on the global structure of a directed graph, so this
model produce high-quality similarity results. *All-pairs* SimRank calculation is
one of the classic SimRank problems, which is calculating the SimRank scores
of all the node pairs in a directed graph G with a given decay factor of c.

In the past decade, *all-pairs* SimRank calculation has been extensively studied [3–6,9,12–14,19,21–28]. However, all these *all-pairs* SimRank algorithms suffer from two major deficiencies.

– **Efficiency issue.** Some *all-pairs* SimRank calculation algorithms [3,6,12–
 14,21,22,25,27] cost $O(n^2)$ memory or more [4,5,9], which is not suitable for

© The Author(s), under exclusive license to Springer Nature Switzerland AG 2023
X. Wang et al. (Eds.): DASFAA 2023, LNCS 13945, pp. 252–268, 2023.
https://doi.org/10.1007/978-3-031-30675-4_17

large-scaled graphs. Because these algorithms are based on matrix multiplication, and the SimRank matrix is always very sparse, there will be a lot of
redundant calculations in the calculation process.
- **Accuracy issue.** A part of methods [3–5, 9, 23–25] are equivalent to random
  walk *without* the first-meeting constraint, but the first encounter constraints
  is defined in the original definition of SimRank [6]. Therefore, the accuracy
  of the solutions of these algorithms is not guaranteed.

Motivations. The state-of-the-art *all-pairs* algorithms are RLP(Reduced Local
Push, also known as opt-FLP) [19] and IncSR [24]. IncSR [24] follow the *approximate* SimRank definition to accelerate computation, therefore, the maximum
error of algorithm IncSR cannot be guaranteed. RLP updates the SimRank
matrix by pushing the residuals to the neighbors. Theoretically, the average
time complexity of RLP is $O(\frac{\bar{d}^2}{c((1-c)^2)\epsilon})$, in which, $\bar{d}$ is the average degree of
the graph. However, there still be a lot of redundant residual calculation, which
won't be added to the SimRank matrix, when the threshold is very small. For
example, when we set $\epsilon = 0.001$, on large scale graphs, the efficiency of RLP
will be greatly affected. We observe that the SimRank values are mainly concentrated from the local structures. So we divide the calculating process into two
stage, locally coarse calculation and globally fine calculation. In this paper, we
aim to *quickly* and *correctly* calculate all-pairs SimRank matrix on large graphs.

Contributions. We first derive a simpler closed formulation of linear system from
the original SimRank definition in work [6], and develop a efficient method to
solve the linear system. The computations are divided into two stages. Firstly,
we make locally coarse calculation through the local structures of the graph.
All the calculations of the first stage can be done entirely in a limited RAM,
which greatly improve the updating efficiency of residuals, and we'll drop some
redundant residuals to improve the efficiency. Then we refine the SimRank scores
on the full graph from the residuals generated in the first stage. Based on the
symmetry of SimRank matrix, we also have done some optimization work during
the calculation. In summary, we have made the following contributions:

- We derived a simple closed formulation of linear system from the original
  SimRank definition, and we come up with an algorithm to efficiently solve
  the linear system.
- We propose a hierarchical framework to reduce the scale of the residuals.
- We present some optimizations to improve the efficiency of our algorithms.
- Our methods outperform the state-of-the-art algorithms on real graphs.

2   Preliminaries

Table 1 shows the notations that are frequently used in the remainder of the
paper, and in Subsect. 2.1 we will first introduce some basic concepts that will
be used later. Based on the conceptions introduced in Subsect. 2.1, we derive the
closed formulation of linear system for SimRank computation in Subsect. 2.2.

Table 1. Table of Notations

Notation	Description
c	the decay factor
ϵ	the maximum absolute error
$G = (V, E)$	directed graph G with vertex set V and edge set E
$G^2 = (V^2, E^2)$	node pair graph of G
n, m	the number of vertices and edges in G
$d_{out}(i), d_{out}(i, j)$	out-degree of node i in G and node pair (i, j) in G^2
$d_{in}(i), d_{in}(i, j)$	in-degree of node i in G and node pair (i, j) in G^2
A	Adjacency matrix
Q	the column normalization matrix of A, $Q[i, j] = \frac{1}{d_{in}(j)}$
$I(v), O(v)$	the set of in-neighbors and out-neighbors of vertex v
$I(u, v), O(u, v)$	$I(u, v) = I(u) \times I(v)$, $O(u, v) = O(u) \times O(v)$
$S, s(u, v)$	SimRank matrix and SimRank score of u and v

2.1   Background of SimRank

Let $G = (V, E)$ be an unweighted directed graph with $|V| = n$ and $|E| = m$. We aim to calculate the similarity matrix S with a maximum error is less than ϵ, i.e. $\max(|S - S'|) \leq \epsilon$. Given two nodes u and v in a directed graph $G = (V, E)$, the SimRank score of u and v is defined as follows:

$$s(u, v) = \begin{cases} 0, & I(u) = \emptyset \text{ or } I(v) = \emptyset \\ 1, & u = v \\ \frac{c}{|I(u)| \cdot |I(v)|} \sum_{x \in I(u), y \in I(v)} s(x, y), & u \neq v \end{cases} \tag{1}$$

where $I(u)$ donates the set of in-neighbors of u, and $c \in (0, 1)$ is a decay factor, which typically set to 0.6 or 0.8 [6,12].

In the original paper [6] of SimRank, Jeh and Widom show that SimRank score can be thought as *flowing* from nodes to its neighbors on G^2 graph, which is a linear system, so we can make a closed form of SimRank without non-linear symbols. Before making a closed form of linear system derivation, we first introduce some basic concepts that will be used later.

Definition 1. (G^2 Graph). *Let $G = (V, E)$ be an unweighted directed graph, its node pair graph $G^2 = (V^2, E^2)$ is defined as follows: $V^2 = \{(x, y)|x, y \in V\}$ and $E^2 = \{((x_1, x_2), (y_1, y_2)) | (x_1, y_1), (x_2, y_2) \in E\}$*

Definition 2 (Kronecker Product). *The Kronecker product of two matrices $A \in \mathbb{R}^{m \times n}$ and $B \in \mathbb{R}^{p \times q}$ is a matrix of $\mathbb{R}^{mp \times nq}$, $A \otimes B = \begin{bmatrix} a_{11}B & \cdots & a_{1n}B \\ \vdots & \ddots & \vdots \\ a_{m1}B & \cdots & a_{mn}B \end{bmatrix}$*

Definition 3 (Vec-Operator). *$S \in \mathbb{R}^{n \times n}$ is a SimRank matrix, $\boldsymbol{Vec}(S) = [s_{11}, \cdots, s_{n1}, s_{12}, \cdots, s_{n2}, \cdots, s_{nn}]^{\top}$, $S[i, j] = \boldsymbol{Vec}(S)[i + (j - 1) \times n]$*

Theorem 1. *Let the new id of node pair (i, j) in G^2 graph is $(i-1)n + j$, in which i and j are the node id in graph G, $A_{G^2} = A_G \otimes A_G$ and $Q_{G^2} = Q_G \otimes Q_G$*

Proof. It's easy to derive this from Definitions 1 and 2.

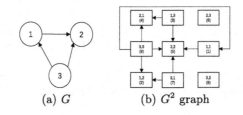

(a) G (b) G^2 graph

Fig. 1. Joy Graph and Its Node Pair Graph

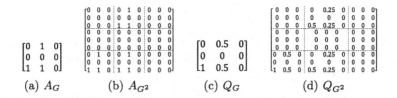

(a) A_G (b) A_{G^2} (c) Q_G (d) Q_{G^2}

Fig. 2. Adjacency Matrices of G and G^2

Figure 1(a) is an unweighted directed graph G, and Fig. 1(b) is the node pair graph G^2. The adjacency matrices of Fig. 1 are shown in Fig. 2, the adjacency matrix of directed graph G is A_G, the adjacency matrix of node pair graph G^2 is $A_{G^2} = A_G \otimes A_G$. The column normalized matrix of A_G is Q_G, and the column normalized matrix of A_{G^2} is $Q_{G^2} = Q_G \otimes Q_G$.

2.2   Linear System for SimRank

As mentioned above, SimRank scores are *flowing* in G^2. The SimRank value $s(u, u)$ is constant of 1.0, so we call node pair (u, u) the *source node* of SimRank *flow*. And the SimRank scores of (u, v) are the amount of SimRank *flows* that propagated into (u, v) from the *source nodes*. According to Theorem 2, we rebuild a SimRank graph G_S^2. As is shown in Fig. 3(a), SimRank graph G_S^2 is transformed from G^2.

Theorem 2. *Let $S \in \mathbb{R}^{n \times n}$ be the SimRank matrix, the closed linear formulation of SimRank is*

$$\mathbf{Vec}(S) = P \cdot \mathbf{Vec}(S)$$

in which $P[k, :] = \begin{cases} e_i \otimes e_j & i = j \\ c \cdot Q_{G^2}[:, k]^\top & i \neq j \end{cases}$, $k = i + (j-1)n$, $\mathbf{Vec}(S)[k] = 1.0$

when $k = i + (i-1)n$.

Proof. $S[i,j]$ is the SimRank score between vertex i and j. Since the SimRank matrix is a symmetric matrix, so

$$S[i,j] = S[j,i] = \boldsymbol{Vec}(S)[(i-1)\cdot n + j] = \boldsymbol{Vec}(S)[i + (j-1)\cdot n]$$

According Eq. (1) we can get that

$$\boldsymbol{Vec}(S)[i + (j-1)n] = \begin{cases} 1 & i = j \\ c \cdot Q_G[:,i]^\top \otimes Q_G[:,j]^\top \cdot \boldsymbol{Vec}(S) & i \neq j \end{cases}$$

in which $c \in (0,1)$, $i,j = 1,2,3\cdots n$, $Q_G[:,i]^\top$ is the transition of the i-th column of matrix Q_G, we could get the following equation.

$$\boldsymbol{Vec}(S) = P \cdot \boldsymbol{Vec}(S) \tag{2}$$

$$P[k,:] = \begin{cases} e_i \otimes e_j & i = j \\ c \cdot Q_{G^2}[:,k]^\top & i \neq j \end{cases} \tag{3}$$

in which $k = i + (j-1)n$, and $Q_{G^2}[:,k]^\top = Q_G[:,i]^\top \otimes Q_G[:,j]^\top$.

So the all-pairs SimRank scores are the solution of the linear system. Since $S[i,i] = 1.0$, so $\boldsymbol{Vec}(S)[k] = 1.0$, when $k = i + (i-1)n$.

Definition 4 (SimRank Graph). *Given a directed graph $G(V,E)$, and Sim-Rank Graph G_S^2 is a weighted directed graph with an adjacency matrix of $P^\top$, in which P is defined in Eq. (3).*

When $c = 0.6$, P is shown in Fig. 3(b), and the SimRank Graph G_S^2 is shown in fig.3(a).

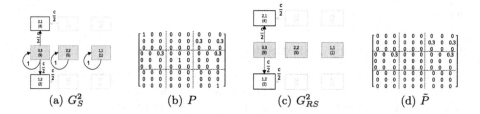

(a) G_S^2 (b) P (c) G_{RS}^2 (d) $\tilde{P}$

Fig. 3. SimRank Graph and Residual Graph

2.3 Numerical Solution

In this subsection, we come up with an algorithm to get the numerical solution of the linear system of Eq. (2). We first give the definitions.

Definition 5 (Numerical Solution). *Let S be the theoretical solution of the linear system, given an absolute error threshold ϵ, the numerical solution returns an estimated matrix S_k, such that $\max(|S_k - S|) < \epsilon$*

Suppose we have an estimate solution $Vec(S_k)$ of $Vec(S)$ for Eq. (2). The increment of S_k is the residuals, namely R_k. And $Vec(R_k) = Vec(S_k - S_{k-1})$. A more accurate $Vec(S_k)$ can be obtained by each iteration.

$$Vec(S_k) = P \cdot Vec(S_{k-1})$$
$$= Vec(S_{k-1}) + Vec(R_k) \tag{4}$$
$$Vec(R_k) = Vec(S_k) - Vec(S_{k-1}) = P \cdot Vec(R_{k-1}) \tag{5}$$

Since the SimRank values of the source nodes are always 1, the residuals of the source nodes are always 0. So Eq. (5) can be rewrite as follows,

$$Vec(R_k) = \tilde{P} \cdot Vec(R_{k-1}) \tag{6}$$

in which $\tilde{P} = P - Diag(Vec(I))$.

Equation(2) means that the SimRank *flow* is propagated in SimRank Graphs and, Eq. (6) means that the residuals *flow* is propagated in Residual Graphs G_{RS}^2 which is shown in Fig. 3(c). Equations (4) and (6) shows how the numerical solution $Vec(S_{k-1})$ and the residual $Vec(R_{k-1})$ are updated to $Vec(S_k)$ and $Vec(R_k)$: the *flow* is aggregated from its neighbors and scaled by the the weight of the edge. From Eq. (5), when $\|Vec(R_k)\|_\infty$ is small, $Vec(S_k)$ would converge to $Vec(S)$.

Algorithm 1. ForwardFlow(G, c, ϵ)

Require: Directed Graph G; Maximum error ϵ and Decay factor c
Ensure: Approximately Solution S_k
1: Set node pair QUEUE Qu, HASH TABLE $S_k = R_k = I$
2: **for all** $v \in V$ **do**
3:　　$Qu.push(v, v)$
4: **while** $Qu \neq \emptyset$ **do**
5:　　$(u, v) = Qu.pop()$
6:　　$tmp = R_k[u, v]$
7:　　$R_k[u, v] = 0$
8:　　**for all** $(ou, ov) \in O(u, v)$ and $ou \neq ov$ **do**
9:　　　　$res = \frac{c \cdot tmp}{I(ou, ov)}$
10:　　　$S_k[ou, ov] = S_k[ou, ov] + res$
11:　　　$R_k[ou, ov] = R_k[ou, ov] + res$
12:　　　**if** $R_k[ou, ov] \geq \frac{(1-c)\epsilon}{c}$ and (ou, ov) not in Qu **then**
13:　　　　$Qu.push(ou, ov)$
14: **return** S_k

The basic algorithm of solving the linear system are shown in Algorithm 1. Since $S_k[i, i] = 1$, we first set $Vec(S_0) = Vec(R_0) = Vec(I)$ (line 1), and we restore the active nodes in an queue Qu. Then we update the residual matrix R_k (in line 6, line 7, line 9 and line 11) and the SimRank matrix S_k (in line 9-10).

The calculation process of the joy graph in Fig. 1 are shown in Fig. 4. At first, initialize $S_k = R_k = I$, and the active node pairs are the source nodes.

Then update the residuals of the active nodes' neighbors, since only source node $(3,3)$ has out neighbors, so we update $R_k[2,1] = R_k[1,2] = 0.3$ and $S_k[2,1] = S_k[1,2] = 0.3$, and the active node pairs are $(1,2)$ and $(2,1)$. Since $(1,2)$ and $(2,1)$ have got no out neighbors, the update process is terminated.

$$
R_k \quad
\begin{array}{c} R_0 \\ \begin{bmatrix} 1 & 0 & 0 \\ 0 & 1 & 0 \\ 0 & 0 & 1 \end{bmatrix} \end{array}
\rightarrow
\begin{array}{c} R_1 \\ \begin{bmatrix} 0 & 0.3 & 0 \\ 0.3 & 0 & 0 \\ 0 & 0 & 0 \end{bmatrix} \end{array}
\rightarrow
\begin{array}{c} R_2 \\ \begin{bmatrix} 0 & 0 & 0 \\ 0 & 0 & 0 \\ 0 & 0 & 0 \end{bmatrix} \end{array}
$$

$$
S_k \quad
\begin{bmatrix} 1 & 0 & 0 \\ 0 & 1 & 0 \\ 0 & 0 & 1 \end{bmatrix}
\rightarrow
\begin{bmatrix} 1 & 0.3 & 0 \\ 0.3 & 1 & 0 \\ 0 & 0 & 1 \end{bmatrix}
\rightarrow
\begin{bmatrix} 1 & 0.3 & 0 \\ 0.3 & 1 & 0 \\ 0 & 0 & 1 \end{bmatrix}
$$

$$S_0 \qquad S_1 = S_0 + R_1 \qquad S_2 = S_1 + R_2$$

Fig. 4. Calculation Process, $c = 0.6$, $\epsilon = 0.01$

Theorem 3. *When $\|Vec(R_k)\|_\infty < \frac{(1-c)\epsilon}{c}$, $\max(|S_k - S|) < \epsilon$*

Proof. Assuming that the algorithm terminates at the k-th iteration, so $\|Vec(R_k)\|_\infty < \frac{(1-c)\epsilon}{c}$. $\|Vec(R_{k+1})\|_\infty = \|\tilde{P} \cdot Vec(R_k)\|_\infty$. $\|\tilde{P} \cdot Vec(R_k)\|_\infty \le \|\tilde{P}\|_\infty \cdot \|Vec(R_k)\|_\infty$. According to Eqs. (3) and (6), $\|\tilde{P}\|_\infty \le c$, so $\|Vec(R_{k+1})\|_\infty \le c \cdot \|Vec(R_k)\|_\infty$. The maximum absolute error is $\sum_{i=k+1}^{\infty} \|Vec(R_i)\|_\infty < \|Vec(R_k)\|_\infty (\sum_{i=1}^{\infty} c^i) \rightarrow \frac{c \cdot \|Vec(R_k)\|_\infty}{1-c} < \epsilon$.

Only non-zero scores will be stored and updated in S_k and R_k, and the scale of active nodes queue Qu is less than $nnz(R_k)$, so the space cost of algorithm 1 is $O(nnz(R_k) + m + n)$, in which $O(nnz(R_k))$ is non-zero SimRank scores. Each iteration reduce the maximum $\|VecR_k\|_\infty$ to $c \cdot \|Vec(R_k)\|_\infty$ in the worst cases, so the time complexity is $O\left(\log_c(1-c)\epsilon \cdot nnz(R_k) \cdot \max(d_{out})^2\right)$.

3 A Hierarchical All-Pairs SimRank Calculation Framework

With the increase of the scale of the graph and the decrease of ϵ, the number of non-zero values in S_k and R_k will increase sharply, which will greatly affect the updating efficiency. Meanwhile, we find that, 1) the update order in S_k and R_k will not change the calculation results, 2) and the SimRank score of each node pair is mainly affected by the local structures, 3) and the non-zero SimRank values are densely distributed around source nodes. So we can first solve the linear system of subgraphs, and then refine the results in full graph.

In this section, we propose a hierarchical framework, shown in Fig. 5, to solve the linear system. We first introduce the hierarchical calculation framework in Subsect. 3.1, and then propose some optimizations in Subsect. 3.2.

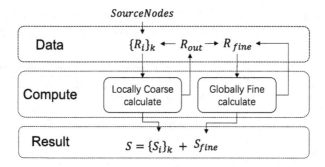

Fig. 5. Hierarchical Framework

3.1   Hierarchical Framework

We first employ a graph partitioning algorithm (e.g., METIS[1]) that decompose the large graph G into several small blocks such that the number of the edges with endpoints in different blocks is minimised. In Fig. 6, we partition G into 2 blocks, $G_1 \cup G_2$, along with 1 edge $\{(6,3)\}$ across the blocks. Then SimRank matrix and residuals matrix are divided into 3 kinds of blocks, $S = \begin{bmatrix} S_1 & S_{crs} \\ S_{crs}^\top & S_2 \end{bmatrix}$, $R = \begin{bmatrix} R_1 & R_{crs} \\ R_{crs}^\top & R_2 \end{bmatrix}$, in which, the SimRank values and residuals of the subgraphs are restored in S_i and R_i respectively, and the values of cross parts G_{crs}^2 are restored in S_{crs} and R_{crs} respectively.

Algorithm 2. Hierarchical Framework

Require: Graph G and $\{G_1, G_2, ..., G_k\}$; Absolute error ϵ and Decay factor c
Ensure: All-pairs SimRank Score matrix S
1: Initialize S_{coarse} and HASH TABLE S_{fine}
2: Initialize R_{coarse} as identity matrices and HASH TABLE R_{fine}
3: **while** not $R_{coarse}.empty()$ **do**
4:     **for all** G_i **do**
5:         $R_{out} = \text{SubForwardFlow}(G_i \cup Outer(G_i), R_i, c, \epsilon)$
6:         update R_{coarse} and R_{fine} From R_{out}
7: **while** not $R_{fine}.empty()$ **do**
8:     $\text{SubForwardFlow}(G, R_{fine}, c, \epsilon)$
9: **return** $S = S_{coarse} + S_{fine}$

As is shown in Fig. 5, we present a hierarchical framework to calculate the All-Pairs SimRank values: locally coarse calculate and globally refine. The algorithm of the hierarchical framework is shown in Algorithm 2. We first initialize $S_{coarse} = S_1 \cup ... \cup S_k$ and R_{coarse} (line 1-2). In the first stage (line 5), we locally

---

[1] http://glaros.dtc.umn.edu/gkhome/views/metis.

update S_i from R_i, and return the residuals for the next stage (line 6). In the second stage (line 7-8), we update the sparse hash table S_{fine} from R_{fine}.

Algorithm 3. SubForwardFlow(G^*, R, c, ϵ)

Require: $G^* = (G \cup Outer(G))$; residuals R; Maximum error ϵ and Decay factor c
Ensure: Outer residuals R_{out}
1: Set node pair QUEUE Qu, and load SimRank score matrix S
2: **for all** non-zero coordinates (u, v) in R **do**
3:     $Qu.push(u, v)$
4:     $S[u, v] = S[u, v] + R[u, v]$
5: **while** $Qu \neq \emptyset$ **do**
6:     $(u, v) = Qu.pop()$
7:     $tmp = R[u, v]$
8:     $R[u, v] = 0$
9:     **for all** $(ou, ov) \in O(u, v)$ and $ou \neq ov$ **do**
10:       $res = \frac{c \cdot R[u,v]}{I(ou,ov)}$
11:       **if** $ou \in G$ and $ov \in G$ **then**
12:         $S[ou, ov] = S[ou, ov] + res$
13:         $R[ou, ov] = R[ou, ov] + res$
14:         **if** $R[ou, ov] \geq \frac{(1-c)\epsilon}{2c}$ and (ou, ov) not in Qu **then**
15:           $Qu.push(ou, ov)$
16:       **else**
17:         $R_{out}[ou, ov] = R_{out}[ou, ov] + res$
18: save(S)
19: **return** R_{out}

We use SubForwardFlow to update S_i in the first stage and S_{fine} in the second stage. The SubForwardFlow are shown in Algorithm 3. Algorithm Sub-ForwardFlow and ForwardFlow are very similar. But the SubForwardFlow will update and restore the outer SimRank values in R_{out} (line 17) and return R_{out} (line 19) in the first stage.

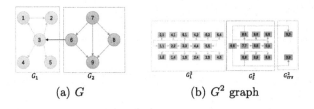

(a) G (b) G^2 graph

Fig. 6. Another Joy Graph

In Fig. 6, for example, after the locally coarse calculation stage, we can get $S_{coarse} = \begin{bmatrix} \tilde{S}_1 & \\ \hline & \tilde{S}_2 \end{bmatrix}$, and $R_{fine} = \begin{bmatrix} & R_{crs} \\ \hline R_{crs}^\top & \end{bmatrix}$. After the globally fine calculation stage, we can get $S_{fine} = \left[\begin{array}{c|c} \Delta S_1 & S_{crs} \\ \hline S_{crs}^{\top} & \Delta S_2 \end{array} \right]$. $\tilde{S}_1$ and $\tilde{S}_2$ are all dense, and

S_{fine} is sparse.

Theorem 4. *The results obtained from Algorithm 2 are the numerical solutions.*

Proof. From Theorem 3, we can get that the residuals of the locally coarse calculation are less than $\frac{\epsilon}{2}$, and the residuals of the globally fine calculation are also less than $\frac{\epsilon}{2}$. So the global residuals are less than $\frac{\epsilon}{2} + \frac{\epsilon}{2} = \epsilon$.

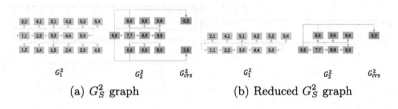

(a) G_S^2 graph   (b) Reduced G_S^2 graph

Fig. 7. Reduced SimRank Graph

3.2 Optimization

The SimRank matrix is symmetrical, $s(u,v) = s(v,u)$, and the SimRank graphs are also symmetrical. So we can just conduct the *SubForwardFlow* algorithm on reduced SimRank graphs, e.g., Fig. 7(b). The pruning operation is conducted at the first step (line 7-8), and the reduction operation is conducted at other steps (line 9-10). In Fig. 7(b), for example, after the locally coarse calculation stage, we can get S_{coarse}, in which we only need half of the space to store $\tilde{S}_1$ and $\tilde{S}_2$, and R_{fine}. After the globally fine calculation stage, we can get S_{fine}, in which we only need half of the space to store ΔS_1 and ΔS_2.

Algorithm 4. optSubForwardFlow(G^*, R, c, ϵ)

Require: $G^* = (G \cup Outer(G))$; residuals R; Maximum error ϵ and Decay factor c
Ensure: Outer residuals R_{out}
1: the same as line 1-4 of Algorithm 3
2: **while** $Q \neq \emptyset$ **do**
3:    $(u,v) = Q.pop()$
4:    $tmp = R[u,v]$
5:    $R[u,v] = 0$
6:    **for all** $(ou, ov) \in O(u,v)$ and $ou \neq ov$ **do**
7:       **if** $u == v$ and $ou < ov$ **then**
8:          continue
9:       **if** $ou < ov$ **then**
10:          swap(ou, ov)
11:       the same as line 11-17 of Algorithm 3
12: save(S)
13: **return** R_{out}

---

4   Related Work

The SimRank matrix can be calculated by *all-pairs* algorithms, or by traversing all nodes using the *single-source* algorithms. The *all-pairs* algorithms can be divided into Iterative methods and Non-iterative methods, and the single source algorithms can be divided into index based methods and index free methods.

4.1   All-Pairs SimRank Algorithms

Iterative Methods. There are lots of research on static all pairs SimRank calculations [3,6,9,12,14,19,21,22,24,27]. The naive iterative method [6] is the first All-Pairs SimRank algorithm, which requires $O(n^2)$ space and $O(km^2)$ time. By using *patial sum* in [12], the time complexity is improved to $O(kdn^2)$, Yu. [22] further improved the time complexity to $O(kd'n^2)$ ($d' < d$) by constructing a minimum spanning tree. Yu. [21,27] also improve the time complexity to $O(k \min\{mn, n^{\log_2^7}\})$ by CSR. For the iterative methods, k should be set at least $\log_c \epsilon$ for the given ϵ [12].

Non-Iterative. In paper [9], Li. come up with NI-SR with a linear formulation for the all-pairs SimRank calculations. Through SVD decomposition and the Kronecher production, they can quickly calculate and update the SimRank matrix. The time complexity of NI-SR is $O(r^4 n^2)$. Based on the linear formulation in [9], Yasuhiro. [3] come up with another non-iterative method, namely SimMat, and SimMat has a better time complexity of $O(rn)$. Based on G^2 graph, Lu. [14] come up with another non-iterative method BiPCG [14]. The time complexity of BiPCG is $O(kmn)$ and the space cost is $O(n^2)$. The state-of-the-art all-pairs SimRank calculating method is FLP(RLP) [19]. FLP improve the efficiency by only calculating the SimRank values which is larger than $(1 - c)\epsilon$ with local push methods. The space complexity of FLP(RLP) is $O(nonzero(R) + m + n)$, in which $nonzero(R)$ is the non-zero values of residual matrix, and the time complexity is $O\left(\frac{d^2}{c(1-c)^2 \epsilon}\right)$.

4.2   Other SimRank Algorithms

Index-Based. Index-Based Algorithms are Monte Carlo based algorithms, and the are generally divided into two phases, preprocessing phase and query phase. In the preprocessing phase, the algorithms will build an index to restore the random walks. Different algorithms have different methods of index construction and random walk strategies. Daniel F. firstly come up with a index based Monte Carlo algorithm for computing single source SimRank in MC [2]. They first build FingerPrint graphs to restore the random walks with a preprocessing time of $O(\frac{n \log(n)}{\epsilon^2})$. TSF [16] is another index-based single-source algorithm, and they compress the random walks in R_g one-way graphs. SLING [18] firstly use $\sqrt{c}$ walks to build the index with a preprocessing time of $O(\frac{m}{\epsilon} + \frac{n \log \frac{n}{\epsilon}}{\epsilon^2})$. READS [7] is an optimized version of $\sqrt{c}$ walk based algorithm. They build an index of

nR compressed $\sqrt{c}$ walks such that the algorithms only need to generate a few more walks in the query phase.

Index-Free. The index-free algorithms can be divide into into two phases, reversed walks and forward walks. The reversed walks to find the meeting nodes, and the forward walks to reach the nodes of non-zero similarity. Uniwalk [15] generates unidirectional random walks from the source node on undirected graphs, and Uniwalk regard the midpoint of the random walks as the meeting point. TopSim is an index-free algorithm based on local exploitation, but TopSim is only able to exploit a few levels on large graphs, which leads to a low precision. ProbeSim [11] apply the $\sqrt{c}$ walks in the reversed and forward walks, and use reverse reachability tree to compactly store the reversed $\sqrt{c}$ walks. Wei. [20] build a partial index by precomputing the $l - hopPPR$ of a small subset of v_k to further improve the efficiency of ProbeSim.

5 Experiments

In this section, we conduct experiments to verify the effectiveness of the proposed algorithms. All experiments are conducted on a machine with a Intel(R) Xeon(R) CPU E7-4820 v2 @ 2.00GHz CPU and 128 GB memory. All algorithms are implemented in C++ and compiled by g++ 5.4.0 with the -O3 option.

Table 2. Datasets

	Name	n	m	Average degree	Partitions
small	ca-GrQc(CG)	5242	14496	2.76	5
	wiki-Vote(WV)	7115	103689	14.57	5
	ca-HepTh(CH)	9877	25998	2.63	5
median	email-Enron(EN)	36692	367662	10.02	5
	loc-Brightkite(LB)	58228	214078	7	5
	soc-sign-Slashdot(SL)	77357	516575	6.7	5
large	Youtube(YT)	1134890	5975248	5.27	10
	soc-LiveJournal(LJ)	4847571	68993773	14.23	20

We use 8 datasets, which are commonly used in the literature [19,28] for all-pairs calculation. Our datasets used in experiments are from Stanford Large Network Dataset Collection[2]. The statistical information of these datasets is shown in Table 2. The major competitor is the state-of-the-art *all-pairs* solution *RLP* [19] (Reduced Forward Local Push). We set $c = 0.6$, and $\epsilon = 0.01$ by default. In our algorithm, namely *HF*, the default number of partitions we devide the graph is shown in Table 2. We divide the graph into blocks as shown in the table. The standard of our graph division is mainly to put each sub-graph into RAM for calculation, while taking into account the calculation efficiency.

---

[2] http://snap.stanford.edu/data/index.html.

Metrics. We use the Naive method [6] with 55 iterations to compute the ground-truth SimRank values of each node pair. Since the naive method can only be conducted in small and medium graphs. Therefore, we only verify the correctness of the algorithms on small and medium graphs. We compute the maximum error with respect to the ground-truth.

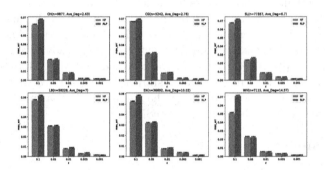

Fig. 8. Maximum Error

Effectness. We conduct *RLP* and *HF* on small and medium datasets, with ϵ changes from 0.001 to 0.1. As is shown in Fig. 8, the maximum errors of *RLP* and *HF* are marginally below the set *epsilon*, which not only satisfied the error requirements, but also reduces some unnecessary calculations. Meanwhile, we can find that almost all the maximum errors of algorithm *HF* are lower than the maximum errors of *RLP*. This is because the termination condition in *HF* is half of *RLP*. So there are more pushes in *HF*, which leads to smaller absolute errors.

Time Efficiency

Varying ϵ The CPU time costs are shown in Fig. 9 with the varying of ϵ. We can see that the CPU time of *RLP* increase sharply with the decrease of ϵ, and *HF* is at most 100× faster than *RLP* when $\epsilon = 0.001$. The reason is that the updating operations in RAM is much faster than the operations in hash tables when the residuals matrix is very large. Even there are more updating operations in *HF*, *HF* still much faster overall, especially when ϵ is very small. From Fig. 9 we can see when $\epsilon \leq 0.01$, *HF* performs better on most graphs.

Varying Average Degree. In addition, the efficiency of *RLP* will be greatly affected by the increasing of the average degrees. This is because the scale of residuals matrix R is increasing exponentially. In *HF*, however, most of the updates are conducted in sub-models, and the sub-models are updated in RAM. As is shown in Fig. 9, with the increase of accuracy and average degrees, *HF* performs better than *RLP*.

We also find that when $\epsilon \leq 0.01$ and $avg\_deg \leq 7$(graph CH, CG and LB), the efficiency of *HF* is a little worse. That's because the scale of nonzero data of *RLP* is not very large, and the initialization conductions of *HF* costs too much time. However, *HF* is less sensitive to the scale of nonzero data during the calculations. For example, when the scale of nonzero number is large, e.g. $\epsilon < 0.01$, and the average degrees are very high ($avg\_deg \geq 7$), the efficiency of *RLP* will be greatly affected.

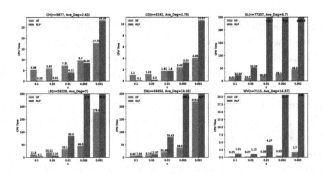

Fig. 9. Time Efficient

Memory Efficiency

*Varying ϵ*As is shown in Fig. 10, *HF* is in better performance when $\epsilon \leq 0.01$. However, the space consumption of *HF* is a little larger when $\epsilon \geq 0.01$. Because in *HF* most of the update operations are conducted in RAM, when $\epsilon > 0.01$, some of the pre-allocated space is unused.

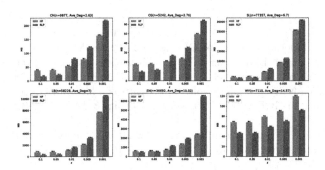

Fig. 10. Memory Efficient

Large Graphs

We also conduct experiments on large-scale graphs, YT and LJ. When we set $\epsilon = 0.001$, the memory consumption of *RLP* exceeds the limit. But algorithm *HF* can still do calculations, because we can limit the memory consumption of each sub-model by increasing the number of sub models.

6   Conclusions

In this paper, we derived a simpler closed form of SimRank, and proposed a Hierarchical Framework to calculate the *all-pairs* SimRank. We further optimize the proposed algorithm by reducing half unnecessary update operations. The experiments shows that *HF* performs better than the state of the art algorithm *RLP*.

In the Hierarchical Framework, the sub-modules can be small, most of the updating in *HF* can be conducted in RAM completely. Through the experiment, even the *HF* makes more 'pushes' than *RLP*, however, most of the operations are conducted in RAM rather than hash tables, the time efficiency of a single update operation is 5 to 10 times that of *RLP*, therefore, the overall speed of *HF* is much faster. Meanwhile, as we can limit the memory consumption of each sub-model by increasing the number of sub-models, so the *HF* can be applied on a large scale graph with high accuracy. Our experiments also show that *HF* outperforms the state-of-the-art *all-pairs* SimRank algorithm *RLP*.

Acknowledgements. This work is supported by National Natural Science Foundation of China under the grant No. 62072460, 62076245, 62172424, 62276270, and Beijing Natural Science Foundation (4212022).

References

1. Abbassi, Z., Mirrokni, V.S.: A recommender system based on local random walks and spectral methods. In: WebKDD, pp. 102–108. ACM (2007)
2. Fogaras, D., Rácz, B.: Scaling link-based similarity search. In: Proceedings of the 14th International Conference on World Wide Web, pp. 641–650 (2005)
3. Fujiwara, Y., Nakatsuji, M., Shiokawa, H., Onizuka, M.: Efficient search algorithm for simrank. In: ICDE 2013, Brisbane, Australia, 8–12 April 2013, pp. 589–600 (2013)
4. He, G., Feng, H., Li, C., Chen, H.: Parallel SimRank computation on large graphs with iterative aggregation. In: SIGKDD, Washington, DC, USA, 25–28 July 2010, pp. 543–552 (2010)
5. He, G., Li, C., Chen, H., Du, X., Feng, H.: Using graphics processors for high performance SimRank computation. IEEE Trans. Knowl. Data Eng. **24**(9), 1711–1725 (2012)

6. Jeh, G., Widom, J.: Simrank: a measure of structural-context similarity. In: Proceedings of the Eighth ACM SIGKDD, 23–26 July 2002, Edmonton, Alberta, Canada, pp. 538–543 (2002)
7. Jiang, M., Fu, A.W.C., Wong, R.C.W.: Reads: a random walk approach for efficient and accurate dynamic SimRank. PVLDB **10**(9), 937–948 (2017)
8. Jin, R., Lee, V.E., Hong, H.: Axiomatic ranking of network role similarity. In: Proceedings of the 17th ACM SIGKDD, pp. 922–930. ACM (2011)
9. Li, C., et al.: Fast computation of SimRank for static and dynamic information networks. In: EDBT 2010, 13th International Conference on Extending Database Technology, Lausanne, Switzerland, 22–26 March 2010, Proceedings, pp. 465–476 (2010)
10. Lin, Z., Lyu, M.R., King, I.: MatchSim: a novel similarity measure based on maximum neighborhood matching. Knowl. Inf. Syst. **32**(1), 141–166 (2012)
11. Liu, Y., et al.: ProbeSim: scalable single-source and top-k SimRank computations on dynamic graphs. Proceed. VLDB Endow. **11**(1), 14–26 (2017)
12. Lizorkin, D., Velikhov, P., Grinev, M.N., Turdakov, D.: Accuracy estimate and optimization techniques for SimRank computation. PVLDB **1**(1), 422–433 (2008)
13. Lizorkin, D., Velikhov, P., Grinev, M.N., Turdakov, D.: Accuracy estimate and optimization techniques for SimRank computation. VLDB J. **19**(1), 45–66 (2010)
14. Lu, J., Gong, Z., Lin, X.: A novel and fast SimRank algorithm. IEEE Trans. Knowl. Data Eng. **29**(3), 572–585 (2017)
15. Luo, X., Gao, J., Zhou, C., Yu, J.X.: UniWalk: unidirectional random walk based scalable SimRank computation over large graph. In: 2017 IEEE 33rd International Conference on Data Engineering (ICDE), pp. 325–336. IEEE (2017)
16. Shao, Y., Cui, B., Chen, L., Liu, M., Xie, X.: An efficient similarity search framework for SimRank over large dynamic graphs. Proceed. VLDB Endow. **8**(8), 838–849 (2015)
17. Spirin, N., Han, J.: Survey on web spam detection: principles and algorithms. ACM SIGKDD Explorations Newsl **13**(2), 50–64 (2012)
18. Tian, B., Xiao, X.: SLING: a near-optimal index structure for SimRank. In: Proceedings of the 2016 International Conference on Management of Data, SIGMOD Conference 2016, San Francisco, CA, USA, 26 June - 01 July 2016, pp. 1859–1874 (2016)
19. Wang, Y., Lian, X., Chen, L.: Efficient SimRank tracking in dynamic graphs. In: 2018 IEEE 34th International Conference on Data Engineering (ICDE), pp. 545–556. IEEE (2018)
20. Wei, Z., et al.: PRSim: Sublinear time SimRank computation on large power-law graphs. In: Proceedings of the 2019 International Conference on Management of Data, pp. 1042–1059 (2019)
21. Yu, W., Lin, X., Le, J.: A space and time efficient algorithm for SimRank computation. In: Advances in Web Technologies and Applications, Proceedings of the 12th Asia-Pacific Web Conference, APWeb 2010, Busan, Korea, 6–8 April 2010, pp. 164–170 (2010)
22. Yu, W., Lin, X., Zhang, W.: Towards efficient SimRank computation on large networks. In: 29th IEEE International Conference on Data Engineering, ICDE 2013, Brisbane, Australia, 8–12 April 2013, pp. 601–612 (2013)
23. Yu, W., Lin, X., Zhang, W.: Fast incremental SimRank on link-evolving graphs. In: IEEE 30th International Conference on Data Engineering, Chicago, ICDE 2014, IL, USA, 31 March - 4 April 2014, pp. 304–315 (2014)
24. Yu, W., Lin, X., Zhang, W., McCann, J.A.: Dynamical SimRank search on time-varying networks. VLDB J.-Int. J. Very Large Data Bases **27**(1), 79–104 (2018)

25. Yu, W., McCann, J.A.: Sig-SR: SimRank search over singular graphs. In: The 37th International ACM SIGIR Conference on Research and Development in Information Retrieval, SIGIR 2014, Gold Coast, QLD, Australia, 06-11 July 2014, pp. 859–862 (2014)

26. Yu, W., McCann, J.A.: High quality graph-based similarity search. In: Proceedings of the 38th International ACM SIGIR Conference on Research and Development in Information Retrieval, Santiago, Chile, 9–13 August 2015, pp. 83–92 (2015)

27. Yu, W., Zhang, W., Lin, X., Zhang, Q., Le, J.: A space and time efficient algorithm for SimRank computation. World Wide Web **15**(3), 327–353 (2012)

28. Zhang, Y., Li, C., Chen, H., Sheng, L.: Fast SimRank computation over disk-resident graphs. In: Meng, W., Feng, L., Bressan, S., Winiwarter, W., Song, W. (eds.) DASFAA 2013. LNCS, vol. 7826, pp. 16–30. Springer, Heidelberg (2013). https://doi.org/10.1007/978-3-642-37450-0_2

Contraction Hierarchies with Label Restrictions Maintenance in Dynamic Road Networks

Bo Feng[1], Zi Chen[1(✉)], Long Yuan[2], Xuemin Lin[3], and Liping Wang[1]

[1] East China Normal University, Shanghai, China
bfeng@stu.ecnu.edu.cn, {zchen,lipingwang}@sei.ecnu.edu.cn
[2] Nanjing University of Science and Technology, Nanjing, China
longyuan@njust.edu.cn
[3] The Shanghai Jiao Tong University, Shanghai, China
xuemin.lin@sjtu.edu.cn

Abstract. In the real world, road networks with weight and label on edges can be applied in several application domains. The shortest path query with label restrictions has been receiving increasing attention recently. To efficiently answer such kind of queries, a novel index, namely, Contraction Hierarchies with Label Restrictions (CHLR), is proposed in the literature. However, existing studies mainly focus on the static road networks and do not support the CHLR maintenance when the road networks are dynamically changed. Motivated by this, in this paper, we investigate the CHLR maintenance problem in dynamic road networks. We first devise a baseline approach to update CHLR by recomputing the potential affected shortcuts. However, many shortcuts recomputed in baseline do not change in fact, which leads to unnecessary overhead of the baseline. To overcome the drawbacks of baseline, we further propose a novel CHLR maintenance algorithm which can only travel little shortcuts through an update propagate chain with accuracy guarantee. Moreover, an optimization strategy is presented to further improve the efficiency of index maintenance. Extensive and comprehensive experiments are conducted on real road networks. The experimental results demonstrate the efficiency and effectiveness of our proposed algorithms.

Keywords: Contraction Hierarchy with Label Restrictions · Dynamic Road Networks · Shortest Path Query

1 Introduction

Shortest path query (SPQ) in road networks is a fundamental problem in graph analysis and has been widely studied due to its applications in various fields. Recently, with the proliferation of applications, the information of road networks becomes more complex and diversified, such as edge weight (e.g., transit time, toll fee) and edge label (e.g., toll road, trucks prohibited road). Existing works mainly aim to address the SPQ with minimum weight, i.e., find the shortest path with minimum distance or toll fee between two locations [1,3,6,12,14,28]. However, the label information of edges can service more

© The Author(s), under exclusive license to Springer Nature Switzerland AG 2023
X. Wang et al. (Eds.): DASFAA 2023, LNCS 13945, pp. 269–285, 2023.
https://doi.org/10.1007/978-3-031-30675-4_18

personalized scenarios. For example, if the drivers do not want to pay the toll fee, all toll roads should be neglected in the route planning [5]. Moreover, if a truck driver requires path service, the path returned should avoid all trucks prohibited roads [21]. The SPQ with considering the label constraints has many practical applications, such as personalized location-based services [29], logistics and commercial transportation [13]. [24] formalizes this problem as $\underline{L}$abel $\underline{C}$onstrained $\underline{S}$hortest $\underline{P}$ath (LCSP) query. Formally, given a restricted label set R, LCSP query aims to find the shortest path with minimum weight avoiding the edges with labels in R. To efficiently answer the LCSP query, [24] proposes a novel index which is based on Contraction Hierarchies, namely, $\underline{C}$ontraction $\underline{H}$ierarchies with $\underline{L}$abel $\underline{R}$estrictions (CHLR). In CHLR, a *shortcut* $e = ((u, v), w, l)$ denotes a path from u to v with weight w and label l. By shortening the path traversal, CHLR can significantly improve the LCSP query processing performance.

Motivation. In the real world, road networks can be dynamically changed. For each edge, the toll fee can be changed at any time by traffic department and the transit time is time-dependent due to the change of traffic conditions [23,31]. Specifically, the traffic conditions can be updated by collecting the travel information of taxi-hailing app users and drivers (e.g., Didi, Uber). However, the number of app users and drivers is huge. In 2022, Didi has 15 millions drivers and 493 millions users [11], Uber has 3.5 millions drivers and 93 millions users [9]. They can generate massive updates for traffic conditions per second. Accordingly, the edge weight can be dynamically changed at a high frequency. Moreover, the label of edges can be changed as well. For example, roads can be labelled as light traffic road, medium traffic road and heavy traffic road depending on the traffic conditions which will change in real time [19]. Unfortunately, most existing works for CHLR focus on static road networks and do not support the CHLR maintenance in dynamic road networks, which makes them unpractical for real applications. Motivated by this, in this paper, we study the CHLR maintenance problem in dynamic road networks.

Our Approach. To solve the CHLR maintenance problem in dynamic road networks, we first propose a Baseline approach. When an edge e_o changes its weight and label, a naive approach is to find all potential affected shortcuts due to the change of e_o, and then recompute the affected shortcuts by invoking the existing CHLR construction algorithm [24]. However, the Baseline would recompute lots of shortcuts that actually do not change, which leads to expensive overhead of Baseline. To overcome the drawbacks of Baseline, in this paper, we devise a more efficient CHLR maintenance algorithm. Specifically, we design a novel *update propagate chain* to avoid massive invalid updates in Baseline. In the update propagate chain, the *neighborhood of shortcuts* are defined as *parents*, *child* and *partner*. We observe that a shortcut only affects its child, and the weight of child can be obtained by its parents. We give an *update order* for the update propagate chain following the observation. Moreover, to further improve the efficiency, we propose an optimization strategy based on *weight count* of shortcuts, which can guarantee that shortcuts will only be recomputed when necessary. We prove that the optimized algorithm has a tightly theoretical boundedness in terms of the number of changed shortcuts E_Δ, i.e., $O(|E_\Delta| \cdot log|E_\Delta| + |E_\Delta| \cdot d_{\max})$, where $d_{\max}$ is the maximum degree.

Contributions. This paper has the following contributions.

(1) *The first work for CHLR maintenance in dynamic road networks.* In this paper we aim to incrementally maintain the CHLR when the road networks are dynamically changed. To the best of our knowledge, this is the first work to address the CHLR maintenance problem in dynamic road networks.

(2) *Efficient CHLR maintenance algorithms.* In this paper, we devise an efficient CHLR maintenance algorithm. Specifically, we design a novel update propagate chain to only propagate the updated shortcuts, which can avoid massive invalid updates in Baseline. Moreover, we propose an optimization strategy based on weight count of shortcuts, which can guarantee that shortcuts will only be recomputed when necessary. We prove that the optimized algorithm has a tightly theoretical boundedness.

(3) *Extensive experiments on the real road networks.* We conduct extensive and comprehensive experiments on real road networks. As shown in our experiments, the optimized CHLR maintenance algorithm has good efficiency and effectiveness. It can achieve up to 2 orders of magnitude speedup compared with Baseline.

2 Related Work

Shortest Path Indexes in Road Networks. Graph analysis has been receiving increasing attention recently [7,8,18,20,26,27,32]. In the literature, many shortest path query (SPQ) algorithms are proposed based on traditional methods, such as Dijkstra [12], they compute the shortest path on the weighted road networks directly. Recently, many index-based SPQ algorithms are proposed achieving remarkable results on speeding up the query process. Hub Labeling (HL) [1,3,15] is one of the most important shortest path indexes , which can improve the efficiency of SPQ by reducing the points on the shortest path. Pruned Landmark Labeling (PLL) [2] and Parallel Shortest-distance Labeling (PSL) [16] are further proposed based on HL. Arterial Hierarchy (AH) [33] index can improve the query time efficiently by splitting a graph into grid structure. Contraction Hierarchies (CH) [4,10,14] is a fundamental index that can reduce the number of iterations during the SPQ process by introducing shortcuts. Moreover, Hierarchical 2-hop labelling(H2H) [22] and P2H [6] combine the advantages of 2-hop labelling and hierarchy among all vertices. Based on CH, [24] proposes an important shortest path index under the label restrictions (CHLR) in labelled road networks. [17,25] experimentally investigate the performance of different SPQ algorithms in multiple dimensions.

Shortest Path Index Maintenance in Dynamic Road Networks. In real world, road networks often change in practice. Hence, it is necessary to study the maintenance of shortest path indexes. [30] gives a solution for maintaining HL index in dynamic road networks. [6] introduces mechanisms for P2H index maintenance for edge weight updating. In order to maintain CH index, a number of methods [23] are proposed. Moreover, for the theoretical boundedness of the dynamic maintenance of CH and H2H, [31] presents the theoretical analysis according to the weight increase and decrease scenario separately. However, the CHLR maintenance problem in dynamic road networks remains to be investigated.

3   Preliminaries

Let $G = (V, E, w, \Sigma, \ell)$ be a directed, weighted and labelled road network, where V and E are the set of vertices and edges in the road network, respectively. w represents the weight function $w : E \rightarrow \mathbb{R}^+$ mapping each edge to a positive weight number. Σ is a finite alphabet used for labeling edges in G and ℓ is a function $\ell : E \rightarrow \Sigma$ mapping edges to labels in Σ. Besides, a vertex order function $\phi : V \rightarrow \{1, \cdots, |V|\}$ is used to sort vertices in G. The neighbors of a vertex v are denoted as $N(v)$. The neighbors with higher/lower order than v are denoted as $N^{+/-}(v)$. We use n and m to denote the number of vertices and edges in the road network, respectively. A path in G is denoted as $P_{s,t} = \{e_1, e_2, \cdots, e_k\}$ passing from s to t. The length of the path is expressed as $w(P_{s,t}) = \sum_{1 \leq i \leq k} w(e_i)$. The label of path is denoted as $\ell(P_{s,t}) = \ell(e_1) \cup \ell(e_2) \cdots \cup \ell(e_k)$. The order of P is the minimum order of vertices in the path, denoted as $\phi(P)$. For simplicity, G is denoted as $G = (V, E)$ if the context is self-evident. Note that the approach proposed in this paper can be smoothly adapted to undirected road networks as an undirected edge can be considered as two directed edges.

Definition 1. *(**L**abel **C**onstrained **S**hortest **P**ath (LCSP)) Given a road network $G = (V, E)$, a restricted label set R and two vertices s and t, the LCSP under restricted labels R from s to t is a path $P_{s,t}^R = \{e_1, e_2, \cdots, e_k\}$, such that, (1) for $1 \leq i \leq k$, $\ell(e_i) \cap R = \emptyset$, i.e., the path avoids all restricted labels in R; (2) $w(P_{s,t}^R)$ is minimum, i.e., $P_{s,t}^R$ is the shortest path satisfying condition (1).*

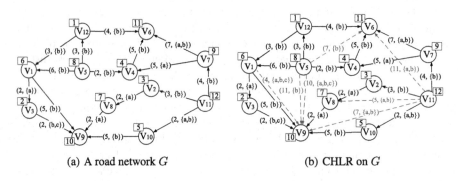

(a) A road network G (b) CHLR on G

Fig. 1. An example for road network G and CHLR on G

Example 1. Consider the road network G shown in Fig. 1, for a LCSP query with $s = v_{11}$, $t = v_9$, $R = \{c\}$, there are two LCSPs $P_1 = \{(v_{11}, v_2), (v_2, v_8), (v_8, v_9)\}$ and $P_2 = \{(v_{11}, v_{10}), (v_{10}, v_9)\}$. For both paths, the length is 7, the restricted label c is avoided, and there is no other path shorter than them without label c.

3.1 Contraction Hierarchies with Label Restrictions

To address LCSP query efficiently, [24] proposes a novel shortest path index based on Contraction Hierarchies (CH) with label restrictions.

Definition 2. (_C_ontraction _H_ierarchies with _L_abel _R_estrictions (CHLR) *Given a road network $G = (V, E)$ and a vertex order function ϕ, for two edges $e' = (u, q)$ and $e'' = (q, v)$ where $\phi(q) < \phi(u)$, $\phi(q) < \phi(v)$, a **shortcut** $e = ((u, v), w, l)$ is created with $w = w(e') + w(e'')$ and $l = \ell(e') \cup \ell(e'')$ if there is no path P' shorter than $P = \{e', e''\}$ satisfying $\phi(P') > \phi(P)$ and $\ell(P') \subseteq \ell(P)$. The CHLR index on G is the set of shortcuts, denoted as E'. The road network with CHLR is called **augmented graph**, denoted as $G' = (V, E \cup E')$.*

Example 2. Considering the road network G in Fig. 1(a), the order of each vertex is marked close the vertex. The CHLR index on G is shown in Fig. 1(b), which is represented in blue dotted lines. For instance, $e = ((v_1, v_9), 4, \{a, b, c\})$ is a shortcut for edges (v_1, v_3) and (v_3, v_9), as $\phi(v_3) < \phi(v_1)$, $\phi(v_3) < \phi(v_9)$ and there is no path P' from v_1 to v_9 shorter than $P = \{(v_1, v_3), (v_3, v_9)\}$ satisfying $\phi(P') > \phi(P)$ and $\ell(P') \subseteq \ell(P)$.

CHLR-Based LCSP Query. Given a road network G, and G' with CHLR, to address a LCSP query for $P_{s,t}^R$, the result can be retrieved by using a bidirectional Dijkstra query variant which performs a simultaneous forward query from s in upward graph $G'^{\uparrow}$ and backward query from t in the downward graph $G'^{\downarrow}$, where $E(G'^{\uparrow}) = \{e = (u, v) | \phi(u) < \phi(v) \wedge \ell(e) \cap R = \emptyset\}$ and $E(G'^{\downarrow}) = \{e = (u, v) | \phi(u) > \phi(v) \wedge \ell(e) \cap R = \emptyset\}$. The query terminates once the minimum key for the priority queue of Dijkstra's algorithm exceeds the tentative shortest distance and $P_{s,t}^R$ on G is obtained.

CHLR Construction. Given a road network G and a vertex order function ϕ, CHLR can be constructed through contracting vertices in the increasing order of ϕ. When contracting a vertex q, based on Definition 2, for each pair of neighbors (u, v) in $N^+(q)$, a potential shortcut $e = (u, v)$ is considered with $\ell(e) = \ell((u, q)) \cup \ell((q, v))$ and $w(e) = w((u, q)) + w((q, v))$. If there is a path P' shorter than $P = \{(u, q), (q, v)\}$ satisfying $\phi(P') > \phi(P)$ and $\ell(P') \subseteq \ell(P)$, e is unnecessary and can be omitted in G'. Otherwise, e is created as a shortcut and inserted into G'. P' can be retrieved by Dijkstra's algorithm from u to v through vertices with higher order than q and edges with labels in $\ell(P)$. When all vertices are contracted, all shortcuts in CHLR are obtained.

3.2 Problem Statement

In this paper, we investigate the CHLR maintenance problem on dynamic road networks. Given a road network G and edge change including weight increase/decrease and label change, we incrementally update the CHLR index on G after the edge change.

4 Baseline Approach

To address the CHLR maintenance problem, in this section, we first explore a Baseline approach. Consider a changed edge $e = (q, v)$ with new weight or label, the main idea of Baseline is to explore the affected shortcuts and recompute the affected shortcuts. There are two steps: 1) *Explore the affected shortcuts.* Due to the change of $e = (q, v)$, assume $\phi(q) < \phi(v)$, the shortcuts like $e' = (u, v)$ produced by contracting q and incident to e may be changed. As e' can be updated by recontracting q who can produce e', q is marked as the vertex to be recontracted and stored into Θ. Recursively, the change of e' may further affect more shortcuts. Similarly, the vertices supporting the affected shortcuts are stored into Θ. 2) *Recompute the affected shortcuts.* When all effected shortcuts are explored, they can be updated by recontracting the vertex in Θ by invoking the existing CHLR construction algorithm [24].

Example 3. Consider the road network G shown in Fig. 1. If an edge $e = ((v_4, v_6), 5, \{b\})$ is changed to $e = ((v_4, v_6), 1, \{c\})$, following the Baseline, at step 1, the vertices to be recontracted in Θ is $\{v_4, \cdots, v_7, v_{11}, v_{12}\}$. Then, we recontract them in the increasing order of ϕ at step 2. For instance, when recontracting v_4, it will produce two new shortcuts $e_1 = ((v_5, v_6), 3, \{b, c\})$ and $e_2 = ((v_7, v_6), 6, \{a, c\})$.

Theorem 1. *The time complexity of* Baseline *is* $O(|\Theta| \cdot d_{\max}^2 \cdot m \cdot \log n)$, *where* $d_{\max}$ *is the maximum degree of vertices in* G'.

Proof. At step1, for each affected shortcut e, exploring the shortcuts that e further affects and finding the vertices that can update e consume $O(d_{\max})$ time, and the number of the affected shortcuts is $O(|\Theta|)$. Next, at step 2, for each vertex in Θ, it can produce $O(d_{\max}^2)$ shortcuts. Furthermore, to generate a shortcut, [24] invokes Dijkstra's algorithm with time complexity $O(m \cdot \log n)$. Therefore, the total time complexity of Baseline is $O(|\Theta| \cdot d_{\max}^2 \cdot m \cdot \log n)$.

Drawbacks of Baseline. Unfortunately, the Baseline approach may introduce unnecessary overhead. First, the affected shortcuts explored at step 1 may do not change their weights and labels. Second, at step 2, all shortcuts incident to vertices in Θ will be recomputed by vertex contraction. However, most of them are unchanged in fact. All of these lead to unnecessary overhead of Baseline.

5   CHLR Maintenance Algorithm

In this section, we explore a new CHLR maintenance algorithm to update index more efficiently by avoiding the overhead of Baseline for unchanged shortcuts.

Framework. Consider a road network G and a changed edge e, due to the weight or label change of e, we can recursively update shortcuts in a BFS-like way. In our new CHLR maintenance algorithm, we aim to reveal an update propagate chain among shortcuts. Only when current shortcut e' is changed, it will extend the update propagate chain to update more shortcuts incident to e'. In this way, the unnecessary overhead of Baseline for updating unchanged shortcuts can be reduced sharply.

Definition 3. (Neighborhood of Shortcuts) *Given the CHLR index on the road network* G, *for a shortcut* $e = (u, v)$ *and two shortcuts* $e' = (u, q)$ *and* $e'' = (q, v)$, *if* $\phi(q) < \phi(u)$, $\phi(q) < \phi(v)$ *and* $\ell(e) = \ell(e') \cup \ell(e'')$, *then,* e' *and* e'' *are the* **parents** *of* e, *denoted as* $(e', e'') \in N^-(e)$. e *is the* **child** *of* e' *and* e'', *denoted as* $e = N^+(e', e'')$. e' *is a* **partner** *of* e'', *denoted as* $e' \in N^=(e'')$.

Algorithm 1. CalWeight

Input: shortcut: e
Output: the weight of e: k

1: $k \leftarrow \infty$; $cnt(e) \leftarrow 0$;
2: **if** $e \in G$ and $w(e) \neq \infty$ **then** $k \leftarrow w(e)$; $cnt(e) \leftarrow 1$;
3: **for each** $(e', e'') \in N^-(e)$ **do**
4:     **if** $w(e') + w(e'') < k$ **then** $k \leftarrow w(e') + w(e'')$; $cnt(e) \leftarrow 1$;
5:     **else if** $w(e') + w(e'') = k$ **then** $cnt(e) \leftarrow cnt(e) + 1$;
6: **return** k;

Lemma 1. *Given a road network G and a shortcut e, the weight of e can be computed by $w(e) = min\{w, w(e') + w(e'') | \forall (e', e'') \in N^-(e)\}$, where w is the weight of e in G if e is an original edge in G.*

Algorithm of CalWeight. Based on Lemma 1, we first propose a weight calculation algorithm to compute the weights of shortcuts on the augmented graph G', which is named CalWeight. The pseudocode is shown in Algorithm 1. Since the pseudocode is self-explanatory, we omit the detailed description for brevity. Note that cnt of Algorithm 1 can be ignored here which will be introduced later.

After obtaining the weights of shortcuts, based on Definition 2, the shortcuts with same (u, v) but different labels may conflict with each other in G'. For instance, there are two shortcuts $e = ((u, v), w_1, \{a, b\})$ and $e' = ((u, v), w_2, \{a, b, c\})$, $w_1 < w_2$. Then, for any LCSP query with restricted label set R, if the target $P_{s,t}^R$ passes through u and v, e has higher priority than e' and $P_{s,t}^R$ will never pass through e'. It's because e' has heavier weight and stricter label constraint than e. To distinguish the conflict relationship, we propose the formal definition of shortcut dominance below.

Definition 4. (Shortcut Dominance) *Given a road network G and CHLR on G, consider two shortcuts e and e' for vertices u and v with different weights or labels, if $w(e) \leq w(e')$ and $\ell(e) \subset \ell(e')$, then e dominates e', e' is dominated by e. $\Phi(e)$ is the dominant shortcut set of e, $e' \in \Phi(e)$.*

Lemma 2. *Given a road network G and CHLR on G, considering two shortcuts e and e', if e dominates e', then e' can be removed from CHLR safely.*

Algorithm of KeepSCDom. Following Lemma 2, we propose a KeepSCDom algorithm to keep the shortcut dominance, the pseudocode is shown in Algorithm 2. Given a shortcut e, type is used to mark the weight increase or decrease of e, Q is a priority queue to store the changed shortcuts. Based on Definition 4, if e can be dominated by e' in the augmented graph G', e is marked as invalid shortcut in G', the weight of e is assigned as ∞ (line 1–2). Otherwise, if the weight of e increases, we recompute the weight of shortcut e' in $\Phi(e)$, if $w(e') < w(e)$, e' is not dominated by e anymore. In this case, e' becomes valid in G' (line 4–9). In another hand, if the weight of e decreases, it can dominate more shortcuts who have heavier weight and longer label than e. Such dominated shortcuts become invalid in G' (line 10–14). Since the weights of shortcuts

---

Algorithm 2. KeepSCDom

Input: shortcut: e; type
Output: the queue: Q
1: **if** $\exists e' \in G'$, s.t., $e'.u = e.u$ and $e'.v = e.v$ and $\ell(e') \subset \ell(e)$ and $w(e') \leq w(e)$ **then**
2:     $w(e) \leftarrow \infty$; $\text{cnt}(e) \leftarrow 0$; $\Phi(e') \leftarrow \Phi(e') \cup e$;
3: **else**
4:     **if** type $=$'inc' **then**
5:         **for each** $e' \in \Phi(e)$ **do**
6:             $k \leftarrow$ CalWeight (e');
7:             **if** $k < w(e)$ **then**
8:                 **if** $(e',$'dec'$) \notin Q$ **then** Q.push$(e',$'dec'$)$;
9:                 $w(e') \leftarrow k$; $\Phi(e) \leftarrow \Phi(e) \setminus e'$; $G' \leftarrow G' \cup \{e'\}$;
10:     **if** type $=$'dec' **then**
11:         **for each** $e' \in G'$, s.t., $e'.u = e.u$ and $e'.v = e.v$ and $\ell(e') \supset \ell(e)$ **do**
12:             **if** $w(e') \geq w(e)$ **then**
13:                 **if** $(e',$'inc'$) \notin Q$ **then** Q.push$(e',$'inc'$)$;
14:                 $w(e') \leftarrow \infty$; $\text{cnt}(e') \leftarrow 0$; $\Phi(e') \leftarrow \Phi(e') \cup e'$;
15: **return** Q;

---

in line 9 and line 14 are updated, they are pushed into Q to further update more short-cuts.

Now, we analyse the update propagate chain among shortcuts.

Definition 5. (Update Propagate Chain) *When the weight of a shortcut e decreases or increases, e only changes its child shortcuts e'' in $N^+(e, e')$, where e' is a partner of e in $N^=(e)$. Specifically,*

– *when $w(e)$ decreases, update $w(e'') = w(e) + w(e')$ if $w(e'') > w(e) + w(e')$;*
– *when $w(e)$ increases, recompute $w(e'')$ by invoking* CalWeight *algorithm.*

If $w(e'')$ changes, KeepSCDom *algorithm is utilized to keep the shortcut dominance and e'' is pushed into Q to be the next shortcut of the update propagate chain.*

Definition 6. (Update Order) *Given the augmented graph G', the update order of edges $e = ((u, v), w, l)$ is the order to be processed in the priority queue Q, denoted as $\phi(e)$. For other arbitrary edge $e' = ((u', v'), w', l')$, $\phi(e) < \phi(e')$ if $min\{\phi(u), \phi(v)\} < min\{\phi(u'), \phi(v')\}$ or $min\{\phi(u), \phi(v)\} = min\{\phi(u'), \phi(v')\}$ and $|l| < |l'|$.*

Theorem 2. *Based on Definition 5 and Definition 6, when an edge change for e arrives, performing the update propagate chain from e and propagating the update following the update order can update CHLR correctly.*

Proof. Theorem 2 holds based on two significant facts. (1) Given an updated shortcut e, e only affects its child shortcuts e''. The weight of e'' can be updated following Definition 5. Specifically, if $w(e)$ decreases, it only needs to determine whether e will produce a shorter path for e'', if so, $w(e'') = w(e) + w(e')$. Otherwise, if $w(e)$ increases,

Algorithm 3. CHLRMaintenance

Input: augmented graph: G'; changed edge: $e_o = ((u, v), w_o, l_o)$; new weight: w_n; new label: l_n

Output: the updated augmented graph: G'

1: Q: priority queue; $Q \leftarrow \emptyset$;
2: **if** $l_n \neq l_o$ **then**
3:     $e_o \leftarrow ((u, v), \infty, l_o)$; $e_n \leftarrow ((u, v), w(e_n), l_n)$; $w(e_o) \leftarrow$ CalWeight (e_o);
4:     **if** $w(e_o) > w_o$ **then** Q.push(e_o,'inc'); KeepSCDom $(e_o,$'inc'$)$;
5:     **if** $w(e_n) > w_n$ **then** $w(e_n) \leftarrow w_n$; Q.push(e_n,'dec'); KeepSCDom $(e_n,$'dec'$)$;
6: **else**
7:     $e_o \leftarrow ((u, v), w_n, l_o)$; $w(e_o) \leftarrow$ CalWeight (e_o);
8:     Q.push($e_o, w(e_o) > w_o$?'inc':'dec'); KeepSCDom $(e_o, w(e_o) > w_o$?'inc':'dec'$)$;
9: **while** !Q.empty() **do**
10:     $(e,\text{type}) \leftarrow Q$.pop();
11:     * **update weight for child shortcut** *\
12:     **for each** $e' \in N^=(e)$ **do**
13:         $e'' \leftarrow N^+(e, e')$;
14:         **if** type =‘inc’ **and** $w(e'') \neq \infty$ **then**
15:             * **recompute weight of** e'' *\
16:             $k \leftarrow$ CalWeight (e'');
17:             **if** $w(e'') < k$ **and** $(e'',$‘inc’$) \notin Q$ **then**
18:                 $w(e'') \leftarrow k$; Q.push(e'',‘inc’); KeepSCDom $(e'',$‘inc’$)$;
19:         **if** type =‘dec’ **then**
20:             * **directly update weight of** e'' *\
21:             **if** $w(e'') > w(e) + w(e')$ **then**
22:                 $w(e'') \leftarrow w(e) + w(e')$;
23:                 **if** $(e'',$‘dec’$) \notin Q$ **then** Q.push(e'',‘dec’); KeepSCDom $(e'',$‘dec’$)$;

---

it's necessary to recompute $w(e'')$ by invoking CalWeight. This is because due to the weight increase of e, the shortest path for e'' may be destroyed by e and a new shortest path should be generated by other parents of e''. Moreover, for other shortcuts incident to e but with lower order than e, since e will only affect the shortcuts with higher order than it based on the contracting process of CHLR, they will not be affected by e. (2) For accuracy guarantee, parents shortcuts should be updated before their child shortcut. By Definition 3 and Definition 6, for shortcut $e'' = N^+(e, e')$, we have $\phi(e) < \phi(e'')$ and $\phi(e') < \phi(e'')$. Hence, updating the shortcut with lowest order in Q at each time ensures that their child shortcut will be updated correctly as the weights of their parents shortcuts are up to date now. In this way, CHLR can be updated correctly.

Algorithm of CHLRMaintenance. Based on Theorem 2, we propose our CHLR maintenance algorithm, the pseudocode is shown in Algorithm 3. Given the augmented graph G', a changed edge e_o with new weight w_n and new label l_n, we first update e_o and create e_n with new weight and label (line 2–8). If the label of e_o is changed, we compute the weight of e_o with old label, if the weight increases, e_o is pushed into Q with type=‘inc’, meanwhile, if weight of e_n with new label decreases, e_n is pushed into Q with type=‘dec’ (line 3–5). Otherwise, e_o is pushed into Q depending on the weight increase or decrease (line 7–8). After that, the update propagate chain starts. At each time, Q selects the shortcut e with lowest order $\phi(e)$ to propagate update. Following the update rules in Definition 5, we update the weight of each child e'' of e. If the weight of e increases, the weight of e'' is recomputed by invoking CalWeight algorithm. And if $w(e'')$ increases due to e, e'' is pushed into Q with type='inc' and invokes KeepSCDom algorithm (line 14–18). On the other hand, if the weight of e decreases, we just update the weight of e'' directly if $w(e'') > w(e) + w(e')$. Then, e'' is pushed into Q and invokes KeepSCDom algorithm if $w(e'')$ decreases (line 19–23). When all shortcuts in Q finish their update propagation, we obtain the new CHLR on G' correctly.

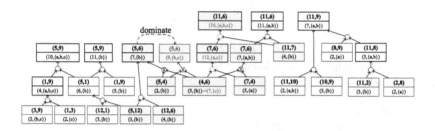

Fig. 2. Example for update process of CHLRMaintenance algorithm

Example 4. Consider the road network G and CHLR on G shown in Fig. 1, the neighborhood of shortcuts is shown in Fig. 2. Given an updated edge $e_o = ((v_4, v_6), 5, \{b\})$, and new weight 7 and new label $\{c\}$ for e_o, following Algorithm 3, as the label of e_o changes, e_o becomes $((v_4, v_6), \infty, \{b\})$ and is pushed into Q with type='inc', while $e_n = ((v_4, v_6), 7, \{c\})$ is created and pushed into Q with type='dec'. Then, e_o first updates its child shortcuts $e_1 = ((v_5, v_6), 7, \{b\})$ and $e_2 = ((v_7, v_6), 7, \{a, b\})$. As the weight of e_o is ∞ now, we recompute $w(e_1)$ and $w(e_2)$, their weights are unchanged. Next, e_n produces two new child shortcuts $e_3 = ((v_5, v_6), 9, \{b, c\})$ and $e_4 = ((v_7, v_6), 12, \{a, c\})$. Here, based on Definition 4, e_1 can dominate e_3 as $w(e_1) < w(e_3)$ and $\ell(e_1) \subset \ell(e_3)$, therefore, e_3 is invalid in G'. Subsequently, e_4 further produces a new child shortcut $e_5 = ((v_{11}, v_6), 16, \{a, b, c\})$. Now, Q is empty, the algorithm finishes.

5.1 Optimization Strategy

To further improve the efficiency of CHLR maintenance, we explore an optimization strategy based on the following observation. Reconsider the update operation of Algorithm 3, for a shortcut e from Q, when the weight of a shortcut e increases, we have to recompute the weight of each child shortcut e'' of e (line 16 of Algorithm 3) whenever $w(e'')$ will change or not. However, if e does not support the current weight of e'', i.e., $w(e'') < w(e) + w(e')$ where e and e' are the parents of e'', $w(e'')$ will not be affected by e. Moreover, even e and e' can support the current weight of e'', i.e., $w(e'') = w(e) + w(e')$, if there exists another couple of parents e_1 and e_1' that can support $w(e'')$, $w(e'')$ will not change as well. To avoid the invalid computation for weight,

---

Algorithm 4. CHLRMaintenance*

Input: augmented graph: G'; changed edge: $e_o = ((u,v), w_o, l_o)$; new weight: w_n; new label: l_n

Output: the updated augmented graph: G'

1:   Q: priority queue; $Q \leftarrow \emptyset$; $e_n \leftarrow ((u,v), w(e_n), l_n)$;

2:   **if** $l_n \neq l_o$ **then**

3:      **if** $w(e_o) = w_o$ and $\text{cnt}(e_o) < 2$ **then** $\text{cnt}(e_o) \leftarrow 0$; Q.push(e_o,'inc');

4:      **if** $w(e_n) = w_n$ **then** $\text{cnt}(e_n) \leftarrow \text{cnt}(e_n) + 1$;

5:      **if** $w(e_n) > w_n$ **then** $w(e_n) \leftarrow w_n$; $\text{cnt}(e_n) \leftarrow 1$; Q.push(e_n,'dec');

6:   **else**

7:      **if** $w(e_o) = w_o$ **then** $\text{cnt}(e_o) \leftarrow \text{cnt}(e_o) - 1$;

8:      **if** $w(e_o) = w_n$ **then** $cnt(e_o) \leftarrow \text{cnt}(e_o) + 1$;

9:      **if** $w(e_o) > w_n$ **then** $w(e_o) \leftarrow w_n$; $\text{cnt}(e_o) \leftarrow 1$; Q.push(e_o,'dec');

10:     **if** $w(e_o) < w_n$ and $\text{cnt}(e_o) < 1$ **then** $\text{cnt}(e_o) \leftarrow 1$; Q.push(e_o,'inc');

11: **while** $!Q$.empty() **do**

12:     $(e,\text{type}) \leftarrow Q$.pop();

13:     **for each** $e' \in N^=(e)$ **do**

14:       $e'' \leftarrow N^+(e,e')$

15:       **if** type $=$'inc' and $w(e'') \neq \infty$ **then**

16:         **if** $\text{cnt}(e'') \geq 1$ and $w(e) + w(e') = w(e'')$ **then**

17:           $\text{cnt}(e'') \leftarrow \text{cnt}(e'') - 1$;

18:           **if** $\text{cnt}(e'') < 1$ and $(e'',\text{'inc'}) \notin Q$ **then** Q.push(e'','inc');

19:       **if** type $=$'dec' **then**

20:         **if** $w(e'') = w(e) + w(e')$ **then** $\text{cnt}(e'') \leftarrow \text{cnt}(e'') + 1$;

21:         **else if** $w(e'') > w(e) + w(e')$ **then**

22:           $w(e'') \leftarrow w(e) + w(e')$; $\text{cnt}(e'') \leftarrow 1$;

23:           **if** $(e'',\text{'dec'}) \notin Q$ **then** Q.push(e'','dec');

24:     **if** type $=$'inc' **then** $w(e) \leftarrow$ CalWeight (e); KeepSCDom $(e,$'inc'$)$;

25:     **else**   KeepSCDom $(e,$'dec'$)$;

---

we propose an optimized CHLR maintenance algorithm based on weight count. We first give the definition of weight count.

Definition 7. (Weight Count) *Given a road network G and the CHLR on G, for each shortcut e in CHLR, the weight count of e is the number of parents that can support the weight of e, denoted as* $\text{cnt}(e)$, *i.e.,* $\text{cnt}(e) = |\{(e_1, e_2) | w(e) = w(e_1) + w(e_2), (e_1, e_2) \in N^-(e)\}|$. *Note that if e is an original edge in G with weight $w(e)$, $\text{cnt}(e)$ is increased by one.*

Lemma 3. *In the update propagate chain, when the weight of a shortcut e increases, for e's each child shortcut e'', $e'' = N^+(e, e')$, if $w(e'') < w(e) + w(e')$ or $\text{cnt}(e'') > 1$, $w(e'')$ will not change, the weight update for e'' can be skipped safely.*

Algorithm of CHLRMaintenance*. Following Lemma 3, we devise our optimized algorithm CHLRMaintenance*. The pseudocode is shown in Algorithm 4. Given the augmented graph G', a changed edge e_o, new weight w_n and new label l_n for e_o, we first update the weight count cnt for e_o and e_n. The weight $w(e_o)$ is destroyed, cnt(e_o) is assigned as 0 and e_o is pushed into Q with type ='inc' (line 3). For shortcut e_n, if $w(e_n) = w_n$, we increase cnt(e_n) by one (line 4). Besides, if $w(e_n)$ decreases due to w_n, cnt(e_n) is assigned as 1 and e_n is pushed into Q with type ='dec' (line 5). In another hand, if the label of e_n is unchanged, cnt(e_o) can be updated similarly depending on the change of $w(e_o)$ (line 7–10). After that, the update propagate chain starts from the shortcuts in Q. At each time, Q selects the shortcut e with lowest order $\phi(e)$ to propagate update. If e will increase its weight, for each child shortcut e'', $e'' = N^+(e, e')$, based on Lemma 3, only if cnt$(e'') \geq 1$ and $w(e) + w(e') = w(e'')$, cnt(e'') is decreased by 1. Then, if cnt$(e'') < 1$, e'' is pushed into Q with type ='inc' (line 15–18). Otherwise, if e has decreased its weight and $w(e'') = w(e) + w(e')$, then we add cnt(e'') by 1. If $w(e'') > w(e) + w(e')$, we directly update $w(e'')$ by $w(e) + w(e')$ and assign 1 to cnt(e''). Meanwhile, e'' is pushed into Q with type ='dec' (line 19–23). Finally, we update $w(e)$ if type='inc' and invoke KeepSCDom algorithm (line 24–25). Note that, in line 14 of KeepSCDom algorithm, we omit the update for $w(e')$ as it will be updated later in line 24 of CHLRMaintenance* algorithm. When all shortcuts in Q finish their update propagation, we obtain the new CHLR on G' correctly.

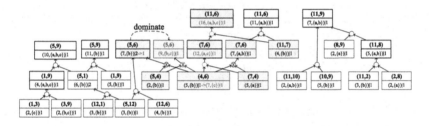

Fig. 3. Example for update process of CHLRMaintenance* algorithm

Example 5. Consider the road network G and CHLR on G shown in Fig. 1, Fig. 3 shows the update process of Algorithm 4. Given the same updated edge $e_o = ((v_4, v_6), 5, \{b\})$, and new weight 7 and new label $\{c\}$ for e_o, when process the child shortcuts of e_o, compared with Algorithm 3, the update operation for $e_1 = ((v_5, v_6), 7, \{b\})$ is skipped as cnt$(e_1) = 2$, Algorithm 4 just needs to decrease cnt(e_1) by 1. Hence, Algorithm 4 can improve the efficiency of CHLR maintenance.

Theorem 3. *The time complexity* CHLRMaintenance* *algorithm is* $O(|E_\Delta| \cdot log|E_\Delta| + |E_\Delta| \cdot d_{max})$, *where* E_Δ *denotes the changed shortcuts in CHLR.*

Proof. In CHLRMaintenance* algorithm, only the changed shortcuts will be pushed into the priority queue Q. Q needs to sort shortcuts based on their update order which consumes $O(|E_\Delta| \cdot log|E_\Delta|)$ time. Moreover, when update a shortcut $e = (u, v)$ in Q, if type ='dec', e can be updated in $O(1)$ time (line 22 of Algorithm 4), otherwise, if cnt$(e) < 1$, we need to recompute the weight of e by visiting its parents which costs $O(d_{max})$ time. Meanwhile, to keep the shortcut dominance, in the worst case, the weights of all shortcuts incident to (u, v) should be computed (line 6 of Algorithm 2) with $O(d_{max})$ time. Therefore, the total time complexity of CHLRMaintenance* is $O(|E_\Delta| \cdot log|E_\Delta| + |E_\Delta| \cdot d_{max})$.

6 Performance Studies

Datasets. We evaluate our algorithms on 8 real road networks, in which, the weight of edges is the length of each road and the label shows the category of different roads. The datasets are downloaded from DIMACS[1]. Table 1 gives the statistic of datasets, where $|\Sigma|$ is the number of labels and d_{max} represents the maximum degree of vertices in the corresponding augmented graphs. All the experiments are conducted on a machine with Xeon(R) Gold 6258R CPU 2.70GHz and 1TB RAM running Linux.

Table 1. Statistic of the real datasets

Dataset	n	m	$|\Sigma|$	d_{max}	Dataset	n	m	$|\Sigma|$	d_{max}
NV	261155	311043	15	145	FL	1,048,506	1,330,551	23	133
LA	413,574	499,254	23	147	CA	1,613,325	1,989,149	28	139
OH	676,058	842,872	24	142	TX	2,073,870	2,584,159	25	152
IL	793,336	1,012,817	23	129	US	2,353,8407	57,131,384	32	258

Algorithms. We compare the performance among these algorithms: (1) Rebuild: the state-of-the-art algorithm to construct CHLR index [24]; (2) Baseline: the Baseline approach proposed in Sect. 4; (3) CHLRM: the CHLRMaintenance algorithm proposed in Sect. 5; (4) CHLRM*: the CHLRMaintenance* algorithm with optimization strategy proposed in Sect. 5.1.

For the vertex order in the road networks, we generate vertex order following the method used in [14,24] with the same setting.

Exp-1: Efficiency of Algorithms When Varying the Percent of Weight Decrease /Increase. In this experiment, we evaluate the efficiency of Baseline, CHLRM and CHLRM* with varying the percent of weight decrease/increase on all datasets. On each dataset, we randomly select 1000 edges as changed edges. We use $\mathcal{W}_o$ to denote the original weights of the changed edges. We first change edges' labels and decrease their weights as $20\%\mathcal{W}_o$–$80\%\mathcal{W}_o$. The running time of three algorithms with different percent of weight decrease is recorded in Fig. 4. Then, we increase their weights as $120\%\mathcal{W}_o$–$200\%\mathcal{W}_o$. The corresponding running time is shown at Fig. 5.

On all datasets, CHLRM* and CHLRM can achieve up to 2 orders of magnitude speedup compared with Baseline. It's because that Baseline often performs the update for edges whose weights are not changed, which leads to unnecessary overhead of Baseline. However, our CHLRM* and CHLRM can avoid the unnecessary overhead by

---

[1] https://www.diag.uniroma1.it/challenge9/data/tiger/.

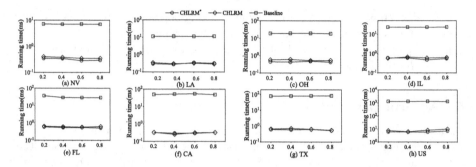

Fig. 4. Running time of different algorithms varying the percent of weight decrease

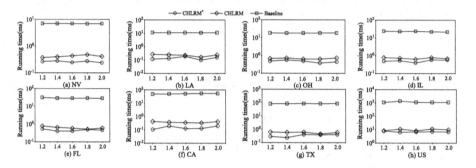

Fig. 5. Running time of different algorithms varying the percent of weight increase

a novel update propagate chain, in which, only the changed shortcuts can further affect other shortcuts. For the efficiency of algorithms when varying the percent of weight increase, Fig. 5 shows similar phenomena with Fig. 4. CHLRM* is fastest among three algorithms. Furthermore, CHLRM* outperforms CHLRM on all datasets. It's because CHLRM* utilizes the optimization strategy based on weight count, which can further avoid the invalid update. Moreover, the running time of three algorithms changed with varying the percent of weight decrease/increase is inapparent, which demonstrates our algorithms have good scalability with the percent of weight change.

Exp-2: Efficiency of Algorithms When Varying the Number of Changed Edges. In this experiment, we compare the efficiency of Rebuild, Baseline, CHLRM and CHLRM* algorithms with changing $0.1\%|E|$–$100\%|E|$ edges. The results are shown in Fig. 6.

CHLRM and CHLRM* are always faster than Baseline. Compared with Rebuild, CHLRM and CHLRM* spend less time updating the CHLR when 10% edges change. Meanwhile, when 100% edges are changed, the running time of CHLRM and CHLRM* is near to the time of rebuilding the CHLR by Rebuild algorithm. The experiment results indicate the significant efficiency of our CHLRM and CHLRM* algorithm.

Exp-3: Effectiveness of CHLR Maintenance Algorithms. In this experiment, we evaluate the effectiveness of different algorithms by comparing the hit rate of them.

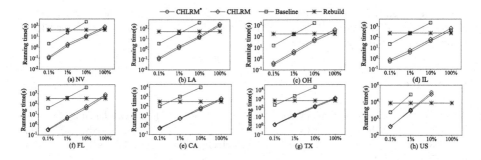

Fig. 6. Running time of different algorithms varying the number of changed edges

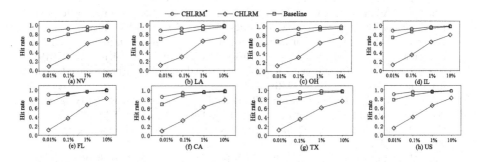

Fig. 7. Hit rate of different algorithms varying the number of changed edges

The hit rate is denoted as $\frac{|E_\Delta|}{|E_U|}$, where E_Δ denotes the shortcuts whose weights are changed and E_U denotes the shortcuts recomputed by algorithms.

As shown in Fig. 7, with the increasing of the number of changed edges, the hit rate of three algorithms increases as well due to more shortcuts will be further changed. Besides, CHLRM performs better than Baseline, CHLRM* achieves the highest hit rate among them on all update cases. It verifies that our CHLRM can avoid the unnecessary update compared with Baseline, and CHLRM* can further reduce the invalid update benefitted from the optimization strategy, which also confirms the reasons for the efficiency of our optimized algorithm at Exp-1 and Exp-2.

7 Conclusion

In this paper, we propose a novel CHLR maintenance algorithm, in which an update propagate chain is designed to propagate the shortcuts whose weights are actually changed without losing the accuracy guarantee. Moreover, an optimization strategy is presented to make sure that shortcuts are updated when necessary, it can further improve the efficiency of index maintenance. Extensive and comprehensive experiments are conducted on real road networks. The experimental results demonstrate the efficiency and effectiveness of our optimized algorithms which can achieve up to 2 orders of magnitude speedup compared with Baseline.

Acknowledgments. Zi Chen is supported by CPSF 2021M701214. Long Yuan is supported by National Key RD Program of China 2022YFF0712100, NSFC61902184, and Science and Technology on Information Systems Engineering Laboratory WDZC20205 250411.

References

1. Abraham, I., Delling, D., Goldberg, A.V., Werneck, R.F.: A hub-based labeling algorithm for shortest paths in road networks. In: Pardalos, P.M., Rebennack, S. (eds.) SEA 2011. LNCS, vol. 6630, pp. 230–241. Springer, Heidelberg (2011). https://doi.org/10.1007/978-3-642-20662-7_20
2. Akiba, T., Iwata, Y., Yoshida, Y.: Fast exact shortest-path distance queries on large networks by pruned landmark labeling. In: SIGMOD, pp. 349–360 (2013)
3. Anirban, S., Wang, J., Islam, M.S., Kayesh, H., Li, J., Huang, M.L.: Compression techniques for 2-hop labeling for shortest distance queries. WWW **25**(1), 151–174 (2022)
4. Buchhold, V., Sanders, P., Wagner, D.: Real-time traffic assignment using engineered customizable contraction hierarchies. ACM J. Exp. Algorithmics **24**(1), 2.4:1–2.4:28 (2019)
5. Chen, Y., Perera, L., Thompson, R.G.: An advanced method to provide best route information in city logistics with toll roads. ATRF (2018)
6. Chen, Z., Fu, A.W., Jiang, M., Lo, E., Zhang, P.: P2H: efficient distance querying on road networks by projected vertex separators. In: SIGMOD, pp. 313–325. ACM (2021)
7. Chen, Z., Yuan, L., Han, L., Qian, Z.: Higher-order truss decomposition in graphs. IEEE TKDE, p. 1 (2021)
8. Chen, Z., Yuan, L., Lin, X., Qin, L., Yang, J.: Efficient maximal balanced clique enumeration in signed networks. In: WWW, pp. 339–349. ACM/IW3C2 (2020)
9. Dean, B.: Uber statistics 2022: How many people ride with uber? https://backlinko.com/uber-users (2022)
10. Dibbelt, J., Strasser, B., Wagner, D.: Customizable contraction hierarchies. ACM J. Exp. Algorithmics **21**(1), 1.5:1–1.5:49 (2016)
11. Didi. Didi statistics and facts (2022). https://expandedramblings.com/index.php/didi-chuxing-facts-statistics/
12. Dijkstra, E.W.: A note on two problems in connexion with graphs. Numer. Math. **1**, 269–271 (1959)
13. Engström, R.: The roads' role in the freight transport system. Transp. Res. Procedia **14**, 1443–1452 (2016)
14. Geisberger, R., Sanders, P., Schultes, D., Delling, D.: Contraction hierarchies: faster and simpler hierarchical routing in road networks. In: McGeoch, C.C. (ed.) WEA 2008. LNCS, vol. 5038, pp. 319–333. Springer, Heidelberg (2008). https://doi.org/10.1007/978-3-540-68552-4_24
15. Lakhotia, K., Kannan, R., Dong, Q., Prasanna, V.K.: Planting trees for scalable and efficient canonical hub labeling. Proc. VLDB Endow. **13**(4), 492–505 (2019)
16. Li, W., Qiao, M., Qin, L., Zhang, Y., Chang, L., Lin, X.: Scaling distance labeling on small-world networks. In: SIGMOD, pp. 1060–1077 (2019)
17. Li, Y., L.H. U, Yiu, M.L., Kou, N.M.: An experimental study on hub labeling based shortest path algorithms. Proc. VLDB Endow. **11**(4), 445–457 (2017)
18. Liu, B., Yuan, L., Lin, X., Qin, L., Zhang, W., Zhou, J.: Efficient (α, β)-core computation in bipartite graphs. VLDB J. **29**(5), 1075–1099 (2020). https://doi.org/10.1007/s00778-020-00606-9
19. Mahajan, B.: Classification of roads in india. https://civiconcepts.com/blog/classification-of-roads

20. Meng, L., Yuan, L., Chen, Z., Lin, X., Yang, S.: Index-based structural clustering on directed graphs. In: ICDE, pp. 2831–2844. IEEE (2022)
21. Mwakalonge, J.L., Moses, R.: Evaluation of truck lane restriction on non-limited access urban arterials. IJTST 1(2), 191–204 (2012)
22. Ouyang, D., Qin, L., Chang, L., Lin, X., Zhang, Y., Zhu, Q.: When hierarchy meets 2-hop-labeling: Efficient shortest distance queries on road networks. In: SIGMOD, pp. 709–724 (2018)
23. Ouyang, D., Yuan, L., Qin, L., Chang, L., Zhang, Y., Lin, X.: Efficient shortest path index maintenance on dynamic road networks with theoretical guarantees. Proc. VLDB Endow. 13(5), 602–615 (2020)
24. Rice, M.N., Tsotras, V.J.: Graph indexing of road networks for shortest path queries with label restrictions. Proc. VLDB Endow. 4(2), 69–80 (2010)
25. Wu, L., Xiao, X., Deng, D., Cong, G., Zhu, A.D., Zhou, S.: Shortest path and distance queries on road networks: an experimental evaluation. Proc. VLDB Endow. 5(5), 406–417 (2012)
26. Yuan, L., Qin, L., Lin, X., Chang, L., Zhang, W.: Diversified top-k clique search. VLDB J. 25(2), 171–196 (2016)
27. Yuan, L., Qin, L., Zhang, W., Chang, L., Yang, J.: Index-based densest clique percolation community search in networks. IEEE Trans. Knowl. Data Eng. 30(5), 922–935 (2018)
28. Zhang, J., Li, W., Yuan, L., Qin, L., Zhang, Y., Chang, L.: Shortest-path queries on complex networks: experiments, analyses, and improvement. Proc. VLDB Endow. 15(11), 2640–2652 (2022)
29. Zhang, J., Yuan, L., Li, W., Qin, L., Zhang, Y.: Efficient label-constrained shortest path queries on road networks: a tree decomposition approach. Proc. VLDB Endow. 15(3), 686–698 (2021)
30. Zhang, M., Li, L., Hua, W., Mao, R., Chao, P., Zhou, X.: Dynamic hub labeling for road networks. In: ICDE, pp. 336–347 (2021)
31. Zhang, Y., Yu, J.X.: Relative subboundedness of contraction hierarchy and hierarchical 2-hop index in dynamic road networks. In: SIGMOD, pp. 1992–2005 (2022)
32. Zhao, Y., Chen, Z., Chen, C., Wang, X., Lin, X., Zhang, W.: Finding the maximum k-balanced biclique on weighted bipartite graphs. IEEE TKDE, pp. 1–14 (2022)
33. Zhu, A.D., Ma, H., Xiao, X., Luo, S., Tang, Y., Zhou, S.: Shortest path and distance queries on road networks: towards bridging theory and practice. In: SIGMOD, pp. 857–868 (2013)

GRMI: Graph Representation Learning of Multimodal Data with Incompleteness

Xian Xu, Xiao Xu$^{(\boxtimes)}$, Xiang Li, and Guotong Xie

Ping An Health Technology, Beijing, China
{xuxian969,xuxiao780,lixiang453,xieguotong}@pingan.com.cn

Abstract. Multimodal data can provide supplementary information of the subjects, which is of great potential for exploring the data-driven insights in various application scenarios. A large amount of researches focus on modal fusion to deriving quality representations of multimodal data. However, missing modality is a common issue, i.e. a sample may not contain full modalities, bringing difficulties to apply existing modal fusion methods on the incomplete multimodal data. In this paper, we present GRMI, a graph-based framework for representation learning of multimodal data with incompleteness. GRMI constructs a bipartite graph for multimodal data, where samples and modalities are viewed as two types of nodes, and the observed modality values as edges. GRMI leverages Graph Neural Network (GNN) to derive edge embeddings and sample node embeddings on the graph, which can be respectively used for missing modality imputation and modal fusion. A self-supervised strategy is utilized to pretrain the GNN by fully exploiting the multimodal data. Extensive experiment results show the superiority of the proposed framework over existing state-ofthe-art methods for both modality imputation task and modal fusion task.(The source code has been anonymously uploaded to https://github.com/GRMI2022/GRMI).

Keywords: multimodal data · incompleteness · graph representation learning · self-supervised learning

1 Introduction

Most of existing works on multimodal data assume that the data are complete, i.e. each sample contains full modalities [5,7,9,15,18,21]. However, in real-world, missing modality is a common issue due to data corruption, equipment damage, budget limitation or human mistake. The incompleteness brings challenges, such as inconsistent feature dimensions, to explore these multimodal learning methods on the original dataset. Excluding the data with missing modalities is a simple and widely-used way to tackle the challenges. While it may bring new problems: low-data-volume and data-bias.

X. Xu and X. Xu—These authors contributed equally to this work.

© The Author(s), under exclusive license to Springer Nature Switzerland AG 2023
X. Wang et al. (Eds.): DASFAA 2023, LNCS 13945, pp. 286–296, 2023.
https://doi.org/10.1007/978-3-031-30675-4_19

There are two popular strategies to solve the modality incompleteness problem. One is to try to impute the missing modality, leveraging the seen modalities of a give sample to reconstruct the missing ones, through expressive deep neural networks, e.g. autoencoders, Generative Adversarial Networks (GANs) [2,13]. While this strategy fails to make full use of modality values from other samples. Another strategy attempts to directly accomplish a modality fusion to get quality sample representations for downstream tasks, with the missing modalities present in the input data [4,10,11,17]. However, due to the complexity of missing patterns in real-world multimodal data, the strategy not only requires large amount of groups (exponentially related with the number of modalities), but also can hardly handle unseen missing patterns during training time. Furthermore, accurate modality imputation is also an important task for multimodal learning.

To address the problems, we present a new representation learning framework (**GRMI**) for multimodal data with incompleteness. The core characteristic of GRMI is to leverage graph structure to share the complementary information among samples, with well-designed self-supervised learning strategy (the pipeline is shown in Fig. 1). Particularly, GRMI models the multimodal data as a bipartite graph, where samples and modalities are encoded into two types of nodes, and the attributed edges represent the available modality values between the sample nodes and modality nodes (an example is illustrated in the left part of Fig. 1). Here, a modality node plays a role of bridging the samples that hold this modality. It facilitates the information passing among these samples, which are of various conditions of missing modalities. Based on the graph structure, good edge embeddings and sample node embeddings can be respectively used for modal imputation and modal fusion. To achieve this goal, GRMI utilizes GNNs with a self-supervised learning approach, which is composed by three learning objectives. The pretrained GNN model can be further used for other tasks, e.g. modal imputation, sample classification, by fine-tuning. Experiments show that GRMI advanced the state-of-the-art performances on multiple datasets.

2   Problem Formulation

A multimodal dataset is denoted as $\mathcal{D} = \{x_{i1}, x_{i2}, \cdots, x_{im}\}_{i=1}^{n}$, where m is the number of modalities and n is the number of samples. The j-th modality value of the i-th sample is denoted as x_{ij}. For multimodal data with missing modalities, a mask matrix is denoted as $\mathcal{M} = \{0,1\}^{n \times m}$, where the value of the vector x_{ij} can be observed only if $\mathcal{M}_{ij} = 1$. This paper aims to design a model that can learn from an incomplete multimodal dataset, deriving quality representations for modal imputation and modal fusion.

Multimodal data as a bipartite graph: Multimodal data $\mathcal{D}$ with mask matrix $\mathcal{M}$ can be represented as an undirected bipartite graph $\mathcal{G} = (\mathcal{V}, \mathcal{E})$, where $\mathcal{V} = \mathcal{V}_S \cup \mathcal{V}_M$ is the node set that consists of two types of nodes. $\mathcal{V}_S = \{u_1, u_2, \ldots, u_n\}$ is the sample node, each node represents a sample; $\mathcal{V}_M = \{v_1, v_2, \ldots, v_m\}$ is the modality node. $\mathcal{E}$ is the edge set, where edges only exist between nodes in different partitions: $\mathcal{E} = \{e_{u_i v_j} | u_i \in \mathcal{V}_S, v_j \in \mathcal{V}_M, \mathcal{M}_{ij} = 1\}$, where the edge attributes $e_{u_i v_j}$ takes the values of the corresponding data x_{ij}. To simplify the notation $e_{u_i v_j}$, we use e_{ij} in the context of multimodal dataset $\mathcal{D}$, and e_{uv} in the context of graph $\mathcal{G}$.

Modal imputation as edge embedding learning: Given a sample i with missing modality j ($\mathcal{M}_{ij} = 0$), the goal of modal imputation is to estimate the edge value $\hat{e}_{ij}$, which is as similar as possible to the true value e_{ij}. It relies on deriving an expressive edge embedding to support the estimation.

Modal fusion as node embedding learning: Modal fusion aims to generate quality representations for samples, i.e. sample node embeddings, which can be used for downstream tasks. Here, we pick sample classification as the example task. Suppose $Y \in \mathbb{R}^n$ is the classification labels and $V \in \{0, 1\}^n$ the train/test partition, where $V_i = 0$ refers to the label of sample i is to be predicted. The goal of sample classification is to minimize the difference between $\hat{Y}_i$ and Y_i ($\forall V_i = 0$), where $\hat{Y}$ is the predicted labels.

3   Methodology

In this section, we introduce the proposed framework for representation learning of multimodal data with missing modality. Inspired by GRAPE [19], the framework first modelled incomplete multimodal data in a bipartite graph structure, which has been detailed in the previous section. Then, GNN is used to accomplish the information passing among multimodal samples in the graph. Meanwhile, we design a self-supervised learning strategy to pretrain the GNN, which could bring benefits for both modal imputation and modal fusion task. Figure 1 shows the pipeline of GRMI.

3.1   GNN Architecture

We first project the original values of different modalities into a unified dimension to constitute the initial edge embedding, via a modal-specific linear transformation: $e_{ij} = f_j(x_{ij})$, where $f_j(\cdot)$ is composed of a transformation matrix with a bias vector for j-th modal. Then, at each GNN layer l, we update the embeddings for nodes and edges in turn. In detail, it first aggregates the neighboring edge information for a node as follows:

$$n_u^{(l)} = \text{AGG}(\sigma(P^{(l)}[h_u^{(l-1)}||e_{uv}^{(l-1)}])|\forall v \in \mathcal{N}(u, \mathcal{E})) \tag{1}$$

Then, node embedding $h_u^{(l)}$ is updated by fusing the information from both the pre-layer $h_u^{(l-1)}$ and the neighbours $n_u^{(l)}$:

$$h_u^{(l)} = \sigma(Q^{(l)}[h_u^{(l-1)}||n_u^{(l)}]) \tag{2}$$

Next, we combine the information of the pre-layer edge embedding $e_{uv}^{(l)}$ and the head/tail node embeddings as the updated edge embedding:

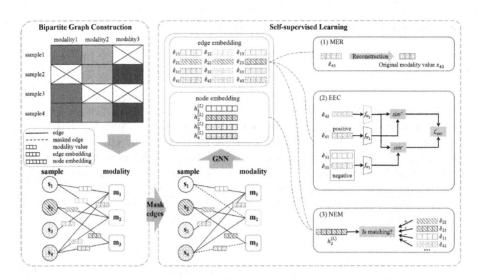

Fig. 1. The framework of GRMI. The left part illustrates the process of construction the bipartite graph from the multimodal data with incompleteness.

$$e_{uv}^{(l)} = \sigma(W^{(l)}[e_{uv}^{(l-1)}||h_u^{(l)}||h_v^{(l)}]) \tag{3}$$

We denote the final edge embedding as the concatenation of the embeddings of head and tail node, with a feedforward neural networks f_{edge}:

$$\hat{e}_{uv} = f_{edge}(h_u^{(L)}||h_v^{(L)}) \tag{4}$$

3.2 Embeddings for Modal Imputation and Modal Fusion

Here, we introduce how to utilize the embeddings, generated by GNN on the bipartite graph, to support the key application tasks on multimodal data with incompleteness. (1) **Edge embedding for modal imputation**. Given a sample with missing modality, there is no edge between the the sample and the modality node in the original bipartite graph. While we can derive its edge embeddings $\hat{e}_{uv}$ from the two nodes. As e_{uv} is linearly transformed from the original modality value, we can directly set the objective of modal imputation as minimizing the distance between $\hat{e}_{uv}$ and e_{uv}. (2) **Node embedding for modal fusion**. With L-layers GNN, the massive information from multiple modalities and other samples have been encoded into the embedding of each sample node $h_u^{(L)}$, i.e. the result of modal fusion. We can use the embeddings of the labeled samples ($V_i = 1$) as the input to train a classifier, and then apply it on the unlabeled samples to obtain their classification results.

3.3 Self-supervised Learning Strategy

We propose a self-supervised learning strategy to pretrain the GNN model, which contains three training tasks.

Masked Edge Reconstruction. MER mimics the modal imputation process that making an estimation for the modality values on masked edges through an autoencoder way. In particular, we randomly mask out edges $\mathcal{E}$ with the rate r_{mask} to generate $\mathcal{G}' = (\mathcal{V}, \mathcal{E}'|\mathcal{E}' = \mathcal{E} \setminus \mathcal{E}_{mask})$ as the model input. Let e denotes the initial embedding of a masked edge that linearly transformed from the corresponding original modality value, and $\hat{e}$ refers to the output embedding that encoded by the L-layers GNN. We use a simple fully-connected (FC) layer as the decoder to generate the reconstruction result. The mean-square error is used as the loss function.

$$\mathcal{L}_{mer} = \frac{1}{|\mathcal{E}_{mask}|} \sum_{e \in \mathcal{E}_{mask}} (e - f_{mer}(\hat{e}))^2 \tag{5}$$

Edge-edge Contrasting. EEC aims to promote the representational ability of the edge embeddings of unimodal. It adopts a contrastive learning way to identify the edge embedding pairs extracted from the same sample. Particularly, given two different modalities v and v', namely modality pairwise combination $v2v'$, we randomly select edges connected to the corresponding nodes, e.g. e_{uv} and $e_{u'v'}$. Then, we make up edge pairs from the two modalities, that the edges from the same sample are denoted as positive pairs $(u = u')$, and the others the negative ones $(u \neq u')$. The learning objective is to learn such an edge embedding space in which positive pairs stay close to each other while negative ones are far apart. InfoNCE (denoted as $\mathcal{L}^{v2v'}$) is used as the contrastive loss function for modality pairwise combination $v2v'$. For a dataset of M-modalities, there are $M(M-1)/2$ different modality pairwise combination. For each combination, we calculate two contrastive loss $\mathcal{L}^{v2v'}$ and $\mathcal{L}^{v'2v}$, and hence the loss of EEC task is defined as:

$$\mathcal{L}_{eec} = \frac{1}{M(M-1)} \sum_{v=1}^{M} \sum_{v'=1}^{M} (\mathcal{L}^{v2v'}|v \neq v') \tag{6}$$

Node-edge Matching. NEM focuses on the quality of sample embeddings, i.e. the ability of fully fusing the multimodal information. Here, it is achieved by enabling the model to distinguish the matched sample node embedding and edge embedding in a contrastive way. In particular, we randomly select a sample node embedding $h_u^{(L)}$ and an edge embedding $\hat{e}_{u'v}$ as the input to a fully connected layer, followed by a sigmoid function to predict a two-class probability p^{nem}. If the edge is connected to the node, i.e. $u = u'$, it is a positive node-edge pair, and vice versa. To enhance the difficulty of the distinguishing task, we select the edges whose embeddings are similar to the positive ones as the negative set. The loss function is defined as:

$$\mathcal{L}_{nem} = -\frac{1}{|\mathcal{Z}|} \sum CE(y^{nem}, p^{nem}) \tag{7}$$

where $CE(\cdot, \cdot)$ is the cross-entropy function, y^{nem} is the ground-truth label, $|\mathcal{Z}|$ is the normalization factor.

We aggregate the loss of the three tasks as the whole objective of the self-supervised learning approach. With the pretrained GNN model, it can be further used for modal imputation and modal fusion tasks via a fine-tuning way.

4 Experiment

4.1 Data

Three popular datasets are used [1,3,14]. The statistics is shown in Fig. 1).

Incompleteness of datasets: We evaluate the performance of our algorithm and the baseline models on different missing ratio. Given a incompleteness ratio ρ, we randomly delete a part of data from the original complete datasets to ensure that $\rho\%$ of the data have missing modality. For example, given a M-modal dataset, we randomly extract $\rho/(M-1)\%$ samples to remove their one modality, $\rho/(M-1)\%$ samples two modalities, and so on. Not that, we guarantee that each samples retains at least one modality (Table 1).

Table 1. Statistics of Datasets (M: # of modalities; $|C|$: # of classes)

Dataset	Train	Valid	Test	M	$|C|$
POM	600	100	203	3	9
IEMOCAP	2717	798	938	3	2
NTU	1393	246	373	2	67

4.2 Baseline Methods

We compare our model with three categories of approaches, one for modal imputation and two for modal fusion. **Imputation**: K Nearest Neighbors (KNN), Autoencoder based generator (AE) [8], and Multimodal Factorization Model (MFM) [16]. **Fusion with disregarding missing modality**: Tensor Fusion Network (TFN) [20], Low-rank Multimodal Fusion (LMF) [12], and Hypergraph Neural Network (HGNN) [6]. **Fusion with consider missing modality**: Heterogeneous Graph-based Multimodal Fusion (HGMF) [4], and SMIL [13].

4.3   Experimental Setup

Here, we introduce the tasks, measurements and implementation details.

Imputation task and evaluation measurements: In this tasks, we attempt to estimate the values of the missing modalities. The experiments are conducted on POM dataset. For each method, only the two observed modalities are used as the input to reconstruct the missing one. Root mean square error (RMSE) is adopted to evaluate the imputation performance.

Sample classification task and evaluation measurements: We conduct experiments on datasets from two application domains, namely Multimodal Emotion Recognition and 3D Object Recognition. For binary classification datasets, we utilize F1 score to evaluate the model performance. For multi-class datasets, we report the classification accuracy Acc-k, where k denotes the number of classes.

4.4   Result Analysis

Imputation. We conduct the imputation task on POM dataset with various incompleteness rate. The performance of the methods, which can support the imputation task, are illustrated in Fig. 2. We can observe that GRMI outperforms both the generative and discriminative baselines on this task. The incompleteness inevitably brings adverse effect on the performance, because it would decrease the available data for model training. While GRMI is the most robust one to the incompleteness.

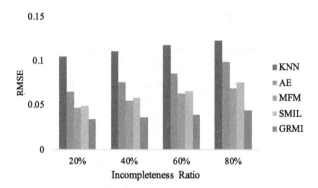

Fig. 2. RMSE for different methods under various multimodal incompleteness rate on POM dataset

Sample Classification. We evaluate the impact of missing modality by changing the multimodal incompleteness rate from 0% to 80% intermittently by 20%. The prediction performance of the different methods on the three datasets is shown in Table 2 and Table 3.

Table 2. Acc (%) for different methods under various multimodal incompleteness rate on POM ($M{=}3$) and NTU ($M{=}2$) datasets.

Method	POM					NTU				
	0%	20%	40%	60%	80%	0%	20%	40%	60%	80%
TFN	36.55	35.60	34.42	32.74	30.71	83.59	82.19	80.86	79.28	77.07
LMF	36.47	35.83	34.73	33.18	32.01	83.04	81.80	80.39	78.41	76.21
HGNN	37.17	36.23	35.26	34.27	32.37	84.28	82.43	81.38	79.73	77.87
HGMF	36.88	36.17	35.15	34.76	33.81	83.94	82.95	81.74	80.35	78.68
SMIL	36.97	36.36	35.28	34.56	33.63	84.10	83.02	81.83	80.24	78.50
GRMI	**37.39**	**36.98**	**36.36**	**35.89**	**35.22**	**84.39**	**83.21**	**82.23**	**81.48**	**80.32**

Table 3. F1 scores for different methods under various multimodal incompleteness rate (ρ) on IEMOCAP (M=3) datasets.

Emotion	ρ	TFN	LMF	HGNN	HGMF	SMIL	GRMI
Happy	0%	84.31	84.41	85.30	85.55	85.40	**85.63**
	20%	83.34	83.42	84.26	84.99	84.87	**85.16**
	40%	82.27	82.34	82.87	84.04	83.92	**84.42**
	60%	81.08	80.94	81.77	82.75	82.81	**83.79**
	80%	79.74	79.34	80.46	81.49	81.66	**82.82**
Sad	0%	83.25	83.69	85.39	**86.12**	85.76	85.92
	20%	82.54	82.70	84.11	**85.30**	85.02	85.24
	40%	81.45	81.27	82.69	84.18	84.11	**84.51**
	60%	80.20	79.81	81.32	82.80	82.86	**83.59**
	80%	78.29	77.83	79.68	81.37	81.50	**82.52**
Angry	0%	85.55	85.24	85.85	86.11	86.04	**86.25**
	20%	84.85	84.59	84.97	85.59	85.47	**85.85**
	40%	84.06	83.70	84.14	85.01	84.84	**85.34**
	60%	83.06	82.57	83.08	84.17	84.08	**84.81**
	80%	81.86	81.30	81.90	82.96	83.07	**84.16**
Neutral	0%	68.37	68.45	69.54	69.68	69.79	**69.81**
	20%	67.61	67.78	68.98	69.21	69.24	**69.41**
	40%	66.76	66.91	68.01	68.45	68.35	**68.68**
	60%	65.15	65.26	66.46	67.25	67.03	**67.76**
	80%	63.58	63.75	64.96	65.86	65.79	**66.70**

Our approach significantly outperforms all the state-of-the-art competitors among most ratios of modality missing, which display the efficiency of our approach in the sample classification problem. The improvement over the strongest baselines w.r.t Accuracy achieves 4.2% on POM ($\rho = 80\%$) and 2.1% on NTU ($\rho = 80\%$), F1 score achieves 1.4% in average on IEMOCAP ($\rho = 80\%$).

On datasets without missing ($\rho = 0\%$), most of the methods achieve competitive performance. It demonstrates the effectiveness of design of the modal fusion in these methods. While GRMI perform slightly better than the other methods on most of datasets. It demonstrates that the graph structure and three pre-training tasks enable the model to obtain high-quality multimodal representations, which can effectively deal with the downstream task.

All methods suffer from the increasing of missing rate. The higher missing rate gets worse performance. It proves that the missing issue would bring significantly adverse impact on downstream tasks of multimodal data. However, with the consideration of missing modalities, the decline proportions of the performance of SMIL, HGMF and GRMI are usually smaller than the others. It shows the necessity of handling the missing issue for multimodal fusion.

Among the methods without considering missing modality, the performance of HGNN is better than TFN and LMF. It may benefit from the design of the hyperedge convolution operation in HGNN, which can alleviate the impact of missing data to a certain extent. However, it can be seen that all these methods perform poorly under high missing rate.

Compare to SMIL, the incorporation of graph techniques brings performance improvement for HGMF, which demonstrates the ability of GNN of passing information among samples. GRMI shows the superiority among the methods with considering missing issue. Its performance is the closest one to the model learned from a full modality dataset. It indicates that GRMI can better leverage the rich information among samples and modalities for modal fusion with incompleteness.

5   Conclusion

In this paper, we introduces a novel framework for representation learning of multimodal data with incompleteness. The key idea is to model the data as a bipartite graph, and leverage GNN with a self-supervised learning strategy to accomplish the information interaction among massive samples and modalities. Based on the graph, the two common tasks, modal imputation and modal fusion, can be naturally convert to the graph representation learning. The graph structure also promises the flexibility on data with large amount of modalities, and the ability of tackling unseen missing patterns. The proposed model demonstrates significant performance improvement on multiple datasets beyond the state-of-the-art methods.

References

1. Busso, C., et al.: IEMOCAP: interactive emotional dyadic motion capture database. Lang. Resour. Eval. **42**(4), 335–359 (2008)
2. Cai, L., Wang, Z., Gao, H., Shen, D., Ji, S.: Deep adversarial learning for multi-modality missing data completion. In: Proceedings of the 24th ACM SIGKDD International Conference on Knowledge Discovery & Data Mining, pp. 1158–1166 (2018)
3. Chen, D.Y., Tian, X.P., Shen, Y.T., Ouhyoung, M.: On visual similarity based 3D model retrieval. In: Computer Graphics Forum, vol. 22, pp. 223–232. Wiley Online Library (2003)
4. Chen, J., Zhang, A.: HGMF: heterogeneous graph-based fusion for multimodal data with incompleteness. In: Proceedings of the 26th ACM SIGKDD International Conference on Knowledge Discovery & Data Mining, pp. 1295–1305 (2020)
5. Chen, Y.-C., et al.: UNITER: universal image-text representation learning. In: Vedaldi, A., Bischof, H., Brox, T., Frahm, J.-M. (eds.) ECCV 2020. LNCS, vol. 12375, pp. 104–120. Springer, Cham (2020). https://doi.org/10.1007/978-3-030-58577-8_7
6. Feng, Y., You, H., Zhang, Z., Ji, R., Gao, Y.: Hypergraph neural networks. In: Proceedings of the AAAI Conference on Artificial Intelligence, vol. 33, pp. 3558–3565 (2019)
7. Kim, W., Son, B., Kim, I.: ViLT: vision-and-language transformer without convolution or region supervision. In: International Conference on Machine Learning, pp. 5583–5594. PMLR (2021)
8. Lee, H.C., Lin, C.Y., Hsu, P.C., Hsu, W.H.: Audio feature generation for missing modality problem in video action recognition. In: ICASSP 2019–2019 IEEE International Conference on Acoustics, Speech and Signal Processing (ICASSP), pp. 3956–3960. IEEE (2019)
9. Li, J., Selvaraju, R., Gotmare, A., Joty, S., Xiong, C., Hoi, S.C.H.: Align before fuse: vision and language representation learning with momentum distillation. In: Advances in Neural Information Processing Systems, vol. 34 (2021)
10. Liu, M., Gao, Y., Yap, P.T., Shen, D.: Multi-hypergraph learning for incomplete multimodality data. IEEE J. Biomed. Health Inform. **22**(4), 1197–1208 (2017)
11. Liu, M., Gao, Y., Yap, P.T., Shen, D.: Multi-hypergraph learning for incomplete multimodality data. IEEE J. Biomed. Health Inform. **22**(4), 1197–1208 (2018)
12. Liu, Z., Shen, Y., Lakshminarasimhan, V.B., Liang, P.P., Zadeh, A., Morency, L.P.: Efficient low-rank multimodal fusion with modality-specific factors. In: Proceedings of the 56th Annual Meeting of the Association for Computational Linguistics, pp. 2247–2256 (2018)
13. Ma, M., Ren, J., Zhao, L., Tulyakov, S., Wu, C., Peng, X.: SMIL: multimodal learning with severely missing modality. In: Proceedings of the AAAI Conference on Artificial Intelligence, vol. 35, pp. 2302–2310 (2021)
14. Park, S., Shim, H.S., Chatterjee, M., Sagae, K., Morency, L.P.: Computational analysis of persuasiveness in social multimedia: a novel dataset and multimodal prediction approach. In: Proceedings of the 16th International Conference on Multimodal Interaction, pp. 50–57 (2014)
15. Radford, A., et al.: Learning transferable visual models from natural language supervision. In: International Conference on Machine Learning, pp. 8748–8763. PMLR (2021)

16. Tsai, Y.H.H., Liang, P.P., Zadeh, A., Morency, L.P., Salakhutdinov, R.: Learning factorized multimodal representations. arXiv preprint arXiv:1806.06176 (2018)
17. Wang, Q., Zhan, L., Thompson, P., Zhou, J.: Multimodal learning with incomplete modalities by knowledge distillation. In: Proceedings of the 26th ACM SIGKDD International Conference on Knowledge Discovery & Data Mining, pp. 1828–1838 (2020)
18. Wang, X., et al.: Heterogeneous graph attention network. In: The World Wide Web Conference, pp. 2022–2032 (2019)
19. You, J., Ma, X., Ding, Y., Kochenderfer, M.J., Leskovec, J.: Handling missing data with graph representation learning. Adv. Neural Inform. Process. Syst. **33**, 19075–19087 (2020)
20. Zadeh, A., Chen, M., Poria, S., Cambria, E., Morency, L.P.: Tensor fusion network for multimodal sentiment analysis. In: Proceedings of the 2017 Conference on Empirical Methods in Natural Language Processing, EMNLP, pp. 1103–1114 (2017)
21. Zhang, C., Song, D., Huang, C., Swami, A., Chawla, N.V.: Heterogeneous graph neural network. In: Proceedings of the 25th ACM SIGKDD International Conference on Knowledge Discovery & Data Mining, pp. 793–803 (2019)

Efficient Network Representation Learning via Cluster Similarity

Yasuhiro Fujiwara[1]([✉]), Yasutoshi Ida[2], Atsutoshi Kumagai[2],
Masahiro Nakano[1,2], Akisato Kimura[1,2], and Naonori Ueda[1,2]

[1] NTT Communication Science Laboratories, Atsugi, Kanagawa, Japan
{yasuhiro.fujiwara.kh,masahiro.nakano.pr,naonori.ueda.fr}@hco.ntt.co.jp,
akisato@ieee.org
[2] NTT Computer and Data Science Laboratories, Musashino, Tokyo, Japan
atsutoshi.kumagai.ht@hco.ntt.co.jp, yasutoshi.ida@ieee.org

Abstract. Network representation learning is a de-facto tool for graph analytics. The mainstream of the previous approaches is to factorize the proximity matrix between nodes. However, if n is the number of nodes, since the size of the proximity matrix is $n \times n$, it needs $O(n^3)$ time and $O(n^2)$ space to perform network representation learning. The proposed approach computes the representations of the clusters from similarities between clusters and computes the representations of nodes by referring to them. If l is the number of clusters, since $l \ll n$, we can efficiently obtain the representations of clusters from a small $l \times l$ similarity matrix. Experiments show that our approach can perform network representation learning more efficiently and effectively than existing approaches.

Keywords: Efficient · Algorithm · Network representation learning

1 Introduction

Many real-world applications can be naturally modeled as graphs [4,5,7]. Network representation learning converts each graph node to a fixed-length vector such that the representation vectors preserve the inherent properties and structures of the graph. Since it is easy to subject the representation vectors to feature-based machine learning methods such as LIBLINEAR, network representation learning has become a fundamental task in graph analytics. NetMF is a matrix factorization-based network representation learning approach [10]. It realizes higher accuracy in performing node classification than DeepWalk and its variants. However, NetMF needs to factorize a dense $n \times n$ proximity matrix between nodes where n is the number of nodes, and thus it needs $O(n^3)$ time and $O(n^2)$ space. This paper proposes G̲raph C̲lustering-based N̲etwork R̲epresentation L̲earning, *GC-NRL*. It performs graph clustering for a given graph and computes a similarity matrix between clusters. It then computes the representation vectors of the clusters by factorizing their similarity matrix with sparse matrices. It determines the representation vectors of the nodes by referring

© The Author(s), under exclusive license to Springer Nature Switzerland AG 2023
X. Wang et al. (Eds.): DASFAA 2023, LNCS 13945, pp. 297–307, 2023.
https://doi.org/10.1007/978-3-031-30675-4_20

to those of clusters. Since it uses the similarity matrix between clusters instead of the proximity matrix between nodes, it can efficiently compute the representation vectors. In the remainder of this paper, Sect. 2 gives an overview of the background, Sect. 3 introduces our approach, Sect. 4 reviews our experimental results, and Sect. 5 provides conclusions.

2   Preliminaries

Given graph $G = (\mathbb{V}, \mathbb{E})$ with $\mathbb{V}$ being the set of n nodes and $\mathbb{E}$ being the set of m edges, network representation learning computes a low-dimensional representation vector $\mathbf{x}_v$ of dimension d for each node $v \in \mathbb{V}$ where $d \ll n$ is the predefined number of dimensions. Representation vector $\mathbf{x}_v$ is set to capture the structural property of node v. NetMF is a factorization-based approach that computes the following $n \times n$ high-order proximity matrix between nodes [10]:

$$\mathbf{M} = \log(\max(\mathbf{M}', 1)), \ \mathbf{M}' = \frac{vol(G)}{bT}(\sum_{r=1}^{T}(\mathbf{D}^{-1}\mathbf{A})^r)\mathbf{D}^{-1} \qquad (1)$$

In this equation, $\log(\cdot)$ is an element-wise logarithm, $vol(G) = \sum_{1 \le i,j \le n} A[i,j]$ is the volume of the graph where $A[i,j]$ is the $[i,j]$ element of the adjacency matrix $\mathbf{A}$ corresponding to the edge weight from the j-th node to the i-th node, b is the number of negative samples [10], T is the window size, and $\mathbf{D}$ is the diagonal matrix $\mathbf{D} = \text{diag}(\mathbf{A}\mathbf{1}_n)$ where $\mathbf{1}_n$ is a vector of length n with all ones. NetMF obtains the representation vectors by using the d left singular vectors and the first d singular values after computing the SVD of matrix $\mathbf{M}$. However, it needs $O(n^3)$ time and $O(n^2)$ space to compute matrix $\mathbf{M}$ of $n \times n$ size. Furthermore, NetMF degrades the quality of representation vectors by truncating the small nonzero elements from the proximity matrix.

3   Proposed Method

Instead of proximities between nodes, we exploit similarities between clusters to reduce the computational cost since the number of clusters is much smaller than that of nodes [11]. Since similarities between clusters represent the structural property of the graph, we can effectively generate the representation vectors by using them. Let c_i be the i-th cluster, we perform graph clustering and compute representation vectors of c_i by using a similarity matrix between clusters. We compute the representation vectors of the nodes from the representation vectors of the clusters. Let $\mathbf{W}$ be the row normalized adjacency matrix and $\mathbf{r}_i$ be the representation vector of c_i, we compute the representation vectors of node v as $\mathbf{x}_v = \sum_{i=1}^{l} \sum_{w \in \mathbb{C}_i} W[v,w]\mathbf{r}_i$. In this equation, l is the number of clusters, $\mathbb{C}_i$ is a node set included in c_i, and $W[v,w]$ is the edge weight between node v and w.

3.1   Similarity Between Clusters

To compute representation vector $\mathbf{r}_i$, we factorize the similarity matrix between clusters. Our approach uses the IncMod method for graph clustering since it can compute clusters efficiently [11]. Note that, since the IncMod method can handle undirected and directed graphs, our approach can handle both graphs. The IncMod method automatically sets number of clusters l based on graph structure; we cannot specify the number of clusters.

Our approach determines the similarity matrix between clusters from the difference from a random graph. Specifically, if $\mathbf{S}$ is the $l \times l$ similarity matrix between clusters, we define its elements as follows:

$$S[i,j] = \sum_{v \in \mathbb{C}_i} \sum_{w \in \mathbb{C}_j} A[v,w] - \frac{1}{vol[G]} \sum_{v \in \mathbb{C}_i} \sum_{w \in \mathbb{V}} A[v,w] \sum_{v \in \mathbb{V}} \sum_{w \in \mathbb{C}_j} A[v,w]$$
(2)

Since $\frac{1}{vol[G]} \sum_{v \in \mathbb{C}_j} \sum_{w \in \mathbb{V}} A[v,w]$ is the ratio of edge weights connected to cluster c_i, if we assume that G is a random graph, the second term of the right side in Eq. (2), $\frac{1}{vol[G]} \sum_{v \in \mathbb{C}_i} \sum_{w \in \mathbb{V}} A[v,w] \sum_{v \in \mathbb{V}} \sum_{w \in \mathbb{C}_j} A[v,w]$, corresponds to the expectation of the sum of edge weights connected to cluster c_i from c_j. On the other hand, the first term, $\sum_{v \in \mathbb{C}_i} \sum_{w \in \mathbb{C}_j} A[v,w]$, corresponds to the sum of edge weights actually connected to cluster c_i from c_j. Therefore, $S[i,j]$ would be positive if c_i and c_j are well-connected compared to a random graph; otherwise, it would be negative. As a result, $\mathbf{S}$ effectively represents the structural relationships between the clusters. Therefore, even if nodes are included in different clusters, our approach can place the nodes closely in the representation space if their clusters have high similarity.

Our approach uses SVD on $\mathbf{S}$ to compute the representation vectors of the clusters. Specifically, we decompose $\mathbf{S}$ as $\mathbf{S} = \mathbf{U} \mathbf{\Sigma} \mathbf{V}^{\top}$ and compute representation matrix $\mathbf{R}$ as $\mathbf{R} = \mathbf{U} \mathbf{\Sigma}^{\frac{1}{2}}$. If $\mathbf{r}_i$ is the i-th row vector of $\mathbf{R}$, we use $\mathbf{r}_i$ as the representation vector of the i-th cluster. However, using row vectors in $\mathbf{R}$ has a problem in computing the representation vectors of the clusters. Since the size of $\mathbf{S}$ is $l \times l$, the length of $\mathbf{r}_i$ is l. Therefore, if $l < d$, the representation vectors of the clusters would be shorter than the representation vectors of the nodes with length d. As a result, it is difficult to effectively use the representation vectors of the clusters to compute the representation vectors of the nodes if $l < d$.

3.2   Dimensionality Expansion

If $l < d$, our approach expands the dimensionality of the representation vectors of the clusters by exploiting a sparse matrix. Let $\mathbf{E}$ be the $l \times d$ expansion matrix and $\mathbf{e}_i$ be the i-th column vector of $\mathbf{E}$, we set the elements of $\mathbf{e}_i$ as follows:

$$e_i[j] = \begin{cases} \sqrt{\frac{l}{\log l}} & \text{with probability } \frac{\log l}{2l} \\ 0 & \text{with probability } 1 - \frac{\log l}{l} \\ -\sqrt{\frac{l}{\log l}} & \text{with probability } \frac{\log l}{2l} \end{cases}$$
(3)

To obtain $\mathbf{R}$, we project low-dimensional matrix $\mathbf{U} \mathbf{\Sigma}^{\frac{1}{2}}$ into a high-dimensional space by using $\mathbf{E}$. Specifically, we expand the dimensionality of the representation vectors by computing $\mathbf{R} = \mathbf{U} \mathbf{\Sigma}^{\frac{1}{2}} \mathbf{E}$ as shown in Algorithm 1. Let $\mathbf{r}_i$ be the i-th column vector of matrix $\mathbf{R}$, we have the following property for $\mathbf{R}$:

Lemma 1. *Let $E(\cdot)$ represent expectation, if $i \neq j$ holds, we have $E(\mathbf{r}_i^\top \mathbf{r}_j) = 0$.*

Proof. Let $\mathbf{U}' = \mathbf{U}\boldsymbol{\Sigma}^{\frac{1}{2}}$, since $\mathbf{r}_i = \mathbf{U}'\mathbf{e}_i$ from Algorithm 1, we have

$$
\begin{aligned}
\mathbf{r}_i^\top \mathbf{r}_j &= \sum_{k=1}^l \left(\sum_{i'=1}^l U'[k, i']e_i[i']\right)\left(\sum_{j'=1}^l U'[k, j']e_j[j']\right) \\
&= \sum_{k=1}^l \sum_{i'=1}^l \sum_{j'=1}^l U'[k, i']U'[k, j']e_i[i']e_j[j']
\end{aligned}
\tag{4}
$$

If $i \neq j$, we have $E(e_i[i']e_j[j']) = 0$. As a result, if $i \neq j$, we have

$$
E\left(\sum_{k=1}^l \sum_{i'=1}^l \sum_{j'=1}^l U'[k, i']U'[k, j']e_i[i']e_j[j']\right) = 0
\tag{5}
$$

Therefore, we have $E(\mathbf{r}_i^\top \mathbf{r}_j) = 0$ from Eq. (4).   □

The column vectors in $\mathbf{U}\boldsymbol{\Sigma}^{\frac{1}{2}}$ are orthogonal to each other since SVD produces orthogonal matrices. Specifically, let $\mathbf{u}_i'$ be the i-th column vector of $\mathbf{U}' = \mathbf{U}\boldsymbol{\Sigma}^{\frac{1}{2}}$, we have $(\mathbf{u}_i')^\top \mathbf{u}_j' = 0$ such that $i \neq j$, which is a necessary condition for preserving pairwise similarities between vectors [1]. On the other hand, Lemma 1 shows that each column of matrix $\mathbf{R}$ would be orthogonal to each other, the same as matrix $\mathbf{U}'$. Consequently, Lemma 1 indicates that we can preserve the preferable property for the representation vectors even after dimensionality expansion. In terms of the quality of dimensionality expansion, we have the following lemma:

Lemma 2. *Let $V(\cdot)$ represent variance, the following equation holds if we have $i \neq j$: $V(\mathbf{r}_i^\top \mathbf{r}_j) = \sum_{k=1}^l \sum_{i'=1}^l \sum_{j'=1}^l (U'[k, i'])^2 (U'[k, j'])^2$.*

Proof. From Eq. (4), we have

$$
V(\mathbf{r}_i^\top \mathbf{r}_j) = \sum_{k=1}^l V\left(\sum_{i'=1}^l \sum_{j'=1}^l U'[k, i']U'[k, j']e_i[i']e_j[j']\right)
\tag{6}
$$

Since $i \neq j$ holds, we have

$$
\begin{aligned}
&\left(\sum_{i'=1}^l \sum_{j'=1}^l U'[k, i']U'[k, j']e_i[i']e_j[j']\right)^2 \\
&= \sum_{i'=1}^l \sum_{j'=1}^l (U'[k, i'])^2 (U'[k, j'])^2 (e_i[i'])^2 (e_j[j'])^2 \\
&\quad + 2\sum_{i'=1}^l \sum_{j'=1}^l \sum_{i''<i'} \sum_{j''<j'} U'[k, i']U'[k, j']U'[k, i'']U'[k, j'']e_i[i']e_j[j']e_i[i'']e_j[j'']
\end{aligned}
\tag{7}
$$

Since $E((e_i[i'])^2(e_j[j'])^2) = \frac{l}{\log l}\frac{\log l}{l} \cdot \frac{l}{\log l}\frac{\log l}{l} = 1$ and $E(e_i[i']e_j[j']e_i[i'']e_j[j'']) = 0$,

$$
E\left(\left(\sum_{i'=1}^l \sum_{j'=1}^l U'[k, i']U'[k, j']e_i[i']e_j[j']\right)^2\right) = \sum_{i'=1}^l \sum_{j'=1}^l (U'[k, i']U'[k, j'])^2
\tag{8}
$$

As a result, from Eq. (5), (6), and (8), if $i \neq j$, we have

$$
\begin{aligned}
V(\mathbf{r}_i^\top \mathbf{r}_j) &= E\left(\left(\sum_{i'=1}^l \sum_{j'=1}^l U'[k, i']U'[k, j']e_i[i']e_j[j']\right)^2\right) \\
&\quad - \left(E\left(\sum_{k=1}^l \sum_{i'=1}^l \sum_{j'=1}^l U'[k, i']U'[k, j']e_i[i']e_j[j']\right)\right)^2 \\
&= \sum_{k=1}^l \sum_{i'=1}^l \sum_{j'=1}^l (U'[k, i'])^2 (U'[k, j'])^2
\end{aligned}
\tag{9}
$$

which completes the proof.   □

Algorithm 1 Effective Vector Computation

Input: d such that $l < d$
Output: $l \times d$ representation matrix **R**
1: **for** $i = 1$ to l **do**
2: **for** $j = 1$ to l **do**
3:   compute $S[i, j]$ from Equation (2);
4: compute SVD of rank l on $\mathbf{S} = \mathbf{U\Sigma V}^{\top}$;
5: compute **E** from Equation (3);
6: $\mathbf{R} = \mathbf{U\Sigma}^{\frac{1}{2}}\mathbf{E}$;

Algorithm 2 Efficient SVD Computation

Input: d such that $d \leq l$
Output: $l \times d$ representation matrix **R**
1: compute **B** from Equation (10);
2: **for** $i = 1$ to l **do**
3: compute $\mathbf{s}_i$ from Equation (2);
4: compute $\mathbf{s}'_i = \mathbf{s}_i\mathbf{B}$;
5: compute SVD of rank d on $\mathbf{S}' = \mathbf{U\Sigma V}^{\top}$;
6: $\mathbf{R} = \mathbf{U\Sigma}^{\frac{1}{2}}$;

As shown in Lemma 2, since $V(\mathbf{r}_i^{\top}\mathbf{r}_j)$ is represented as the cumulative summation of l elements, it would have a small value as number of clusters l is small. Besides, as shown in Lemma 2, $V(\mathbf{r}_i^{\top}\mathbf{r}_j)$ is independent from number of dimensions d. This indicates that we have the preferable property of the column vectors in **R** for the representation vectors regardless of the expanded dimensionality. Algorithm 1 has the following property:

Lemma 3. *Algorithm 1 takes $O(d \log l + l^3)$ time and $O(ld)$ space for computing representation matrix* **R**.

Proof. It would take $O(d \log l)$ time to compute **E**. It takes $O(l^2)$ time to compute **S**. Besides, it needs $O(l^3)$ time to compute SVD on **S**. It requires $O(l^2 \log l)$ time to compute $\mathbf{R} = \mathbf{U\Sigma}^{\frac{1}{2}}\mathbf{E}$ since $\mathbf{e}_i$ has $\log l$ nonzero elements. It needs $O(l^2)$, $O(d \log l)$, and $O(ld)$ spaces to hold **S**, **E**, and **R**, respectively. Therefore, Algorithm 1 needs $O(d \log l + l^3)$ time and $O(ld)$ space. □

3.3 SVD Computation

The previous section described the approach for the case of $l < d$. If we have $l \geq d$, we can obtain the representation vectors of the clusters with length d by computing the SVD of rank d for **S**. However, since the computation cost of SVD is $O(l^3)$, it is impractical to compute SVD if we have a large number of clusters. To efficiently compute SVD, we use $l \times d$ basic matrix **B** whose i-th row vector, $\mathbf{b}_i$, is set as follows:

$$b_i[j] = \begin{cases} \sqrt{\frac{1}{\log d}} & \text{with probability } \frac{\log d}{2d} \\ 0 & \text{with probability } 1 - \frac{\log d}{d} \\ -\sqrt{\frac{1}{\log d}} & \text{with probability } \frac{\log d}{2d} \end{cases} \tag{10}$$

Algorithm 2 details the procedure. It uses basic matrix **B** to project $l \times l$ large matrix **S** into an $l \times d$ low-dimensional space corresponding to matrix **S'** as a form of $\mathbf{S}' = \mathbf{SB}$. However, since the size of **S** is $l \times l$, it requires high memory cost if we directly hold **S**. To reduce the memory cost, our approach processes row vectors of **S** one by one. Specifically, let $\mathbf{s}_i$ be the i-th row vector of **S** and $\mathbf{s}'_i$ be the i-th row vector of **S'**, we compute row vectors as $\mathbf{s}'_i = \mathbf{s}_i\mathbf{B}$, as shown in Algorithm 2. Since it does not directly use **S**, we can reduce the memory cost in computing SVD. Let $\mathbf{V}'^\top = \mathbf{V}^\top \mathbf{B}^\top$, Algorithm 2 computes the representation matrix by factorizing the following $l \times l$ matrix $\tilde{\mathbf{S}}$:

$$\tilde{\mathbf{S}} = \mathbf{SBB}^\top = \mathbf{S}'\mathbf{B}^\top = \mathbf{U}\Sigma\mathbf{V}'^\top \tag{11}$$

We have the following property for matrix $\tilde{\mathbf{S}}$:

Lemma 4. *For matrix $\tilde{\mathbf{S}}$, $E(\tilde{S}[i,j]) = S[i,j]$ holds.*

Proof. From Eq. (11), we have

$$\tilde{S}[i,j] = \mathbf{s}_i \mathbf{Bb}_j^\top = \mathbf{s}_i [\mathbf{b}_1 \mathbf{b}_j^\top \ \mathbf{b}_2 \mathbf{b}_j^\top \ \dots \ \mathbf{b}_l \mathbf{b}_j^\top]^\top = \sum_{j'=1}^l S[i,j'] \left(\sum_{k=1}^d b_{j'}[k] b_j[k] \right) \tag{12}$$

If $j' = j$, we have $E\left(\sum_{k=1}^d b_{j'}[k] b_j[k]\right) = 1$. Otherwise, $E\left(\sum_{k=1}^d b_{j'}[k] b_j[k]\right) = 0$ holds. As a result, we have $E(\tilde{S}[i,j]) = E\left(\sum_{j'=1}^l S[i,j'] \left(\sum_{k=1}^d b_{j'}[k] b_j[k]\right)\right) = S[i,j]$, which completes the proof. $\square$

As shown in Eq. (11), we can exactly compute matrix $\tilde{\mathbf{S}}$ by using Algorithm 2. Therefore, this lemma indicates that we can effectively approximate $\mathbf{S}$ as $\tilde{\mathbf{S}}$. Concerning the approximation quality, we have the following property;

Lemma 5. *We have $V(\tilde{S}[i,j]) = \left(\frac{1}{\log d} - \frac{1}{d}\right)(S[i,j])^2 + \frac{1}{d} \sum_{j' \neq j} (S[i,j'])^2$.*

Proof. From Eq. (12), we have

$$V(\tilde{S}[i,j]) = \sum_{j'=1}^l (S[i,j'])^2 \, V\left(\sum_{k=1}^d b_{j'}[k] b_j[k]\right) \tag{13}$$

If $j' = j$, we have

$$\left(\sum_{k=1}^d b_{j'}[k] b_j[k]\right)^2 = \sum_{k=1}^d (b_j[k])^4 + 2\sum_{k=1}^d \sum_{k' < k} (b_j[k])^2 (b_j[k'])^2 \tag{14}$$

In this equation, we have $E\left(\sum_{k=1}^d (b_j[k])^4\right) = d \frac{1}{(\log d)^2} \frac{\log d}{d} = \frac{1}{\log d}$. Besides, since $k' < k$, we have $E((b_j[k])^2 (b_j[k'])^2) = \frac{1}{(\log d)^2} \frac{(\log d)^2}{d^2} = \frac{1}{d^2}$. As a result, we have $E(2\sum_{k=1}^d \sum_{k' < k} (b_j[k])^2 (b_j[k'])^2) = d(d-1)\frac{1}{d^2} = 1 - \frac{1}{d}$. Therefore, if $j' = j$, we have $E\left(\left(\sum_{k=1}^d b_{j'}[k] b_j[k]\right)^2\right) = \frac{1}{\log d} + 1 - \frac{1}{d}$. As a result, if $j' = j$, we have

$$V\left(\sum_{k=1}^d b_{j'}[k] b_j[k]\right) = E\left(\left(\sum_{k=1}^d b_{j'}[k] b_j[k]\right)^2\right) - \left(E\left(\sum_{k=1}^d b_{j'}[k] b_j[k]\right)\right)^2 = \frac{1}{\log d} - \frac{1}{d} \tag{15}$$

If $j' \neq j$, the following equation holds:

$$\left(\sum_{k=1}^d b_{j'}[k] b_j[k]\right)^2 = \sum_{k=1}^d (b_{j'}[k])^2 (b_j[k])^2 + 2\sum_{k=1}^d \sum_{k' < k} b_{j'}[k] b_j[k] b_{j'}[k'] b_j[k'] \tag{16}$$

In this equation, we have $E(b_{j'}[k] b_j[k] b_{j'}[k'] b_j[k']) = 0$. Besides, $(b_{j'}[k])^2 (b_j[k])^2 = \frac{1}{(\log d)^2}$ holds with probability $\frac{(\log d)^2}{d^2}$; $(b_{j'}[k])^2 (b_j[k])^2 = 0$, otherwise. Therefore, if $j' \neq j$, we have

$$E\left(\left(\sum_{k=1}^d b_{j'}[k] b_j[k]\right)^2\right) = d \frac{1}{(\log d)^2} \frac{(\log d)^2}{d^2} = \frac{1}{d} \tag{17}$$

As a result, if $j' \neq j$, we have

$$V\left(\sum_{k=1}^d b_{j'}[k]b_j[k]\right) = E\left(\left(\sum_{k=1}^d b_{j'}[k]b_j[k]\right)^2\right) - \left(E\left(\sum_{k=1}^d b_{j'}[k]b_j[k]\right)\right)^2 = \frac{1}{d} \tag{18}$$

Therefore, we have

$$V(\tilde{S}[i,j]) = \left(\frac{1}{\log d} - \frac{1}{d}\right)(S[i,j])^2 + \frac{1}{d}\sum_{j'\neq j}(S[i,j'])^2 \tag{19}$$

which completes the proof. □

Lemma 5 indicates that $V(\tilde{S}[i,j])$ would be small as the dimensions of the representation d increase. Therefore, Algorithm 2 can effectively compute the representation vectors as d increases. The computational and memory costs of Algorithm 2 are as follows:

Lemma 6. *Algorithm 2 needs $O(l^2 + ld^2)$ time and $O(ld)$ space for computing representation matrix* **R**.

Proof. Since $\mathbf{b}_i$ would have $\log d$ nonzero elements, it would take $O(l\log d)$ time to compute **B**. It needs $O(l^2)$ time to compute **S**. Since each column of **B** has $\log d$ nonzero elements, it needs $O(l\log d)$ time to compute $\mathbf{S}' = \mathbf{SB}$. It takes $O(ld^2)$ time to compute SVD on $\mathbf{S}'$ and $O(ld)$ time to compute $\mathbf{R} = \mathbf{U}\boldsymbol{\Sigma}^{\frac{1}{2}}$. Besides, it needs $O(l)$ space to hold $\mathbf{s}_i$ and $O(ld)$ space to hold **B**, $\mathbf{S}'$, and **R**. As a result, Algorithm 2 takes $O(l^2 + ld^2)$ time and $O(ld)$ space. □

3.4   Representation Learning Algorithm

Algorithm 3 gives a full description of our algorithm. It first identifies the clusters using the IncMod method (line 1). If the number of clusters, l, is smaller than the number of dimensions, d, it computes the clusters' representation vectors from Algorithm 1 (line 2–3). Otherwise, it computes the representation vectors using Algorithm 2 (line 4–5). It then computes the representation vectors for the nodes from the obtained representation vectors of the clusters (line 6–7). The computational and memory costs of Algorithm 3 are given as follows:

Theorem 1. *Our approach takes $O(m + d\log l + l^3)$ time and $O(nd + m)$ space if $l < d$ holds. Otherwise, it requires $(m + l^2 + ld^2)$ time and $O(nd + m)$ space.*

Proof. The IncMod method needs $O(m)$ time and $O(m)$ space [11]. If $l < d$, as shown in Lemma 3, it takes $O(d\log l + l^3)$ time and $O(ld)$ space to compute the representation vectors of the clusters. Otherwise, it needs $O(l^2 + ld^2)$ time and $O(ld)$ space, as shown in Lemma 6. It needs $O(m)$ time to compute the representation vectors of the nodes. It needs $O(nd)$ space to hold the representation vectors. As a result, it needs $O(m + d\log l + l^3)$ time and $O(nd + m)$ space if $l < d$ holds. Otherwise, it takes $(m + l^2 + ld^2)$ time and $O(nd + m)$ space. □

Algorithm 3. GC-NRL

Input: graph $G = (\mathbb{V}, \mathbb{E})$ and number of dimen-
     sion d
Output: representation vector $\mathbf{x}_v$ for each node
1: compute the clusters in G by using IncMod
     method;
2: **if** $l < d$ **then**
3:     compute matrix $\mathbf{R}$ from Algorithm 1;
4: **else**
5:     compute matrix $\mathbf{R}$ from Algorithm 2;
6: **for** each $v \in \mathbb{V}$ **do**
7:     compute $\mathbf{x}_v = \sum_{i=1}^{l} \sum_{w \in \mathbb{C}_i} W[v, w] \mathbf{r}_i$;

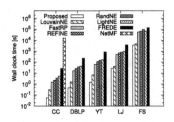

Fig. 1. Processing time.

Table 1. Characteristics of the experimental graphs.

	CC	DBLP	YT	LJ	YT
#Nodes	44,034	317,080	1,138,499	3,997,962	65,608,366
#Edges	390,722	2,099,732	5,980,886	69,362,378	3,612,134,270
#Labels	15	50	47	50	50

4   Experimental Evaluation

This section compared our approach to the previous approaches; FastRP [3], REFINE [14], RandNE [13], FREDE [12], LightNE [9], LouvainNE [2], and NetMF [10]. As shown in Table 1, we used five real-world graphs; *CoCit (CC)*, *com-DBLP (DBLP)*, *YouTube (YT)*, *com-LiveJournal (LJ)*, and *com-Friendster (FS)*. For NetMF, we set the target rank of eigendecomposition to 1,024 as in [10]. We set negative sampling to 20, as shown in [6]. We set the window size used in NetMF, as well as FastRP, REFINE, RandNE, and LightNE to ten by following [10]. We set the number of nodes from which we compute personalized PageRank to 1,000 for FREDE. For RandNE and FastRP, we set weights used in the high-order proximity matrices to one, the same as in [10]. For REFINE and LightNE, we set the number of diffusion steps to two by following the previous paper [14]. For LouvainNE, we set the damping parameter to 0.01 following [2]. We used the same programming language, C++, to implement the approaches examined. We conducted the experiments on a Linux server using an Intel Xeon Platinum 8280 CPU with a 2.70 GHz processor and 1.5 TB memory.

4.1   Network Representation Learning Time

We evaluated the network representation learning time of each approach. Figure 1 plots the processing time to compute the representation vectors from the given graphs. This experiment set the number of dimensions to $d = 128$.

As shown in Fig. 1, our approach offers higher efficiency than the previous approaches; it is up to 5.5, 43.7, 77.3, 99.9, 111.5, 601.0, and 293071.0 times faster than LouvainNE, FastRP, REFINE, RandNE, LightNE, FREDE, and NetMF,

Table 2. Node classification performance of each approach.

Approach	Micro-F1 [%]					Macro-F1 [%]				
	CC	DBLP	YT	LJ	FS	CC	DBLP	YT	LJ	FS
Proposed	**37.127**	**16.823**	**2.894**	**10.599**	**0.145**	**23.179**	**17.539**	**12.933**	**9.801**	**0.836**
FastRP	14.268	4.970	0.666	1.250	0.034	5.072	4.692	3.136	2.109	0.463
REFINE	29.310	12.236	2.645	9.902	0.069	20.935	17.419	3.627	7.583	0.202
RandNE	16.659	5.686	2.128	9.449	0.037	1.904	2.766	5.403	3.908	0.587
FREDE	30.150	9.868	2.662	9.790	0.049	19.189	13.367	8.685	7.062	0.833
LightNE	14.651	5.642	2.367	9.461	0.035	4.963	2.456	5.758	4.109	0.575
LouvainNE	36.969	16.613	2.783	10.462	0.144	22.642	17.161	8.235	9.273	0.828
NetMF	13.879	–	–	–	–	8.119	–	–	–	–

respectively. NetMF incurs a high computational cost to apply eigendecomposition to the proximity matrix since the matrix has $O(n^2)$ number of nonzero elements. RandNE and REFINE incur high computation costs to obtain the orthogonal matrix used in the iterative projection procedure. LightNE also incurs high computation costs to perform orthonormalization for the basic matrix used in SVD. FastRP incurs high computation costs since it recursively performs expensive matrix computations to obtain the representation vectors. FREDE needs a high computational cost to compute personalized PageRank and SVD iteratively. The computation cost of LouvainNE is high as the Louvain method is iteratively performed to obtain the hierarchical structure. On the other hand, the proposed approach factorizes the small $l \times l$ similarity matrix by performing graph clustering only once to efficiently generate the representation vectors.

4.2   Multi-label Node Classification

This experiment performed node classification. We used the one-vs-rest logistic regression model implemented by LIBLINEAR. In the test phase, the one-vs-rest model yielded a ranking of labels rather than an exact label assignment. We took the assumption that was made in DeepWalk; the number of labels for nodes in the test data is given [8]. Table 2 shows the Micro-F1 and Macro-F1 scores where we set the training ratio to 5%. We set the number of dimensions to $d = 128$. For DBLP, YT, LJ, and FS, we omit the results of NetMF since it failed to compute the representation vectors due to the lack of memory space.

Table 2 indicates that our approach yields higher Micro-F1 and Macro-F1 scores than the previous approaches. This is because, as described in Sect. 3.1, we exploit the structural similarity matrix to capture the relationships between clusters and compute the representation vectors of the clusters by factorizing their similarity matrix. NetMF applies the element-wise matrix logarithm to the proximity matrix. However, it harms the quality of representations by cutting small nonzero elements. Even though the base matrices used of FastRP, REFINE, and RandNE would be orthogonal, they do not accurately capture the structural property of nodes since the obtained representation vectors are not orthogonal after performing the iterative projection procedure. Although FREDE uses personalized PageRank, it fails to capture the structural property of nodes. Since the path-sampling approach used in LightNE yields a sparse proximity matrix where at most m node pairs can have nonzero elements, it has difficulty in effectively representing the proximities between nodes. Since LouvainNE does not exploit relationships between clusters, it separates nodes independently in the representation space according to the clusters.

5 Conclusions

This paper addressed the problem of improving the efficiency and accuracy of network representation learning. We perform graph clustering just once and factorize the similarity matrix between clusters to capture the structural property of the graph. Experiments show that our approach is more efficient than existing approaches with greater accuracy.

Acknowledgment. This work was supported by JSPS KAKENHI Grant Number 22H03596.

References

1. Arriaga, R.I., Vempala, S.S.: An algorithmic theory of learning: robust concepts and random projection. Mach. Learn. **63**(2), 161–182 (2006)
2. Bhowmick, A.K., Meneni, K., Danisch, M., Guillaume, J., Mitra, B.: Louvainne: hierarchical Louvain method for high quality and scalable network embedding. In: WSDM, pp. 43–51 (2020)
3. Chen, H., Sultan, S.F., Tian, Y., Chen, M., Skiena, S.: Fast and accurate network embeddings via very sparse random projection. In: CIKM, pp. 399–408 (2019)
4. Fujiwara, Y., Irie, G., Kuroyama, S., Onizuka, M.: Scaling manifold ranking based image retrieval. Proc. VLDB Endow. **8**(4), 341–352 (2014)
5. Ida, Y., Fujiwara, Y., Kashima, H.: Fast sparse group lasso. In: NeurIPS, pp. 1700–1708 (2019)
6. Mikolov, T., Sutskever, I., Chen, K., Corrado, G.S., Dean, J.: Distributed representations of words and phrases and their compositionality. In: NIPS, pp. 3111–3119 (2013)
7. Nakatsuji, M., Fujiwara, Y., Toda, H., Sawada, H., Zheng, J., Hendler, J.A.: Semantic data representation for improving tensor factorization. In: AAAI, pp. 2004–2012 (2014)
8. Perozzi, B., Al-Rfou, R., Skiena, S.: Deepwalk: online learning of social representations. In: KDD, pp. 701–710 (2014)
9. Qiu, J., Dhulipala, L., Tang, J., Peng, R., Wang, C.: Lightne: a lightweight graph processing system for network embedding. In: Li, G., Li, Z., Idreos, S., Srivastava, D. (eds.) SIGMOD, pp. 2281–2289. ACM (2021)
10. Qiu, J., Dong, Y., Ma, H., Li, J., Wang, K., Tang, J.: Network embedding as matrix factorization: Unifying deepwalk, line, pte, and node2vec. In: WSDM, pp. 459–467 (2018)

11. Shiokawa, H., Fujiwara, Y., Onizuka, M.: Fast algorithm for modularity-based graph clustering. In: AAAI (2013)
12. Tsitsulin, A., Munkhoeva, M., Mottin, D., Karras, P., Oseledets, I.V., Müller, E.: FREDE: anytime graph embeddings. Proc. VLDB Endow. **14**(6), 1102–1110 (2021)
13. Zhang, Z., Cui, P., Li, H., Wang, X., Zhu, W.: Billion-scale network embedding with iterative random projection. In: ICDM, pp. 787–796 (2018)
14. Zhu, H., Koniusz, P.: REFINE: random range finder for network embedding. In: Demartini, G., Zuccon, G., Culpepper, J.S., Huang, Z., Tong, H. (eds.) CIKM, pp. 3682–3686. ACM (2021)

Learning with Small Data: Subgraph Counting Queries

Kangfei Zhao[1]([✉]), Jeffrey Xu Yu[2], Zongyan He[2], and Yu Rong[3]

[1] Beijing Institute of Technology, Beijing, China
zkf1105@gmail.com
[2] The Chinese University of Hong Kong, Sha Tin, Hong Kong
{yu,zyhe}@se.cuhk.edu.hk
[3] Tencent AI Lab, Shenzhen, China

Abstract. Deep Learning (DL) has been widely used in many applications, and its success is achieved with large training data. A key issue is how to provide a DL solution when there is no efficient training data to learn initially. In this paper, we explore a meta learning approach for a specific problem, subgraph isomorphism counting, which is a fundamental problem in graph analysis to count the number of a given pattern graph, p, in a data graph, g, that matches p. This problem is NP-hard, and needs large training data to learn by DL in nature. To solve this problem, we design a Gaussian Process (GP) model which combines graph neural network with Bayesian nonparametric, and we train the GP by a meta learning algorithm on a small set of training data. By meta learning, we obtain a generalized meta-model to better encode the information of data and pattern graphs and capture the prior of small tasks. We handle a collection of pairs (g, p), as a task, where some pairs may be associated with the ground-truth, and some pairs are the queries to answer. There are two cases. One is there are some with ground-truth (few-shot), and one is there is none with ground-truth (zero-shot). We provide our solutions for both. We conduct substantial experiments to confirm that our approach is robust to model degeneration on small training data, and our meta model can fast adapt to new queries by few/zero-shot learning.

1 Introduction

Deep Learning (DL) has achieved remarkable success in database systems to support estimation tasks [12,13]. The success lies in not only end-to-end modeling but also learning from a large number of training data. Almost all the work focus on DL techniques assuming that it is possible to collect enough training data to learn a model. A natural question that arises is what a system can do if there is only a few training data to learn a model that can be effectively used. The solution rules out learning a model until the training dataset is large. To alleviate this data insufficient issue, a new learning paradigm called meta learning [6] is developed by the machine learning community. The target of meta learning is to learn a model and refine the model with limited or even no training data

© The Author(s), under exclusive license to Springer Nature Switzerland AG 2023
X. Wang et al. (Eds.): DASFAA 2023, LNCS 13945, pp. 308–319, 2023.
https://doi.org/10.1007/978-3-031-30675-4_21

if any from time to time. We will discuss it with a specific problem – subgraph isomorphism counting, as it is difficult to come up with a general solution to deal with the requirement of sufficient training at this stage.

In this paper, we study subgraph isomorphism counting queries that support a variety of applications. A subgraph isomorphism counting query is specified by a pair of data graph g and a pattern graph p, aiming to find the number of matches of p in g. As subgraph isomorphism problem is NP-complete [1], the counting problem is also difficult to solve. In general, given a set of data graphs, $\mathcal{G} = \{g_1, \cdots, g_n\}$ and a set of pattern graphs $\mathcal{P} = \cup_{1 \leq i \leq n} \mathcal{P}(g_i)$, where $\mathcal{P}(g_i)$ is the set of pattern graphs associated with g_i, Liu et al. [4] propose DL models by feeding the queries (g, p) in $\mathcal{G}$ and $\mathcal{P}$ with their true counts as the training data. The models are used to predict unseen queries. The DL models [4] need hundreds of thousands of training queries, and for each graph g_i, its $\mathcal{P}(g_i)$ should also be large enough. In real applications, the data graphs may come from different domains, and the pattern graphs may be diverse regarding the sizes, node/edge labels, and structures. It is infeasible to exhaustively collect a sufficiently large training set to synthesise comprehensive features to be learned.

Hence, for subgraph counting, learning a DL model from limited training pairs of $\mathcal{G}$ and $\mathcal{P}$ is an inevitable and challenging task. To deal with the problem, we construct a meta model to learn the prior knowledge of subgraph counting across multiple tasks. Here, a task is a batch of queries which may be subject to underlying distribution within the task. For a new task where a small number of training queries, a.k.a, shots, is possibly available, the meta model can swiftly adapt to answer new queries that are subject to the similar distribution of the task. Inspired by deep kernel learning [9] and deep kernel transfer [6], we devise a new meta model that warps a Graph Neural Network (GNN) as a special Gaussian Process (GP). For one thing, the GNN preserves the powerful modeling capability of DL for subgraph counting. For the other thing, Bayesian nonparametric, inherited by GP, enables learning from scratch over small samples with a distribution-free assumption. Furthermore, as the new task may not provide new training data, we adapt the kernel-based meta learning algorithm to support this zero-shot case in a data-driven fashion.

The contributions of this paper are summarized as follows: ① We study subgraph isomorphism counting in a paradigm of meta-learning. We propose a GP model, called RGIN-GP, that combines GNN and kernel method, aiming to learn over limited training data. We employ a Bayesian meta-learning algorithm to train the meta model. ② We provide solutions for both few-shot and zero-shot cases to deal with a new subgraph counting task. In particular, for zero-shot, we propose a new data-driven approach to predict the count values for a new task without any ground-truth. ③ We conduct extensive experiments on real and synthetic graph datasets for different task configurations. The experimental results verify the superiority of meta learned RGIN-GP on small training data and its effectiveness for few/zero-shot learning.

2 Preliminaries

We model both data graph g and pattern graph p as a labeled undirected graph, i.e., a tuple $G = (V, E, L_V, L_E, \Sigma_V, \Sigma_E)$. Here, V is a set of nodes, E is a set of undirected edges, and L_V (L_E) is a mapping function that maps a node $u \in V$ (edge $e \in E$) to a node label (edge label) in Σ_V (Σ_E). We denote neighbors of node u in G as $N(u) = \{v | (u, v) \in E\}$.

Subgraph Isomorphism: Given a data graph $g = (V_g, E_g, L_V, L_E, \Sigma_V, \Sigma_E)$ and a pattern graph $p = (V_p, E_p, L_V, L_E, \Sigma_V, \Sigma_E)$, subgraph isomorphism p to g is an *injective* function $f \colon V_p \mapsto V_g$ such that (1) for every $u \in V_p$, $L_V(u) = L_V(f(u))$, (2) for every $(u, v) \in E_p$, $(f(u), f(v)) \in E_g$, and (3) for every $e = (u, v) \in E_p$ and $e' = (f(u), f(v)) \in E_g$, $L_E(e) = L_E(e')$.

Subgraph Isomorphism Counting Query: A graph database is a set of small/medium sized graphs, $\mathcal{G} = \{g_1, g_2, \cdots, g_n\}$, with a set of pattern graphs $\mathcal{P} = \cup_{1 \leq i \leq n} \mathcal{P}(g_i)$, where $\mathcal{P}(g_i)$ is a set of pattern graphs associated with g_i. For simplicity, we use $\mathcal{P} = \{p_1, p_2, \cdots, p_m\}$ to denote the whole possible set of patterns. Given a data graph $g \in \mathcal{G}$, and a pattern graph $p \in \mathcal{P}(g)$, a subgraph isomorphism counting query, (g, p), is to find the total number of subgraph isomorphism matchings of p to g, denoted as $c(g, p)$. Here, a node (edge) in p is allowed to be unlabeled, indicating its label can be any one in L_V (L_E).

A regression model for subgraph counting can be built from a collection of training queries, $X = \{x_1, x_2, \cdots, x_{|X|}\}$, where $x_i = (g_i, p_i)$ is a query, associated with the true count $c(x_i)$ (or $c(g_i, p_i)$). The model will predict the count $\hat{c}(g^*, p^*)$ for an unseen test query (g^*, p^*). Here, either g^* or p^*, or both g^* and p^* does not appear in the training data X. We use the absolute error as Eq. (1) to evaluate the accuracy of the estimated count.

$$\text{abs-error}(g^*, p^*) = |c(g^*, p^*) - \hat{c}(g^*, p^*)| \tag{1}$$

Note that the regression model can answer the subgraph isomorphism query, i.e., whether p^* is subgraph isomorphism to g^* by $\hat{c}(g^*, p^*) > 0.5$.

Problem Statement: Our problem is to build a meta model $\mathcal{M}$ to support subgraph isomorphism counting *tasks*, where a task $\mathcal{T}$ is a batch of queries $\{(g_i, p_i)\}_{i=1}^{b}$ that may be subject to underlying distribution within the task. The model $\mathcal{M}$ is trained by a set of training task $\mathcal{D} = \{\mathcal{T}_1, \cdots, \mathcal{T}_n\}$ where all the queries have the ground-truth. Specifically, a test task $\mathcal{T}^*$ is the union of two subsets of queries, $\mathcal{S}^*$ and $\mathcal{Q}^*$, denoted as $\mathcal{T}^* = (\mathcal{S}^*, \mathcal{Q}^*)$. Here, $\mathcal{S}^* = \{(g_i^*, p_i^*)\}_{i=1}^{k}$ is called the *support set*, where the ground-truth count $c(g_i^*, p_i^*)$ for each $i \in [1, \cdots k]$ is given. And $\mathcal{Q}^* = \{(g_j^*, p_j^*)\}_{j=k+1}^{b}$ is called the *query set* where each query (g_j^*, p_j^*) is to be answered by model $\mathcal{M}$. Note that $\mathcal{S}^*$ may be empty, i.e., $|\mathcal{S}^*| = 0$ (called *zero-shot*), and is small in size when it is non-empty (called *few-shot*). The problem is how to build and exploit $\mathcal{M}$ to answer queries in $\mathcal{Q}^*$ on-demand with the assistant of $\mathcal{S}^*$ which may be empty.

In this paper, we explore the following 5 task configurations where a single variable (e.g., a data or pattern graph) is controlled. ① Same Graph Tasks

(SameG): Data graphs are from a single domain. The data graphs that appear in training tasks will not appear in any testing task. The pattern graphs $\mathcal{P}(g_i)$ that are associated with a data graph g_i will appear together with g_i in a task where g_i appears. ② Same Pattern Tasks (SameP). Data graphs are from a single domain. The pattern graphs that appear in training tasks will not appear in any testing task. The data graphs g_i will appear in a task together with p_j if $p_j \in \mathcal{P}(g_i)$ appears in the task. ③ Hybrid Domains with Same Graph Tasks (HySameG). Data graphs are from multiple domains whereas data graphs in one task are from the same domain. For one domain, training and testing tasks follow SameG. ④ Hybrid Domains with Same Pattern Tasks (HySameP). Data graphs are from multiple domains whereas data graphs in one task are from the same domain. For one domain, training and testing tasks follow SameP. ⑤ Random Tasks (Random). Data graphs are from a single domain. Pairs are randomly and disjointly distributed in all the training and testing task sets.

2.1 GNN-Based Encoder for Subgraph Counting

Recently, a learning framework has been proposed for subgraph isomorphism counting of a pair of data and pattern graphs in [4]. This neural network framework is composed of graph representation layers, interaction layers, and Multilayer perceptron (MLP), to learn $\mathcal{M}$ with large training data. [4] explores different options for the graph representation layers and the interaction layers, where a GNN variant, Relational Graph Isomorphism Network (RGIN) coupled with a sum pooling interaction layer achieves the best trade-off between prediction accuracy and efficiency. Below, we introduce RGIN and sum pooling interaction which we deploy as the encoder of our meta model.

RGIN Graph Representation: The K-layer GNN [3] follows a neighborhood aggregation paradigm to update the representation of each node by alternatively applying an aggregation function and a combine function in K iterations. Take the RGIN layer as an example, let $e_v^{(k)}$ denote the representation of node v generated in the k-th iteration. For each node v, the aggregate function in Eq. (2) distinguishes its neighbors by the edge label, and aggregates the $|\Sigma_E|$ types of neighbors respectively. Here, $W_l^{(k)}$ is the weight matrix for the neighbors with edge label l in the k-th layer. Then, the representations for $|L_E|$ types are further summed to one representation $a_v^{(k)}$. In the combine function of Eq. (3), the aggregated $a_v^{(k)}$ is summed up with the $(k-1)$-layer representation $e_v^{(k-1)}$, which is transformed by the weight $W_0^{(k)}$, and finally is transformed by an MLP layer.

$$a_v^{(k)} = \sum_{l \in \Sigma_E} \sum_{u \in N(v), L_E((u,v))=l} W_l^{(k)} e_u^{(k-1)} \tag{2}$$

$$e_v^{(k)} = \mathsf{MLP}(W_0^{(k)} e_v^{(k-1)} + a_v^{(k)}) \tag{3}$$

For the data graph g and pattern graph p, two independent RGIN models generate the data graph and pattern graph node embedding, respectively.

Sum Pooling Interaction. The interaction layer is to combine the data graph and pattern graph embeddings into one pair-wise embedding. The sum pooling interaction sums up the node embedding of the data and pattern graphs, respectively, and concatenates the two vectors to a long vector, as shown in Eq. (4).

$$h = \mathsf{Concat}\left(\sum_{v \in V_g} e_v^{(K)}, \sum_{v' \in V_p} e_{v'}^{(K)} \right) \tag{4}$$

The concatenated embedding h will be used to predict $\hat{c}(g, p)$ for query (g, p).

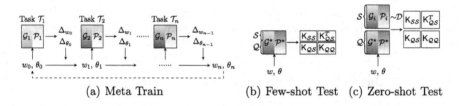

(a) Meta Train   (b) Few-shot Test   (c) Zero-shot Test

Fig. 1. Meta Model Train & Test

3   A Meta Learning Approach

Based on the neural network encoder, we introduce how to build a meta subgraph counting model $\mathcal{M}$. The model is trained on a collection of training tasks $\mathcal{D} = \{\mathcal{T}_1, \cdots, \mathcal{T}_n\}$, where task-common parameters are learned to capture the prior knowledge of the subgraph counting tasks. For a test task $\mathcal{T}^* = (\mathcal{S}^*, \mathcal{Q}^*)$, the meta model $\mathcal{M}$ will adapt to $\mathcal{T}^*$ by few-shot learning if $|\mathcal{S}^*| > 0$ or by zero-shot learning if $|\mathcal{S}^*| = 0$ to perform task-specific estimations.

Our basic idea is inspired by Deep Kernel Transfer [6] which learns a task-common Gaussian Process (GP) model shared by multiple query tasks. We propose a model called RGIN-GP that warps the RGIN with sum pooling as a feature transfer layer in a kernel function. The neural network parameter and the kernel hyperparameter are jointly optimized by deep kernel learning [9]. Concretely, in the training process illustrated in Fig. 1(a), RGIN-GP computes the kernel function for each training task $\mathcal{T}_i$ as a batch, where the parameters are optimized by minimizing the negative marginal (log) likelihood of task $\mathcal{T}_i$. In testing, for a test task $\mathcal{T}^* = (\mathcal{S}^*, \mathcal{Q}^*)$, the model leverages the kernel matrix of the task $\mathcal{T}^*$ to predict the queries in $\mathcal{Q}^*$ conditioned on the support set $\mathcal{S}^*$ by Bayesian inference (Fig. 1(b)). In the case of zero-shot, i.e., $|\mathcal{S}^*| = 0$, we take a data-driven approach to build the kernel by making use of one task, $\mathcal{T}_i$, drawn from the training data (Fig. 1(c)).

3.1 RGIN Gaussian Process (RGIN-GP)

To learn over a small set of samples, nonparametric modeling is an effective method in Bayesian learning. Inspired by this, we construct a kernel function for RGIN with a sum pooling layer by deep kernel learning [9]. Given an input $x = (g, p)$ as a pair of data and pattern graphs, the deep kernel function K measures the similarity of a pair of inputs x_i, x_j as

$$\mathsf{K}(x_i, x_j; w, \theta) = \mathcal{K}(\mathcal{F}(x_i; w), \mathcal{F}(x_j; w); \theta) \tag{5}$$

Here, $\mathcal{F}(x; w)$ is a non-linear transformation specified by a deep neural network with parameters w, i.e., the RGIN together with the sum pooling interaction layer. And the function $\mathcal{K}(h_i, h_j; \theta)$ is a stationary kernel function that is invariant to input transformation with the hyperparameter θ, e.g., the RBF kernel.

Given n training inputs, $X = \{x_1, \cdots, x_n\}$, the deep kernel K defined in Eq. (5), the model $f(X)$ is a Gaussian Process as Eq. (6) [7] that we call RGIN-GP, where $\mu_X = [\mu]^n$ is an assumed constant mean and $\mathsf{K}_{X,X} = [\mathsf{K}(x_i, x_j; w, \theta)]^{n \times n}$ is the covariance function.

$$f(X) = [f(x_1), \cdots, f(x_n)]^T \sim \mathcal{N}(\mu_X, \mathsf{K}_{X,X}) \tag{6}$$

To make prediction for the testing inputs $X^* = \{x_1^*, \cdots, x_m^*\}$, we need to compute the conditional distribution $p(f(X^*)|f(X))$ as the prediction, assuming the output is disturbed by a Gaussian noise $\mathcal{N}(0, \sigma^2)$. It is also proved to be a Gaussian distribution as Eq. (7), where the expectation and covariance of the predictive distribution can be solved in closed form in Eq. (8)–(9).

$$f(X^*)|f(X) \sim \mathcal{N}(\mathbb{E}(c^*), C) \tag{7}$$

$$\mathbb{E}(c^*) = \mu_X + \mathsf{K}_{X,X^*}^T [\mathsf{K}_{X,X} + \sigma^2 \mathcal{I}]^{-1}(c - \mu_X) \tag{8}$$

$$C = \mathsf{K}_{X^*,X^*} - \mathsf{K}_{X,X^*}^T \cdot [\mathsf{K}_{X,X} + \sigma^2 \mathcal{I}]^{-1} \mathsf{K}_{X,X^*} \tag{9}$$

Here, $c = [c(x_i)]^n$ is the ground-truth of the input X. $\mathsf{K}_{X,X} = [\mathsf{K}(x_i, x_j; w, \theta)]^{n \times n}$, $\mathsf{K}_{X,X^*} = [\mathsf{K}(x_i, x_j^*; w, \theta)]^{n \times m}$ and $\mathsf{K}_{X^*,X^*} = [\mathsf{K}(x_i^*, x_j^*; w, \theta)]^{m \times m}$ are the train-train, train-test, test-test kernel matrices, respectively. The expectation $\mathbb{E}(c^*)$ will be treated as the explicit prediction counts $\hat{c}$, and the diagonal element of matrix C in Eq. (9) measures the variance of the prediction.

Training the GP is to fit the neural network weight w and the kernel hyperparameter θ by minimizing the negative (log) likelihood of X as Eq. (10).

$$\mathcal{L}_{\mathsf{mll}} = -\log p(c|X) \propto c^T [\mathsf{K}_{X,X} + \sigma^2 \mathcal{I}]^{-1} c + \log |\mathsf{K}_{X,X} + \sigma^2 \mathcal{I}| \tag{10}$$

To train an RGIN-GP, the neural network weights w and the kernel hyperparameter θ are jointly optimized by stochastic gradient descent.

3.2 Meta Learning for RGIN-GP

Given training task set $\mathcal{D} = \{\mathcal{T}_1, \cdots, \mathcal{T}_n\}$, we discuss how to train an RGIN-GP as a meta model and test it in the few-shot and zero-shot scenarios.

Meta Training: The meta training process is to learn (w, θ) of the kernel K that minimizes the negative marginal likelihood across all the training tasks. As shown in Fig. 1(a), for each gradient step, a task $\mathcal{T}$ is sampled from the training tasks, then the marginal likelihood $\mathcal{L}_{\text{mll}}$ (Eq. (10)) is computed over all the pairs in the task, i.e., $\mathcal{S} \cup \mathcal{Q}$, and the parameters (w, θ) are updated for that task. The meta training algorithm is different from training the kernel from scratch, where marginalization of the likelihood is computed on all data instead of a distinct task. The learned parameters (w, θ) better leverage the structure of the tasks, which are shared across all tasks as the task-common parameters.

Table 1. Profile of Datasets

Dataset	Data Graphs					Pattern Graphs					# (g, p)	$c(g, p)$
	$\|V_g\|$	$\|E_g\|$	$\|\Sigma_V\|$	$\|\Sigma_E\|$	# g	$\|V_p\|$	$\|E_p\|$	$\|\Sigma_V\|$	$\|\Sigma_E\|$	# p		
MUTAG	[10, 28]	[20, 66]	[3, 7]	[3, 4]	188	[3, 4]	[2, 3]	[1, 2]	[1, 2]	24	4,512	[0, 156]
SYN-S	[10, 28]	[20, 66]	[3, 7]	[3, 4]	30,681	[3, 4]	[2, 3]	[1, 2]	[1, 2]	240	30,681	[0, 126]
SYN-M	[10, 56]	[22, 132]	[3, 7]	[3, 4]	102,057	[3, 8]	[2, 12]	[1, 2]	[1, 2]	1,680	102,057	[0, 128]
SYN-L	[64, 512]	[64, 2,048]	[16, 64]	[16, 64]	127,897	[3, 16]	[2, 16]	[2, 16]	[2, 16]	100	127,897	[0, 512]

Few-shot Testing: Given a testing task $\mathcal{T}^* = (\mathcal{S}^*, \mathcal{Q}^*)$ where $|\mathcal{S}^*| \neq 0$, the meta model will adapt to the task based on its support set and the task-common parameters learned. As shown in Fig. 1(b), the predictive distribution of Eq. (7) for the query set $\mathcal{Q}^*$ is computed by conditioning on the support set $\mathcal{S}^*$, which analytical solution is given in Eq. (8)–(9).

Zero-shot Testing: Given a testing task $\mathcal{T}^* = (\mathcal{S}^*, \mathcal{Q}^*)$ where $|\mathcal{S}^*| = 0$, the meta model cannot adapt to the task based on its support set. In the literature [10], zero-shot learning is mainly done for classification where the classes are limited. Different from classification, To make predictions for the regression task, we utilize training tasks. The basic idea is to borrow some training task $\mathcal{T}_i$ as the support set for the new coming task $\mathcal{T}^*$ as shown in Fig. 1(c). The assumptions are that the training tasks and test task may be similar regarding data/pattern graphs, and they share some specific task structures. The kernel K leverages the similarity. First, a training task is sampled randomly from the training data. Then, a set of auxiliary data (X, c) is randomly drawn from the task to serve as the support set and is used to compute the posterior of the parameters $p(\rho_{\mathcal{T}^*} | c, X)$.

3.3 Feature Encoding

Encoding initial node representation $\mathbf{e}_v^{(0)}$ for RGIN-GP in the neural network mapping $\mathcal{F}$ is important in learning. For subgraph isomorphism, the node/edge labels of a pattern node serve as the predicates of the pattern, and are used to filter nodes in the data graph. However, the widely used one-hot encoding of the labels is lack insight for the analytical subgraph counting. We utilize frequency-based encoding and pre-trained embedding-based encoding to encode label information and topological structure, which are specifically proposed for subgraph counting tasks [13].

4 Experimental Studies

In this section, we give the test setting and report our experimental results.

Implementation and Setting: We give the settings of RGIN-GP. For the neural network transformation $\mathcal{F}$, the number of RGIN layers is 3, where each hidden layer has 64 units and a Dropout probability of 0.2. For the stationary kernel function $\mathcal{K}$, we use the spectral mixtures based kernels [8]. For the embedding based encoding, we try 4 scalable task-independent node embedding approaches, and finally choose *ProNE* [11] as the embedding algorithm for the label-augmented graph. Following the setting in [11], the dimension of the embedding is 128. The learning framework is built on PyTorch. We use the Adam optimizer with a decaying learning rate to train our models via 200 epochs. The initial learning rates α_w and α_θ are set to 5e-4 and 1e-3 empirically, respectively.

Datasets: We use one real graph dataset MUTAG, and three synthetic graph datasets SYN-S, SYN-M and SYN-L. MUTAG collection has 188 unique compounds where nodes represent atoms and edges represent bonds. The 24 patterns are from [4]. The three synthetic datasets are generated by the generator of [4]. SYN-S follows the same scale as MUTAG, and SYN-M enlarges the scale of the MUTAG data and pattern graphs two times. SYN-L follows the largest scale of the data and pattern graphs in [4]. Table 1 lists the profile of the four datasets.

Baseline Approaches: We compare meta learned RGIN-GP (RGIN-GP) with the neural network baselines RGIN+SumPool and RGIN+DIAMNet [4].

Evaluation Metrics: We use the mean of abs-error (Eq. (1)), i.e., MAE, of the counts and the accuracy of the subgraph isomorphism query to evaluate the model performance.

Exp-1: RGIN-GP vs. Neural Network Models. We first compare our RGIN-GP with its neural network counterpart RGIN+SumPool, and a more powerful model RGIN+DIAMNet on the MUTAG dataset. For RGIN-GP, we organize the training and testing pairs as Random tasks with 128 pairs in each task. Testing is conducted in the zero-shot mode with 128 auxiliary pairs drawn from the training data. For the two neural network models, they are trained by standard supervised learning. Table 2 shows the testing performance on 20% testing pairs when the training pairs are set to 60%, 40%, and 20% of the overall pairs, respectively. In general, the 3 RGIN-GP variants remarkably outperform the two neural network models w.r.t. MAE. The neural network RGIN+SumPool and RGIN+DIAMNet suffer from model degradation. The implication of this experiment is our RGIN-GP is robust, data-efficient and much easier to train than its neural network counterpart.

Table 2. MAE (↓)/Accuracy (↑) on MUTAG and SYN-L

Train Ratio	Model	MUTAG		SYN-L	
		MAE	**Acc.**	**MAE**	**Acc.**
0.6	RGIN+SumPool	10.41 ± 6.25	0.89 ± 0.02	14.47 ± 4.32	0.95 ± 0.01
	RGIN+DIAMNet	3.29 ± 0.82	0.83 ± 0.06	5.55 ± 3.58	0.97 ± 0.00
	RGIN-GP(onehot)	0.96 ± 0.15	0.92 ± 0.07	2.03 ± 0.02	0.93 ± 0.00
	RGIN-GP(freq)	0.92 ± 0.18	0.94 ± 0.04	2.05 ± 0.01	0.93 ± 0.00
	RGIN-GP(prone)	0.87 ± 0.13	0.93 ± 0.02	2.03 ± 0.01	0.93 ± 0.00
0.4	RGIN+SumPool	8.78 ± 9.14	0.88 ± 0.03	6.04 ± 3.78	0.93 ± 0.04
	RGIN+DIAMNet	7.72 ± 1.21	0.85 ± 0.36	5.62 ± 2.30	0.95 ± 0.03
	RGIN-GP(onehot)	0.94 ± 0.14	0.92 ± 0.06	2.21 ± 0.06	0.93 ± 0.00
	RGIN-GP(freq)	0.87 ± 0.12	0.91 ± 0.05	2.21 ± 0.05	0.93 ± 0.00
	RGIN-GP(prone)	0.82 ± 0.13	0.89 ± 0.04	2.19 ± 0.05	0.93 ± 0.00
0.2	RGIN+SumPool	8.65 ± 5.12	0.87 ± 0.02	7.07 ± 2.63	0.96 ± 0.01
	RGIN+DIAMNet	8.66 ± 5.11	0.87 ± 0.02	5.30 ± 3.34	0.98 ± 0.01
	RGIN-GP(onehot)	1.19 ± 0.23	0.84 ± 0.02	2.27 ± 0.07	0.93 ± 0.00
	RGIN-GP(freq)	1.21 ± 0.07	0.87 ± 0.05	2.19 ± 0.03	0.93 ± 0.00
	RGIN-GP(prone)	1.21 ± 0.08	0.84 ± 0.01	2.20 ± 0.04	0.93 ± 0.00

Exp-2: Sampled Data for Few/Zero-Shot Testing. We investigate the effect of adding sampled data from the training tasks to the support set of the test task in the zero-shot and few-shot scenarios for the 5 task types. For SameG, SameP and Random, 7 tasks from MUTAG are used for training. For HySameG and SameP, 239 tasks from SYN-S are added to the training tasks. We test 28 MUTAG tasks by varying the number of shots and the auxiliary pairs in $0 \sim 64$ and $0 \sim 128$, respectively. The size of all the tasks is 128. The testing performance over the 5 task configurations is shown in Table 3. For SameG, HySameG, and Random, the upper-left cell of the table is the worst performance for zero-shot without auxiliary data and the lower-right cell is the best performance for 64 shots with 128 auxiliary. As the number of shots or auxiliary pairs increases, the test performance improves from the upper-left to the upper-right, lower-left and upper-right. For a fixed size support set, the more data from the shot, the better the performance. However, we find for tasks with the type SameP and HySameP, adding auxiliary data from training tasks will degrade the MAE and accuracy. Recall that SameP and HySameP task is one new pattern p^* for different data graphs $\mathcal{G}$ in a database. We observe that for different patterns, p_1 and p_2, their true count distributions of $\mathcal{G}$ are rather different, because of the different topology between p_1 and p_2. A large discrepancy between the ground-truth distribution makes the zero-shot transfer difficult.

Exp-3: Comparison with Algorithmic Approaches. We compare our meta learned RGIN-GP with traditional subgraph counting algorithms, including 7 approximate algorithms in the GCARE benchmark [5], and an exact counting algorithm VF2 [2] implemented by NetworkX. The MAE, quantiles of the error and the total counting time are presented in Table 4. The prediction results of RGIN-GP are collected by 5-fold cross-validation where one model is trained over 20% MUTAG pairs that are organized in 7 tasks with type of Random and size of

Table 3. Test MAE (↓)/Accuracy (↑) for Zero-shot

MAE/Acc.		# auxiliary data				
		0	16	32	64	128
SameG # shots	0	8.06/0.72	4.01/0.82	3.13/0.84	2.37/0.84	2.24/0.86
	1	6.97/0.74	3.5/0.83	2.84/0.82	2.29/0.84	1.73/0.87
	4	6.22/0.77	3.14/0.83	2.71/0.85	2.28/0.85	1.81/0.87
	16	3.65/0.82	2.68/0.83	2.24/0.85	2.04/0.86	1.79/0.87
	64	1.80/0.85	1.67/0.86	1.67/0.86	1.49/0.87	1.48/0.87
SameP # shots	0	7.08/0.75	15.02/0.57	15.78/0.50	15.00/0.54	14.92/0.48
	1	6.10/0.77	11.11/0.52	12.45/0.54	16.45/0.40	14.54/0.42
	4	4.54/0.80	11.84/0.51	11.65/0.52	11.27/0.55	11.01/0.53
	16	2.51/0.84	6.48/0.54	6.68/0.63	6.51/0.64	10.09/0.57
	64	1.34/0.88	2.61/0.76	2.57/0.76	3.89/0.70	3.68/0.69
HySameG # shots	0	10.92/0.28	3.68/0.86	3.02/0.88	2.23/0.89	1.67/0.91
	1	8.51/0.49	3.76/0.86	2.69/0.87	2.01/0.90	1.81/0.90
	4	4.91/0.81	3.61/0.86	2.44/0.89	2.13/0.90	1.71/0.91
	16	3.71/0.87	2.89/0.88	1.88/0.89	1.77/0.90	1.59/0.91
	64	1.49/0.89	1.35/0.90	1.29/0.90	1.21/0.91	1.11/0.92
HySameP # shots	0	10.22/0.23	11.35/0.27	10.60/0.30	10.93/0.32	9.62/0.37
	1	7.67/0.32	7.77/0.36	9.54/0.36	8.32/0.36	9.34/0.42
	4	4.67/0.49	5.94/0.48	5.96/0.46	6.79/0.48	5.41/0.50
	16	1.95/0.78	2.50/0.72	2.45/0.71	2.54/0.69	2.62/0.70
	64	0.86/0.90	1.00/0.84	1.08/0.85	1.05/0.85	1.02/0.88
Random # shots	0	8.05/0.73	4.38/0.81	2.72/0.84	1.79/0.85	1.41/0.85
	1	7.17/0.76	3.83/0.82	3.45/0.83	1.74/0.84	1.43/0.87
	4	6.00/0.78	3.49/0.82	2.39/0.84	1.84/0.85	1.46/0.86
	16	3.98/0.79	2.21/0.82	2.06/0.81	1.45/0.84	1.26/0.85
	64	1.81/0.86	1.61/0.86	1.51/0.86	1.35/0.86	1.23/0.88

128. The prediction is conducted by zero-shot testing with 128 auxiliary pairs. In Table 4, `RGIN-GP` achieves the lowest MAE among the 8 approximate approaches and its prediction is 6× faster than the exact algorithm `VF2`.

Table 4. Comparison with Subgraph Counting Algorithms

Method	RGIN-GP	WJ	CS	CSET	IMPR	JSUB	BSK	SumRDF	VF2
MAE	1.32	8.05	25.35	7.87	40.28	7.69	156.15	6.48	0
5%	−4.66	−48	−56	−56	−24	−48	0	−45	0
25%	0.01	−4	−4	−4	0	−4	0	0	0
50%	0.01	0	0	0	0	0	0	0	0
75%	0.35	0	0	0	0	0	66	0	0
95%	3.68	0	0	0	252.96	0	932	2.21	0
Time (s)	0.14	0.59	0.98	0.55	0.82	0.61	693.02	40.55	0.89

5   Conclusion

In this paper, we study an NP-complete problem, subgraph isomorphism counting, by DL techniques. To alleviate the reliance on a large volume of training data, we devise a GP, called RGIN-GP. The model is trained end-to-end by a meta learning algorithm, which aims to exploit the knowledge prior of training tasks. Compared with the baseline approach, the meta trained RGIN-GP reduces the MAE from 8 to 1, with only one thousand training samples.

Acknowledgement. This work was supported by the Research Grants Council of Hong Kong, China, under No. 14203618, No. 14202919 and No. 14205520.

References

1. S. A. Cook. The complexity of theorem-proving procedures. In: Proceedings of the STOC, pp. 151–158 (1971)
2. Cordella, L.P., Foggia, P., Sansone, C., Vento, M.: A (sub)graph isomorphism algorithm for matching large graphs. IEEE Trans. Pattern Anal. Mach. Intell. **26**(10), 1367–1372 (2004)
3. Hamilton, W.L., Ying, Z., Leskovec, J.: Inductive representation learning on large graphs. In: Proceedings of the NeurIPS 2017, pp. 1024–1034 (2017)
4. Liu, X., Pan, H., He, M., Song, Y., Jiang, X., Shang, L.: Neural subgraph isomorphism counting. In: Proceedings of the KDD 2020, pp. 1959–1969 (2020)
5. Park, Y., Ko, S., Bhowmick, S.S., Kim, K., Hong, K., Han, W.: G-CARE: a framework for performance benchmarking of cardinality estimation techniques for subgraph matching. In: Proceedings of the SIGMOD 2020, pp. 1099–1114 (2020)
6. Patacchiola, M., Turner, J., Crowley, E.J., Storkey, A.: Bayesian meta-learning for the few-shot setting via deep kernels. In: Proceedings of NeurIPS (2020)
7. Rasmussen, C.E., Williams, C.K.I.: Gaussian Processes for Machine Learning. MIT Press (2006)
8. Wilson, A.G., Adams, R.P.: Gaussian process kernels for pattern discovery and extrapolation. In: Proceedings of ICML, vol. 28, 1067–1075 (2013)
9. Wilson, A.G., Hu, Z., Salakhutdinov, R., Xing, E.P.: Deep kernel learning. In: Proceedings of AISTATS, vol. 51, pp. 370–378 (2016)

10. Yang, Q., Zhang, Y., Dai, W., Pan, S.J.: Transfer Learning. Cambridge University Press, Cambridge (2020)
11. Zhang, J., Dong, Y., Wang, Y., Tang, J., Ding, M.: Prone: fast and scalable network representation learning. In: Proceedings of the IJCAI 2019, pp. 4278–4284 (2019)
12. Zhao, K., Yu, J.X., He, Z., Li, R., Zhang, H.: Lightweight and accurate cardinality estimation by neural network gaussian process. In: SIGMOD 2022, pp. 973–987. ACM (2022)
13. Zhao, K., Yu, J.X., Zhang, H., Li, Q., Rong, Y.: A learned sketch for subgraph counting. In: Proceedings of SIGMOD 2021 (2021)

FairHELP: Fairness-Aware Heterogeneous Information Network Embedding for Link Prediction

Meng Cao, Jianqing Song, Jinliang Yuan, Baoming Zhang, and Chongjun Wang[✉]

Department of Computer Science and Technology, Nanjing University, Nanjing, China
{caomeng,sjq,yuanjl19,zhangbm}@smail.nju.edu.cn, chjwang@nju.edu.cn

Abstract. Heterogeneous information networks (HINs) are ubiquitous in real-world social systems. To effectively learn representations of HINs, Graph Neural Networks (GNNs) have been widely studied as a powerful tool. Nevertheless, there is growing concern that GNNs are prone to make biased predictions in critical decision-making scenarios such as link prediction and social recommendation. Despite recent progress on fair graph learning, few attempts have been made toward promoting fairness in HIN embedding models. In this paper, we study the problem of mitigating link prediction bias in HINs. First, we formalize the definition of fairness in link prediction in HINs, and design fairness measures for the link prediction task. Second, we propose a flexible and model-agnostic debiasing framework named FairHELP for learning fair embeddings in HINs. Third, we conduct extensive experiments on three real-world datasets. The results validate the effectiveness of the proposed fairness measures and the FairHELP framework in achieving fair and accurate link prediction results.

Keywords: Heterogenegous Information Networks · Graph Neural Networks · Link Prediction · Algorithmic Fairness

1 Introduction

Heterogeneous information networks (HINs) contain abundant information with various types of nodes and multi-typed structural relations. To better process and analyze HINs, heterogeneous network embedding has emerged as a fundamental technique for various downstream network analysis tasks, such as node classification, link prediction, clustering, etc. Among the tasks, link prediction, aiming at inferring missing relations or future interactions in the network, plays a vital role in various social network mining applications, such as social recommendations [14].

© The Author(s), under exclusive license to Springer Nature Switzerland AG 2023
X. Wang et al. (Eds.): DASFAA 2023, LNCS 13945, pp. 320–330, 2023.
https://doi.org/10.1007/978-3-031-30675-4_22

Recently, Graph Neural Networks (GNNs) [10] have achieved significant progress in network representation learning. The key idea of GNNs is to learn node embeddings via neighborhood message passing and information aggregation, which has achieved superior performance compared to traditional network embedding techniques in various network analysis tasks [11,12]. Inspired by the strength of GNNs in learning effective network representations, a myriad of heterogeneous graph neural network models have been proposed to deal with the more complex information in HINs, such as RGCN [15], HAN [18], HGT [8], etc.

Despite the success of the heterogeneous GNN models in achieving state-of-the-art performance, little attention has been paid to understanding the fairness and trustworthiness of HINs. Recently, some research has revealed that GNN models are prone to inheriting and exacerbating the data bias during the message-passing process [4]. Although there have been progress in developing fair GNNs [1,4,13], very few works have concentrated on fairness issues in HINs. For example, in the link prediction task for recommending jobs to applicants, the embedding algorithms may be more likely to recommend high-salary jobs to males, exacerbating discrimination against females with similar qualifications.

In this paper, we focus on learning fair HIN embeddings for link prediction via Graph Neural Networks. Different from the existing fair GNN models that focus on homogeneous networks, we aim to mitigate the semantic disparity caused by the heterogeneity of demographic groups (race, gender, etc.) in HINs. Specifically, we seek to solve two main challenges: *1) how to measure fairness in HINs with heterogeneous sensitive attributes; 2) how to mitigate bias in heterogeneous graph neural networks for link prediction.* To address these above problems, we propose a novel HIN embedding framework named **FairHELP**. First, we formally provide appropriate fairness measures for link prediction in HINs. Then, we propose an adversarial learning based HIN embedding framework by encouraging the model to learn de-biased link embeddings that are independent of the intrinsic link semantics. The experiments demonstrate that FairHELP produces fair and accurate link prediction performance compared with state-of-the-art baselines.

2   Related Works

2.1   Heterogeneous Information Network Embedding

Over the past few years, significant progress has been made toward heterogeneous information network embedding. Metapath2vec [5] and HIN2VEC [6] adopt meta-paths constrained random walks to extract complex semantics in HINs. Recently, inspired by the success of GNNs, RGCN [15] and CompGCN [17] design the relation-specific feature aggregation mechanisms for embedding learning in HINs. HAN [18] and HGT [8] adopt different attention mechanisms for heterogeneous node feature aggregation. In summary, the heterogeneous GNN models have become a powerful tool in learning HIN embeddings for various network analysis tasks.

2.2 Fairness in Machine Learning

In the research area of network representation learning, increasing research attention has been focused on the fairness of network representations [2,13]. Recently, FairDrop [16] proposes a pre-processing method which reduces the homogeneity of the network to promote fairness. FairGNN [4] proposes a GNN-based fairness-aware learning framework for fair node classification. FairHIN [19] employs several bias-mitigating techniques including data pre-processing, model in-processing, and post-processing to improve fairness in HIN embeddings. In summary, most existing methods are designed for homogeneous networks, and the fair embedding methods for heterogeneous networks are highly demanded.

3   Preliminaries

Definition 1 (Heterogeneous Information Network Embedding). *A **heterogeneous information network (HIN)** is defined as a graph $\mathcal{G} = (\mathcal{V}, \mathcal{E}, \mathcal{A}, \mathcal{R})$ with a node mapping function $\phi : \mathcal{V} \to \mathcal{A}$, and an edge mapping function $\psi : \mathcal{E} \to \mathcal{R}$, where $\mathcal{V}$ and $\mathcal{E}$ denote nodes and edges in $\mathcal{G}$. $\mathcal{A}$ and $\mathcal{R}$ denote the sets of node types and edge types, and $|\mathcal{A}| + |\mathcal{R}| > 2$. **HIN embedding** aims to learn a low-dimensional vector $z_i \in \mathbb{R}^d$ for each node $v_i \in \mathcal{V}$, where d is the embedding dimension, and $d \ll |\mathcal{V}|$.*

Definition 2 (Semantic Disparity for Heterogeneous Link Prediction). *In the HIN $\mathcal{G}$, each node $v \in \mathcal{V}$ is mapped to a node type $\phi(v) \in \mathcal{A}$, and is associated with a binary sensitive attribute $s_{\phi(v)} \in \{0, 1\}$. For the link prediction task, we aim at predicting the existence of links with type $r(\phi(v), \phi(u)) \in \mathcal{R}$. We define the **semantic subgroup** s_{r_ρ} for link type $r(\phi(v), \phi(u))$ as the composition of the demographic groups of the node pair (v, u), where $S = \{s_{r_\rho}1, s_{r_\rho}2, ..., s_{r_\rho}T\}$ is the semantic subgroup set with T types of semantics. The **semantic disparity for heterogeneous link prediction** refers to the discrepancy of the model in favoring different link semantic subgroups in S by making positive predictions.*

Definition 3 (Demographic Parity Discrepancy (DPd)). *Demographic Parity [3] requires the positive predictions of link $\hat{y}$ to be independent of the semantic subgroups, i.e., $\hat{y} \perp S$. Specifically, the Demographic Parity Discrepancy (DPd) is defined as:*

$$DPd = \max_{s_{r_\rho} \in S} E[\hat{y} = 1 | s = s_{r_\rho}] - \min_{s_{r_\rho} \in S} E[\hat{y} = 1 | s = s_{r_\rho}]. \tag{1}$$

Definition 4 (Equalized Odds Discrepancy (EOd)). *To consider the positive predictions with ground truth in evaluation, we follow the definition of Equalized Odds [3] and propose Equalized Odds Discrepancy (EOd) as:*

$$EOd = TPRd + FPRd, \tag{2}$$

where TPRd and FPRd is defined as:

$$TPRd = \max_{s_{r_\rho} \in S} E[\hat{y} = 1|y = 1, s = s_{r_\rho}] - \min_{s_{r_\rho} \in S} E[\hat{y} = 1|y = 1, s = s_{r_\rho}], \quad (3)$$

$$FPRd = \max_{s_{r_\rho} \in S} E[\hat{y} = 1|y = 0, s = s_{r_\rho}] - \min_{s_{r_\rho} \in S} E[\hat{y} = 1|y = 0, s = s_{r_\rho}]. \quad (4)$$

Definition 5 (Demographic Parity Variance (DPv) and Equalized Odds Variance (EOv)). *As the DPd and EOd neglect the distribution difference among the intermediate groups, we propose Demographic Parity Variance (DPv) and Equalized Odds Variance (EOv) as:*

$$DPv = \underset{s_{r_\rho} \in S}{Var} (E[\hat{y} = 1|s = s_{r_\rho}]), \quad (5)$$

$$EOv = TPRv + FPRv. \quad (6)$$

where Var(·) is the variance function, TPRv and FPRv are the variances of TPR and FPR across all link semantic subgroups in S.

4 Methods

In this section, we propose a novel **Fair**ness-aware **H**eterogeneous information network **E**mbedding framework for **L**ink **P**rediction named **FairHELP**. We provide an overall framework of FairHELP in Fig. 1, which consists of a GNN-based network embedding generator $f_{\mathcal{G}}$, a link predictor $f_{\mathcal{L}}$, and a semantic subgroup discriminator $f_{\mathcal{D}}$. We elaborate on each part in detail as follows.

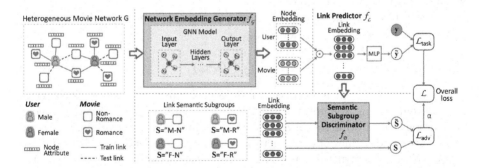

Fig. 1. The overall framework of the proposed FairHELP. In the example movie network, *user* nodes have gender (*Male, Female*) as the sensitive attribute, and *movie* nodes have genre (*Non-Romance, Romance*) as the sensitive attribute. The framework contains the following three parts. (1) Network embedding generator, (2) link predictor, and (3) semantic subgroup discriminator.

4.1   Network Embedding Generator

In this section, we introduce the basic structure of the GNN model as the network embedding generator $f_{\mathcal{G}}$. The core idea of GNNs is to aggregate attribute message from the neighboring nodes iteratively. For a target node v, the message at the l-th layer is aggregated as:

$$\mathbf{MSG}_v^{(l)} = f_{AGG}^{(l-1)}\left(\left\{\mathbf{h}_u^{(l-1)} : u \in \mathcal{N}(v)\right\}\right), \tag{7}$$

where $\mathcal{N}(v)$ is the neighboring nodes of v, and $\mathbf{h}_u^{(l-1)}$ is the feature vector of node u at the $(l\text{-}1)$-th layer, and f_{AGG} is the aggregation function. For heterogeneous GNN models, without loss of generality, the aggregated message in Eq. (7) can be extended as:

$$\mathbf{MSG}_v^{(l)} = \left\{f_{AGG(\phi_1)}^{(l-1)}\left(\left\{\mathbf{h}_u^{(l-1)} : u \in \mathcal{N}^{\phi_1}(v)\right\}\right), ..., f_{AGG(\phi_N)}^{(l-1)}\left(\left\{\mathbf{h}_u^{(l-1)} : u \in \mathcal{N}^{\phi_N}(v)\right\}\right)\right\}, \tag{8}$$

where $\phi_1, ..., \phi_N \in \mathcal{A}$ are the node types in $\mathcal{G}$. Notably, f_{AGG} can take various formats based on heterogeneous information types. For example, the HAN model [18] performs attention-based dual-level aggregations based on both node-level and semantic-level information, while HGT [8] performs different aggregations according to both node types and link types, etc.

After the aggregation process, the target node v's feature at the l-th layer is updated based on the message at this layer and its feature $\mathbf{h}_v$ at the previous layer:

$$\mathbf{h}_v^{(l)} = f_{UPD}^{(l)}\left(\mathbf{h}_v^{(l-1)}, \mathbf{MSG}_v^{(l)}\right), \tag{9}$$

where f_{UPD} is the update function.

After L layers of message passing and feature update, node v's embedding $\mathbf{h}_v^{(L)}$ at the output layer is denoted as:

$$\mathbf{h}_v^{(L)} = f_{\mathcal{G}}^{(L)}\left(\mathcal{G}, \mathbf{x}_v\right), \tag{10}$$

where $\mathbf{x}_v$ is the initial feature vector of node v, and $f_{\mathcal{G}}^{(L)}$ is the GNN-based network embedding generator with L layers. In this way, we obtain the node embeddings with both structural and attribute similarity in the original network. In our proposed framework, the network embedding generator $f_{\mathcal{G}}$ is flexible as any GNNs with a structure that follows Eq. (8)–(9) can be utilized. With the generated network embedding, we then employ a link predictor for task-specific model training.

4.2   Link Predictor

In the link prediction settings, the model takes node embedding pairs to compute link probabilities via a scoring function, such as dot product, cosine similarity, etc. For a node pair (v, u) in HIN $\mathcal{G}$, instead of directly computing the link probability scores, we first obtain the link embedding from the node embeddings via the Hadamard product operator [7][1]:

$$\mathbf{h}_{vu} = \mathbf{h}_v \odot \mathbf{h}_u. \tag{11}$$

Then we adopt a linear neural network to compute the link probability score:

$$\hat{y}_{vu} = \sigma \left(\mathbf{h}_{vu}^{\top} \cdot \mathbf{w} \right), \tag{12}$$

where $\mathbf{w} \in \mathbb{R}^d$ is the weight parameter, and σ is the sigmoid activation function.

For model training, we follow the contrastive learning methods [2] to maximize the likelihood of true links compared to the negative link samples. The binary cross entropy loss for the link prediction task is formulated as follows:

$$\min_{\theta_{\mathcal{G}}, \theta_{\mathcal{L}}} \mathcal{L}_{task} = -\frac{1}{|\mathcal{E}_r^+ \cup \mathcal{E}_r^-|} \sum_{(v,u) \in \mathcal{E}_r^+ \cup \mathcal{E}_r^-} [y_{vu} \log (\hat{y}_{vu}) + (1 - y_{vu}) \log (1 - \hat{y}_{vu'})],$$
$$\tag{13}$$

where $\mathcal{E}_r^+$ and $\mathcal{E}_r^-$ represent the true links and negative link samples of link type r, respectively. $\hat{y}_{vu}$ is the output of the link prediction results for node pair (v, u), and y_{vu} is the ground truth link label. $\theta_{\mathcal{G}}$ and $\theta_{\mathcal{L}}$ are model parameters of $f_{\mathcal{G}}$ and $f_{\mathcal{L}}$.

4.3   Semantic Subgroup Discriminator

It has been demonstrated that network embedding generated by GNN models may inherit bias from the data [1,4], which makes the link predictor produce unfair predictions. In this section, we propose to eliminate the bias from the link embeddings. Specifically, we aim to answer the following question: *will the algorithm make decisions regardless of the links' semantic subgroups?* In other words, will the model be biased against specific semantic subgroups?

Ideally, for a link (v, u), given the link's semantic subgroup as the sensitive attribute s_{vu} and the link embedding $\mathbf{h}_{vu}$, we make the following fairness assumption of independence:

$$f_{\mathcal{L}}(\mathbf{h}_{vu}) \perp s_{vu} \quad \forall (v, u) \in \mathcal{E}, \tag{14}$$

where $f_{\mathcal{L}}(\mathbf{h}_{vu})$ is the model's output from the link predictor. Equation (14) indicates that, for a fair prediction model, the predictions of link existence would be the same regardless of the link semantics.

To achieve the above independence assumption, we propose an adversarial learning framework as follows. Specifically, we adopt a link semantic subgroup discriminator $f_{\mathcal{D}}$ which tries to predict s_{vu} for each link $(v, u) \in \mathcal{E}$ given the

---

[1] Other operators such as concatenation are also applicable here, we leave the analysis of different operators for future studies.

link embedding $\mathbf{h}_{vu}$. Meanwhile, the embedding generator $f_{\mathcal{G}}$ aims to learn link embeddings that fool the discriminator $f_{\mathcal{D}}$ to make inaccurate predictions on the link semantic subgroup s_{vu}. The above training process can be written as a min-max game:

$$\min_{\theta_{\mathcal{G}}} \max_{\theta_{D}} \mathcal{L}_{adv} = - \sum_{(v,u)\in\mathcal{E}} \mathcal{L}_s(f_{\mathcal{D}}(f_{\mathcal{G}}(\mathcal{G}, \mathbf{x}_u) \odot f_{\mathcal{G}}(\mathcal{G}, \mathbf{x}_v), s_{vu})), \qquad (15)$$

where $f_{\mathcal{G}}(\mathcal{G}, \mathbf{x}_v) \odot f_{\mathcal{G}}(\mathcal{G}, \mathbf{x}_u)$ generates the link embedding $\mathbf{h}_{vu}$, and $\mathcal{L}_s$ is the cross-entropy classification loss for predicting sensitive attribute s_{vu} from $\mathbf{h}_{vu}$.

4.4   Model Training

We define the overall loss as a combination of link prediction loss and adversarial loss:

$$\mathcal{L} = \min_{\theta_{\mathcal{G}}, \theta_{\mathcal{L}}} \max_{\theta_{\mathcal{D}}} \left(\mathcal{L}_{task} + \alpha \mathcal{L}_{adv} \right), \qquad (16)$$

where $\theta_{\mathcal{G}}$, $\theta_{\mathcal{L}}$, and $\theta_{\mathcal{D}}$ are the parameters of $f_{\mathcal{G}}$, $f_{\mathcal{L}}$, and $f_{\mathcal{D}}$, respectively. α is a balancing parameter that controls the contribution of the adversarial debiasing framework.

For model training, we first train the network embedding generator and the link predictor to optimize on $\theta_{\mathcal{G}}$ and $\theta_{\mathcal{L}}$, then we fix $\theta_{\mathcal{G}}$ and $\theta_{\mathcal{L}}$ to train the discriminator and update $\theta_{\mathcal{D}}$. We optimize the overall loss of the model in Eq. (16) via the ADAM optimizer [9].

5   Experiments

5.1   Experimental Settings

Datasets. We conduct experiments on three real-world HIN datasets: Movie-Lens[2], DBLP[3], and LastFM[4]. We summarize the dataset statistics in Table 1.

Evaluation Metrics. We adopt the prediction Accuracy and Area Under ROC (AUROC) scores to evaluate the link prediction performance as utility metrics. For fairness evaluation, we employ the Demographic Parity Discrepancy (DPd), Equalized Odds Discrepancy (EOd), Demographic Parity Variance (DPv), and Equalized Odds Variance (EOv) defined in Sect. 3 as fairness metrics.

---

[2] http://www.movielens.org/.
[3] http://arnetminer.org/billboard/citation.
[4] http://ir.ii.uam.es/hetrec2011.

Table 1. The statistics of the datasets.

Datasets	A-B	#A	#B	#A-B	Target X-Y	Sens. X	Sens. Y
MovieLens	User-Movie	943	1,682	100,000	User-Movie	gender	genre
DBLP	Paper-Author	14,376	14,475	41,794	Paper-Author	topic	area
	Paper-Conf	14,376	20	14,376			
LastFM	Artist-User	17,632	1,892	92,834	Artist-User	popularity	activeness
	Artist-Tag	17,632	1,088	23,253			
	User-User	1,892	1,892	25434			

*Sens.: sensitive attribute.

Baselines. We compare FairHELP with the following state-of-the-art methods, including random walk-based methods, fairness-aware methods, and GNN-based methods. The random walk-based methods include **Metapath2vec** [5]. For fairness-aware methods, we adopt **FairHIN** [19] with its two variations: **FairHIN-dp** and **FairHIN-eo**. For the GNN-based methods, we choose **RGCN** [15], **CompGCN** [17], **HAN** [18], and **HGT** [8]. We compare the above GNN-based models with the FairHELP integrated adaptations, denoted as **RGCN-fair**, **CompGCN-fair**, **HAN-fair**, and **HGT-fair**, respectively.

5.2 Bias Mitigation Performance in Link Prediction

We report the utility and fairness results in Table 2. From the results, we can observe that, **1) the FairHELP framework achieves the fairest results** across all datasets compared to the vanilla versions of the GNN models. **2) FairHELP can maintain a good link prediction performance in terms of Accuracy and AUROC** compared to Metapath2vec and FairHIN. Although Metapath2vec and FairHIN has relatively fair link prediction results compared with other baselines, they perform poor in prediction utility, especially for FairHIN. A possible explanation is that since FairHIN directly adds demographic parity and equal opportunity to the loss function, it encourages the model to give uniform predictions for all links. Hence the fairness may look "well ensured", but the model utility is not guaranteed in this circumstance. In summary, the results validate that our proposed FairHELP can achieve an excellent fairness-accuracy trade-off among all baselines.

5.3 Parameter Analysis

To better analyze the contribution of the adversarial loss in mitigating bias, we vary the balancing parameter α in $\{0.001, 0.005, 0.01, 0.05, 0.1\}$. Figure 2 shows the performance of four GNN-based FairHELP models on three datasets in terms of DPd and accuracy. We observe that, with an increasing value of α, the DPd

Table 2. The comparison results of the proposed FairHELP with the baselines. We show the average performance (%) over five independent runs. Arrows (↑,↓) indicate the direction of better performance. The shaded area shows that FairHELP improves the link prediction fairness, meanwhile keeping the model utility.

Dataset	Models	Accuracy(↑)	AUROC(↑)	DPd(↓)	EOd(↓)	DPv(↓)	EOv(↓)
MovieLens	Metapath2vec	67.97±0.24	66.92±0.22	10.28±1.31	15.39±1.96	0.21±0.04	0.30±0.10
	FairHIN-dp	54.50±6.19	55.67±8.44	41.14±11.10	81.98±20.56	3.20±1.81	6.38±3.47
	FairHIN-eo	55.94±5.53	57.97±6.83	35.49±19.33	71.61±35.33	2.81±2.27	5.59±4.33
	RGCN	70.60±0.46	77.47±0.47	18.83±2.10	32.27±3.87	0.79±0.17	1.10±0.27
	RGCN-fair	69.49±1.47	77.32±0.43	**16.28±1.04**	**27.84±2.51**	**0.58±0.07**	**0.83±0.10**
	CompGCN	72.74±2.43	79.28±2.91	20.03±3.09	30.85±6.96	0.62±0.15	0.82±0.28
	CompGCN-fair	67.82±2.60	75.64±2.52	**13.33±1.15**	**22.20±2.25**	**0.37±0.08**	**0.57±0.09**
	HAN	71.34±0.17	77.65±0.14	19.14±1.86	32.76±3.43	0.79±0.13	1.09±0.21
	HAN-fair	70.88±0.79	77.14±0.39	**17.88±2.50**	**29.94±4.79**	**0.66±0.18**	**0.86±0.26**
	HGT	78.01±0.17	85.25±0.20	16.99±0.75	20.00±2.43	0.46±0.05	0.41±0.10
	HGT-fair	75.51±2.90	82.54±3.67	**14.79±3.38**	**19.53±7.93**	**0.37±0.10**	**0.39±0.23**
DBLP	Metapath2vec	60.51±0.12	64.55±0.12	19.59±0.65	32.77±1.50	0.53±0.03	0.96±0.11
	FairHIN-dp	82.12±0.88	89.96±0.45	42.38±1.96	21.32±3.92	3.49±0.32	0.41±0.09
	FairHIN-eo	48.43±22.29	46.46±28.07	34.36±9.60	37.83±21.28	2.26±1.25	1.40±1.34
	RGCN	74.10±0.80	82.40±1.07	48.21±1.22	48.74±5.67	3.79±0.18	2.16±0.36
	RGCN-fair	74.34±0.56	82.99±0.79	**45.73±2.04**	**42.55±4.29**	**3.52±0.13**	**1.62±0.25**
	CompGCN	69.22±0.93	70.31±3.11	34.37±5.15	44.50±7.71	1.85±0.60	1.60±0.59
	CompGCN-fair	68.96±1.31	69.71±3.09	**29.75±6.06**	**43.25±4.50**	**1.39±0.56**	1.77±0.36
	HAN	66.80±3.77	71.86±3.72	38.77±16.02	51.43±15.92	2.73±1.78	2.91±1.69
	HAN-fair	63.65±2.89	68.56±3.02	**26.39±9.45**	**48.07±16.35**	**1.08±0.58**	2.32±0.98
	HGT	72.67±0.71	79.59±0.80	39.48±1.72	39.84±8.40	2.53±0.13	1.23±0.49
	HGT-fair	72.49±0.31	78.38±0.42	**35.31±1.68**	**34.14±5.91**	**2.05±0.17**	**0.93±0.22**
LastFM	Metapath2vec	53.07±0.13	53.10±0.13	9.74±0.09	16.85±0.84	0.15±0.00	0.25±0.01
	FairHIN-dp	49.87±0.32	49.90±0.29	1.44±0.74	5.37±1.54	0.00±0.00	0.03±0.01
	FairHIN-eo	50.00±0.21	50.11±0.18	1.19±0.33	4.49±1.39	0.00±0.00	0.02±0.01
	RGCN	71.05±1.08	77.05±0.79	58.87±1.21	106.95±3.07	7.90±0.45	14.08±0.92
	RGCN-fair	67.30±0.55	74.97±0.53	**46.18±1.92**	**80.49±4.21**	**4.83±0.35**	**8.35±0.68**
	CompGCN	74.43±1.23	78.40±0.56	64.75±5.75	113.47±12.12	10.17±1.58	16.91±2.93
	CompGCN-fair	73.45±1.94	77.52±0.95	**62.26±4.26**	**107.62±7.29**	**8.98±1.17**	14.8±1.96
	HAN	75.91±2.69	79.0±1.89	68.57±8.09	121.91±15.71	11.70±2.80	19.76±4.62
	HAN-fair	71.10±1.06	76.83±1.88	**57.20±6.45**	**99.08±14.49**	**7.36±1.61**	**12.45±3.01**
	HGT	72.17±0.96	78.96±0.53	52.94±3.12	86.99±6.11	6.80±0.68	10.96±1.22
	HGT-fair	71.14±0.80	78.61±0.37	**50.81±1.50**	**83.45±2.41**	**6.18±0.32**	**9.90±0.49**

results generally decrease, indicating that the models are becoming fairer. Meanwhile, the prediction accuracy also drops as α increases, which shows a fairness-accuracy trade-off in our framework, and similar observations are also discovered in [4,16]. Besides, a larger α will make the model harder to train properly, which may be a possible reason for the large variance in the results with $\alpha > 0.01$. In summary, $\alpha = 0.01$ maintains a good fairness-accuracy trade-off.

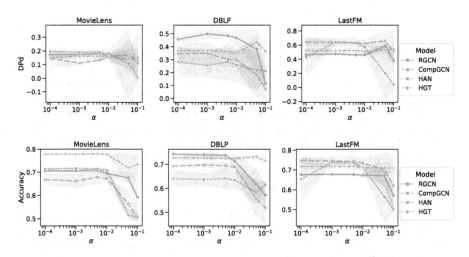

Fig. 2. Parameter analysis of FairHELP on four GNN models with different balancing parameters α. We evaluate the link prediction fairness regarding DPd (top row) and the link prediction accuracy (bottom row). The results show that increasing the value of α will improve model performance in fairness and result in a drop in model utility as a trade-off.

6   Conclusion

In this paper, we propose FairHELP, a novel fairness-aware heterogeneous information network embedding framework for link prediction. First, we formalize the fairness definitions for link prediction in HINs based on semantic disparity. Second, we design effective fairness measures for the semantic disparity in link prediction with heterogeneous sensitive attributes. Third, we propose a bias mitigation framework, which employs adversarial learning to encourage the model to learn link embeddings independent of the link semantic subgroups. Our experimental results on three real-world heterogeneous network datasets demonstrate that FairHELP achieves satisfying results with a good trade-off between fairness and utility compared with state-of-the-art baselines.

Acknowledgement. This paper is supported by the National Natural Science Foundation of China (Grant No. 62192783, U1811462), the Collaborative Innovation Center of Novel Software Technology and Industrialization at Nanjing University.

References

1. Agarwal, C., Lakkaraju, H., Zitnik, M.: Towards a unified framework for fair and stable graph representation learning. In: Uncertainty in Artificial Intelligence, pp. 2114–2124. PMLR (2021)
2. Bose, A., Hamilton, W.: Compositional fairness constraints for graph embeddings. In: International Conference on Machine Learning, pp. 715–724. PMLR (2019)

3. Caton, S., Haas, C.: Fairness in machine learning: a survey. arXiv preprint arXiv:2010.04053 (2020)
4. Dai, E., Wang, S.: Say no to the discrimination: learning fair graph neural networks with limited sensitive attribute information. In: Proceedings of the 14th ACM International Conference on Web Search and Data Mining, pp. 680–688 (2021)
5. Dong, Y., Chawla, N.V., Swami, A.: metapath2vec: scalable representation learning for heterogeneous networks. In: Proceedings of the 23rd ACM SIGKDD International Conference on Knowledge Discovery and Data Mining, pp. 135–144 (2017)
6. Fu, T., Lee, W., Lei, Z.: Hin2vec: explore meta-paths in heterogeneous information networks for representation learning. In: Proceedings of the 2017 ACM on Conference on Information and Knowledge Management, pp. 1797–1806. ACM (2017)
7. Horn, R.A., Johnson, C.R.: Matrix Analysis. Cambridge University Press, Cambridge (2012)
8. Hu, Z., Dong, Y., Wang, K., Sun, Y.: Heterogeneous graph transformer. In: Proceedings of the Web Conference 2020, pp. 2704–2710 (2020)
9. Kingma, D.P., Ba, J.: Adam: a method for stochastic optimization. arXiv preprint arXiv:1412.6980 (2014)
10. Kipf, T.N., Welling, M.: Semi-supervised classification with graph convolutional networks. arXiv preprint arXiv:1609.02907 (2016)
11. Li, X., et al.: Braingnn: interpretable brain graph neural network for FMRI analysis. Med. Image Anal. **74**, 102233 (2021)
12. Lv, L., Cheng, J., Peng, N., Fan, M., Zhao, D., Zhang, J.: Auto-encoder based graph convolutional networks for online financial anti-fraud. In: 2019 IEEE Conference on Computational Intelligence for Financial Engineering & Economics (CIFEr), pp. 1–6. IEEE (2019)
13. Rahman, T., Surma, B., Backes, M., Zhang, Y.: Fairwalk: towards fair graph embedding. In: International Joint Conference on Artificial Intelligence (2019). https://publications.cispa.saarland/2933/
14. Sanz-Cruzado, J., Castells, P.: Contact recommendations in social networks. In: Collaborative Recommendations: Algorithms, Practical Challenges and Applications, pp. 519–569. World Scientific (2019)
15. Schlichtkrull, M., Kipf, T.N., Bloem, P., van den Berg, R., Titov, I., Welling, M.: Modeling relational data with graph convolutional networks. In: Gangemi, A., et al. (eds.) ESWC 2018. LNCS, vol. 10843, pp. 593–607. Springer, Cham (2018). https://doi.org/10.1007/978-3-319-93417-4_38
16. Spinelli, I., Scardapane, S., Hussain, A., Uncini, A.: Fairdrop: biased edge dropout for enhancing fairness in graph representation learning. IEEE Trans. Artif. Intell. **3**(3), 344–354 (2021)
17. Vashishth, S., Sanyal, S., Nitin, V., Talukdar, P.: Composition-based multi-relational graph convolutional networks. arXiv preprint arXiv:1911.03082 (2019)
18. Wang, X., et al.: Heterogeneous graph attention network. In: The World Wide Web Conference, pp. 2022–2032 (2019)
19. Zeng, Z., Islam, R., Keya, K.N., Foulds, J., Song, Y., Pan, S.: Fair representation learning for heterogeneous information networks. In: Proceedings of the International AAAI Conference on Weblogs and Social Media, vol. 15 (2021)

HAEP: Heterogeneous Environment Aware Edge Partitioning for Power-Law Graphs

Xian Zhang[1], Junchang Xin[1,3(✉)], Jinyi Chen[1], Beibei Wang[1],
and Zhiqiong Wang[2,3]

[1] School of Computer Science and Engineering, Northeastern University,
Shenyang 110819, China
{zhangxian,chenjinyi}@stumail.neu.edu.cn,xinjunchang@mail.neu.edu.cn
[2] College of Medicine and Biological Information Engineering, Northeastern
University, Shenyang 110819, China
wangzq@bmie.neu.edu.cn
[3] Key Laboratory of Big Data Management and Analytics (Liaoning Province),
A Northeastern University, Shenyang 110819, China

Abstract. Graph partitioning is an important preprocessing step for
distributed processing of large-scale graph data. By balancing workloads and reducing communication costs among nodes, graph partitioning methods improve the efficiency of homogeneous clusters for processing power-law graphs. However, a real cluster usually consists of heterogeneous nodes, each with different computing and communication
ability. Nodes handle the same workload with different time cost, and
the slowest node is the bottleneck. Therefore, a Heterogeneous environment Aware Edge Partitioning method (HAEP) is proposed to balance
graph processing time by skewing the workload. HAEP can adapt to the
challenge of unbalanced performance among nodes. First, a k-time balanced graph partitioning problem is defined to balance the expected time
cost of graph processing in heterogeneous environments. Then, a neighborhood heuristic expansion is performed according to the node performance, minimizing the communication time among nodes and assigning
an appropriate workload for each node. Further, a distributed method
of HAEP, DHAEP, is proposed to improve the efficiency of graph partitioning. The performance evaluation shows that HAEP and DHAEP can
improve graph processing efficiency by up to 41% compared to state-of-
the-art partitioning methods, and the graph partition time of DHAEP
is 15% of HAEP.

Keywords: Graph partitioning · Heterogeneous environments ·
Distributed computing

1 Introduction

As the scale of graph data increases, many distributed graph processing systems have been developed for large-scale graph processing, such as Pregel [13],

© The Author(s), under exclusive license to Springer Nature Switzerland AG 2023
X. Wang et al. (Eds.): DASFAA 2023, LNCS 13945, pp. 331–340, 2023.
https://doi.org/10.1007/978-3-031-30675-4_23

GraphLab [12], PowerGraph [7] and PowerLayer [6]. Graph partitioning is an important preprocessing step in distributed graph processing systems, balancing workloads among cluster nodes while reducing communication costs.

Edge partitioning can partition a power-law graph more efficiently than traditional vertex partitioning in homogeneous clusters. Skew-degree distributions [7] often appear in many large-scale real-world graphs, such as web or social graphs. However, the computing and communication ability among physical nodes are often unbalanced in real cluster environments. For example, in a real EC2 cluster with 128 nodes [5], the highest network bandwidth of nodes is 500 MB/s and the lowest is only 37.5 MB/s. When homogeneous graph partition methods allocate the same workload to the cluster nodes, the graph processing time of each node is different, and the slowest node becomes the system bottleneck.

In this paper, a Heterogeneous environment Aware Edge Partitioning method (HAEP) is proposed to balance graph processing time by skewing the workload. HAEP can adapt to the challenge of uneven node performance in heterogeneous clusters. First, the k-time balanced graph partitioning problem is defined, and the goal is to balance the expected graph processing time cost of heterogeneous clusters. Then, the neighborhood heuristic expansion is performed according to the node performance, minimizing the communication time among nodes and assigning an appropriate workload to the nodes. Further, a Distributed Heterogeneous environment Aware Edge Partitioning method (DHAEP) is proposed to improve the efficiency of graph partitioning. The main contributions in this paper are as follows:

- The time balanced graph paritioning method HAEP is proposed to solve the challenge of unbalanced performance among nodes. HAEP balances the expected time cost of graph processing for each node by heuristic neighborhood expansion and reduces communication time among nodes by assigning center boundary vertices. HAEP effectively partitions power-law graphs in heterogeneous computing environments.
- DHAEP is proposed to solve the scalability issue in heterogeneous computing environments. DHAEP can speed up graph partitioning time for HAEP while keeping the quality.
- An extensive evaluation of multiple real-world graphs shows that, in most cases, HAEP and DHAEP handle power-law graphs more efficiently compared to seven graph partitioning methods in heterogeneous computing environments.

2   Related Work

Graph partitioning is an unavoidable and challenging problem in distributed graph computing, which has been studied for decades. Most of the existing graph partitioning methods assume that the partitioning is performed in a homogeneous computing environment. Existing homogeneous methods include vertex partitioning and edge partitioning.Vertex partitioning is a traditional graph partitioning method that replicates edges across partitions by assigning vertices to partitions. Many vertex partitioning methods [9] use the multi-level heuristic scheme. A recent study [1] finds that edge partitioning methods are more efficient in most real-world large graphs. Because these real-world graphs (e.g., network graphs and social networks) typically have skewed power-law degree distributions, edge partitioning methods are able to provide better workload balancing in power-law graphs. Bourse et al. [1] propose that the vertex partitioning method can be converted into an edge partitioning method.

The above graph partitioning algorithms always assume that the hardware environment is homogeneous. However, heterogeneous computing environments [3,4] are ubiquitous in distributed computing. The clusters often use hardware computing units with different types of instruction sets and architectures, and use various programming frameworks compatible with heterogeneous hardware platforms [2]. The computational efficiency of graph processing systems is often affected by the performance of the cluster nodes. Michael et al. [10] offer some graph partitioning strategies to improve data-ingress on heterogeneous clusters. Chen et al. propose a multi-layer graph partitioning algorithm [5] considering the bandwidth difference of cluster networks. However, the algorithm is not fast enough for the current scale of graph data. HeAPS [17] is a streaming graph partitioning algorithm in a heterogeneous environment, but the algorithm is a vertex partitioning method for sparse graphs.

3 Problem Definition

In the problem of graph partitioning, an undirected graph is represented by $G = (V, E)$, where V is the set of all vertices in the graph G, $n = |V|$ is the size of the vertex set, and E represents the set of all edges in the graph G, $m = |E|$ is the size of the edge set. The symbol $Comp_i$ represents the computing time of graph processing in the partitioning P_i. $f_p(E_i, V_i)$ represents the computing amount of the partition P_i, and $comp_i$ represents the computing ability of the partition P_i, then the computing time is defined as

$$Comp_i = \frac{f_p(E_i, V_i)}{comp_i} \tag{1}$$

The computational cost depends on the number of edges rather than the number of vertices, so the partition computation amount of edge partitioning can be represented by the function $f_p(E_i)$.

The cut type of the graph affects the communication mode among partitions. Edge partitions synchronize information by copying vertices, so the communication time $Comm_i$ can be defined as

$$Comm_i = \sum_{j=1, i \neq j}^{k} \frac{f_m(V_i, V_j)}{comm_{i,j}} \tag{2}$$

The function $f_m(V_i, V_j)$ represents the total amount of communication in the partition P_i, and the communication ability between the cluster nodes $Worker_i$

and $Worker_j$ is represented by the symbol $comm_{i,j}$. The time cost of graph processing in each partition P_i is consist of computing time and communication time. In addition, considering that the time cost is also affected by the system I/O mode, the time cost of graph processing in the partition is defined as

$$T_i = \begin{cases} Comp_i + Comm_i & if \ I/O = SISO \\ Max(Comp_i + Comm_i) & otherwise \end{cases} \tag{3}$$

Based on the above discussion, our k-time balanced graph partitioning tailored in heterogeneous computing environments is defined as follows.

Definition 1. k-time balanced graph partition. Given a graph $G = (V, E)$ and a partition number k, k-time balanced partition aims to find a partitioning scheme $G = \{G_1, \cdots, G_k\}$ such that

- The time cost of graph processing in each partition is balanced, i.e., $T_i \leq (1 + \lambda) \sum_{j=1}^{k} T_j / k$;
- The sum of computing time of each partition is minimized, i.e., $\min \sum_{i=1}^{k} f_p(E_i)/comp_i$;
- The
  communication time of each partition is minimized, i.e., $\min \sum_{i=1}^{k} Comm_i$, where $Comm_i = \sum_{j=1, j \neq i}^{k} f_m(V_i, V_j)/comm_{i,j}$.

4   HAEP

4.1   Neighbor Expansion

The main idea of neighborhood extension is to balance the computation time of each partition by heuristically selecting edges for each partition.

Heuristic Selection of Edges. The vertex set $B_s = \{v \mid v \in V(E_i) \land \exists e_{v,u} \in E \backslash E_i\}$ is represented as E_i boundary. Based on the edge partitioning of expansion, the graph G can be divided into k balanced partitions by using the boundary set B_s. The vertices associated with at least one partition are called boundary vertices in an expansion process. In each partition, the edge set is expanded according to heuristic rules. When the size of the edge set reaches a limit or there are no edges suitable for expansion, the expansion will stop.

Through the above process, an edge partition can be obtained. Each expansion selects a vertex from the boundary set to minimize increase in vertex replication, and the selected vertex is added to the vertex set C_s as a marker. The basic heuristic for selecting a vertex [6,7,18] is defined as

$$v_{min} = argmin_{v \in B_s \backslash C_s} |N(v) \backslash B_s| \tag{4}$$

Algorithm 1. Heuristic Neighbor Expansion

1: **function** EXPAND$(E, comp)$
2:     $C_s \leftarrow \emptyset$ ▷ $comp = \{comp_1, ..., comp_k\}$
3:     **for** each $P_i \in P$ **do**
4:         $B_s, E_i \leftarrow \emptyset$
5:         $\alpha_i = comp_i / \sum_{j=1}^{k} comp_j$
6:         **while** $|E_i| \leq (1+\lambda)|E| * \alpha_i / k$ **do**
7:             **if** $B_s \backslash C_s = \emptyset$ **then**
8:                 $v =$ random vertex in $V \backslash C_s$
9:             **else**
10:                $v \leftarrow argmin_{v \in V_i \backslash C_s} |N(v) \backslash B_s|$
11:            **end if**
12:            ALLOCEDGES(C_s, B_s, E_i, v)
13:        **end while**
14:    **end for**
15: **end function**

Balance Setting. Neighbor expansion requires an explicit termination condition. First, the computational cost of each partition is estimated as follows.

$$f_p(E_i) = |E_i| \tag{5}$$

When all partitions are divided, the balance condition for the computation cost of each partition is as follows.

$$|E_i|/C_i \leq \frac{((1+\lambda)\sum_{j=1}^{k} |E_j|/C_j)}{k} \tag{6}$$

The imbalance factor, $\lambda \geq 0$, is a constant parameter. The relative expansion of each partition α_i is as follows.

$$\alpha_i = \frac{comp_i}{\sum_{j=1}^{k} comp_j} \tag{7}$$

The judgment condition for the end partitioning of each partition can be expressed as follows.

$$|E_i| \leq \frac{(1+\lambda)|E| * \alpha_i}{k} \tag{8}$$

The Algorithm 1 summarizes the whole process of neighbor expansion. The heuristic expansion process requires p rounds of iteration. In each round i, the edge set E_i will be obtained from the graph. Initially, the core set C_s is empty (line 1). Then the expansion operation is performed (lines 3–13). Figure 1 shows the expansion process of HEAP in a heterogeneous cluster. Each color represents a node in the cluster. First, the boundary set B_s and edge set E_i are initialized (line 4). Line 5 estimates the expansion ability α_i of each partition. The ratio of α in Fig. 1 is 3:2:1. In each round (lines 6–13), E_i is expanded when it satisfies Eq. 8. If $B_s \backslash C_s = \emptyset$, a vertex is randomly selected from $V \backslash C_s$, otherwise vertex is selected according to the Eq. 4. Finally, v is added to the core set C_s and the adjacent edges of v are added to the E_i.

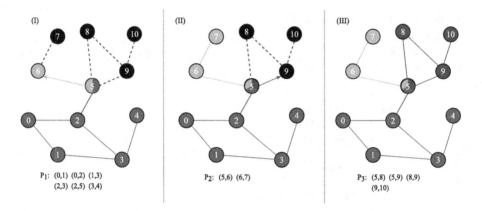

Fig. 1. Expansion of HEAP. The ratio of the performance of each color (blue, red, green) in the cluster is $3:2:1$. (Color figure online)

4.2 Center Boundary Vertices

In distributed graph processing systems, border nodes usually incur communication costs. Reconsidering $Comp_i$ and $Comm_i$ for each partition, HAEP also needs to reallocate center boundary nodes in order to further reduce the expected time cost of partitions. Note that setting the center vertex does not increase the computation time of the partition, as it only adjusts the center node mapping. The center of boundary vertex v is assigned to the partition with minimal cost. v is assigned to P_i, such that

$$i = argmin_{j \in \{i_1, \cdots, i_k\}} Comp_j + Comm_j + \Delta Comm_j. \tag{9}$$

Once the center of v is assigned to a partition P_i, the time cost of that partition will include the communication time of vertex v.

4.3 Distributed HAEP

In most cases, the HAEP algorithm is efficient enough to process data sequentially. But the graph of the real world is getting bigger, and the target scale of graph analysis has gone from a billion edges to a trillion edges. In order to partition a graph using HAEP, a distributed partitioning algorithm is required.

The core of the distributed algorithm is the use of two distributed processes: an expansion process and an allocation process. The expansion process computes and manages the boundary vertices for each partition. The allocation process distributedly manages the allocation of vertices and edges. First, the input graph is randomly distributed into the allocation process. Each expansion process randomly selects a vertex to start expanding. The algorithm is performed iteratively. Although the partition results of DHAEP are consistent with the final partition results in Fig. 1, the partition results of DHEAP are not always consistent in real-world graphs. The edges allocated during the algorithm iteration are copied and sent from the allocation process to the expansion process, and at the end of the calculation, the entire edge is allocated to the extension process.

5 Experiments

5.1 Experiment Settings

In order to verify the correctness of the algorithm, four different graph datasets will be used, namely, HT, AS, LJ and OK from SNAP [11]. All experimental datasets are shown in Table 1. DBH [16], HDRF [15], SNE [18], NE [18], DNE [8], HEP [14] and HASH partitioning methods are chosen to compare with HAEP and DHAEP.

Table 1. Datasets

Graph	Vertices	Edges	Type
higgs-twitter (HT)	456,631	14,855,875	Social Network
com-livejournal (LJ)	3,997,962	34,681,189	Social Network
as-skitter (AS)	1,696,415	11,095,298	Topology Graph
com-orkut (OK)	3,072,441	117,185,083	Social Network

Experiments use graph processing algorithms, PageRank (PR), Single Source Shortest Path (SSSP) and Connected Component (CC) running time as metrics to evaluate the performance of different partitions. The partitioning efficiency of HAEP and DHAEP is evaluated by measuring replication factor and partitioning time.

Table 2. Cluster Topologies

Topo	CPU(Number)	Network(Number)
T_1	3.4 GHz(32)	1 Gbps(32)
T_2	1.6 GHz(16),3.4 GHz(16)	500 bps(16), 1 Gbps(16)
T_3	1.6 GHz(2),3.4 GHz(30)	500 bps(2), 1 Gbps(30)

A homogeneous cluster topology (T_1) and two heterogeneous cluster topologies (T_2, T_3) are used in comparative experiments. Table 2 lists topologies and their symbols.

5.2   Experiment Results

To test the performance of HAEP and DHAEP methods, these methods are compared with other state-of-the-art partitioning methods in different environments. These graph partitioning methods divide the input graph into 8 partitions. After partitioning, the PageRank, BFS and CC algorithms are performed on the topology shown in Table 2.

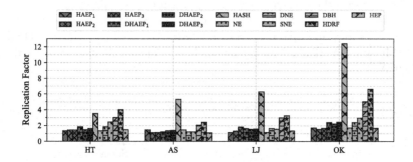

Fig. 2. Replication factor of real-world graphs.

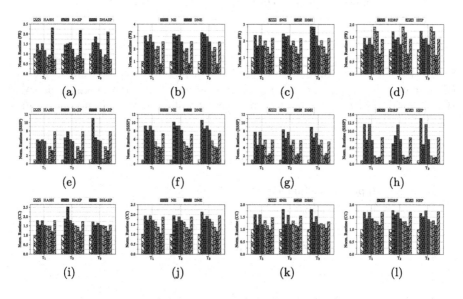

Fig. 3. Performance results on real-world graphs.

Figure 2 shows the comparison results of replication factors of HEAP, DHAEP and other partitioning methods. In the heterogeneous environment, HAEP can still approximate the replicator of NE and is the partition with the best replicator effect in the experiment. To more clearly compare the experimental results, the normalized runtime results of PageRank, BFS and CC are plotted

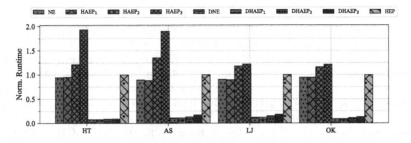

Fig. 4. Runtime for partitioning real-world graphs.

against hash partitioning. Figure 3 shows a normalized runtime comparison on three plots. HEAP and DHAEP usually have a better running performance. Figure 3(e) shows that HAEP and DHAEP can improve graph processing efficiency by up to 41% compared to state-of-the-art partitioning methods. However, in Fig. 3(a) and Fig. 3(d), the runtime of HAEP in T_2 is still lower than that of the NE algorithm. In partitions with weak communication ability, the bottleneck of execution time at this time is determined by communication time. The skewed workload may increase the communication processing time of nodes. Figure 4 measures the graph partitioning time of different graph partitioning algorithms with the HEP algorithm as the baseline in the T_1 cluster. DHAEP reduces the partition time by 85%–90% compared to HAEP.

6   Conclusion

Traditional graph partitioning methods fail to handle power-law graphs efficiently in heterogeneous clusters. Based on the idea of balancing expected graph processing time cost, HAEP and DHAEP algorithms are proposed to adapt to the challenge of performance imbalance among nodes. DHAEP solves the issue of scalability in heterogeneous environments whiling keeping the quality. The performance evaluation shows that HAEP and DHAEP can greatly imporve the performance of graph processing and are quite efficient to partition large-scale power-law graphs.

Acknowledgement. This work was supported by National Natural Science Foundation of China (62072089); Fundamental Research Funds for the Central Universities of China (N2116016, N2104001, N2019007).

References

1. Bourse, F., Lelarge, M., Vojnovic, M.: Balanced graph edge partition. In: KDD, pp. 1456–1465. ACM (2014)
2. Cao, J., Simonin, M., Cooperman, G., Morin, C.: Checkpointing as a service in heterogeneous cloud environments. In: CCGRID, pp. 61–70. IEEE Computer Society (2015)

3. Chan, Y., Fathoni, H., Yen, H., Yang, C.: Implementation of a cluster-based heterogeneous edge computing system for resource monitoring and performance evaluation. IEEE Access **10**, 38458–38471 (2022)
4. Chen, C., Nagel, L., Cui, L., Tso, F.P.: Distributed federated service chaining for heterogeneous network environments. In: UCC, pp. 2:1–2:10. ACM (2021)
5. Chen, R., Yang, M., Weng, X., Choi, B., He, B., Li, X.: Improving large graph processing on partitioned graphs in the cloud. In: SoCC, p. 3. ACM (2012)
6. Chen, R., Shi, J., Chen, Y., Chen, H.: Powerlyra: differentiated graph computation and partitioning on skewed graphs. In: EuroSys, pp. 1:1–1:15. ACM (2019)
7. Gonzalez, J.E., Low, Y., Gu, H., Bickson, D., Guestrin, C.: Powergraph: distributed graph-parallel computation on natural graphs. In: OSDI, pp. 17–30. USENIX Association (2012)
8. Hanai, M., Suzumura, T., Tan, W.J., Liu, E.S., Theodoropoulos, G., Cai, W.: Distributed edge partitioning for trillion-edge graphs. Proc. VLDB Endow. **12**(13), 2379–2392 (2019)
9. Karypis, G., Kumar, V.: A fast and high quality multilevel scheme for partitioning irregular graphs. SIAM J. Sci. Comput. **20**(1), 359–392 (1998)
10. LeBeane, M., Song, S., Panda, R., Ryoo, J.H., John, L.K.: Data partitioning strategies for graph workloads on heterogeneous clusters. In: SC, pp. 56:1–56:12. ACM (2015)
11. Leskovec, J., Krevl, A.: SNAP Datasets: Stanford large network dataset collection (2014). http://snap.stanford.edu/data
12. Low, Y., Gonzalez, J., Kyrola, A., Bickson, D., Guestrin, C., Hellerstein, J.M.: Distributed graphlab: a framework for machine learning in the cloud. Proc. VLDB Endow. **5**(8), 716–727 (2012)
13. Malewicz, G., et al.: Pregel: a system for large-scale graph processing. In: SIGMOD, pp. 135–146. ACM (2010)
14. Mayer, R., Jacobsen, H.: Hybrid edge partitioner: partitioning large power-law graphs under memory constraints. In: SIGMOD, pp. 1289–1302. ACM (2021)
15. Petroni, F., Querzoni, L., Daudjee, K., Kamali, S., Iacoboni, G.: HDRF: stream-based partitioning for power-law graphs. In: CIKM, pp. 243–252. ACM (2015)
16. Xie, C., Yan, L., Li, W., Zhang, Z.: Distributed power-law graph computing: theoretical and empirical analysis. In: NIPS, pp. 1673–1681 (2014)
17. Xu, N., Cui, B., Chen, L., Huang, Z., Shao, Y.: Heterogeneous environment aware streaming graph partitioning. IEEE Trans. Knowl. Data Eng. **27**(6), 1560–1572 (2015)
18. Zhang, C., Wei, F., Liu, Q., Tang, Z.G., Li, Z.: Graph edge partitioning via neighborhood heuristic. In: KDD, pp. 605–614. ACM (2017)

A Graph Embedding Approach for Link Prediction via Triadic Closure Based Direct Aggregation and Weighted Concatenation

Yahui Chai[1], Xiaobin Rui[1], Jie Yang[1], Philip Yu[2],
and Zhixiao Wang[1(✉)]

[1] School of Computer Science, China University of Mining and Technology,
Xuzhou 221116, Jiangsu, China
{yhchai,ruixiaobin,jyang_1,zhixwang}@cumt.edu.cn
[2] Department of Computer Science, University of Illinois at Chicago,
Chicago, IL 60607, USA
psyu@uic.edu

Abstract. Graph embedding based on deep learning is an effective approach for link prediction. However, there still remain some unsolved problems. Firstly, existing methods iteratively aggregate node embeddings from the neighborhood, which cannot retain the global structural information at lower node aggregation cost. Secondly, existing methods can hardly retain rich local structural information for edge embeddings obtained from node embeddings by Hadamard product, summation, or direct concatenation. To tackle these challenges, this paper proposes a novel graph embedding approach for link prediction via triadic closure based direct aggregation and weighted concatenation. To aggregate neighborhood information with high efficiency, our proposed approach directly aggregates multi-order neighbors to the central node and utilizes the triadic closure structure to assign different aggregating weights. To well retain the local structural information, our proposed approach generates edge embeddings through weighted summation. Extensive experiments demonstrate that the triadic closure based direct aggregation and weighted concatenation enable our proposed approach to efficiently learn more accurate embeddings for link prediction, outperforming state-of-the-art methods.

Keywords: link prediction · graph embedding · triadic closure

Supported by the National Natural Science Foundation of China (No. 61876186, No. 61977061), the Xuzhou Science and Technology Project (No. KC21300), the Graduate Innovation Program of China University of Mining and Technology (No. 2022WLJCR-CZL26), and the Postgraduate Research & Practice Innovation Program of Jiangsu Province (No. KYCX22_2568).

© The Author(s), under exclusive license to Springer Nature Switzerland AG 2023
X. Wang et al. (Eds.): DASFAA 2023, LNCS 13945, pp. 341–350, 2023.
https://doi.org/10.1007/978-3-031-30675-4_24

1   Introduction

Graph is an important data representation model to reflect the structure of social networks, in which each node corresponds to a person or a social entity, and the link between two nodes shows their interaction. When studying the evolution of social networks, link prediction is of great significance. It refers to the task of predicting missing links or links that are likely to occur in the future [5].

Recently, numerous methods of link prediction have been implemented. As an effective and popular method, network embedding maps high-dimensional nodes of graphs to low-dimensional vector spaces. Embedding-based techniques mainly include matrix factorization [12], random walk [7] and deep learning [3]. In this paper, we mainly concentrate on the deep learning methods. Generally speaking, this kind of methods first obtains node embeddings and then extracts edge embeddings by combining the corresponding node representations to predict likely but unobserved links finally. Goyal et al. [6] demonstrated the excellent performance of deep learning based embedding in link prediction. However, there still remain some unsolved problems for this kind of methods:

(1) When obtaining node embeddings, traditional methods iteratively aggregate nodes from neighborhood. Based on such an inefficient aggregation strategy, complicated neighborhood weight settings will lead to a high node aggregation cost, while simple neighborhood weight settings (e.g. the same weight for the same order nodes) will result in the loss of global structural information.
(2) When generating an edge embedding from two node embeddings, existing methods regard the two nodes equivalent without considering their different contributions to the formation of the edge. In fact, the topology-dependent formation of an edge between two nodes depends on their respective degrees. Ignoring such information will result in the loss of local structural information and affect the performance of link prediction.

In this paper, we propose a novel embedding approach for link prediction named **Tri**adic closure based direct **A**ggeragtion and weighted **C**oncatenation (TriAC for short), which can effectively improve aggregation efficiency while preserving the global structural information. Moreover, emphasizing on such triadic closure structure will better capture important nodes in the formation of edges, and thus improve the final quality of edge embeddings.

The main contributions of this work are summarized as follows.

- When obtaining node embeddings, this paper directly aggregates multi-order neighbors to the central node and utilizes the triadic closure structure to assign different aggregating weights. Larger weight will be assigned to neighbors with richer triadic structures. These neighbors are usually aggregated more times in traditional iterative aggregation. Thus, node embeddings can be efficiently obtained with direct aggregation, rather than complicated iterative aggregation of traditional methods.

- When obtaining edge embeddings, this paper proposes a novel approach to generate edge embeddings through the weighted summation of the corresponding node embeddings. The weight is based on the number of triadic neighbors of each node, where more triadic neighbors implies higher importance of a node, hence higher weight. Weighted summation of node embeddings can better preserve the local structural information of the two ends. By associating the rank of matrix, we show the superiority of summation, compared with Hadamard product and direct concatenation.

2   Related Work

Embedding-based methods for link prediction mainly include matrix factorization [12], random walk [7] and deep learning [3]. Matrix factorization represents attributes of a graph in the form of sparse matrix. It decomposes the sparse matrix to obtain node embeddings. However, when graph scale increases, it becomes difficult to calculate the eigenvalues corresponding to the Eigen matrix [12]. Given a starting node, random walk repeatedly samples one of its neighbors as the next visiting. After obtaining a sufficient number of random walk sequences, the SkipGram [2] model is employed for vector learning to obtain the node embedding. Though random walk methods involve higher-order neighbors, nodes in the same window share a uniform weight [7], ignoring structural difference in neighborhood.

The growing research on deep learning has led to a deluge of Deep Neural Networks (DNN) based methods applied to graphs. Graph Convolutional Networks (GCN) [11] iteratively aggregates the embeddings of neighbors for a node to obtain its new embedding. GraphSAGE [8] generates the node embedding through feature aggregation within a fixed-size neighbor set. GAT [15] introduces an attention mechanism into the propagation process and assigns different attention weights to each neighbor to identify important neighbors. SEAL [16] utilizes Node2Vec [7] to obtain node embedding, then takes node embedding as part of the subgraph representation. DeepEdge [1], an edge-based embedding method, models edges as function of nodes, then jointly optimizes the edge function and node representations with a new objective-graph likelihood. CensNet [10] co-embeds both nodes and edges to a latent feature space by using a line graph of the original undirected graph.

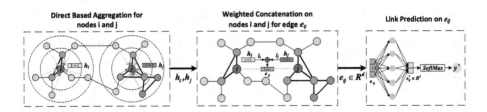

Fig. 1. Framework of TriAC.

3   Method

In this section, we introduce our proposed approach – **Tri**adic closure based direct **A**ggregation and weighted **C**oncatenation (TriAC), which consist of three parts: triadic closure based direct aggregation, weighted concatenation and link prediction (Fig. 1).

3.1   Triadic Closure Based Direct Aggregation

The social triad, a group of three people, is one of the simplest and most fundamental social groups [9]. Close social network connections consist of the social triad relationship [14]. Let $G = (V, E)$ denotes a network, where V refers to the set of nodes and $E \in V \times V$ represents the set of relationships connecting those nodes. If for any two nodes in a triad Δ, i.e., $\forall v_i, v_j \in \Delta$, there exists $e_{ij} \in E$, then we call Δ a closed triad or triadic closure.

Take Fig. 2 as an example, v_0 is the central node, and its first-order neighbors are $\{v_1, v_2, v_3, v_4, v_5, v_6, v_7\}$. TriAC first samples these first-order neighbors. If two first-order neighbors have a connecting edge, such as (v_1, v_7), they will form a triadic closure structure with their "common friend" (v_0). Thus, the first-order triadic neighbors are $\{v_1, v_5, v_6, v_7\}$. Similarly, the second-order triadic neighbors can be obtained as $\{v_8, v_9, v_{13}, v_{14}\}$. Finally, we can obtain the triadic set as $\{v_1, v_5, v_6, v_7, v_8, v_9, v_{13}, v_{14}\}$.

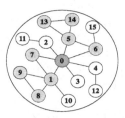

Fig. 2. Example for the triadic set identification.

After neighbor grouping, the neighbors are directly aggregated to obtain the embedding of the central node. Different from previous methods [8,11,15], TriAC assigns two types of weights, one for triadic neighbors, the other for non-triadic neighbors. The non-triadic aggregating weight is $1 - \alpha/n$ while the triadic aggregating weight is $1 + \theta/\Theta$, where n denotes the number of sampled neighbors, α refers to the number of the triadic neighbors, θ represents the number of triaidc closures that a node belongs to and Θ refers to the maximal number of triaidc closures that the node has in the aggregated neighnorhood. This weighting scheme implies that if a central node has more triadic neighbors, those non-triadic neighbors would be less important to the central node compared with the triadic neighbors during aggregating. Thus, TriAC makes the embedding of the central node biased towards the triadic neighbors.

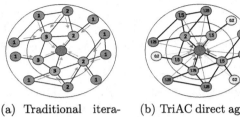

(a) Traditional itera-
tive aggregation

(b) TriAC direct aggre-
gation

Fig. 3. Different node aggregation strategies.

Figure 3 shows the difference between traditional iterative aggregation and TriAC direct aggregation. Figure 3(a) refers to traditional iterative aggregation. Nodes are iteratively aggregated to the central node. The labeled number implies the times that a node would be aggregated. The central node has five first-order neighbors and ten second-order neighbors. The five first-order neighbors need to aggregate their own neighborhood (the second-order neighbors of the target node) to obtain corresponding embeddings, and then aggregate their embeddings to the target node. Figure 3(b) shows the aggregating process of TriAC. The triadic set and non-triadic set are first identified. There are 12 triadic neighbors and 3 non-triadic neighbors. Then, these neighbors are directly aggregated to the central node with different aggregating weights (as labeled in Fig. 3(b)). The non-triadic aggregating weight is $3/15 = 0.2$ and the triadic aggregating weight is $1 + \theta/4$ (the maixmal number of triadic closures for a node is 4 in Fig. 3(b)) where θ is the number of triadic closures that a node belongs to. Figure 3 shows that the triadic closure based direct aggregation can obtain node embeddings efficiently, avoiding complicated iterative aggregations of traditional methods.

3.2 Weighted Concatenation for Edge Embedding

In order to obtain an edge embedding, existing methods combine two node embeddings by Hadamard product, summation or direct concatenation. On one hand, these methods regard the two nodes equivalent without considering their different contributions to the formation of the edge. On the other hand, to the best of our knowledge, there is no existing research that theoretically analyzes the pros and cons of these concatenating methods. To tackle these problems, TriAC introduces a weight scheme for edge concatenation.

Supposing there are two nodes v_i, v_j, the corresponding node embeddings $h_i = [h_{i1}, h_{i2}, ..., h_{id}]$, $h_j = [h_{j1}, h_{j2}, ..., h_{jd}]$, TriAC defines the edge embedding e_{ij} as

$$e_{ij} = L_i h_i + L_j h_j = [L_i h_{i1} + L_j h_{j2}, ..., L_i h_{id} + L_j h_{jd}],$$

$$L_i = \frac{2l_i}{l_i + l_j}; L_j = \frac{2l_j}{l_i + l_j}, \tag{1}$$

where L_i denotes the concatenation weight of v_i, l_i refers to the number of triadic neighbors of v_i. According to the handshake lemma [4], each edge in a graph has two ends, *i.e.*, each edge provides $2°$ for the graph. Therefore, our proposed TriAC sets the total weights of the two nodes to 2, *i.e.*, $L_i + L_j = 2$.

Here, we present an example to show the superiority of weighted concatenation. After obtaining node embeddings, Fig. 4(a) directly mixes them to get the edge embedding, ignoring their structural differences. In contrast, Fig. 4(b) highlights the structural difference that v_j has 5 triadic neighbors while v_i only has 2 triadic neighbors. This is to say, $l_i = 2, l_j = 5, L_i = (2l_i)/(l_i + l_j) = 4/7, L_j = (2l_j)/(l_i + l_j) = 10/7$. Obviously, v_j contains more information due to its rich triadic structure. Therefore, in the concatenating process, v_j has a greater contribution to the formation of the edge.

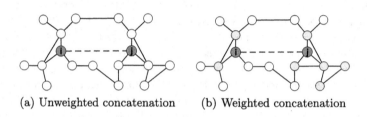

(a) Unweighted concatenation　　　(b) Weighted concatenation

Fig. 4. Two concatenation schemes.

Compared with Hadamard product and direct concatenation, the edge embedding obtained from node embeddings summation is of higher possibility to contain more structural information. We explain this by associating the linear correlation between edge embeddings with the rank of a matrix.

Assuming there is a batch of node embedding matrices where each of them consists of n random node embeddings, we can select two of these matrices and combine them to obtain a new matrix. The new matrix is thus the corresponding edge embedding matrix and each row represents an edge representation. The edge embeddings should reflect the structural difference between edge, which can be characterized by the linear correlation between these embeddings. Therefore, the rank (denoted as $r(\cdot)$) of a edge embedding matrix is able to evaluate whether the edge embedding can better reflect the structural differences between edges.

Let $A = (a_{ij})_{m*n}$ and $B = (b_{ij})_{m*n}$ ($n \leq m$) denote two node embedding matrices. Let $(A, B)_{m*2n}$, $(A \odot B)_{m*n}$, and $(A + B)_{m*n}$ be the result of the matrix obtained by the direct concatenation, the Hadamard product, and the summation, respectively. It is obviously that $r(A \odot B) \leq min(r(A), r(B)) \leq r(A, B) \leq r(A)+r(B)$, and $r(A+B) \leq r(A)+r(B)$, which means the rank of $(A \odot B)$ will only decrease while the rank of (A, B) and $(A + B)$ is likely to increase. However, the direct concatenation involves a DNN to complete dimensionality reduction, which can further results in a loss of its rank. To sum up, the rank of a matrix obtained by summation is likely to increase without involving any loss, which is superior to Hadamard product and direct concatenation.

3.3   Link Prediction

In this paper, the link prediction problem is transformed into a binary classification problem. The edges are classified according to their embedding representations, which can be implemented by using a deep neural network. The pseudocode of TriAC is shown in Algorithm 1.

Algorithm 1. Triadic Closure Based Direct Aggregation and Weighted Concatenation (TriAC)

Require: network G;
      triadic sequence T; non-triadic sequence $G - T$;
      activation function σ;
      concatenating weight W;
      neural network f;
      the number of sampled neighbors n;
      the number of triadic closures that a node belongs to θ ;
      the maximal number of triadic closures that a node has Θ;
Ensure: Link prediction result $y_{i,j}$ for $i = 1, 2, ..., N; j = 1, 2, ..., N; i \neq j$
 1: **for** $i = 1, 2, ..., N$ **do**
 2:     $l_i = length(T_i)$;
 3:     $h_i^0 = (1 - l_i/n_i)T_i + (1 + \theta_i/\Theta_i)(G - T)_i$;
 4: **end for**
 5: **for** $i = 1, 2, ..., N$ **do**
 6:     $\overline{h_i} = \sigma(W_0 h_i^0 + b_0) + \sigma(W_1 h_i^0 + b_1)$;
 7: **end for**
 8: **for** $i = 1, 2, ..., N; j = 1, 2, ..., N; i \neq j$ **do**
 9:     $L_i = \frac{2l_i}{l_i + l_j}, L_j = \frac{2l_j}{l_j + l_i}$;
10:     $e_{ij} = L_i \overline{h_i} + L_j \overline{h_j} = [L_i \overline{h_{i1}} + L_j \overline{h_{j2}}, ..., L_i \overline{h_{id}} + L_j \overline{h_{jd}}]$;
11: **end for**
12: **for** $i = 1, 2, ..., N; j = 1, 2, ..., N; i \neq j$ **do**
13:     $e_{ij}^* = f(e_{ij})$;
14:     $y_{ij} = softmax(e_{ij}^*)$;
15: **end for**

---

4   Experiments

4.1   Datasets and Settings

Seven state-of-the-art graph embedding methods are selected as baselines, including DeepWalk [13], GCN [11], GAT [15], GraphSAGE [8], DeepEdge [1], CensNet-VAE [10], and SEAL [16]. Besides, we also design **ITER** (adopting our proposed direct aggregation approach but weighted by the aggregation times of traditional **ITER**ative aggregation methods, rather than the number of triadic closures) as one of baselines. Table 1 shows the networks used in experiments.

Table 1. Datasets.

Networks	$|V|$	$|E|$
Hamster	1858	12534
Yeast	2375	11693
Facebook	4039	88234
Douban	13786	214391
CS4	22499	43858

4.2 Performance of Link Prediction

The experimental results are shown in Table 2. All in all, our approach has the best or occasionally 2nd best performance for all metrics in all datasets. In contrast, ITER is not competitive among those baselines. It implies that weighted with aggregation times of iterative aggregation is not enough to yield high-quality node embeddings. The local topology structure should be involved.

Table 2. Comparison with state-of-the-art methods (the best one in bold and the 2nd best with underline).

Networks	Metric	Deep Walk	GCN	GAT	Graph SAGE	DeepEdge	CensNet	SEAL	TriAC	ITER
	PRE	0.8034	0.7328	0.8317	0.8150	0.6950	0.8359	**0.9105**	0.8991	0.8678
Hamster	ACC	0.9158	0.7310	0.8344	0.7780	0.8860	0.8357	0.8901	**0.9231**	0.8735
	ROC	0.8962	0.8357	0.9099	0.8900	0.8530	0.9081	0.9452	0.9465	0.9257
	PRE	0.8192	0.7930	0.7670	0.8480	0.8300	0.8567	0.8946	**0.9145**	0.8881
Yeast	ACC	0.9427	0.8260	0.8590	0.8100	0.8910	0.8564	0.8984	**0.9466**	0.9119
	ROC	0.9133	0.8340	0.8390	0.9260	0.9160	0.9190	**0.9585**	0.9440	0.9258
	PRE	0.8311	0.7573	0.9325	0.8169	0.8520	0.8965	0.9456	**0.9647**	0.9610
Facebook	ACC	**0.9581**	0.8234	0.9543	0.8545	0.8760	0.9041	0.9523	0.9552	0.9546
	ROC	0.9389	0.8697	0.9711	0.9223	0.9310	0.9421	0.9831	**0.9892**	0.9820
	PRE	0.6136	**0.8060**	0.7076	0.7123	0.7211	0.7456	0.7722	0.7967	0.7662
Douban	ACC	0.6344	0.8166	0.7259	0.7015	0.7376	0.7488	0.7874	**0.8300**	0.7819
	ROC	0.6504	0.8244	0.7386	0.7354	0.7431	0.7514	0.7853	**0.8598**	0.7327
	PRE	0.8774	0.8524	0.9127	0.9056	0.9171	0.9225	0.9330	**0.9763**	0.6762
CS4	ACC	0.8426	0.8223	0.9025	0.9018	0.8825	0.9027	0.8474	**0.9775**	0.6819
	ROC	0.8567	0.7961	0.8712	0.8892	0.9017	0.8914	0.9079	**0.9946**	0.7327

4.3 Performance of Node Direct Aggregation

There are three common iterative aggregation schemes: (1) Assigning the same weight to all neighbors (typically used in GCN); (2) Assigning different weights to each neighbor according to the self-attention mechanism (typically used in GAT); (3) Sampling a fixed number of neighbors and assigning the same weight (typically used in SAGE). We compare our proposed direct aggregation

Table 3. Comparison of link prediction results (ROC) with state-of-the-art aggregation methods for node embedding.

Networks	SUM (summation)				HM (hadamard product)				CAT (direct concatenation)			
	GCN	GAT	SAGE	TriAC	GCN	GAT	SAGE	TriAC	GCN	GAT	SAGE	TriAC
Hamster	0.8435	0.9157	0.9100	**0.9414**	0.8347	0.8271	0.8900	**0.9412**	0.8315	0.9132	0.8546	**0.9375**
Yeast	0.8421	0.8490	0.8473	**0.9371**	0.8206	0.8366	0.8260	**0.9353**	0.885	0.8398	0.8573	**0.9321**
Facebook	0.8607	0.9741	0.9334	**0.9853**	0.8605	0.9609	0.9223	**0.9848**	0.8602	0.9711	0.9257	**0.9869**
Douban	0.8244	0.7386	0.7354	**0.8424**	0.8110	0.7255	0.7249	**0.8311**	0.8206	0.7321	0.7301	**0.8400**
CS4	0.7843	0.8652	0.8816	**0.9884**	0.7518	0.8592	0.8638	**0.9803**	0.7672	0.8661	0.8785	**0.9816**
Mean	0.8310	0.8685	0.8615	**0.9389**	0.8157	0.8419	0.8454	**0.9345**	0.8329	0.8665	0.8492	**0.9356**

with these aggregation schemes, and then the different concatenation methods are employed to generate edge embeddings from node embeddings. It should be noted that here these concatenation methods do not involve any weight.

Table 3 shows that our proposed direct aggregation approach has competitive and stable performance over diverse approaches (GCN, GAT and SAGE) under different concatenation methods, *i.e.* summation (SUM), Hadamard product (HM) and direct concatenation (CAT). Besides, SUM shows superior performance than HM and CAT, which is in accordance with our previous claims.

4.4 Performance of Edge Concatenation

Table 4. Comparison of link prediction results (ROC) with different concatenation methods for edge embedding.

Networks	HM	HM-W	CAT	CAT-W	SUM	SUM-W
Hamster	**0.9412**	0.9245	0.9375	**0.9438**	0.9414	**0.9465**
Yeast	**0.9353**	0.9312	0.9321	**0.9555**	0.9371	**0.9440**
Facebook	0.9848	**0.9914**	0.9869	**0.9908**	0.9853	**0.9892**
Douban	0.8311	**0.8404**	0.8400	**0.8436**	0.8424	**0.8598**
CS4	0.9803	**0.9904**	0.9816	**0.9931**	0.9884	**0.9946**

Table 4 presents the results of three different concatenation methods with or without weight. Obviously, the performance of link prediction is significantly improved with weighted concatenation. Especially, SUM-W is still better than HM-W and CAT-W, which again verifies our previous claims. Therefore, both the proposed direct aggregation and the proposed weighted edge concatenation contribute to the performance improvements of link prediction.

5 Conclusion

In this paper, we propose a deep learning based embedding approach, *i.e.*, TriAC, to learn structure-preserving embeddings for link prediction in social networks.

Different from the existing methods, TriAC directly aggregates node embeddings by adopting the triadic closure structure in social networks. Also, TriAC generates edge embeddings with weight that in line with the number of triadic closures. Experimental results on five datasets show that the proposed TriAC can better predict links in social network, outperforming state-of-the-art methods.

References

1. Abu-El-Haija, S., Perozzi, B., Al-Rfou, R.: Learning edge representations via low-rank asymmetric projections. In: Proceedings of the 2017 ACM on Conference on Information and Knowledge Management, pp. 1787–1796 (2017)
2. Cai, H., Zheng, V.W., Chang, K.C.C.: A comprehensive survey of graph embedding: problems, techniques, and applications. IEEE Trans. Knowl. Data Eng. **30**(9), 1616–1637 (2018)
3. Cao, S., Lu, W., Xu, Q.: Deep neural networks for learning graph representations. In: Proceedings of the Thirtieth AAAI Conference on Artificial Intelligence, pp. 1145–1152 (2016)
4. Dalfo, C., Fiol, M.A.: Graphs, friends and acquaintances. arXiv preprint arXiv:1611.07462 (2016)
5. Ghorbanzadeh, H., Sheikhahmadi, A., Jalili, M., Sulaimany, S.: A hybrid method of link prediction in directed graphs. Expert Syst. Appl. **165**, 113896 (2021)
6. Goyal, P., Ferrara, E.: Graph embedding techniques, applications, and performance: a survey. Knowl.-Based Syst. **151**, 78–94 (2018)
7. Grover, A., Leskovec, J.: node2vec: scalable feature learning for networks. In: Proceedings of the 22nd ACM SIGKDD International Conference on Knowledge Discovery and Data Mining, pp. 855–864 (2016)
8. Hamilton, W.L., Ying, R., Leskovec, J.: Inductive representation learning on large graphs. In: Proceedings of the 31st International Conference on Neural Information Processing Systems, pp. 1025–1035 (2017)
9. Huang, H., Dong, Y., Tang, J., Yang, H., Chawla, N.V., Fu, X.: Will triadic closure strengthen ties in social networks? ACM Trans. Knowl. Discovery Data (TKDD) **12**(3), 1–25 (2018)
10. Jiang, X., Zhu, R., Li, S., Ji, P.: Co-embedding of nodes and edges with graph neural networks. IEEE Trans. Pattern Anal. Mach. Intell., 1–1 (2020)
11. Kipf, T.N., Welling, M.: Semi-supervised classification with graph convolutional networks. arXiv preprint arXiv:1609.02907 (2016)
12. Ma, X., Sun, P., Qin, G.: Nonnegative matrix factorization algorithms for link prediction in temporal networks using graph communicability. Pattern Recogn. **71**, 361–374 (2017)
13. Perozzi, B., Al-Rfou, R., Skiena, S.: Deepwalk: online learning of social representations. In: Proceedings of the 20th ACM SIGKDD International Conference on Knowledge Discovery and Data Mining, pp. 701–710 (2014)
14. Simmel, G.: The sociology of Georg Simmel. Simon and Schuster, New York (1950)
15. Veličković, P., Cucurull, G., Casanova, A., Romero, A., Lio, P., Bengio, Y.: Graph attention networks. arXiv preprint arXiv:1710.10903 (2017)
16. Zhang, M., Chen, Y.: Link prediction based on graph neural networks. Adv. Neural. Inf. Process. Syst. **31**, 5165–5175 (2018)

MPGCL: Multi-perspective Graph Contrastive Learning

Miao Zhang and Yan Yang[✉]

Heilongjiang University, Harbin, Heilongjiang, China
yangyan@hlju.edu.cn

Abstract. In recent years, contrastive learning has emerged as a successful method for unsupervised graph representation learning. It generates two or more different views by data augmentation and maximizes the mutual information between the views. Prior approaches usually adopt naive data augmentation strategies or ignore the rich global information of the graph structure, leading to sub-optimal performance. This paper proposes a contrast-based unsupervised graph representation learning framework, MPGCL. Since data augmentation is the key to contrastive learning, this paper proposes constructing higher-order networks by injecting similarity-based global information into the original graph. Then, adaptive and random augmentation strategies are combined to generate two views with complementary semantic information, which preserve important semantic information while not being too similar. In addition, the previous methods only consider the same nodes as positive samples. In this paper, the positive samples are identified by capturing global information. In extensive experiments on eight real benchmark datasets, MPGCL outperforms both the SOTA unsupervised competitors and the fully supervised methods on the downstream task of node classification. The code is available at: https://github.com/asfdd3/-miao/tree/src/MPGCL.

Keyword: Graph Neural Networks · Graph Representation Learning · Unsupervised Learning · Self-Supervised Learning · Contrastive learning

1 Introduction

Graph representation learning aims to obtain a low-dimensional embedding of the nodes in the graph to encode the attributes and structural features of the nodes. However, existing methods are primarily established in a supervised manner [1, 2], which requires abundant labeled nodes for training. Recently, contrastive learning [3] addressed the problem of label dependency. Contrastive learning maximizes the mutual information [30] between similar instances and minimizes the mutual information between dissimilar instances by following the InfoMax [4] principle, enabling the learning of differentiated embeddings even in unsupervised case.

First, data augmentation has been shown to be a key component of contrastive learning [37]. Existing methods such as DGI [3], inspired by DIM [8], using randomly shuffling features to generate negative samples is difficult to provide a powerful supervised

© The Author(s), under exclusive license to Springer Nature Switzerland AG 2023
X. Wang et al. (Eds.): DASFAA 2023, LNCS 13945, pp. 351–366, 2023.
https://doi.org/10.1007/978-3-031-30675-4_25

signal in the case of sparse feature matrices. MVGRL [9] performs graph diffusion to enhance the original graph, which enriches the global information but injects noise at the same time. GRACE [10] and BGRL [18] adopt a simple augmentation strategy that results in the disruption of important semantic information. Second, how to define positive and negative samples is still rarely explored. For example, BGRL only regards the same nodes as positive samples, which obviously ignores the rich global information of the graph. GRACE and GCA [11] require a large number of negative samples to improve the performance of downstream tasks, requiring high computational and memory costs, which is unrealistic in reality.

To address the problems of existing graph contrastive learning, we propose a simple and efficient framework to learn node representations, denoted as **M**ulti-**P**erspective **G**raph **C**ontrastive **L**earning (MPGCL). Specifically, in order to enable the original view to perceive more global information, the original graph is diffused. However, only a simple graph diffusion will unavoidably introduce a large amount of noise. We first construct a k-NN graph by node features to filter out these noisy edges and combine them with the original adjacency matrix to generate a higher-order network. Then augmentation is performed on the higher-order network. It is known that an appropriate increase in the inconsistency of the two views can further facilitate the learning of differentiated embeddings, so we perform a combination of adaptive and random augmentation strategies for the higher-order networks. In addition, it is crucial to construct positive and negative samples to provide self-supervised signals. Previous approaches [10, 11] have required introducing a large number of negative samples to prevent model collapse. Recently, BYOL [13] in computer vision proposed using momentum updating Siamese networks as an architecture to maximize the mutual information of identical instances without explicit negative samples for contrastive learning. However, optimizing only the above objectives in the graph domain ignores the rich structural information of the graph. Since different nodes may also have similar semantic information, the same nodes should not simply be defined as positive samples. To alleviate this problem, we search from a global perspective to capture potential positive samples while injecting a small number of negative samples to enrich node contextual information.

In summary, our contributions are as follows:

- We propose an efficient framework, MPGCL, for data augmentation from multiple perspectives. First, we inject global information into the original graph to construct a higher-order network. Then further augmentation is performed on the higher-order network enabling the information of the two views to complement each other after augmentation, which provides a powerful self-supervised signal for cross-view contrastive learning.
- We consider that the graph contains rich structural information, from a global perspective instead of considering only the same nodes as positive samples, we mine the representation space for potential positive samples, while injecting a small number of negative samples to further enrich the self-supervised signal.
- We conduct extensive experiments on eight real-world datasets, and compared with twelve existing methods, MPGCL consistently achieves better performance.

2   Related Work

In recent years, graph representation learning algorithms have attracted much attention in the scientific community. Most successful graph representation methods combine neural networks with graph structured data. Most of the existing graph neural network methods follow message passing mechanisms, such as GCN [1], GAT [2], GraphSAGE [13] and their various variants. Despite the great success of these methods, most of them follow a semi-supervised learning paradigm, i.e., they require a large number of node labels or graph labels. Firstly, labeling labels depends on human intervention and requires expertise in the relevant domain, making it costly [14]. Secondly, the generalization ability of the model in supervised scenarios is poor, and it is easy to overfit [15]. Therefore, learning how to perform graph representation without relying on labels is crucial.

Contrastive learning, a manner of unsupervised learning, learns representations by contrasting positive and negative pairs and has succeeded in vision [12, 31–33] and natural language processing [34, 35]. Here we focus on graph-related contrastive learning methods. Initially, many traditional unsupervised representation learning [5–7, 26, 36] methods also imply the idea of contrastive learning behind them, e.g., node2vec [5] and DeepWalk [6] generate sequences of nodes based on random wandering, forcing neighboring nodes to have similar representations. Recent works DGI [3] and InfoGraph [16] combine contrastive learning with neural networks and propose to maximize the mutual information of node and graph level representations. MVGRL [9] proposes multi-view contrastive learning and uses an augmentation strategy of graph diffusion to optimize a similar objective as DGI. However, MVGRL has the same drawback as DGI, i.e., the readout function often has difficulty satisfying the requirements of the injective function, leading to a graph-level representation that does not contain sufficient global information as the size of the input graph increases. GMI [17] builds on DGI to maximize the mutual information of nodes and edges using implicit data augmentation. However, the above methods suffer from the same drawback that they require mutual information estimators with parameterization to score positive and negative sample pairs, which severely increases the time and space overhead. GRACE [10] and GCA [11] adopt a node-level contrastive strategy by considering the same nodes in the two augmented views as positive sample pairs and all other nodes as negative samples, avoiding the negative impact of adopting the readout function and without parameterized mutual information estimators. Although achieving significant success, GRACE and GCA introduce a large number of negative samples to prevent model collapse to trivial solutions, and the hyperparameters of the augmentation need to be carefully chosen to generate two high-quality views. BGRL [18] and SelfGNN [19], inspired by BYOL [13], proposed unsupervised representation learning of graphs without using negative samples. However, regard only the same nodes as positive samples ignores the rich structural information in the graph and requires the design of complex asymmetric structures.

3   Method

3.1   Problem Formulation and Graph Neural Network

Problem formulation. let $\mathcal{G} = (\mathcal{V}, \mathcal{E})$ denote a graph, where $\mathcal{V} = \{v_1, v_2, \cdots, v_N\}$, $\mathcal{E} \subseteq \mathcal{V} \times \mathcal{V}$ represent the node set and the edge set respectively. We denote the feature matrix and the adjacency matrix as $X \in \mathbb{R}^{N \times F}$ and $\in \{0, 1\}^{N \times N}$, where $x_i \in \mathbb{R}^F$ is the feature of v_i, and $A_{ij} = 1$ iff $(v_i, v_j) \in \mathcal{E}$. There is no given class information of nodes in $\mathcal{G}$ during training in the unsupervised setting. Our objective is to learn a GNN encoder $f(X, A) \in \mathbb{R}^{N \times F'}$ receiving the graph features and structure as input, that produces node embeddings in low dimensionality, i.e., $F' \ll F$. We denote $H = f(X, A)$ as the learned representations of nodes, where h_i is the embedding of node v_i. These representations can be used in downstream tasks, such as node classification.

Graph neural network. Denote a graph as $\mathcal{G} = (\mathcal{V}, \mathcal{E})$ where the node features are x_v for $v \in V$. In this paper, we focus on the node classification task using Graph Neural Networks (GNNs). GNNs generate node-level embedding h_v through aggregating the node features x_v of its neighbors. Each layer of GNNs serves as an iteration of aggregation, such that the node embedding after the k-th layers aggregates the information within its k-hop neighborhood. The k-th layer of GNNs can be formulated as:

$$a_v^{(k)} = \text{AGGREGATE}^{(k)}\left(\left\{h_v^{(k-1)} : u \in \mathcal{N}(v)\right\}\right)$$
$$h_v^{(k)} = \text{COMBINE}^{(k)}\left(h_v^{(k-1)}, a_v^{(k)}\right) \tag{1}$$

Different GNNs use different formulations of the COMBINE and AGGREGATE functions. In this work, we will focus on GCN encoders. Formally, the GCN propagation rule for a single layer is as follows:

$$\text{GCN}_i(X, A) = \sigma\left(\hat{D}^{-\frac{1}{2}}\hat{A}\hat{D}^{-\frac{1}{2}}XW_i\right) \tag{2}$$

where $\hat{A} = A + I$ is the adjacency matrix with self-loops, $\hat{D}$ is the degree matrix, σ is a non-linearity such as ReLU, and W_i is a learned weight matrix for the k-th layer.

3.2 Overall Framework

As shown in Fig. 1, the proposed framework MPGCL in this paper follows the graph contrastive learning paradigm. The goal is to maximize the mutual information of positive sample pairs in different views to learn node representations. The framework consists of three main parts, data augmentation, Siamese networks for momentum updating, and cross-network contrastive learning. We first perform data augmentation on the original graph to generate two views and feed the two views into the Siamese networks to obtain the representation of the two views. Finally, cross-view contrastive learning is performed to pull closer to the representation of the positive sample pairs while distinguishing the embeddings of different nodes.

3.3 Graph Data Augmentation

The success of contrastive learning in computer vision relies heavily on data augmentation [20], e.g., by rotation, cropping, and scaling. It is challenging to apply methods from the vision domain directly to graphs due to their non-Euclidean nature. Therefore, designing augmentation strategies on graph-structured data remains a challenge. In this paper, we propose to augment the original graph at both the node attribute and topology levels. However, unlike previous methods, we inject high-quality global information into the original graph first to construct the higher-order network. Then the combination of adaptive and random augmentation strategies is adopted for the topology and attribute information of the higher-order network to ensure that the two generated views preserve adequate semantic information while not being too similar. The data augmentation techniques used are described in detail below.

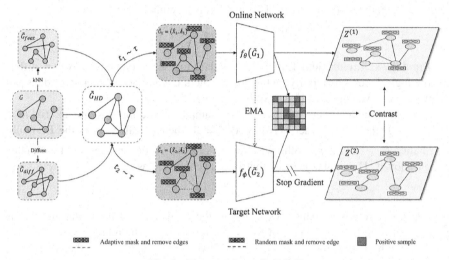

Fig. 1. The overview of MPGCL

High-order network construction. We transform a graph via diffusion to generate a congruent view. The effectiveness of this method attributed to the extra global information provided by the diffused view. This process is formulated as follows:

$$S = \sum_{k=0}^{\infty} \theta_k T^k \in \mathbb{R}^{N \times N} \tag{3}$$

where θ is a parameter to control the distribution of local and global signals, $T \in \mathbb{R}^{N \times N}$ is the transition matrix to transfer the adjacency matrix. In this paper, we adopt the Personalized PageRank (PPR) kernel to power the graph diffusion. Formally, given the adjacency matrix A, the identity matrix I, and the degree matrix D, the equation can be reformulated as:

$$S^{Diff} = \alpha \left(I - (1 - \alpha) D^{-\frac{1}{2}} A D^{-\frac{1}{2}} \right)^{-1} \tag{4}$$

where α is a tunable parameter for the random walk teleport probability. However, simply constructing the diffusion graph introduces noisy edges, which is not beneficial for message passing. We construct the feature similarity graph based on the cosine similarity of node features. The low-quality edges are filtered by computing the intersection of the diffusion graph and the feature similarity graph. The rules for feature similarity graph construction are as follows.

$$S_{ij}^{Feat} = \frac{x_i^\top x_j}{\|x_i\| \, \|x_j\|} \tag{5}$$

Then we choose 6 nearest neighbors following the above cosine similarity for each node and obtain the feature similarity graph. Finally, we combine the original adjacency matrix with the intersection of the diffusion graph and the feature similarity graph to generate the higher-order network. This process is formulated as:

$$A^{HD} = A \cup \left(S^{Feat} \cap S^{Diff} \right) \tag{6}$$

Topology augmentation. Topology-level data augmentation aims to explore different topological views by modifying the graph structure. We randomly remove edges and adaptively remove edges for higher-order network.

For randomly removing edges, we first sample a random masking matrix $\tilde{R} \in \{0, 1\}^{N \times N}$. If the higher-order network $A_{ij}^{HD} = 1$, then its value obeys the Bernoulli distribution $\tilde{R}_{ij} \sim \mathcal{B}(1 - p_r)$, otherwise $\tilde{R}_{ij} = 0$. Here p_r is the probability of each edge being removed. The final adjacency matrix can be denoted as:

$$\tilde{A} = A^{HD} \circ \tilde{R} \tag{7}$$

For adaptively removing edges, we consider the probability of removing an edge is closely related to the importance of the edge. In this paper, the importance of an edge is calculated by measuring the importance of two nodes connected by the edge. First, given the node centrality measure $\varphi_c(\cdot) : \mathcal{V} \rightarrow \mathbb{R}^+$, we use the average of the degree centrality of the two nodes connected the edge to measure the importance of the edge, i.e. $s_{i,j} = \log \frac{\varphi_c(v_i) + \varphi_c(v_j)}{2}$.

We normalize the importance scores of the edges to obtain the probability values:

$$p_{i,j} = min \left(\frac{s_{max} - s_{i,j}}{s_{max} - \mu_s} \cdot p_e, \, p_\tau \right) \tag{8}$$

where s_{max} and μ_s are the maximum and average values of the importance of the edge. p_e is the hyperparameter controlling the overall edge removal probability. p_τ is the cutoff probability, which prevents the probability of removing an edge from being excessively high and thus destroying the graph structure.

Feature augmentation. Most augmentation strategies focus mainly on the topology and do not consider the importance of attribute-level augmentation. We adopt adaptive masking and random masking for node features.

For randomly masking, we randomly mask a fraction of dimensions with zeros in node features. Formally, we first sample a random vector $\tilde{m} \in \{0, 1\}^F$ where each dimension of it independently is drawn from a Bernoulli distribution with probability $1 - p_m$, i.e., $\tilde{m}_i \sim \mathcal{B}(1 - p_i)$, $\forall i$. Then, the generated node features $\tilde{X}$ can be denoted as:

$$\tilde{X} = \left[x_1 \circ \tilde{m}; x_2 \circ \tilde{m}; \cdots ; x_N \circ \tilde{m} \right]^T \tag{9}$$

For adaptively masking, we expect to p_i reflect the importance of the i-th dimension of node features. Define that if a dimension frequently occurs in important nodes, this dimension is usually considered to be necessary, where the importance of the i-th dimension can be denoted as:

$$w_i^f = \sum_{u \in V} x_{ui} \cdot \varphi_c(u) \tag{10}$$

where x_{ui} is the i-th dimensional feature of node u. $\varphi_c(u)$ is the importance of node u.

Similar to the topology-level augmentation, we normalize the weights. The probability of the i-th dimensional importance of the final node attribute can be formulated as:

$$p_i^f = min\left(\frac{s_{max}^f - s_i^f}{s_{max}^f - \mu_s^f} \cdot p_f, p_\tau \right) \tag{11}$$

where $s_i^f = logw_i^f$, s_{max}^f and μ_s^f is the maximum and the average value of s_i^f respectively, and p_f is a hyperparameter that controls the overall magnitude of feature augmentation. p_τ is the cutoff probability, which prevents the probability of masking from being excessively high.

3.4   Network Structure

Siamese networks [21] is a neural architecture that contains two or more identical structures to make multi-class prediction or entity comparison. Traditionally, it has been used on supervised tasks such as signature verification and face matching [22]. Recently, Siamese networks appeared in the self-supervised approach to computer vision BYOL [18] for representation learning. In this paper, we use an architecture similar to the Siamese network, as shown in Fig. 1, consisting of an online network encoder and a target network encoder. Both encoders have the same internal architecture. The difference between them is that the online network updates its parameters by backpropagation, while the target network updates its parameters by learning the content of the online network. This process is implemented using momentum updates. The updated rules can be formulated as:

$$\phi \leftarrow \tau\phi + (1 - \tau)\theta \tag{12}$$

where τ, ϕ, and θ are the momentum, target network parameters and online network parameters, respectively. Since the target network adopts the strategy of momentum update, it makes the network structure asymmetric. The model does not collapse to a trivial solution without using negative samples.

Prior methods [13, 19] relied on MLPs as predictors connected to the encoder to increase the asymmetry of the network to prevent model collapse [21]. We empirically found that introducing a small number of negative samples in the cross-view contrastive learning can replace the prediction head and enrich the supervised signal. Therefore, we drop the prediction head.

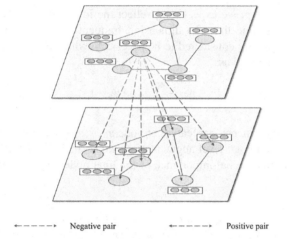

←– – – –→   Negative pair   ←– – – –→   Positive pair

Fig. 2. The red dashed lines are positive sample pairs and the blue dashed lines are negative sample pairs. We pull the same representation in both views close by the objective function and push the representation of different nodes far away. (Color figure online)

3.5 Cross-View Contrastive Learning

Contrastive learning can be successful in the graph domain in large part because of following the classical InfoMax [4] principle which maximizes the mutual information between similar instances. Specifically, in cross-view contrast learning, an objective function is defined to pull the distance between positive samples while pushing away the distance between negative samples. In this paper, we define the objective function for contrastive learning from two perspectives.

In this paper, we define the objective function from two perspectives. One perspective is to capture potential positive samples.. For example, in an academic collaboration network whose nodes denote authors and edges denote collaborations between authors, even if two authors work on the same research topic (same label), they may not be connected in the graph because they have neither collaborated nor shared any collaborators in the past. We believe that such semantically similar but disconnected entities can be mined by searching global information. As in Fig. 1, in each iteration, the original graph data $G = (X, A)$ performs augmentation to generate two semantically informative complementary views $\tilde{G}_1$ and $\tilde{G}_2$, which are then fed to the online network f_θ and the target network f_ϕ to obtain the node representations $Z^{(1)}$ and $Z^{(2)}$. Then, we construct the similarity matrix by calculating the cosine similarity of the node representations and selecting top-k similar nodes for each node as positive samples, such as the nodes corresponding to the red dashed lines in Fig. 2. The objective is to minimize the distance between positive sample pairs. The above process can be represented by:

$$\mathcal{L}_{sim} = -\frac{1}{N} \sum_{i=1}^{N} \sum_{v_j \in \mathbb{P}_i} \frac{z_i^{(1)} \left(z_j^{(2)} \right)^T}{\| z_i^{(1)} \| \| z_j^{(2)} \|} \tag{13}$$

where z_i is the embedding representation of the node v_i and $\mathbb{P}_i$ is the set of positive samples of v_i. The above objective does not introduce negative samples, and although it achieves better results [18], it ignores the rich contextual information of the graph topology. In addition, using the above objective function alone requires a prediction head to further increase the asymmetric structure of the model. Therefore, we enriched the contextual information by introducing a small number of negative samples, while discarding the prediction head. As shown in the blue dashed line in Fig. 2, this paper further pushes the distances of different nodes by introducing a small number of negative samples between views. The loss function with the introduced negative samples can be represented by:

$$\mathcal{L}_{cl} = -\frac{1}{N} \sum_{i=1}^{N} \log \frac{\exp\left(\text{sim}\left(z_i^{(1)}, z_i^{(2)}\right)/\tau\right)}{\sum_{k \in \{\mathbb{P}_i \cup \mathbb{N}_i\}} \exp\left(\text{sim}\left(z_i^{(1)}, z_i^{(2)}\right)/\tau\right)} \tag{14}$$

where $\text{sim}(\cdot)$ denotes the cosine similarity and τ is a temperature hyperparameter. Finally, the final objective function is obtained by integrating Eq. (13) and Eq. (14) as follows:

$$\mathcal{L} = \lambda \cdot \mathcal{L}_{\text{sim}} + (1 - \lambda) \cdot \mathcal{L}_{cl} \tag{15}$$

where λ is the influence factor to balance the two loss functions, and the two losses are regularization terms for each other. Loss function $\mathcal{L}_{\text{sim}}$ captures positive samples from a global perspective, and $\mathcal{L}_{cl}$ further introduces negative samples to enrich self-supervised signals. We balance the strength of the two regularization terms $\mathcal{L}_{\text{sim}}$ and $\mathcal{L}_{cl}$ by tuning λ.

4 Experiment

In this section, we conduct node classification tasks on eight widely used benchmark datasets to verify the effectiveness of MPGCL. In the following, we introduce the datasets, describe the experimental setup, and then report our experimental results.

4.1 Dataset

We use eight datasets provided by the PyTorch Geometric library, including Cora [23], Citeseer [23], Pubmed [23], Wiki-CS [24], Amazon-computers [25], Amazon-photos [25], Coauthor-CS [25], and Coauthor-Physics [25]. Note that these are benchmark datasets that are widely used in the evaluation of node classification.

- Cora, Citeseer and Pubmed are citation networks where each node represents an article, and the edges connected by two nodes represent the citation relationship. For example, Cora consists of seven papers in machine learning, and the nodes have a total of seven categories.
- Wiki-CS is a reference network constructed based on Wikipedia; the nodes represent articles about computer science, the edges represent links between the articles, and the Nodes are labeled with ten classes, each representing a field branch.

- Amazon-computers and Amazon-photos are two networks of co-purchase relation-ships constructed from Amazon. The nodes are commodities, and if two kinds of commodities are frequently purchased together, an edge exists between the two nodes. Nodes have 10 and 8 classes, respectively.
- Coauthor-CS and Coauthor-Physics are extracted from Microsoft Academic Graph. Nodes represent authors, and edges represent the collaboration between two authors. Each node has a sparse bag-of-words feature based on the paper keywords of the author. The label of an author corresponds to their most active research field. Nodes have 10 and 8 classes, respectively.

We adopt a public split for the Wiki-CS dataset [24]. Since no public split is available for Coauthor-CS, Coauthor-Physics, Amazon-computers and Amazon-photo, we follow the split of most methods, i.e., 10%, 10%, and 80% for the training set, validation set and test set, respectively. We perform 20 training sessions for different splitting methods and report the average performance for each dataset in Table 1. It is known that although the three citation networks mentioned have standard fixed splitting methods, they are not reliable in evaluating GNN methods [25]. Therefore, we report in Table 2 the average results of 20 random splits. Note that the baseline for their comparison also uses random splits.

4.2 Baselines

We consider the following three representative classes of methods as a baseline, includ-ing two traditional unsupervised methods node2vec [5] and DeepWalk [6], two semi-supervised methods GCN [1] and GAT [2] and eight self-supervised learning methods GAE [26], DGI [3], MVGRL [9], GRACE [10], GCA [11], BGRL [18], COSTA [38] and AFGRL [39], where they are trained in an end-to-end fashion.

4.3 Experimental Setup

Evaluation protocol. We follow the linear evaluation protocol introduced in DGI [3] for the node classification task. All nodes in the original graph are first trained without supervision. After that, we freeze the parameters of the encoder and get the embedding of all nodes. We provide the learned embeddings across the training set to the logistic regression classifier [27] and give the results on the test nodes. We repeat the experiments in Tables 1 and Table 2 twenty times and report the average accuracy with the standard deviation.

Setting-up. We used Pytorch-Geometric 2.0.4 and Pytorch 1.11.0 to implement the methods in this paper, experimenting on a server with two NVIDIA GeForce3090 (with 24 GB memory each). Our model was initialized with Glorot [28] using the AdamW [29] optimizer with the initial learning rate and weight decay rate set to 10^{-5}. We used the same encoder architecture in all experiments. The target and online networks use two layers of standard GCN as our encoders, each followed by a BatchNorm layer. We use PreLU activation in all experiments.

Table 1. Node classification accuracy for the Wiki-cs, Amazon and Coauthor datasets, where X, A, Y denote node features, adjacency matrix and labels, respectively. Bold indicates the best performing method. Baseline results are from BGRL [18] and AFGRL [39].

Method	Input	Wiki-cs	Computers	Photo	CS	Physics
Node2vec	A	71.79 ± 0.05	84.39 ± 0.08	89.67 ± 0.12	85.08 ± 0.03	91.19 ± 0.04
DeepWalk	A	74.35 ± 0.06	85.68 ± 0.06	89.44 ± 0.11	84.61 ± 0.22	91.77 ± 0.15
GAE	X, A	70.15 ± 0.01	85.27 ± 0.19	91.62 ± 0.13	90.01 ± 0.71	94.92 ± 0.07
DGI	X, A	75.35 ± 0.14	75.35 ± 0.14	91.61 ± 0.22	92.15 ± 0.63	94.51 ± 0.52
MVGRL	X, A	77.52 ± 0.08	87.52 ± 0.11	91.74 ± 0.07	92.11 ± 0.12	95.33 ± 0.03
GRACE	X, A	78.19 ± 0.01	87.46 ± 0.22	92.15 ± 0.24	92.93 ± 0.01	95.26 ± 0.02
GCA	X, A	78.35 ± 0.05	88.94 ± 0.15	92.53 ± 0.16	93.10 ± 0.01	95.73 ± 0.03
BGRL	X, A	79.36 ± 0.53	89.68 ± 0.31	92.87 ± 0.27	93.21 ± 0.18	95.56 ± 0.12
COSTA	X, A	79.12 ± 0.02	88.32 ± 0.03	92.56 ± 0.45	92.95 ± 0.12	95.74 ± 0.02
AFGRL	X, A	77.62 ± 0.49	89.88 ± 0.33	93.22 ± 0.28	93.27 ± 0.17	95.69 ± 0.10
MPGCL	X, A	**80.42 ± 0.22**	**90.87 ± 0.13**	**93.85 ± 0.15**	**93.69 ± 0.09**	**96.05 ± 0.08**
GCN	X, A, Y	77.19 ± 0.12	86.51 ± 0.54	92.42 ± 0.22	93.03 ± 0.31	95.65 ± 0.16
GAT	X, A, Y	77.65 ± 0.11	86.93 ± 0.29	92.56 ± 0.35	92.31 ± 0.24	95.47 ± 0.15

Table 2. Classification accuracy on the three citation network datasets. Bold indicates the best performing algorithm. Note that we use the random split followed by GRACE [10] and the baseline method of comparison is also random.

Method	Cora	Citeseer	Pubmed
Node2vec	74.8	52.3	80.3
DeepWalk	75.7	50.5	80.5
GAE	76.9	60.6	82.9
DGI	82.6 ± 0.4	68.8 ± 0.7	86.0 ± 0.1
GRACE	83.3 ± 0.4	72.1 ± 0.5	86.7 ± 0.1
BGRL	83.83 ± 1.61	72.32 ± 0.89	86.03 ± 0.33
MPGCL	**85.81 ± 0.56**	**74.51 ± 0.21**	**87.04 ± 0.12**

4.4 Result and Analysis

Tables 1 and 2 summarize the node classification accuracy of all methods on the eight real graph benchmark datasets. Compared with other supervised and self-supervised methods, MPGCL achieves the best classification accuracy on all eight benchmark datasets, demonstrating the robust performance of our proposed method. Note that the existing baseline has achieved high classification accuracy in the Coauthor and Amazon datasets; MPGCL can achieve further significant improvements. The reasons for our success are as follows, (1) We propose a novel data augmentation strategy to construct a higher-order network by injecting high-quality global information into the original view, and then adopt a combination of adaptive and random augmentation strategies for the higher-order network topology and attribute information. (2) Capture potential positive samples from a global perspective instead of simply regarding the same nodes as positive samples, and introduce a small number of negative samples to enrich the self-supervised signals further.

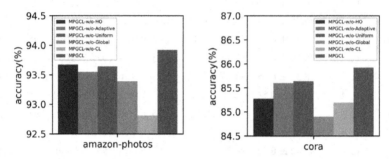

Fig. 3. Comparison of MPGCL and its variants.

4.5 Ablation Study

In this section, to verify the effectiveness of each of our proposed components, we perform ablation experiments on the Amazon-photos and Cora datasets. The experimental results are shown in Fig. 3, for the variant MPGCL-w/o-HD, we investigate the effectiveness of constructing a higher-order network. It can be seen that by constructing the higher-order network for the Cora dataset improves 0.7%, which proves that it is meaningful to introduce high-quality global information to the original view. For the variants MPGCL-w/o-Uniform and MPGCL-w/o-Adaptive, we adopt a single data augmentation for the higher-order network. From Fig. 3, we can find that the performance of a single augmentation strategy for the original view only decreases by 0.3% and 0.4%, respectively, in the Cora dataset. This proves the significance of our multi-perspective augmentation.

To verify the effectiveness of searching positive samples from a global perspective and introducing negative samples reasonably in cross-view comparisons, we report the variants MPGCL-w/o-Global and MPGCL-w/o-CL. As can be seen from Fig. 3, on Amazon-photos, compared with MPGCL, MPGCL-w/o-Global and MPGCL-w/o-CL decreases the performance by 0.5% and 1.1%, respectively. The effectiveness of capturing positive samples from the global and introducing a small number of negative samples is verified.

4.6 Analysis of Hyper-parameters

In this section, we perform a parameter-sensitive analysis of MPGCL, as shown in Fig. 4, where we demonstrate the stability of our model by modifying the probability of

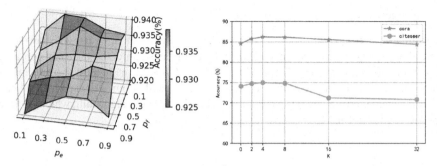

Fig. 4. Classification results of our method on different dataset with different hyperparameters.

Table 3. Hyperparameter settings.

Method	Wiki-cs	Computers	Photo	CS	Physics
$p_{f,1}$	0.1	0.1	0.1	0.4	0.1
$p_{f,2}$	0.1	0.1	0.1	0.3	0.1
$p_{e,1}$	0.2	0.3	0.5	0.3	0.4
$p_{e,2}$	0.4	0.3	0.5	0.2	0.4
α	0.05	0.05	0.05	0.05	0.05
λ	0.7	0.7	0.7	0.8	0.8
τ	0.7	0.7	0.7	0.7	0.7
K	3	4	3	4	4
embedding sizes	256	128	256	256	128
hidden sizes	512	256	512	512	256

removing edges and the probability of masking. For Amazon-photos, we propose that the augmentation strategy performs better in most cases. Existing approaches [10, 11] rely on carefully designed hyperparameters to guarantee the generation of two different views to obtain the best performance. Note that we propose multiple angle augmentation to obtain better performance even when both have the same probability. We also explore the effect of the number of positive samples on the performance of MPGCL. As shown in Fig. 4, the best performance is obtained when k = 4 (i.e., each node has four as positive samples in addition to itself). Note that the performance of the model also does not fluctuate significantly by modifying the k value, which confirms that our method is insensitive to hyperparameters compared to other methods while outperforming them in most cases. We also notice a slight decrease in performance when the value of k is too large, due to the fact that node embeddings tend to be similar when the value of k is too large. In addition, the hyperparameter configurations for all datasets are summarized in Table 3.

4.7 Visualization

To further intuit the performance of our model, as shown in Fig. 5, we use a visualization algorithm to project the node embedding of the Cora dataset into 2D space and compare it with the node representation obtained using the GRACE methods. Where different colors of nodes indicate different categories, we can see that our proposed method can present a clearer outline in 2D space compared to the GRACE algorithm, which is more conducive to handling downstream tasks.

RAW   GRACE   MPGCL

Fig. 5. Visualization of the learned node embedding on the Cora dataset.

5 Conclusion

In this paper, we propose an efficient contrast-based unsupervised learning framework, MPGCL, which first introduces global information to the original view to construct a higher-order network, then performs data augmentation on the higher-order network to generate two views with complementary semantic information. We search for potentially positive samples from the global representation. Experiments are conducted on eight publicly available datasets, MPGCL consistently outperforms other SOTA methods, demonstrating the effectiveness of our approach.

References

1. Kipf, T.N., Welling, M.: Semi-supervised classification with graph convolutional net-works. In: ICLR, pp. 1–14 (2017)
2. Veličković, P., Cucurull, G., Casanova, A., Romero, A., Lio, P., Bengio, Y.: Graph at-tention networks. In: ICLR, pp. 1–12 (2018)
3. Veličković, P., Fedus, W., Hamilton, W.L., Liò, P., Bengio, Y., Hjelm, R.D.: Deep graph infomax. In: ICLR 2(3), 4 (2019)
4. Linsker, R.: Self-organization in a perceptual network. Computer 21(3), 105–117 (1988)
5. Grover, A., Leskovec, J.: node2vec: scalable feature learning for networks. In: SIGKDD, pp. 855- 864 (2016)
6. Perozzi, B., Al-Rfou, R., Skiena, S.: Deepwalk: online learning of social representations. In: Proceedings of the 20th ACM SIGKDD International Conference on Knowledge Discovery and Data Mining, pp. 701–710. KDD 2014 (2014)
7. Church, K. W.: Word2Vec.Natural Language Engineering, vol. 23, no. 1, pp. 155–162, Jan (2017)

8. Hjelm, R. D., et al.: Learning deep representations by mutual information estimation and maximization. In: ICLR (2019)
9. Hassani, K., Khasahmadi, A.H.: Contrastive multi-view representation learning on graphs. In: ICML, pp. 4116–4126 (2020)
10. Zhu, Y., Xu, Y., Yu, F., Liu, Q., Wu, S., Wang, L.: Deep graph contrastive representation learning. arXiv preprint arXiv:2006.04131 (2020)
11. Zhu, Y., Xu, Y., Yu, F., Liu, Q., Wu, S., Wang, L.: Graph contrastive learning with adaptive augmentation. In: WWW, pp. 2069–2080 (2021)
12. Chen, T., Kornblith, S., Norouzi, M., and Hinton, G.: A simple framework for contrastive learning of visual representations. In: ICML, pp. 1597–1607 (2020)
13. Grill, J.-B., et al.: Bootstrap your own latent-a new approach to self-supervised learning. NeurIPS **33**, 21271–21284 (2020)
14. Rong, Y., et al.: Self-supervised graph transformer on large-scale molecular data. Adv. NeurIPS **33**, 12559–12571 (2020)
15. Rong, Y., Huang, W., Xu, T., Huang, J.: Dropedge: towards deep graph convolutional networks on node classification. In ICLR, pp. 1–17 (2020)
16. Sun, F.-Y., Hoffmann, J., Verma, V., Tang, J.: Infograph: Unsupervised and semi-supervised graph-level representation learning via mutual information maximization. In ICLR (2020)
17. Peng, Z., et al.: Graph representation learning via graphical mutual information maximization. In: WWW, pp. 259–270 (2020)
18. Thakoor, S., Tallec, C., Azar, M.G., Munos, R., Veličković, P., Valko, M.: Boot-strapped representation learning on graphs. In: ICLR Workshop (2021)
19. Kefato, Z.T., Girdzijauskas, S.: SelfGNN: self-supervised graph neural networks without explicit negative sampling. In WWW Work-shop (2021)
20. Chen, T., Kornblith, S., Norouzi, M., Hinton, G.: A simple framework for contrastive learning of visual representations. In: ICML, 1597–1607 (2020)
21. Chen, X., He, K.: Exploring simple Siamese representation learning. In: CVPR, pp. 15750–15758 (2021)
22. Bromley, J., Guyon, I., LeCun, Y., Säckinger, E., Shah, R.: Signature verification using a "Siamese" time delay neural network. In: NeurIPS (1993)
23. Sen, P., Namata, G., Bilgic, M., Getoor, L., Galligher, B., EliassiRad, T.: Collective classification in network data. AI Mag. **29**, 93 (2008)
24. Mernyei, P., Cangea, C.: Wiki-cs: a wikipedia-based benchmark for graph neural net-works. arXiv preprint arXiv:2007.02901 (2020)
25. Shchur, O., Mumme, M., Bojchevski, A., Günnemann, S.: Pitfalls of graph neural net-work evaluation. In: NeurIPS (2018)
26. Kipf, T.N., Welling, M.: Variational graph auto-encoders. arXiv preprint arXiv:1611.07308 (2016)
27. Pedregosa, F., et al.: Scikit-learn: machine learning in Python. J. Mach. Learn. Res. **12**, 2825–2830 (2011)
28. Glorot, X., Bengio, Y.: Understanding the difficulty of training deep feedforward neural networks. In: AISTATS, pp. 249–256 (2010)
29. Kingma, D.P., Ba, J.: Adam: a method for stochastic optimization. In: ICML (2015)
30. Oord, A. van den, Li, Y., Vinyals, O.: Representation learning with contrastive predictive coding. arXiv preprint arXiv:1807.03748 (2018)
31. Bachman, P., Hjelm, R.D., Buchwalter, W.: Learning representations by maximizing mutual information across views. In: NeurIPS (2019)
32. Fey, M., Lenssen, J.E.: Fast graph representation learning with PyTorch Geometric. arXiv preprint arXiv:1903.02428 (2019)

33. Tian, Y., Krishnan, D., Isola, P.: Contrastive Multiview Coding. In: Vedaldi, A., Bischof, H., Brox, T., Frahm, J.-M. (eds.) ECCV 2020. LNCS, vol. 12356, pp. 776–794. Springer, Cham (2020). https://doi.org/10.1007/978-3-030-58621-8_45
34. Collobert, R., Weston, J.: A unified architecture for natural language processing: Deep neural networks with multitask learning. In: ICML, pp. 160–167 (2008)
35. Mnih, A., Kavukcuoglu, K.: Learning word embeddings efficiently with noise-contrastive estimation. In: NeurIPS, p. 26 (2013)
36. Hamilton, W.L., Ying, R., Leskovec, J.: Representation learning on graphs: methods and applications. arXiv preprint arXiv:1709.05584 (2017)
37. Wu, M., Zhuang, C., Mosse, M., Yamins, D., Goodman, N.: On mutual information in contrastive learning for visual representations. arXiv preprint arXiv:2005.13149 (2020)
38. Lee, N., Lee, J., Park, C.: Augmentation-free self-supervised learning on graphs. Proc. AAAI Conf. Artif. Intell. **36**, 7372–7380 (2022)
39. Zhang, Y., Zhu, H., Song, Z., Koniusz, P., King, I.: COSTA: covariance-preserving feature augmentation for graph contrastive learning. In: SIGKDD, pp. 2524–2534 (2022)

Retrieval

A Joint Link-Retrieve Framework for Open Table-and-Text Question Answering

Jian Zou[1], Jiaan Wang[1], Ying He[3], Jianfeng Qu[1(✉)], Zhixu Li[2],
Pengpeng Zhao[1], An Liu[1], and Lei Zhao[1]

[1] School of Computer Science and Technology, Soochow University, Suzhou, China
`{jzou2,jawang1}@stu.suda.edu.cn`
`{jfqu,ppzhao,anliu,zhaol}@suda.edu.cn`
[2] School of Computer Science, Fudan University, Shanghai, China
`zhixuli@fudan.edu.cn`
[3] IFLYTEK Research, Suzhou, China
`yinghe@iflytek.com`

Abstract. Open Table-and-Text Question Answering (OTTQA) task aims at answering questions that require retrieving and combining information from unstructured passages and semi-structured tables. Despite the great success achieved by the existing models, they still flaw in several drawbacks, including lacking valuable global or contextual information at the linking stage, ignoring potential error propagation and noise distraction from linking to retrieval, and neglecting the huge structural gap between table and passage. To address these problems, in this paper we propose a novel joint link-retrieve OTTQA framework, where the table block embedding and passage embedding are shared in the unified framework, and all modules are trained jointly. More specifically, we encode the table block and passage by taking full advantage of their global and contextual information and then leverage a contrastive learning approach to deeply interact with their embedding and link them together. Meanwhile, we establish a novel retrieval method to retrieve fusion blocks after linking, which does not require explicitly generating or encoding the fusion block but only focuses on the passage most relevant to the question. Furthermore, considering the structural gap between the table and text, we introduce the Table-to-Text as an auxiliary task to help the model better understand the structural difference and capture the inner semantic correlations between them. Our empirical study demonstrates that the proposed framework achieves new state-of-the-art performance.

Keywords: Open Table-and-Text Question Answering · Joint Link-Retrieve Framework · Table-to-Text

© The Author(s), under exclusive license to Springer Nature Switzerland AG 2023
X. Wang et al. (Eds.): DASFAA 2023, LNCS 13945, pp. 369–384, 2023.
https://doi.org/10.1007/978-3-031-30675-4_26

1   Introduction

Open Table-and-Text Question Answering(OTTQA) is a question-answering task, which requires retrieving and combining information from unstructured passages and semi-structured tables to answer a specific question. Compared with traditional Open-Domain Question Answering(ODQA), which only needs to retrieve unstructured text [12,16,26], OTTQA is more challenging and in line with the real world where a large amount of world's knowledge is stored not only in unstructured passages but also in semi-structured tables. Chen et al. [5] construct a large dataset OTT-QA for this new task. In particular, questions in OTT-QA are multi-hop and require aggregating information from both tables and texts to answer. As illustrated in Fig. 1, to answer the question *"How many points per game did Lebron James get in the NBA Season suspended by COVID?"*, an OTTQA system first needs to retrieve that the NBA Season suspended by COVID was the 19-20 Season and James got 25.3 points per game in the 19-20 Season from the passage and table corpora respectively; and then aggregates this information to get the answer *25.3*.

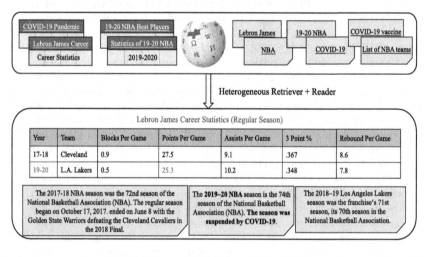

Fig. 1. An example of OTT-QA, an OTT-QA model needs to retrieve evidence from tables and passages and then perform multi-hop reasoning to find answers

To retrieve tables and passages at the same time, Chen et al. [5] propose an early fusion mechanism, which links table blocks and passages as fusion blocks first, and then retrieves these obtained fusion blocks. In particular, they input a row of the table as the table block into GPT-2 [23], a generative language model, to generate titles of passages and calculate the BM25 [8] score of the generated titles with the ground-truth titles to link table block and passages. After linking, they adopt the DPR [12] as the retriever which needs to concatenate the table block and linked passages to explicitly generate the fusion block and encode the fusion block and question to retrieve question-related fusion blocks as evidence. Similarly, Zhong et al. [31] use BLINK [30] which is an entity linking model and can link against all Wikipedia entities as the linker and DPR as the retriever. They take the cell of the table block as the input of BLINK to link relevant passages for each cell to form the fusion block.

Despite the above Linker-Retriever-Reader methods have been proven to be more efficient than the Retriever-Reader models that retrieve tables and passages separately [5] [31], there still exist the following shortages: **Firstly**, both Chen et al. [5] and Zhong et al. [31] only focus on local information and ignore global information. Between them, Chen et al. [5] link the table block and passage only through the title of the passage, ignoring the entire content information of the passage. Zhong et al. [31] only input the cell instead of the entire table block into the BLINK, which results in the linker lacking sufficient contextual information. **Secondly**, the DPR retriever requires to generate and encode the fusion block explicitly. Under this mechanism, passages in the fusion block are treated equally, even though there are incorrectly linked passages and passages irrelevant to the question in the fusion block. Error propagation and noise interference brought by this mechanism will cause a serious impact on retrieval, degrading the performance of the entire model. **Thirdly**, the table block is encoded by transformer-based [27] encoder like BERT [7] and RoBERTa [18]. However, this type of encoder is suitable for the input of unstructured plain texts rather than semi-structured tables. And there is a significant structural gap between the table and the passage that prevents the encoder from understanding the table well.

To address these above limitations, we propose an improved Joint Link-Retrieve framework(JLRF). Specifically, **for the first challenge**, we utilize a linker based on contrastive learning. We employ dual encoders to separately encode table blocks and passages. Based on that, we then calculate the score of the table block and passage to optimize their representations, so that the linked table block and passage have a higher score to link each other. Our linker takes full advantage of the contextual information of table block and passage and links them together effectively. **For the second challenge**, we propose a novel retrieval method to compute the similarity of fusion blocks after linking, which does not require explicitly generating and encoding fusion blocks. We encode the question and compute similarity for question embedding with table block embedding and passage embedding in the linking step. The similarity of the fusion block is obtained by considering the score of the table block with the highest score of passage in the fusion block. In this way, we share the table block embedding and passage embedding to alleviate the error propagation, and only focus on the passage that is most relevant to the question to reduce noise interference. **For the third one**, we introduce the Table-to-Text task, reformulating the table block fluently in natural language as an auxiliary task. The task can promote the table block encoder to understand the content together with the structure of the table block, to narrow the structural gap between the table and the passage.

As shown in Fig. 2, our framework consists of Linker, Retriever, and Table-to-Text modules. Since embedding is shared in these modules and these modules are trained simultaneously, the framework is called Joint Link-Retrieve Framework(JLRF). Finally, We feed the retrieval results of JLRF to the reader to answer questions. Our contributions can be summarized as follows:

- We propose a novel Joint Link-Retrieve Framework(JLRF) for Open Table-and-Text Question Answering where the table block embedding and passage embedding are shared in the framework and all modules in the framework are trained simultaneously.
- We leverage a contrastive learning-based linker that can take full advantage of the global information of the table block and passage, and propose a crafted retriever that can mitigate error propagation and reduce noise interference without explicitly generating or encoding fused blocks.
- We introduce an auxiliary table-to-text task to alleviate the structural gap between table and text, helping the encoder better understand the table.
- Experiments show that JLRF can significantly improve retrieval performance. Moreover, we achieve new state-of-the-art results in linking, retrieval, and question-answering evaluations.

2 Related Work

2.1 Open Domain Question Answering

Given questions, open domain question answering(ODQA) retrieves evidence candidates from external resources and provides their answers. Previous ODQA works typically follow the two-stage retriever-reader pipeline where a retriever first gathers relevant passages as evidence candidates, then a reader extracts answers from the retrieved candidates [4]. To retrieve evidence, traditional retrieval methods (e.g., TF-IDF [11] and BM25 [8]) generally retrieve passages based on the lexical overlap between the given question and candidate passages. Though great success has been achieved, these methods can not capture the semantics similarity, resulting in sub-optimal solutions. Further, representation-based retrieval models which can capture semantic information have been developed in the past few years [32]. Among them, DPR [12] employs two independent BERT encoders to encode the given question and passages, respectively and then calculates their relevance based on the similarity of these two encoded representations. Nogueira et al. [20] input both the question and the candidate passage into one single BERT to score their relevance. These two model structures are called two-tower and one-tower architectures, respectively. The former is faster to retrieve relevant passages due to the representation of questions and passages are calculated separately, thus the representations of all candidate passages could be pre-processed and one can utilize efficient similarity search toolkits (e.g., FAISS [10]) when deploying the retrieval model. The Latter has better performance because of its strong interactivity, but limited retrieval speed, and it hardly works for large-scale candidate documents/passages. In this paper, considering the efficiency and sharing representations, we focus on the two-tower structure.

After retrieving evidence, the reader is further adapted to extract answers. In general, existing readers can be classified into the following two types: (1) extractive readers that predict an answer span from the evidence candidates, and (2) generative readers directly generate answers in the text-generation paradigm. For example, Extended Transformer Construction (ETC) [1] is an extractive reader that supports long-sequence texts (up to 4,096 tokens) as input through a carefully designed global-local sparse attention mechanism. Fusion-in-Decoder (FiD) [9] is a T5-based [24] generative reader whose encoder embeds the given question and the retrieved passages separately. Then the decoder is aware of the attention over the concatenation of the embedding and generates the corresponding answer. Grounding the truth that generative readers are more flexible than extractive ones (e.g., the generative readers can provide answers not limited to the text span from retrieved candidates), we equip our retrieval model with generative readers.

2.2   Open Table and Text Question Answering

Semi-structured tables are essential knowledge sources storing a significant amount of real-world knowledge. Open table and text question answering (OTTQA) extends ODQA to a more realistic scene where not only the textual passages but also semi-structured tables are provided for answering questions. Chen et al. [5] pioneer this task and construct the OTT-QA dataset for this task. Specifically, they collect OTT-QA samples based on the HybridQA [6] dataset, a close-domain table-and-text question-answering dataset where each question is aligned with a Wikipedia table and multiple entity-centric passages. To answer questions in HybridQA, models should reason between tabular and textual information. To adapt HybridQA to the open domain, Chen et al. [5] first remove hyperlinks between tables and passages, and then decontextualize the original context-dependent questions via crowd-sourcing. As a result, the proposed OTT-QA contains about 45.8K context-independent questions.

To build OTTQA systems, Chen et al. [5] propose an early fusion mechanism to improve the retrieval performance, which groups each table segment with its relevant passages to a fusion block before retrieval. The fusion blocks could serve as the basic retrieval units to facilitate the retrieval of relevant evidence across different modalities. Then, a DPR model is adopted to retrieve relevant fusion blocks which are further fed into a reader to provide answers. Similarly, Zhong et al. [31] utilize the BLINK, an effective entity linking model, to link each cell of the table block with passages. Besides, they propose a chain-centric reasoning and pre-training framework (CARP), which employs a hybrid chain to model the explicit intermediate reasoning process across tables and passages for question answering and leverages a chain-centric pre-training strategy. Finally, the retrieved evidence and the reasoning chain are used to get the final answers.

3　Methodology

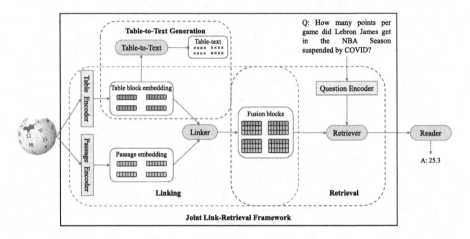

Fig. 2. An overview of JLRF containing three modules: Linker, Retriever, and Table-to-Text generator. All three modules are trained end-to-end.

The structure of our proposed Joint Link-Retrieve Framework (JLRF) is illustrated in Fig. 2, which contains three modules, i.e., linker, retriever, and table-to-text generator. Specifically, the linker (Sect. 3.1) that can take full advantage of the global information of table blocks and passages is utilized to link each table block with its relevant passages to form a fusion block. The retriever (Sect. 3.2) uses fusion blocks as the basic retrieval units and retrieves the evidence candidates relevant to the given question without explicit generation and encoding of fusion blocks. To narrow the semantic gap between tables and passages, a table-to-text generator (Sect. 3.3) is adopted as an auxiliary module to generate textual descriptions for table blocks. The table block embedding and passage embedding are shared among these modules and these modules are trained at the same time. Finally, the given question as well as the retrieved evidence candidates are fed into a generative reader (Sect. 3.4) to reason and generate the corresponding answer.

3.1　Linking Table Blocks with Passages

To incorporate the global information of the table blocks and passages, we employ a contrastive learning-based linker. Specifically, we first input a table block t and a passage p separately into dual BERT encoders to obtain their representations $\mathbf{E}_t$ and $\mathbf{E}_p$, which could be formally defined as:

$$t = [\text{CLS}] \langle title \rangle [\text{T}] \langle section\_title \rangle [\text{ST}] \langle header_i \rangle [\text{H}] \langle cell_i ||_{i=1}^n \rangle [\text{C}] \tag{1}$$

$$p = [\text{CLS}] \langle title \rangle [\text{T}] \langle passage \rangle \tag{2}$$

$$\mathbf{E}_t, \mathbf{E}_p = \text{BERT}_1(t), \text{BERT}_2(p) \tag{3}$$

where n is the number of cells in the table block, $\langle title \rangle$, $\langle section\_title \rangle$, $\langle header \rangle$, $\langle cell \rangle$ and $\langle passage \rangle$ are contents of the table block (i.e., title, section title, header, and cells) and passage. [CLS], [T], [ST], [H] and [C] are special tokens to indicate the different information sources. In this way, we can not only fully utilize the global information of table blocks and passages but also easily share the representation of table blocks and passages in subsequent modules.

Then, we adopt the representation of the first token (i.e., [CLS]) as the overall representations of the input table block t and passage p, respectively. The similarity between t and p is calculated as the dot product of their overall representations:

$$sim(t,p) = \mathbf{E}_{t,0}^{\top} \cdot \mathbf{E}_{p,0} \tag{4}$$

where $\mathbf{E}_{t,0}$ and $\mathbf{E}_{p,0}$ denote the representation of the first token in t and p, respectively.

After calculating the similarity between table blocks and passages, we use the Multi-Similarity-Loss [29] to optimize the linker, which is a contrastive learning-based loss function:

$$\mathcal{L}_l = \frac{1}{m}\sum_{i=1}^{m}\left\{\frac{1}{\alpha}\log\left[1+\sum_{k\in P_i}e^{-\alpha(s_{jk}-\lambda)}\right]+\frac{1}{\beta}\log\left[1+\sum_{k\in N_i}e^{\beta(s_{jk}-\lambda)}\right]\right\} \tag{5}$$

where m is the number of table blocks in each batch, s_{jk} is the similarity between table block b_{tj} and passage b_{pk}, P_i and N_i are the positive and negative passage sets of batch i respectively, α, β are the scaling factor hyper-parameters, λ is a pre-defined penalty coefficient, which punishes positive samples if their scores below λ and negative samples if their scores above λ.

During linking process, for each table block t, we group t with its relevant passages to a fusion block $\mathcal{F}_t = \{t, p_{t,1}, p_{t,2} \ldots, p_{t,|\mathcal{F}_t|-1}\}$, where the similarity between each $p_{t,i}$ and t exceeds λ.

3.2 Retrieving Evidence Candidates

To alleviate error propagation and reduce noise interference, our retriever shares the representations of table blocks and passages with the linker. In this way, the linker and retriever can interact with each other in an end-to-end manner to mitigate the error propagation issue.

In detail, we encode the question q to obtain its representation $\mathbf{E}_q$ via a BERT encoder, and then calculate the similarity between q and each fusion block $\mathcal{F} = \{t, p_{t,1}, p_{t,2} \ldots, p_{t,|\mathcal{F}_t|-1}\}$ as:

$$\mathbf{E}_q = \mathrm{BERT}_1(q) \tag{6}$$

$$sim'(q,\mathcal{F}) = \mathbf{E}_{q,0}^{\top} \cdot \mathbf{E}_{t,0}^{\top} + \max\{0, \mathbf{E}_{q,0}^{\top} \cdot \mathbf{E}_{p_t,1,0}^{\top}, \mathbf{E}_{q,0}^{\top} \cdot \mathbf{E}_{p_t,2,0}^{\top}, \ldots, \mathbf{E}_{q,0}^{\top} \cdot \mathbf{E}_{p_t,|\mathcal{F}_t|-1,0}^{\top}\} \tag{7}$$

Therefore, the similarity between q and $\mathcal{F}$ is calculated based on the similarity between q and t as well as the similarity between q and its most relevant passage in $\mathcal{F}$. Since most of the questions in OTTQA are two-hop questions, the way we compute the similarity of fusion blocks enables the retriever to focus only on the table block and passage most relevant to the question, thereby reducing the noise of irrelevant passages. This approach is simple yet effective since it does not require explicitly generating and encoding each fusion block like Chen et al. [5].

Next, the InfoNCE loss [21] is adopted as supervised signals to optimize the retriever:

$$\mathcal{L}_r = -\frac{1}{m}\sum_{i=1}^{m}\frac{e^{s'_i}}{\sum_{j=1}^{n}e^{s'_{ij}}}, \tag{8}$$

where m and n are the number of questions and fusion blocks in each batch, s'_{ij} denotes the similarity between question q_i and fusion block $\mathcal{F}_j$. s'_i indicates the similarity between question q_i and the ground-truth fusion block.

During the inference process, we retrieve top-k fusion blocks with the highest similarity with the given question, which are further fed into a reader (Sect. 3.4) as evidence candidates to reason and provide its answer.

3.3   Table-to-Text Generation

To fully understand the structure and content of table blocks, we utilize table-to-text generation as an auxiliary task to reformulate each table block into its textual description. In this manner, the table-to-text generator could help the table block encoder (i.e., BERT$_1$) better comprehend and encode table blocks, thus bridging the structural gap between table and text.

Textual Description Collection. ToTTo [22] is a widely-used table-to-text dataset based on Wikipedia tables that are the same source and form as the table in OTTQA, whose goal is to generate the final text given the table and table metadata(such as the title), and set of highlighted cells. Since there are no ground-truth textual descriptions for the table blocks in OTT-QA, similar to An et al. [2] who use the T5 model based on prompt-tuning strategies [13,15,17] to generate table-text, we train a T5 model on the ToTTo to obtain a table-to-text annotator. Then, the annotator is used to generate/label textual descriptions for the table blocks of OTT-QA, serving as the pseudo labels to supervise our table-to-text generator.

Table-to-Text Generator. We pair the table block encoder (i.e., BERT$_1$) with a transformer decoder (which is also initialized by the weights of BERT in a BERT-to-BERT structure [25]) to form our table-to-text generator. We calculate cross-entropy loss(i.e., $\text{CE}(\cdot,\cdot)$) between the generated textual description y and the pseudo label $\hat{y}$:

$$\mathcal{L}_t = \text{CE}(y, \hat{y}) \tag{9}$$

with the help of the table-to-text auxiliary task, the table block encoder is able to better understand the table block content to reduce the structural gap between the table and the passage. Since the table-to-text module is only used in the training process, during inference, the decoder is dropped.

The summation loss of the above three modules (i.e., linker, retriever, and graph-to-text generator) is used to jointly train JLRF in an end-to-end manner:

$$\mathcal{L} = \mathcal{L}_l + \mathcal{L}_r + \mathcal{L}_t, \tag{10}$$

3.4 Reasoning and Answering

For the given question, our JLRF retrieves top-k relevant fusion blocks. We not only sort the retrieved fusion blocks but also sort and filter the passages inside the fusion blocks(i.e., passages with a similarity to the question below the threshold μ will be discarded.). Lastly, a reader is leveraged to reason the information across the retrieved fusion blocks and provide the final answer. Here, we directly utilize FiD [9] as our reader. In detail, FiD adopts the backbone of transformer encoder-decoder architecture [28] and concatenates the given question and fusion blocks as input. Then, the decoder performs attention over the whole input sequences to aggregate and reason information across the fusion blocks and generate the corresponding answer. To train the FiD reader, cross-entropy loss between the generated answer a and the ground truth answer $\hat{a}$ is calculated:

$$\mathcal{L}_{FiD} = \text{CE}(a, \hat{a}) \tag{11}$$

4 Experiments

4.1 Experiment Setup

We set α, β, and λ in our linker to 1, 1, and 0, respectively, and set the threshold for filtering passages μ to 0. There are 8 questions, 16 table blocks, and 50 passages in each batch when training the JLRF. During training the reader, we use top-15 ($k = 15$) retrieved fusion blocks as evidence for each question and there are 16 questions in each batch. We use the AdamW optimizer [19], linear warmup of 1,000 steps, and set the learning rate to 3e-5 to train both JLRF and FiD reader.

In order to train our framework quickly and smoothly, we first use the linker-predicted labels directly and then use the ground-truth labels to generate fusion blocks. During the inference process, we use FAISS [10] to store the pre-processed embeddings and build the index for fast search. The BERT encoder used in JLRF is implemented based on the bert-base model with default settings (768 hidden size, 12 multi-head attention, and 12 hidden layers).

4.2   Dataset and Evaluation

We evaluate the performance of our approach on the OTT-QA dataset [5]. OTT-QA is a large-scale table-and-text open-domain question-answering benchmark that aims to evaluate open-domain question answering over both tabular and textual knowledge. Table 1 lists the statistics of OTT-QA, as we can see, OTT-QA contains over 40 K questions and it also provides a corpus collected from Wikipedia with over 400 K tables and 6M passages.

Table 1. Data statistics of OTT-QA dataset

Type	Numbers
Questions in the training set	41,469
Questions in the development set	2,214
Questions in the test set	2,158
Total tables	410,740
Total passages	6,342,314

Following Chen et al. [5], we adopt exact match (EM) and F1 scores to evaluate QA model performance. Precision, recall, and F1 are used to evaluate the linking performance. In addition, we follow the more fine-grained and challenging metric of Zhong et al. [31] to evaluate the retrieval performance, fused block recall at top-k ranks (R@k) is adopted, where a fused block is considered as a correct match when it comes from the ground truth table and contains the correct answer.

4.3   Baseline Methods

We compare our model with several typical baselines and the state-of-the-art baselines as follows:

- **HYBRIDER** is originally designed for the closed-domain HybridQA [6] which is a two-stage model dealing with the heterogeneous information across table and text, respectively. Since this model requires a ground truth table with its hyperlinks to do modularized reasoning, Chen et al. [5] use BM25 to retrieve the most relevant table and passages to reconstruct an approximated input for this model.
- **GPT+DPR+ETC** is proposed by Chen et al. [5] with the early fusion mechanism. This method uses GPT-2 as the linker, DPR as the retriever, and ETC as the reader. The ETC reader receives the 4096 token space with the top-k retrieval results and generates answers for the given questions.
- **CAPR** [31] is chain-centric reasoning and pre-training framework, which utilizes a hybrid chain to model the explicit intermediate reasoning process across table and text for question answering. This method uses BLINK as the linker, DPR as the retriever, and CAPR as the reader.

- **DUREPA** [14] is a dual reader-parser framework that takes both textual and tabular data as input, and generates either direct answers or SQL statements based on the context. In detail, if the answer is listed in the passages, DUREPA generates the answer directly. Otherwise, it generates a SQL statement and executes the statement to extract the answer from the table. The model is equipped with BM25 to retrieve tables and passages separately.

Table 2. Overall linking and retrieving performances. The **bold** denotes the best performance. † indicates the results are re-implemented by us. Hit@4K is used to measure the retrieval recall on 4096 tokens.

Methods	Linking			Retrieval			
	Precision	Recall	F1	R@1	R@10	R@100	Hit@4K
GPT + DPR	50.7	50.1	50.4	–	–	–	52.4
BLINK + DPR	**68.7**†	52.6†	59.6†	16.3	46.7	75.5	–
JLRF(ours)	60.6	**79.8**	**68.9**	**30.2**	**61.7**	**86.3**	**68.2**

Table 3. Question Answering results on the dev set and blind test set. The **bold** denotes the best performance.

Models	Dev		Test	
	EM	F1	EM	F1
BM25+HYBRIDER	10.3	13.0	9.7	12.8
DUREPA	15.8	–	–	–
GPT+DPR+ETC	28.1	32.5	27.2	31.5
CAPR	33.2	38.6	32.5	38.5
JLRF + FiD(ours)	**36.3**	**43.0**	**35.4**	**40.9**

4.4 Main Results

Table 2 shows the experimental results of linking and retrieving. As the only reported retrieval performance of Chen et al. [5] is Hit@4K which means the retrieval recall for table blocks on the retrieved 4096 tokens, we also report the result of our JLRF with Hit@4K for fair. Our model significantly outperforms all the baselines in most of the metrics. Specifically, since our linker focuses on the global information of table blocks and passages, and can better understand the table with the help of table-to-text auxiliary tasks, we achieve an improvement of 18.5% / 9.3% over GPT+DPR / BLINK+DPR in terms of F1 score in linking evaluation. For retrieval evaluation, along with the mitigation of error propagation and the reduction of noise interference, our model outperforms GPT+DPR by 15.8% in terms of Hit@4K and outperforms BLINK+DPR by 15.9%, 15.0%, and 10.8% in terms of R@1, R@10, and R@100, respectively.

Table 4. Ablation experiments of JLRF on the OTT-QA dataset. The **bold** denotes the best performance.

Methods	Linking			Retrieval		
	Precision	Recall	F1	R@1	R@10	R@100
JLRF	**60.6**	**79.8**	**68.9**	**30.2**	**61.7**	**86.3**
w/o table-to-text	56.6	69.3	62.2	25.5	54.8	79.2
w/o joint training	51.5	63.6	56.9	22.1	50.8	75.8
w/o embedding sharing	51.5	63.6	56.9	12.5	42.6	68.2

Table 5. Experiments on the effect of retrieval results on reading. We input 4096 tokens and top-1 top-5 fusion blocks of JLRF into Longformer respectively to compare with the GPT+DPR+ETC baseline on the dev set. The **bold** denotes the best performance.

Models	EM	F1
GPT+DPR+ETC		
with 4096 tokens	28.1	32.5
JLRF+Longformer		
with top-1 fusion block	22.8	26.2
with top-2 fusion blocks	24.9	26.5
with top-3 fusion blocks	26.3	27.9
with top-5 fusion blocks	28.7	32.6
with 4096 tokens	**29.4**	**34.1**

Table 3 shows the results of question answering evaluation, with significant improvement in the retrieval performance, our reader also achieves the new state-of-the-art (SOTA) performance in terms of both EM and F1 scores. In detail, compared with CAPR, the previous SOTA method, we achieve an improvement of 3.1% / 4.4% in terms of EM / F1 metric on the dev set and an improvement of 2.9% / 2.4% in terms of EM / F1 metric on the blind test set.

4.5   Ablation Study

Effectiveness of Joint Framework. In order to explore the effectiveness of each module in our framework, We sequentially add the following settings for ablation experiments:

- **JLRF (w/o table-to-text)** removes the Table-to-Text module.

- **JLRF (w/o joint training)** trains the linker and retriever separately.

- **JLRF (w/o embedding sharing)** breaks the joint framework and adopts DPR as the retriever which requires the explicit generation and encoding of fusion blocks.

The effects of these ablations are shown in Table 4. In each case, the evaluation scores are lower than our vanilla JLRF, justifying the rationality of our framework. Specifically, compared with vanilla JLRF, JLRF (w/o table-to-text) drops 6.7% in terms of F1 score on linking performance and 4.7% in terms of R@1 score on retrieving performance. On this basis, JLRF (w/o joint training) drops another 5.3% and 3.4% respectively. In particular, JLRF (w/o embedding sharing) dramatically drops another 9.6% on the R@1 score of retrieval. From the above results, it can be concluded that our joint framework can significantly improve the performances of linking and retrieval, the Table-to-Text module can narrow the gap between tables and passages, and joint training can mitigate error propagation in lining and retrieval.

Effect of Retrieval on Reading. Since the FiD reader we adopted is different from those readers in baseline methods, in order to explore the effect of retrieval on reading, we also replace the FiD reader with the longformer [3] reader (which is the same route as the ETC reader). We attempt the following strategies to input the longformer reader with the retrieved fusion blocks: (1) the concatenation of the most relevant fusion blocks until up to 4,096 tokens (denoted as "longformer with 4,096 tokens"), and (2) top-k ($k \in \{1, 2, 3, 4, 5\}$) relevant fusion blocks (denoted as "longformer with top-k fusion blocks"). We compare the above-modified models with GPT+DPR+ETC baseline [5] on the development set. As shown in Table 5, the retrieval performance of the modified JLRF is better than GPT+DPR, and the Longformer (with 4096 tokens) outperforms the ETC model. In addition, as the increase of input length, the performance of Longformer also increases, and our Longformer reader achieves competitive results with ETC as long as top-5 fusion blocks are input. Based on the above analysis, we conclude that the improvement of retrieval can significantly improve the QA performance of the reader.

5   Conclusion

In this paper, we study Open Table-and-Text Question Answering(OTTQA) and propose a Joint Link-Retrieve Framework(JLRF) where table block embedding and passage embedding are shared, and the linker and retriever are jointly trained in an end-to-end-manner. We leverage a contrastive learning-based linker to fully exploit the global and contextual information of table blocks and passages. In the meanwhile, we propose a novel retrieval method to compute the similarity of the fusion block which does not require explicitly generating and encoding the fusion block after linking. Besides, considering the structural gap between the table and passage, we utilize table-to-text as an auxiliary task to help the table block encoder better comprehend tables. Extensive experiments show that our framework has the ability to take full advantage of global information, mitigate error propagation, reduce noise interference, and bridge the gap between table and passage. We achieve new state-of-the-art performances on the OTT-QA dataset.

Acknowledgement. This research is supported by the National Natural Science Foundation of China (Grant No. 62072323, 62102276), Shanghai Science and Technology Innovation Action Plan (No. 22511104700), Natural Science Foundation of Jiangsu Province (Grant No. BK20210705, BK20211307), the Major Program of Natural Science Foundation of Educational Commission of Jiangsu Province, China (Grant No. 21KJD520005), the Priority Academic Program Development of Jiangsu Higher Education Institutions, and the Collaborative Innovation Center of Novel Software Technology and Industrialization.

References

1. Ainslie, J., et al.: ETC: Encoding long and structured inputs in transformers. In: Conference on Empirical Methods in Natural Language Processing (EMNLP), pp. 268–284. Association for Computational Linguistics (2020). https://aclanthology.org/2020.emnlp-main.19
2. An, S., et al.: Input-tuning: adapting unfamiliar inputs to frozen pretrained models. arXiv preprint arXiv:2203.03131 (2022). https://arxiv.org/abs/2203.03131
3. Beltagy, I., Peters, M.E., Cohan, A.: Longformer: the long-document transformer. arXiv preprint arXiv:2004.05150 (2020). https://arxiv.org/abs/2004.05150
4. Chen, D., Fisch, A., Weston, J., Bordes, A.: Reading Wikipedia to answer open-domain questions. In: Meeting of the Association for Computational Linguistics(ACL), pp. 1870–1879. Association for Computational Linguistics, Vancouver, Canada (2017). https://aclanthology.org/P17-1171
5. Chen, W., Chang, M.W., Schlinger, E., Wang, W.Y., Cohen, W.W.: Open question answering over tables and text. In: International Conference on Learning Representations(ICLR) (2021). https://openreview.net/forum?id=MmCRswl1UYl
6. Chen, W., Zha, H., Chen, Z., Xiong, W., Wang, H., Wang, W.Y.: HybridQA: a dataset of multi-hop question answering over tabular and textual data. In: Findings of the Association for Computational Linguistics(EMNLP). pp. 1026–1036. Association for Computational Linguistics (2020). https://aclanthology.org/2020.findings-emnlp.91
7. Devlin, J., Chang, M.W., Lee, K., Toutanova, K.: BERT: Pre-training of deep bidirectional transformers for language understanding. In: Conference of the North American Chapter of the Association for Computational Linguistics(NACL), pp. 4171–4186. Association for Computational Linguistics, Minneapolis, Minnesota (2019). https://aclanthology.org/N19-1423

8. Harman, D.K.: Overview of the third text retrieval conference (TREC) (1995)
9. Izacard, G., Grave, E.: Leveraging passage retrieval with generative models for open domain question answering. In: Conference of the European Chapter of the Association for Computational Linguistics(EACL), pp. 874–880. Association for Computational Linguistics (2021). https://aclanthology.org/2021.eacl-main.74
10. Johnson, J., Douze, M., Jégou, H.: Billion-scale similarity search with GPUs. IEEE Trans. Big Data (TBD) **7**(3), 535–547 (2021). https://doi.org/10.1109/TBDATA.2019.2921572
11. Jones, K.S.: A statistical interpretation of term specificity and its application in retrieval. J. Documentation **28**, 11–21 (1972). https://doi.org/10.1108/eb026526
12. Karpukhin, V., et al.: Dense passage retrieval for open-domain question answering. In: Conference on Empirical Methods in Natural Language Processing (EMNLP), pp. 6769–6781. Association for Computational Linguistics (2020). https://aclanthology.org/2020.emnlp-main.550
13. Lester, B., Al-Rfou, R., Constant, N.: The power of scale for parameter-efficient prompt tuning. In: Conference on Empirical Methods in Natural Language Processing(EMNLP), pp. 3045–3059. Association for Computational Linguistics, Online and Punta Cana, Dominican Republic (2021). https://aclanthology.org/2021.emnlp-main.243
14. Li, A.H., Ng, P., Xu, P., Zhu, H., Wang, Z., Xiang, B.: Dual reader-parser on hybrid textual and tabular evidence for open domain question answering. In: Meeting of the Association for Computational Linguistics and International Joint Conference on Natural Language Processing (ACL&IJCNLP), pp. 4078–4088. Association for Computational Linguistics (2021). https://aclanthology.org/2021.acl-long.315
15. Li, X.L., Liang, P.: Prefix-tuning: Optimizing continuous prompts for generation. In: Meeting of the Association for Computational Linguistics and International Joint Conference on Natural Language Processing (ACL&IJCNLP), pp. 4582–4597. Association for Computational Linguistics (2021). https://aclanthology.org/2021.acl-long.353
16. Li, Y., Li, W., Nie, L.: Dynamic graph reasoning for conversational open-domain question answering. ACM Trans. Inf. Syst. (TOIS) **40**(4), 1–24 (2022). https://doi.org/10.1145/3498557
17. Liu, X., et al.: GPT understands, too. arXiv preprint arXiv:2103.10385 (2021). https://arxiv.org/abs/2103.10385
18. Liu, Y., et al.: Roberta: a robustly optimized bert pretraining approach. arXiv preprint arXiv:1907.11692 (2019). https://arxiv.org/abs/1907.11692
19. Loshchilov, I., Hutter, F.: Decoupled weight decay regularization. In: International Conference on Learning Representations(ICLR) (2019). https://openreview.net/forum?id=Bkg6RiCqY7
20. Nogueira, R., Cho, K.: Passage re-ranking with BERT. arXiv preprint arXiv:1901.04085 (2019). https://arxiv.org/abs/1901.04085
21. van den Oord, A., Li, Y., Vinyals, O.: Representation learning with contrastive predictive coding. arXiv preprint arXiv:1807.03748 (2018). https://arxiv.org/pdf/1807.03748.pdf
22. Parikh, A., et al.: ToTTo: a controlled table-to-text generation dataset. In: Conference on Empirical Methods in Natural Language Processing (EMNLP), pp. 1173–1186. Association for Computational Linguistics (2020). https://aclanthology.org/2020.emnlp-main.89
23. Radford, A., Wu, J., Child, R., Luan, D., Amodei, D., Sutskever, I., et al.: Language models are unsupervised multitask learners. OpenAI blog **1**(8), 9 (2019). https://github.com/openai/gpt-2

24. Raffel, C., et al.: Exploring the limits of transfer learning with a unified text-to-text transformer. J. Mach. Learn. Res. (JMLR) **21**(140), 1–67 (2020). https://jmlr.org/papers/v21/20-074.html
25. Rothe, S., Narayan, S., Severyn, A.: Leveraging pre-trained checkpoints for sequence generation tasks. Trans. Assoc. Comput. Linguist. **8**, 264–280 (2020). https://aclanthology.org/2020.tacl-1.18
26. Sachan, D., et al.: End-to-end training of neural retrievers for open-domain question answering. In: Meeting of the Association for Computational Linguistics and International Joint Conference on Natural Language Processing (ACL&IJCNLP), pp. 6648–6662. Association for Computational Linguistics (2021). https://aclanthology.org/2021.acl-long.519
27. Vaswani, A., et al.: Attention is all you need. In: Guyon, I., et al. (eds.) Advances in Neural Information Processing Systems(NeurIPS), vol. 30. Curran Associates, Inc. (2017). https://proceedings.neurips.cc/paper/2017/file/3f5ee243547dee91fbd053c1c4a845aa-Paper.pdf
28. Vaswani, A., et al.: Attention is all you need. In: Conference on Neural Information Processing Systems (2017). https://proceedings.neurips.cc/paper/2017/hash/3f5ee243547dee91fbd053c1c4a845aa-Abstract.html
29. Wang, X., Han, X., Huang, W., Dong, D., Scott, M.R.: Multi-similarity loss with general pair weighting for deep metric learning. In: Proceedings of the IEEE/CVF Conference on Computer Vision and Pattern Recognition(CVPR) (2019). https://arxiv.org/abs/1904.06627
30. Wu, L., Petroni, F., Josifoski, M., Riedel, S., Zettlemoyer, L.: Scalable zero-shot entity linking with dense entity retrieval. In: Conference on Empirical Methods in Natural Language Processing (EMNLP), pp. 6397–6407. Association for Computational Linguistics (2020). https://aclanthology.org/2020.emnlp-main.519
31. Zhong, W., et al.: Reasoning over hybrid chain for table-and-text open domain question answering. In: Raedt, L.D. (ed.) International Joint Conference on Artificial Intelligence(IJCAI), pp. 4531–4537. International Joint Conferences on Artificial Intelligence Organization (2022). https://doi.org/10.24963/ijcai.2022/629
32. Zhu, F., Lei, W., Wang, C., Zheng, J., Poria, S., Chua, T.S.: Retrieving and reading: a comprehensive survey on open-domain question answering. arXiv preprint arXiv:2101.00774 (2021). https://arxiv.org/abs/2101.00774

L2QA: Long Legal Article Question Answering with Cascaded Key Segment Learning

Shugui Xie[1], Lin Li[1(✉)], Jingling Yuan[1], Qing Xie[1], and Xiaohui Tao[2]

[1] Wuhan University of Technology, Wuhan, China
{xieshugui,cathylilin,yjl,felixxq}@whut.edu.cn
[2] University of Southern Queensland, Toowoomba, Australia
xiaohui.tao@usq.edu.au

Abstract. Evidences in Legal Question Answering (LQA) help infer accurate answers. Current sentence-level evidence extraction based methods may lose the discourse coherence of legal articles since they tend to make the extracted sentences scattered over an article. To this end, this paper proposes a cascaded key segment learning enhanced framework for **L**ong **L**egal article **Q**uestion **A**nswering, namely **L2QA**. The framework consists of three cascaded modules: *Sifter*, *Reader*, and *Responder*, which first transfers a long legal article into segments and each segment is inherent in the discourse coherence from consecutive sentences. And then, the *Sifter* is trained by automatically sifting out key segments in an iterative answer-guided coarse-to-fine way. The *Reader* utilizes a range of co-attention and self-attention mechanisms to obtain the semantic representations of the question and key segments. Finally, the *Responder* predicts final answers in a cascaded manner, identifying where the answer is located. Conducted on CAIL 2021 Law MRC dataset, our L2QA achieves 83.1 Macro-F1 and 65.8 EM and outperforms a state-of-the-art legal QA model by 4.1% and 9.1%.

Keywords: Legal Question Answering · Key Segment · Coarse-to-fine

1 Introduction

Legal Question Answering (LQA) refers to finding answers to a given question by reading and understanding a set of legal articles [1]. To assist the research community in LQA with multiple question types, several datasets have been constructed, such as JEC-QA with 28,641 multiple-choice and multiple-answer questions [2], COLIEE-2021 with 800 yes/no questions [3], the United States Multistate Bar Examination(MBE) corpus with 600 yes/no [4], OAB exam data [5], and so on. The CJRC-2021[1] has four kinds of question types, including single-span, yes/no, unanswerable and multi-span answers [6], which is publicly released

---

[1] http://cail.cipsc.org.cn/.

© The Author(s), under exclusive license to Springer Nature Switzerland AG 2023
X. Wang et al. (Eds.): DASFAA 2023, LNCS 13945, pp. 385–394, 2023.
https://doi.org/10.1007/978-3-031-30675-4_27

with the greatest variety of question types in LQA. The various question types in LQA expand its applicability, however, brings further challenges to QA models.

Legal articles are rigorously structured and logical. A densely connected encoder stack is designed to obtain multi-scale semantic features of legal articles [7]. But, when dealing with various answer types, it is highly demanded to accurately locate evidence sentences for answer prediction. Current studies demonstrate the effectiveness of extracting evidence like key sentences or paragraphs to guide answer prediction in long article question answering [8–12]. Recently, self-training [10] method is employed to supervise the evidence extractor with auto-generated evidence sentence labels in an iterative process. Few evidence annotations (strong semi-supervision) combining with abundant document-level labels (weak supervision) are also adopted for evidence extraction [11]. However, the above methods are mainly based on sentence-level or span-level extraction. They tend to make the extracted sentences scattered over the article, which results in unsatisfied answering for consecutive evidence sentences. On the other hand, at paragraph-level evidence extraction, reinforcement-based methods are used to jointly train the model for answer prediction [13,14], where the evidence is obtained through a ranker that learns a weight distribution from the retrieved passage-level text and assigns probabilities based on the relevance of each passage to the question. However, the passage-level extraction method is limited by the input size of the language model. To meet this size limit, the popular hard truncation method may not guarantee contextual semantics (discourse coherence). Their performance becomes unstable when some answers need to be reasoned through multiple, consecutive sentences.

It is observed that preserving the contextual semantics (discourse coherence) of evidence sentences can help outputting accurate answers. Tackling the problem, this paper proposes L2QA, a *Sifter, Reader, Responder* cascaded framework with answer-guided key segment learning. The *Sifter* first divides a long legal article into several segments in a dynamic programming method, to guarantee each of which contains one or more consecutive sentences. We then design an answer-guided learning on the *Sifter*, in a coarse-to-fine manner to select more key segments from the article and go through several iterations. The *Reader* utilizes co-attention and segment-level self-attention mechanisms to learn semantic representation. The *Responder* maps the semantic representation to the input of a multi-label classification task and obtains the final answer in a cascaded manner. Conducted on a legal QA dataset, our L2QA framework achieves 83.1 in Macro-F1 and 65.8 in EM and outperformed the state-of-the-art model by 4.1% and 9.1%.

2   Methodology

This study is focused on the following legal question answering task:

- Input: a long legal article x, a question q
- Output: y (span(s), yes, no, unknown)

The answer is an element of the set containing *single-span, multi-span, yes, no,* and *unknown* As an extractive reading comprehension task, the L2QA is modeled as a multi-label classification task with the above four answer types.

Our L2QA is a cascaded two-stage training, answer-guided key segment learning framework for long legal question answering, as shown in Fig. 1.

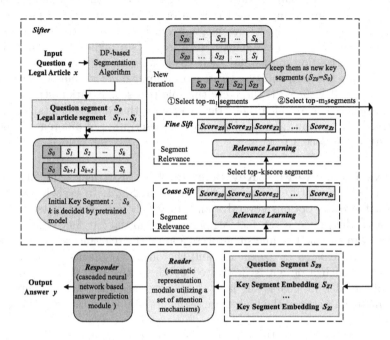

Fig. 1. Overview of L2QA framework.

2.1 Sifter: Answer-Guided Key Segment Selection Module

As shown in Fig. 1, the *Sifter* aims to maintain the discourse coherence of the input legal article x and question q to guide answer prediction while satisfying the input size of text representation learning model. Based on the Dynamic Programming (DP) algorithm, the *Sifter* first segments the legal article x and question q into several segments, each of which contains one or more consecutive sentences. A coarse-to-fine relevance calculation is then performed between legal segments and key fact segments. It is an iterative process based on an answer-guided training method. Segments with the highest relevance scores will be merged with the existing key fact segments, forming a new set of key fact segments. The segments sifted out through several iterations are fed into the semantic representation module, *Reader*, in the order of the original article.

Dynamic Programming Algorithm Based Segmentation. A legal article is divided to several segments based on a dynamic programming algorithm that pre-defines the cost of all punctuation marks in a legal article. Here, if a segmentation is ended with a punctuation mark, its cost in DP is lower than a hard truncation that usually loses discourse conference. We traverse the entire legal article, record all punctuation marks and the corresponding cost, and when the distance between two punctuation marks is greater than the maximum segment length (e.g., 64), recording a hard truncation. The ending with a punctuation mark, possibly a full stop ".", a question mark "?", or a comma mark ",", deserves a lower cost and the hard truncation gets a higher cost. Then, based on all the punctuation marks and their costs, we traverse forward in a reverse order to obtain the segmentation strategy of minimizing the sum of all segments costs. The smaller the cost, the more suitable this segmentation strategy is for the input legal article, which theoretically ensures that our segments will not be hard truncation as much as possible. See the description of different segmentation methods in Sect. 3.1.

Iterative Coarse-to-Fine Key Segment Sift. Key fact segments will be sifted out via several iterations, as illustrated in Fig. 1. In each iteration, the coarse sift process obtains several segments which have the highest relevance scores with the previous key fact segments. The initial key fact segment is the question segment s_0. The fine sift process adopts the relevance score calculation again between existed key segments and the output segments from the coarse sift process. Such a design is motivated by the realization that the relevance scores obtained from coarse sift lack of sufficient accuracy without interaction and comparison between the high score segments, similar to the re-ranking problem [15].

In each fine sift step, the top-m_1 relevance score segments will be merged with key segments and form the new key fact segments (evidence). m_1 is the parameter. After iterations, the question segment S_{Z_0}(the same as S_0) and key fact segments S_{Z_1}-S_{Z_l} will be obtained, and labeled as x_l. The size of $S_{Z_0}+S_{Z_1}$-S_{Z_l} is m. The accumulated quantity of key segments is determined by the input size of the downstream module *reader*.

Relevance Learning. Relevance learning is trained in a supervised manner:

$$loss_{sifter}(S) = CrossEntropy(R, relv\_label(S)) \tag{1}$$

$$relv\_label(S) \in [0, 1]^{len(S)} = [\underbrace{1, \cdots, 1}_{S_{Z_0}}, \underbrace{0, \cdots, 0}_{S_{Z_1}\text{irrelevant}}, \underbrace{1, \cdots, 1}_{S_{Z_2}\text{relevant}}, \cdots] \tag{2}$$

where the training sample S is either a sequence of continuous segments x_{rand} sampled from the legal article x or a mixture of all relevant and randomly selected irrelevant segments x_{relv}.

The relevance learning obtains the embedding of all input tokens through a pretrained model. Then MLP (sigmoid) performs a multi-label classification task on the embedding, aiming to obtain the token logits. We regard the classification result of the input sequence as its confidence on the question. The average of all tokens in the range of each segment as the confidence of this segment, that is, the relevance with the input question. Then the segment relevance is produced by the mean pooling result of the token logits.

2.2   Reader: Semantic Representation Module

The *Reader* is a cascaded structure with a range of co-attention [16,17] and segment-level self-attention.

Encoding Question and Key Segments. We firstly employ a pretrained language model such as Electra to obtain initial embedding of question segment and key legal segments $E_1^Q \in \mathbb{R}^{n \times h}$, $E_1^x \in \mathbb{R}^{m \times h}$, where h is the hidden size.

We concatenate three tokens of "[CLS][UNK][YES][NO]" token, following those common Machine Reading Comprehension tasks, the tokenized question q and a final "[SEP]" token. The tokenized key segments S_{Z_i} and a final "[SEP]" token are concatenated in the same way. Then segments from S_{Z_0} to S_{Z_l} are concatenated and fed into the pretrained language model.

The co-attention mechanism has been shown effective in other MRC models [16,17], outputting a co-dependent representation of the question and key fact segments. It takes E_1^Q, E_1^x as inputs and respectively outputs $Q_2^x \in \mathbb{R}^{n \times h}$ and $x_2^Q \in \mathbb{R}^{m \times h}$ and feed them to the segment-level self-attention layer.

Segment-Level Self-attention. This layer works for gathering information on key tokens in each key segment. The token-level representation is given by:

$$x_3^Q = \text{Transformer}(x_2^Q) \in \mathbb{R}^{m \times h} \tag{3}$$

x_3^Q is the concatenation from segment $S_{Z_1}^x$ to segment $S_{Z_l}^x$ and the question self-attention is calculated in the same way. Finally, a mean pooling is used to obtain the segment representation $x_4^Q \in \mathbb{R}^{l \times h}$ as:

$$x_4^Q[i,:] = MeanPooling(S_{Z_i}^x) \tag{4}$$

where $S_{z_i}^x$ is the i-th segment representation in x_3^Q.

The difference between the original multi-head self-attention and our segment-level self-attention layer is that we incorporate an attention mask that is the sum of consists of two masks as introduced in equations below.

$$M_s[i,j] = \begin{cases} 0 & i,j \in \text{same segment} \\ -\infty & \text{otherwise} \end{cases} \tag{5}$$

$$M_{rand}[i,j] = \begin{cases} 0 & 128 \text{ random tokens} \\ -\infty & \text{otherwise} \end{cases} \tag{6}$$

where $M_s \in \mathbb{R}^{m \times m}$ is a segment-level attention mask, as each segment representation only focus on the token information inside the segment. $M_{rand} \in \mathbb{R}^{m \times m}$ is a random mask for each segment to pay attention to some tokens of other segments. The 0 will not be overwritten by negative infinity.

Q_3^x will be concatenated with x_3^Q as token-level representation $E_{QS} \in \mathbb{R}^{m+n}$ and fed into the *Responder*. Also, the legal article and question segment representation $H \in \mathbb{R}^{(l+1) \times h}$ is obtained through the concatenation of Q_4^x and x_4^Q.

2.3   Responder: Answer Predictor

Since our task is an extractive reading comprehension task, the final answer will be obtained through the start and end positions. In *Responder*, a dense layer F^S with Tanh activation function is first employed as segment learning layer. It takes $H \in \mathbb{R}^{(l+1) \times h}$ as inputs and outputs the segment representation $H_S \in \mathbb{R}^{(l+1) \times h}$.

Answer Start and End Locator. The segment representation H_S and token-level representation E_{QS} are used as the inputs to predict the start position of the answer. Since the row-dimension of $E_{QS} \in \mathbb{R}^{(m+n) \times h}$ is different from that of $H_S \in \mathbb{R}^{(l+1) \times h}$, the H_S cannot directly concatenate to E_{QS}. Then $H_S \in \mathbb{R}^{(l+1) \times h}$ is tiled with $H_S^2 \in \mathbb{R}^{(m+n) \times h}$ along row-dimension: $H_S^2[i,:] = H_S[\mathbb{L}_i,:] \in \mathbb{R}^h$. Note that $\mathbb{L}_i$ indicates the index of the segment where the i-th token is located. Thus, the model can consider the information of the segment learning layer when predicting the answer start. The start logits of the answer are predicted by:

$$H_{start} = F_{start}([H_S^2; E_{QS}]) \in \mathbb{R}^{(m+n) \times h} \tag{7}$$

$$O_{start} = H_{start} W_{start} \in \mathbb{R}^{m+n} \tag{8}$$

where o_{start} is the output logit vector of the start positions of the answer, F_{start} is a dense layer with Tanh activation function, and $W_{start} \in \mathbb{R}^{(h \times 1)}$ is a trainable parameter. H_{start} will be used in answer end locator.

Similarly, the end logits of the answer are predicted by:

$$H_{end} = F_{end}([H_{start}; E_{QS}]) \in \mathbb{R}^{(m+n) \times h} \tag{9}$$

$$O_{end} = H_{end} W_{end} \in \mathbb{R}^{m+n} \tag{10}$$

where o_{end} is the logit vector of the end positions of the answer, F_{end} is a dense layer with Tanh activation function, and $W_{end} \in \mathbb{R}^{(h \times 1)}$ is trainable.

We map the LQA task into a multi-label classification task. The cross-entropy loss is computed over the aforementioned output start and end logits, and jointly minimizes these two cross-entropy losses.

$$loss_{start} = CrossEntropy(o_{start}, start\_label) \tag{11}$$

$$loss_{end} = CrossEntropy(o_{end}, end\_label) \tag{12}$$

$$Minimize(loss_{start} + loss_{end}) \tag{13}$$

3   Experiment

3.1   Dataset and Experimental Setup

The CAIL 2021 Law MRC dataset (CJRC) [6] contains data released in 2019 and 2021(39333+4200 examples). Following the official splitting of CAIL 2021 Law MRC contest, we get the train set(43133), dev set(1000) and test set(1000). Note that Dev(Test) has 3 sets and each question type is 200 samples.

In Train set, 56.38% articles' length are longer than the input size of pre-trained models(512). The commonly used data segmentation method, sliding window, results in hard truncated segments accounting for 97.1%, which can't guarantee the semantic coherence of sentences since a sentence is cutoff directly. Our DP-based segmentation do much better with 6.9% hard truncated segments.

Our L2QA adopts a two-stage training method, that is, the training of *Sifter* and the training of *Reader, Responder* are separated. It takes approximately 10 h and 30 h to train the two modules respectively with a TITANX GPU. The HFL released legal-electra-base [18] is used as our pretrained model.

The number of segments selected in the first iteration m_1, and the number in the second iteration m_2 in *Sifter* are set as 3 and 5. The single segment size is set to 64 tokens. The learning rate and batch size of *Sifter* and *Reader* are 7×10^{-5}, 4. The gradient accumulation of *Sifter* and *Reader* are set to 4, 8 respectively. And the train epoch is set to 2,4 respectively. The Adamw is used for optimizer.

Our baselines are some existing methods which enhance QA performance from different perspectives, i.e., **Sliding Window** method [19], **CogLTX** [20], **RikiNet** [17], **MTMSN** [21] and **MacBERT** [22]. Our evaluation measures are extract match(EM) and macro-average F1(Macro-F1), the same as SQuAD [23].

3.2   Main Results

Overall performance comparisons are conducted between previously published works and our L2QA. The average results of Exact Match(EM), Macro-F1 on both 3 dev sets and 3 test sets are listed in Table 1.

Table 1. Overall performance comparisons.

	Average Dev		Average Test	
	EM	F1	EM	F1
1 - Sliding Window [19]	59.8	76.2	61.9	77.2
2 - CogLTX [20]	60.2	77.9	60.7	78.3
3 - CogLTX+12 hidden layers attention	59.8	77.5	60.4	77.9
4 - CogLTX+4,8,12 hidden layers attention	60.2	78.2	61.3	78.8
5 - RikiNet [17]	60.1	77.5	61.6	78.2
6 - MTMSN [21]	44.2	67.3	44.8	67.1
7 - MacBERT-base [22]	60.4	80.3	60.3	79.8
L2QA(Ours)	**64.0**	**81.9**	**65.8**	**83.1**

The CogLTX used as Sifter module in L2QA has achieved improvement in Macro-F1 over the sliding window method on both the dev set and test set (Dev set: Macro-F1 from 76.2 to 77.9; Test set: Macro-F1 from 77.2 to 78.3). It demonstrates the effectiveness of extracting evidence like legal segments. The effects of the hidden layers of pretrained models are also widely studied [24]. The attention over the 4th, 8th and 12th hidden layers achieve better performance than the other two methods (dev set:Macro-F1 78.2; test set:Macro-F1 78.8). The result shows that it is worth adding an attention to the hidden layers of the pretrained model to improve the model's answer prediction ability.

RikiNet focuses on semantic understanding, utilizing a set of attention mechanisms and sliding window. For the reason of using dynamic paragraph dual-attention, the training of the RikiNet takes more than 80 h, which is three times the training time of CogLTX. It brings improvements over the single CogLTX model (Dev set: Macro-F1 from 76.2 to 77.5; Test set: Macro-F1 from 77.2 to 78.2). MTMSN aims to deal with questions of multi-type answers. The result shows 44.2 EM and 67.3 Macro-F1 on dev set, 44.8 EM and 67.1 Macro-F1 on test set. Predicting the number of answers and the type of answer in MTMSN brings a negative effect to the result in this long legal article dataset.

For our L2QA, the accuracy of the trained *Sifter* sifting out key sentences exceeds 99.5%. And, L2QA outperforms the other six baselines on both dev set and test set (dev set: 64.0 EM, 81.9 Macro-F1; test set: 65.8 EM, 83.1 Macro-F1). The result demonstrates that the cascaded answer-guided key segment learning framework is capable of accurately answering multiple type questions.

3.3   Ablation Study

We conduct an in-depth ablation study on probing three key modules with results shown in Table 2. All three modules show their importance to the L2QA performance and cascading the three key modules significantly improves the framework's performance in answering legal questions.

Table 2. Ablation study of our L2QA.

Setting	EM	F1	Setting	EM	F1
L2QA	**64.9**	**82.5**	**L2QA**	**64.9**	**82.5**
(a) - Sifter	63.3	80.2	(e) - Rand Mask(64 token)	64.7	81.3
(b) - Reader	52.6	79.4	(f) - Responder	63.7	80.8
(c) - Co-attention	63.6	81.3	(g) - Reader & Responder	60.4	78.1
(d) - Key segment Self-attention	64.8	81.9			

(1) Ablations of Sifter: In (a), we remove the entire Sifter and replace it with the sliding window method. Since our Reader is designed corresponding to the segments involved in Sifter, after removing the Sifter, we make some difference to Reader, replacing the segment-level self-attention with token-level self-attention. We can see that after removing the Sifter, the performance drops sharply, for example, EM is reduce to 63.3 from 64.9.

(2) Ablations of Reader: In (b), we keep the Sifter, the Responder and remove the Reader. In (c), and (d), we remove the co-attention layer, and key segment self-attention layer respectively. In (e), we remove the random mask and keep the segment mask as introduced in Sect. 2.2. It shows that after removing each component of Reader, the performance drops accordingly. Moreover, the question self-attention layer and co-attention layer both enhance the performance.

(3) Ablations of Responer: In (f), components in the Responder are further replaced. We have simplified models through two parallel calculations on the start_logits and end_logits, which means canceling the cascaded manner for answer prediction. The result of Macro-F1 drops from 82.5 to 80.8. In (g), we further remove the Sifter and Reader, which results in additional performance decreasing, suggesting the essence of Reader and Responder in L2QA.

4 Conclusions

This study proposes a new Legal Question Answering framework namely L2QA that reads legal articles to answer multi-type questions. The L2QA is constituted by three modules, Sifter, Reader and Responder, working in a cascaded manner to predict the final answers. Experiments conducted on the CAIL 2021 Law MRC dataset show the promising performance of the proposed L2QA and its superior to state-of-the-art QA models.

Acknowledgments. This work is supported in part by the National Natural Science Foundation of China (62276196) and the Key Research and Development Program of Hubei Province (2021BAA030).

References

1. Zhong, H., Xiao, C., Tu, C., Zhang, T., Liu, Z., Sun, M.: How does NLP benefit legal system: a summary of legal artificial intelligence. In: ACL, pp. 5218–5230 (2020)
2. Zhong, H., et al.: JEC-QA: a legal-domain question answering dataset. In: AAAI, pp. 9701–9708 (2020)
3. Rabelo, J., et al.: Overview and discussion of the competition on legal information extraction/entailment (coliee) 2021. In: The Review of Socionetwork Strategies, pp. 1–23(2022)
4. Fawei, B., Wyner, A., Pan, J.: Passing a USA national bar exam: a first corpus for experimentation. In: LREC, pp. 3373–3378 (2016)
5. Delfino, P., Cuconato, B., Haeusler, E.H., Rademaker, A.: Passing the Brazilian OAB exam: data preparation and some experiments. In: Legal Knowledge and Information Systems-JURIX, pp. 89–94 (2017)
6. Duan, X., et al.: CJRC: a reliable human-annotated benchmark DataSet for Chinese judicial reading comprehension. In: Sun, M., Huang, X., Ji, H., Liu, Z., Liu, Y. (eds.) CCL 2019. LNCS (LNAI), vol. 11856, pp. 439–451. Springer, Cham (2019). https://doi.org/10.1007/978-3-030-32381-3_36

7. Nai, P., Li, L., Tao, X.: A densely connected encoder stack approach for multi-type legal machine reading comprehension. In: WISE, pp. 167–181 (2020)
8. Ji, D., Tao, P., Fei, H., Ren, Y.: An end-to-end joint model for evidence information extraction from court record document. Inf. Process. Manag. **57**(6), 102305(2020)
9. Li, X., Burns, G., Peng, N.: Scientific discourse tagging for evidence extraction. In: EACL, pp. 2550–2562 (2021)
10. Niu, Y., Jiao, F., Zhou, M., Yao, T., Xu, J., Huang, M.: A self-training method for machine reading comprehension with soft evidence extraction. In: ACL, pp. 3916–3927 (2020)
11. Pruthi, D., Dhingra, B., Neubig, G., Lipton, Z.C.: Weakly-and semi-supervised evidence extraction. In: EMNLP (Findings) (2020)
12. Xie, S., Li, L., Yuan, J., Xie, Q., Tao, X.: Long legal article question answering via cascaded key segment learning (Student Abstract). In: AAAI (2023)
13. Wang, S., et al.: R^3: reinforced ranker-reader for open-domain question answering. In: AAAI, pp. 5981–5988 (2018)
14. Choi, E., Hewlett, D., Uszkoreit, J., Polosukhin, I., Lacoste, A., Berant, J.: Coarse-to-fine question answering for long documents. In: ACL (Volume 1: Long Papers), pp. 209–220 (2017)
15. Collins, M., Koo, T.: Discriminative reranking for natural language parsing. Comput. Linguist. **31**(1), 25–70 (2005)
16. Xiong, C., Zhong, V., Socher, R.: DCN+: mixed objective and deep residual coat-tention for question answering. In: ICLR (2018)
17. Liu, D., et al.: RikiNet: reading Wikipedia pages for natural question answering. In: ACL, pp. 6762–6771 (2020)
18. Cui, Y., Che, W., Liu, T., Qin, B., Wang, S., Hu, G.: Revisiting pre-trained models for Chinese natural language processing. In: EMNLP (Fingddings), pp. 657–668 (2020)
19. Wang, Z., Ng, P., Ma, X., Nallapati, R., Xiang, B.: Multi-passage BERT: a globally normalized BERT model for open-domain question answering. In: EMNLP-IJCNLP, pp. 5878–5882 (2019)
20. Ding, M., Zhou, C., Yang, H., Tang, J.: Cogltx: applying BERT to long texts. In: NeurIPS (2020)
21. Hu, M., Peng, Y., Huang, Z., Li, D.: A multi-type multi-span network for reading comprehension that requires discrete reasoning. In: EMNLP-IJCNLP, pp. 1596–1606 (2019)
22. Cui, Y., Che, W., Liu, T., Qin, B., Yang, Z.: Pre-training with whole word masking for Chinese Bert. In: IEEE/ACM Trans on Audio, Speech and Language Processing, vol. 29, pp. 3504–3514 (2021)
23. Rajpurkar, P., Zhang, J., Lopyrev, K., Liang, P.: SQuAD: 100,000+ questions for machine comprehension of text. In: EMNLP (2016)
24. Ganesh Jawahar, Benoît Sagot, and Djamé Seddah.: What does BERT learn about the structure of language? In: ACL, pages 3651–3657(2019)

An Adaptive Video Clip Sampling Approach for Enhancing Query-Based Moment Retrieval in Videos

Lingdu Kong, Tieying Li, Xiaochun Yang$^{(\boxtimes)}$, Shengzhi Han, and Bin Wang

School of Computer Science and Engineering, Northeastern University, Shenyang, China
2210707@stu.neu.edu.cn, {yangxc,binwang}@mail.neu.edu.cn

Abstract. Query-based moment retrieval aims to localize the most relevant moment in an untrimmed video according to the given natural language query. Existing retrieval models require the same length for easy training and use. Therefore, videos with different lengths are pre-processed using the fixed sampling method. As a result, the longer the video, the more video clips are lost, thus affecting the accuracy of retrieval. We observed the fixed sampling method causes two accuracy issues, including missing clips and sparse clips. In this paper, we propose an adaptive video clip sampling method including resampling missing clips and enhancing sparse sampled clips to increase the retrieval accuracy. Resampling missing clips is used to address situations in which annotated clips are completely lost during fixed sampling. Enhancing sparse sampled clips aims to prevent the clips containing the same semantics from being too sparse. Our approach first obtains multiple video features through the adaptive sampling methods based on the backbone networks. Then we propose a consistency loss maintenance method to learn the semantics of adaptive sampled features. The extensive experiments on three real datasets demonstrate the effectiveness of our proposed method, especially for long videos.

Keywords: Query-based moment retrieval · Multi-modal · Consistency · Video sampling

1 Introduction

Query-based moment retrieval (a.k.a. natural language video localization) has drawn increasing attention over the last years, which aims to locate the start and end boundaries of the most relevant video segment from an untrimmed video according to a given natural language query. For example, given an untrimmed video in Fig. 2 and a query "fire sparks erupt," we aim to locate the best matching segment for the query.

Most existing moment retrieval works [2–5, 10, 11, 13] focus on different aspects of this emerging task, such as the query representation learning [7], video context modeling [1,9], and cross-modal fusion [11,13], while ignoring the video segment representations. Figure 1 shows the recent methods can be classified into unfixed size and fixed size sampling methods, where fixed size sampling methods (i.e. videos of different lengths are sampled into the same length for easy training and use) perform better.

However, the longer the video, the more video clips are lost, thus affecting the accuracy of retrieval. Generally, a video contains different segments, each consisting of clips

© The Author(s), under exclusive license to Springer Nature Switzerland AG 2023
X. Wang et al. (Eds.): DASFAA 2023, LNCS 13945, pp. 395–404, 2023.
https://doi.org/10.1007/978-3-031-30675-4_28

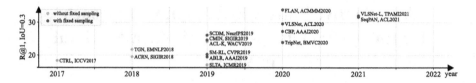

Fig. 1. Comparison of the performance of existing methods on TACoS dataset in recent years.

(a) Original video.

(b) Target video after fixed sampling.

Fig. 2. An example of query-based moment retrieval in videos based on fix-sampling.

and semantically linked to a sentence. A clip is made up of continuous frames. In the example shown in Fig. 2 with 9 clips, a clip "fire sparks erupt" may be missed if the fixed sampling size is set to 5, causing inaccurate information learning in models.

Fixed sampling size leads to limited adaptability in existing models. It is challenging to determine a suitable sampling size due to varying video lengths affecting representation accuracy differently. We find that longer videos lose more clips, and short segments become sparse after sampling. Therefore, an adaptive video sampling is desired to reduce performance loss.

In this work, we propose an adaptive video clip sampling to improve accuracy of existing moment retrieval models to avoid missing and sparse clips from fixed sampling. The training phase adaptively samples videos by resampling missing clips and enhancing sparse clips. We adopt three widely used backbone networks [11, 13, 14] to obtain fusion representations, and design a consistency loss function to learn more accurate semantic information and reduce the loss caused by fixed sampling.

The contributions of our work are as follows:

- We propose an adaptive sampling method to avoid missing clip and sparse clip caused by fixed sampling.
- We propose consistency loss maintenance on multiple fusion representation base on backbone networks. The maintenance is used to learn the semantics of adaptive sampled features, which reduces the loss caused by fixed sampling.
- We demonstrate the effectiveness of our proposed approach on three real moment retrieval datasets. Our approach shows better improvement, especially for long videos.

2   Analysis of the Existing Fixed Size Sampling Methods

In this section, we formally introduce the problem of query-based moment retrieval. We then analyze why the existing fixed size sampling methods are unable to achieve high accuracy especially on long videos.

2.1   Problem Formulation

Given an untrimmed video $V^u = \{v_1^u, \ldots, v_n^u\}$ with n frames and a sentence query $Q = \{w_1, \ldots, w_m\}$ with m words, our task aims to retrieve the best semantic matching M in V, i.e. the moment $M = V[i, j]$ from frame v_i to v_j that delivers the same semantic meaning as the input query sentence Q.

In order to answer Q on V semantically, the goal is to train a retrieval model by building potential relationships between given video-sentence pairs. More specifically, given a training video stream $S = \{s_1, \ldots, s_z\}$, let $\{X, T, A\}$ be the training set of the video-sentence pairs. The k-th video segment $x_k (= S[i, j]) \in X$ is specified by its annotation $a_k = (i, j) \in A$ (i.e. labeling the start and end frame of the video segment $S[i, j]$), and can be semantically described by its corresponding text $t_k \in T$.

2.2   Problems of Existing Video Representations with Fixed Sampling Size

The existing fixed sized sampling approaches first segment the input video stream S into a number of video clips. Each video clip c_i consists of T frames. For each video clip c_i, its feature is represented as v_i^o. Therefore, we could use features of all video clips to represent the whole video feature, denoted $V^o = \{v_1^o, \ldots, v_p^o\}$, where $v_i^o \in \mathbb{R}^d$, $p = \lfloor \frac{z}{T} \rfloor$, and d is the feature dimension. In order to feed different number of video clips into the same training model, the existing methods transform different lengths of videos into fix number of video clips. Such transformation is called fixed size sampling. To be specific, given a fix number N, the video is transformed into a sampled video with N clips, where each $v_i^r = v_{\lfloor p \times i/N \rfloor}^o$. Then the feature of the sampled video is denoted as $V^S = \{v_1^S, \ldots, v_N^S\}, v_i^S \in \mathbb{R}^d$.

Figure 3(a) shows a video with 17 clips, and Fig. 3(b) shows a sampled video with 9 clips. Let x_1, x_2, and x_3 be three segments in the original video with corresponding texts t_1, t_2, and t_3. Other segments have no text annotation.

Figure 3(b) illustrates the two issues caused by fixed size sampling.

(i) *Missing clips.* Some video clip (e.g. x_2) is not sampled (i.e. the only clip's feature representing x_2 does not exist in the sampled video), which results in some important clips are missing after fixed size sampling.

(ii) *Sparse clips.* When a segment contains a small number of clips, the number of clips to be sampled may be very sparse, thus affecting the accuracy of model training. For example, the segment x_1 only contains two clips. After fixed sampling, only one clip is sampled, which means 50% of clips are lost.

Table 1 shows the statistic of the above two issues and the total number of video-sentence pairs in three datasets. We observe that the missing clip issue happens frequently in TACoS dataset. It is mainly because TACoS contains more number of long

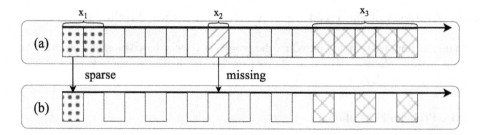

Fig. 3. Examples of sampled video clips using fix-interval sampling. (a) Original video, and (b) sampled clips.

Table 1. The statistics of (i) *missing clips*, (ii) *sparse clips*, and the total number of video-sentence pairs (denoted ALL) in three datasets with the fixed sampling size $N = 200$.

Video length (Sec.)	ActivityNet dataset			TACoS dataset			TACoS$_{2DTAN}$ dataset		
	issue (i)	issue (ii)	All	issue (i)	issue (ii)	All	issue (i)	issue (ii)	All
0–100	0	271	28,272	0	0	2,096	0	0	2,329
100–200	1	26,030	32,083	0	1,729	5,647	0	1,894	5,171
200–300	1	11,289	11,505	24	3,372	3,556	4	3,368	3,396
300–400	0	33	33	31	1,663	1,678	0	1,606	1,607
400–500	0	32	32	104	1,775	1,775	7	1,701	1,702
500–600	0	5	5	61	933	934	1	913	913
600–700	0	9	9	193	1,539	1,546	12	1,523	1,523
700–+∞	0	18	18	478	1,586	1,586	40	1,586	1,586

videos than ActivityNet, which results in many short segments in TACoS dataset have a high probability of not being sampled. Meanwhile, the sparse issue occurs frequently in the videos within $100 \sim +\infty$ seconds in the three datasets, because the fixed sampling makes the sampling interval large.

3   An Adaptive Sampling

Based on the analysis in Sect. 2.2, we propose an adaptive sampling approach to avoid missing clips and sparse clips.

3.1   Framework Overview

Figure 4 shows the overall framework of our method which consists three parts: backbone network, adaptive sampling, and consistency maintenance. (i) Backbone network is adopted by most of the existing approaches to train the retrieval model. (ii) Adaptive sampling includes resampling missing clips and enhancing sparse sampled clips. Resampling missing clips aims to adjust feature representation by substituting features corresponding to the important but unsampled clips for those unimportant or highly repetitive clips due to fixed size sampling. Enhancing sparse clips aims to enhance the

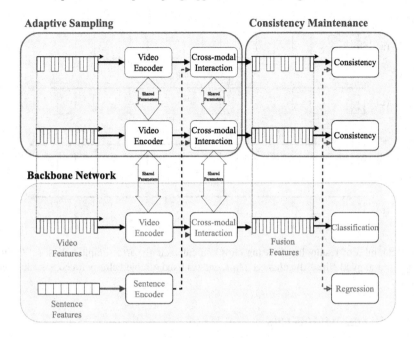

Fig. 4. The basic framework of our model.

feature representation for sparse clips. (iii) Consistency maintenance is used for generating the fusion of adjusted features and enhanced features by making multiple feature representations close to each other in the vector space.

3.2   Backbone Network

For completeness of presentation, we first briefly introduce the widely adopted backbone network in query-based movement retrieval. Generally, the backbone network consists of three parts: video encoder, sentence encoder, and cross-modal interaction. Specifically, given a video feature V^S, the video encoder computes the hidden representation H^v of this video. Similarly, for the text t_k, sentence encoder computes the hidden representation H^s of t_k. Cross modal interaction calculates the fusion representation of video and text, denoted $F = \{f_1, \ldots, f_N\}$, $f_i \in \mathbb{R}^d$, as follows:

$$H^v = VideoEncoder(V^S), \quad H^s = SentenceEncoder(t_k)$$
$$F = CrossModalInteraction(H^v, H^s). \tag{1}$$

After obtaining F, the backbone network will get a time segment through classification or regression. Since different backbone networks define different loss functions, we use L_{bone} to denote the loss function of a backbone network.

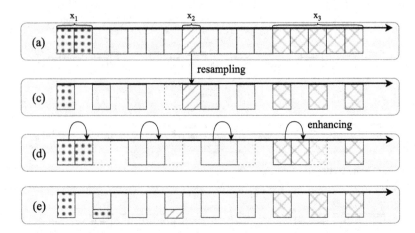

Fig. 5. Examples of resampling missing clips and enhancing sparse sampled clips. (a) Original video, (c) resampled clips, (d) enhanced clips, and (e) fixed sampled clips with consistency maintenance.

3.3 Resampling Missing Clips

For missing clips issue, we propose a solution to resample missing clips. This solution adjusts feature representation by replacing features of annotated and lost clips with those of unannotated clips.

We first uniformly select N clips from the original p clips. Suppose there is a sentence describing a moment with the annotation $a = (s, e) \in A$. Moreover, clip number $e - s$ is less than the interval $\frac{p}{N}$. The training model cannot learn the feature information of video clips that matches the sentence. Therefore, for each sentence in the training set, if its corresponding video segment clips are not selected, we replace the last clip's feature with the middle of the corresponding clip's feature of this segment to obtain a new video representation. The definition of obtaining each feature v_i^r in a new video representation $V^r = \{v_1^r, v_2^r, \ldots, v_N^r\}$, $v_i^r \in \mathbb{R}^d$ is given below:

$$
v_i^r = \begin{cases} v^o_{\lfloor \frac{s+e}{2} \rfloor}, & if \ \lfloor s \ mod \ r \rfloor \neq 0 \ \& \ \lfloor e \ mod \ r \rfloor \neq 0 \ \& \ \lfloor \frac{s}{r} \rfloor = \lfloor \frac{e}{r} \rfloor = i \\ v^o_{\lfloor r \times i \rfloor}, & others \end{cases}
\tag{2}
$$

where $r = \frac{p}{N}$ is the number of clips separated by the fixed sampling clips. It is noteworthy that since there is no annotated segment information in the test phase, this sampling method is only used as auxiliary information to learn in the training phase. Specific auxiliary learning operations will be introduced in Sect. 3.5.

3.4 Enhancing Sparse Sampled Clips

For spare clips issue, we propose an enhancing solution for sparse clips by repeating a set of video clips, to achieve an enhancement effect. This creates a new video representation, which also serves as auxiliary information.

Specifically, we define a set of sampling rates $R = \{r_1, r_2, \ldots, r_w\}, r_j \in \mathbb{R}^1$, which is used to select the degree of enhancement for a single clip. The more the rate, the more the enhancement. For each sampling rate r_j, we get the enhanced video representation $V_j^e = \{v_1^e, v_2^e, \ldots, v_N^e\}, v_i^e \in \mathbb{R}^d$ from V^S as:

$$v_i^e = v_{\lfloor \lfloor \frac{i}{r_j} \rfloor \times r_j \rfloor}^S,\tag{3}$$

where Eq. 3 is the enhancing function, and r_j is the sampling rate.

Then each video V_j^e is considered as the input of the backbone network. We take the union of $\{V^r\}$ and $\{V_1^e, \ldots, V_w^e\}$ as $\{V_1', \ldots, V_{w+1}'\}$. For each V_i', a fusion representation set is obtained as $\{F_1', \ldots, F_{w+1}'\}$ by the backbone.

3.5 Consistency Loss Maintenance

In this section, we propose consistency loss maintenance method to use F_i' to assist the learning of the original fusion representation F. Since F_i' and F are the fusion of the same video and sentence pair, the representation F_i' should be similar to the representation F. Therefore, to reduce the distance of F_i' and F in the vector space, F is maintained by the consistency loss function defined in Eq. 4.

$$L_{con} = \sum_{i=1}^{w+1} SmoothL1(F, F_i'),\tag{4}$$

where L_{con} is a smooth L1 loss function and $w + 1$ is the number of sampled features.

The representation of fusion feature in Fig. 5(e) obtained by the consistency maintenance is closer to the representations in Fig. 5(c) and Fig. 5(d). Then we can retrieve the time boundary by a better fusion representation $F = \{f_1, \ldots, f_N\}, f_i \in \mathbb{R}^d$ than that without the adaptive sampling and consistency loss maintenance method. Combining with the backbone loss fuction L_{bone}, the final loss function is defined as:

$$L = L_{bone} + \lambda_1 L_{con},\tag{5}$$

where λ_1 are hyper-parameter to control the balance of the two losses.

4 Experiments

4.1 Datasets and Evaluation Metrics

In order to validate our adaptive sampling approach, we conduct experiments on three real datasets, ActivityNet Captions [6][1], TACoS [8][2], and TACoS$_{2DTAN}$ [12]. Table 2 shows the details of these datasets.

We adopt "R@n and IoU=m", the two evaluation metrics proposed in [3] to measure our approach. "R@n" means the top-n retrieved moment results. IoU is calculated

---

[1] http://activity-net.org/challenges/2020/tasks/anet_captioning.html.
[2] https://www.coli.uni-saarland.de/projects/smile/page.php?id=tacos.

Table 2. Dataset details.

Datasets	Number of video-sentence pairs			Video lengths (Sec.)			Segment lengths (Sec.)		
	Training set	Validating set	Test set	Min	Average	Max	Min	Average	Max
ActivityNet	37, 417	17, 505	17, 031	2	124	755	1	37	409
TACoS	10, 146	4, 589	4, 083	49	332	1, 403	1	6	167
TACoS$_{2DTAN}$	9, 790	4, 436	4, 001	49	333	1, 403	1	28	843

Table 3. Performance comparison on TACoS using the C3D features. ("-" indicates that CPN cannot provide corresponding results.)

Methods	R@1 IoU0.1	R@1 IoU0.3	R@1 IoU0.5	R@5 IoU0.1	R@5 IoU0.3	R@5 IoU0.5	Time Per Pair	Train Video Mem	Num Para
CPN	33.01	24.66	15.04	–	–	–	19.750 ms	13873 MB	20.82 MB
CPN+Adaptive	**33.23**	**26.25**	**16.68**	–	–	–	25.781 ms	14917 MB	20.82 MB
CMIN	32.48	24.64	18.05	62.13	38.46	27.02	9.016 ms	6959 MB	18.96 MB
CMIN+Adaptive	**38.97**	**28.46**	**18.35**	**66.19**	**41.10**	**29.90**	14.078 ms	13251 MB	18.96 MB

Table 4. Performance comparison on ActivityNet using the C3D features.

Methods	R@1 IoU0.3	R@1 IoU0.5	R@1 IoU0.7	R@5 IoU0.3	R@5 IoU0.5	R@5 IoU0.7	Time Per Pair	Train Video Mem	Num Para
CPN	62.81	45.10	**28.10**	–	–	–	19.437 ms	16713 MB	14.82 MB
CPN+Adaptive	**63.71**	**45.63**	26.85	–	–	–	20.187 ms	18947 MB	14.82 MB
CMIN	**63.61**	43.40	23.88	80.54	67.95	50.73	8.750 ms	6963 MB	18.97 MB
CMIN+Adaptive	63.43	**43.55**	**24.25**	**81.51**	**68.43**	**50.93**	14.687 ms	13265 MB	18.97 MB

as IoU $= \frac{min(j,\hat{j})-max(i,\hat{i})}{max(j,\hat{j})-min(i,\hat{i})}$, where $S[i,j]$ is the ground truth segment, and $\hat{S}[\hat{i},\hat{j}]$ is the retrived segment. "R@n and IoU=m" means the percentage of queries having at least one result whose IoU is larger than m in top-n results. And the "mIoU" means the average IoU of all top-1 retrieved results.

4.2   Performance Comparisons

Firstly, we show the improvement of our approach. We used CPN and CMIN as the backbone network, and conducted experiments using C3D features to test the improvement of using adaptive sampling. The experimental results on the TACoS and ActivityNet datasets are shown in Tables 3 and 4. We can see that our adaptive sampling improved both CPN and CMIN on TACoS dataset. It demonstrates that our adaptive sampling method can adapt to long videos.

Secondly, we validate the ability to solve the problems in Sect. 2.2 through experiments with varying video lengths using CMIN. Figure 6 shows improvement in mIoU with our approach (green lines) compared to CMIN with fixed size sampling (blue lines). Our approach improves retrieval accuracy. Long videos show better improvements than since long videos have more missing and sparse clips than short videos.

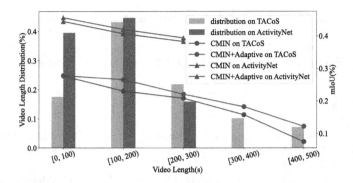

Fig. 6. Videos with different lengths in TACoS and ActivityNet using the CMIN backbone. (Color figure online)

Table 5. Performance comparison on ActivityNet and TACoS$_{2DTAN}$ using the I3D features.

Methods	ActivityNet			TACoS$_{2DTAN}$		
	R@1 IoU0.3	R@1 IoU0.5	R@1 IoU0.7	R@1 IoU0.3	R@1 IoU0.5	R@1 IoU0.7
CMIN	58.70	36.61	16.03	47.66	34.51	16.57
CMIN+Adaptive	**58.96**	**36.99**	**16.51**	**49.84**	**35.47**	**17.67**
CPN	63.20	44.89	27.16	47.69	**36.33**	21.58
CPN+Adaptive	**64.49**	**45.87**	**27.52**	**48.53**	36.11	**21.99**
VSLNet	**63.16**	43.22	26.16	47.11	36.34	26.42
VSLNet+Adaptive	60.81	**43.78**	**27.00**	**49.54**	**39.19**	**26.84**

Thirdly, we illustrate the generality of our approach using different backbone networks. We perform experiments on ActivityNet and TACoS$_{2DTAN}$ using the I3D [1] features. The experimental results are shown in Table 5. We can see that adaptive sampling improve the effectiveness of three backbone networks (i.e. CMIN, CPN, and VSLNet) on different datasets.

5 Conclusion

In this paper, we proposed a novel adaptive video frames sampling approach for query-based moment retrieval in videos. Specifically, we resample the missing clips and enhance the sparse clips to reduce the loss caused by most recent fixed sampling methods. We further proposed consistency loss maintenance to learn the semantic of the missing clips and the sparse clips. As a result, our approach improves the retrieval accuracy on three real datasets especially for long videos under different backbone networks.

Acknowlegments. The work is partially supported by the National Key Research and Development Program of China (No. 2020YFB1707901), National Natural Science Foundation of China (Nos. U22A2025, 62072088, 62232007), Ten Thousand Talent Program (No. ZX20200035), Liaoning Distinguished Professor (No. XLYC1902057), and 111 Project (B16009).

References

1. Carreira, J., Zisserman, A.: Quo vadis, action recognition? a new model and the kinetics dataset. In: 2017 IEEE Conference on Computer Vision and Pattern Recognition (CVPR), pp. 4724–4733 (2017)
2. Chen, J., Chen, X., Ma, L., Jie, Z., Chua, T.S.: Temporally grounding natural sentence in video. In: EMNLP (2018)
3. Gao, J., Sun, C., Yang, Z., Nevatia, R.: Tall: Temporal activity localization via language query. In: 2017 IEEE International Conference on Computer Vision (ICCV), pp. 5277–5285 (2017)
4. Hahn, M., Kadav, A., Rehg, J.M., Graf, H.P.: Tripping through time: efficient localization of activities in videos. arXiv:abs/1904.09936 (2020)
5. Hendricks, L.A., Wang, O., Shechtman, E., Sivic, J., Darrell, T., Russell, B.C.: Localizing moments in video with natural language. In: 2017 IEEE International Conference on Computer Vision (ICCV), pp. 5804–5813 (2017)
6. Krishna, R., Hata, K., Ren, F., Fei-Fei, L., Niebles, J.C.: Dense-captioning events in videos. In: 2017 IEEE International Conference on Computer Vision (ICCV), pp. 706–715 (2017)
7. Pennington, J., Socher, R., Manning, C.D.: Glove: global vectors for word representation. In: EMNLP (2014)
8. Regneri, M., Rohrbach, M., Wetzel, D., Thater, S., Schiele, B., Pinkal, M.: Grounding action descriptions in videos. Trans. Assoc. Comput. Linguist. **1**, 25–36 (2013)
9. Tran, D., Bourdev, L.D., Fergus, R., Torresani, L., Paluri, M.: Learning spatiotemporal features with 3d convolutional networks. In: 2015 IEEE International Conference on Computer Vision (ICCV), pp. 4489–4497 (2015)
10. Wang, W., Huang, Y., Wang, L.: Language-driven temporal activity localization: a semantic matching reinforcement learning model. In: 2019 IEEE/CVF Conference on Computer Vision and Pattern Recognition (CVPR), pp. 334–343 (2019)
11. Zhang, H., Sun, A., Jing, W., Zhen, L., Zhou, J.T., Goh, R.: Natural language video localization: a revisit in span-based question answering framework. IEEE Trans. Pattern Anal. Mach. Intell. **44**, 4252–4266 (2022)
12. Zhang, S., Peng, H., Fu, J., Luo, J.: Learning 2D temporal adjacent networks for moment localization with natural language. In: AAAI (2020)
13. Zhang, Z., Lin, Z., Zhao, Z., Xiao, Z.: Cross-modal interaction networks for query-based moment retrieval in videos. In: Proceedings of the 42nd International ACM SIGIR Conference on Research and Development in Information Retrieval (2019)
14. Zhao, Y., Zhao, Z., Zhang, Z., Lin, Z.: Cascaded prediction network via segment tree for temporal video grounding. In: 2021 IEEE/CVF Conference on Computer Vision and Pattern Recognition (CVPR), pp. 4195–4204 (2021)

Video Retrieval with Tree-Based Video Segmentation

Seong-Min Kang, Dongin Jung, and Yoon-Sik Cho[✉]

Chung-Ang University, Seoul 06974, South Korea
{kang7734,dongin1009,yoonsik}@cau.ac.kr

Abstract. Text-to-video retrieval aims to find relevant videos from text queries. The recently introduced Contrastive Language Image Pretraining (CLIP), a pretrained vision-language model trained on large-scale image and caption pairs, has been extensively used in the literature. Existing studies have focused on directly applying CLIP to learn the temporal dependency. While leveraging the dynamics of the video intuitively sounds reasonable, learning temporal dynamics has demonstrated no advantage or only small improvements. When temporal dynamics are not incorporated, most studies focus on constructing representative images from a video. However, we found these images tend to be noisy, degrading the performance of text-to-video task. This observation is the intuition for designing the proposed model, we introduce a novel tree-based frame division method to focus on the most relevant image for learning.

Keywords: Text-Video Retrieval · CLIP · Video Segmentation

1 Introduction

With the growth of online video-sharing platforms, text-video retrieval (TVR) has attracted significant attention from industry and academia. In these video-sharing platforms, searching for videos through a text-based query has become one of the main functionalities. In TVR, the model tries to learn a similarity function that performs feature learning both from videos and texts. The retrieval model needs to understand cross-modal relations and their semantics independently. Therefore, reducing the semantic and modality gaps in video-text is a significant and challenging task for TVR.

Over the past years, various TVR models have been proposed [1,4,10,12], which usually formulate the task as learning and matching based on the multi-modality encoder's output to the same embedding space. Recently, Contrastive Language Image Pretraining(CLIP) [14] introduced a contrastive learning-based matching method for images and text and gained a lot of attention by outperforming existing image models in image classification. The CLIP [14] is trained on image-text pairs to align image and text data into a joint embedding space. CLIP4Clip [12] adapted pre-trained weight of CLIP for video and performed both text-to-video (T2V) and video-to-text (V2T) retrieval tasks. In CLIP4Clip [12], the video representation is obtained by performing a frame sampling on a given video. A set of multiple images is encoded by ViT [3], and aggregated into an image representation. CLIP4Clip [12] proposed several frame aggregating methods. The best performing models obtain the representative image from

© The Author(s), under exclusive license to Springer Nature Switzerland AG 2023
X. Wang et al. (Eds.): DASFAA 2023, LNCS 13945, pp. 405–414, 2023.
https://doi.org/10.1007/978-3-031-30675-4_29

a video by aggregating video frames using mean-pooling. This image is later compared with the text embeddings for computing similarities.

The CLIP4Clip outperformed existing pretrained models including Frozen [1] and ClipBERT [10]. However, the visual backbone which extends image-language pre-trained model to video-language tends to over-smooth the video frames when fusing the video representation. This over-smoothing becomes more problematic when a video clip involves frequent scene changes. To this end, we propose a novel framework based on the *multiple choice learning* (MCL) scheme to consider *multiple scenes* from a video. Rather than aggregating all frames into a single image in CLIP4Clip, we propose the non-parametric video segmentation method named MCL-C4C (MCL-CLIP4Clip) that splits the clip into multiple segments for better matching with the given text by relaxing the problem to MCL. Thus, a text query in a video-text pair can better attend to the video encoding focusing on the target sub-frames. Our contributions are as follows:

- We introduce a novel framework that splits a video into multiple video segments in recursive fashion for obtaining multiple *clean* average frames.
- We construct each average frame from the segments, and resort to MCL scheme for choosing the most relevant image that matches with the given text queries.
- We show how our framework can be easily applied to existing work, and show performance improvement over CLIP4Clip.

2   Related Work

2.1   Video and Language Understanding for T2V Retrieval

Text-to-video (T2V) retrieval, along with video-to-text (V2T), is one of the most popular tasks for video-language understanding. Since then, various methods [1, 4, 10, 12, 16] have been proposed with the spotlight on T2V retrieval tasks. ClipBERT [10] and Frozen [1] employed sparse sampling and single uniform sampling extraction from the raw video, respectively. Another line of work is based on visual representation learning from text paired with images. CLIP [14] used natural language supervision for image representation learning and had shown significant improvements over various existing computer vision models. CLIP4Clip [12], which first introduced how CLIP can be applied to videos, successfully extracted representative images from videos by simply taking the average of frames and applying CLIP for T2V and V2T tasks. CLIP4Clip outperformed previous approaches in the T2V task with a large margin, and has been further extended in [4, 16]. These works [4, 16] focused on a better alignment of word-video tokens by splitting videos into fixed-length segments. We further study on video segmentation with variable lengths while maintaining the benefit of CLIP pretraining.

2.2   Multiple Choice Learning

Multiple choice learning (MCL) [7] is an algorithm that produces multiple structured outputs. In MCL [7], data examples are put into a multiple output model and trained by multiple output loss. MCL has been extended in many ways [8, 9, 11, 15], mainly focusing on image classification and segmentation. In this study, we adopt the MCL scheme into TVR, treating video segments as multiple models. We aim to select the most likely model (or sub-video in our context) from the multiple models. Backpropagation only flows to the selected model in the training step. In the following section, we show how our proposed approach segments a video into multiple video subsets.

3   Proposed Methods

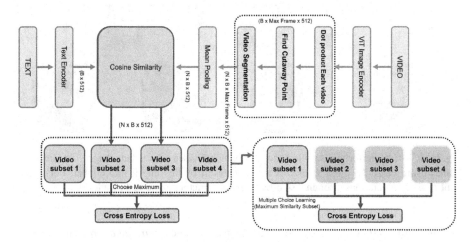

Fig. 1. The overall structure of proposed model based on CLIP4Clip [12]. Our methods are highlighted in a box with dotted lines, where other modules are borrowed from CLIP4Clip [12]. The content in parentheses is the dimension of the video. The output for final learning uses the concept of multiple choice learning to become a specific subset of video judged to have the most significant similarity to the text within the batch size.

We propose a novel framework based on the MCL scheme to address the aforementioned effects caused by scene changes in video clips. Instead of learning a pair between a whole single clip and its corresponding text, we consider multiple scenes in a given clip. We treat multiple scenes as an ensemble (multiple) of models, with texts as labels. Through MCL, we select the most relevant scene from *multiple* scenes for better matching between a video and its corresponding text. With the best match obtained through MCL, we can better learn the video representations by focusing on the exact match scene instead of the whole video with redundant scenes attached. Our method is based upon the backbone from CLIP4Clip [12] which adopts CLIP [14] for the encoders. The overall structure of our method is presented in Fig. 1. We use the same text encoder, a 12-layer 512-wide model with eight attention heads, from CLIP4Clip [12] to extract text features. The video encoder is ViT [3] which was used in CLIP4Clip [12]. In CLIP4Clip [12] the *parameter-free* approach uses cosine similarity between the text encoding and the average frame obtained from mean-pooling. The *sequential transformer* approach uses a transformer with trainable parameter to obtain video representation.

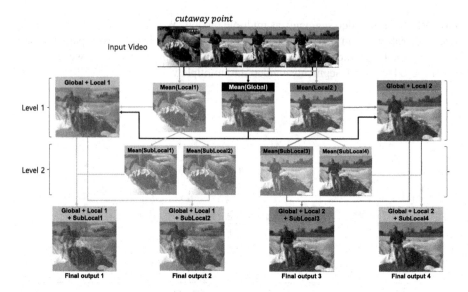

Fig. 2. MCL-C4C performed on selected video, which divides a 3-seconds range video into four frames using Frozen [1] and visualizes the final result by dividing it into Global, Local, and SubLocal representations through a step-by-step tree structure expressed by a level.

We further study the best performing two approaches: *parameter-free* and *sequential transformer* in CLIP4Clip. We also consider the video representation using the encoding for each frame as $V_i = \{F_1, F_2, ..., F_N\}$ and text caption representation T_j. We extend the existing framework by proposing the *Tree-based Video Segmentatio* algorithm and resort to MCL for finding better representative images, where we apply mean-pooling on each segment.

3.1 Tree-Based Video Segmentation

The input data consist of video-text pairs. Each video $V_i \in \mathcal{V}$ matches each text caption $T_i \in \mathcal{T}$, where a video V_i can be further represented as a collection of N frames: $V_i = \{F_1, F_2, F_3, ..., F_N\}$. Our proposed method performs video segmentation using binary splits in a binary tree. As shown in Fig. 2, we start with a video clip as a root node. A video is split into two segments, which are represented as two nodes (Local1 and Local2) at level 1 in a binary tree context. From level 2 and onward, we examine the consecutive frames and recursively perform a binary split on tree nodes to the next level only when necessary using the threshold. While our proposed algorithm is general enough to handle long video clips, we only consider up to level 2 (or four leaves at most). Each tree node performs its own "mean-pooling" on its corresponding video frames. This mechanism has two benefits. First, we can alleviate the over-smoothing effect by aggregating the sub-frames that are relevant each other. Second, the text queries can better find the relevant videos through comparing with all tree nodes with MCL.

3.2 Global-Local Video Representation

Global. We treat the existing video representation in CLIP4Clip as the Global representation. We therefore use the root node as the Global representation.

We denote the Global representation of video i as G_i which is expressed as below:

$$G_i = \text{mean-pooling}(V_i) = \text{mean-pooling}(F_1, F_2, ..., F_N), \tag{1}$$

where F_i is the generated representation of each frame and N is the index of last frame.

Algorithm 1. Tree-based Video Segmentation

Require: Video representation $V \in \mathbb{R}^{(B \times N \times 512)}$ ▷ B and N are batch size and frame number
Ensure: Video frames $V_i = \{F_1, F_2, ..., F_N\}$ ▷ F_n is each frame of video
   $G_i \leftarrow Mean(V_i)$ ▷ Global representation
   $M \in \mathbb{R}^{1 \times N} \leftarrow 0$ ▷ Initialize video frame mask
   $S \in \mathbb{R}^{N \times N} \leftarrow V_i \cdot V_i^\top$ ▷ S is a score matrix
   $S_0 \leftarrow S[0]$
   $S_0' \leftarrow S_0[1:].\text{append}(0)$ ▷ Shift
   $S_{Global} \leftarrow S_0 - S_0'$ ▷ Computes forward difference
   $cp' \leftarrow \text{argmax}(S_{Global})$ ▷ After first largest point has empty score
   $cp \leftarrow \text{argmax}(S_{Global}[:cp'])$ ▷ Second largest point is actual *cutaway point*
   $\text{Mask}_{Loc_1} \leftarrow M[:cp] = 1$
   $\text{Mask}_{Loc_2} \leftarrow M[cp:] = 1$
   $Loc_j = V_i * \text{Mask}_{Loc_j}$
   for $j = 1, 2$ **do**
     $S_{Loc} = Loc_j \cdot Loc_j^\top$
     $S_{Loc_k} = S[k]$ ▷ k is the First nonzero frame index in Loc_j
     $S_{Loc_k}' = S_{Loc_k}[k:].\text{append}(0)$
     $S_{Local} \leftarrow S_{Loc_k} - S_{Loc_k}'$
     $S_{Local}[S_{Local} \geq \lambda] = 1$ ▷ λ is a meaningful score threshold
     $S_{Local}[S_{Local} < \lambda] = 0$
     $scp' = \text{argmax}(S_{Local})$ ▷ After first largest point has empty score
     $scp = \text{argmax}(S_{Local}[:scp'])$ ▷ *cutaway point* in Loc_j
     $\text{Mask}_{SubLoc_u}, \text{Mask}_{SubLoc_{u+1}} = \text{Mask}_{Loc_j}[:scp], \text{Mask}_{Loc_j}[scp:]$
   end for
   $SubLoc_u = V_i * \text{Mask}_{SubLoc_u}$
   $Out = (G + Mean(Loc_j) + (Mean(SubLoc_u) \times \frac{\# \text{SubLoc frame}}{\# \text{Global frame}}))$

Local. In Fig. 2, mean-pooling is applied to each of the child nodes, and Local representations can be obtained. From the root node (input video) to level 1, we divide the video into two segments based on the scene transition point. We recursively repeat the process to the child nodes following Algorithm 1. At each leaf node, we compute the "path sum" from the root, which is used in MCL.

We first compute the cosine similarity between the frames of each video to find the scene transition point and make a similarity matrix S. The created S has number of frame N by N size, and each value s_{ij} represents the similarity between the i-th frame and the j-th frame. We apply the forward difference to the similarity between the first frame and the rest of the frame to find the *cutaway point* of the video. In this case, since one dimension is reduced by applying the forward difference, we append a zero-column vector at the end of S to match the dimension with the existing matrix. Using this, we can obtain the difference in similarity between the first frame and the other frame. In this matrix, the *cutaway point* is the index of the value with the second largest difference in similarity with the first frame. By applying the forward difference, it is possible to see how much the similarity of each frame changes sequentially, and this value increases as there is a rapid change between frames. Since the similarity between the first frame and the previously added zero will have a meager value, the index with a zero value will have the largest value in the matrix obtained through the forward difference. Similarly, when the frame of a video is less than N, the last index of the actual video frame will have the greatest value because the similarity is calculated with nonexistent value in a video. Thus, cp is the value with the second largest similarity difference.

Then, Loc_1 and Loc_2 are generated based on cp having the largest similarity difference. One video is divided based on cp through the generated mask Loc_i, and the similarity is calculated separately through text and cosine similarity between the two videos. The Local representation is expressed as Loc_i. As shown in Algorithm 1, the Local representations on level 1 can be expressed as follows:

$$Loc_1 = \text{mean-pooling}(F_1, ..., F_{cp}), \tag{2}$$

$$Loc_2 = \text{mean-pooling}(F_{cp+1}, ..., F_N), \tag{3}$$

where cp denotes the *cutaway point* in Local area.

Likewise, we can further define named SubLocal representations for the nodes on level 2. SubLocal representations are split into two nodes from each Local representation. For each SubLocal *cutaway point*, as indicated scp in Algorithm 1. However, all videos have not necessarily multiple *cutaway point*; thus, SubLocal representation is extracted only if there is a difference in similarity beyond a specific threshold λ. The method for extracting the SubLocal representation is divided into a tree structure in the Local representation, as shown in Fig. 2. From level 1, we have Local representations with mask based on the *cutaway point*. Accordingly, Loc_1, masked to the right by the *cutaway point*, determines the *cutaway point* based on its similarity to the first frame. However, for Loc_2 masked from the left based on the *cutaway point*, we determine the *cutaway point* to obtain it using similarity with the frame immediately following it. Further, suppose all the values in the video do not exceed the threshold. In that case, the division of the video stops there, and each SubLocal representation becomes the same as the Local representation. Hence, SubLocal representation exists only when there is a frame difference exceeding a threshold from the frames in a node from the previous level.

$$SubLoc_1 = \text{mean-pooling}(F_1, ..., F_{scp_1}), \tag{4}$$

$$SubLoc_2 = \text{mean-pooling}(F_{scp_1+1}, .., F_{cp}), \tag{5}$$

where the scp_1 is the cp which is obtained in a similar way when finding cp in level 1. We can find other SubLocal representation from another Local representation below:

$$SubLoc_3 = \text{mean-pooling}(F_{cp+1}, ..., F_{scp_2}), \tag{6}$$

$$SubLoc_4 = \text{mean-pooling}(F_{scp_2+1}, .., F_N). \tag{7}$$

Global-Local Aggregation. When Global representation is solely used, it is a video representation from the previous model, CLIP4Clip [12]. Local representations obtained from MCL-C4C focus on insignificant segmented fragment information by losing entire video content. The model cannot easily distinguish videos containing similar content when using the Local representation only. So we combine Global and Local representations, achieving better than Global-only and Local-only methods. From the obtained binary tree-based structure, we add up the Global and Local representations. If we divide the video into two (up to level 1), we get the final output by summing the tree structure at the first level, namely Global representation which is the root node, and Local representation which is the child node of the root node. Suppose the video is divided into four or fewer (up to level 2). In that case, the final output is multiplied by a slight weight to reduce the influence of the SubLocal representation obtained in level 2. In short, the purpose of the summation of each representation is to increase the proportion of Local representation in the final output.

3.3 MCL-CLIP4Clip

The mean-pooling method and sequential transformer method in CLIP4Clip [12] generate an image by performing just frame-wise mean-pooling over sampled frames with each method. However, when a video involves scene changes, the mean-pooling on a whole frame only obtains noisy video representation. Based on this observation, the fundamental assumption of **MCL-C4C** is that the given text query does not reasonably match the whole video, but better matches directly to a sub-frame.

In Fig. 2, we show the results through recursive tree structure algorithm in the areas marked by dotted lines after using cosine similarity in Fig. 1. In the first level, a video is divided into a subset of videos called Loc_1 and Loc_2 according to the cp. The output for each level is given by Algorithm 1. In the second level, if at least one of the frame differences is over the threshold, Loc_2 is split again through the dividing point, such as $SubLoc_3$ and $SubLoc_4$, creating a new video subset. The proposed method has a complete binary tree structure, and we obtain the final output by a weighted sum of edges connecting from root to leaf, rather than simply using the leaf node value.

The similarity between j-th text representation T_j and i-th video representation V_i in our methods is measured by the cosine similarity as below:

$$\text{sim}(V_i, T_j) = \max_m \frac{T_j^\top \bar{V}_i^m}{\|T_j\|\|\bar{V}_i^m\|}, \tag{8}$$

where $\bar{V}_i^m$ denotes Global-Local aggregated representation m-th segment of V_i. The Global-Local aggregation is the sum of the path from the root to the given node in

Fig. 2. On training, we use cross-entropy loss on each T2V and V2T task.

$$\mathcal{L}_{v2t} = -\frac{1}{B}\sum_B \log \frac{\exp(\text{sim}(V_i, T_i))}{\Sigma_{j=1}^{B}\exp(\text{sim}(V_i, T_j))} \tag{9}$$

$$\mathcal{L}_{t2v} = -\frac{1}{B}\sum_B \log \frac{\exp(\text{sim}(V_i, T_i))}{\Sigma_{j=1}^{B}\exp(\text{sim}(V_j, T_i))} \tag{10}$$

$$\mathcal{L} = \mathcal{L}_{v2t} + \mathcal{L}_{t2v}, \tag{11}$$

where B denotes the batch size and $\mathcal{L}$ denotes training loss.

Table 1. Result on MSR-VTT [17] dataset with [5], [13], and MSVD [2] dataset.

Methods	meanP				seqTransf			
	R@1↑	R@5↑	R@10↑	MnR↓	R@1↑	R@5↑	R@10↑	MnR↓
MSR-VTT 7k-training								
CLIP4Clip [12]	**43.6**	68.5	78.9	17.2	42.5	<u>69.0</u>	78.9	17.9
Ours (Div2)	**43.6**	<u>69.4</u>	<u>79.3</u>	<u>16.5</u>	<u>42.9</u>	**69.5**	<u>80.1</u>	**15.8**
Ours (Div4)	**43.6**	69.8	79.7	15.5	43.7	69.5	80.4	<u>16.5</u>
MSR-VTT 9k-training								
CLIP4Clip [12]	42.4	69.7	**80.8**	15.7	43.7	71.0	81.2	14.7
Ours (Div2)	**44.1**	<u>70.8</u>	80.6	**13.9**	**44.4**	<u>72.5</u>	81.8	<u>13.4</u>
Ours (Div4)	<u>43.4</u>	72.2	80.2	<u>14.4</u>	<u>44.3</u>	72.8	82.9	13.3
MSVD								
CLIP4Clip [12]	46.2	<u>76.1</u>	85.2	**9.9**	<u>45.3</u>	<u>75.8</u>	84.8	<u>10.1</u>
Ours (Div2)	**46.6**	76.3	85.3	**9.9**	**45.6**	75.9	84.8	**10.0**
Ours (Div4)	<u>46.5</u>	<u>76.1</u>	85.3	**9.9**	45.2	<u>75.8</u>	84.8	<u>10.1</u>

4   Experiments

To verify performance improvement by applying our method on [12], we only set the baseline as methods of [12] and follow the same evaluation criteria.

4.1   Datasets

MSR-VTT. MSR-VTT [17] is a dataset consisting of 10,000 videos and 20 sentence captions per video. Each video has a duration between about 10 and 30 s. According to the previous dataset split variant, we evaluate our model on a 7k-training [13] set and 9k-training [5] set, respectively, having 7,180 and 9,000 training videos. The testing dataset samely uses 1,000 clip-text pairs provided by [17].

MSVD. MSVD [2] contains 1,970 videos of various lengths from 1 to 62 s. On average, each video is paired with a set of 40 sentences. The split was suggested by [6] consisting of 1,200 videos for training, 100 for validation, and 670 for testing.

4.2   Experimental Details

Our proposed model relies on a text encoder and a video encoder with CLIP (ViT-B/32) [14]. For the experiments, we use four NVIDIA GeForce RTX 2080 Ti GPUs, with initial learning rates set to $1e-4$. The batch size is set to 64. The maximum number of frames is set to 12 same as [12]. So, we train our model with only two different approaches, where one of the settings uses 2-split videos and the other setting uses up to 4-split videos. Because, splitting to 2 or up to 4 segments is enough in this study. However, our proposed scheme is general enough to handle multiple splits if necessary.

4.3   Experimental Results

We compare the performance of our approach to the baselines from CLIP4Clip [12]. We apply our MCL schemes on two approaches to verify the effectiveness of our proposed model. Two approaches are expressed as **meanP** and **seqTransf** in Table 1, which correspond to *parameter-free* and *sequential transformer*. Table 1 summarizes the model performance across different datasets: [13], [5] and [2]. For the baselines, we conduct experiments in our environment using the code provided in [12]. Our method achieves the better or same performance as most datasets and metrics, excluding R@10 on meanP in MSR-VTT 9k [5].

5   Conclusion

In this paper, we propose MCL-C4C, which matches the text embedding with the most likely subset videos, based on MCL scheme. Two best performing models in CLIP4Clip use mean-pooling to aggregate frames, the average frame tends to be noisy when video frames have high dynamics. Our approach addresses this problem by segmenting a video through our proposed mechanism. Our video segmenting algorithm examines the difference of each frame representation, and dynamically split the video if necessary.

Acknowledgement. This work was supported by Institute of Information & Communications Technology Planning & Evaluation (IITP) grant funded by the Korean government (MSIT) (No. 2021-0-01341, Artificial Intelligence Graduate School Program of Chung-Ang Univ.), and (No. 2021-0-02067, Next Generation AI for Multi-purpose Video Search).

References

1. Bain, M., Nagrani, A., Varol, G., Zisserman, A.: Frozen in time: A joint video and image encoder for end-to-end retrieval. In: Proceedings of the IEEE/CVF International Conference on Computer Vision, pp. 1728–1738 (2021)
2. Chen, D., Dolan, W.: Collecting highly parallel data for paraphrase evaluation. In: Proceedings of the 49th Annual Meeting of the Association for Computational Linguistics: Human Language Technologies, pp. 190–200. Association for Computational Linguistics, Portland, Oregon, USA, June 2011
3. Dosovitskiy, A., et al.: An image is worth 16x16 words: transformers for image recognition at scale. arXiv preprint arXiv:2010.11929 (2020)
4. Fang, H., Xiong, P., Xu, L., Chen, Y.: Clip2video: mastering video-text retrieval via image clip. arXiv preprint arXiv:2106.11097 (2021)
5. Gabeur, V., Sun, C., Alahari, K., Schmid, C.: Multi-modal transformer for video retrieval. In: Vedaldi, A., Bischof, H., Brox, T., Frahm, J.-M. (eds.) ECCV 2020. LNCS, vol. 12349, pp. 214–229. Springer, Cham (2020). https://doi.org/10.1007/978-3-030-58548-8_13
6. Guadarrama, S., et al.: Youtube2text: recognizing and describing arbitrary activities using semantic hierarchies and zero-shot recognition. In: 2013 IEEE International Conference on Computer Vision, pp. 2712–2719 (2013)
7. Guzman-Rivera, A., Batra, D., Kohli, P.: Multiple choice learning: Learning to produce multiple structured outputs. In: Advances in Neural Information Processing Systems, vol. 25 (2012)
8. Lee, K., Hwang, C., Park, K., Shin, J.: Confident multiple choice learning. In: Precup, D., Teh, Y.W. (eds.) Proceedings of the 34th International Conference on Machine Learning. Proceedings of Machine Learning Research, vol. 70, pp. 2014–2023. PMLR, 06–11 August 2017
9. Lee, S., et al.: Stochastic multiple choice learning for training diverse deep ensembles. In: Lee, D., Sugiyama, M., Luxburg, U., Guyon, I., Garnett, R. (eds.) Advances in Neural Information Processing Systems, vol. 29. Curran Associates, Inc. (2016)
10. Lei, J., et al.: Less is more: clipbert for video-and-language learning via sparse sampling. In: Proceedings of the IEEE/CVF Conference on Computer Vision and Pattern Recognition, pp. 7331–7341 (2021)
11. Li, Z., Chen, Q., Koltun, V.: Interactive image segmentation with latent diversity. In: Proceedings of the IEEE Conference on Computer Vision and Pattern Recognition (CVPR), June 2018
12. Luo, H., et al.: CLIP4Clip: an empirical study of clip for end to end video clip retrieval. arXiv preprint arXiv:2104.08860 (2021)
13. Miech, A., Zhukov, D., Alayrac, J.B., Tapaswi, M., Laptev, I., Sivic, J.: Howto100m: Learning a text-video embedding by watching hundred million narrated video clips. In: Proceedings of the IEEE/CVF International Conference on Computer Vision, pp. 2630–2640 (2019)
14. Radford, A., et al.: Learning transferable visual models from natural language supervision. In: International Conference on Machine Learning, pp. 8748–8763. PMLR (2021)
15. Tian, K., Xu, Y., Zhou, S., Guan, J.: Versatile multiple choice learning and its application to vision computing. In: Proceedings of the IEEE/CVF Conference on Computer Vision and Pattern Recognition, pp. 6349–6357 (2019)
16. Wang, Q., Zhang, Y., Zheng, Y., Pan, P., Hua, X.S.: Disentangled representation learning for text-video retrieval. arXiv preprint arXiv:2203.07111 (2022)
17. Xu, J., Mei, T., Yao, T., Rui, Y.: MSR-VTT: a large video description dataset for bridging video and language. In: 2016 IEEE Conference on Computer Vision and Pattern Recognition (CVPR), pp. 5288–5296 (2016)

CMT: Cross-modal Memory Transformer for Medical Image Report Generation

Yiming Cao[1,2], Lizhen Cui[1,2(✉)], Lei Zhang[1,2], Fuqiang Yu[1,2],
Ziheng Cheng[1,2], Zhen Li[3], Yonghui Xu[2(✉)], and Chunyan Miao[4]

[1] School of Software, Shandong University, Jinan, China
[2] Joint SDU-NTU Centre for Artificial Intelligence Research (C-FAIR),
Shandong University, Jinan, China
clz@sdu.edu.cn, xu.yonghui@hotmail.com
[3] Department of Gastroenterology, Qilu Hospital of Shandong University,
Jinan, China
[4] Joint NTU-UBC Research Centre of Excellence in Active Living for the Elderly,
Nanyang Technological University, Singapore, Singapore

Abstract. Automatic medical image report generation has attracted extensive research interest in medical data mining, which effectively alleviates doctors' workload and improves report standardization. The mainstream approaches adopt the Transformer-based Encoder-Decoder architecture to align the visual and linguistic features. However, they rarely consider the importance of cross-modal interaction (*e.g.,* the interaction between images and reports) and do not adequately explore the relations between multi-modal medical data, leading to inaccurate and incoherent reports. To address these issues, we propose a **C**ross-modal **M**emory **T**ransformer model (CMT) to process multi-modal medical data (*i.e.,* medical images, medical terminology knowledge, and medical report text), and leverage the relations between multi-modal medical data to generate accurate medical reports. To explore the interaction of cross-modal information, we design a novel cross-modal feature memory decoder to memorize the relations between image and report features. Furthermore, the multi-modal feature fusion module in CMT exploits the multi-modal medical data to adaptively measure the contribution of multi-modal features for word generation, which improves the accuracy of generated reports. Extensive experiments on three real datasets demonstrate that our proposed CMT outperforms benchmark methods on automatic metrics.

Keywords: Medical Data Mining · Medical Report Generation · Encoder-Decoder Architecture

1 Introduction

Medical image reports serve an essential role in medical diagnosis and treatment [7]. In clinical practice, doctors describe medical images in free text to compose medical image reports as diagnostic evidence. However, the quality of reports

© The Author(s), under exclusive license to Springer Nature Switzerland AG 2023
X. Wang et al. (Eds.): DASFAA 2023, LNCS 13945, pp. 415–424, 2023.
https://doi.org/10.1007/978-3-031-30675-4_30

depends significantly on the levels and experience of doctors, which leads to a tremendous workload for doctors and uneven quality of reports. Therefore, automatic medical report generation [1,13,18] has attracted extensive research interest in medical data mining to alleviate doctors' workload while improving the quality and automation of medical image reports.

Most existing approaches follow the Encoder-Decoder paradigm to automatically generate medical image reports. Although these works can generate textual narratives for medical images, there are several issues that need to be further addressed. *Firstly*, the interaction between cross-modal features is not fully utilized. A few approaches employ memory units to record the relation between one modality, *i.e.*, the relation between images [5] and that between texts [4]. They do not sufficiently consider the interaction between cross-modality features in the word generation process, leading to the problem of image-report inconsistency. *Secondly*, the relationships between multi-modal medical data need to be further exploited. Most works [1,2,18] only use knowledge as an intermediate medium and do not explore the associations between different modal data.

To tackle the above limitations, in this paper, we propose a Cross-model Memory Transformer model (CMT) to generate medical image reports. CMT exploits the multi-modal nature of medical data to simultaneously incorporate three modalities of medical data, improving the quality of medical report generation. In order to enhance the interaction of cross-modal features, we design a memory module in the CMT decoder to memorize the interaction between the visual features of images and the linguistic features of reports. Furthermore, to exploit the characteristics of multi-modal medical data, a multi-modal feature fusion module is employed to perform sequence prediction by adaptively computing the contribution of multi-modal medical features to report sequences, and to generate accurate medical reports. Experimental results on three real-world datasets demonstrate the effectiveness of our proposed CMT.

2 Cross-modal Memory Transformer

To generate accurate and consistent medical image reports, we propose a cross-modal memory Transformer, which consists of four components, *i.e.*, a visual encoder, a medical term enhanced module, a cross-modal feature memory decoder, and a multi-modal feature fusion module. The overview of our proposed CMT is illustrated in Fig. 1.

2.1 Visual Encoder

Given a medical image I, the visual encoder is adopted to extract visual features $\mathbf{f}^v$ for I. The visual encoder is composed of a pre-trained CNN and a Transformer-based encoder [15], which can be formalized as:

$$\mathbf{f}^p = \{\mathbf{f}_1^p, \mathbf{f}_2^p, ..., \mathbf{f}_i^p, ..., \mathbf{f}_N^p\} = CNN(\{I_1, I_2, ..., I_i, ..., I_N\}), \tag{1}$$

$$\mathbf{f}^m = AddNorm(Attention(\mathbf{f}^p \mathbf{W}_q, \mathbf{f}^p \mathbf{W}_k, \mathbf{f}^p \mathbf{W}_v)), \tag{2}$$

$$\mathbf{f}^v = AddNorm(FFN(\mathbf{f}^m)), \tag{3}$$

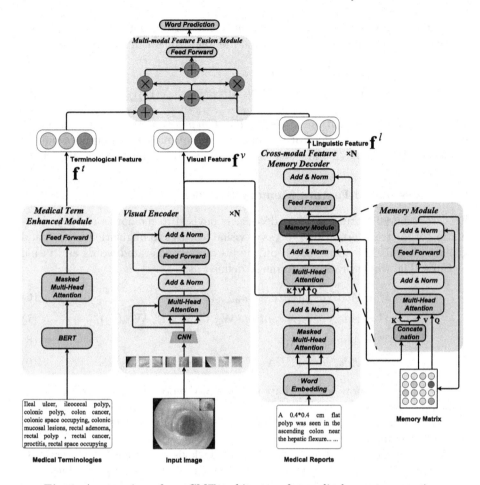

Fig. 1. An overview of our CMT architecture for medical report generation.

where N is the number of patches, $AddNorm$ includes a residual connection and a normalization layer, $Attention$ denotes a multi-head attention, and FFN represents the fully connected feed-forward network.

2.2 Medical Term Enhanced Module

The medical term enhanced module is employed to learn the contextual information representation of medical terminologies related to medical reports, which helps to improve the accuracy of the report generation.

Following the works [1] and [17], two medical terminology corpora are created to store descriptions of medical terminologies and findings that frequently appear in medical reports for gastrointestinal and thoracic diseases. Subsequently, the medical term enhanced module is adopted to process the two corpora mentioned above for contextual information learning. The terminological features $\mathbf{f}^t$ are extracted by the medical term enhanced module, which can expressed as:

$$\mathbf{f}^b = BERT(Corpus), \tag{4}$$

$$\mathbf{f}^t = Softmax(FFN(Attention_{mask}(\mathbf{f}^b))), \tag{5}$$

where $\mathbf{f}^b$ denotes the output of the BERT, $Corpus$ represents the sequence in the corpus, $\mathbf{f}^t$ is terminological features from a softmax distribution of output sequence, $Attention_{mask}$ indicates the masked multi-head attention operation.

2.3   Cross-modal Feature Memory Decoder

The cross-modal feature memory decoder is devised to memorize the key information of the transformation process of visual features to linguistic features, and generate linguistic features based on the previously generated words and visual features. The whole process is formally defined as:

$$\mathbf{f}^s = AddNorm(Attention_{mask}(\mathbf{f}^e)), \tag{6}$$

$$\mathbf{f}^i = AddNorm(Attention(\mathbf{W}_{qs}\mathbf{f}^s, \mathbf{W}_{ks}\mathbf{f}^v, \mathbf{W}_{vs}\mathbf{f}^v)), \tag{7}$$

$$\mathbf{f}^m = Memory(\mathbf{f}^i), \tag{8}$$

$$\mathbf{f}^l = AddNorm(FFN(\mathbf{f}^m)), \tag{9}$$

where $\mathbf{f}^s$ and $\mathbf{f}^i$ denote intermediate outputs of the decoder, $\mathbf{f}^e$ represents word embedding vectors, $\mathbf{f}^v$ is visual features, $\mathbf{f}^m$ is the memory features, $\mathbf{W}_{qs}$, $\mathbf{W}_{ks}$, and $\mathbf{W}_{vs}$ are matrices of learnable weights, $Memory$ indicates the memory module, and $\mathbf{f}^l$ is the linguistic features.

As mentioned above, the memory module is devised to record the interaction and mapping between visual features and linguistic hidden states. Specifically, it employs a memory matrix to pass mnemonic information during the generation process of the decoder, where the mnemonic information memorizes the interaction and mapping of visual features and word vectors. The memory matrix is updated simultaneously with the decoding process, which is formulated as:

$$\mathbf{Q}_m = \mathbf{M}_{t-1}^m\mathbf{W}_{qm}, \tag{10}$$

$$\mathbf{K}_m = Concat[\mathbf{M}_{t-1}^m, \mathbf{f}_{t-1}^i]\mathbf{W}_{km}, \tag{11}$$

$$\mathbf{V}_m = Concat[\mathbf{M}_{t-1}^m, \mathbf{f}_{t-1}^i]\mathbf{W}_{vm}, \tag{12}$$

$$\mathbf{f}^o = Attention(\mathbf{Q}_m, \mathbf{K}_m, \mathbf{V}_m), \tag{13}$$

$$\mathbf{f}^m = AddNorm(FFN(AddNorm(\mathbf{f}^o))), \tag{14}$$

$$\mathbf{M}_t^m = FFN(Add(\mathbf{M}_{t-1}^m, \mathbf{f}^m)), \tag{15}$$

where $\mathbf{Q}_m$, $\mathbf{K}_m$, and $\mathbf{V}_m$ denote inputs of the multi-head attention layer, $\mathbf{W}_{qm}$, $\mathbf{W}_{km}$, and $\mathbf{W}_{vm}$ are weight matrices, $Concat$ indicates the concatenation operation, $\mathbf{f}^o$ is the output from the multi-head attention layer, $\mathbf{M}_t^m$ represents the memory matirx of time step t, and Add is the matrix add operation.

2.4   Multi-modal Feature Fusion Module

In order to take full advantage of the association of medical multi-modal data, a multi-modal feature fusion module is designed to adaptively fuse the medical features of these three modalities. The multi-modal feature fusion module adaptively fuses three medical features using three operations: the element-wise summation $\boxplus$, element-wise multiplication $\otimes$, and vector summation $\oplus$ [6]. Formally, the multi-modal feature fusion module can be expressed as:

$$\mathbf{f}^f = \mathbf{f}^t \boxplus \mathbf{f}^v, \tag{16}$$

$$\mathbf{f}^a = w(\mathbf{f}^f \boxplus \mathbf{f}^l) \otimes \mathbf{f}^f \oplus ((1 - w)(\mathbf{f}^f \boxplus \mathbf{f}^l)) \otimes \mathbf{f}^l, \tag{17}$$

$$Output = FFN(\mathbf{f}^a), \tag{18}$$

where $\mathbf{f}^f$ is the combination of terminological features and visual features, ensuring that our CMT reconsiders the effect of terminology context before word prediction, and w denotes the weight of attention.

2.5   Training

For each sample (I, r), where I is a group of images and r is the corresponding report composed of the ground truth sequence, the loss of report generation $\mathcal{L}$ is minimized by the cross-entropy loss:

$$\mathcal{L}(\theta) = -\sum_{i=1}^{N} \log(p_\theta(s_i|s_{1:i-1})), \tag{19}$$

where θ is the parameters of our CMT model, $s_{1:i-1}$ represents the ground truth sequence of the report r.

3   Experiment

3.1   Experimental Settings

Dataset. Three real-world medical image report datasets are used in our experiment, which are described as follows.

GE [1]. The GE dataset consists 15,345 white light images of gastrointestinal endoscopy and 3,069 corresponding Chinese reports. For medical terminologies in GE, the gastroenterologists provide 126 medical terminologies frequently used in the report, containing 37 normal and 89 abnormal terminologies.

IU-CX [8]. The IU-CX is a widely used chest X-ray dataset. 2896 radiology reports with both frontal and lateral view chest X-ray images are extracted from the original dataset. For the construction of medical terminology corpus, we automatically extracted the medical terminologies for IU-CX from the "Abstract" field in the report, including 80 abnormal findings and 17 normal terminologies.

MIMIC-CXR [10]. The MIMIC-CXR is the largest public chest X-ray dataset with 473,057 images and 206,563 reports. Following the same data extraction criterion as IU-CX, we extract 142,772 images and 71,386 reports. The medical terminologies for MIMIC-CXR are the same as that for IU-CX.

Parameter Settings. The three datasets are randomly split into the training, validation and testing set in the ratio of 7:1:2. A pre-trained DenseNet-121 is adopted to extract image features, where the patch size is set to $7 * 7$. The number of layers and heads in Transformer-based modules are set to 3 and 8. The number of rows and dimension in Memory Matrix are set to 3 and 512 with random initialization. The hidden dimension in our CMT is set to 512.

Evaluation Metrics. We employ automatic metrics to evaluate the performance, including evaluation metrics widely-used in medical report generation models, *i.e.,* BLEU, CIDEr, ROUGE-L, and METEOR.

Baselines. The state-of-the-art report generation and image captioning approaches are used as baselines to compare with our CMT, including SaT [16], AAtt [14], CoAtt [9], Transformer [4], R2GEN [4], RGKG [18], PPKED [13], CMN [3], and KdTNet [1]. For IU-CX, the template retrieval methods such as HRGRA [12] and KEP [11] are also compared with our CMT for thoracic diseases. Note that due to the lack of predefined templates on GE, these two methods are not used for comparison on GE.

3.2 Results on Report Generation

To demonstrate the effectiveness of our proposed CMT, we compared the performance of CMT and baseline approaches on three datasets for automatic metrics, with the results presented in Table 1, where the **best** and second best results are highlighted.

It is observed that our CMT outperforms all baseline approaches on the CIDEr (or METEOR) and BLEU-n scores for all three datasets, which demonstrates the consistency and accuracy of our CMT in generating medical reports. For CIDEr, our CMT improves the performance by 0.85%–6.04% compared to the current state-of-the-art results, which indicates that our CMT is capable of generating more coherent reports since CIDEr is widely used to evaluate the similarity between generated medical reports and ground-truth reports. In terms of METEOR and BLEU, our CMT also achieves optimal results, which demonstrates that our CMT can generate accurate medical reports because METEOR considers both the precision and recall of the generated reports on the ground truth, and BLEU measures the consensus. As for ROUGR-L, our CMT performance is slightly worse than PPKED. One possible explanation is that PPKED employs additional medical knowledge (*e.g.,* retrieved radiology reports) to address the issue of data bias, making it easier to generate the same subsequence in generated reports as the ground-truth report.

Qualitative Analysis. To further investigate the effectiveness of our CMT, the qualitative analysis on three datasets is performed. Figure 2 presented the

Table 1. Comparison of report generation models on automatic metrics

Dataset	Methods	CIDEr	ROUGE-L	BLEU-1	BLEU-2	BLEU-3	BLEU-4
GE	SaT	0.557	0.613	0.643	0.552	0.506	0.414
	AAtt	0.579	0.617	0.649	0.549	0.491	0.419
	CoAtt	0.674	0.748	0.774	0.654	0.618	0.575
	RGKG	0.684	0.726	0.752	0.676	0.609	0.554
	Transformer	0.604	0.691	0.689	0.572	0.584	0.521
	R2GEN	0.679	0.736	0.779	0.677	0.619	0.574
	PPKED	0.691	**0.749**	0.791	0.684	0.624	0.579
	CMN	0.686	0.742	0.782	0.679	0.621	0.572
	KdTNet	0.692	0.748	0.792	0.686	0.624	0.583
	CMT(ours)	**0.698**	0.748	**0.795**	**0.689**	**0.628**	**0.586**
IU-CX	SaT	0.294	0.307	0.216	0.124	0.087	0.066
	AAtt	0.295	0.308	0.220	0.127	0.089	0.068
	CoAtt	0.277	0.369	0.455	0.288	0.205	0.154
	HRGRA	0.343	0.322	0.438	0.298	0.208	0.151
	KER	0.277	0.369	0.455	0.288	0.205	0.154
	RGKG	0.304	0.367	0.441	0.291	0.203	0.147
	Transformer	–	0.342	0.396	0.254	0.179	0.135
	R2GEN	–	0.371	0.470	0.304	0.219	0.165
	PPKED	0.351	**0.376**	0.483	0.315	0.224	0.168
	CMN	–	0.375	0.475	0.309	0.222	0.17
	KdTNet	0.341	0.375	0.474	0.316	0.225	0.169
	CMT(ours)	**0.354**	0.374	**0.485**	**0.321**	**0.229**	**0.175**

Dataset	Methods	METEOR	ROUGE-L	BLEU-1	BLEU-2	BLEU-3	BLEU-4
MIMIC-CXR	SaT	0.124	0.263	0.299	0.184	0.121	0.084
	AAtt	0.118	0.266	0.299	0.185	0.124	0.088
	CoAtt	0.138	0.274	0.410	0.267	0.189	0.144
	Transformer	0.125	0.265	0.314	0.192	0.127	0.090
	R2GEN	0.142	0.277	0.353	0.218	0.145	0.103
	PPKED	0.149	**0.284**	0.360	0.224	0.149	0.106
	CMN	0.142	0.278	0.353	0.218	0.148	0.106
	KdTNet	0.148	0.281	0.358	0.226	0.151	0.108
	CMT(ours)	**0.158**	0.283	**0.372**	**0.241**	**0.156**	**0.113**

visualization results of CMT on three datasets, including the image with ground-truth report and attention maps, the reports generated by our CMT, and the mappings from visual (image regions) and textual features (words and terminologies), where the attention maps and mapping areas are highlighted with different colors. It can be observed that our CMT is able to cover the normal medical descriptions and abnormalities for all three datasets. For example, the regions (*i.e.*, sigmoid colon) and types (*i.e.*, Yamada polyp) of the lesion in the first sample of GE are correctly reported in the generated report. Similarly, our CMT also precisely generates locations or types of the organ and lesion for IU-CX and MIMIC-CXR samples. In addition, reports generated by CMT cover essential medical terminologies describing normal conditions and abnormal symptoms, such as "Yamada polyp", "smooth mucosa", and "clear vascular texture" in the GE samples, and "mediastinal contour", "vascular engorgement", and "no pleural effusion" in the IU-CX and MIMIC-CXR samples. Meanwhile, the report generated by CMT is consistent with the abnormal regions in the

422   Y. Cao et al.

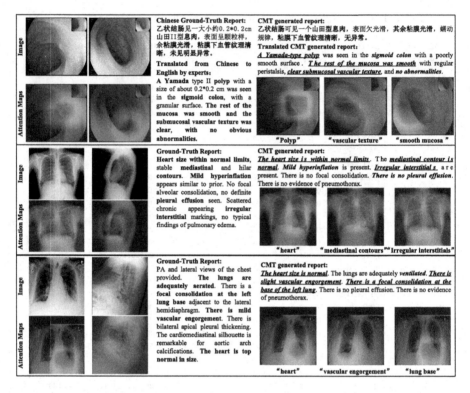

Fig. 2. Sample cases of CMT on GE (the first part), IU-CX (the middle part) and MIMIC (the last part). The left column is the image with generated attention maps by CMT, the middle column is the ground-truth report, and the right column is the CMT generated reports and the mappings of image region and medical terminologies. **Bold text** indicates consistency between the generated reports and ground truth. Underlined text indicates the correspondence between the generated reports and the attention maps.

attention maps. The result explicates that CMT is capable of generating reports consistent with the reports written by doctors.

To study the manner in which the CMT aligns the visual features with the terminological features, the image-text attention mappings of several terminologies from the multi-head attention layer of the cross-modal feature memory decoder are visualized, with the results exhibited in the lower right of each sample in Fig. 2. It is observed that the CMT is capable of aligning the locations in the image with the medical terminologies of the disease or organs. This observation indicates that our model not only improves the accuracy of medical report generation, but also better aligns images and important medical terminologies.

3.3   Ablation Studies

Effect of Components. The ablation studies are conducted on three datasets to explore the utility of each module in CMT, which are demonstrated in Fig. 3.

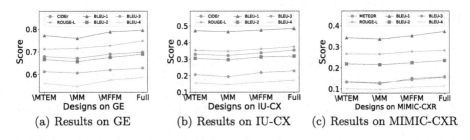

(a) Results on GE   (b) Results on IU-CX   (c) Results on MIMIC-CXR

Fig. 3. Ablation study for different designs

\MTEM excludes the medical term enhanced module from CMT, \MM and \MFFM drops the memory module in the cross-modal feature memory decoder and the multi-modal feature fusion module, respectively. CMT\MM performs the worst, indicating that memorizing the relation between visual and linguistic features will bring gains in accuracy for automatic report generation tasks. CMT\MTEM also has poor results, which suggests that introducing critical terminology knowledge into the report generation task is effective in enhancing the quality of reports. The performance of CMT\MFFM is worse than that of CMT, indicating that considering medical features of different modalities before word prediction can improve prediction performance.

4   Conclusion

In this paper, we propose a cross-modal memory Transformer model for accurate and coherent medical reports generation, which is able to process three modalities of medical data simultaneously and exploit the interaction between the different modalities of data. The cross-modal feature memory decoder is designed on the top of visual encoder to memorize the interaction between medical image and report features. A medical term enhanced module is adopted to provide the terminological information. Furthermore, the multi-modal feature fusion module exploits the cross-modal data to measure the contribution of multi-modal features for sequence generation. Extensive experiments on three real datasets show that our CMT outperforms baseline approaches.

Acknowledgements. This work is partially supported by the NSFC No. 62202279; National Key R&D Program of China No. 2021YFF0900800; Shandong Provincial Key Research and Development Program (Major Scientific and Technological Innovation Project) (No. 2021CXGC010108 and No. 2021CXGC010506); Shandong Provincial Natural Science Foundation (No. ZR202111180007); the Fundamental Research Funds of Shandong University; the State Scholarship Fund by the China Scholarship Council (CSC).

References

1. Cao, Y., et al.: Kdtnet: medical image report generation via knowledge-driven transformer. In: DASFAA 2022, pp. 117–132 (2022)
2. Cao, Y., et al.: Kdinet: knowledge-driven interpretable network for medical imaging diagnosis. In: BIBM 2022, pp. 1457–1460. IEEE (2022)
3. Chen, Z., Shen, Y., Song, Y., Wan, X.: Cross-modal memory networks for radiology report generation. In: ACL/IJCNLP 2021. pp. 5904–5914 (2021)
4. Chen, Z., Song, Y., Chang, T., Wan, X.: Generating radiology reports via memory-driven transformer. In: EMNLP 2020, pp. 1439–1449 (2020)
5. Cornia, M., Stefanini, M., Baraldi, L., Cucchiara, R.: Meshed-memory transformer for image captioning. In: CVPR 2020, pp. 10575–10584 (2020)
6. Dai, Y., Gieseke, F., Oehmcke, S., Wu, Y., Barnard, K.: Attentional feature fusion. In: IEEE WACV 2021, pp. 3559–3568. IEEE (2021)
7. Delrue, L., et al.: Difficulties in the interpretation of chest radiography. In: Comparative In- terpretation of CT and Standard Radiography of the Chest, pp. 27–49 (2011)
8. Demner-Fushman: Preparing a collection of radiology examinations for distribution and retrieval. MIA **23**(2), 304–310 (2016)
9. Jing, B., Xie, P., Xing, E.P.: On the automatic generation of medical imaging reports. In: Proceedings of the 56th Annual Meeting of the Association for Computational Linguistics, ACL 2018, pp. 2577–2586 (2018)
10. Johnson, A.E.W.: MIMIC-CXR: a large publicly available database of labeled chest radiographs. CoRR abs/1901.07042 (2019)
11. Li, C.Y., Liang, X., Hu, Z., Xing, E.P.: Knowledge-driven encode, retrieve, paraphrase for medical image report generation. In: AAAI 2019, pp. 6666–6673 (2019)
12. Li, Y., Liang, X., Hu, Z., Xing, E.P.: Hybrid retrieval-generation reinforced agent for medical image report generation. In: NeurIPS 2018, pp. 1537–1547 (2018)
13. Liu, F., Wu, X., Ge, S., Fan, W., Zou, Y.: Exploring and distilling posterior and prior knowledge for radiology report generation. In: IEEE, CVPR 2021, pp. 13753–13762 (2021)
14. Lu, J., Xiong, C., Parikh, D., Socher, R.: Knowing when to look: Adaptive attention via a visual sentinel for image captioning. In: 2017 IEEE Conference on Computer Vision and Pattern Recognition, CVPR 2017, pp. 3242–3250 (2017)
15. Vaswani, A., et al.: Attention is all you need. In: Advances in Neural Information Processing Systems 30: Annual Conference on Neural Information Processing Systems, NIPS 2017. pp. 5998–6008 (2017)
16. Vinyals, O., Toshev, A., Bengio, S.: Show and tell: a neural image caption generator. In: IEEE CVPR, pp. 3156–3164 (2015)
17. Zhang, X., et al.: Rstnet: captioning with adaptive attention on visual and non-visual words. In: IEEE, CVPR 2021. pp. 15465–15474 (2021)
18. Zhang, Y., Wang, X., Xu, Z., Yu, Q., Yuille, A.L., Xu, D.: When radiology report generation meets knowledge graph. In: The Thirty-Fourth AAAI Conference on Artificial Intelligence, AAAI 2020, pp. 12910–12917 (2020)

Fintech Key-Phrase: A New Chinese Financial High-Tech Dataset Accelerating Expression-Level Information Retrieval

Weiqiang Jin[(✉)] [iD], Biao Zhao [iD], and Chenxing Liu

School of Information and Communications Engineering, Xi'an Jiaotong University,
Innovation Harbour, Xi'an, Shaanxi, China
`{weiqiangjin,lcx459455791}@stu.xjtu.edu.cn, biaozhao@xjtu.edu.cn`

Abstract. Expression-Level Information Extraction is a challenging Natural Language Processing (NLP) task that aims to retrieve important information from the linguistic documents. However, there still lacks the up-to-date data sources for accelerating the Expression-Level Information Extraction, especially in the field of Chinese financial high technology. To fill this gap, we present Fintech Key-Phrase: a human-annotated Chinese financial high technology field related key-phrase dataset, which contains more than 12K paragraphs together with the annotated domain-specific key-phrases. We extract the publicly released reports on Chinese Management's Discussion and Analysis (CMD&A) from the well-known Chinese Research Data Services Platform (CNRDS) and then filter the Financial High-Tech related reports. The Financial High-Tech key-phrases are annotated through pre-defined philosophy guidelines to control the annotation quality. To demonstrate that our released Fintech Key-Phrase helps retrieve valuable information in the field of Chinese financial high technology, we adopt several superior Information Retrieval systems as representative baselines to validate its significance and report the performance statistics correspondingly. We hope this dataset can facilitate the scientific research and further exploration in the Chinese Financial High-Tech domain. We have made our Fintech Key-Phrase dataset and experimental code of the adopted baselines accessible at Github (https://github.com/albert-jin/Fintech-Key-Phrase/). To motivate newcomers to get involved in the Information Retrieval of the Chinese financial high technology field, we have built an open website (https://albert-jin.github.io/FintechKP-frontend/) and a real-time information retrieval API tool (https://31863ew564.zicp.fun/information_retrieval/).

Keywords: Information Retrieval · Data Mining and knowledge discovery · Expression-Level Information Extraction · Financial High Technology Field

© The Author(s), under exclusive license to Springer Nature Switzerland AG 2023
X. Wang et al. (Eds.): DASFAA 2023, LNCS 13945, pp. 425–440, 2023.
https://doi.org/10.1007/978-3-031-30675-4_31

1 Introduction

In Natural Language Processing (NLP), Information Retrieval remains crucial as it includes various research directions, such as Named Entity Recognition [22,23] and Relation Extraction [24]. Expression-Level Information Extraction is a sub-task of Information Retrieval that focuses on retrieving interesting words, phrases, or even paragraphs from large-scale textual documents [16], and is similar to Named Entity Recognition. It requires the machine to obtain a comprehensive semantic understanding and accurately identify whether each atomic expression (Minimum Semantic Unit) in a sentence contains interesting information. Expression-Level Information Extraction can facilitate many other downstream NLP tasks, such as the Passage Retrieval-based Question Answering [1,12] and Document Topic Classification [20]. Thus, Expression-Level Information Extraction continues to be one of the major linguistic processing problems in recent years.

Despite the recent success in Expression-Level Information Extraction, there is still a lack of specific-domain related information extraction datasets with which to extrapolate its potential in research and application, which limits the further development of textual semantic analysis. In particular, to our knowledge, there are few disclosed data sources for the Expression-Level Information Extraction task in the Chinese Financial High-Tech field.

To fill these gaps and stimulate in-depth research, we present a new dataset, named Chinese Financial High-Tech Based Key-Phrase (Fintech Key-Phrase), which can be regarded as the newest Chinese domain-specific Expression-Level Information Extraction benchmark.

1) First, the original corpus consisting of nearly one hundred thousand annual financial corporation reports is extracted from the well-known Chinese Research Data Services Platform (CNRDS)[1].
2) Then, we design an enterprise category mapping strategy to filter the high technology enterprise parts from all annual financial reports. The filtered enterprise annual reports are the original documents used to generate our Fintech Key-Phrase.
3) Considering the dataset reliability, the annual reports are manually annotated by crowd-sourced annotators based on pre-defined annotation rules rather than automatic labelling based on fixed matching patterns.

In this work, we make the following major contributions:

- We propose a new Chinese Financial High-Tech dataset (Fintech Key-Phrase) for Information Retrieval, which is derived from the publicly released Chinese Management's Discussion and Analysis (CMD&A). To the best of our knowledge, together with more than 1.2K human-annotated instances, Fintech Key-Phrase is the largest reliable Chinese benchmark for the Expression-Level Information Extraction task;

---

[1] https://www.cnrds.com/.

- To highlight the importance of our Fintech Key-Phrase in Expression-Level Information Extraction, we conduct comprehensive experiments by utilizing several (state-of-the-art) SOTA approaches. Experimental results demonstrate that our dataset can serve as solid baseline for future Information Extraction related research;
- The original corpus (Excel format), the annotated Key-Phrase dataset (JSON format), and the experimental evaluation scripts (Python) are publicly released[2] for reproducibility;
- To motivate newcomers to get involved in the Chinese Financial High-Tech Information Retrieval field, we have built a series of tools, including an open website[3] and the corresponding real-time information retrieval APIs[4].

2   Related Work

2.1   Expression-Level Information Extraction

Expression-Level Information Extraction, which acts as a popular text mining technique, has recently received much attention due to humans' continuously increasing need for valuable knowledge retrieved from numerous digital documents [3]. It requires automatically discovering valuable domain-specific information (e.g., phrases, proper nouns, and event descriptions) from large-scale unstructured documents [25]. The extracted results can effectively facilitate many NLP tasks, such as Question Answering [14] and Sentiment Analysis [13,15].

Recently, diverse studies have yielded significant performances in the Expression level Information Extraction. Zhou et al. [26] present the *PowerBioNE*, an *Hidden Markov Models* HMM & K-Nearest Neighbor algorithm (KNN) based system, which enhances the post-processing patterns to extract the interesting entities from a sentence automatically. Zhiheng et al. [11] introduce the BiLSTM-CRF network, which transfers the initial information extraction to a simple sequence tagging problem. Fei et al. [7] introduce a novel multitask learning method based on dispatched attention module. Nguyen et al. [19] propose a joint extraction framework for Information Extraction, named *Multi-Stage Attentional U-Net*, which extracts the valuable information from large-scale unstructured texts. Shen et al. [22] propose *Parallel Instance Query Network* (PIQN), which sets up global and learnable instance queries to extract interest phrases from a sentence in a parallel manner. Also, inspired by the Object Detection task [13], Shen et al. [23] propose *Locate and Label*, a two-stage information retrieval method, to find the domain-specific named entities from a sentence.

2.2   Domain-Specific Benchmarks

Despite numerous model innovations made in the Expression-Level Information Extraction, multiple widely used benchmarks also have been put forward.

---

[2] https://github.com/albert-jin/Fintech-Key-Phrase/.
[3] https://albert-jin.github.io/FintechKP-frontend/.
[4] https://31863ew564.zicp.fun/information_retrieval/.

There exists a vast number of widely used datasets for open-domain information extraction. *NNE* [21] is an English language dataset that is designed to extract the people, locations, and organisation from natural language texts. GermEval 2014 [4] is a well-known German NER benchmark which is built on top of the German News and Wikipedia Corpora. It involves 12 named entity classes, including *Person, Location, Organisation,* and *Others.*

Furthermore, many domain-specific benchmark datasets have been constructed by researchers to promote the development of the domain-specific information retrieval task. Li et al. [16] propose a large-scale chemical patent analysis benchmark for the ChEMU 2022 campaign, which focuses on promoting the drug design techniques and material manufacturing analysis in the Biological Chemical domain. In the mental health field, Soumitra et al. [9] re-annotate the *CEASE-v2.0* dataset [8] and propose a new benchmark dataset about emotion-cause suicide.

3   Fintech Key-Phrase: A New Dataset for Expression-Level Information Retrieval

3.1   Main Motivation

With the advent of technology and industrialization, the 21st century has continued the inevitable trend of technological revolutions and economic globalization. The financial high technology field has recently received much attention from both industries and academic communities. An intelligent financial analysis system can effectively help investors to conduct various investment activities (e.g., *stock price forecasting* and *corporate risk aversion*). The solutions about how to efficiently extract interesting information from the financial high technology domain remain an important research direction. For the mainstream supervised Neural Network-based (NN-based) approaches, the quality of annotated corpora directly affects the model's final performance. However, since domain-specific information extraction is still in its infancy, there is no available data sources in the financial high technology domain. This severe data deficiency has caused a considerable gap in the specific domain-related research. The TENCENT AI & NLP team built a natural language processing & analysis tool, *TexSmart*[5] [17], to conduct various NLP tasks (e.g., Nested Named Entity Recognition, and Fine-Grained Semantic Expression). Nevertheless, due to its open-domain oriented feature, *TexSmart* handles poorly in some domain-specific scenarios (Financial High-Tech domain).

We make a detailed Information Retrieval comparison between the Information Retrieval performance of *TexSmart* and the financial experts' agreed standard. As shown in Fig. 1, the *TexSmart* mispredicts multiple *Missing Items* (that should be included in the interesting keywords) and *Redundant Items* (that should not be included in the interesting keywords), which is an apparent discrepancy compared with the financial experts' professional annotations.

---

[5] https://texsmart.qq.com/en.

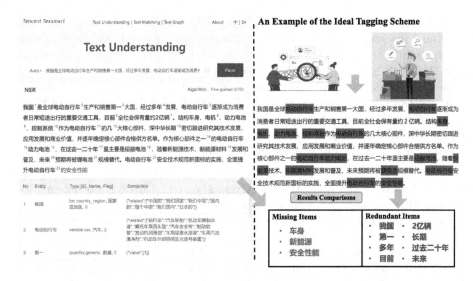

Fig. 1. Information extraction results comparison between the open-domain text understanding interface *TexSmart* and our ideal tagging results annotated by the financial experts.

This incorrect key-phrase identification affects the downstream NLP tasks. Thus, it is essential to develop an high-quality Financial High-Tech domain Information Retrieval dataset to generate a better information extraction model. To fill these gaps, we contribute a sizeable human-annotated dataset, named *Fintech Key-Phrase*, for the Expression-Level Information Extraction. We hope that our proposed *Fintech Key-Phrase* plays an important role in discovering the valuable knowledge in the Chinese financial high technology field.

3.2 Dataset Construction Guidelines

The construction of our presented dataset contains a series of preparation procedure.

Corpus Source Collecting. Before manual annotations, the corpora collection strategy is a crucial step that directly affects the annotating procedures and the final dataset qualities. The corpus collection can be divided into two primary steps: *Extraction from CMD&A*, and *Fintech Domain Filtering*.

Extraction from CMD&A. The Internet stores a massive quantity of Chinese financial market data, such as company trade records and corporate annual reports. On the Internet, the well-known Chinese Research Data Services Platform (CNRDS) has recorded a large amount of historical information on most Chinese public companies. Specifically, the CNRDS platform stores hundreds of millions of pieces of multi-dimension textual information for comprehensive analysis in all aspects, such as research reports, stock volatility records, credit or annual reports, and legal documents. We pay attention to the Chinese annual enterprise reports, which are also called Management's Discussion and Analysis (CMD&A)[6]. CMD&As primarily include both the business information (e.g., Financial Statements) and non-business information (e.g., Corporate Administrations, Core Techniques) together with the professional perspectives of these corporations. We request the official API and download the CMD&A documents of Chinese financial enterprises from 2008 to 2022 (14-years span until the present). The downloaded corpus contains up to millions of annual enterprise reports.

Fintech Domain Filtering.   After acquiring abundant CMD&A documents, we need to filter the high technology companies from the original documents accurately. We found that China Securities Regulatory Commission (CNSRC)[7] has formulated the China National Economic Industry Classification Guidelines[8]. The Guidelines indicate the pre-defined industrial classification for all Chinese public companies, including the industrial categories and their corresponding indexes (e.g., 核燃料加工   *Nuclear-Fuel Processing* [No. 253], and 航空器制造   *Spacecraft Manufacturing* [No. 3726]). We utilize these classification guidelines to filter the high technology-related CMD&A documents from the original corpus.

Here, we briefly introduce the data statistics of the filtered CMD&A documents:

– The High-Tech CMD&A annual reports comprise more than 16,600 documents, and the documents have recorded up to 2692 different companies' annual business reports;
– The High-Tech CMD&A documents contain about 11171 words on average. The maximum length and the minimum length in the documents are 115 and 32,006 words, respectively;
– The statistics on the document lengths and the document released time in the different intervals are shown in Fig. 2.

Key-Phrase Annotation.   Different from automatic construction, human annotation can endow datasets with higher recall and precision while decreasing unexpected noise. We formulate several primary labelling schemes as follows to circumscribe the scope of the interesting phrases that need to be annotated. In other words, given a set of specific-domain documents, crowd-sourced annotators were recruited to mark the target texts following these regulations. The pre-annotated documents is assigned to the students which are now the Masters of Finance from Nanjing University Business School. These crowd-sourcing annotators have a wide financial related background knowledge, which is crucial and essential to the overall quality of our Fintech Key-Phrase.

---

[6] https://www.cnrds.com/Home/index#/FeaturedDatabase/DB/CMDA.

[7] http://www.csrc.gov.cn/.

[8] http://www.csrc.gov.cn/csrc/c100103/c1452025/content.shtml.

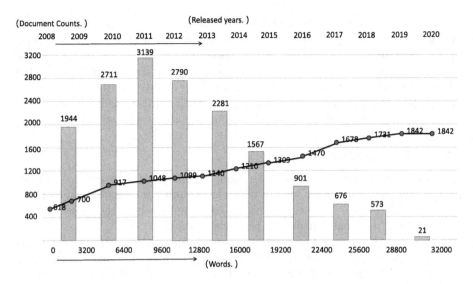

Fig. 2. Statistics of the High-Tech enterprises CMD&A reports.

- **Word Meaning Rules.** Annotators should pay more attention to the phrases related to the enterprise products, techniques, attributes, and events that lead to better insights into the enterprise nature or its recent condition, and the theme of the whole document;
- **Suffixes Exclusion Rules.** Annotators should delete the common-used suffixes of each key-phrase because of the semantic redundancy situation;
- **Conjunction Exclusion Rules.** Regarding to the conjunctions between two key-phrases, annotators should drop the conjunctions because the conjunctions have no semantics;
- **Universal Semantic Rules.** Annotators should skip labelling the micro-aspect words, such as the proprietary phrases created by the enterprise itself; Annotators are recommended to extract the explicit phrases which contain universal semantic information rather than infrequently appearing phrases.

Table 1 gives several annotation instances from our annotations, from which we can better understand the pre-defined tagging scheme.

Table 1. Examples of key-phrase annotations that follow the labelling scheme above.

	Sentence	Key-Phrases
S1	完成"面向下一代互联网感知-智能流量控制设备产业化项目"	互联网感知　智能流量控制
S2	天融信开展永磁环保柜、永磁环网柜及环网柜等开发项目	永磁环保柜　永磁环网柜　环网柜
S3	哈工易科加强物联网市场基础建设,加大研发人工智能机器人	物联网　人工智能机器人

After the annotation steps above, we generate more than 12,000 labelled instances. Notice that the distraction of annotators might lead annotators themselves to the omission of some key-phrase labelling, which may be harmful to the quality of our dataset. Considering these possible situations, we integrate all the labelled phrases and back-label them into the previously annotated dataset to ensure the key-phrases coverage of a certain quality. After the back-label step, the dataset is additionally populated with up to 165 key-phrases.

3.3 Dataset Statistics

After crowd-sourcing annotations, we randomly split our dataset into a subset for training and a subset for test, in which the training set contains up to 11579 samples, and the test set contains up to 610 samples where the ratio is about 19:1 following the same schedule of previous works [24].

Figure 3 statistics the key-phrases counts in different length segment intervals. We can clearly observe that the majority of Fin-tech domain key-phrases are scattered in the length range from 1 to 6, together with a smart part of key-phrases whose length is more than 7. This observation indicates that, generally, the key-phrases in which the financial experts are interested are short and simple.

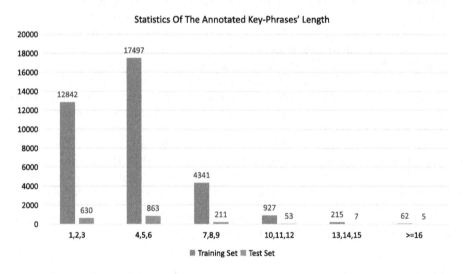

Fig. 3. The length scattering of the split train set (Blue) and the test set (Orange). (Color figure online)

4   Experiments

To provide follow-up researchers with explicit baseline comparisons towards our Fintech Key-Phrase, we conduct extensive experiments on a series of Neural

Network driven SOTA methods. Owing to our experiments' exhaustiveness and comprehensiveness, we hope that the experimental results reported in this work are sufficient to qualify as meanwhile representative baselines for future domain-specific information retrieval research.

4.1 Baseline Models

The baseline models which we choose are all rely on the Pre-trained Language Models (PLMs), including Chinese-BERT [2,5] and Chinese-RoBERTa [18]. Based on the publicly released code[9] and its experimental settings, we adopted six information extraction models to evaluate their different performances. These baseline models' details are listed as follows:

- **BERT-Linear** is a simple but effective approach which directly adds a single label decoding classifier after the feature encoder BERT;
- **RoBERTa-Linear** replaces the Transformer BERT with the enhanced Transformer *A Robustly Optimized BERT (RoBERTa)* to better utilize the large-scale PLMs' superiority compared with the *BERT-Linear*;
- **BERT-CRF** is also a BERT-based labelling model which replaces the simple *Full-Connected* classifier with a complex Conditional Random Field (CRF) [6] classifier compared with the model *BERT-Linear*;
- **RoBERTa-CRF** replaces the *BERT-CRF*'s BERT encoder with the advanced RoBERTa encoder to gain better information retrieval performances, which is similar to *BERT-Linear*;
- **BERT-BiLSTM-CRF** adds an extra Bidirectional Long Short Term Memory (Bi-LSTM) network among the BERT encoder and the CRF classifier of the *BERT-CRF* to better capture the dependencies on observation sequences;
- **RoBERTa-BiLSTM-CRF** replaces the *BERT-BiLSTM-CRF*'s BERT encoder with the enhanced Transformer RoBERTa to achieve superior performance (Fig. 4).

4.2 Performance Results

Following previous works [10,11], we adopt the *F1-score* as evaluation metric, which is also a typical classification metric in multi-label classification tasks. Formally, *F1-score* could be interpreted as a weighted balance of the *Recall* metrics and *Precision* metrics. Generally, a powerful information extraction model always leads to a higher *F1-score* performance.

As we know, experimental random fluctuations are crucial factors that might affect the model's final performance. Since one single experiment for each baseline has certain random fluctuations, these experimental results lack enough stability and representative conclusions. The unexpected issues include the CPU or GPU's random seeds, which are used to initialize model parameters (e.g., Linear

---

[9] https://github.com/hemingkx/CLUENER2020.

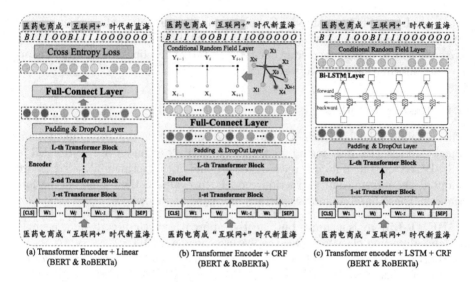

Fig. 4. Overall architectures of all the baselines we adopted. The feature encoders can be substituted with various Transformer-based PLMs.

Layer, Dropout Layer), the trivial difference of hyperparameter setting, and so on.

To alleviate the perturbation from various random factors, we perform tenfold Information Extraction cross-validation on our Fintech Key-Phrase instead of conducting model training and evaluation just once. The average *F1-score* and *Standard Deviation* on the test set are reported in Table 2, which can represent the final capabilities of these baselines. We hope these extensive experimental results can act as standard baseline performances for research on the Expression-Level Information Retrieval task in the cross-field of Chinese financial high technology.

From the Table 2 overall F1-score performance comparisons on our *Fintech Key-Phrase* dataset, we clearly find the following capability rankings of these baselines: *RoBERTa-BiLSTM-CRF > BERT-BiLSTM-CRF > RoBERTa-CRF > BERT-CRF > RoBERTa-Linear > BERT-Linear*. The performance gain of *BERT-BiLSTM-CRF* is trivial when compared with the *BERT-CRF* (similar with *RoBERTa-CRF* and *RoBERTa-CRF* counterparts). We speculate that this is because stacking multiple layers could bring a small performance gain, but sometimes the performance is reduced when the network depth is increased. Disregarding the trivial factors, these results further demonstrate the powerful capabilities of the Transformer-based encoders: BERT and its variant, RoBERTa, the superiority of CRF layer over other simple classifiers, and the effectiveness of *BiLSTM* for the long-term language sequence modelling. Based on the empirical explorations on the *Fintech Key-Phrase* dataset, we expect a wide range of powerful SOTA methods for the Expression-Level Information Retrieval task will be proposed in future.

Table 2. Performance statistics of the ten-fold evaluation experiments (under different random seeds) on *Fintech Key-Phrase* dataset towards the adopted six baselines, from which the *Min* and *Max* columns respectively denote the Minimum *F1-score* and Maximum *F1-score*. The column *Avg. ± Std.* denotes the average performances within the standard deviation range of all adopted baselines.

Baseline/Experiments	1	2	3	4	5	6	7	8	9	10	Min	Max	(Avg. ± Std.)
BERT-Linear	75.45	76.29	**77.92**	77.19	76.94	77.32	75.72	77.47	75.29	76.38	75.29	77.92	(76.596 ± 0.864)
RoBERTa-Linear	79.44	**80.57**	78.37	80.16	78.92	78.79	78.37	80.16	78.71	79.32	78.37	80.57	(79.278 ± 0.752)
BERT-CRF	81.25	80.22	80.78	**81.51**	80.31	79.92	81.41	80.75	80.49	79.41	79.41	81.51	(80.645 ± 0.689)
RoBERTa-CRF	81.93	80.57	81.77	**82.31**	81.93	80.57	81.77	81.65	**82.31**	80.96	80.57	82.31	(81.475 ± 0.66)
BERT-BiLSTM-CRF	80.96	81.62	81.24	80.27	81.48	81.24	80.96	**82.07**	81.60	80.48	80.96	82.07	(81.202 ± 0.526)
RoBERTa-BiLSTM-CRF	81.77	81.62	**82.31**	80.75	80.96	81.25	81.93	81.41	**82.31**	80.75	80.75	82.31	(81.512 ± 0.551)

5   Case Study

To understand the in-depth significance of our proposed *Fintech Key-Phrase*, we collect some information retrieval comparison instances between our well-trained baseline *RoBERTa-BiLSTM-CRF* and the *TENCENT TextSmart*. The *TENCENT TextSmart* is one of the most advanced open-domain information retrieval tools in China, which is well-known and convenient. As shown in Table 3, four specific examples are listed. For each case, the standard results of expert annotations, our trained baseline's extracted phrases, and the *TexSmart*'s extracted phrases are shown in different columns, respectively. As we expected, all the samples clearly demonstrate the effectiveness of *RoBERTa-BiLSTM-CRF*, which is fully trained on *Fintech Key-Phrase* benchmark. Before the analysis, it should also be emphasized that the experts mainly pay attention to the concept-level phrases of Financial High-Tech domain rather than the reified things.

Specifically, we could have the following observations. (1) In the first case, our trained baseline correctly predicts all the annotated phrases which are annotated by experts, while the *TexSmart* API makes several wrong predictions such as the redundant extracted expressions (长期 , 50%) and the missed expressions (天然气). (2) For the second case, our model performs very well while the *TexSmart* makes several minor errors. Compared with the expert annotation phrases, the minor errors includes: missing the words 研究 for the phrase 聚氨酯研究 ; over-predicting the word 所 for the phrases, 醇胺工程研究 and 节能新材料研究 ; and missing the words 分析 for the phrase 分析检测中心 . The examples shown in the third and fourth rows are similar to the above cases.

All the listed cases prove sufficient evidence that our fully-trained *RoBERTa-BiLSTM-CRF* is far superior to the other methods for the Financial & High-Tech domain Expression-Level Information Retrieval tasks. In conclusion, the case analysis demonstrates that it is practical and helpful to provide a high-quality dataset to solve the inaccuracy problem for the domain-specific information extraction task.

Table 3. Case analysis of the information extraction performance comparisons between the baseline *RoBERTa-BiLSTM-CRF* and the *TENCENT TextSmart* official released API. Referring to the expert annotations as standards, the words which are crossed out denote the incorrect predictions.

Original Text	Expert Annotations	Our RoBERTa BiLSTM-CRF	TexSmart Preds
从长期来看，全球天然气的需求将增长 50%以上，是化石燃料中增长最快的。	天然气 化石燃料	天然气 化石燃料	~~长期~~ ~~50%~~ 化石燃料
研究院下设聚氨酯研究所、醇胺工程研究所、节能新材料研究所、分析检测中心等部门。	聚氨酯研究 醇胺工程研究 节能新材料研究 分析检测中心	聚氨酯研究 醇胺工程研究 节能新材料研究 分析检测中心	~~聚氨酯~~ 醇胺工程研究~~所~~ 节能新材料研究~~所~~ ~~检测中心~~
植物提取物、人参提取物、人参皂苷单体分离方法的研究、产品的开发取得了阶段性成果。	植物提取物 人参提取物 人参皂苷单体分离	植物提取物 人参提取物 人参皂苷单体分离	植物提取物 人参提取物 ~~人参皂苷~~
单抗行业现状及未来发展趋势单克隆抗体药物是当今国际医药界的前沿领域。	单抗 单克隆抗体药物	单抗 单克隆抗体药物	~~未来~~ ~~当今~~

6 Released Tools for Financial High-Tech Domain Information Retrieval

To motivate newcomers to get involved in the Financial High-Tech specific-domain Information Retrieval field, we have built a series of functional tools, including a website [10] which gives brief introductions on our Fintech Key-Phrase together with six real-time user interactive planes[11] for information extraction (corresponding to six adopted baselines), and the released information extraction APIs[12]. The tools allow the user to know more about our *Fintech Key-Phrase* benchmark and query the key expressions from a piece of Chinese Financial High-Tech domain text.

6.1 Website of Our Fintech Key-Phrase

As shown in Fig. 5, when accessing the homepage, you will see that the screen displays four major sections:

- **Introduction** mainly gives a brief introduction of the *Fintech Key-Phrase* benchmark. The main plane shows the original statistics and the train/test split details of the *Fintech Key-Phrase* dataset. Mouse-clicking on the *Learn More* button, the specific information including the *DataSet Construction Motivation,* and *DataSet Construction Procedures* is displayed;

---

[10] Online Website: https://albert-jin.github.io/FintechKP-frontend/.

[11] https://albert-jin.github.io/FintechKP-frontend/model_prediction_rblc.html.

[12] https://31863ew564.zicp.fun/information_retrieval.

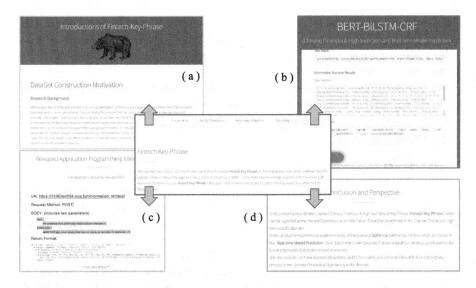

Fig. 5. The functional partitions of our released website. The part (a), (b), (c), and (d) respectively depict the four pages: Introduction, Model Prediction, Application Interface and Summary.

- **Model Prediction** mainly provides convenient user experiences of the well-trained baselines' performances for all possible users. In *Real-time Model Prediction* section, mouse-clicking on one of the six sub-planes (e.g., the *BERT-BiLSTM-CRF* or *RoBERTa-BiLSTM-CRF*), it redirects to a new user interface in which you can make Chinese fin-tech domain information extraction in real-time;
- **Application Interface** introduces the detailed API requesting and calling approaches for each well-trained baseline, including various calling parameters (e.g., headers, body, and standard returned formats);
- **Summary** Last but not least, we summarize our major contributions towards our proposed *Fintech Key-Phrase* in the last part of the website.

6.2 Released APIs

To further accelerate future research on domain-specific information retrieval, we release a series of online APIs[13]. As depicted in Fig. 5 (c), users should prepare all needed request parameters in advance, including:

- (1) The request method only allows the POST way. The API querying process may affect a certain network delay, wait for a few seconds patiently;
- (2) The request headers should contain *content-type: application/json* to ensure that the returned data belongs to the standard JSON format;

---

[13] https://31863ew564.zicp.fun/information_retrieval.

- (3) It needs to attach the complete Chinese sentence into the body parameter *sent*. Also, please select a baseline from the list: [bert_linear, roberta_linear, bert_crf, roberta_crf, bert_lstm_crf, roberta_lstm_crf] and attach it into the body parameter, *model_type*;
- (4) Furthermore, the real-time APIs are under long term maintenance. We encourage users who are interested in to contact with us proactively and report their questions or issues.

7   Conclusion

In the current work, we propose a high-quality human-annotated dataset, Fintech Key-Phrase. To the best of our knowledge, the benchmark is the first also the largest Chinese fin-tech domain information extraction dataset. The dataset aims to provide the missing training data sources in the Chinese financial high technology field information retrieval task. The professional also careful annotation procedures under a series of pre-defined annotation guidelines ensure the dataset's overall quality. We hope our proposed corpus extraction methodology and the key-phrase annotation guidelines will be helpful to future research in domain-specific Information Retrieval. Furthermore, we conduct comprehensive experiments by utilizing several SOTA approaches, where the experimental results can serve as solid baselines for future research studies. The Fintech Key-Phrase benchmark and corresponding experimental code are publicly released[14]. To motivate subsequent related works, we build a customized website and real-time information retrieval APIs for the Fintech Key-Phrase benchmark. In the future, we plan to complete the following works: (1) Annotate more Financial High-Tech domain corpus from other possible information channels (e.g., Chinese Business Website, Innovation China Website); (2) Conduct more experiments on other advanced information retrieval approaches on our Fintech Key-Phrase to improve the performances fully.

Acknowledgments. This work was carried out by the first author (Weiqiang Jin) during his research at Xi'an Jiaotong University. Weiqiang Jin is the corresponding author. We would like to thank Guizhong Liu (Xi'an Jiaotong University) and Bryce Guo (Nanjing University) for providing helpful discussions and valuable recommendations. We thank the anonymous reviewers, and the action editor for their insightful feedback that helped improve the paper.

References

1. Albarede, L., Mulhem, P., Goeuriot, L., Le Pape-Gardeux, C., Marie, S., Chardin-Segui, T.: Passage retrieval on structured documents using graph attention networks. In: Hagen, M., et al. (eds.) ECIR 2022. LNCS, vol. 13186, pp. 13–21. Springer, Cham (2022). https://doi.org/10.1007/978-3-030-99739-7_2
2. Ashish, V., et al.: Attention is all you need (2017)

---

[14] https://github.com/albert-jin/Fintech-Key-Phrase/.

3. Bai, C.: E-commerce knowledge extraction via multi-modal machine reading comprehension. In: Bhattacharya, A., et al. (eds.) DASFAA 2022. LNCS, vol. 13247, pp. 272–280. Springer, Cham (2022). https://doi.org/10.1007/978-3-031-00129-1_21
4. Benikova, D., Biemann, C., Reznicek, M.: Nosta-d named entity annotation for German: guidelines and dataset. In: LREC (2014)
5. Devlin, J., Chang, M.W., Lee, K., Toutanova, K.: BERT: pre-training of deep bidirectional transformers for language understanding (2018). https://doi.org/10.48550/ARXIV.1810.04805
6. Eddy, S.R.: Hidden Markov models. Curr. Opin. Struct. Biol. **6**(3), 361–365 (1996). https://doi.org/10.1016/S0959-440X(96)80056-X
7. Fei, H., Ren, Y., Ji, D.: Dispatched attention with multi-task learning for nested mention recognition. Inf. Sci. **513**, 241–251 (2020). https://doi.org/10.1016/j.ins.2019.10.065
8. Ghosh, S., Ekbal, A., Bhattacharyya, P.: A multitask framework to detect depression, sentiment and multi-label emotion from suicide notes. Cogn. Comput. **14**(1), 110–129 (2021). https://doi.org/10.1007/s12559-021-09828-7
9. Ghosh, S., Roy, S., Ekbal, A., Bhattacharyya, P.: CARES: CAuse recognition for emotion in suicide notes. In: Hagen, M., et al. (eds.) ECIR 2022. LNCS, vol. 13186, pp. 128–136. Springer, Cham (2022). https://doi.org/10.1007/978-3-030-99739-7_15
10. Hochreiter, S., Schmidhuber, J.: Long short-term memory. Neural Comput. **9**, 1735–80 (1997). https://doi.org/10.1162/neco.1997.9.8.1735
11. Huang, Z., Xu, W., Yu, K.: Bidirectional LSTM-CRF models for sequence tagging (2015). https://doi.org/10.48550/ARXIV.1508.01991
12. Izacard, G., Grave, E.: Leveraging passage retrieval with generative models for open domain question answering. In: Proceedings of the 16th Conference of the European Chapter of the Association for Computational Linguistics: Main Volume, pp. 874–880. Association for Computational Linguistics, Online, April 2021. https://doi.org/10.18653/v1/2021.eacl-main.74
13. Jin, W., Yu, H., Luo, X.: CVT-ASSD: convolutional vision-transformer based attentive single shot multibox detector. In: 2021 IEEE 33rd International Conference on Tools with Artificial Intelligence (ICTAI), pp. 736–744 (2021). https://doi.org/10.1109/ICTAI52525.2021.00117
14. Jin, W., Zhao, B., Yu, H., Tao, X., Yin, R., Liu, G.: Improving embedded knowledge graph multi-hop question answering by introducing relational chain reasoning. Data Min. Knowl. Discove. (2022). https://doi.org/10.1007/s10618-022-00891-8
15. Jin, W., Zhao, B., Zhang, L., Liu, C., Yu, H.: Back to common sense: Oxford dictionary descriptive knowledge augmentation for aspect-based sentiment analysis. Inf. Process. Manag. **60**(3), 103260 (2023). https://doi.org/10.1016/j.ipm.2022.103260, https://www.sciencedirect.com/science/article/pii/S0306457322003612
16. Li, Y., et al.: The ChEMU 2022 evaluation campaign: information extraction in chemical patents. In: Hagen, M., et al. (eds.) ECIR 2022. LNCS, vol. 13186, pp. 400–407. Springer, Cham (2022). https://doi.org/10.1007/978-3-030-99739-7_50
17. Liu, L., et al.: Texsmart: a system for enhanced natural language understanding. In: The Joint Conference of the 59th Annual Meeting of the Association for Computational Linguistics and the 11th International Joint Conference on Natural Language Processing (ACL-IJCNLP): System Demonstrations (2021)
18. Liu, Y., et al.: RoBERTa: a robustly optimized BERT pretraining approach. arXiv abs/1907.11692 (2019)
19. Nguyen, T.A.D., Thanh, D.N.: End-to-end information extraction by character-level embedding and multi-stage attentional u-net. In: BMVC (2019)

20. Qiang, J., Chen, P., Ding, W., Wang, T., Xie, F., Wu, X.: Heterogeneous-length text topic modeling for reader-aware multi-document summarization. ACM Trans. Knowl. Discov. Data **13**(4) (2019). https://doi.org/10.1145/3333030

21. Ringland, N.: Structured named entities. Ph.D. thesis, The University of Sydney, 30 September 2015. http://hdl.handle.net/2123/14558

22. Shen, Y., Ma, X., Tan, Z., Zhang, S., Wang, W., Lu, W.: Locate and label: a two-stage identifier for nested named entity recognition. In: Proceedings of the 59th Annual Meeting of the Association for Computational Linguistics and the 11th International Joint Conference on Natural Language Processing (Volume 1: Long Papers), pp. 2782–2794. Association for Computational Linguistics, Online, August 2021. https://doi.org/10.18653/v1/2021.acl-long.216, https://aclanthology.org/2021.acl-long.216

23. Shen, Y., et al.: Parallel instance query network for named entity recognition. In: Proceedings of the 60th Annual Meeting of the Association for Computational Linguistics (Volume 1: Long Papers), pp. 947–961. Association for Computational Linguistics, Dublin, May 2022. https://doi.org/10.18653/v1/2022.acl-long.67, https://aclanthology.org/2022.acl-long.67

24. Tao, Q., Luo, X., Wang, H., Xu, R.: Enhancing relation extraction using syntactic indicators and sentential contexts. In: 2019 IEEE 31st International Conference on Tools with Artificial Intelligence (ICTAI), pp. 1574–1580 (2019). https://doi.org/10.1109/ICTAI.2019.00227

25. Wang, Y., Tong, H., Zhu, Z., Li, Y.: Nested named entity recognition: a survey. ACM Trans. Knowl. Discov. Data **16**(6) (2022). https://doi.org/10.1145/3522593

26. Zhou, G., Zhang, J., Su, J., Shen, D., Tan, C.: Recognizing names in biomedical texts: a machine learning approach. Bioinformatics **20**(7), 1178–1190 (2004). https://doi.org/10.1093/bioinformatics/bth060

BACH: Black-Box Attacking on Deep Cross-Modal Hamming Retrieval Models

Jie Zhang[1], Gang Zhou[2,3], Qianyu Guo[4(✉)], Zhiyong Feng[1],
and Xiaohong Li[1(✉)]

[1] College of Intelligence and Computing, Tianjin University, Tianjin, China
xiaohongli@tju.edu.cn
[2] School of Artificial Intelligence, University of Chinese Academy of Sciences,
Beijing, China
[3] Institute of Automation, Chinese Academy of Sciences, Beijing, China
[4] Zhongguancun Laboratory, Beijing, People's Republic of China
guoqy@zgclab.edu.cn

Abstract. The growth of online data has increased the need for retrieving semantically relevant information from data in various modalities, such as images, text, and videos. Thanks to the powerful representation capabilities of deep neural networks (DNNs), deep cross-modal hamming retrieval (i.e., DCMHR) models have become popular in cross-modal retrieval tasks due to their efficiency and low storage cost. However, the vulnerability of DNN models makes them susceptible to small perturbations. Existing attacks on DNN models focus on supervised tasks like classification and recognition, and are not applicable to DCMHR models. To fill this gap, in this paper, we present BACH, an adversarial learning-based attack method for DCMHR models. BACH uses a triplet construction module to learn and generate well-designed adversarial samples in a black-box setting, without prior knowledge of the target models. During the learning process, we estimate the gradient of the objective function by using random gradient-free (RGF) method. To evaluate the effectiveness and efficiency of BACH, we perform thorough experiments on 3 popular cross-modal retrieval dataset and 13 state-of-the-art DCMHR models, including 6 image-to-image retrieval models and 7 image-to-text retrieval models. As a comparison, we select two established adversarial attack methods: CMLA for white-box attack and AACH for black-box attack. The results show that BACH offers comparable attack performance to CMLA while requiring no knowledge of the target models. Furthermore, BACH surpasses AACH on most DCMHR models in terms of attack success rate with limited queries.

Keywords: Cross-modal Retrieval · Hashing · Robustness · Adversarial perturbation

1 Introduction

The rapid advancement in storage and encoding techniques has greatly impacted human life by enabling people to search the internet for what they desire.

G. Zhou and J. Zhang—Contribute equally to this paper.

© The Author(s), under exclusive license to Springer Nature Switzerland AG 2023
X. Wang et al. (Eds.): DASFAA 2023, LNCS 13945, pp. 441–456, 2023.
https://doi.org/10.1007/978-3-031-30675-4_32

While various search techniques have been developed with the growth of social media and multi-modal data, they mostly only work with similarity-based search within a single modality, such as the keyword and tag-based searches, which no longer suffice in the face of diverse multi-modal data [9,32].

To address this limitation, cross-modal retrieval has been proposed and is gaining widespread attention [6,9,18,43]. It maps data from different modalities into a common space with the same dimension, and measures the semantic similarity by comparing samples in this space. However, measuring semantic similarity between data from different modalities is a significant challenge, which is also known as the heterogeneous gap problem. Conventional cross-modal retrieval methods assess semantic similarity by measuring the distance between samples in a common space [18]. Specifically, samples from different modalities with the same semantics are close in the common space [25]. Representational encoding in this space can be either real-valued or binary [37]. While real-valued encoding is often impractical for large dataset, the binary encoding is preferred for large dataset as it reduces storage costs and speeds up retrieval [14], and is used in cross-modal hashing to map data semantics into a binary space and measure semantic similarity using Hamming distance [22].

The quality of semantic feature extraction has a significant impact on the performance of encoding. To minimize the impact of the heterogeneous gap problem, an effective feature extraction method is essential [1]. Existing hash-based cross-modal retrieval methods are based on shallow architectures [13,41], and rely on features extracted by human experts. To date, with the growth of deep learning techniques in computer vision [33], natural language processing [36], and speech analysis [17], deep neural networks (DNNs) have become popular for improving the performance of cross-modal retrieval. DNNs can effectively detect semantic similarities between different modalities, and build cross-modal correlations through their superior representational capabilities. Due to their powerful representational capabilities, DNNs are trained to identify semantic similarities between different modalities and build cross-modal correlations. Research has shown that DNN-based cross-modal retrieval models outperform traditional shallow models [7].

However, it has been well established that even a well-trained deep learning model can be easily misled by inputs with subtle, human-undetectable perturbations, known as the adversarial examples [5,26,34,40]. To date, many effective adversarial methods have been proposed to attack trained DNN models [23]. These attacks can be categorized as white-box or black-box based on whether the attacker has access to the target model's internal information. While these attack methods are primarily designed for supervised tasks such as classification or recognition, little attention has been given to studying the impact of adversarial samples on deep hamming learning in cross-modal retrieval area.

The cross-modal retrieval task differs significantly from tasks like classification and recognition. Firstly, cross-modal retrieval models are trained through unsupervised or semi-supervised methods without ground-truth labels, making them more susceptible to misleading information. Secondly, the objective of attacks on cross-modal retrieval models is to generate semantically unrelated samples rather than incorrect classifications. This makes existing adversarial attacks unsuitable for attacking cross-modal retrieval models. Additionally, there are two major challenges in performing adversarial attacks on deep cross-modal hamming retrieval (DCMHR) models in a black-box setting: 1) the attacker does not have access to information about the target model, including the network architecture, model parameters, and loss functions, and can only obtain the output of the target model through queries; 2) there are often practical constraints on queries, such as a maximum number of queries allowed.

To tackle these challenges, we introduce BACH, a black-box adversarial attacking method for deep cross-Modal hamming retrieval (DCMHR) models. BACH specifically targets DCMHR models and generates adversarial samples by maximizing the hamming distance of semantically similar samples, thereby greatly impairing the performance of DCMHR models. To evaluate the effectiveness of BACH, we conduct experiments on 13 state-of-the-art DCMHR models and 3 popular dataset (i.e., MIRFlickr-25K, NUS-WIDE, and CIFAR-10) in three aspects: 1) attacking DCMHR models on both image-to-image and image-to-text retrieval tasks; 2) investigating the impact of the number of samples in the query dataset used to construct triples on the attack performance; and 3) comparing BACH against the state-of-the-art white-box attack method (i.e., CMLA [20]) and the black-box attack method (i.e., AACH [19]).

To summarize, this paper makes the following contributions:

- We propose BACH, a learning-based approach for adversarial attacks on deep cross-modal hamming retrieval models in a black-box environment. Unlike existing white-box attack methods, BACH does not require any prior knowledge and thus, is more practical in real-world applications. To the best of our knowledge, BACH is the first method designed for attacking cross-modal retrieval models in a black-box setting.
- We select a query-based black-box attacking strategy with performance comparable to white-box attack methods. This is achieved through the use of the random gradient-free (RGF) method and a limited number of target model queries.
- We evaluate the effectiveness and efficiency of BACH by conducting experiments on 13 state-of-the-art cross-modal retrieval models and 3 benchmark dataset. The results show that BACH performs comparably to the white-box attack methods while only requiring a limited number of queries. Our approach can be used to assess the robustness of cross-modal retrieval models.

The rest of the paper is organized as follows. Section 2 briefly introduces deep cross-modal retrieval task and problem formulation. Section 3 presents the technical details of our approach BACH. Section 4 shows our experimental setup as well as the experimental results. Related work is discussed in Sect. 5. Section 6 presents the conclusion and future extensions of this work.

2 Background

2.1 Deep Cross-Modal Retrieval and Problem Formulation

Cross-modal retrieval task refers to using image or text as queries to search for data with another modal in the database, such as using text to search for images or using images to search for text. A well-trained DCMHR model can retrieval semantically relevant data from the database. As shown in Fig. 1(a), using a picture of a flower as a query, the DCMHR model can retrieve some text about the flower, and we define that this picture and the retrieval result (i.e., text) are semantically relevant. In this paper, we use $O = \{O^v, O^t\} = \{o_i\}_{i=1}^{\mathbf{C}}$ to represent a cross-modal database with $\mathbf{C}$ samples. Herein, sample $o_i = \{o_i^v, o_i^t\}$ is an image-text pair, where o_i^v and o_i^t represent the image data and textual data, respectively.

Generally, DCMHR use DNN to extract semantic features $F_v \in \mathbb{R}^{\mathbf{C} \times k_v}$, $F_t \in \mathbb{R}^{\mathbf{C} \times k_t}$ from the original data, where k_v, k_t is the feature-length. After using the feature extraction architecture on the dataset, the semantic features are calculated as:

$$F^v = f_{base}^v \left(O^v, \theta_{base}^v\right), F^t = f_{base}^t \left(O^t, \theta_{base}^t\right), \tag{1}$$

where θ_{base}^v, θ_{base}^t are the parameters that need to be trained for the two feature extraction architectures. Moreover, k_v and k_t are generally set to be the same in order to extract the equipotential features.

Deep cross-modal retrieval model aims to learn two hash functions f_{hash}^v, f_{hash}^t that project image or text samples onto the Hamming space. This process can be formulated as:

$$B^v = sign \left(f_{hash}^v \left(F^v, \theta_{hash}^v\right)\right), B^t = sign \left(f_{hash}^t \left(F^t, \theta_{hash}^t\right)\right), \tag{2}$$

where $B^v, B^t \in \{-1, 1\}^{\mathbf{C} \times d}$ are the binary code, d is the length of the hash space, $F^v, F^t \in [-1, 1]^{\mathbf{C} \times d}$ are the binary-like representation generated by the output layer of a target deep cross-modal network, and θ_{hash}^v and θ_{hash}^t are two parameters that need to be learned for the hash function.

The semantic similarity between samples from different modalities is evaluated by the Hamming distance of the learned binary codes in a hash space:

$$D(X, Y) = \frac{1}{2} \left(K - \langle X, Y \rangle\right), \tag{3}$$

where X and Y are the binary codes of the samples, K is a constant that maintain the distance magnitude. A well-trained cross-modal hash retrieval model should preserve semantic similarity structure between samples of different modalities. Specifically, image sample o_i^v whose Hamming distance from its positive sample $o_{i_P}^t$ (with the shortest Hamming distance) is less than negative sample $o_{i_N}^t$ (with the longest Hamming distance). Here, we can construct a sample's triple $\{o_i^v, o_{i_P}^t, o_{i_N}^t\}$. Hash function is encouraged to satisfy the inequality as follows:

$$D(f_{hash}^v(o_i^v), f_{hash}^t(o_{i_P}^t)) < D(f_{hash}^v(o_i^v), f_{hash}^t(o_{i_N}^t)) \tag{4}$$

Next, we consider generating restricted adversarial perturbations η that can fool the DCMHR models. The attacking goal can be formalized as the following inequality:

$$D(f^v_{hash}(o^v_i + \eta^v), f^t_{hash}(o^t_{i_P})) > D(f^v_{hash}(o^v_i + \eta^v), f^t_{hash}(o^t_{i_N})) \qquad (5)$$

The adversarial images $o^v_i + \eta^v$ obtained by adding well-designed perturbations η can make the retrieval performance of the retrieval model significantly degraded. For example, as shown in Fig. 1(b), given an adversarial flower sample, the retrieval results are some irrelevant textual items with the original flower target.

(a) Original sample Query Results   (b) Adversarial Sample Query Results

Fig. 1. Examples of query results for the original and adversarial image samples

3   Black-Box Attack on DCMHR Models

This part details our proposed black-box adversarial attack named BACH against DCMHR models. Figure 2 shows the overall working pipeline, which mainly consists of three parts. The first part carries out cross-modal querying. The second part constructs a cross-modal triplet for every image based on the query results, and the third generates the adversarial example of an image according to the cross-modal triplet. In this paper, we generate adversarial samples only for images, not text because adding perturbations to text can be easily detected.

3.1   Black-Box Attack Framework

Firstly, we input M image-text pairs samples as cross-modal data queries $(O_q = \{O^v_q, O^t_q\}$, where $O^v_q = \{o^v_i\}^M_{i=1}$ and $O^t_q = \{o^t_i\}^M_{i=1})$ to the target retrieval model. Then, we constructing a triplet $\{o^v_i, o^t_{i_P}, o^t_{i_N}\}$ for each sample by get the hamming distance between M samples. Specifically, for an image-text triplet $\{o^v_i, o^t_{i_P}, o^t_{i_N}\}$, the goal of attacking cross-modal Hamming retrieval model can be formulated as follows:

$$\min_{\eta^v} D(f^v_{hash}(o^v_i + \eta^v), f^t_{hash}(o^t_{i_N})) -$$
$$D(f^v_{hash}(o^v_i + \eta^v), f^t_{hash}(o^t_{i_P})), s.t. \|\eta^v\|_p \le \epsilon^v. \qquad (6)$$

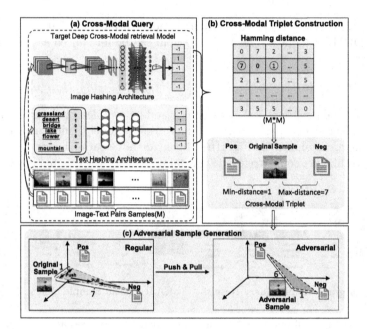

Fig. 2. Overview of BACH

An adversarial image $\hat{o}_i^v = o_i^v + \eta^v$ should satisfy two constraints: 1) the hamming distance between adversarial sample and it's positive sample should be as large as possible, while with negative sample should be as small as possible; Cause the generation of adversarial samples is guided by positive and negative samples to change the pixels of the original image. Specifically, see Fig. 2(c) for example, we continuously push away the hamming distance between the original sample and positive sample (1 to 6), and narrow the hamming distance between the original sample and negative sample (7 to 1), until reaching the preset number of iterations T. The value setting of threshold T is detailed in Sect. 4.2; 2) The perturbation η^v in the attack should be human-imperceptible. To this end, we use $\|\eta^v\|_p \leq \epsilon^v$ to constrain the magnitude of the perturbation. η^v refers to the pixel changes guided by positive and negative samples in this paper. Specifically, let h be the length of perturbation, $\|\cdot\|_p$ be the l_p-norm paradigm, we define the perturbation as $\|\eta\|_p = \sqrt[p]{\frac{1}{h}(|\eta_1|^p + |\eta_2|^p + \cdots + |\eta_h|^p)}$. The dimension of the perturbation is consistent with the raw image in the image dataset. Note that, the l_∞^ϵ bound is the most common way to limit the magnitude of an image, as it strictly limits the maximum image pixel from being perceived. Therefore, we choose the l_∞^ϵ-norm attack in this paper.

However, the optimization problem of Eq. (6) is an NP-hard problem, inspired by the C&W [4] attack, we rewrite the objective loss function as:

$$\min \Gamma(o_i^v, o_{i_P}^t, o_{i_N}^t, \epsilon^v)$$

$$= \min \sum_{i=0}^{M} \max(D(\hat{B}_i^v, B_{i_N}^t) - D(\hat{B}_i^v, B_{i_P}^t) + \kappa, 0), s.t. \|\eta^v\|_\infty \le \epsilon^v, \tag{7}$$

$$\hat{B}_i^v = f_{hash}^v (o_i^v + \eta^v),$$
$$B_{i_N}^t = f_{hash}^t (o_{i_N}^t),$$
$$B_{i_P}^t = f_{hash}^t (o_{i_P}^t), \tag{8}$$

and $\kappa \ge 0$ is a tuning parameter for attack transferability.

In the white-box setting, the optimization problem of Eq. (7) can be solved by back-propagating the loss function gradient. In the black-box setting, however, we cannot get the network information of target model and only get the model output (i.e., $B_i^*, * \in \{v, t\}$ in Eq. (8)). Gradient-based estimation is the most effective method in black-box attacks. Inspired by [29], we use the random gradient-free (RGF) method to estimate the gradient of the loss function in Eq. (7), and Ilyas et al. [11] have proved this method is optimal to estimate the gradient. Specifically, the gradient $\frac{\partial \Gamma}{\partial o_i^v}$ (defined as $\hat{g}_i$) of an image sample o_i^v can be estimated by the following equation:

$$\hat{g} = \frac{1}{q}\sum_{i=1}^{q} \hat{g}_i, \text{ with } \hat{g}_i = \frac{f(x + \sigma u_i, y) - f(x, y)}{\sigma} \cdot u_i, \tag{9}$$

where $\{u_i\}_{i=1}^{q}$ are the random vectors sampled independently from a uniform distribution $\mathcal{P}$ on $\mathbb{R}^D$, q is the number of the random direction, σ is the sampling variance and D is the dimension of original image. We set $\sigma = 0.01$, and $q = 50$ in this paper. However, it is a box-constraint problem for Eq. (7) that cannot be solved directly based on the commonly-used optimizers. Therefore, we used the following treatment to perturbation ϵ^v to solve this problem:

$$\epsilon^v = \frac{1}{2}(\tanh(\epsilon^v) + 1) - o^v. \tag{10}$$

Then, we choose the Adam [16] optimizer to solve Eq. (7). Finally, we learn the following adversarial perturbations for cross-modal retrieval:

$$\eta^v = \underset{\epsilon^v}{\arg\min}\, \Gamma(o_i^v, o_{i_P}^t, o_{i_N}^t, \epsilon^v). \tag{11}$$

Now that we have detailed the attack method's whole process, specifically, we attack the target retrieval model by inputting the adversarial sample. The entire process of adversarial sample generation is shown in Algorithm 1. Line 1 to Line 4 of the algorithm is the querying part. Line 5 to Line 6 describe the triplet construction. Line 7 to Line 11 illustrate the adversarial sample generation, where Line 9 corresponds to the gradient estimation.

Algorithm 1: Black-box Adversarial Perturbation Generation Method for Deep Cross-modal Hash Retrieval Models (BACH)

Input : Target deep cross-modal retrieval model: $f^*_{hash}(o^*_i)$, $* \in \{v, t\}$, data $O = \{o^v_i, o^t_i\}^C_{i=1}$, iteration T, adversarial queries M

Output: A adversarial sample of query image: $\hat{o}^v_i = o^v_i + \eta^v$

---

1 initialize $iter = 0$;

2 Random select query data $\{o^v_i, o^t_i\}^M_{i=1}$;

3 Compute $B^t = sign(f^t_{hash}(O^t_q))$;

4 Compute $B^v = sign(f^v_{hash}(O^v_q))$;

5 Compute Hamming distance matrix according to Equation (3) based on $\{B^v, B^t\} = \{B^v_i, B^t_i\}^M_{i=1}$;

6 Create cross-modal triplets $\{o^v_i, o^t_{i_P}, o^t_{i_N}\}$ for every image o^v_i;

7 Select η^v:**while** $iter \leq T$ **do**

8      $\eta^v = \arg\min_{\epsilon^v} \Gamma(o^v_i, o^t_{i_P}, o^t_{i_N}, \epsilon^v)$;

9   Estimate $\hat{g}_i$ using Equation (9);

10   Using Adam optimizer;

11      $iter = iter + 1$;

12 **return** $\hat{o}^v_i$;

---

4 Experiment

This section evaluates the performance of BACH on several commonly-used deep cross-modal hamming retrieval models and dataset. We assess the attack on image-to-text retrieval task and image-to-image retrieval task.

4.1 Dataset

The dataset of image-to-text retrieval task include MIRFlickr-25K and NUS-WIDE. The dataset of image-to-image retrieval task include CIFAR10 and NUS-WIDE. We use these three dataset to train several deep cross-modal hamming retrieval models. In all of our attack experiments below, dataset are divided into three-part, including train, query, and gallery parts. Note that the attacks does not use the train set.

MIRFlickr-25K contains 25,000 images from the Flickr website, each image with a corresponding text description constituting an image-text pair. According to [42], we randomly divided the dataset into a training dataset with 5000 samples and a test dataset with 20000 samples. There are M samples as query dataset in the test dataset, while remaining samples as a gallery set.

NUS-WIDE is a multi-label dataset containing 81 labels. There are 269,648 image-text pairs. We select a total of 195834 samples from the most commonly used 21 labels as the image retrieval dataset according to [12]. We select 500

Table 1. The attack performance in term of mAP on the state-of-the-art DCMHR models for image-to-text retrieval task, based on MIRFlickr-25K and NUS-WIDE sets

Dataset	MIRFLICKR-25k								NUS-WIDE							
CL	16		32		64		128		16		32		64		128	
Method	REG	ATK	REG	ATK	REG	ATK	REG	ATK	REG	ATK	REG	ATK	REG	ATK	REG	ATK
DJSRH	0.66	0.60	0.66	0.61	0.67	0.61	0.68	0.62	0.46	0.41	0.49	0.43	0.45	0.41	0.53	0.44
AGAH	0.74	0.59	0.77	0.60	0.78	0.62	0.77	0.63	0.62	0.43	0.64	0.45	0.65	0.46	0.65	0.46
SSAH	0.64	0.60	0.68	0.61	0.69	0.61	0.71	0.61	0.48	0.41	0.51	0.41	0.53	0.41	0.53	0.42
DCMH	0.71	0.62	0.74	0.62	0.72	0.63	0.73	0.64	0.64	0.43	0.66	0.44	0.65	0.46	0.67	0.46
DSAH	0.69	0.60	0.70	0.60	0.71	0.61	0.71	0.61	0.56	0.42	0.60	0.42	0.61	0.43	0.62	0.44

[a] REG is the abbreviation of regular that used to represent regular retrieval performance, and ATK is the abbreviation of attack that used to represent attack performance.

[b] CL refers to code length. Here, M is set to 500 and T is set to 800.

pairs for each label to construct the training dataset randomly, with 100 pairs of each label randomly selected to query, and the rest are used as the gallery dataset. In addition, this paper uses NUS-WIDE as a dataset for attacking the deep cross-modal hamming retrieval models for image-to-image retrieval task. Following [8], a total of 5000 samples are selected randomly as the query dataset and the remaining samples as a gallery set.

CIFAR10 dataset consists of 60,000 images whose sizes are 32×32 and belong to 10 categories. Each category has 6,000 images. There are 50,000 training images and 10,000 testing images. We extract 100 samples for each category from the testing dataset for querying, and the remaining samples are as a gallery set.

To evaluate BACH, we use two commonly used evaluation criteria for cross-modal retrieval tasks in the field of information retrieval, namely, mean Average Precision (mAP) and Normalized Discounted Cumulative Gain (NDCG).

Table 2. The attack performance in term of NDCG on the state-of-the-art DCMHR models for image-to-text retrieval task, based on MIRFlickr-25K and NUS-WIDE sets

Dataset	MIRFLICKR-25k								NUS-WIDE							
CL	16		32		64		128		16		32		64		128	
Method	REG	ATK	REG	ATK	REG	ATK	REG	ATK	REG	ATK	REG	ATK	REG	ATK	REG	ATK
DJSRH	0.63	0.59	0.66	0.60	0.66	0.61	0.68	0.62	0.48	0.41	0.49	0.42	0.53	0.43	0.54	0.44
AGAH	0.76	0.61	0.79	0.62	0.80	0.62	0.81	0.63	0.64	0.44	0.67	0.44	0.68	0.45	0.68	0.46
SSAH	0.64	0.59	0.67	0.60	0.67	0.52	0.69	0.62	0.49	0.42	0.50	0.42	0.53	0.43	0.54	0.44
DCMH	0.75	0.62	0.75	0.62	0.76	0.63	0.77	0.63	0.63	0.43	0.64	0.44	0.65	0.44	0.66	0.45
DSAH	0.68	0.59	0.70	0.60	0.71	0.61	0.72	0.62	0.58	0.42	0.60	0.42	0.61	0.43	0.61	0.44

[a] CL refers to code length. Here, M is set to 500 and T is set to 800.

4.2 Evaluation

BACH is a black-box adversarial attack on DCMHR models. To evaluate the effectiveness and efficiency of BACH, we design experiments to answer the following three research questions:

- **RQ1:** Is BACH effective to attack classical deep cross-modal hamming retrieval models for image-to-text and image-to-image retrieval tasks?
- **RQ2:** Does the number of samples of the query dataset used to construct triples affect the attack performance?
- **RQ3:** How does BACH perform compared with existing white-box and black-box attacking methods?

Table 3. The attack performance in term of mAP on the state-of-the-art DCMHR models for image-to-image retrieval task, based on CIFAR10 set

Dataset	CIFAR10							
Code Length	12		24		36		48	
Method	REG	ATK	REG	ATK	REG	ATK	REG	ATK
SDH	0.46	0.10	0.64	0.11	0.66	0.12	0.67	0.14
DSH	0.62	0.10	0.66	0.13	0.67	0.14	0.68	0.14
ADSH	0.88	0.15	0.88	0.15	0.87	0.15	0.87	0.15
DSDH	0.73	0.14	0.75	0.15	0.75	0.15	0.75	0.16

[a] Here, M is set to 500, T is set to 800.

Firstly, we show the performance of BACH on several retrieval models. To verify the ability of the attack, we re-produce 7 state-of-the-art DCMHR models for image-to-text retrieval task, including DJSRH [32], AGAH [28], SSAH [18], DCMH [7], DSAH [22], PRDH [38] and CMHH [2]. We construct six state-of-the-art DCMHR models for image-to-image data retrieval task, including DSH [24], DIHN [35], DSDH [21], ADSH [15], HMH [39] and SDH [30]. In addition, to evaluate the attack performance of the adversarial samples with different hash code lengths, for the cross-modal retrieval task, we use 16, 32, 64, and 128 as the length, respectively. Moreover, for image-to-image retrieval tasks, there are two dataset, where the CIFAR10 dataset takes values of 12, 24, 36, and 48 for hash code length, and the NUS-WIDE dataset takes values of 8, 16, 24, and 32 for hash code length. We attacked the above retrieval models, and the comparison between the regular retrieval performance and the performance after being attacked in terms of the mAP/NDCG score. The attack results on the DCMHR models for image-to-text retrieval task are shown in Tables 1 and 2. The attack results of image-to-image retrieval task are in Tables 3 and 4. Furthermore, due to the special requirements of the HMH method for hash code length, 8-bit and 24-bit hash code lengths do not satisfy the HMH requirements, so we do not perform 8-bit and 24-bit attacks on CIFAR10 and NUS-WIDE for HMH. BACH produces adversarial samples that effectively degrade the performance of all the above well-trained retrieval models, which means that all our attacks are successful, demonstrating the lack of robustness of these existing deep retrieval models to small adversarial perturbations.

To construct a triplet of samples, we need to query the hash codes of M samples. Different M will led to different attack performances, so we will take M as 200, 300, 500, and 1000 respectively, to perform the attack. The comparison of the attack performance on MIRFlickr-25K, NUS-WIDE is shown

Table 4. The attack performance in terms of mAP and NDCG on the state-of-the-art DCMHR models for image-to-image retrieval task, based on NUS-WIDE set

Dataset	NUS-WIDE (mAP)								NUS-WIDE (NDCG)							
CL	8		16		24		32		8		16		24		32	
Method	REG	ATK	REG	ATK	REG	ATK	REG	ATK	REG	ATK	REG	ATK	REG	ATK	REG	ATK
DSH	0.66	0.24	0.69	0.26	0.70	0.27	0.71	0.27	0.45	0.26	0.45	0.26	0.45	0.27	0.45	0.27
ADSH	0.80	0.27	0.85	0.28	0.86	0.29	0.87	0.29	0.51	0.28	0.59	0.28	0.61	0.28	0.63	0.29
DIHN	0.74	0.25	0.79	0.27	0.81	0.27	0.80	0.27	0.48	0.26	0.51	0.27	0.58	0.27	0.58	0.29
DSDH	0.77	0.26	0.76	0.26	0.80	0.27	0.80	0.28	0.50	0.28	0.55	0.28	0.59	0.30	0.59	0.23
HMH	-	-	0.74	0.26	-	-	0.78	0.27	-	-	0.53	0.28	-	-	0.52	0.28

[a] CL refers to code length. Here, M is set to 500 and T is set to 800.

Table 5. Comparison of the attack performance for different Adversarial Queries (M) in terms of mAP scores, the code length is set to 32 bits, T is set to 800

Tasks	Adversarial Queries	MIRFlickr-25K				NUS-WIDE			
		DCMH	PRDH	SSAH	CMHH	DCMH	PRDH	SSAH	CMHH
	REG	0.74	0.78	0.68	0.75	0.66	0.64	0.51	0.60
$I \to T$	BACH 200	0.70	0.68	0.66	0.64	0.59	0.52	0.44	0.46
	300	0.65	0.64	0.64	0.63	0.50	0.46	0.42	0.44
	500	0.62	0.60	0.61	0.61	0.44	0.42	0.41	0.43
	1000	0.62	0.61	0.62	0.61	0.45	0.41	0.42	0.43

[a] $I \to T$ denotes retrieval text using an adversarial image query.

in Table 5 (mAP). The mAP scores decreases as M increases, so the attack performance gradually improves. However, we find that the attack performance slightly decreases when the M increases from 500 to 1000, which may be due to some inaccurate information obtained when querying the target model, so high-quality query samples will help to improve the query efficiency and attack performance.

Meanwhile, we compare the impact of different iterative numbers, T, on the attack performance during adversarial sample generation. Here we fix M to 500, and the attack performance comparison on the baseline databases is shown in Table 6. We find the retrieval performance degrades gradually as the number of iterations becomes larger, meaning the attack performance becomes better. However, when T grows from 500 to 800, the attack performance increase is insignificant. Since there is often a limit on the number of queries, we consider T takes 800 as the optimal value.

Last, we compare BACH performance to white-box and black-box attack methods, and the results are shown in Table 7. CMLA [20] is a work to attack DCMHR models in a white-box setting. In contrast, AACH [19] attacks DCMHR models in a black-box setting. Therefore, AACH does not require a priori knowledge, such as the structure of the target network. However, AACH requires constructing a surrogate model, which we do not need. We attack by directly estimating the gradient of the loss function. We compare the attack performance of the three methods on top of two different dataset according to [19]. CMLA achieves the best performance, which is attributed to the fact that CMLA has

Table 6. Comparison of the attack performance for different iteration (T) in terms of mAP scores, the code length is set to 32 bits, M is set to 500.

Tasks	Iteration		MIRFlickr-25K				NUS-WIDE			
			DCMH	PRDH	SSAH	CMHH	DCMH	PRDH	SSAH	CMHH
	REG		0.74	0.78	0.68	0.75	0.66	0.64	0.51	0.60
$I \rightarrow T$	BACH	300	0.66	0.63	0.60	0.69	0.46	0.50	0.44	0.55
		500	0.63	0.60	0.62	0.62	0.44	0.45	0.41	0.45
		800	0.62	0.60	0.61	0.61	0.44	0.42	0.41	0.43

[a] $I \rightarrow T$ denotes retrieval text using an adversarial image query.

Table 7. Comparison of the attack performance of BACH, CMLA and AACH in terms of mAP scores on different dataset, the code length is 32 bits.

Tasks	Methods	MIRFlickr-25K				NUS-WIDE			
		DCMH	PRDH	SSAH	CMHH	DCMH	PRDH	SSAH	CMHH
	REG	0.74	0.78	0.68	0.75	0.66	0.64	0.51	0.60
$I \rightarrow T$	CMLA	0.52	0.60	0.60	0.56	0.46	0.40	0.36	0.33
	AACH	0.63	0.62	0.56	0.65	0.44	0.50	0.40	0.41
	BACH	0.62	0.61	0.61	0.58	0.44	0.49	0.41	0.40

[a] $I \rightarrow T$ denotes retrieval text using an adversarial image query.

all the prior knowledge of the target model as a white-box attack. However, the attack performance of our BACH is more potent than AACH on both benchmark dataset. It validates the effectiveness of our approach.

5   Related Work

5.1   Deep Cross-Modal Hashing

In order to measure the semantic similarity between samples of different modalities and maintain the similarity between data samples, the features of data samples belonging to different modalities are often mapped into a common subspace. As shown in Fig. 3, Hash codes learning and retrieval tasks are all based on this common subspace. For example, Inter-Media Hashing [31] uses inter-modal and intra-modal consistency as benchmarks to construct a common Hamming space and introduces regularized linear regression into the hashing process. Latent Semantic Sparse Hashing [43] the latent space through sparse coding and matrix factorization and then fuses the features of different modal data into a unified hash code. The Composite Correlation Quantization [27] uses the maximum mapping method to construct a common subspace. Collective Matrix Factorization Hashing [5] and Supervised Collective Matrix Factorization Hashing [6] exploit collaborative matrix factorization to learn hash codes from different modalities. Based on common subspace learning, adding label information can effectively improve the performance of cross-modal retrieval models, and such supervised models can generate hash codes that preserve semantics.

The Semantic Correlation Maximization [42] method proposed earlier attempts to integrate label information into the learning process to obtain a similarity matrix. Semantics Preserving Hashing [23] first used a distribution function to formulate the hashing process, converted the semantic relationship contained in the label information into a probability distribution, then trained the model by minimizing the Kullback-Leibler divergence.

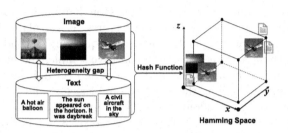

Fig. 3. Regular cross-modal Hamming retrieval

Meanwhile, DNN can enhance the feature learning capability for different modal data, which yields deep cross-modal hashing using DNN as a feature extraction network. Typical approaches are Deep Cross-Modal Hashing [14] and Pairwise Relation Guided Deep Hashing [38], both of which use deep convolutional networks and fully connected networks to extract the image and text features, respectively, while adding semantic labeling information to maintain the original semantic similarity between samples. The Deep Visual-Semantic Hashing [3] method further uses Long Short Term Memory (LSTM) [10] to learn textual information in the form of sentences. Self-Supervised Adversarial Hashing [18] proposes to capture semantic features from different modalities further using generative adversarial networks and proposes labeling networks to generate hash codes of label vectors.

5.2 Adversarial Attacks

Szegedy et al. were the first to propose the concept of adversarial sample [5], and they found that small perturbations that are not sensitive to the human visual system can make the neural network too sensitive to produce false recognition. Subsequent researchers have proposed many more powerful and effective methods for attack generation. The existing adversarial attacks can be divided into two main categories: white-box attacks and black-box attacks. White-box attacks refer to the information of the target model is fully accessible, and the most commonly used white-box attacks are fast gradient symbolic method (FGSM) [8] and projected gradient descent method (PGD) [23]. Although the performance of white-box attacks is relatively high, obtaining specific information about the target model in the real world is complicated. The black-box attack can only obtain the model's output or even the information about the model is completely unknown. This setting increases the difficulty of attacks, but it is more practical than white-box. Research shows that, black-box attacks based on gradient estimation are already close to the performance of the best white-box attacks.

6   Conclusion

This paper presents BACH, a learning-based adversarial attack method aimed at fooling deep cross-modal retrieval models on hamming space in a black-box setting. BACH consists of three parts: first, it calculates the hamming distance between samples through cross-modal querying; second, it constructs cross-modal triplets (i.e., original sample, positive sample, and negative sample) for each image based on the hamming distance; and third, it learns to generate adversarial samples by pulling the negative samples close and pushing away the positive sample, using a random gradient-free gradient estimation method to reduce the number of queries. BACH was tested on 3 popular dataset and 13 state-of-the-art deep cross-modal hamming retrieval models, including 6 models for image-to-image retrieval and 7 models for image-to-text retrieval. The experiments show that BACH can effectively attack existing retrieval models and has comparable attack performance to the white-box attack method (i.e., CMLA) and the black-box attack method (i.e., AACH). The results highlight the unreliability of current cross-modal hamming retrieval models, as well-designed perturbations can easily mislead them in practice. Thus, BACH can serve as a baseline for evaluating the robustness of cross-modal hamming retrieval models, and call for advanced method to enhance the robustness of cross-modal retrieval models in the future.

Acknowledgements. This paper was supported by the Ministry of Science and Technology of China under Grant No. 2020AAA0108401, and the Natural Science Foundation of China under Grant Nos. 72225011 and 71621002.

References

1. Andoni, A., Indyk, P.: Near-optimal hashing algorithms for approximate nearest neighbor in high dimensions. In: 2006 47th Annual IEEE Symposium on Foundations of Computer Science (FOCS 2006), pp. 459–468. IEEE (2006)
2. Cao, Y., Liu, B., Long, M., Wang, J.: Cross-modal hamming hashing. In: Ferrari, V., Hebert, M., Sminchisescu, C., Weiss, Y. (eds.) ECCV 2018. LNCS, vol. 11205, pp. 207–223. Springer, Cham (2018). https://doi.org/10.1007/978-3-030-01246-5_13
3. Cao, Y., Long, M., Wang, J., Yang, Q., Yu, P.S.: Deep visual-semantic hashing for cross-modal retrieval. In: Proceedings of the 22nd ACM SIGKDD International Conference on Knowledge Discovery and Data Mining, pp. 1445–1454 (2016)
4. Carlini, N., Wagner, D.: Towards evaluating the robustness of neural networks. In: 2017 IEEE Symposium on Security and Privacy (SP), pp. 39–57. IEEE (2017)
5. Ding, G., Guo, Y., Zhou, J.: Collective matrix factorization hashing for multimodal data. In: Proceedings of the IEEE Conference on Computer Vision and Pattern Recognition, pp. 2075–2082 (2014)

6. Ding, G., Guo, Y., Zhou, J., Gao, Y.: Large-scale cross-modality search via collective matrix factorization hashing. IEEE Trans. Image Process. **25**(11), 5427–5440 (2016)

7. Gong, Y., Lazebnik, S., Gordo, A., Perronnin, F.: Iterative quantization: a procrustean approach to learning binary codes for large-scale image retrieval. IEEE Trans. Pattern Anal. Mach. Intell. **35**(12), 2916–2929 (2012)

8. Goodfellow, I.J., Shlens, J., Szegedy, C.: Explaining and harnessing adversarial examples. arXiv preprint arXiv:1412.6572 (2014)

9. Gu, W., Gu, X., Gu, J., Li, B., Xiong, Z., Wang, W.: Adversary guided asymmetric hashing for cross-modal retrieval. In: Proceedings of the 2019 on International Conference on Multimedia Retrieval, pp. 159–167 (2019)

10. Hochreiter, S., Schmidhuber, J.: Long short-term memory. Neural Comput. **9**(8), 1735–1780 (1997)

11. Ilyas, A., Engstrom, L., Madry, A.: Prior convictions: black-box adversarial attacks with bandits and priors. In: International Conference on Learning Representations (2018)

12. Ilyas, A., Santurkar, S., Tsipras, D., Engstrom, L., Tran, B., Madry, A.: Adversarial examples are not bugs, they are features. In; Advances in Neural Information Processing Systems, vol. 32 (2019)

13. Indyk, P., Motwani, R.: Approximate nearest neighbors: towards removing the curse of dimensionality. In: Proceedings of the Thirtieth Annual ACM Symposium on Theory of Computing, pp. 604–613 (1998)

14. Jiang, Q.Y., Li, W.J.: Deep cross-modal hashing. In: Proceedings of the IEEE Conference on Computer Vision and Pattern Recognition, pp. 3232–3240 (2017)

15. Jiang, Q.Y., Li, W.J.: Asymmetric deep supervised hashing. In: Proceedings of the AAAI Conference on Artificial Intelligence, vol. 32 (2018)

16. Kingma, D.P., Ba, J.: Adam: a method for stochastic optimization. arXiv preprint arXiv:1412.6980 (2014)

17. LeCun, Y., Bengio, Y., Hinton, G.: Deep learning. Nature **521**(7553), 436–444 (2015)

18. Li, C., Deng, C., Li, N., Liu, W., Gao, X., Tao, D.: Self-supervised adversarial hashing networks for cross-modal retrieval. In: Proceedings of the IEEE Conference on Computer Vision and Pattern Recognition, pp. 4242–4251 (2018)

19. Li, C., Gao, S., Deng, C., Liu, W., Huang, H.: Adversarial attack on deep cross-modal hamming retrieval. In: Proceedings of the IEEE/CVF International Conference on Computer Vision, pp. 2218–2227 (2021)

20. Li, C., Gao, S., Deng, C., Xie, D., Liu, W.: Cross-modal learning with adversarial samples. In: Advances in Neural Information Processing Systems, vol. 32 (2019)

21. Li, Q., Sun, Z., He, R., Tan, T.: Deep supervised discrete hashing. In: Advances in Neural Information Processing Systems, vol. 30 (2017)

22. Li, Y., van Gemert, J.: Deep unsupervised image hashing by maximizing bit entropy. In: Proceedings of the AAAI Conference on Artificial Intelligence, vol. 35, pp. 2002–2010 (2021)

23. Lin, Z., Ding, G., Hu, M., Wang, J.: Semantics-preserving hashing for cross-view retrieval. In: Proceedings of the IEEE Conference on Computer Vision and Pattern Recognition, pp. 3864–3872 (2015)

24. Liu, H., Wang, R., Shan, S., Chen, X.: Deep supervised hashing for fast image retrieval. In: Proceedings of the IEEE Conference on Computer Vision and Pattern Recognition, pp. 2064–2072 (2016)

25. Liu, J., Xu, C., Lu, H.: Cross-media retrieval: state-of-the-art and open issues. Int. J. Multimedia Intell. Secur. **1**(1), 33–52 (2010)

26. Liu, X., Huang, L., Deng, C., Lang, B., Tao, D.: Query-adaptive hash code ranking for large-scale multi-view visual search. IEEE Trans. Image Process. **25**(10), 4514–4524 (2016)
27. Long, M., Cao, Y., Wang, J., Yu, P.S.: Composite correlation quantization for efficient multimodal retrieval. In: Proceedings of the 39th International ACM SIGIR conference on Research and Development in Information Retrieval, pp. 579–588 (2016)
28. Nakkiran, P.: Adversarial robustness may be at odds with simplicity. arXiv preprint arXiv:1901.00532 (2019)
29. Nesterov, Y., Spokoiny, V.: Random gradient-free minimization of convex functions. Found. Comput. Math. **17**(2), 527–566 (2017)
30. Shen, F., Shen, C., Liu, W., Tao Shen, H.: Supervised discrete hashing. In: Proceedings of the IEEE Conference on Computer Vision and Pattern Recognition, pp. 37–45 (2015)
31. Song, J., Yang, Y., Yang, Y., Huang, Z., Shen, H.T.: Inter-media hashing for large-scale retrieval from heterogeneous data sources. In: Proceedings of the 2013 ACM SIGMOD International Conference on Management of Data, pp. 785–796 (2013)
32. Su, S., Zhong, Z., Zhang, C.: Deep joint-semantics reconstructing hashing for large-scale unsupervised cross-modal retrieval. In: Proceedings of the IEEE/CVF International Conference on Computer Vision, pp. 3027–3035 (2019)
33. Sun, Y., Chen, Y., Wang, X., Tang, X.: Deep learning face representation by joint identification-verification. In: Advances in Neural Information Processing Systems, vol. 27 (2014)
34. Szegedy, C., et al.: Intriguing properties of neural networks. arXiv preprint arXiv:1312.6199 (2013)
35. Wu, D., Dai, Q., Liu, J., Li, B., Wang, W.: Deep incremental hashing network for efficient image retrieval. In: Proceedings of the IEEE/CVF Conference on Computer Vision and Pattern Recognition, pp. 9069–9077 (2019)
36. Wu, Y., et al.: Google's neural machine translation system: bridging the gap between human and machine translation. arXiv preprint arXiv:1609.08144 (2016)
37. Xu, C., Tao, D., Xu, C.: A survey on multi-view learning. arXiv preprint arXiv:1304.5634 (2013)
38. Yang, E., Deng, C., Liu, W., Liu, X., Tao, D., Gao, X.: Pairwise relationship guided deep hashing for cross-modal retrieval. In: proceedings of the AAAI Conference on Artificial Intelligence, vol. 31 (2017)
39. Yuan, L., et al.: Central similarity quantization for efficient image and video retrieval. In: Proceedings of the IEEE/CVF Conference on Computer Vision and Pattern Recognition, pp. 3083–3092 (2020)
40. Yuan, X., He, P., Zhu, Q., Li, X.: Adversarial examples: attacks and defenses for deep learning. IEEE Trans. Neural Netw. Learn. Syst. **30**(9), 2805–2824 (2019)
41. Zhai, X., Peng, Y., Xiao, J.: Heterogeneous metric learning with joint graph regularization for cross-media retrieval. In: Twenty-Seventh AAAI Conference on Artificial Intelligence (2013)
42. Zhang, D., Li, W.J.: Large-scale supervised multimodal hashing with semantic correlation maximization. In: Proceedings of the AAAI Conference on Artificial Intelligence, vol. 28 (2014)
43. Zhou, J., Ding, G., Guo, Y.: Latent semantic sparse hashing for cross-modal similarity search. In: Proceedings of the 37th International ACM SIGIR Conference on Research and Development in Information Retrieval, pp. 415–424 (2014)

Category-Highlighting Transformer Network for Question Retrieval

Denghao Ma[1], Li Chong[2], Yueguo Chen[2(✉)], and Liang Shen[1]

[1] Meituan, Beijing, China
{madenghao,shenliang03}@meituan.com
[2] DEKE Lab, Renmin University of China, Beijing, China
{chongli,chenyueguo}@ruc.edu.cn

Abstract. Question retrieval aims to find the semantically equivalent questions from question archives for a user question. Recently, Transformer-based models have significantly advanced the progress of question retrieval, which mainly focus on capturing the content-based semantic relations of two questions. However, they can not well capture the category-based semantic relations of two questions, even question categories are very important to identify the semantic equivalence of two questions. To capture both the content-based and category-based semantic relations, we study the issue of improving Transformer by highlighting and incorporating the category information. To this end, we innovatively propose the Category-Highlighting Transformer Network (CHT). Because questions are not equipped with explicit categories, CHT first uses a category identification unit to construct category-based semantic representations for the question and its embedded words. Second, to "deeply" capture the category-based and content-based semantic relations, we develop the category-highlighting Transformer by improving the self-attention unit with the category-based representations. The cascaded category highlighting Transformers are used for modelling "individual" semantics of a question and "joint" semantics of two questions. Extensive experiments on three public datasets show that the category-highlighting Transformer network outperforms the state-of-the-art solutions.

Keywords: Question answering · Question retrieval · Transformer

1 Introduction

Question answering (QA) has been a popular platform where users can seek relevant answers for their questions. As a key component of QA systems, question retrieval aims to retrieve semantically equivalent questions from the question and answer archives. By the equivalent questions, a QA system will aggregate, rank and show their answers to users. Due to the importance of question retrieval, it has attracted much attention from the communities of both academia and industry [8,12,13]. Meanwhile, a variety of models have been proposed for addressing the challenge of question retrieval, i.e., lexical gap— semantically equivalent

---

D. Ma and L. Chong—Equal contribution.

© The Author(s), under exclusive license to Springer Nature Switzerland AG 2023
X. Wang et al. (Eds.): DASFAA 2023, LNCS 13945, pp. 457–467, 2023.
https://doi.org/10.1007/978-3-031-30675-4_33

questions have different words. The work [12,24] takes the translation probability from an archived question to the user question as their semantic similarity. In work [2,8,13], the topic distributions of two questions are constructed and used for identifying their semantic equivalence. With the development of deep learning techniques, various neural network architectures have been proposed for deep encoding two questions [18,23]. Recently, pre-trained representation models, such as BERT [5] and ERNIE [17], have been proposed to model the semantics of questions, and have achieved significant performance gains.

These neural network architectures and pre-trained models focus on modelling the content-based semantic relations of two questions. Different from them, we propose new insights and solution to capture the semantic relations of both content-level and category-level. *Data insight: questions with different categories can not be semantically equivalent.* The question categories represent the important semantics of a question [10,11], and are critical for the semantic equivalence identification of questions. This is because questions with different categories can not be semantically equivalent. For example, given two questions "how to rescue a phone that fell into the water" and "where to rescue a phone that fell into the water", they have high text similarity but are not semantically equivalent, because they belong to different question categories, i.e., "solution" and "location". *Technical insight: Transformer can not well capture the category-based semantic relations of questions.* Many Transformer-based solutions have been developed and achieved new state-of-the-art results of question retrieval. The self-attention in Transformer mainly focuses on modelling the contextual dependencies of words, while don't specially model the semantics on question categories. So Transformer-based solutions can not well capture the category-based semantic relations of two questions.

Solution: Category-Highlighting Transformer Network. According to the above insights, we propose to improve Transformer by incorporating the category information, for capturing both content-based and category-based semantic relations of two questions. To realize this idea, we develop the category-highlighting Transformer network (CHT) with two cascaded units:

Category Identification Unit (CIU). In the datasets of question retrieval, questions are not equipped with explicit categories. CIU is developed to construct category-based semantic representations for a question and its embedded words. We define some question categories for a question dataset, e.g., "time", "location" and "people". Based on the categories, CIU estimates the relevance between a question and each category, and uses the relevance as weights to sum the embeddings of categories. The summed embedding is used for representing the category-based semantics of the question. The relevance between a word and the summed embedding as well as the word and the categories is estimated and used for constructing the category-based semantic representation of the word.

Category-Highlighting Transformer Unit. For deeply modelling both category-based and content-based semantic relations of two questions, we develop the cascaded category-highlighting Transformers to model "individual" semantics of a question and the "joint" semantics of two questions. In each Transformer, we improve the self-attention unit by incorporating the category-based attentive similarities that are estimated based on the category-based representations of words, so that both content-based and category-based semantics are captured.

Our contributions are concluded as follows:

- We propose new insights for question retrieval from data and technical views.
- We develop the category-highlighting Transformer network to model both content-based and category-based semantic relations of two questions, without the supervision of question-to-category labelling data.
- We are the first to improve the self-attention unit in Transformer by incorporating the question category information.
- We conduct extensive experiments on three public datasets and validate the effectiveness of the category-highlighting Transformer network.

2 Preliminaries

Question retrieval aims to find semantically equivalent questions from a large question repository D for a user question q. It can be formulated: $\mathcal{D} = f(q, D)$, where f is a pipeline of selecting candidate questions $\mathcal{D}$ from D. A classical pipeline consists of two stages, i.e., retrieval and identification. The retrieval stage is to find relevant candidates, and the identification stage is to select the semantically equivalent questions from the candidates. The retrieval stage needs to be completed in a limited time. So it typically uses some term-matching solutions to find candidate questions, e.g., language model [16] and BM25 [15].

2.1 Related Work of the Identification Stage

Many solutions have been proposed to identify the semantic equivalence of two questions, which can be grouped into four classes:

Term-Matching Solutions. Many traditional techniques of information retrieval are used for estimating the semantic similarity of two questions, such as BM25 [15] and language model [16]. These solutions are based on an assumption that two questions with higher text similarity are more likely to be semantically equivalent, suffering from the lexical gap challenge.

Translation-Model Solutions. These solutions use the translation probability from one question to another question to identify the semantic equivalence of two questions [7,22]. The word-to-word translation probabilities are first learned from different parallel corpora, and then are applied to some retrieval models.

Topic-Model Solutions. These solutions [2,8] identify the semantic relations of two questions in the latent topic space. The work [2] proposes a topic model that incorporates category information into the process of discovering latent topics. The work [8] proposes a Question-Answer Topic Model (QATM) that learns the latent topics from the question-answer pairs.

Deep Learning Solutions. The deep learning techniques have been widely applied in question retrieval, and achieve better performance than traditional solutions. The work [20] uses a bidirectional LSTM to generate multiple positional

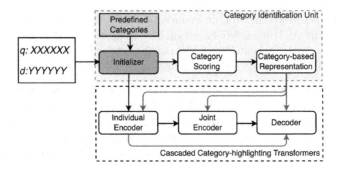

Fig. 1. The architecture of category-highlighting Transformer network.

sentence representations for each question. In work [4], a Siamese CNN is proposed to estimate the semantic similarity of two questions. Recently, the pre-training language models have been demonstrated strong performance on text representations in many NLP tasks, such as BERT [5] and Sentence-BERT [14].

3   Category-Highlighting Transformer Network

To accurately identify the semantically equivalent questions, we propose new insights: 1) *questions with different categories can not be semantically equivalent*; 2) *Transformer can not well capture the category-based semantic relations of questions*. Accordingly, we design the category-highlighting Transformer network (CHT) to model both content-based and category-based semantic relations of two questions, and show its architecture in Fig. 1. CHT first exploits a category identification unit to automatically construct the category-based semantic representations for a question and its embedded words. Second, we develop the cascaded category-highlighting Transformers by incorporating the category-based representations of words into the self-attention unit, so that both content-based and category-based relations of questions can be captured and deeply modelled. Moreover, the category-based representations of two questions and the deeply encoded results of the category-highlighting Transformers are combined in a decoder function to estimate the semantic equivalence of two questions. We formulate the category-highlighting Transformer network as follows:

$$eq(q, d) = \sigma_1(w_1[c(q); c(d); che(q, d)], b_1) \qquad (1)$$

where $c(q)$ and $c(d)$ are the category-based representations of question q and d, learned by the category identification unit. The $che(q, d)$ is the deeply encoded results over q and d, which contain both content-based semantics and category-based semantics. It is constructed by our cascaded category-highlighting Transformers. The $[;]$ is the concatenation operation. The σ_1 is a $m \times n \times 1$ MLP with activation function ReLU, ReLU and Sigmoid. The w_1 and b_1 are the weight vectors and bias vectors, respectively.

3.1   Category Identification Unit

As motivated in Sect. 1, in the datasets of question retrieval, both user questions and archived questions are not equipped with explicit question categories. To capture the category-based semantics, we propose to automatically construct the category-based representations for a question and its embedded words. Initially, we need to define some question categories for a dataset. Inspired by the work [6] that uses the 5W1H ("what", "when", "where", "why", "who", "how") to extract main events from news articles, we define question categories as "time", "location", "people", "solution", "cause", "object" and "other". Note that fine-grained categories can also be defined and applied to our model.

Given a set of predefined question categories C, we construct the category-based semantic representation of a question q as follows:

$$c(q) = \sum_{c \in C} w(c, q)e(c), \quad \sum_{c \in C} w(c, q) = 1. \tag{2}$$

where $e(c)$ is the embedding of c. We use BERT to encode the category c and take the embedding of $[CLS]$ as $e(c)$. To well construct the semantic representation over a set of categories, $w(c, q)$ is used for adjusting the weight of $e(c)$. If c is more relevant to q $w(c, q)$ is higher. So we estimate $w(c, q)$ as follows:

$$w(c, q) = \sigma_2(w_2[e(c); e(q); CoA(e(c), e(q))] + b_2) \tag{3}$$

where σ_2 with parameters w_2 and b_2 denotes a three-layer MLP with activation functions ReLU, ReLU and Sigmoid. The CoA is a function of performing some interactions over $e(c)$ and $e(q)$, such as $e(c) - e(q)$, $|e(c) - e(q)|$ and $e(c) * e(q)$.

Inspired by the topic model where both sentences and words are represented as latent topics [1], we argue that in a question q, a word t_q should have a category-based representation. Since the word is embedded by a question, its category-based representation can be partly derived from the categories of the question, i.e., question-specific category representation. Besides, a word may appear in many questions, and thus it should have a global category representation. Combining the question-specific and global category representations, we construct the category-based representation for a word as follows:

$$c(t_q, q) = w(e(t_q), c(q))c(q) + \sum_{c \in C} w(e(t_q), e(c))e(c), \quad \sum_{c \in C} w(e(t_q), e(c)) = 1. \tag{4}$$

where $c(q)$ is the category-based representation of q. The $w(e(t_q), c(q))$ is the weight of $c(q)$, and $w(e(t_q), e(c))$ is the weight of $e(c)$, which are estimated by Eq. 3. The $w(e(t_q), c(q))c(q)$ is to capture the question-specific category semantics, and $\sum_{c \in C} w(e(t_q), e(c))e(c)$ is to model the global category semantics.

3.2   Category-Highlighting Transformer

To deeply model the content-based and category-based relations of two questions, we develop the cascaded category-highlighting Transformers as encoders, i.e., individual encoder and joint encoder (see Fig. 1). Given a question pair q and d, the individual encoder models q and d individually, and derives the "individual" semantic representation for each question. The joint encoder first concatenates the individual semantic representations of q and d, and then encodes the concatenated results to derive the "joint" semantic representation. We formulate the cascaded category-highlighting Transformers as follows:

$$che(q, d) = [IT(q); IT(d); JT(IT(q), IT(d))] \tag{5}$$

where IT and JT are two category-highlighting Transformers, denoting the individual encoder and joint encoder respectively. For a question q (or d), IT first encodes the words in q to get their embeddings, and then averages the embeddings as the individual semantics of q. To get the joint semantics of q and d, JT first concatenates $IT(q)$ and $IT(d)$ with a start token [CLS] and separated token [SEP], and then encodes the concatenated result. The embedding of [CLS] is taken as the joint semantics of q and d.

We design the category-highlighting Transformer by improving the self attention unit with the category-based representations of words (IT and JT). Specifically, we first apply the category-based representations of words to estimate the category-based attentive similarities between a word and other words. Second, the context-based attentive similarities are estimated by the scaled Dot-Product [19]. The category-based attentive similarities and context-based attentive similarities are combined as the final attentive similarities. The category-highlighting attention unit is formulated as follows:

$$Attention(Q, K, V, G, H) = softmax(\frac{QK^T + \lambda GH^T}{\sqrt{d_k}})V, \quad \lambda \in (0, 1) \tag{6}$$

where $Q = E_q W^Q$, $K = E_q W^K$, $V = E_q W^V$, $G = C_q W^G$, $H = C_q W^H$. The $C_q = \{c(t_q^1, q), \cdots, c(t_q^n, q)\}$ is the category-based representations of words $\{t_q^1, \cdots, t_q^n\}$ in a question q, and $c(t_q^n, q)$ is estimated by Eq. 4. The E_q is the context-based representations of words. The W^Q, W^K, W^V, W^G and W^H are projection matrices. The GH^T is to model the category-based attentive similarity, and QK^T is to model the context-based attentive similarity. The parameter λ is the weight of GH^T, to balance the importance of QK^T and GH^T.

4   Experiments

4.1   Experimental Setup

Experimental Objectives. Experimental objectives are designed as follows:

- O1: Can CHT better accomplish the question retrieval task than baselines?
- O2: Can CHT effectively model the category-based relations of two questions?
- O3: Ablation Study. What is the effectiveness of components in CHT?

Table 1. Performance comparisons.

Model	Acc. (%)	AUC. (%)	Acc. (%)	AUC. (%)	Acc. (%)	AUC. (%)
	BankQ		LCQMC		Quora	
BiMPM	79.48	87.50	83.59	93.75	85.82	93.46
RE2	81.07	88.94	84.74	94.78	89.20	95.56
BERT	84.06	89.34	86.70	94.84	91.07	96.96
ERNIE	84.66	92.31	87.17	95.89	91.08	96.84
Sentence-BERT	83.53	90.95	84.07	94.62	88.69	94.84
tBERT	80.65	88.82	85.12	93.13	89.31	95.69
CHT	**85.24**	**92.43**	**88.98**	**96.23**	**91.67**	**97.11**

Datasets Description. We conduct experiments on three public datasets.

- BankQ [3]: It is the largest QA dataset in the financial domain and sampled from the session logs of an online bank custom service system.
- Quora[1]: The dataset is sampled from a Q&A website Quora.com. Each question pair is labeled with a binary value that indicates equivalent or not.
- LCQMC [9]: It is a large-scale Chinese question retrieval dataset and sampled from the largest online Chinese question answering platform, i.e., Baidu Knows.

Comparison Solutions. According to the survey in Sect. 2.1, we select state-of-the-art solutions as baselines to verify the above experimental objectives:

- BiMPM [21]: BiLSTM model is used for encoding questions, and multiple matching methods are aggregated to model the relations of two questions.
- RE2 [23]: The aligned features, original point-wise features, and contextual features are applied to residual networks.
- BERT [5]: It is a well-known pre-training representation model, widely applied to many NLP tasks and achieves new state-of-the-art results.
- ERNIE [17]: It is a pre-training framework where the lexical, syntactic, and semantic information are learnt by using multi-task learning strategy.
- Sentence-BERT [14]: It is a framework with twin networks, and individually generates an embedding for every sentence.
- tBERT [13]: It first learns topic-based representations for words, and then use them to improve BERT for semantically equivalent question detection.

Performance Metrics. Similar to the studies [13,17,23], we use the accuracy and AUC metrics to measure the effectiveness of all models.

Reproducibility. The parameters of all models are assigned the default values for fair comparisons. Specifically, the batch sizes of models on the BankQ and LCQMC datasets are 32 and that of models on Quora datasets is 64. All models are developed in Python 3.8 and Pytorch 1.10 development environment. We set the learning rate to $2e-5$ and use the warm-up learning rate method.

4.2 Experimental Results

To verify the experimental objective O1, we perform all models over the three public datasets, and show their performances in Table 1. BERT and ERNIE

---

[1] https://data.quora.com/First-Quora-Dataset-ReleaseQuestion-Pairs.

Table 2. Effectiveness of modelling the category-based relations of two questions.

Model	Acc. (%)	Acc. (%)	Acc. (%)
	ND	*PS*	*other*
BiMPM	84.52	82.16	74.48
RE2	84.59	85.82	75.61
BERT	90.48	83.25	80.52
ERNIE	89.24	86.77	80.31
Sentence-BERT	88.12	83.77	80.45
tBERT	83.47	84.59	76.19
CHT	**90.81**	**86.88**	**80.59**

Table 3. Ablation study: effectiveness of components in CHT.

Model	Acc. (%)	AUC. (%)
CHT-*che*+BERT	84.30	92.16
CHT-*cqd*	84.57	92.31
CHT-*gh*	85.04	92.27
CHT-*ind*	84.72	92.41
CHT-*joint*	84.71	92.31
CHT	**85.24**	**92.43**

perform better than other baselines. tBERT uses the topic distributions of words to improve BERT. But it is hard to construct the accurate topic distributions for a word because its context (question) is very short. Sentence-BERT is a modification of BERT and is to reduce the computation cost of BERT not improve the accuracy of BERT. Besides, we see that CHT performs better than all baseline models on three datasets. Compared with BERT and ERNIE, CHT not only benefits from the pre-trained embeddings but also benefits from the incorporation of question categories in Transformer. The above comparisons positively verify O1: CHT better accomplishes the question retrieval task than baseline models.

To verify experimental objective O2, we make an analysis based on BankQ. Specifically, we first divide the test examples into three groups by manually labelling, i.e., negative examples with different categories (ND), positive examples with the same categories(PS), and *other*. Secondly, we investigate the performances of models over the three groups and show the results in Table 2. It can be seen that 1) the advantage of CHT on the ND group is larger than that on the PS group. This is because examples with different categories can not be semantically equivalent, but examples with the same categories may be positive examples or negative examples. CHT can better capture the data insight than baseline models; 2) The advantage of CHT on PS group is larger than that on the *other* group, since the examples with the same categories are more likely to be positive examples. These metric comparisons verify the ability of CHT on modelling category-based semantic relations of two questions.

To verify the effectiveness of components in CHT (O3), we conduct the ablation study and present the results in Table 3. The notation CHT-cqd denotes the category-based representations are not applied, i.e., $c(q)$ and $c(d)$ in Eq. 1 are not applied. The CHT-che+BERT denotes the cascaded category-highlighting Transformers are not applied, and BERT is used for generating the deep encoding result $che(q, d)$. It can be seen that the metrics of CHT are higher than those of CHT-che+BERT and CHT-cqd. The above comparisons verify the effectiveness of both the category identification unit and the cascaded category-highlighting Transformers. In the cascaded category-highlighting Transformers, the first one is to model the individual semantics of a question, and the second one is to model the joint semantics of two questions. We test the two category-highlighting Transformers, and show their results in Table 3 where the CHT-ind denotes the individual semantics are not applied, and CHT-$joint$ denotes the joint semantics are not applied. Comparing the metrics of CHT, CHT-ind and CHT-$joint$, we find that CHT performs better than CHT-ind and CHT-$joint$. This illustrates that the combination of the individual semantics and joint semantics can better capture the relations of two questions than any single one. Besides, on AUC metric, CHT-ind performs better than CHT-$joint$. This illustrates that the joint semantics are more important for modelling the relations of two questions than the individual semantics.

In the category-highlighting Transformer, we use the category-based representations of words to improve the self-attention unit of the original Transformer. Specifically, in Eq. 6, both the category-based attentive similarity (GH^T) and context-based attentive similarity (QK^T) are used for updating the embedding of words. Comparing the original Transformer, we incorporate the category-based attentive similarity GH^T into the self-attention unit. To verify the effectiveness of this incorporation, we perform an experiment where GH^T is not applied to CHT and denote the results as CHT-gh in Table 3. Comparing the metrics between CHT and CHT-gh, we find that CHT performs better than CHT-gh. The comparisons illustrate that the incorporation of category-based attentive similarity helps to identify the semantic relations of two questions.

5 Conclusion

In this paper, we propose new insights from data and technical perspectives, to address the lexical gap challenge of question retrieval. To capture these insights, we develop a category-highlighting transformer network (CHT), which models both content-based and category-based semantic relations of two questions. Experiments demonstrate 1) the category-highlighting transformer network can better accomplish the question retrieval task than baseline models; 2) The incorporation of implicit question categories can effectively identify the semantic distinctions of two questions with high text similarity yet different categories.

Acknowledgements. This work is supported by National Key Research and Development Program (No. 2020YFB1710004) and the National Science Foundation of China under the grant 62272466.

References

1. Blei, D.M., Ng, A.Y., Jordan, M.I.: Latent Dirichlet allocation. In: NIPS, pp. 601–608 (2001)
2. Cai, L., Zhou, G., Liu, K., Zhao, J.: Learning the latent topics for question retrieval in community QA. In: IJCNLP, pp. 273–281 (2011)
3. Chen, J., Chen, Q., Liu, X., Yang, H., Lu, D., Tang, B.: The BQ corpus: a large-scale domain-specific chinese corpus for sentence semantic equivalence identification. In: EMNLP, pp. 4946–4951 (2018)
4. Das, A., Yenala, H., Chinnakotla, M.K., Shrivastava, M.: Together we stand: Siamese networks for similar question retrieval. In: ACL (2016)
5. Devlin, J., Chang, M.W., Lee, K., Toutanova, K.: BERT: pre-training of deep bidirectional transformers for language understanding. In: NAACL-HLT, pp. 4171–4186 (2019)
6. Hamborg, F., Breitinger, C., Gipp, B.: Giveme5W1H: a universal system for extracting main events from news articles. In: INRA@RecSys, pp. 35–43 (2019)
7. Jeon, J., Croft, W.B., Lee, J.H.: Finding similar questions in large question and answer archives. In: CIKM, pp. 84–90 (2005)
8. Ji, Z., Xu, F., Wang, B., He, B.: Question-answer topic model for question retrieval in community question answering. In: CIKM, pp. 2471–2474 (2012)
9. Liu, X., et al.: LCQMC: a large-scale Chinese question matching corpus. In: COLING, pp. 1952–1962 (2018)
10. Ma, D., Chen, Y., Chang, K.C.C., Du, X., Xu, C., Chang, Y.: Leveraging fine-grained wikipedia categories for entity search. In: WWW, pp. 1623–1632 (2018)
11. Ma, D., Chen, Y., Du, X., Hao, Y.: Interpreting fine-grained categories from natural language queries of entity search. In: DASFAA, pp. 861–877 (2018)
12. Murdock, V., Croft, W.B.: A translation model for sentence retrieval. In: EMNLP, pp. 684–691 (2005)
13. Peinelt, N., Nguyen, D., Liakata, M.: tBERT: topic models and BERT joining forces for semantic similarity detection. In: ACL, pp. 7047–7055 (2020)
14. Reimers, N., Gurevych, I.: Sentence-BERT: sentence embeddings using siamese BERT-networks. In: EMNLP-IJCNLP, pp. 3980–3990 (2019)
15. Robertson, S.E., Zaragoza, H.: The probabilistic relevance framework: BM25 and beyond. Found. Trends Inf. Retr. **3**(4), 333–389 (2009)
16. Song, F., Croft, W.B.: A general language model for information retrieval. In: CIKM, pp. 316–321 (1999)
17. Sun, Y., et al./: ERNIE 2.0: a continual pre-training framework for language understanding. In: AAAI, pp. 8968–8975 (2020)
18. Tan, C., Wei, F., Wang, W., Lv, W., Zhou, M.: Multiway attention networks for modeling sentence pairs. In: IJCAI, pp. 4411–4417 (2018)
19. Vaswani, A., et al.: Attention is all you need. In: NIPS, pp. 5998–6008 (2017)
20. Wan, S., Lan, Y., Guo, J., Xu, J., Pang, L., Cheng, X.: A deep architecture for semantic matching with multiple positional sentence representations. In: AAAI, pp. 2835–2841 (2016)
21. Wang, Z., Hamza, W., Florian, R.: Bilateral multi-perspective matching for natural language sentences. In: IJCAI, pp. 4144–4150 (2017)

22. Xue, X., Jeon, J., Croft, W.B.: Retrieval models for question and answer archives. In: SIGIR, pp. 475–482 (2008)
23. Yang, R., Zhang, J., Gao, X., Ji, F., Chen, H.: Simple and effective text matching with richer alignment features. In: ACL, pp. 4699–4709 (2019)
24. Zhou, G., Cai, L., Zhao, J., Liu, K.: Phrase-based translation model for question retrieval in community question answer archives. In: ACL, pp. 653–662 (2011)

Text Processing

HANOIT: Enhancing Context-aware Translation via Selective Context

Jian Yang[1], Yuwei Yin[2], Shuming Ma[3], Liqun Yang[1(✉)], Hongcheng Guo[1],
Haoyang Huang[3], Dongdong Zhang[3], Yutao Zeng[1], Zhoujun Li[1], and Furu Wei[2]

[1] State Key Lab of Software Development Environment,
Beihang University, Beijing, China
{jiaya,lqyang,hongchengguo,zengyutao,lizj}@buaa.edu.cn
[2] The University of Hong Kong, Hong Kong, China
yuweiyin@hku.hk
[3] Microsoft Research Asia, Beijing, China
{shumma,haohua,dozhang,fuwei}@microsoft.com

Abstract. Context-aware neural machine translation aims to use the document-level context to improve translation quality. However, not all words in the context are helpful. The irrelevant or trivial words may bring some noise and distract the model from learning the relationship between the current sentence and auxiliary context. To mitigate this problem, we propose a novel end-to-end encoder-decoder model with a layer-wise selection mechanism to sift and refine the long document context. To verify the effectiveness of our method, extensive experiments and extra quantitative analysis are conducted on four document-level machine translation benchmarks. The experimental results demonstrate that our model significantly outperforms previous models on all datasets via the soft selection mechanism.

Keywords: Neural Machine Translation · Context-aware Translation · Soft Selection Mechanism

1 Introduction

Recently, neural machine translation (NMT) based on the encoder-decoder framework has achieved state-of-the-art performance on the sentence-level translation [2,5,6,23,24,26,31–33]. However, the sentence-level translation solely considers single isolated sentence in the document and ignores the semantic knowledge and relationship among them, causing difficulty in dealing with the discourse phenomenon such as lexis, ellipsis, and lexical cohesion [27,30].

To model the document-level context, there are two main context-aware neural machine translation schemes. One approach introduces an additional context encoder to construct dual-encoder structure, which encodes the current source sentence and context sentences separately and then incorporates them via the gate mechanism [3,4,9,16,28,29]. The other one directly concatenates the current source sentence and context sentences as a whole input to the standard Transformer architecture, though the input sequence might be quite long

© The Author(s), under exclusive license to Springer Nature Switzerland AG 2023
X. Wang et al. (Eds.): DASFAA 2023, LNCS 13945, pp. 471–486, 2023.
https://doi.org/10.1007/978-3-031-30675-4_34

> Source: Many foreigners like bass .

> Context: The bass is a bowed instrument like the violin and cello . This explanation about instrument bass guitar is provided by Wikipedia , which is a free encyclopedia hosted by Wikimedia Foundation .

Fig. 1. An example of the source and the context sentence. Above is a source sentence to be translated, and below is its context in the same document. The underlined words are useful to disambiguate the source sentence, while the rest is less important.

[1,3,20,25]. The previous works [1,13] conclude that the Transformer model has the capability to capture long-range dependencies, where the self-attention mechanism enables the simple concatenation method to have competitive performance with multi-encoder approaches.

Most aforementioned previous methods use the whole context sentences and assume that all words in the context have a positive effect on the final translation. Despite the benefits of part of the context, not all context words are useful to the current translation. In Fig. 1, the underlined words provide supplementary information for disambiguation, while the others are less important. The irrelevant words may bring some noise and redundant content, increasing the difficulty for the model to learn the relationship between the context and the translation. Therefore, these useless words should be discarded so that the model can focus on the relevant information of the current sentence.

In this work, we propose an end-to-end model to translate the source document based on layer-wise context selection over encoder. In our model, the context is concatenated with current source sentence as external knowledge to be fed into the unified self-attention, where they are precisely selected among multiple layers to gradually discard useless information. The criteria on context selection is based on context-to-source attention score which are recursively calculated layer-by-layer. Ultimately, the context on the top layer is expected to be the most useful knowledge to help current source sentence translation. The architecture of our model looks like a Tower of **Hanoi** over the **T**ransformer structure (HANOIT). Our proposed model captures all context words at the bottom layer and focuses more on the essential parts at the top layer via the soft selection mechanism.

To verify the effectiveness of our method, we conduct main experiments and quantitative analysis on four popular benchmarks, including IWSLT-2017, NC-2016, WMT-2019, and Europarl datasets. Experimental results demonstrate that our method significantly outperforms previous baselines on these four popular benchmarks and can be further enhanced by the sequence-to-sequence pretrained model, such as BART [12]. Analytic experiments and attention visualization illustrate our proposed selection mechanism for avoiding the negative interference introduced by noisy context words and focusing more on advantageous context pieces.

2   Our Approach

In this section, we will describe the architecture of our HANOIT, and apply HANOIT to context-aware machine translation.

2.1   Problem Statement

Formally, let $X = \{x^{(1)}, .., x^{(k)}, .., x^{(K)}\}$ denote a source language document composed of K source sentences, and $Y = \{y^{(1)}, .., y^{(k)}, .., y^{(K)}\}$ is the corresponding target language document. $\{x^{(k)}, y^{(k)}\}$ forms a parallel sentence, where $x^{(k)}$ denotes the kth source sentence and $y^{(k)}$ is the translation of $x^{(k)}$. $X_{<k} = \{x^{(1)}, .., x^{(k-1)}\}$ denotes the historical context of $x^{(k)}$ and $X_{>k} = \{x^{(k+1)}, .., x^{(K)}\}$ represents the future context. Given the current source sentence $x^{(k)}$, the historical context $X_{<k}$, and the future context $X_{>k}$, the translation probability is calculated by:

$$P(y^{(k)}|X; \theta) = \prod_{i=1}^{N} P(y_i^{(k)}|X, y_{<i}^{(k)}; \theta) \tag{1}$$

where $y_i^{(k)}$ is the ith word of the kth target sentence and $y_{<i}^{(k)}$ are the previously generated words of the target sentence $y^{(k)}$ before i^{th} position. $y^{(k)}$ has N words. In this work, we use one previous and one next sentence as the context.

2.2   HANOIT

Figure 2 shows the overall structure of our HANOIT model. At the bottom of the encoder, it models the concatenation of the source sentence and the context with unified self-attention layers. At the top of the encoder, it gradually selects the context words according to the attention weights.

Embedding. We use the segment embedding to distinguish the current sentence, source, and target context sentences. In Fig. 2, we concatenate the current sentence and the source context as a whole. To model the positions of the different parts, we also reset the positions of the current source sentence and source context sentences. Therefore, the final embedding of the input words is the sum of the word embedding, position embedding, and segment embedding, which can be described as:

$$E = E_w + E_p + E_s \tag{2}$$

where E_w is the word embedding, E_s is the segment embedding from the learned parameter matrix, and E_p is the position embedding.

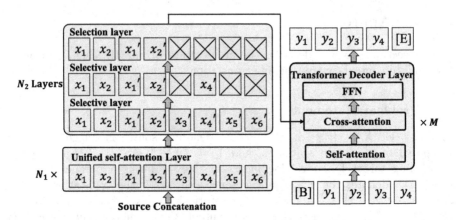

Fig. 2. Overview of our proposed model HANOIT. For simplicity, layer normalization and other components of the Transformer architecture are omitted in the picture. Cross symbols denote dropped words. (x_1, x_2) is the current source sentence and $(x_1', x_2', x_3', x_4', x_5', x_6')$ is the source context. N_1 and N_2 denote the number of unified self-attention layers and selection layers. $(x_1', x_2', x_3', x_4', x_5', x_6') \rightarrow (x_1', x_2', x_4') \rightarrow (x_1', x_2')$ is the selective procedure, where important words are selected gradually by multiple selection layers.

Encoder. Since the inputs of context-aware neural machine translation are composed of several sentences, we build our model based on the multi-head attention to capture long-range dependencies and compute the representation of the document-level context. Our encoder consists of two groups of layers: unified self-attention layers and selection layers. The unified self-attention layers is to compute a joint representation of the source sentence and the context, while the selection layer is to select the context for the next layer.

Unified Self-attention Layer. Given the concatenation of the source sentence and the source context, we obtain the document representation $s^0 = \{s_1^0, .., s_p^0, ..., s_m^0\}$ after the embedding layer, where p is the length of $x^{(k)}$ and m is the length of source concatenation. Then, we feed the s^0 into N_1 unified self-attention layers to compute their representations.

$$s^l = \text{FFN}(\text{MultiHeadAttn}(s^{l-1}; \theta_{N_1})) \tag{3}$$

where the l is the number of the unified self-attention layer and $l \in [1, N_1]$.

Selection Layer. After N_1 unified self-attention layers, we get representations of source concatenation $s^{N_1} = \{s_1^{N_1}, .., s_p^{N_1}, ..., s_m^{N_1}\}$, which can be used to select important context words. In the selection layer, we apply multi-head attention to s^{N_1}, and then average attention scores across different heads, which can be described as below:

$$a_{i,j} = \frac{1}{h} \sum_{1 \leq i \leq h} \text{MultiHeadAttn}(s^{N_1}) \tag{4}$$

where h is the number of attention heads. $a_{i,j}$ represents the average attention score between the ith token and the jth token.

Then, we calculate the average attention score between i^{th} word and other tokens in the source current sentence $x^{(k)}$:

$$a_{i,\neq i} = \frac{1}{p} \sum_{j \in [1,p], j \neq i} a_{i,j} \tag{5}$$

where $a_{i,\neq i}$ represents the average correlation between i^{th} word and the other words, and p is the number of tokens in the source sentence.

In order to decide which context words should be selected, we compute the correlation scores s between each context word and the whole source sentence. For the kth context word, we count how many words in the current sentence have a higher attention score with it compared to the average attention score $a_{i,\neq i}$:

$$s_k = \sum_{i \in [p+1, m]} \delta_{a_{i,k} \geq a_{i,\neq i}} \tag{6}$$

where $\delta_{a_{i,k} \geq a_{i,\neq i}}$ equals 1 if $a_{i,k} \geq a_{i,\neq i}$ else 0, p is the number of tokens in the source sentence, and m is the total number of tokens in the concatenation of the source sentence and the source context.

Finally, we can select the context words with top correlation scores s_k. We use v_k to denote whether the kth word is selected:

$$v_k = \delta_{s_k \geq q*p} \tag{7}$$

where $\delta_{s_k \geq q*p}$ equals to 1 if $s_k \geq q*p$ else 0, p is the number of tokens in the source sentence, and $q \leq 1$ is a hyper-parameter to control the percentage of the selective context. In this work, we set $q \in [0.1, 0.5]$ according to the performance in the validation set.

Decoder. The source selective concatenation $s^{N_2} = \{s_1^{N_2}, .., s_p^{N_2}, .., s_{m_1}^{N_2}\}$ is fed into the standard Transformer decoder to predict the final translation.

2.3   Bi-lingual Context Integration

Section 2.2 only considers the mono-lingual context, i.e. the source context. In practice, when translating a document, we can also obtain the target context by sentence-level translating the document before context-aware translation [27]. In this section, we extend our HANOIT model to integrate the bi-lingual context, i.e. the source context and the target context.

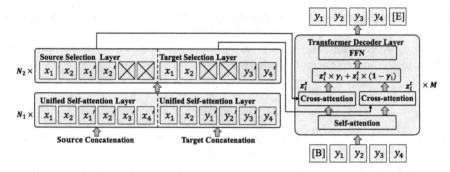

Fig. 3. Overview of the extended HANOIT to integrate the bilingual context. Cross symbols denote masked words. Source concatenation consists of the current source sentence (x_1, x_2) and source context (x'_1, x'_2, x'_3, x'_4). Target concatenation is composed of the source sentence (x_1, x_2) and the target context (y'_1, y'_2, y'_3, y'_4). Then the source and target selective concatenations are incorporated by the gate mechanism to predict the final translation.

Formally, let $X = \{x^{(1)}, .., x^{(k)}, .., x^{(K)}\}$ denote a source language document composed of K source sentences and $Y = \{y^{(1)}, .., y^{(k)}, .., y^{(K)}\}$ denotes the sentence-level translation of X. $X_{<k}$ is the historical source context and $X_{>k}$ is the future source context. Similarity, we denote historical target context $\{y^{(1)}, .., y^{(k-1)}\}$ as $Y_{<k}$ and future target context $\{y^{(k+1)}, .., y^{(K)}\}$ as $Y_{>k}$. We model the translation probability that is conditioned on the bi-lingual source context $X_{\neq k}$ and target context $Y_{\neq k}$ as:

$$P(y^{(k)}|X; \theta) = \prod_{i=1}^{N} P(y_i^{(k)}|X, Y_{\neq k}, y_{<i}^{(k)}; \theta) \tag{8}$$

where $y_i^{(k)}$ is the ith word of the kth target sentence and $y_{<i}^{(k)}$ are the previously generated words of the target sentence $y^{(k)}$ before i^{th} position.

Encoder. As shown in Fig. 3, the current source sentence and the source context are merged as the source concatenation. Besides, the current source sentence and the target context are also merged as the target concatenation. Both concatenations are fed into unified self-attention and selection layers to compute representations of source concatenation s^{N_2} and target concatenation t^{N_2}.

Decoder. With the above encoder, we obtain the representations of the selective source concatenation $s^{N_2} = \{s_1^{N_2}, .., s_p^{N_2}, .., s_{m_1}^{N_2}\}$ and the selective target concatenation $t^{N_2} = \{t_1^{N_2}, .., t_p^{N_2}, .., t_{n_1}^{N_2}\}$, where m_1 and n_1 are lengths of selective source and target concatenation. Given both selective concatenations, we deploy the multi-head attention by two attention components. Using query, key, value parameters (W_s^Q, W_s^V, W_s^K), the decoder gets the hidden state z_i^s. Similarly, another hidden state z_i^t is generated by the additional attention component with parameters (W_t^Q, W_t^V, W_t^K). Considering the previous insight [10] that the gate network is a good component for bi-lingual context setting, we employ the gate mechanism to incorporate the source and target context.

Gate Mechanism. Given the ith hidden states z_i^s and z_i^t, the gate mechanism can be described as:

$$\gamma_i = c\sigma(W_s z_i^s + U_t z_i^t + b) \tag{9}$$

where W_s and U_t are parameters matrices and b is a bias. $c \in [0,1]$ is a hyper-parameter to control range of the gate weight. $\sigma(\cdot)$ is the sigmoid function.

$$z_i = (1 - \gamma_i)z_i^s + \gamma_i z_i^t \tag{10}$$

where z_i is the ith decoder final hidden state derived from the source context and target context.

2.4 Training

Given the mono-lingual context only, the training objective is a cross-entropy loss function on the top of Eq. 1. The objective $\mathcal{L}_m$ is written as:

$$\mathcal{L}_m = - \sum_{X,y^{(k)} \in D} \log P_\theta(y^{(k)}|X) \tag{11}$$

where θ are model parameters.

Considering the bi-lingual context, the training objective $\mathcal{L}_b$ is calculated as:

$$\mathcal{L}_b = - \sum_{X,y^{(k)},Y_{\neq k} \in D} \log P_\theta(y^{(k)}|X, Y_{\neq K}) \tag{12}$$

where θ are model parameters.

The quality of the target context depends on the sentence-level translation model, which may bring additional errors. To reduce the possible harm by these errors and make the training stable, our model optimizes a combination of the mono-lingual objective $\mathcal{L}_m$ and the bi-lingual objective $\mathcal{L}_b$:

$$\mathcal{L}_{all} = \alpha \mathcal{L}_m + (1 - \alpha)\mathcal{L}_b \tag{13}$$

where α is a scaling factor to balance two objectives between $\mathcal{L}_m$ and $\mathcal{L}_b$. We find when the value of α equals 0.5, our model gets the optimal performance by balancing two objectives. We adopt Eq. 11 to train the model with mono-lingual context, and Eq. 13 to train the model with bi-lingual context.

3 Experiments

To prove the efficiency of our method, we conduct experiments on four public benchmarks.

Table 1. Sentence-level evaluation results on four tasks with BLEU% metric using the source context. Bold numbers denote the best BLEU points. RNN and Transformer are context-agnostic baselines and others are context-aware baselines. The results with the symbol "†" are directly reported from the previous work. BLEU points with the symbol "*" are re-implemented by ourselves. "‡" denotes our proposed method.

Mono-lingual Context	IWSLT-2017	NC-2016	Europarl	WMT-2019
RNN [2]	$19.24^{\dagger}$	$16.51^{\dagger}$	$26.26^{\dagger}$	–
Transformer [26]	$23.28^{\dagger}$	$22.78^{\dagger}$	$28.72^{\dagger}$	–
Transformer (our re-implementation)	24.52*	24.45*	29.98*	38.02*
ECT [25]	24.32*	24.40*	30.08*	38.14*
Dual Encoder [9]	24.14*	24.36*	30.12*	38.12*
DCL [41]	$24.00^{\dagger}$	$23.08^{\dagger}$	$29.32^{\dagger}$	–
HAN [30]	$24.58^{\dagger}$	$25.03^{\dagger}$	$28.60^{\dagger}$	–
Transformer + QCN [40]	$24.41^{\dagger}$	$22.22^{\dagger}$	$29.48^{\dagger}$	–
SAN [16]	$24.55^{\dagger}$	$24.78^{\dagger}$	$29.75^{\dagger}$	–
Flat Transformer [13]	$24.87^{\dagger}$	$23.55^{\dagger}$	$30.09^{\dagger}$	38.34*
HanoiT (our method)	$\textbf{24.94}^{\ddagger}$	$\textbf{25.22}^{\ddagger}$	$\textbf{30.49}^{\ddagger}$	$\textbf{38.52}^{\ddagger}$

3.1 Datasets

To evaluate our method, we use the same dataset as previous work, including IWSLT-2017, NC-2016, Europarl, and WMT-2019 En-De translation [16].

IWSLT-2017. This corpus is from IWSLT-2017 MT track and contains transcripts of TED talks aligned at the sentence level.

NC-2016. NC-2016 dataset is from Commentary v9 corpus. Newstest2015 and newstest2016 are used as the valid and the test set.

Europarl. The dataset from Europarl v7 is split into training, valid and test sets according to the previous work [16]. Europarl is extracted from the European Parliament website.

WMT-2019. The WMT-2019 dataset comes from the WMT-2019 news translation shared task for English-German. Newstest2016, newstest2017, and newstest2018 are concatenated as the valid set. Newstest2019 is used as the test set.[1]

3.2 Implementation Details

Considering the model performance and computation cost, we use one previous and one next sentence as the source and target context for all our experiments.

---

[1] https://www.statmt.org/wmt19/.

Table 2. Sentence-level evaluation results on four tasks with BLEU% metric using the bi-lingual context. Bold numbers represent the best BLEU points. The results with the symbol "†" are directly reported from the previous work. BLEU points with the symbol "*" are re-implemented by ourselves. "‡" represents our proposed method.

Bi-lingual Context	IWSLT-2017	NC-2016	Europarl	WMT-2019
ECT [25]	24.38*	24.55*	30.24*	38.16*
Dual Encoder [9]	24.26*	24.46*	30.25*	38.24*
DCL [41]	23.82†	22.78†	29.35†	–
HAN [30]	24.39†	24.38†	29.58†	–
CADec [27]	24.45*	24.30*	29.88*	–
SAN [16]	24.62†	24.36†	29.80†	–
HanoiT (our method)	**25.04‡**	**25.28‡**	**30.89‡**	**38.55‡**

The evaluation metric is case-sensitive tokenized BLEU [18]. For different benchmarks, we adapt the batch size, the beam size, the length penalty, the number of unified self-attention layers N_1, and the number of selection layers N_2 to get better performance. For all experiments, we use a dropout of 0.1 and cross-entropy loss with a smoothing rate of 0.1 for sentence-level and context-aware baselines except notification. All sentences are tokenized with Moses [11] and encoded by BPE [21] with a shared vocabulary of 40K symbols. The batch size is limited to 2048 target tokens by default. For the **IWSLT-2017** dataset, we deploy the small setting of the Transformer model, which has 6 layers with 512 embedding units, 1024 feedforward units, 4 attention heads, a dropout of 0.3, a l_2 weight decay of 1e-4. For the **NC-2016** dataset, we use the base setting of Transformer [26], in which both the encoder and the decoder have 6 layers, with the embedding size of 512, feedforward size of 2048, and 8 attention heads. We set both dropout and attention dropout as 0.2 for our method. For the **Europarl** and the **WMT-2019** dataset, the base setting of the Transformer model with 4000 warming-up steps is used.

3.3   Baselines

For the mono-lingual and the bi-lingual context setting, we compare our method with other baselines.

Mono-lingual Context: **RNN** [2] and **Transformer** [26] are backbone models. **ECT** [25] simply concatenates the source sentence and context into the standard Transformer model. Besides, **Dual Encoder** [9] uses two encoders to incorporate the source sentence and context sentences to predict the translation. Moreover, **DCL** [41] incorporates context hidden states into both the source encoder and target decoder. **Flat Transformer** [13] focus on the current self-attention at the top. Furthermore, **HAN** [30] and **SAN** [16] introduce the hierarchical and selection attention mechanism. **QCN** [40] is a query-guided capsule networks.

Table 3. Sentence-level evaluation results on four benchmarks with BLEU% metric under the mono-lingual context setting. The architecture $N_1 + N_2$ represents our HANOIT consists of N_1 unified self-attention layers and N_2 selection layers. The architecture ($N_1 = 6$, $N_2 = 0$) only uses six unified self-attention layers with the segment embedding, which select all context words to generate the final translation.

Architecture	IWSLT-2017	NC-2016	Europarl	WMT-2019	Average
6 + 0	24.48	24.64	30.22	38.22	29.39
5 + 1	24.32	24.95	30.52	38.46	29.74
4 + 2	24.55	24.52	30.72	38.37	29.54
3 + 3	24.64	24.88	30.65	38.12	29.58
2 + 4	24.74	24.60	30.62	**38.62**	29.65
1 + 5	**24.94**	**25.22**	30.49	38.40	**29.85**
0 + 6	24.56	24.75	30.66	37.98	29.49

Bi-lingual Context: **CADec** [27] is composed of identical multi-head attention layers, of which the decoder has two multi-head encoder-decoder attention with encoder outputs and first-pass decoder outputs. Also, **Dual Encoder**, **ECT**, **DCL**, **HAN** and **SAN** can also use the bi-lingual context to improve the performance.

3.4 Main Results

Mono-lingual Context. We present the results of our proposed method, sentence-level baselines, and other context-aware baselines in Table 1, which all only use the mono-lingual source context. The context-aware baselines include ECT, Dual Encoder, DCL, HAN, SAN, and Flat Transformer. The sentence-level Transformer model gets 24.52, 24.55, 29.98, and 38.02 BLEU points on four benchmarks. Compared to this strong baseline, our model also significantly gains an improvement of +0.42, +0.77, +0.81, and +0.51 BLEU points respectively on four benchmarks. Furthermore, our method outperforms SAN by +0.39, +0.44, +0.74 BLEU points on IWSLT-2017, NC-2016, and Europarl datasets. We also observe that most context-aware models gain better performance than the sentence-level model Transformer, especially on IWSLT-2017, NC-2016, and Europarl datasets. We conjecture these three datasets are suitable for evaluating context-aware models, where the current sentence needs to learn longer dependencies.

Bi-lingual Context. Under the bi-lingual context setting, our method also outperforms other baselines, including Dual Encoder, ECT, DCL HAN, CADec,

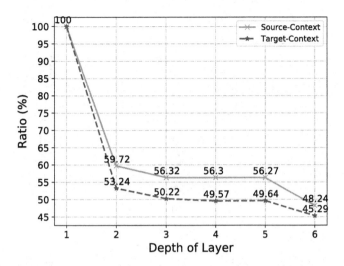

Fig. 4. Ratio of the source and target context selected words on the NC-2016 dataset. Our model selects useful words from both the source and the target context gradually layer by layer. Therefore, the number of context words reduces as the depth of selection layer increasing.

and SAN. HANOIT can achieve improvements of +1.02, +1.33, +0.91, +0.53 BLEU points than the sentence-level Transformer baseline. It proves that HANOIT also can be compatible with the target context to select useful words. Besides, HANOIT can significantly outperform the related baseline SAN model by +0.49, +0.50, +1.14 BLEU points, achieving better performance on three benchmarks. We also observe that the bi-lingual context provides marginal improvements over the mono-lingual context. According to these results, we infer that whether the context-aware model benefits from the bi-lingual context setting is dependent on the specific dataset.

4   Analysis

Attention Visualization. Our model encodes the concatenation of the source words and all context words by the unified attention layers at the bottom layers. As shown in Fig. 5(a), the model focuses on the source sentence "What do you do when you have a headache ?" and all context words "You swallow an aspir@@ in ." using the self-attention mechanism, which ensures that all context words can provide the external guidance and implicitly contribute to the translation. The context words with higher attention weights tend to be selected. In Fig. 5(b), the model only focuses on the source sentence and selected context word "aspir@@". The source word "headache" has a correlation with the context word "aspir@@". In this way, our method pays more attention to the current sentences and the selected words, while the other context words also provide the supplemental semantics for the current sentences at the bottom.

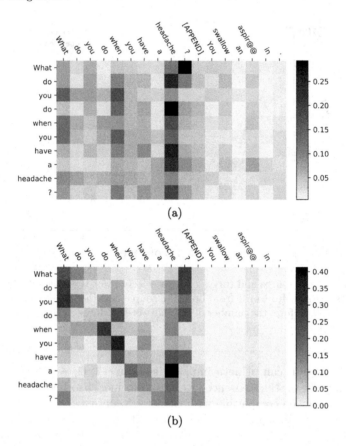

Fig. 5. Attention visualization of the encoder self-attention weights of the bottom unified attention layer (a) and top layer after the selection operation (b).

Number of Selection Layers. To better understand the impact of the selection layers for the translation performance, we tune different numbers of concatenation self-attention layers (N_1 layers) and selection layers (N_2 layers) to get the better performance under the mono-lingual setting. For the fair comparison, we keep the $N_1 + N_2 = 6$, which equals the number of the base setting Transformer layers. As shown in Table 3, we find that the architecture "$N_1 = 1$, $N_2 = 5$" gains the best average performance on four benchmarks. Besides, stacking too many selection layers also leads to worse performance, which may be caused by wrongly discarding too many context words. In summary, our proposed model uses all context words by unified self-attention layers, and focuses those important context words at the top of encoder blocks.

Ratio of Selected Words. We investigate how many words in the context are selected on the NC-2016 dataset. Figure 4 shows that the first selection layer reserves only 60% words from the context. After multiple selection layers, the

Table 4. Results of our method with the offline (1 previous + 1 next) and the online (1 previous) setting.

Context	IWSLT-2017	NC-2016	Avg.
Online (1 previous)	24.62	24.78	24.70
cre Offline (1 previous + 1 next)	**24.94**	**25.12**	25.07

ratio gradually reduces to 48.24%. Another obvious phenomenon is that the ratio of the target context is less than that of the source context. An intuitive explanation is that we use the source sentence words to select source or target context words, where source-source attention has a higher score compared with the source-target attention. Representations of the same language have a closer relationship than those of different languages on average [19].

Online vs. Offline Setting. Table 4 lists results of our method with different context settings. "1 previous" denotes the online setting where the context only includes one previous sentence. "1 previous + 1 next" denotes the offline setting where the context includes one previous and one next sentence. From the table, we can find that our proposed method has the similar performance with online and offline settings on the IWSLT-2017 and NC-2016 datasets. We also try the longer context including "2 previous + 2 next" and "3 previous + 3 next", but find no significant improvement.

Mono-lingual vs. Bi-lingual Context. For the source and target mono-lingual context setting, our method gets 24.94 and 24.98 BLEU points on the IWSLT-2017 dataset. Furthermore, we conduct experiments with the bi-lingual context sentences and get 25.04 BLEU points, where the target context sentences are the translation of the source context sentences. The bi-lingual context setting of our method has limited improvement over the mono-lingual context setting. The reason is that the target-side context shares similar information to the source-side, which also has been found by the previous work [16].

Leveraging Pre-trained Model. Since the parameters of our model are the same as the standard Transformer, our model can be initialized with the pre-trained model to enhance our method. The pre-trained model BART-large [12] is used for initialization under the mono-lingual context setting. We extract 12 bottom layers of the BART encoder and 6 bottom layers of the BART decoder to initialize our model. On IWSLT-2017 dataset, our model gains +1.91 BLEU improvement (24.94 → 26.85) with pre-trained model BART.

5 Related Work

Sentence-level Machine Translation. Sentence-level neural machine translation has developed immensely in the past few years, from RNN-based [2,6,24,32,33,

36], CNN-based [5], to the self-attention-based architecture [23,26,31,34,35,37–39]. However, these models always performed in a sentence-by-sentence manner, ignoring the long-distance dependencies. The past or future context can be important when it refers to using discourse features to translate the source sentence to the target sentence.

Context-aware Machine Translation. Context-aware machine translation aims to incorporate the source or target context to help translation. Previous works [3,7,9,15,22,27–29] have proven the importance of context in capturing different types of discourse phenomena such as Deixis, Ellipsis, and Lexical Cohesion. Others [3,9,16,28,29] explore the dual encoders and concatenation-based context-aware models.

Recently, a promising line of research to improve the performance of the context-aware NMT is to select useful words of the whole context, which can be used to enhance the positive use of context [8,17,42]. Other researchers propose selective attention mechanism by introducing sparsemax function [14,16].

6   Conclusion

In this work, we explore the solution to select useful words from the context. We propose a novel model called HANOIT, consisting of unified self-attention layers and selection layers. The experiments on both mono-lingual and bi-lingual context settings further prove the effectiveness of our method Experimental results demonstrate that our proposed method can select useful words to yield better performance.

References

1. Agrawal, R.R., Turchi, M., Negri, M.: Contextual handling in neural machine translation: look behind, ahead and on both sides. In: EAMT 2018, pp. 11–20 (2018)
2. Bahdanau, D., Cho, K., Bengio, Y.: Neural machine translation by jointly learning to align and translate. In: ICLR 2015 (2015)
3. Bawden, R., Sennrich, R., Birch, A., Haddow, B.: Evaluating discourse phenomena in neural machine translation. In: NAACL 2018, pp. 1304–1313 (2018)
4. Chen, L., et al.: Improving context-aware neural machine translation with source-side monolingual documents. In: IJCAI 2021, pp. 3794–3800. https://www.ijcai.org/ (2021)
5. Gehring, J., Auli, M., Grangier, D., Yarats, D., Dauphin, Y.N.: Convolutional sequence to sequence learning. In: ICML 2017, pp. 1243–1252 (2017)
6. Geng, X., Feng, X., Qin, B., Liu, T.: Adaptive multi-pass decoder for neural machine translation. In: EMNLP 2018, pp. 523–532 (2018)
7. Gonzales, A.R., Mascarell, L., Sennrich, R.: Improving word sense disambiguation in neural machine translation with sense embeddings. In: WMT 2017, pp. 11–19 (2017)
8. Jean, S., Cho, K.: Context-aware learning for neural machine translation. CoRR abs/1903.04715 (2019)

9. Jean, S., Lauly, S., Firat, O., Cho, K.: Does neural machine translation benefit from larger context? CoRR abs/1704.05135 (2017)
10. Junczys-Dowmunt, M.: Microsoft translator at WMT 2019: towards large-scale document-level neural machine translation. In: WMT 2019, pp. 225–233 (2019)
11. Koehn, P., et al.: Moses: open source toolkit for statistical machine translation. In: ACL 2007, pp. 177–180 (2007)
12. Lewis, M., et al.: BART: denoising sequence-to-sequence pre-training for natural language generation, translation, and comprehension. In: ACL 2020, pp. 7871–7880 (2020)
13. Ma, S., Zhang, D., Zhou, M.: A simple and effective unified encoder for document-level machine translation. In: ACL 2020 (2020)
14. Martins, A.F.T., Astudillo, R.F.: From softmax to sparsemax: a sparse model of attention and multi-label classification. In: ICML 2016, pp. 1614–1623 (2016)
15. Maruf, S., Haffari, G.: Document context neural machine translation with memory networks. In: ACL 2018, pp. 1275–1284 (2018)
16. Maruf, S., Martins, A.F.T., Haffari, G.: Selective attention for context-aware neural machine translation. In: NAACL 2019, pp. 3092–3102 (2019)
17. Maruf, S., Saleh, F., Haffari, G.: A survey on document-level machine translation: methods and evaluation. CoRR abs/1912.08494 (2019)
18. Papineni, K., Roukos, S., Ward, T., Zhu, W.: BLEU: a method for automatic evaluation of machine translation. In: ACL 2002, pp. 311–318 (2002)
19. Qin, L., Ni, M., Zhang, Y., Che, W.: CoSDA-ML: multi-lingual code-switching data augmentation for zero-shot cross-lingual NLP. In: IJCAI 2020, pp. 3853–3860 (2020)
20. Scherrer, Y., Tiedemann, J., Loáiciga, S.: Analysing concatenation approaches to document-level NMT in two different domains. In: EMNLP 2019, pp. 51–61 (2019)
21. Sennrich, R., Haddow, B., Birch, A.: Neural machine translation of rare words with subword units. In: ACL 2016, pp. 1715–1725 (2016)
22. Smith, K.S., Aziz, W., Specia, L.: The trouble with machine translation coherence. In: EAMT 2016, pp. 178–189 (2016)
23. Stern, M., Chan, W., Kiros, J., Uszkoreit, J.: Insertion transformer: flexible sequence generation via insertion operations. In: ICML 2019, pp. 5976–5985 (2019)
24. Sutskever, I., Vinyals, O., Le, Q.V.: Sequence to sequence learning with neural networks. In: NIPS 2014, pp. 3104–3112 (2014)
25. Tiedemann, J., Scherrer, Y.: Neural machine translation with extended context. In: EMNLP 2017, pp. 82–92 (2017)
26. Vaswani, A., et al.: Attention is all you need. In: NIPS 2017, pp. 5998–6008 (2017)
27. Voita, E., Sennrich, R., Titov, I.: When a good translation is wrong in context: context-aware machine translation improves on deixis, ellipsis, and lexical cohesion. In: ACL 2019, pp. 1198–1212 (2019)
28. Voita, E., Serdyukov, P., Sennrich, R., Titov, I.: Context-aware neural machine translation learns anaphora resolution. In: ACL 2018, pp. 1264–1274 (2018)
29. Wang, L., Tu, Z., Way, A., Liu, Q.: Exploiting cross-sentence context for neural machine translation. In: EMNLP 2017, pp. 2826–2831 (2017)
30. Werlen, L.M., Ram, D., Pappas, N., Henderson, J.: Document-level neural machine translation with hierarchical attention networks. In: EMNLP 2018, pp. 2947–2954 (2018)
31. Wu, F., Fan, A., Baevski, A., Dauphin, Y.N., Auli, M.: Pay less attention with lightweight and dynamic convolutions. In: ICLR 2019 (2019)
32. Wu, Y., et al.: Google's neural machine translation system: Bridging the gap between human and machine translation. CoRR abs/1609.08144 (2016)

33. Xia, Y., et al.: Deliberation networks: sequence generation beyond one-pass decoding. In: NIPS 2017, pp. 1784–1794 (2017)
34. Yang, J., et al.: GanLM: encoder-decoder pre-training with an auxiliary discriminator. CoRR **abs/2212.10218** (2022). https://doi.org/10.48550/arXiv.2212.10218
35. Yang, J., et al.: Multilingual machine translation systems from microsoft for WMT21 shared task. In: WMT@EMNLP 2021, pp. 446–455 (2021)
36. Yang, J., Ma, S., Zhang, D., Li, Z., Zhou, M.: Improving neural machine translation with soft template prediction. In: ACL 2020, pp. 5979–5989 (2020)
37. Yang, J., et al.: Learning to select relevant knowledge for neural machine translation. In: Wang, L., Feng, Y., Hong, Yu., He, R. (eds.) NLPCC 2021. LNCS (LNAI), vol. 13028, pp. 79–91. Springer, Cham (2021). https://doi.org/10.1007/978-3-030-88480-2_7
38. Yang, J., Yin, Y., Ma, S., Zhang, D., Li, Z., Wei, F.: High-resource language-specific training for multilingual neural machine translation. In: Raedt, L.D. (ed.) Proceedings of the Thirty-First International Joint Conference on Artificial Intelligence, IJCAI 2022, Vienna, Austria, 23-29 July 2022, pp. 4461–4467. ijcai.org (2022). https://doi.org/10.24963/ijcai.2022/619
39. Yang, J., et al.: UM4: unified multilingual multiple teacher-student model for zero-resource neural machine translation. In: Raedt, L.D. (ed.) Proceedings of the Thirty-First International Joint Conference on Artificial Intelligence, IJCAI 2022, Vienna, Austria, 23-29 July 2022, pp. 4454–4460. ijcai.org (2022). https://doi.org/10.24963/ijcai.2022/618
40. Yang, Z., Zhang, J., Meng, F., Gu, S., Feng, Y., Zhou, J.: Enhancing context modeling with a query-guided capsule network for document-level translation. In: EMNLP 2019, pp. 1527–1537 (2019)
41. Zhang, J., et al.: Improving the transformer translation model with document-level context. In: EMNLP 2018, pp. 533–542 (2018)
42. Zheng, Z., Huang, S., Sun, Z., Weng, R., Dai, X., Chen, J.: Learning to discriminate noises for incorporating external information in neural machine translation. CoRR abs/1810.10317 (2018)

Speculation and Negation Scope Resolution via Machine Reading Comprehension Formulation with Data Augmentation

Zhong Qian[1]([✉]), Tiening Sun[1], Ting Zou[1], Peifeng Li[1,2], Qiaoming Zhu[1,2], and Guodong Zhou[1,2]

[1] School of Computer Science and Technology, Soochow University, Suzhou, China
qianzhong@suda.edu.cn, {tnsun,tzou}@stu.suda.edu.cn
[2] AI Research Institute, Soochow University, Suzhou, China
{pfli,qmzhu,gdzhou}@suda.edu.cn

Abstract. As a key sub-task in the field of speculation and negation extraction, Speculation and Negation Scope Resolution (SpNeSR) focuses on extracting speculative and negative texts within sentences, i.e., distinguishing between factual and non-factual information, which means it is an important and fundamental task in Natural Language Processing (NLP) community. Previous work utilized various methods for SpNeSR that are quite domain-specific, and failed to build a unified framework with good generalization. In addition, they were limited by the sizes of datasets and ignored producing more samples for training, since SpNeSR is a data-hungry task. With consideration of the above problems, we not only propose a unified Machine Reading Comprehension (MRC) formulation for SpNeSR, but also design a Data Augmentation (DA) method fit for scopes. Experimental results on several English and Chinese corpora manifest that both MRC and DA mechanism are effective, and our MRC model with DA is superior to several state-of-the-arts.

Keywords: Speculation and negation scope · Machine reading comprehension · Data augmentation

1 Introduction

Speculation and Negation Scope Resolution (SpNeSR) aims to identify the scope of a cue, where a cue is a word or phrase with speculative/uncertain or negative/counterfactual semantics, e.g., speculative cues "may"/"indicate that" in the sentences S1.1/S1.2, and negative cues "not"/"no longer" in S1.3/S1.4. And a scope is a continuous span governed by the corresponding cue, e.g., scope "may be liquid water on the planet" of the cue "may", just as exemplified by Fig. 1.

Early studies relied on rules [1,18,19] and traditional machine learning methods [11,26,27]. Neural networks on SpNeSR can learn lexical [6] and syntactic knowledge from parse trees [20,21]. But these work have the following limitations: 1) They depend on rules difficult to summarize, or syntactic features relying on NLP tools; 2) They did not treat multi-token cues as complete phrases,

© The Author(s), under exclusive license to Springer Nature Switzerland AG 2023
X. Wang et al. (Eds.): DASFAA 2023, LNCS 13945, pp. 487–496, 2023.
https://doi.org/10.1007/978-3-031-30675-4_35

> (S1) Scientists think there {may be liquid water on the planet} since space tele-scopes have observed plumes of smoke on it.
>
> (S2) Estimates {indicate that the aquifer contains enough water to fill Lake Huron} ...
>
> (S3) He does {not participate in the game} due to injury caused by the accident.
>
> (S4) {Superhigh scores like VOS Savant's are no longer possible}, because scoring is now based on a statistical population distribution among age peers ...

Fig. 1. Examples of scopes, where speculative and negative cues are colored, and {scopes} are in {curly braces} with corresponding colors (the same below). (Color figure online)

and considered the first token for simplification [20, 21], which may lead to incomplete semantics; 3) They are confined to datasets of specific domain, e.g., biomedicine [22], detective fictions [6], and did not build a unified paradigm, since structures of scopes of several cues are similar across different datasets; 4) They failed to produce more instances since it is a data-driven task.

To address the aforementioned challenges, we propose a novel method termed Scope Resolution via Machine Reading Comprehension Formulation with Data Augmentation, i.e., SR-MRC-DA. Our main contributions are three-fold: 1) We formulate SpNeSR as a unified MRC paradigm applied to several datasets and compatible with various cues, including multi-token ones; 2) We leverage data augmentation for SpNeSR; 3) Experimental results on several English and Chinese corpora indicate that our model is superior to state-of-the-arts.

2   Approach

Task Definition. Given a set of cues $\{Cue_0, \ldots, Cue_{I-1}\}$ in a sentence Sen, SpNeSR is required to determine a scope Scp_i for each Cue_i, where $i = 0, \ldots, I - 1$, and each cue is regarded an individual sample. A cue is a sub-string of its scope, i.e., $Cue_i \subseteq Scp_i \subseteq$ Sen. We use golden cues for SpNeSR. Therefore, a sample can be represented as $\mathbb{S} = \{Cue_i, Scp_i\}$, and MRC version of a sample is a triple group $\mathbb{S} = \{\mathbb{Q}, \mathbb{C}, \mathbb{A}\}$, where

- $\mathbb{Q}$ is the *Question* that integrates the type and position of Cue_i. Specially, English and Chinese versions of $\mathbb{Q}$ are:

    What is the scope of the $[Typ_i]$ cue $[Cue_i]$ at the position $[Loc_i]$?
    位置为 $[Loc_i]$ 的 $[Typ_i]$ 线索词 $[Cue_i]$ 的作用范围是什么?

    where $[Typ_i] \in \{$speculative/不确定 , negative/否定 $\}$ denotes the type of cue, $[Loc_i]$ is the location/position of $[Cue_i]$ in the current sentence.
- $\mathbb{C}$ is the *Context* that refers to the current sentence Sen.
- $\mathbb{A}$ is the *Answer* that is a sub-string (i.e., Scp_i) of $\mathbb{C}$ (i.e., Sen).

Data Augmentation. For each sample $\mathbb{S}$, we perform the following automatic data augmentation operations for $\mathbb{C}$ during the training phase:

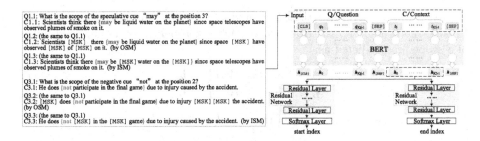

Fig. 2. Overall architecture of SR-MRC-DA model, where curly braces are just used to highlight scopes, instead of tokens of input.

1) Out-of-Scope Mask (OSM) is designed for the token t_j that is OUT of the scope, i.e., $\{t_j | t_j \in \mathbb{C} \& t_j \notin \mathbb{A}\}$, and t_j is replaced by the token [MSK] with the probability of p_{OSM}. OSM generates n_{OSM} more samples for the current $\mathbb{S}$;

2) In-Scope Mask (ISM) is launched for t_j that is IN the scope and is NOT the CUE, i.e., $\{t_j | t_j \in \mathbb{C} \& t_j \in \mathbb{A} \& t_j \neq Cue_i\}$. Then, t_j is replaced by [MSK] with the probability of p_{ISM}. ISM reduplicates n_{ISM} more samples for $\mathbb{S}$.

Since the least frequent cues can hardly influence the performance, we choose $\alpha_d \in (0,1)$ of the original training samples and apply data augmentation operations, instead of using all the samples. Then we can obtain a set of produced samples whose size is not too large, and the distribution of cues is almost consistent with the original set.

Model Structure. Figure 2 shows the architecture of SR-MRC-DA model. Firstly, we concatenate $\mathbb{Q}$ and $\mathbb{C}$ to obtain the input $\mathbb{X}$, and then use Bert-base [4] to encode $\mathbb{X}$ as the hidden representation $\boldsymbol{H}_0 \in \mathbb{R}^{d \times L}$, where d is the dimension of hidden states in Bert, and L is the length of $\mathbb{X}$:

$$\mathbb{X} = \{[\text{CLS}], \mathbb{Q}, [\text{SEP}], \mathbb{C}, [\text{SEP}]\} \tag{1}$$

$$\boldsymbol{H}_0 = \text{Bert}(\mathbb{X}) \tag{2}$$

To learn more meaningful semantics for scopes, $\boldsymbol{H}_0$ is fed into Residual Networks (RNet) [8] to compute high-level information of start and end indices of the scope separately:

$$\boldsymbol{H}_s = \text{RNet}_s(\boldsymbol{H}_0) \tag{3}$$

$$\boldsymbol{H}_e = \text{RNet}_e(\boldsymbol{H}_0) \tag{4}$$

where each RNet is comprised of a stack of several identical Residual Layers, i.e., $\text{RNet} = \{\text{RLayer}_i\}$. Formally, each residual layer RLayer_i with the input $\boldsymbol{H}_0$ is calculated as follows:

$$\boldsymbol{H}_r = \text{FC}_2(\text{GELU}(\text{FC}_1(\text{LN}(\boldsymbol{H}_0)))) \tag{5}$$

$$\text{RLayer}_i(\boldsymbol{H}_0) = \boldsymbol{H}_0 + \boldsymbol{H}_r \tag{6}$$

where FCs are Fully-Connected layers, and LN is Layer Normalization operation. Finally, the probability distributions of each token being a start and end index of a scope are predicted as:

$$p_s = \text{softmax}(W_s H_s + b_s) \tag{7}$$

$$p_e = \text{softmax}(W_e H_e + b_e) \tag{8}$$

The objective function $\mathcal{L}(\theta)$ of the model is designed as:

$$\mathcal{L}(\theta) = -\frac{1}{N}[\epsilon \sum_{i=0}^{N-1} \log P_s(y_s^i|\theta) + (1 - \epsilon) \sum_{i=0}^{N-1} \log P_e(y_e^i|\theta)] \tag{9}$$

where N is the number of samples, y_s^i and y_e^i are annotated start and end indices of the i-th sample, ϵ is the trade-off coefficient, and we set $\epsilon = 0.5$.

3 Experimentation

In this section, firstly, we introduce experimental settings, including corpus, evaluation details, and baselines. Then we give detailed experimental analysis.

3.1 Experimental Settings

Corpus. To evaluate our model, we select these datasets: 1) BioScope [22] consists of three sub-corpora: biological scientific abstracts (Abs), clinical radiology reports (Cli), full papers (Ful); 2) CoNLL-2010 Shared Task [7] proposed Task2 (Tk2) for biological speculation scope resolution; 3) CD-Sco (CD) [17] is Conan Doyle stories with negation scopes, and contains two test sets: the Adventure of the Cardboard Box (Cdb) and Red Circle (Cir); 4) Sfu [10] consists of reviews from Epinions.com; 5) CNeSp [28] is a Chinese corpus includes three sub-corpora: financial articles (Fin), product reviews (Rev), scientific literature (Sci).

Evaluation Details. 1) In-domain evaluation. We launch 10-fold cross validation on Abs, Sfu, Fin, Rev, Sci [20,21,27,28], and report the performance on the test set of CD [17]; 2) Cross-domain evaluation. Results of Cli, Ful, and Tk2 are predicted by models trained on Abs [7,20,21,27]. *Percentage of Correct Scopes (PCS)* [20,21,27] is the main evaluation metric, where a scope is resolved correctly if all the tokens are determined whether in the scope or not. *Percentage of Correct Left/Right Boundaries (PCLB/PCRB)* are partial measurements.

Baselines. We consider models with complex syntactic features: 1) CNN-Path [21] and BiLSTM-Path [20] regards SpNeSR as a token classification problem, using syntactic paths from cues to tokens as the syntactic features. Depending on constituency/dependency trees, CNN-Path and BiLSTM-Path are classified as CNN-CPath/CNN-DPath and BiLSTM-CPath/ BiLSTM-DPath. We also employ MRC baselines: 2) QANet [24] consists of convolution and self-attention; 3) Bert-MRC uses Bert-base [4] and employs MRC framework for SpNeSR defined in §2, but considers neither data augmentation nor residual networks.

Table 1. Performance of speculation scope resolution. Format: PCS (%).

Models	Abs	Cli	Ful	Tk2	Sfu	Fin	Rev	Sci
CNN-CPath	85.75	73.92	59.82	58.86	65.49	71.63	53.87	54.47
CNN-DPath	74.43	64.39	52.98	53.63	62.95	67.84	51.47	49.72
BiLSTM-CPath	86.20	73.11	63.10	**65.34**	67.86	74.09	54.15	54.70
BiLSTM-DPath	79.54	62.68	49.85	51.40	66.11	73.68	53.68	52.96
QANet	74.87	65.02	52.53	52.66	64.90	72.54	51.06	51.62
Bert-MRC	84.18	73.56	63.69	62.25	69.06	75.16	57.62	56.83
SR-MRC-DA	**88.41**	**75.09**	**65.48**	64.67	**72.83**	**77.87**	**60.41**	**58.14**

Table 2. Performance of negation scope resolution. Format: PCS (%).

Models	Abs	Cli	Ful	Cdb	Cir	Sfu	Fin	Rev	Sci
CNN-CPath	70.86	89.66	55.32	71.97	58.59	65.86	70.17	53.65	40.68
CNN-DPath	77.14	87.82	53.99	71.21	59.38	67.34	71.23	53.31	42.59
BiLSTM-CPath	80.11	88.74	55.05	71.97	57.81	70.88	70.41	55.84	45.19
BiLSTM-DPath	80.28	92.30	62.50	73.48	57.03	71.49	71.15	57.22	44.93
QANet	74.06	90.11	48.67	72.66	62.50	72.73	72.26	59.28	45.56
Bert-MRC	79.30	91.38	62.77	75.76	64.06	74.04	74.84	62.83	**49.73**
SR-MRC-DA	**82.43**	**93.33**	**64.36**	**77.27**	**65.63**	**77.41**	**77.95**	**64.89**	49.59

3.2 Overall Results

Table 1 and 2 present that SR-MRC-DA is superior to other baselines, proving the validity and usefulness of MRC and DA. We attribute the success to these aspects: 1) The complex structure of the encoder. Based on Bert with multi-head attention layers, SR-MRC-DA is stronger on encoding than simpler models, e.g., CNN-Path; 2) The unified MRC formulation for multi-domain corpora. Compared with models designed for a certain domain (e.g., CNN-Path and BiLSTM-Path for BioScope), our model is fit for all the datasets considered in this paper, and can learn generalized knowledge across datasets; 3) The application of data augmentation (DA), which can produce more samples for training and enable SR-MRC-DA to have better robustness than the models without DA, e.g., Bert-MRC; 4) PCS is a strict evaluation metric requiring exact matching of scopes, and can best reflect the performance of SpNeSR models, rather than token-level F1-score. For example, although SR-MRC-DA can obtain higher F1 (89.64) than BiLSTM-CPath (89.15), BiLSTM-CPath is superior to SR-MRC-DA in term of PCS on Tk2 corpus. Therefore, we do not consider F1.

3.3 Detailed Analysis

Detailed Analysis involves those aspects, i.e., light-weighted MRC model, data augmentation, error analysis, and top cues.

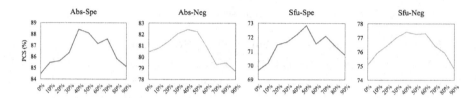

Fig. 3. The performance of SR-MRC-DA model with regard to α_d defined in §2. "Spe" and "Neg" mean the task of Speculation and Negation scope resolution, respectively.

Light-Weighted MRC Model. To eliminate the influence from large-scale pre-trained models and complicated syntactic features, we consider QANet, which is a light-weighted MRC model and significantly outperform CNN-Path that uses syntactic features on several datasets. Therefore, these performance can validate the effectiveness of MRC formulation.

Scale of Data Augmentation. We present Fig. 3 where α_d is the main argument, and select Abs and Sfu as the evaluation sets. We focus on α_d because produced training samples should have similar distributions of cues with the original set, where infrequent cues have little impact on results. Figure 3 shows that appropriately increasing the size of training set can boost the results on Abs and Sfu, and the best PCSs are usually obtained when $\alpha_d = 0.4$ or 0.5. However, performance drops especially when $\alpha_d > 0.5$ due to overfitting and weak generalization. Besides, too many augmented training samples may have different distributions of cues compared with the original set when sampling randomly.

Error Analysis. We dive into each corpus separately.

BioScope. 1) The performance of SR-MRC-DA model on speculation scopes is higher than negation in Abs, since speculation scopes are usually phenomenons or conclusions with complete semantic chunks (S5 in Fig. 4). But the most frequent negative cues "not" and "no" and are flexible in negating words, phrases, and clauses (S6 in Fig. 4); 2) Short sentences and the high proportion of "no" (77.36%) account for the excellent performance of negation scopes in Cli. However, the results of speculation are lower than that of Abs, attributed to the different top cues in Abs (top 3: "may", "suggest", "suggesting") and Cli (top 3: "or", "may", "evaluate for") (S7 in Fig. 4); 3) The performance on Ful and Tk2 is quite low caused by long and complicated sentences and scopes, or by words and phrases in parentheses representing references at the end of a sentence, which hardly appears in Abs (S8 in Fig. 4).

CD-Sco. In addition to the errors of top cues (e.g., "not"/"no", S9 in Fig. 4), some cues with prefix/suffix (e.g., "un-"/"-less") are quite rare (15.97%) and are also the main mistakes (S10 in Fig. 4), whose minority account for low results.

Sfu. Free-style review texts distinguish Sfu from scientific and fiction corpora. The results on Sfu are lower than that on Abs, because the grammar of Sfu is not so standard. Our model achieves better results on Sfu than CD-Sco because of shorter scopes in Sfu. Top cues are analyzed below.

(S5) Interference with the activation or activity of NF-kappa B {⟨may be beneficial in suppressing toxic/septic shock, graft-vs-host reactions, acute inflammatory reactions⟩, acute phase response, and radiation damage}. (Abs-Spe)

(S6) ... since inhibiting cell proliferation did {⟨not alter the level of MNDA⟩ mRNA and cell cycle} variation in MNDA mRNA levels were not observed. (Abs-Neg)

(S7) These findings are likely chronic and may ⟨represent {reactive airways disease or viral airways disease}⟩. (Cli-Spe)

(S8) ... that pupariated usually showed typical GFP expectoration, ⟨{indicating the presence of a high premetamorphic peak of ecdysteroids} (Figure⟩ 7E) . (Ful-Spe)

(S9) ... he solved it in an original fashion , and so effectively that her presence was {⟨not even known⟩ to the landlady who supplies her with food}. (CD)

(S10) "He is in ⟨{unoccupied premises} under suspicious circumstances⟩, " ... (CD)

(S11) ..., so please do give me a heads up {⟨if there is any vital info you want⟩ in addition to what I have put here}. (Sfu-Spe)

(S12) It's ⟨{not convenient} to spontaneously add the caller ID received phone call⟩ to both directories. (Sfu-Neg)

(S13) 总体而言{⟨市场或维持⟩弱势震荡的基调}。 (Overall, the market may maintain the tone of weak shocks.) (Fin-Spe)

(S14) ...，客厅⟨里居然⟨窗户都没有⟩}， ... (..., and there are even no windows in the living room, ...) (Rev-Neg)

(S15) ...该方法需要构建系统枚举索引关键词词典并在用户端存储，⟨{这无疑不符合实际应用}且造成较大客户端⟩负担。 (... this method needs to build a dictionary of system enumeration index keywords and store it on the client, which is undoubtedly not in line with the practical application and causes a large burden on the client.) (Sci-Neg)

Fig. 4. Some typical error cases, where {annotated scopes} are in {curly brace} and ⟨predicted scopes⟩ are in ⟨angle brackets⟩. (Color figure online)

CNeSp. Table 1 and 2 show that some token-based models (e.g., CNN-Path and BiLSTM-Path) get unsatisfactory results on CNeSp because of tokenization. But SR-MRC-DA is a character-based model without errors of Chinese word segmentation. Our model obtains excellent performance on Fin whose texts are grammatical, and scopes are usually opinions of stock trends, which is the main source of erros. Besides, polysemy may also cause mistaken scopes, e.g., "或" has the meanings of "may/possibly" (S13 in Fig. 4) and "or". The performance on Rev is lower than Fin because of free-style texts, e.g., error scopes because cues may be at the end of the scope (S14 in Fig. 4). All the models gain worse results on Sci owing to the fewer samples and longer sentences containing terminology of scientific literature (S15 in Fig. 4).

Top Cues. Figure 5 displays PCSs of the most frequent cues in Abs and Sfu that represent two typical language styles (i.e., scientific literature and free reviews) in SpNgSR, since performance depends largely on top cues. 1) Speculative Cues. Several top cues, e.g., "may", "suggest", "suggesting", "indicate that", occupy 42.87% among Abs, and generally state opinions or situations with complete semantic chunk. The performance of "or" is lower than others because of multiple usage for linking tokens, phrases, and clauses. Top cues of Sfu are different from that of Abs, i.e., "if" and "or" account for 39.40% in total, where "if" are

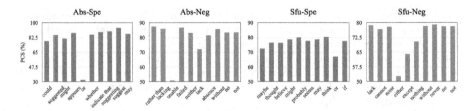

Fig. 5. The performance of top 10 cues on Abs and Sfu corpus using SR-MRC-DA model. In terms of morphology, different forms of one word are considered as different cues, since the structure of scopes may vary among them.

primarily used to lead conditional (S11 in Fig. 4) or object sentences (i.e., yes or no). The polysemy of top cues results in the lower performance on Sfu-Spe than Abs-Spe; 2) Negative Cues are concentrated on "not" & "no" that occupy 72.77%/79.47%, where "not" make up 60.73%/66.67% in Abs/Sfu. Their performance in Abs is a bit higher than that in Sfu. Abs contains scientific texts with correct grammar. But "not" is flexible Sfu, e.g., negating a word or verbal phrase (S12 in Fig. 4), leading to the difficulty of detecting right boundaries of scopes.

Additionally, we can draw more conclusions: 1) Multi-token cues (e.g., "indicate that", "rather than") achieves excellent performance, which means SR-MRC-DA is suitable for both single- and multi-token cues; 2) Some top cues get unsatisfactory results, e.g., "unable" and "either", owing to their small proportions (1.75% and 1.12%) and different structures of scopes from others. Unlike most negative cues in Sfu, "either" is often used after a negative statement or at the end of the sentence, and obtains low PCLB (56.32%).

4   Related Work

Scope Resolution. Early work started from heuristic rules. Özgür et al. [19] exploited part-of-speeches of cues and syntactic structures for speculation. Øvrelid et al. [18] employed rules dervied from dependency trees. Apostolova et al. [1] proposed lexico-syntactic rules. With the release of BioScope [22], Zhu et al. [26] regarded the cue as a predicate and mapped its scope into constituents as arguments. Zou et al. [27,28] considered tree kernel-based method [27] and sub-structure modelling [28] based on syntactic trees. Li et al. [11] devised semi-Markov and latent-variable CRF. Recently, neural networks have been applied to SpNgSR. Fancellu et al. [6] adopted feed forward and BiLSTM networks with word-embedding and lexical features for negation scope detection. Qian et al. [20,21] regarded SpNgSR as a token-level classification task, applying CNN [21] and LSTM [20] working on syntactic paths. With the emergence of large-scale pre-trained models, Zhao et al. [25] directly employed Bert as the encoder.

MRC for NLP. By casting information extraction tasks into MRC frameworks, researchers can integrate discriminative clues into template-based questions. For event trigger and argument extraction, Du et al. [5] generated questions by pre-defined templates. Li et al. [13] framed entity-relation extraction as a multi-turn MRC. Li et al. [12] proposed a unified MRC paradigm for flat and nested NER. Some studies designed more complex paradigms. Mao et al. [16] utilized a joint training dual-MRC framework to handle aspect based sentiment analysis. Liu et al. [14] developed topic-relevant and context-dependent methods for question generation with better generalization.

Data Augmentation For NLP. Token and phrase-level operations, e.g., random insertion, swap, deletion, are used in text classification [23]. NER models [3] employed token, synonym, and mention replacement (or shuffle, dropout). For event argument extraction, Liu et al. [15] used a pre-trained MRC model to label samples according to semantic roles and event types. For sentiment analysis, Chen et al. [2] proposed antonym and synonym replacement. Hsu et al. [9] used a sequence-to-sequence encoding model to produce words and replace unimportant terms.

5   Conclusion

This paper proposes SR-MRC-DA model for Speculation and Negation Scope Resolution. SR-MRC-DA is a unified MRC model for multi-domain evaluation on several datasets, and integrates data augmentation. Experiments on several English and Chinese datasets demonstrated the effectiveness and generalization of our model. In the future, our study will expand to end-to-end framework and more useful DA mechanisms.

Acknowledgements. Authors would like to thank the three anonymous reviewers for their comments. This work was supported by national Natural Science Foundation of China (NSFC) via Grant Nos. 62006167, 62276177 and 61836007.

References

1. Apostolova, E., Tomuro, N., Demner-Fushman, D.: Automatic extraction of lexico-syntactic patterns for detection of negation and speculation scopes. In: ACL 2011, pp. 283–287 (2011)
2. Chen, H., Xia, R., Yu, J.: Reinforced counterfactual data augmentation for dual sentiment classification. In: EMNLP 2021, pp. 269–278 (2021)
3. Chen, S., Aguilar, G., Neves, L., Solorio, T.: Data augmentation for cross-domain named entity recognition. In: EMNLP 2021, pp. 5346–5356 (2021)
4. Devlin, J., Chang, M., Lee, K., Toutanova, K.: BERT: pre-training of deep bidirectional transformers for language understanding. In: NAACL, pp. 4171–4186 (2019)
5. Du, X., Cardie, C.: Event extraction by answering (almost) natural questions. In: EMNLP 2020, pp. 671–683 (2020)
6. Fancellu, F., Lopez, A., Webber, B.L.: Neural networks for negation scope detection. In: ACL 2016 (2016)

7. Farkas, R., Vincze, V., Móra, G., Csirik, J., Szarvas, G.: The CoNLL-2010 shared task: Learning to detect hedges and their scope in natural language text. In: CoNLL 2010, pp. 1–12 (2010)
8. He, K., Zhang, X., Ren, S., Sun, J.: Deep residual learning for image recognition. In: Proceedings of CVPR 2016, pp. 770–778 (2016)
9. Hsu, T., Chen, C., Huang, H., Chen, H.: Semantics-preserved data augmentation for aspect-based sentiment analysis. In: EMNLP 2021, pp. 4417–4422 (2021)
10. Konstantinova, N., de Sousa, S.C.M., Díaz, N.P.C., López, M.J.M., Taboada, M., Mitkov, R.: A review corpus annotated for negation, speculation and their scope. In: LREC 2012, pp. 3190–3195 (2012)
11. Li, H., Lu, W.: Learning with structured representations for negation scope extraction. In: ACL 2018, pp. 533–539 (2018)
12. Li, X., Feng, J., Meng, Y., Han, Q., Wu, F., Li, J.: A unified MRC framework for named entity recognition. In: ACL 2020, pp. 5849–5859 (2020)
13. Li, X., et al.: Entity-relation extraction as multi-turn question answering. In: ACL 2019, pp. 1340–1350 (2019)
14. Liu, J., Chen, Y., Liu, K., Bi, W., Liu, X.: Event extraction as machine reading comprehension. In: EMNLP 2020, pp. 1641–1651 (2020)
15. Liu, J., Chen, Y., Xu, J.: Machine reading comprehension as data augmentation: a case study on implicit event argument extraction. In: EMNLP, pp. 2716–2725 (2021)
16. Mao, Y., Shen, Y., Yu, C., Cai, L.: A joint training dual-mrc framework for aspect based sentiment analysis. In: AAAI 2021, pp. 13543–13551 (2021)
17. Morante, R., Blanco, E.: *SEM 2012 shared task: Resolving the scope and focus of negation. In: *SEM 2012, pp. 265–274 (2012)
18. Øvrelid, L., Velldal, E., Oepen, S.: Syntactic scope resolution in uncertainty analysis. In: COLING 2010. pp. 1379–1387 (2010)
19. Özgür, A., Radev, D.R.: Detecting speculations and their scopes in scientific text. In: EMNLP 2009, pp. 1398–1407 (2009)
20. Qian, Z., Li, P., Zhou, G., Zhu, Q.: Speculation and negation scope detection via bidirectional lstm neural networks. J. Software **29**(8) (2018)
21. Qian, Z., Li, P., Zhu, Q., Zhou, G., Luo, Z., Luo, W.: Speculation and negation scope detection via convolutional neural networks. In: EMNLP, pp. 815–825 (2016)
22. Vincze, V., Szarvas, G., Farkas, R., Móra, G., Csirik, J.: The bioscope corpus: biomedical texts annotated for uncertainty, negation and their scopes. BMC Bioinform. **9**(S-11) (2008)
23. Wei, J.W., Zou, K.: EDA: easy data augmentation techniques for boosting performance on text classification tasks. In: EMNLP 2019, pp. 6381–6387 (2019)
24. Yu, A.W., et al.: Qanet: combining local convolution with global self-attention for reading comprehension. In: ICLR 2018 (2018)
25. Zhao, Y., Bethard, S.: How does Bert's attention change when you fine-tune? an analysis methodology and a case study in negation scope. In: ACL 2020, pp. 4729–4747 (2020)
26. Zhu, Q., Li, J., Wang, H., Zhou, G.: A unified framework for scope learning via simplified shallow semantic parsing. In: EMNLP 2010. pp. 714–724 (2010)
27. Zou, B., Zhou, G., Zhu, Q.: Tree kernel-based negation and speculation scope detection with structured syntactic parse features. In: EMNLP, pp. 968–976 (2013)
28. Zou, B., Zhu, Q., Zhou, G.: Negation and speculation identification in Chinese language. In: ACL 2015, pp. 656–665 (2015)

CoDE: Contrastive Learning Method for Document-Level Event Factuality Identification

Zihao Zhang, Zhong Qian$^{(\boxtimes)}$, Xiaoxu Zhu, and Peifeng Li

School of Computer Science and Technology, Soochow University, Suzhou, China
zhzhangpro@stu.suda.edu.cn, {qianzhong,xiaoxzhu,pfli}@suda.edu.cn

Abstract. Document-level event factuality identification (DEFI) aims to assess the veracity degree to which an event mentioned in a document has happened, which is crucial and fundamental for many downstream tasks of Natural Language Processing (NLP). Thus far, studies on DEFI typically regard it as a supervised classification task relying on annotated information, suffering from data scarcity since there is only one existing corpus (DLEF) consisting of two annotated sub-corpora based on news reports. The uneven distribution and data scarcity of DLEF and the existing annotation-relied methods limit the research progress on the DEFI task. To tackle this issue, we introduce a two-stage data augmentation strategy from text to graph via contrastive learning for DEFI (CoDE), which modifies a document at lexical, sentence, and document levels differently. Experiments on two widely used datasets show that our proposed model outperforms the state-of-the-art model, which demonstrates the effectiveness of our method for DEFI.

Keywords: Document-level Event Factuality Identification · Contrastive learning · Data augmentation

1 Introduction

Document-level event factuality identification (DEFI) is defined as the level of information expressing the veracity towards the factual nature of events mentioned in a certain discourse or context [19], namely, the task of determining whether an event is a fact, a possibility, or an impossible situation from the view of document [18]. In general, according to both modality and polarity, event factuality identification (EFI) is a five-label classification task that can be classified into five categories: Certain Positive (CT+), Certain Negative (CT-), Possible Positive (PS+), Possible Negative (PS-) and Underspecified (Uu).

Identifying document-level event factuality is a challenging task. As illustrated by Fig. 1, the factuality from document to sentence may vary. Sentence-level event factuality values are often varied due to the influence of speculation words and negation words, e.g., S4 is defined as CT- under the impact of negation ***don't***, S5 is PS+ according to speculation ***if*** and ***will***, S7 is CT+ under the impact of modal ***officially*** and ***declare***. However, the document-level event factuality value of the example given in Fig. 1 is uniquely determined as CT- based

© The Author(s), under exclusive license to Springer Nature Switzerland AG 2023
X. Wang et al. (Eds.): DASFAA 2023, LNCS 13945, pp. 497–512, 2023.
https://doi.org/10.1007/978-3-031-30675-4_36

Event: Biden Says He Doesn't Believe There's a Looming Recession.	
The Document-level factuality value: CT-	
[S1]	The president's remarks come after many analysts and leading banks have warned of a global recession (CT+) amid the Ukraine war, rising inflation, interest rate hikes and other uncertainties.
[S2]	Despite many analysts and banks projecting a looming U.S. recession (PS+), President Joe Biden says he doesn't believe the country is heading in that direction.
[S3]	The president acknowledged that economic projections warning of a recession (CT-) are periodic — but he doesn't necessarily agree with the assessment.
[S4]	He continued: "It hadn't happened yet. It hadn't... I don't think there will be a recession (CT-).
[S5]	If it is, it'll be a very slight recession (PS+). That is, we'll move down slightly."
[S6]	Biden's remarks about a recession (CT-) came just one day after JPMorgan Chase CEO Jamie Dimon warned on Monday that a U.S. recession (PS+) is likely within the next six to nine months.
[S7]	Officially, the National Bureau of Economic Research declares recessions (CT+).
[S10]	"When you are creating almost 400,000 jobs a month, that is not a recession (CT-)."

Fig. 1. An example of document-level event factuality, where event factuality varies from sentences to document.

on the full-text semantic content. Such variation and conflict between document and sentences are highly likely to happen in real scenarios. When assessing document-level event factuality, in addition to the inconsistencies of event factuality between sentences and document, there are often conflicts between the truth values of the sentences within a document, as S1, S2, and S3 shown in Fig. 1. Such inter-sentence conflicts can increase the complexity and difficulty of the DEFI task.

Various studies exploited document-level event factuality with different approaches. Qian et al. [18] proposed an adversarial neural network to embark on the task of DEFI. Zhang et al. [29] facilitated DEFI by using cross-domain negation and speculation scope features. Cao et al. [3] proposed a graph-based model by using graph convolutional networks [11] relying on event triggers. All of which are annotation-dependent methods focusing on exploiting semantic information, yet the semantic information that can be exploited is limited due to the uneven distribution of data and the scarcity of annotation information. Thus far, there is only one publicly available dataset, i.e., DLEF, constructed from reported news texts with authorized factuality value [18] for this task. Due to the rigorous nature of news reports, documents in DLEF dataset annotated with different event factuality values are highly unevenly distributed and further suffer from the dilemma of data scarcity, thus limiting the research progress on this task.

To tackle the issue analyzed above, we propose a novel two-stage method, i.e., **C**ontrastive Learning based **D**ocument-level **E**vent Factuality Identification framework (**CoDE**). Specifically, we propose four different data augmentation strategies to hierarchically modify a document at the lexical, sentence, and global levels to generate similar representation variants in the text stage, which allows an implicit enhancement of document graph augmentation. More specifically, we use hidden states of some specific layer (e.g., #8, #9, #12) of the pretrained language model, i.e., BERT [6], which contain most of the syntactic and semantic information [10] to ensure maximum usage of a certain document's text.

In summary, our contributions are three-fold as follows.

1. We propose a novel contrastive learning framework, **CoDE**, for document-level event factuality identification. To our best knowledge, this is the first method that optimizes document hierarchically from the perspective of both plain text and graph.
2. We design four data augmentation strategies that modify a document at the lexical, sentence, and global levels from a text perspective to construct similar representations implicitly from a graph perspective on the basis of contrastive learning.
3. Extensive experiments are conducted on two widely used datasets to verify the effectiveness of our model. The experimental results demonstrate that our model achieves state-of-the-art performances, showcasing that our method is an effective way to further improve the performance of DEFI task.

2 Related Work

2.1 Event Factuality Identification

Event factuality identification (EFI) is a fundamental task in event extraction, which is crucial and helpful for many natural language understanding (NLU) applications, e.g., rumor detection [2,15,24], knowledge base construction [26] and fake news detection [1,22]. Currently, studies on EFI of different levels all rely on annotated information.

Sentence-Level EFI. Saurí et al. [19,20] constructed a widely-used sentence-level EFI corpus: FactBank and proposed a rule-based model in the early phase of SEFI studying. On the basis of FactBank, Qian et al. [16,17] first proposed a two-step framework combining rule-based approaches and machine learning and further devised a generative adversarial network with auxiliary classification for SEFI. Current deep learning models have demonstrated the importance of syntactic and semantic structures of sentences to identify important context words for EFI tasks. Based on this, Veyseh et al. [25] proposed a graph-based neural network for SEFI. Le et al. [13] devised a novel model that explicitly considers multi-hop paths with both syntax-based and semantic-based edges among words to obtain sentence structures for representation learning in SEFI.

Document-Level EFI. Existing studies on document-level event factuality are still scarce. Qian et al. [18] constructed the first and only document-level event

Table 1. Event Factuality Category.

Modality	Polarity		
	Positive(+)	Negeative(−)	Underspecified(u)
Certain(CT)	CT+	CT-	NA
Possible(PS)	PS+	PS-	NA
Underspecified(U)	NA	NA	Uu

factuality dataset, DLEF, with two widely used English and Chinese subcorpus, and proposed an LSTM-based adversarial neural network (Att_2+AT) for DEFI. Zhang et al. [30] used a gated convolutional network and self-attention layer to capture feature representation of the overall information for the DEFI task, which outperforms Att_2+AT. Recently, Cao et al. [3] proposed a state-of-the-art graph-based method (ULGN) by utilizing Gaussian distribution to aggregate uncertain local information into a global document structure to assess document-level event factuality.

2.2   Contrastive Learning

Contrastive learning aims to learn effective representation by pulling semantically close neighbors together and pushing apart non-neighbors [9,27], which is popular in both natural language processing [8,21] and computer vision [4,23]. Data augmentation strategies are frequently used in contrastive learning methods to generate positive pairs.

For text representation, positive variants can be generated by some simple text data augmentation methods, e.g., reordering, back-translation, and mix-up. Gao et al. [8] generate similar variants by feeding the same input to the encoder twice with different dropout masks. For vision representation, similar positive variants are usually generated by flipping, rotation, cropping, and distortion from an image [4,23].

While the principles of contrastive learning have been broadly accepted, the implementation is still being explored, with the general guiding principles of alignment and uniformity [27].

3   Methodology

3.1   Task Definition

Let $\mathcal{D}_i \in \mathbb{D}^n$ denotes the i-th document in a dataset which contains n documents, we present a document $\mathcal{D}_i = \{s_1, s_2, \ldots, s_{\mathcal{M}}\}$ as a sequence of $\mathcal{M}$ sentences, where each sentence is composed of $\mathcal{K}$ words that can be denoted as $s_i = \{w_1, w_2, \ldots, w_{\mathcal{K}}\}$. Generally, if the j-th word w_j is an event, it then can be denoted as $w_j = e_j$.

The goal of DEFI seeks to predict a real-valued score in the range of [-2,2] to indicate the occurrence possibility for an event derived throughout a document. Each real-valued score corresponds to a class of event factuality value. Typically,

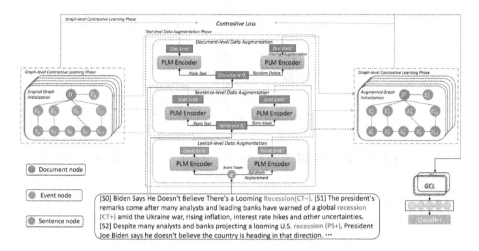

Fig. 2. The architecture of our proposed **CoDE** framework for document-level event factuality identification. Lines with an arrow indicate the flow direction of vectors, dashed lines with a filled arrow represent the initialization of graph/node by original text embedding, dashed lines with a line arrow represent the initialization of augmented graph/node by corresponding augmented variant.

event factuality values are composed of modality and polarity [19,20], where modality depicts the certainty degree of events and polarity coveys whether an event happens. Table 1 shows the detail of event factuality values, where NA means not applicable.

In the following subsections, we will introduce our contrastive learning framework with four different data augmentation strategies from text to graph for document-level event factuality identification based on these notations.

3.2   Overview

Graph-based methods [13,16,30] have shown their superiority in EFI tasks. Thus, we adopt graph convolutional networks into our framework. Existing studies on document-level contrastive learning either generate variants from plain texts or from graph structure, neither of which can fully exploit some specific intrinsic high-level feature of documents. To fully use information from both text and graph, we first optimize the DEFI document from a text perspective at multi-granularities, and then use these information to implicitly generate similar graph variants with contrastive learning for DEFI.

Our approach is schematically illustrated in Fig. 2, which is composed of three major modules: (1) Document Encoding Module, which consists of 4 different data augmentation strategies to modify documents at different text granular levels; (2) Implicit Graph-level Optimization Module, which relies on the generated variants from text to optimize document structure that implicitly constructs augmented graph via contrastive learning to optimize representation; (3) Graph

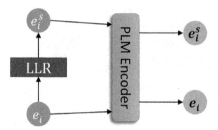

Fig. 3. Illustration of OUR lexical-level replacement (LLR) operation.

Learning Module, which utilizes a simple but efficient graph structure via graph convolutional networks [11] to identify document-level event factuality.

3.3   Document Encoding

Lexical-Level Replacement (LLR). Among the various text augmentation strategies in Natural Language Processing (NLP), Wei et al. [28] proposed four easy data augmentation techniques for boosting performance on text classification tasks, including Synonym Replacement (SR), Random Insertion (RI), Random Swap (RS) and Random Deletion (RD), which can surprisingly enhance the robustness and performance of a proposed model. Figure 3 shows the process of applying LLR data augmentation strategy to event token. Due to our lexical-specific augmentation condition, we use the off-the-shell and effective SR strategy in practice. Precisely, our LLR operation can be defined as follows.

$$e_i^s = \text{SynRep}(e_i) \tag{1}$$

where $\text{SynRep}(\cdot)$ denote a synonym replacement method, e.g., WordNet, e_i denotes event word w_i, e_i^s denotes the synonym of e_i, respectively. We then feed e_i and e_i^s into pretrained language model encoder to get their embedding, which can be defined as follows.

$$\begin{aligned} e_i &= \text{BERT}(e_i) \\ e_i^s &= \text{BERT}(e_i^s) \end{aligned} \tag{2}$$

where $\text{BERT}(\cdot)$ is the BERT base pretrained language model, $\boldsymbol{e}_i$ and $\boldsymbol{e}_i^s$ denote the embedding of e_i and e_i^s, respectively.

Sentence-Level Masking (SLM). Masking-based data augmentation strategy has been widely applied in many NLP applications, e.g., sentiment analysis [5] and question answering [7]. By using masking strategy, the robustness and performance of a model are better improved. Inspired by existing studies, we introduce a term masking strategy that randomly masks tokens over a sentence. Figure 4 shows the process of applying SLM data augmentation strategy to sentences. By operating such augmentation strategy, the generated sentences

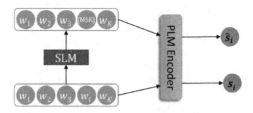

Fig. 4. Illustration of our sentence-level masking (SLM) operation.

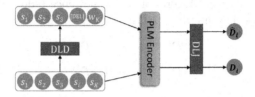

Fig. 5. Illustration of our document-level deletion and document-level jittering (DLDJ) operation.

only have minor differences in some tokens, which are similar to the original sentence. The whole process can be formulated as follows.

$$MP_i = \{idx_1, idx_2, \ldots, idx_k, \ldots, idx_{L_{s_i}}\}$$
$$\hat{s}_i = \{\hat{w}_1, \hat{w}_2, \ldots, \hat{w}_n\}$$
$$= \text{SLM}(\{w_1, w_2, \ldots, w_n\}) \qquad (3)$$
$$\hat{w}_i = \begin{cases} w_i, & i \notin MP_i \\ [\text{MSK}], & i \in MP_i \end{cases}$$

where MP_i denotes the mask proportion set of a sentence, $idx_{L_{s_i}}$ denotes the token index (with the maximum length to L_{s_i}) to be masked out by [MSK], $\text{SLM}(\cdot)$ is the sentence-level masking function. Similar to LLR, the embedding of s_i and $\hat{s}_i$ is defined as follows.

$$s_i = \text{BERT}(s_i)$$
$$\hat{s}_i = \text{BERT}(\hat{s}_i) \qquad (4)$$

where s_i and $\hat{s}_i$ denote the embedding of s_i and $\hat{s}_i$, respectively.

Document-Level Deletion and Jittering (DLDJ). Deletion is a common data augmentation strategy in computer vision and natural language to increase data variety to help models generalize better via creating subsets of images and texts [21,23]. Thus, we formulate our document-level deletion (DLD) strategy as a function on $\mathcal{D}_i$, which is defined as follows.

$$DP_i = \{idx_1, idx_2, \ldots, idx_{L_{\mathcal{D}_i}}\}$$

$$\hat{\mathcal{D}}_i = \{\hat{s}_1, \hat{s}_2, \ldots, \hat{s}_n\}$$

$$= \text{DLD}(\{s_1, s_2, \ldots, s_n\}) \tag{5}$$

$$\hat{s}_i = \begin{cases} s_i, & i \in DP_i \\ [\text{DEL}], & i \notin DP_i \end{cases}$$

where DP_i denotes the delete proportion of sentence index idx, $\text{DLD}(\cdot)$ is the document-level deletion function. [DEL] denotes the deletion token.

$$\mathcal{D}_i^{\text{DLD}} = \text{BERT}(\mathcal{D}_i)$$

$$\hat{\mathcal{D}}_i^{\text{DLD}} = \text{BERT}(\hat{\mathcal{D}}_i) \tag{6}$$

where $\mathcal{D}_i^{\text{DLD}}$ and $\hat{\mathcal{D}}_i^{\text{DLD}}$ denote the embedding of $\mathcal{D}_i$ and $\hat{\mathcal{D}}_i$ encoded by BERT model after DLD operation, respectively.

Similar to the widely used color-jittering method [12], we apply PCA jittering operation to $\hat{\mathcal{D}}_i^{\text{DLD}}$. The document-level jittering operation (DLJ) is defined as follows.

$$\hat{\mathcal{D}}_i^{\text{DLD}} = [h_0, h_1, \ldots, h_d]$$

$$\hat{\mathcal{D}}_i = \text{DLJ}(\hat{\mathcal{D}}_i^{\text{DLD}})$$

$$= [h_0 + \delta, h_1 + \delta, \ldots, h_d + \delta] \tag{7}$$

$$\delta = [p_0, p_1, \ldots, p_d][\alpha\lambda_0, \alpha\lambda_1, \ldots, \alpha\lambda_d]^T$$

where $\hat{\mathcal{D}}_i$ denotes the jittered embedding $\hat{\mathcal{D}}_i^{\text{DLD}}$, $\text{DLJ}(\cdot)$ denotes the document-level jittering function, d is the dimension, $\alpha \sim (0, \sigma^2)$, p_i and λ_i are the i-th eigenvector and eigenvalue, respectively.

The whole process of document-level encoding, i.e., DLDJ (DLD and DLJ), is illustrated in Fig. 5.

3.4   Implicit Graph-Level Optimization

We adopt graph convolutional networks [11] in our framework and construct a simple but efficient graph structure to represent documents with three kinds of nodes, i.e., document node, sentence node, and event node. For a document to be converted into a graph, a document node represents the entire document, sentence nodes are connected adjacently according to context, and event nodes are connected with their corresponding sentence node. We connect all event nodes, including event derived at document-level, with document node. Each node is initialized by its corresponding embedding from Sect. 3.3. The document structure is shown in Fig. 2

Surprisingly, the graph variants similar to the original graph are generated automatically in an easy and implicit way as positive samples for the contrastive learning phase.

3.5 Contrastive Learning Objective

We apply an l-layer GCN [11] to convolve on our constructed DEFI graphs. The optimal l is set to be 2 after extensive experiments on the performance of GCN [11]. The $(l+1)$-th GCN-layer-wise inference is defined as follows.

$$H^{(l+1)} = \sigma(\tilde{D}^{-\frac{1}{2}} \tilde{A} \tilde{D}^{-\frac{1}{2}} H^{(l)} W^{(l)}) \tag{8}$$

where $\tilde{A} = A + I$, A and I denotes the adjacency matrix of the constructed graph and identity matrix, respectively. $\sigma(\cdot)$ denotes an activation function, such as $ReLU(\cdot) = max(0, \cdot)$. $W^{(l)}$ denotes a layer-specific trainable weight matrix.

The i-th element of the $(l+1)$-th GCN-layer-wise inference matrix is defined as follows.

$$h_i^{(l+1)} = \sigma\Big(\sum_{j \in ne(i)} \frac{1}{\sqrt{\tilde{D}_{i,i}\tilde{D}_{j,j}}} h_j^{(l)} W^{(l)} \Big) \tag{9}$$

$$\tilde{D}_{i,i} = \sum_{j \in ne(i)} \tilde{A}_{i,j} \tag{10}$$

where $ne(i)$ denotes the neighbor nodes set of the i-th node.

We take the last layer's hidden state as the final representation of the document node $r^d = h^{d(l+1)}$ to de predicted as $p = \text{softmax}(W r^d + b)$.

A contrastive learning objective is applied to optimize the document node representation. Suppose a minibatch with size N, we obtain set $\{\mathcal{R}^d\}$ with size $2N$ after constructing positive pairs (r_i^d, r_j^d) by hierarchically modifying document at different granular levels from text to graph with strategies proposed in Sect. 3.3 and Sect. 3.4. The contrastive learning loss for a positive pair $\{r_i^d, r_j^d\}$ is defined as follows.

$$l(i,j) = -\log \frac{\exp(\text{sim}(r_i^d, r_j^d)/\tau)}{\sum_{k=1}^{2N} \mathbb{I}_{k \neq i} \exp(\text{sim}(r_i^d, r_j^d)/\tau)}$$

$$\text{sim}(r_i^d, r_j^d) = \frac{r_i^{d\top} r_j^d}{||r_i^d||\,||r_j^d||} \tag{11}$$

where $\text{sim}(\cdot, \cdot)$ is the cosine similarity function, $\mathbb{I}$ is the indicator function that specify whether $k \neq i$, and τ is a temperature hyperparameter.

The overall contrastive loss is defined as all positive pairs in a minibatch as follows.

$$\mathcal{L}_{CLS} = \sum_{i=1}^{2N} \sum_{j=1}^{2N} (l(i,j) + l(j,i)) \tag{12}$$

Following Qian et al. [18], we use cross-entropy as the loss function as follows.

$$\mathcal{L}_D(\theta) = -\frac{1}{M} \sum_{i=0}^{M-1} log\ p(y_j^{(i)}|x^{(i)}; \theta) \tag{13}$$

Table 2. Statistics of DLEF_en and DLEF_zh.

Dataset	Uu	CT-	PS-	PS+	CT+	Total
DLEF_en	12	279	12	274	1150	1727
DLEF_zh	20	1342	36	848	2403	4649

where M is the number of instances, $p(y_j^{(i)}|x^{(i)}; \theta)$ denotes the probability of instance x_i being predicted as the golden label $y^{(i)}$. θ is a hyper-parameter.

Thus, the final loss is defined as follows, where α is a hyperparameter.

$$\mathcal{L} = \mathcal{L}_{\mathcal{D}}(\theta) + \alpha \mathcal{L}_{\mathcal{CLS}} \tag{14}$$

4 Experimentation

4.1 Experimental Settings

To verify the effectiveness of our model, we conduct experiments on two widely-used English and Chinese datasets constructed by Qian et al. [18]. The statistics of these two datasets are shown in Table 2.

We use the AdamW algorithm [14] to optimize model parameters. The optimal dropout rate and the learning rate are set to 0.7 and 2e−5, respectively. The number of graph convolution layers is set to 2. The size of the hidden states of our graph convolution layer is 768. In our implementations, our method uses HuggingFace's Transformers library[1] to implement the BERT Base model.

To be fairly compared with previous studies [3,18,29,30], we focus on the performance of CT+, CT− and PS+, and conduct 10-fold cross-validation on both English and Chinese dataset. F1 score is adopted to evaluate each category of event factuality, micro-/macro-averaged F1 score is also adopted for the overall performance evaluation of event factuality categories.

4.2 Baselines

To verify the effectiveness of our **CoDE** framework, we conduct the following strong baselines for comparison.

- **BERT Base** [6], which utilizes the BERT-base to encode documents, and uses the [CLS] token for prediction.
- **Att_2+LSTM** [18], which utilizes intra-sentence attention to capture the most important information in sentences, and employs the long short-term memory network (LSTM) for DEFI.
- **Att_2+AT** [18], which leverages the intra-sentence and inter-sentence attention to learn the document representation. Adversarial training is adopted to improve the robustness.

---

[1] https://github.com/huggingface/transformers.

Table 3. Experimental results on the document-level event factuality datasets (English and Chinese respectively). The best performance is in **bold**.

Dataset	Methods	CT+	CT-	PS+	Micro-F1	Macro-F1
DLEF_en	MaxEntVote	75.14	58.17	35.89	68.42	56.40
	Att_2+LSTM	79.18	65.25	53.65	73.23	66.03
	SentVote	83.98	70.22	57.85	78.06	70.68
	Att_2+AT	89.84	76.87	62.14	83.56	76.28
	BERT	89.38	71.82	69.09	83.53	76.76
	GCNN	91.19	80.28	70.76	86.37	80.74
	ULGN	92.25	85.53	74.01	87.83	83.26
	BERT_SSF	92.37	83.83	76.15	88.34	84.37
	BERT_MSF	92.50	83.71	76.38	88.64	84.24
	CoDE(Ours)	**93.97**	**86.79**	**84.21**	**91.23**	**88.32**
DLEF_zh	MaxEntVote	72.22	62.44	58.29	67.72	64.32
	Att_2+LSTM	81.89	68.82	49.78	71.12	67.28
	SentVote	80.68	72.66	58.39	74.70	70.58
	Att_2+AT	87.52	83.35	74.06	84.03	81.64
	BERT	84.79	88.71	79.33	85.83	84.28
	GCNN	89.60	85.38	76.81	86.03	83.93
	BERT_SSF	90.94	88.53	85.43	89.20	88.37
	BERT_MSF	92.09	90.08	85.71	90.34	89.35
	ULGN	93.16	94.12	86.78	92.48	91.35
	CoDE(Ours)	**94.26**	**94.96**	**89.53**	**93.77**	**92.92**

- **MaxEntVote** [18], which uses maximum entropy model to identify sentence-level event factuality, and considers voting mechanism, i.e., choose the value committed by the most sentences as the document-level factuality value.
- **SentVote** [18], which is similar to MaxEntVote model, voting mechanism is used to identify document-level event factuality. Inter-sentence is not considered in it.
- **GCNN** [30], which uses a gated convolution network and self-attention layer to capture the feature representation of the overall information to identify the document-level event factuality.
- **BERT_SSF** [29], which utilizes detected negation and speculative scope as a whole to incorporate with BERT.
- **BERT_MSF** [29], which utilizes detected negation and speculative scope separately.
- **ULGN** [3][2], which proposes a graph-based model [11] via graph neural networks relying on event triggers. The original results of ULGN somewhat are far to reach in practice, so we adopt the best implementation results via its publicly available code instead.

---

[2] https://github.com/CPF-NLPR/ULGN4DocEFI.

Table 4. Experimental results on ablation study. The best performance is in **bold**.

Dataset	Methods	CT+	CT-	PS+	Micro-F1	Macro-F1
DLEF_en	**CoDE(Ours)**	**93.97**	86.79	**84.21**	**91.23**	**88.32**
	w/o LLR	93.83	85.11	78.43	90.32	85.79
	w/o SLM	92.57	**89.23**	72.34	89.15	84.72
	w/o DLDJ	90.43	89.80	71.19	86.98	83.81
DLEF_zh	**CoDE(Ours)**	**94.26**	**94.96**	**89.53**	**93.77**	**92.92**
	w/o LLR	93.86	94.36	89.16	93.27	92.46
	w/o SLM	93.68	93.97	88.76	92.93	92.14
	w/o DLDJ	94.07	93.97	85.54	92.47	91.20

4.3   Results and Analysis

Experimental results on the document-level event factuality datasets are shown in Table 3, and we can observe from the experimental results that:

1. Our model outperforms all the baselines on both English and Chinese DLEF datasets. Notably, on the English dataset, our model's micro-/macro-F1 score outperforms the current state-of-the-art model ULGN by 3.4/5.06, and outperforms the previous state-of-the-art model Att_2+AT by 7.67/12.04, on the Chinese dataset, our model's micro-/macro-F1 score outperforms the current state-of-the-art model ULGN by 1.29/1.57, and outperforms the previous state-of-the-art model Att_2+AT by 9.74/11.28, which showcases the robustness and effectiveness of our proposed method for document-level event factuality identification.
2. It can be observed that there is quite an experimental gap between traditional deep learning models, i.e., Att_2+AT and GCNN, and graph-based models, i.e., CoDE and ULGN, we contribute this success to graph structure that is more suitable for processing document-level text-oriented tasks. For tasks like DEFI, it's important and crucial to have a better understanding and global view of given documents. Traditional deep learning models treat document text as a long sequence and then process in a linear way, which can't capture the global structure and information of a document. On the contrary, graph structure can better encapsulate a document and its corresponding data, and better utilizes documents with the higher global feature.
3. The performance of the BERT-based models, i.e., BERT [6], BERT_SFF [29], BERT_MSF [29], is also satisfactory, which showcases BERT as one of the best models that can obtain deeper semantic information. CoDE outperforms BERT on both English and Chinese datasets. We attribute the performance to the effectiveness of our proposed contrastive learning framework with data augmentation strategies at different text granular.
4. CoDE shows a huge nonnegligible improvement in the accuracy performance on **PS+** samples, which varies from [7.83, 48.32] and [2.75, 31.24] on the

Event: Czech President pays respects to victims of Nanjing Massacre.	
The Document-level factuality value: CT+	
[S1]	President Milos Zeman of the Czech Republic visits the Memorial Hall of the Victims in Nanjing Massacre by Japanese Invaders on Tuesday in Nanjing, Jiangsu province, paying (CT+) tribute and bowing to the victims.
[S2]	Zeman was invited by Chinese President Xi Jinping to attend the Belt and Road Forum for International Cooperation in Beijing.
[S3]	Related story: Czech president, first lady pay (CT+) respects at Nanjing memorial.

Fig. 6. An example of document-level event factuality identification prediction made by CoDE on the DLEF dataset correctly.

English and Chinese datasets, respectively, which demonstrate our proposed method for solving the data scarcity problem that the DEFI task suffered.

4.4 Ablation Study

To verify the effectiveness of our proposed method, we design an ablation test to investigate the effectiveness of those proposed strategies separately. As shown in Table 4, we can observe that by removing the proposed data augmentation strategy from different levels separately, both Micro-F1 and Macro-F1 show different degrees of decrease Namely, by removing the data augmentation strategy from different text granular, the micro-/macro-F1 dropped by 0.91/2.53, 2.08/3.6, and 4.25/4.51 on the English dataset, and 0.5/0.46, 0.84/0.78 and 1.3/1.72, respectively. The overall trend is that the decline increases at the lexical level, sentence level, and document level in ascending order.

Specifically, the Macro-F1 values dropped significantly both on the DLEF English and Chinese datasets, which further supports our method for solving the data scarcity problem of DEFI task.

4.5 Case Study

Figure 6 provides an example correctly predicted by CoDE on the test set of DLEF. In this sample, event token *pay* is converted into its synonym *show* and *give*, which enriched the semantic information of the original text without erasing key information. SLM operation randomly masked some words out of sentences to enhance the robustness via adding noise. DLDJ deletes S1 and further adds noise globally, then optimizes representation by contrastive learning. The success correct prediction is a tribute to every stage and every step of our proposed CoDE framework.

4.6 Error Analysis

To better understand the errors made by CoDE for DEFI, we analyze the outputs of our proposed CoDE framework on the test set of DLEF dataset. Figure 7 shows

Event: President Trump says not considering firing special counsel Mueller.	
The Document-level factuality value: CT-	
[S1]	US President Donald Trump said (CT+) on Sunday that he is not considering firing special counsel Robert Mueller.
[S2]	Mueller is investigating allegations of Russian interference in the last year 's US presidential election and collusion between the Trump campaign and Moscow.
[S3]	When asked if (Uu) he was considering dismissing Mueller , Trump responded (CT-), "No, I'm not, " and did not elaborate on the issue.
[S4]	Democratic lawmakers have raised concern that Trump might fire Mueller.
[S5]	The president, however, criticized the fact that Mueller's team had gained access to tens of thousands of emails sent and received by officials from Trump's transition team before the start of the presidency.
[S6]	Trump said it was "not looking good".
[S7]	Mueller's team received those emails from the General Services Administration that had housed Trump's transition team.
[S8]	In a letter to several congressional panels Saturday, a lawyer for the transition accused Mueller of illegally obtaining the emails.

Fig. 7. An example of document-level event factuality identification prediction made by CoDE on the DLEF dataset incorrectly.

a sample of what was incorrectly predicted by CoDE. This sample demonstrates a common problem where multiple different event mention words are annotated in a text uniquely defined by an document-level event. Thus, LLR operation introduces more irrelevant information by converting *if* into *__whether__*. Furthermore, by SLM and DLDJ, some important information may be randomly erased, e.g., S1 being deleted. This example suggests that CoDE may perform better by applying data augmentation strategies at different granular levels in a more efficient and cautious way.

5   Conclusion

In this paper, we propose a novel contrastive learning framework: **CoDE** for document-level event factuality identification, which not only modifies documents at the lexical, sentence, and global levels to generate similar variants hierarchically, allowing us generate graph variants automatically in an implicit and easy way, but also employ contrastive learning at graph phase for better representations. Extensive experiments showed that our framework achieves state-of-the-art performances.

Acknowledgements. The authors would like to thank the three anonymous reviewers for their comments on this paper. This research was supported by the National Natural Science Foundation of China (No. 62006167, 62276177, and 61836007.), and Project Funded by the Priority Academic Program Development of Jiangsu Higher Education Institutions (PAPD).

References

1. Baly, R., Karadzhov, G., Alexandrov, D., Glass, J.R., Nakov, P.: Predicting factuality of reporting and bias of news media sources. In: Proceedings of the 2018 EMNLP, pp. 3528–3539. Association for Computational Linguistics (2018)

2. Bian, T., et al.: Rumor detection on social media with bi-directional graph convolutional networks. In: Proceedings of the 34th AAAI, pp. 549–556. AAAI Press (2020)

3. Cao, P., Chen, Y., Yang, Y., Liu, K., Zhao, J.: Uncertain local-to-global networks for document-level event factuality identification. In: Proceedings of EMNLP 2021, pp. 2636–2645. Association for Computational Linguistics (2021)

4. Chen, T., Kornblith, S., Norouzi, M., Hinton, G.E.: A simple framework for contrastive learning of visual representations. In: Proceedings of the 37th ICML, pp. 1597–1607. PMLR (2020)

5. Dai, A.M., Le, Q.V.: Semi-supervised sequence learning. In: Proceedings of the 2015 NIPS, pp. 3079–3087 (2015)

6. Devlin, J., Chang, M., Lee, K., Toutanova, K.: BERT: pre-training of deep bidirectional transformers for language understanding. In: Proceedings of the 2019 NAACL-HLT, pp. 4171–4186. Association for Computational Linguistics (2019)

7. G, S., N, U., G, K.: LAWBO: a smart lawyer chatbot. In: Proceedings of the ACM India Joint International Conference on Data Science and Management of Data, pp. 348–351. ACM (2018)

8. Gao, T., Yao, X., Chen, D.: SimCSE: simple contrastive learning of sentence embeddings. In: Proceedings of the 2021 EMNLP, pp. 6894–6910. Association for Computational Linguistics (2021)

9. Hadsell, R., Chopra, S., LeCun, Y.: Dimensionality reduction by learning an invariant mapping. In: CVPR 2006, pp. 1735–1742. IEEE Computer Society (2006)

10. Jawahar, G., Sagot, B., Seddah, D.: What does BERT learn about the structure of language? In: Proceedings of the 57th Conference of ACL, pp. 3651–3657. Association for Computational Linguistics (2019)

11. Kipf, T.N., Welling, M.: Semi-supervised classification with graph convolutional networks. In: Proceedings of the 5th ICLR. OpenReview.net (2017)

12. Krizhevsky, A., Sutskever, I., Hinton, G.E.: Imagenet classification with deep convolutional neural networks. In: Proceedings of the 26th NIPS, pp. 1106–1114 (2012)

13. Le, D., Nguyen, T.H.: Does it happen? Multi-hop path structures for event factuality prediction with graph transformer networks. In: Proceedings of the Seventh Workshop on Noisy User-Generated Text, W-NUT 2021, Online, 11 November 2021, pp. 46–55. Association for Computational Linguistics (2021)

14. Loshchilov, I., Hutter, F.: Decoupled weight decay regularization. In: Proceedings of the 7th ICLR. OpenReview.net (2019)

15. Qazvinian, V., Rosengren, E., Radev, D.R., Mei, Q.: Rumor has it: identifying misinformation in microblogs. In: Proceedings of the 2011 EMNLP, pp. 1589–1599. Association for Computational Linguistics (2011)

16. Qian, Z., Li, P., Zhang, Y., Zhou, G., Zhu, Q.: Event factuality identification via generative adversarial networks with auxiliary classification. In: Proceedings of the 27th IJCAI, pp. 4293–4300. ijcai.org (2018)

17. Qian, Z., Li, P., Zhu, Q.: A two-step approach for event factuality identification. In: Proceedings of the 2015 IALP, pp. 103–106. IEEE (2015)

18. Qian, Z., Li, P., Zhu, Q., Zhou, G.: Document-level event factuality identification via adversarial neural network. In: Proceedings of the 2019 NAACL-HLT, pp. 2799–2809. Association for Computational Linguistics (2019)

19. Saurí, R., Pustejovsky, J.: Factbank: a corpus annotated with event factuality. Lang. Resour. Eval. **43**(3), 227–268 (2009)
20. Saurí, R., Pustejovsky, J.: Are you sure that this happened? Assessing the factuality degree of events in text. Comput. Linguist. **38**(2), 261–299 (2012)
21. Shorten, C., Khoshgoftaar, T.M., Furht, B.: Text data augmentation for deep learning. J. Big Data **8**(1), 101 (2021)
22. Singhal, S., Shah, R.R., Chakraborty, T., Kumaraguru, P., Satoh, S.: Spotfake: a multi-modal framework for fake news detection. In: Fifth IEEE International Conference on Multimedia Big Data, pp. 39–47. IEEE (2019)
23. Tian, Y., Krishnan, D., Isola, P.: Contrastive multiview coding. In: Vedaldi, A., Bischof, H., Brox, T., Frahm, J.-M. (eds.) ECCV 2020. LNCS, vol. 12356, pp. 776–794. Springer, Cham (2020). https://doi.org/10.1007/978-3-030-58621-8_45
24. Tu, K., Chen, C., Hou, C., Yuan, J., Li, J., Yuan, X.: Rumor2vec: a rumor detection framework with joint text and propagation structure representation learning. Inf. Sci. **560**, 137–151 (2021)
25. Veyseh, A.P.B., Nguyen, T.H., Dou, D.: Graph based neural networks for event factuality prediction using syntactic and semantic structures. In: Proceedings of the 57th ACL, pp. 4393–4399. Association for Computational Linguistics (2019)
26. Bijl de Vroe, S., Guillou, L., Stanojević, M., McKenna, N., Steedman, M.: Modality and negation in event extraction. In: Proceedings of the 4th Workshop on Challenges and Applications of Automated Extraction of Socio-political Events from Text (CASE), pp. 31–42. Association for Computational Linguistics, Online (2021)
27. Wang, T., Isola, P.: Understanding contrastive representation learning through alignment and uniformity on the hypersphere. In: Proceedings of the 37th ICML, pp. 9929–9939. PMLR (2020)
28. Wei, J.W., Zou, K.: EDA: Easy data augmentation techniques for boosting performance on text classification tasks. In: Proceedings of the 2019 EMNLP-IJCNLP, pp. 6381–6387. Association for Computational Linguistics (2019)
29. Zhang, H., Qian, Z., Zhu, X., Li, P.: Document-level event factuality identification using negation and speculation scope. In: Mantoro, T., Lee, M., Ayu, M.A., Wong, K.W., Hidayanto, A.N. (eds.) ICONIP 2021. LNCS, vol. 13108, pp. 414–425. Springer, Cham (2021). https://doi.org/10.1007/978-3-030-92185-9_34
30. Zhang, Y., Li, P., Zhu, Q.: Document-level event factuality identification method with gated convolution networks. Comput. Sci. **47**(3), 206–210 (2020)

Recovering Missing Key Information: An Aspect-Guided Generator for Abstractive Multi-document Summarization

Haotian Chen[✉], Han Zhang, Houjing Guo, Shuchang Yi, Bingsheng Chen, and Xiangdong Zhou

School of Computer Science, Fudan University, Shanghai 200433, China
{htchen18,hanzhang20,xdzhou}@fudan.edu.cn
{houjingguo21,scyi21,chenbs21}@m.fudan.edu.cn

Abstract. Abstractive multi-document summarization (MDS) paraphrases the salient key information scattered across multiple documents. Due to the large length of the documents, most previous methods opt to first extract salient sentence-level information and then summarize it. However, they neglect the aspect information: documents are often well-organized and written down according to certain aspects. The absence of aspects renders the generated summaries not comprehensive and wastes the prior aspect knowledge. To solve the issue, we propose a novel aspect-guided joint learning framework to detect aspect information for guiding the generating process. Specifically, our proposed method adopts feed-forward networks to detect the aspects in the given context. The detected aspect information serves as both constraints of the objective function and supplement information expressed in the context representations. Aspect information is explicitly discovered and exploited to facilitate generating comprehensive summaries. We conduct extensive experiments on the public dataset. The experimental results demonstrate that our proposed method outperforms previous state-of-the-art (SOTA) baselines, achieving a new SOTA performance on the dataset.

Keywords: Abstractive multi-document summarization ·
Aspect-guided generator · Aspect information

1 Introduction

Multi-document summarization (MDS) paraphrases multiple thematically related documents into a fluent, condensed, and informative summary [14,21,26]. MDS facilitates a wide range of applications, including generating Wikipedia abstracts [18,35], creating news digests [1], and opinion summarization [2]. Compared with single document summarization (SDS), MDS generates more comprehensive summaries from documents, where the given documents comprise various aspects and can overlap and complement with each other [14], and is accordingly

© The Author(s), under exclusive license to Springer Nature Switzerland AG 2023
X. Wang et al. (Eds.): DASFAA 2023, LNCS 13945, pp. 513–522, 2023.
https://doi.org/10.1007/978-3-031-30675-4_37

more complicated as it involves incredibly lengthy input documents and tries to capture and organize information scattered across the long input [8,21].

Traditional methods in MDS are based on feature engineering [8], statistical learning [3], and graph [8,24]. Most of them extract key information, which is represented by most salient textual units, structural dependencies among phrases, keywords, or semantic clusters, to aid in generating summaries [21]. Recently, pre-trained language models (PLMs) have significantly accelerated the development of text summarization. Previous methods based on PLMs can be roughly divided into four popular categories: (1) sparse attention [6,32]; (2) hierarchical model architecture [18,34] focus on improving the capability of models to process all information from multiple documents simultaneously; (3) extract-then-generate methods [31,33]; (4) divide-and-conquer approaches [10,12] aim to shorten the length of input context by extracting salient texts or individually summarizing each part of multiple documents. Previous work reports that extract-then-generate methods, hierarchical models, most traditional methods, and human performance are all based on a common belief: Summaries should be generated in a top-down way. They first detect key information from the input documents explicitly and then use it to guide the summarization [15,23].

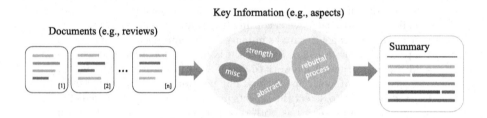

Fig. 1. A brief illustration for multi-document summarization.

However, previous work is hardly aware of aspect information. Summaries are usually considered as plain text even though they summarize various aspects elaborated by multiple documents. As shown in Fig. 1, these aspects compose key information, summarize the input documents, and thereby guide humans in a top-down way to write summaries. A pipeline method [35] is proposed to exploit aspect information for summarization. It first learns to distinguish topics discussed by the input documents and then encodes these topics together with the documents to perform summarization. Despite its outstanding performance, it leaves two problems unsolved: (1) the pipeline method separates aspect detection from summarization, thus suffering from cascade errors; (2) the detected aspect information indirectly aids the generator (merely by inputting the implicit aspect representation). We argue that aspect information is essential and should be explicitly adopted to supervise the generating process of summaries.

In this paper, we propose a novel aspect-guided joint learning framework that captures aspect information as both constraints of the objective function and sufficient expressive power for representations to guide the generating process. Specifically, we construct feed-forward networks to detect the aspect information expressed by the representations of input documents and the generated summaries. Models are required not only to infuse representations with aspect information but also to eliminate the inconsistency between the aspects expressed by documents and summaries, which guides summaries to describe the aspects mentioned in corresponding documents. We evaluate our proposed method on the summarization benchmark. Based on the same backbone, our method outperforms the strong baseline models and achieves a new state-of-the-art performance on the benchmark. Our main contributions are summarized as follows:

- We introduce a novel aspect-guided joint learning framework that detects aspect information to guide the generating process of summaries.
- We exploit the inconsistency between the aspects expressed by documents and summaries to constrain the objective function and give sufficient expressive power to representations, which aids in the generating process of summaries.
- Experimental results show that our framework significantly outperforms previous SOTA methods on the MDS benchmark. We provide a case study to show our recovered missing key information.

2   Related Work

MDS resolves potentially diverse and redundant information in the given multiple documents to distill a coherent, concise, and informative summary [26], which fuels progress in generating Wikipedia abstracts [18,35], creating news digests [1], and opinion summarization [2]. Compared with SDS, MDS is more challenging due to the longer input and more redundant information, which exacerbates the difficulty of abstractive MDS in retaining the most critical contents and paraphrasing them [8].

Traditional MDS techniques, which are based on term frequency-inverse document frequency [27], graphs [22], latent semantic analysis [3], and clustering [11] with manually crafted features [5,8], obtain key information represented by words, sentences, graphs, and semantic clusters to guide the generator. With recent significant improvement in SDS brought by large-scale PLMs [16,28], most researchers tackle MDS based on PLMs in four ways with two underlying purposes: 1) To enhance the long-input processing capability of models, they propose sparse attention [6,32] and hierarchical model architectures [18,34]. The former is proposed for reducing the memory complexity of transformer-based PLMs while the latter is designed for capturing dependency information among sentences and words. 2) To shorten the length of source input, researchers adopt extract-then-generate methods [4,31,33] and divide-and-conquer approaches [10,12]. The former extract salient texts (key information) from the given documents and then summarize them, the latter divide the given documents into sections and then individually summarize them to form a final summary.

As to extract-then-generate methods, traditional methods, and hierarchical model architectures, they try to first collect and merge the information scattered across the source input in a heuristic way, and then summarize the derived key information. Inspired by the paradigm, we focus on exploring and modeling the aspect information, which is crucial and often neglected in previous work, as key information to explicitly supervise the generating process of summaries.

Aspect information is hardly exploited by previous MDS methods in generic summary generation [9,25]. Some aspect-based work focuses on identifying aspects (e.g., words or phrases) in human-written opinions for opinion summarization and sentiment analysis [30]. The aspect-specific context will aid in distinguishing the sentiment polarities of reviews about different aspects of a product. To facilitate identifying aspect-level keywords, a previous work [1] proposes a system to automatically extract aspect-level keywords in a heuristic way without assuming human-annotated training data.

Recently, human-annotated aspect information significantly increases with some held-out aspect-oriented abstractive summarization datasets. They comprise WikiAsp for aspect-oriented Wikipedia summarization [13], meta-review dataset (MRED) for structure-controllable meta-review generation [29], and summaries of popular and aspect-specific customer experience (SPACE) dataset for opinion mining [2]. They facilitate the development of human-annotated aspect information and make it available for generators.

We focus on modeling the human-annotated aspect information instead of the heuristically extracted one since noise information can be introduced in the latter and thus confuses the generator. TWAG [35] proposes a pipeline method to first model the aspect information and then perform abstractive MDS, which we compare with.

3   Methodology

We formulate the task and then introduce our method. In MDS, the input document set $\mathcal{D} = \{D_i\}_{i=1}^n$ consists of multiple documents and is expressed by its concatenate context X. The generated output is their summary y of length T. Given input documents X and the previously generated tokens $y_{<t}$, the learning models in MDS are required to maximize the likelihood of the given optimal summary y^*. The generated summary tokens y can be described by,

$$y^* = \arg\max_y P(y \mid \mathcal{D}, \mathcal{A}) = \arg\max_y \prod_{t=1}^{T} P\left(y_t \mid \mathcal{D}, \mathcal{A}, y_{<t}\right), \qquad (1)$$

where $\mathcal{A}$ comprises all aspects expressed by the input documents and is detected by the feed-forward network FFN,

$$\mathcal{A} = \text{FFN}(\mathcal{D}). \qquad (2)$$

Previous work treats the summary as plain text, neglecting the fact that most documents are well organized and written down according to the underlying aspect information, which guides human-written summaries of multiple documents [35]. That is to say, aspect information is a constraint of the objective function, which is fitted by the generator in the training process. However, such a constraint (aspects $\mathcal{A}$) is missed in previous work, resulting in their increased risk of falling into suboptimal results. To deal with the problem, we construct feed-forward networks to estimate the aspect information expressed by the representations of input documents and generated words, respectively. The imbalance in the amount of two source aspect information supervises the generator, guiding it to recover the missing target aspect information.

4   Experiments

4.1   Dataset

We experiment on MRED [29]. MRED is a highly abstractive dataset focusing on meta-reviews from a peer-reviewing system (ICLR) which contains essential and high-density opinions. It is provided for structure-controllable text generation. The meta-reviews are manually written to summarize the aspects described in different reviews and we use *sent-ctrl* version of MRED.

4.2   Baselines

We compare ETAGE with previous state-of-the-art methods (comprising an extractive model and 4 abstractive models) on the three datasets:

TextRank [24] is a common extractive summarization baseline model which uses vertex scores calculated by a graph-based "random-surfer model" to rank sentences.

TWAG [35] is a two-step abstractive summarization method that first detects the aspects described by the multiple source documents and then performs summarization based on the detected aspects.

BertAbs [19] is an abstractive summarization model with encoder initialized with BERT [7] and transformer decoder randomly initialized.

Longformer [6] is a pre-trained language model tackling long input by sparse attention. Following BertAbs [19], We initialize the encoder with Longformer and randomly initialize the transformer decoder.

BART [16] is a SOTA abstractive summarization model pre-trained with the objective of denoising autoencoding.

We also compare with other baselines mentioned in the work proposing the corresponding dataset.

4.3   Implementation Details

We complete our experiments on a single RTX3090 GPU. We first load the pre-trained models released by Huggingface[1] as the backbones. To keep in line with the basic settings of baselines for fair comparison, we adopt the common hyper-parameters used in the transformer-based baseline models. Specifically, we apply AdamW algorithm [20] to optimize model parameters with a learning rate of 1e–5. We evaluate our generated summaries against the reference

---

[1] https://huggingface.co/models.

manually written ones by calculating the F_1-scores of $ROUGE_1$, $ROUGE_2$, and $ROUGE_L$ [17]. Following previous work, we adopt the Rouge evaluation script[2] provided by Huggingface with "use_stemmer" enabled.

Table 1. Performance on MReD. The signal † denotes that the results of models are quoted in the original paper proposing MReD. The rest of the results are based on our implementation.

Model	R-1	R-2	R-L
MMR[†]	32.37	6.28	17.58
LexRank[†]	32.60	6.66	17.48
TextRank[†]	33.52	7.20	17.75
TWAG	27.82	7.22	19.99
Longformer	23.39	5.63	20.52
BertAbs-BERT$_{base}$	23.57	6.67	18.59
BART$_{large}^{†}$	38.59	10.61	22.93
Ours-BART$_{large}$	**39.04**	**11.00**	**23.65**

4.4   Main Results

We compare our method with all of the previous SOTA methods on MReD. We also further implement several common strong baselines for comparison and deeper analysis. Table 1 shows the main results of the baseline models and our method. We can observe that our method outperforms the existing SOTA baselines on MReD in the BART backbone. Specifically, with BART$_{large}$ as the PLM, our method surpasses BART$_{large}$ by 0.45/0.39/0.72 of ROUGE-1/2/L scores, achieving new SOTA performance on MReD. The experimental results show the effectiveness of the overall framework of our method.

We attribute the improvement to the incorporation of aspect information and our proposed joint learning framework for two reasons. First, aspect information significantly improves the performance of models. We observe the performance gaps (the aforementioned gains of ROUGE-1/2/L scores) between models that adopt our framework and models with the same backbones neglecting aspect information. The difference between the two kinds of models is that the former incorporates aspect information by constraining the objective function, which indicates that by properly infusing models and constraining the objective function with aspect information, models are able to achieve more significant improvements. Second, learning aspect information in a joint way largely enhances the effectiveness of models. Compared to TWAG which models aspect information in a two-step way, our proposed method significantly outperforms TWAG by 11.22/3.78/3.66 on MReD.

---

[2] https://github.com/huggingface/transformers/blob/main/examples/pytorch/summarization/.

4.5   Case Study

We present the case study of our proposed method and the corresponding baseline model. As shown in Fig. 2, the baseline model BART$_{large}$ misses the aspect information of "rebuttal-process", paraphrases too much information about "strength", and wrongly generates the unmentioned aspect "weakness", which ruins the comprehensiveness, faithfulness, and conciseness of the generated summary. Compared with BART$_{large}$, our proposed method recovers the missing aspect and faithfully generates the summary.

Aspects: misc | strength | abstract | rebuttal_process	
Summary	**ROUGE**
Ground Truth : Thank you for submitting you paper to *CONF*. The idea is simple, but easy to implement and effective. The paper examines the performance fairly thoroughly across a number of different scenarios showing that the method consistently reduces variance. How this translates into final performance is complex of course, but faster convergence is demonstrated and the revised experiments in table 2 show that it can lead to improvements in accuracy.	**NA**
BART$_{large}$: Thanks for your detailed feedback to the reviewers, which helped us a lot to better understand your paper. \<sep> This paper addresses an important problem of reducing the variance of the gradient estimator in deep learning. The proposed solution is simple and elegant: a rank-1 random sign matrix averaging of the weights by a sign matrix. This efficient mechanism is only twice as costly as a single random perturbation, and the authors show how to efficiently parallelize it on GPUs, thereby also allowing GPU-ization of evolution strategies. **While the technical contribution of this work is relevant for network learning, several key aspects are yet to be addressed thoroughly, such as the experiments.**	**ROUGE-1: 30.57** **ROUGE-2: 5.29** **ROUGE-3: 18.34**
AGSum-BART$_{large}$: Thank you for submitting you paper to *CONF*. The consensus from the reviewers is that this is a solid, well-written paper. The paper proposes a method to reduce the variance of the gradient estimation in the training of neural networks. The proposed method is simple, and is shown to be competitive with several state-of-the-art baselines. The authors provided a thorough and convincing response to the reviewers' concerns, and promised to include the missing experiments in the final version.	**ROUGE-1: 46.24** **ROUGE-2: 18.48** **ROUGE-3: 34.00**

Fig. 2. The case study of our proposed method.

5   Conclusion

Multi-document summarization (MDS) is a long-standing task and is challenging due to the requirement of paraphrasing the key information scattered across multiple documents. In this paper, we introduce our aspect-guided joint learning framework, which captures aspect information to constrain the optimization and aids representation learning. Our method adopts a multi-task joint learning method to avoid introducing the cascade error and impeding the interaction between aspect detection and generation. The extracted aspect information guides the generating process, improving the comprehensiveness and faithfulness of the generated summaries. The experimental results on three commonly used summarization datasets not only show that our method outperforms the strong baseline models, but also validate the effectiveness of the detected aspects which are accurate and well guide the generating process.

References

1. Ahuja, O., Xu, J., Gupta, A., Horecka, K., Durrett, G.: ASPECTNEWS: aspect-oriented summarization of news documents. In: Proceedings of the 60th Annual Meeting of the Association for Computational Linguistics (Volume 1: Long Papers), pp. 6494–6506. Association for Computational Linguistics, Dublin (2022). https://doi.org/10.18653/v1/2022.acl-long.449
2. Angelidis, S., Amplayo, R.K., Suhara, Y., Wang, X., Lapata, M.: Extractive opinion summarization in quantized transformer spaces. Trans. Assoc. Comput. Linguist. **9**, 277–293 (2021). https://doi.org/10.1162/tacl-a-00366
3. Arora, R., Ravindran, B.: Latent Dirichlet allocation and singular value decomposition based multi-document summarization. In: 2008 Eighth IEEE International Conference on Data Mining, pp. 713–718. IEEE, Pisa, December 2008. https://doi.org/10.1109/ICDM.2008.55
4. Bajaj, A., et al.: Long document summarization in a low resource setting using pre-trained language models. In: Proceedings of the 59th Annual Meeting of the Association for Computational Linguistics and the 11th International Joint Conference on Natural Language Processing: Student Research Workshop, pp. 71–80. Association for Computational Linguistics, Online (2021). https://doi.org/10.18653/v1/2021.acl-srw.7
5. Baxendale, P.B.: Machine-made index for technical literature—an experiment. IBM J. Res. Dev. **2**(4), 354–361 (1958). https://doi.org/10.1147/rd.24.0354
6. Beltagy, I., Peters, M.E., Cohan, A.: Longformer: the long-document transformer (2020). https://doi.org/10.48550/ARXIV.2004.05150
7. Devlin, J., Chang, M.W., Lee, K., Toutanova, K.: BERT: pre-training of deep bidirectional transformers for language understanding. In: Proceedings of the 2019 Conference of the North, pp. 4171–4186. Association for Computational Linguistics, Minneapolis (2019). https://doi.org/10.18653/v1/N19-1423
8. Erkan, G., Radev, D.R.: LexRank: graph-based lexical centrality as salience in text summarization. J. Artif. Intell. Res. **22**, 457–479 (2004). https://doi.org/10.1613/jair.1523
9. Fabbri, A., Li, I., She, T., Li, S., Radev, D.: Multi-news: a large-scale multi-document summarization dataset and abstractive hierarchical model. In: Proceedings of the 57th Annual Meeting of the Association for Computational Linguistics, pp. 1074–1084. Association for Computational Linguistics, Florence (2019). https://doi.org/10.18653/v1/P19-1102
10. Gidiotis, A., Tsoumakas, G.: A divide-and-conquer approach to the summarization of long documents. IEEE/ACM Trans. Audio Speech Lang. Process. **28**, 3029–3040 (2020). https://doi.org/10.1109/TASLP.2020.3037401
11. Goldstein, J., Mittal, V., Carbonell, J., Kantrowitz, M.: Multi-document summarization by sentence extraction. In: NAACL-ANLP 2000 Workshop on Automatic Summarization, vol. 4, pp. 40–48. Association for Computational Linguistics, Seattle (2000). https://doi.org/10.3115/1117575.1117580
12. Grail, Q., Perez, J., Gaussier, E.: Globalizing BERT-based transformer architectures for long document summarization. In: Proceedings of the 16th Conference of the European Chapter of the Association for Computational Linguistics: Main Volume, pp. 1792–1810. Association for Computational Linguistics, Online (2021). https://doi.org/10.18653/v1/2021.eacl-main.154
13. Hayashi, H., Budania, P., Wang, P., Ackerson, C., Neervannan, R., Neubig, G.: WikiAsp: a dataset for multi-domain aspect-based summarization. Trans. Assoc. Comput. Linguist. **9**, 211–225 (2021). https://doi.org/10.1162/tacl-a-00362

14. Jin, H., Wang, T., Wan, X.: Multi-granularity interaction network for extractive and abstractive multi-document summarization. In: Proceedings of the 58th Annual Meeting of the Association for Computational Linguistics, pp. 6244–6254. Association for Computational Linguistics, Online (2020). https://doi.org/10.18653/v1/2020.acl-main.556
15. Kiyoumarsi, F.: Evaluation of automatic text summarizations based on human summaries. Procedia. Soc. Behav. Sci. **192**, 83–91 (2015). https://doi.org/10.1016/j.sbspro.2015.06.013
16. Lewis, M., et al.: BART: denoising sequence-to-sequence pre-training for natural language generation, translation, and comprehension. In: Proceedings of the 58th Annual Meeting of the Association for Computational Linguistics, pp. 7871–7880 (2020). https://doi.org/10.18653/v1/2020.acl-main.703
17. Lin, C.Y.: ROUGE: a package for automatic evaluation of summaries. In: Text Summarization Branches Out, pp. 74–81. Association for Computational Linguistics, Barcelona, July 2004
18. Liu, Y., Lapata, M.: Hierarchical transformers for multi-document summarization. In: Proceedings of the 57th Annual Meeting of the Association for Computational Linguistics, pp. 5070–5081. Association for Computational Linguistics, Florence (2019). https://doi.org/10.18653/v1/P19-1500
19. Liu, Y., Lapata, M.: Text summarization with pretrained encoders. In: Proceedings of the 2019 Conference on Empirical Methods in Natural Language Processing and the 9th International Joint Conference on Natural Language Processing (EMNLP-IJCNLP), pp. 3728–3738. Association for Computational Linguistics, Hong Kong (2019). https://doi.org/10.18653/v1/D19-1387
20. Loshchilov, I., Hutter, F.: Decoupled weight decay regularization. arXiv preprint arXiv:1711.05101 (2017)
21. Ma, C., Zhang, W.E., Guo, M., Wang, H., Sheng, Q.Z.: Multi-document summarization via deep learning techniques: a survey. ACM Comput. Surv. 3529754 (2022). https://doi.org/10.1145/3529754
22. Mani, I., Bloedorn, E.: Multi-document summarization by graph search and matching. In: Proceedings of the Fourteenth National Conference on Artificial Intelligence and Ninth Conference on Innovative Applications of Artificial Intelligence. AAAI'97/IAAI'97, pp. 622–628. AAAI Press, Providence (1997)
23. Mao, Z., et al.: DYLE: dynamic latent extraction for abstractive long-input summarization. In: Proceedings of the 60th Annual Meeting of the Association for Computational Linguistics (Volume 1: Long Papers), pp. 1687–1698. Association for Computational Linguistics, Dublin (2022). https://doi.org/10.18653/v1/2022.acl-long.118
24. Mihalcea, R., Tarau, P.: Textrank: bringing order into text. In: Proceedings of the 2004 Conference on Empirical Methods in Natural Language Processing, pp. 404–411 (2004)
25. Over, P., Yen, J.: An introduction to DUC-2004. National Institute of Standards and Technology (2004)
26. Radev, D.: A common theory of information fusion from multiple text sources step one: cross-document structure. In: 1st SIGdial Workshop on Discourse and Dialogue, pp. 74–83 (2000). https://doi.org/10.3115/1117736.1117745
27. Radev, D.R., Jing, H., Styś, M., Tam, D.: Centroid-based summarization of multiple documents. Inf. Process. Manag. **40**(6), 919–938 (2004). https://doi.org/10.1016/j.ipm.2003.10.006
28. Raffel, C., et al.: Exploring the limits of transfer learning with a unified text-to-text transformer. J. Mach. Learn. Res. **21**(140), 1–67 (2020)

29. Shen, C., Cheng, L., Zhou, R., Bing, L., You, Y., Si, L.: MReD: a meta-review dataset for structure-controllable text generation. In: Findings of the Association for Computational Linguistics: ACL 2022, pp. 2521–2535. Association for Computational Linguistics, Dublin (2022). https://doi.org/10.18653/v1/2022.findings-acl.198

30. Wang, W., Pan, S.J., Dahlmeier, D., Xiao, X.: Recursive neural conditional random fields for aspect-based sentiment analysis. In: Proceedings of the 2016 Conference on Empirical Methods in Natural Language Processing, pp. 616–626 (2016). https://doi.org/10.18653/v1/D16-1059

31. Xu, J., Durrett, G.: Neural extractive text summarization with syntactic compression. In: Proceedings of the 2019 Conference on Empirical Methods in Natural Language Processing and the 9th International Joint Conference on Natural Language Processing (EMNLP-IJCNLP), pp. 3290–3301. Association for Computational Linguistics, Hong Kong (2019). https://doi.org/10.18653/v1/D19-1324

32. Zaheer, M., et al.: Big bird: transformers for longer sequences. In: Larochelle, H., Ranzato, M., Hadsell, R., Balcan, M., Lin, H. (eds.) Advances in Neural Information Processing Systems, vol. 33, pp. 17283–17297. Curran Associates, Inc. (2020)

33. Zhang, Y., et al.: An exploratory study on long dialogue summarization: what works and what's next. In: Findings of the Association for Computational Linguistics: EMNLP 2021, pp. 4426–4433. Association for Computational Linguistics, Punta Cana, Dominican Republic (2021). https://doi.org/10.18653/v1/2021.findings-emnlp.377

34. Zhu, C., Xu, R., Zeng, M., Huang, X.: A hierarchical network for abstractive meeting summarization with cross-domain pretraining. In: Findings of the Association for Computational Linguistics: EMNLP 2020, pp. 194–203. Association for Computational Linguistics, Online (2020). https://doi.org/10.18653/v1/2020.findings-emnlp.19

35. Zhu, F., Tu, S., Shi, J., Li, J., Hou, L., Cui, T.: TWAG: a topic-guided wikipedia abstract generator. In: Proceedings of the 59th Annual Meeting of the Association for Computational Linguistics and the 11th International Joint Conference on Natural Language Processing (Volume 1: Long Papers), pp. 4623–4635. Association for Computational Linguistics, Online (2021). https://doi.org/10.18653/v1/2021.acl-long.356

TETA: Text-Enhanced Tabular Data Annotation with Multi-task Graph Convolutional Network

Chen Ye, Haoshi Zhi, Shihao Jiang, Hua Zhang$^{(\boxtimes)}$, Yifan Wu, and Guojun Dai

School of Computer Science and Technology, Hangzhou Dianzi University,
Hangzhou, China
{chenye,houzss,zhangh,yfwu,daigj}@hdu.edu.cn, shihaojiang@tencent.com

Abstract. Tabular data annotation, which aims to match cells (or columns) to their semantic entities (or types), is crucial to tackling the absence of table content. Recent approaches tend to learn embeddings for tabular data based on deep learning models, but are not conducive to parsing tabular data without metadata. While the metadata may not always be available, entity-related textual information can be easily obtained through external sources such as knowledge bases. Motivated by this, we introduce entity-related textual details in this study to enhance the understanding of tabular data. To obtain better embeddings, we propose a novel model TETA, which adopts the graph convolutional network to refine semantic and structure information from constructed graph features based on tables, entities, types, and text. Meanwhile, we adopt a multi-task learning technique to improve its performance and robustness. We compare TETA with five baselines on five datasets. The results of tabular data annotation and novelty classification demonstrate the effectiveness and promise of TETA.

Keywords: Tabular data annotation · Text data · Graph convolutional network · Multi-task learning

1 Introduction

Nowadays, tabular data is ubiquitous on the web, and it can provide precious information for broad applications such as table question answering [18,20] and semantic parsing [10]. However, incomplete tables (e.g., erroneous structure alignment, ambiguous cells, or missing column names) often exist due to the uneven quality of web data and transmission errors at certain moments. To tackle this issue, tabular data annotation is a fundamental task of crucial importance. The majority of scholars have focused on modifying and fine-tuning pre-training models [4,6,12,18] such as transformers for better embedding representations, while ignoring their inability to fully understand the structural information of tabular data. Furthermore, it requires large artificial corpora and elaborately huge training cost for pre-training. ColNet [2],TabGCN [11], and TCN [14] employ different

© The Author(s), under exclusive license to Springer Nature Switzerland AG 2023
X. Wang et al. (Eds.): DASFAA 2023, LNCS 13945, pp. 523–533, 2023.
https://doi.org/10.1007/978-3-031-30675-4_38

but convolution-related methods. However, most of the models' [4, 14, 18] performance will be greatly compromised without metadata. Fortunately, large quantities of entity-related text exist in the Knowledge Bases (KBs), which may facilitate the semantic understanding of tabular data. In this paper, we propose a novel approach incorporating entity-related text for tabular data annotation. Considering the capability of Graph Convolutional Network (GCN) to capture deep structure information among massive graph data with elaborate relationships [5, 15–17, 19], we use GCN to learn embeddings of mixed data.

Challenges. Modeling tables, entities, types, and text into graph data and leveraging these features to complete the tabular data annotation task is challenging since it raises two significant questions: (1) How to construct a graph based on mixed tabular-textual data? (2) How to design the model architecture to fully utilize the tabular and textual data for the annotation task?

To answer these two questions, we propose a highly effective model TETA by integrating tabular and textual data information. For question (1), we create various types of graph nodes for the crucial features in tables and text, and the relationships between them can be modeled as different types of edges. For question (2), we adopt a two-layer GCN to learn embeddings for tabular and textual data simultaneously. Moreover, we adopt a multi-task learning technique, which designs three classifiers to focus the model's attention on the column-type, cell-entity, and text-type classification tasks. Model parameters are updated inversely through three tasks' weighted joint loss.

Contributions. We summarize our contributions as follows:

- We realize that entity-related text benefits the semantic understanding of tabular data. This idea motivates us to incorporate text extracted from the KBs to enhance the performance of tabular data annotation.
- To encode the relationships between tabular and textual data and capture the syntactic structure of tables, we design a graph construction approach to transform the data into various nodes and edges of a graph.
- We propose a novel model TETA based on a GCN to jointly learn the embeddings of text, table, entity, and type nodes. TETA also adopts a multi-task learning technique to obtain better embeddings.
- To demonstrate the capabilities of TETA, we compare it with five baselines on five datasets and results show that TETA significantly improves the performance of tabular data annotation tasks. The performance of TETA on novelty classification for the downstream task also demonstrates the usefulness of learned embeddings.

2   Problem Definition

Table Definition. Given a table set $T = \{\tau^1, \ldots, \tau^n\}$, the k-th table τ^k includes m rows $R^k = \{r_1^k, \ldots, r_m^k\}$ and o columns $C^k = \{c_1^k, \ldots, c_o^k\}$. The cell set $X^k = \{x_{11}^k, \ldots, x_{1o}^k, \ldots, x_{m1}^k, \ldots, x_{mo}^k\}$ is obtained by traversing the row set R^k (or column set C^k) by row (or by column), where x_{ij}^k denotes the cell belonging to row r_i^k and column c_j^k. For tabular data annotation, each column $c \in C^k$ cor-responds to a single type t^c, and each cell $x \in X^k$ is corresponding to a specific entity e^x. We denote the entity set and type set w.r.t. T as $\mathbb{E}$ and $\mathbb{T}$, respectively.

Text Definition. We assume an entity-related text set $\mathbb{S} = \{S^1, \ldots, S^n\}$ can be obtained for each table $\tau^k \in T$. The text set for τ^k is denoted as S^k, where each sentence $s^x \in S^k$ contains several words that can be auxiliary information for a table cell $x \in X^k$. Concretely, the syntactic structure of sentence s^x is (*entity, copula, proper noun*). Here, *proper noun* is a type generalization of the entity.

Text-Enhanced Tabular Data Annotation. Tabular data annotation [3,11] can be divided into Cell Entity Annotation (CEA), Column Type Annotation (CTA), and Column Pair Annotation (CPA). Following previous work [11], we deal with the first two tasks in a semi-supervised manner. That is, for each table $\tau^k \in T$, we assume that a subset of C^k has been annotated with their corresponding types, and a subset of X^k has been annotated with their corresponding entities. Then, for the cells and columns without annotations, we aim to predict entity e^x that best describes the semantics of cell x (w.r.t. CEA task) and type t^c that best describes the semantics of column c (w.r.t. CTA task) with the help of $\mathbb{S}$.

3   TETA Architecture

This section introduces the model architecture that contains four key components: (1) Text extraction: we extract the entity-related text by using several automatic extraction tools. (2) Graph construction: we construct various types of nodes and edges for different tabular elements, entities, types, and text. (3) Representation learning: we adopt a two-layer GCN to convert features of various graph nodes into text-enhanced embedding. (4) Multi-task learning: we design a multi-task learning module that minimizes a joint loss over three classification tasks to train the node embedding. We detail each component below.

3.1   Text Extraction

Given a table $\tau^k \in T$, its text set S^k can be extracted according to its cell set X^k. Specifically, for a cell $x \in X^k$ annotated with entity e^x, we directly extract its related text s^x from a relevant KB based on e^x. For a cell $x \in X^k$ without entity annotation, we extract its related text s^x from the KB based on x. Then we adopt existing NLP tools to filter text that isn't in the form of (entity, copula, proper noun). The filtered result will be stored in the text set S^k.

3.2   Graph Construction

Given a table set T and a text set $\mathbb{S}$ related to T, we construct a graph for them. Figure 1 shows an example constructed graph, and we will describe the construction methods of nodes and edges below.

Graph Construction of Nodes. To preserve the structure information and capture the crucial features, we construct various types of nodes for each table $\tau^k \in T$. Specifically, we create tabular element nodes (white rounded rectangle) including table node n_τ, row node n_r, column node n_c, and cell node n_x, which comprehensively express the content and structure information of the table. Then, to capture the semantic features of entities and types, we create entity node n_e (blue ellipse) and type node n_t (yellow ellipse) for each entity $e \in \mathbb{E}$ and type $t \in \mathbb{T}$, respectively. Finally, to obtain additional text information, we create text node n_s (red right-angled rectangle) for each text $s \in \mathbb{S}$.

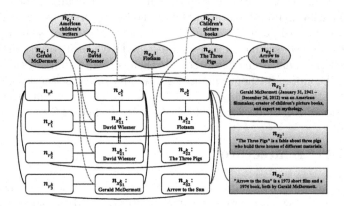

Fig. 1. Graph construction of a web table, entities, types, and entity-related text. (Color figure online)

Graph Construction of Edges. To represent elaborate relationships between nodes and mine deeper semantic and structural information of tabular data, we construct various types of edges, including:

- The tabular basic edge E_t (black solid line) is used for capturing the basic structural relation between tabular elements, including the cell-row edge, the cell-column edge, the row-table edge, and the column-table edge.
- The knowledge edge E_κ (yellow solid line) preserves the relationship between entity e and type t. For instance, if column c is annotated with type t^c and its cell x is annotated with entity e^x, then a knowledge edge E_κ exists between entity node n_e and type node n_t.
- The annotation edge E_a (blue dashed line and yellow dashed line) indicates the annotation information between the cell node n_x and the entity node n_e (or the column node n_c and the type node n_t).

- The auxiliary edge E_u (red solid line) connects the text node n_s and the column node n_c, i.e., the node of column c that includes the cell x corresponding to text s^x. Since text s^x interprets entity e^x in terms of describing type t^c, establishing E_u between n_s and n_c can benefit the CEA task and be more helpful for the CTA task.
- The lexical similarity edge E_l (black double dotted line) aims to connect two lexically similar cell nodes $n_x, n_{x'}$, so that the model's ability to capture the relationships and information among cell nodes can be enhanced. If the cosine similarity between two cell nodes is higher than the threshold ε, then it is considered that there is a lexical similarity between them.

3.3 Representation Learning

To capture the semantic features of tabular data, we learn the embedding of each type of node in the constructed graph, which is depicted in Fig. 2(a).

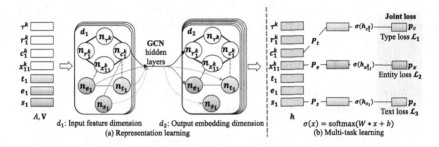

Fig. 2. The process of representation learning and training.

Matrix Construction. We obtain feature matrix $\mathbf{V}$ and adjacent matrix A based on the graph nodes and edges. For nodes n_x, n_e, n_t, n_s with text values, we convert them to the features of the corresponding nodes by means of word vectorization [11,14]. Specifically, the feature of a node is the average of its contained word vectors, i.e., $\mathbf{v}_{x/e/t/s} = \text{mean}(\sum_{\text{word}\in\text{text}(n_{x/e/t/s})} \text{Vectorize(word)})$. For row node n_r and column node n_c, we use the mean of the feature vectors of cells belonging to them as their features, i.e., $\mathbf{v}_{r/c} = \text{mean}(\sum_{x\in r/c} \mathbf{v}_x)$. Similarly, the feature of the table node n_τ is the mean of the feature vectors of all row nodes (or column nodes) in the table, i.e., $\mathbf{v}_\tau = \text{mean}(\sum_{r/c\in\tau} \mathbf{v}_{r/c})$. Therefore, we obtain the feature matrix $\mathbf{V} \in \mathbb{R}^{N\times d_1} = [\mathbf{v}_\tau, \mathbf{v}_r, \mathbf{v}_c, \mathbf{v}_x, \mathbf{v}_e, \mathbf{v}_t, \mathbf{v}_s]^\top$, where N and d_1 represent the total number and feature dimension of all types of nodes, respectively. Meanwhile, we establish adjacency matrix $A \in \mathbb{R}^{N\times N}$ by traversing all nodes and all edges, where each node has a self-connecting edge, and each edge takes the form of an undirected symmetric edge [8,11].

Embedding Learning. Given feature matrix $\mathbf{V}$ and adjacent matrix A, we adopt the GCN [8] to learn text-enhanced embedding representations of tabular data. As the receptive field of the network increases with the number of GCN layers, information from higher-order neighborhoods can be obtained by stacking and convolving through multiple GCN layers. However, if the network is too deep, the node embedding will be affected by irrelevant nodes, and the effect will decrease instead. Thus, we set the number of GCN layers to 2, and the calculation formula for each GCN layer is as follows:

$$L^{(\ell+1)} = \text{ReLU}(\tilde{A}L^{(\ell)}W_\ell) \tag{1}$$

where L is the input or output in each layers and ℓ represents the number of GCN layer ($L^{(0)} = \mathbf{V}$). The output of each subsequent layer is the input of the next layer. $W_\ell \in \mathbb{R}^{d_1' \times d_2'}$ is the weight matrix of the ℓ-th layer. $\tilde{A} = D^{-1/2}AD^{-1/2}$ is a Laplacian regular matrix, which is the result of symmetric normalization of the adjacency matrix A. The element value of the diagonal degree matrix D is the sum of the in and out degrees of the corresponding nodes, i.e., $D_{ii} = \sum_{j=1}^{N} A_{ij} + \sum_{k=1}^{N} A_{ki}$. With a two-layer GCN, we obtain the embeddings $\mathbf{h} \in \mathbb{R}^{N \times d_2} = [\mathbf{h}_\tau, \mathbf{h}_r, \mathbf{h}_c, \mathbf{h}_x, \mathbf{h}_e, \mathbf{h}_t, \mathbf{h}_s]^\top$ of different types of nodes.

3.4 Multi-task Learning

To enhance the performance of TETA on CEA and CTA tasks, we design a multi-task learning module for training. The multi-task learning module consists of three parts: column-type classifier, cell-entity classifier, and text-type classifier, as shown in Fig. 2(b). Take column-type classifier as an example, to predict the type t^c of an unannotated column c, we project the column embedding $\mathbf{h}_c$ to the type space through the projection matrix P_t, and then send the projected vector to the type classifier. Specifically, we use a Fully Connected (FC) layer to vary the dimensions and then use the softmax function to output the predicted value for each category. The cell-entity classifier and text-type classifier are similar. The formula for three classifiers is as follows:

$$p_{c/x/s} = \text{softmax}(W_{t/e/s}(P_{t/e/s}\mathbf{h}_{c/x/s}) + b_{t/e/s}), \tag{2}$$

where $W_{t/e/s} \in \mathbb{R}^{d_2 \times q/d_2 \times g/d_2 \times z}$ and $b_{t/e/s} \in \mathbb{R}^{q/g/z}$ are the parameter matrix and the bias matrix of the FC layer, respectively, d_2 is the dimension of embedding, $q/g/z$ represents the number of type, entity or text-type categories.

Joint Training. We jointly train three tasks via Adam [7] to optimize model parameters. We use cross-entropy as the loss function of three classification tasks. The final loss is the weighted sum of the losses of the three tasks, where the weight ratio is set to 1:2:0.5 (the optimal ratio of manual adjustment):

$$\mathcal{L} = -\left[\frac{1}{N_c}\sum_{i=1}^{N_c}\sum_{j=1}^{q} y_{ij}^t \log p_{c_{ij}} + \frac{2}{N_x}\sum_{i=1}^{N_x}\sum_{j=1}^{g} y_{ij}^e \log p_{x_{ij}} + \frac{0.5}{N_s}\sum_{i=1}^{N_s}\sum_{j=1}^{z} y_{ij}^s \log p_{s_{ij}}\right] \tag{3}$$

where N_c, N_x, N_s represent the number of training samples for annotated columns, cells, and text, respectively. q, g, z represent the number of categories of the corresponding task, and $y_{ij}^t, y_{ij}^e, y_{ij}^s$ represent the symbolic function of the real labels of three tasks. For instance, if the real category of the sample i is j, it outputs 1; otherwise, 0. $p_{c_{ij}}, p_{x_{ij}}, p_{s_{ij}}$ represent the predicted probability that the sample i belongs to the category j in three tasks.

4 Experiments

4.1 Experiment Setup

Datasets. We use five web datasets whose statistics are summarized in Table 1. Note that T2Dv2 and Wikipedia only contain type annotations from DBpedia. Thus, they can only be used for the CTA task. For all the datasets, we randomly select 30% of the entities and types as novel entities and types for the downstream task of novel classification. The cells and columns annotated with these novel entities or types are considered unannotated. In the remaining cells and columns annotated with un-novel entities and un-novel types, respectively, we randomly divide the training set and test set by the ratio of 70%:30%.

Compared Methods. To verify the effect of adding the text-type classification task to loss computation, we divide the proposed model TETA into a base version and a full version, where TETA-base removes the text classification task while retaining the text nodes. We also compare with five baselines: ColNet [2], TaBERT [18], TURL [4], Doduo [12], and TabGCN [11].

Implementation Details. All experiments are implemented in Ubuntu 16.04 LTS operating system and Anaconda-Pytorch environment, using a 2080TI-12G GPU for training. The hyperparameters of baselines are consistent with the corresponding papers, and for comparison, the hyperparameters of TETA are consistent with TabGCN. TETA uses Wikipedia as KB source for entity-related text extraction. We use StanfordOpenIE [1] and Spacy to filter out worthless text. As the full version of TURL uses table and metadata and only T2Dv2 has

Table 1. The statistics of datasets.

Datasets	Table	Row (mean)	Column (mean)	Un-novel entity	Novel entity	Total entity	Un-novel type	Novel type	Total type
Wiki M	40	35.7	4.2	959	410	1369	39	16	55
Web M	404	34.2	3.7	732	313	1045	38	16	54
Limaye	327	31.3	3.8	551	235	786	13	5	18
T2Dv2	367	85.2	4.9	–	–	–	34	14	48
Wikipedia	590	26	5	–	–	–	22	9	31

partial metadata, we implement two variants TURL-v3 (metadata & cells) and TURL-v4 (cells) on T2Dv2 dataset. We use the micro-average F1-score as the evaluation metric and repeat 10 times in each experiment for all the models.

Table 2. Micro-average F1-score comparison on CEA and CTA tasks.

Model	Wiki M		Web M		Limaye		T2Dv2	Wikipedia
	Entity	Type	Entity	Type	Entity	Type	Type	Type
ColNet	–	0.2000	–	0.4700	–	0.4700	0.5900	0.6000
TaBERT	–	0.2000	–	0.4900	–	0.4700	0.5900	0.5900
TURL-v3	–	–	–	–	–	–	0.7604	–
TURL-v4	–	**0.5402**	–	0.8607	–	0.6190	0.6945	0.6951
Doduo	–	0.1333	–	0.8158	–	0.7619	0.6027	0.8915
TabGCN	0.2462	0.3333	0.8110	0.8500	0.7936	0.8571	0.8082	0.8527
TETA-base	0.2554	0.3333	0.8121	0.8500	0.8051	0.8571	0.8438	0.8605
TETA-full	**0.2615**	0.3333	**0.8219**	**0.8684**	**0.8127**	**0.9524**	**0.9062**	**0.9690**

4.2 Experimental Results

Performance Comparison of the CEA and CTA Tasks. Table 2 shows results for all methods on five datasets. Except for TabGCN and TETA-base/full, other methods can only complete the CTA task. ColNet and TaBERT perform worst. ColNet's accuracy largely depends on the coverage of the KB and the informativeness of randomly sampled cells. Although the pre-training method of TaBERT includes the column type prediction, after fine-tuning for the semantic parsing task, TaBERT has lost its advantage on the CTA task. TURL works surprisingly well on the Wiki M dataset, because TURL's pre-training corpus WikiTables contains tables of the Wiki M dataset. However, without metadata, TURL's performance is not as good as TETA-full's on the remaining datasets. Doduo tokenizes the tabular data by column, ignoring the connection between the cells in the same row, so the complete structure information cannot be captured, resulting in tepid performance. TabGCN and TETA-base/full learn the embeddings of the whole table to simultaneously finish the CEA and the CTA tasks, thus outperforming other methods in most cases. Without the loss of text classification, TETA-base also achieves higher micro-average F1-scores than other methods in most cases, demonstrating the helpfulness of textual data to the annotation task. Meanwhile, TETA-full adds the loss of text classification to help the model focus its attention on the entity and type information of text. Therefore, TETA-full can better capture tables' deep semantic information and outperforms TabGCN by up to 12% (i.e., the CTA task on Wikipedia).

Ablation Study. To verify the effectiveness of the graph structure, we analyze the effectiveness of TETA by setting multiple variants. **-T** only focuses on the CEA task by removing all type nodes and type-related edges. Note that the category of the text cannot be obtained when type nodes are removed. Thus, the text classifier is also removed. Similarly, **-E** only focuses on the CTA task by removing all the entity nodes and entity-related edges (the cell-entity edge of the annotation edge, and the knowledge edge), which is divided into **-E$_{\tilde{s}}$** without text nodes and **-E$_S$** with text nodes. **-S** removes the text nodes and their edges. **-A** removes the text classification task during training while reserving text nodes and auxiliary edges. **-L** removes lexical similarity edges. Table 3 shows the results of three datasets with both entity and type annotations. Except for TETA-full, each model has a certain decline in metrics due to the lack of specific components, where TETA-L drops the most in the absence of the lexical similarity edge. Thus, all components of TETA contribute to performance improvements and the lexical similarity edge plays a significant role in the CEA task.

Table 3. Ablation study result of CEA and CTA tasks.

Model	Wiki M		Web M		Limaye	
	Entity	Type	Entity	Type	Entity	Type
TETA-E$_{\tilde{s}}$	–	0.3333	–	0.8333	–	0.8571
TETA-E$_S$	–	0.3333	–	0.8417	–	0.9048
TETA-T	0.2523	–	0.8099	–	0.8043	–
TETA-S	0.2462	0.3333	0.8110	0.8500	0.7936	0.8571
TETA-A	0.2554	0.3333	0.8121	0.8500	0.8051	0.8571
TETA-L	0.2492	0.3333	0.6678	0.8684	0.6055	0.9524
TETA-full	**0.2615**	0.3333	**0.8219**	**0.8684**	**0.8127**	**0.9524**

Downstream Task - Novelty Classification. The novelty classification task is to determine whether an unannotated cell/column belongs to a novel entity/type that has not been seen. In addition to the cells (or columns) belonging to 30% novel entities (or types), we also add cells (or columns) used as test data for the CEA (or CTA) task to explore the effectiveness of learned embeddings in distinguishing novel from un-novel entities (or types). Note that only TabGCN and TETA can accomplish this task. Specifically, we project the embedding $\mathbf{h}_{x/c}$ of an unannotated node $n_{x/c}$ into the entity (or type) space and compare with all un-novel entities (or types), measured by cosine similarity. If the similarities between the projected vector of the node and the projected vectors of all un-novel entities (or types) do not exceed a given threshold, the entity (or type) category of the cell (or column) x/c is determined as novel, otherwise un-novel: $\sigma(\max_{e/t \in \mathbb{E}^u/\mathbb{T}^u} \cos(P_{e/t}\mathbf{h}_{x/c}, P_{e/t}\mathbf{h}_{e/t}) \leq \varphi_{e/t})$, where $\mathbb{E}^u \subset \mathbb{E}$ and $\mathbb{T}^u \subset \mathbb{T}$ represent un-novel entities set and un-novel types set, respectively, σ is the Kronecker delta function, and $\varphi_{e/t}$ is the novelty threshold. The results of novelty classification are shown in Table 4, where the novelty threshold φ_e and φ_t are both manually optimized to 0.85-0.9. TETA-base/full's performance is better than TabGCN on both novel-entity and novel-type classification in most cases, demonstrating the effectiveness of the embeddings learned from TETA.

Table 4. F1-score comparison on the novelty classification task.

Model	Wiki M		Web M		Limaye		T2Dv2	Wikipedia
	Entity	Type	Entity	Type	Entity	Type	Type	Type
TabGCN	**0.8050**	0.8000	0.8850	0.8020	0.7770	0.8890	0.8000	0.7870
TETA-base	0.7720	0.8100	0.8860	0.8170	**0.7960**	0.9230	0.8110	0.7870
TETA-full	0.7720	**0.8240**	**0.8980**	**0.8580**	0.7720	**0.9600**	**0.8310**	**0.8230**

5   Related Work

Tabular Data Annotation. Existing methods for tabular data annotation can
be mainly divided into three categories: graph-based inference methods [9,13],
pre-training methods [4,6,12,18], and convolution-related approach [2,11,14].
The graph-based inference methods use logistic models for prediction, whose
features are manually constructed by domain knowledge. Despite being pre-
trained with a specific large corpus, the pre-training models cannot capture the
complete structured information of tabular data due to the fundamental way of
linearizing cells of tabular data. Several convolution-related models have been
proposed to address the above issues, using automatic feature generation tools
and convolution operations to improve efficiency and better capture structural
information, respectively. Our proposed model TETA enhances the GCN's ability to classify cells and columns by adding entity–related textual information and
multi-task learning module, thus can better capture the semantic and structural
information of the tabular data without metadata.

Representation Learning of Tabular and Textual Data. The popularity of
pre-training language models has promoted the development of the joint learning
of tabular and textual data [10,18,20]. Among these methods, the most related
work to TETA is TaBERT [18], which uses CTA task as a pre-training method
but finetunes on question answering tasks. BRIDGE [10] and MT2Net [20] are
trained for DB-related tasks. None of these are suitable for annotating tables.

6   Conclusion

We present TETA, a novel model for learning text-enhanced representations on
tabular data via graph convolutional network and multi-task learning. Different
from existing methods, TETA (1) utilizes entity-related text extracted from the
knowledge base to assist the model in better understanding the semantic and
structural information of tables, (2) jointly learns enhanced embedding of text,
tables, entities, and types for better representation, (3) uses a multi-task learning
method to improve model performance and robustness. We evaluate TETA on
five datasets for tabular data annotation tasks and a downstream task, the results
of which demonstrate the effectiveness and promise of our model. In the future,
we will explore the potential of using other graph neural networks and test other
downstream tasks of mixing table understanding with text.

Acknowledgments. This paper was partially supported by National Natural Science Foundation of China (No. 62202132), Natural Science Foundation of Zhejiang Province (No. LQ22F020032), and National Key Research and Development Program of China (No. 2022YFE0199300).

References

1. Angeli, G., Premkumar, M.J.J., Manning, C.D.: Leveraging linguistic structure for open domain information extraction. In: Proceedings of ACL, pp. 344–354 (2015)
2. Chen, J., Jiménez-Ruiz, E., Horrocks, I., et al.: ColNet: embedding the semantics of web tables for column type prediction. In: Proceedings of AAAI, pp. 29–36 (2019)
3. Cutrona, V., Chen, J., Efthymiou, V., et al.: Results of semTab 2021. In: Proceedings of ISWC, pp. 1–12 (2021)
4. Deng, X., Sun, H., Lees, A., et al.: TURL: table understanding through representation learning. SIGMOD Rec. **51**(1), 33–40 (2022)
5. He, M., Han, T., Ding, T.: Learning and fusing multiple user interest representations for sequential recommendation. In: Proceedings of DASFAA, pp. 401–412 (2022)
6. Iida, H., Thai, D., Manjunatha, V., et al.: TABBIE: pretrained representations of tabular data. In: Proceedings of NAACL-HLT, pp. 3446–3456 (2021)
7. Kingma, D.P., Ba, J.: Adam: a method for stochastic optimization. In: Proceedings of ICLR (2015)
8. Kipf, T.N., Welling, M.: Semi-supervised classification with graph convolutional networks. In: Proceedings of ICLR (2017)
9. Limaye, G., Sarawagi, S., Chakrabarti, S.: Annotating and searching web tables using entities, types and relationships. Proc. VLDB Endow. **3**(1), 1338–1347 (2010)
10. Lin, X.V., Socher, R., Xiong, C.: Bridging textual and tabular data for cross-domain text-to-SQL semantic parsing. In: Proceedings of EMNLP, pp. 4870–4888 (2020)
11. Pramanick, A., Bhattacharya, I.: Joint learning of representations for web-tables, entities and types using graph convolutional network. In: Proceedings of EACL, pp. 1197–1206 (2021)
12. Suhara, Y., Li, J., Li, Y., et al.: Annotating columns with pre-trained language models. In: Proceedings of SIGMOD, pp. 1493–1503 (2022)
13. Takeoka, K., Oyamada, M., Nakadai, S., et al.: Meimei: an efficient probabilistic approach for semantically annotating tables. In: Proceedings of AAAI, pp. 281–288 (2019)
14. Wang, D., Shiralkar, P., Lockard, C., et al.: TCN: table convolutional network for web table interpretation. In: Proceedings of WWW, pp. 4020–4032 (2021)
15. Wang, H., Zhang, F., Zhang, M., et al.: Knowledge-aware graph neural networks with label smoothness regularization for recommender systems. In: Proceedings of SIGKDD, pp. 968–977 (2019)
16. Wei, X., Yu, R., Sun, J.: View-GCN: view-based graph convolutional network for 3D shape analysis. In: Proceedings of CVPR, pp. 1847–1856 (2020)
17. Yang, L., Gu, X., Shi, H.: A noval satellite network traffic prediction method based on GCN-GRU. In: Proceedings of WCSP, pp. 718–723 (2020)
18. Yin, P., Neubig, G., Yih, W., et al.: TaBERT: pretraining for joint understanding of textual and tabular data. In: Proceedings of ACL, pp. 8413–8426 (2020)
19. Zhang, M., Chen, Y.: Link prediction based on graph neural networks. In: Proceedings of NeurIPS, pp. 5171–5181 (2018)
20. Zhao, Y., Li, Y., Li, C., et al.: MultiHiertt: numerical reasoning over multi hierarchical tabular and textual data. In: Proceedings of ACL, pp. 6588–6600 (2022)

A Two-Stage Label Rectification Framework for Noisy Event Extraction

Zijie Xu, Peng Wang$^{(\boxtimes)}$, Ziyu Shang, and Jiajun Liu

School of Computer Science and Engineering, Southeast University, Nanjing, China
{xuzijie,pwang,ziyus1999,jiajliu}@seu.edu.cn

Abstract. Event extraction aims to identify event triggers and corresponding arguments. Existing methods are mainly devoted to well-designed deep neural networks based on the assumption that training data are high-quality. However, noisy labels are unavoidable in real-world scenarios, which is challenging for current event extraction applications. This paper proposes a novel **Two**-stage **l**abel **r**ectification framework (Tolar) to tackle this problem from explicit and latent aspects. In the first stage, an event schema mapping module is designed to rectify the explicit label inconsistency. Then a self-adaptive iteration module addresses the latent semantic noise in the second stage. In order to cope with extremely noisy labels, we further design a **Co**operative **G**lobal **P**ointer **N**etwork (CoGPN) to train two global pointer networks concurrently and let them filter possibly noisy labels mutually. Extensive experiments on ACE 2005 and MAVEN with synthetic noise demonstrate that our framework Tolar effectively enhances event extraction methods, and our CoGPN achieves state-of-the-art performance in extremely noisy settings.

Keywords: Event extraction · Natural language processing · Noisy label learning · Global pointer network

1 Introduction

Event extraction (EE) [7] is a significant yet challenging task in information extraction (IE), aiming to detect events and extract event arguments from unstructured text. Recent EE is benefited from the rich contextualized representations generated by pretrained language models (PLMs) and elaborate downstream fine-tune strategies [1,8,9]. These approaches focus on improving the performance of PLMs with high quality data labels. However, EE is a data-hungry task and suffers from noisy labels. It is necessary to deal with noisy labels in real-world EE scenarios. We divide noisy labels into explicit label inconsistency and latent semantic noise. The former is the contradiction between the labels of event types and corresponding argument roles. The latter is inconsistent with the latent semantic of the sample when the text spans or types of triggers and arguments are annotated by mistake. Figure 1 illustrates an example of two types of noise. Owing to the misleading word "killing", an Ownership Transfer event

© The Author(s), under exclusive license to Springer Nature Switzerland AG 2023
X. Wang et al. (Eds.): DASFAA 2023, LNCS 13945, pp. 534–543, 2023.
https://doi.org/10.1007/978-3-031-30675-4_39

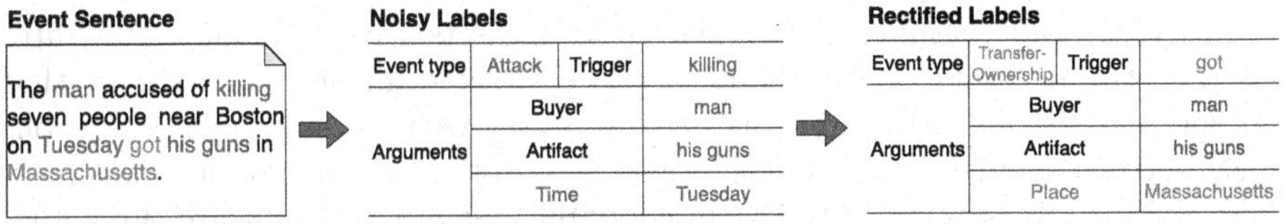

Event Sentence	Noisy Labels				Rectified Labels				

Fig. 1. An example for two types of noisy labels and corresponding rectification. The words in red are noisy labels, the words in green are corresponding revisions, and the words in blue are clean labels. (Color figure online)

is labeled as an Attack event by mistake. However, the event type Attack contradicts the argument roles "Buyer" and "Artifact", since they do not belong to the Attack event according to the event schema. This contradiction is referred to as the explicit label inconsistency. Besides, the word "Tuesday" denotes the time that the man murdered seven people and the word "Massachusetts" was the actual place where the transaction happened. The incorrect label "Time: Tuesday" is regarded as the latent semantic noise, which is more challenging.

To address the above issues, we propose a novel two-stage label rectification framework (Tolar) for noisy event extraction. In the first stage, we focus on the explicit label inconsistency. Since event schema suggests that events with specific types will contain arguments with particular roles, we leverage the pre-defined event schema to rectify the incorrect event type labels in the source data. As shown in Fig. 1, since "Buyer" is an exclusive argument role in the Ownership Transfer event and "Artifact" is not the participant in the Attack event, we correct the event type label "Attack" into "Transfer-Ownership".

In the second stage, to tackle the latent semantic noise, we propose a self-adaptive iteration strategy. The basic idea is that deep neural models tend to give low confidence to noisy labels despite being profoundly influenced. We first train an event extraction model with pre-processed training data from the first stage and then utilize this model to predict the entire training set. According to the confidence of predicted results, the model adaptively corrects the probable noisy labels into proper labels with high confidence. Such self-correction procedure is iterated until obtaining a convergent model.

In addition, it is tough for PLMs to deal with extremely noisy labels when the noise rate is over 50%. To alleviate this problem, we introduce a novel Cooperative Global Pointer Network (CoGPN). We train two independent Global Pointer Networks (GPN) [10] simultaneously. One GPN is span-specific, which focuses on identifying the start and end position of triggers or arguments. Another GPN is cascade [9], which aims to extract the span and types concurrently for candidate triggers or arguments. During the training phase, the two GPNs select the small loss samples as clean ones for each other to update the parameters in each mini-batch. CoGPN takes advantage of different emphases on learning tasks of the two GPNs to filter different types of error introduced by noisy labels. In the inference phase, the *logits* from span-specific GPN additionally enriches the *logits* from the cascade GPN.

We verify our method on two synthetic datasets with different noise rates based on two datasets, ACE 2005 and MAVEN. Experimental results show that our method significantly improves existing event extraction models in various noisy settings, and CoGPN outperforms the state-of-the-art under extremely noisy settings. In summary, the main contributions of this paper are three-fold:

- To the best of our knowledge, we are among the first to address noisy event extraction. To this end, we propose Tolar, a two-stage label rectification framework for noisy event extraction from both explicit and latent aspects based on the event schema mapping mechanism and self-adaptive iteration.
- To ensure the robustness of models encountering extreme noise, we further propose a novel cooperative global pointer network (CoGPN) that utilizes the different learning abilities of two GPNs to filter noisy labels mutually.
- We construct two benchmarks for noisy event extraction task and conduct comprehensive experiments. Experimental results demonstrate that the Tolar framework can efficiently improve the EE models, and CoGPN achieves the state-of-the-art performance in extremely noisy settings.

2   Methodology

2.1   Problem Statements

For noisy event extraction scenarios, an event schema is pre-defined, including k event types $\{t_1, t_2, \cdots, t_k\}$ and corresponding argument roles $\{R_1, R_2, \cdots, R_k\}$, where $R_i = \{r_{i,1}, r_{i,2}, \cdots, r_{i,L}\}$ is the argument role set of specific event type t_i. Consider an event sample (T_p, Y_p), where T_p is the input event text, and Y_p is the corresponding label. The purpose of noisy event extraction is to extract the actual triggers and arguments from T_p while Y_p could be mislabeled.

2.2   Framework Overview

The overview of Tolar framework is illustrated in Fig. 2, composed of two stages: the event schema mapping stage and the self-adaptive iteration stage. In the first stage, the source data is transferred to the event schema mapping module, utilizing the event schema mapping to revise explicit label inconsistency between event types and argument roles. The pre-processed data from the first stage is then fed to the self-adaptive iteration module in the second stage, where the event extraction model corrects the latent noisy labels iteratively according to the confidence score. The event extraction model consists of an PLM encoder and different decoding and inference strategies.

2.3   Event Schema Mapping Stage

We first introduce event schema mapping as a data processing stage. The arguments with particular roles $r_{i,j} \in R_i$ will be involved in the event of specific type t_i, which expresses the mapping from *argument role* to *event type*. Specifically,

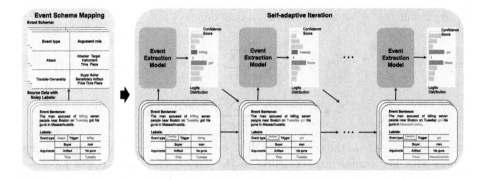

Fig. 2. The architecture of the Tolar framework.

when the labels of argument roles are reliable, any particular argument role $r_{i,j}$ will indicate the event type t_i. Thus, we apply the following logical rule to rectify the explicit label inconsistency:

$$\exists r_{cur,j} \left(r_{cur,j} \in R_i \wedge r_{cur,j} \notin (R - R_i) \right) \wedge (t_{cur} \neq t_i) \Rightarrow t_i \qquad (1)$$

where $r_{cur,j} \in R_{cur}$ and t_{cur} represent the current labels of argument roles and event type within the given event label Y_p. R_i is the argument role set of specific event type t_i while R is the universal set of argument roles.

When the labels of arguments are also noisy, only one particular argument role will be inadequate. We reinforce the additional restriction that all labels of argument roles within the given event label Y_p should be involved in the same specific event type t_i:

$$\exists r_{cur,j} \left(r_{cur,j} \notin (R - R_i) \right) \wedge \forall r_{cur,j} \left(r_{cur,j} \in R_i \right) \wedge (t_{cur} \neq t_i) \Rightarrow t_i \qquad (2)$$

2.4   Self-adaptive Iteration Stage

In the self-adaptive iteration stage, we first train an event extraction model with pre-processed training data from the first stage and then utilize this model to predict the whole training set. For each event sample (T_p, Y_p) from the training set $\mathcal{D}_{train}$, we obtain the *logits* produced by the event extraction model and estimate the span and types for each trigger and argument. The *logits* is denote as $\mathbf{logits}^p_{trg} = \{[l^{1,1}_{trg}, l^{1,2}_{trg}, \cdots , l^{1,k}_{trg}], \cdots , [l^{N,1}_{trg}, l^{N,2}_{trg}, \cdots , l^{N,k}_{trg}]\}$ and $\mathbf{logits}^p_{arg} = \{[l^{1,1}_{arg}, l^{1,2}_{arg}, \cdots , l^{1,L}_{arg}], \cdots , [l^{N,1}_{arg}, l^{N,2}_{arg}, \cdots , l^{N,L}_{arg}]\}$ for trigger detection and argument extraction respectively, where N is the length of T_p, k is the number of event types and L is the size of argument role set R. For each event type t_i or argument role r_j, we operate sigmoid function on the *logits* to estimate the confidence score of predicted result and set average score of each event type t_i

or argument role r_j as the threshold of confidence S :

$$S_{trg}^i = \frac{1}{M \times N} \sum_{p=1}^{M} \sum_{q=1}^{N} \text{sigmoid}\left(l_{trg}^{q,i}\right) \tag{3}$$

$$S_{arg}^j = \frac{1}{M \times N} \sum_{p=1}^{M} \sum_{q=1}^{N} \text{sigmoid}\left(l_{arg}^{q,j}\right) \tag{4}$$

where M is the size of the training set. The model self-corrects the probable noisy labels with low confidence according to the threshold S. For trigger label set $Y_{trg}^p = [y_{trg}^{p,1}, y_{trg}^{p,2}, \cdots, y_{trg}^{p,k}]$ and argument label set $Y_{arg}^p = [y_{arg}^{p,1}, y_{arg}^{p,2}, \cdots, y_{arg}^{p,L}]$ in Y_p of p-th event sample, we estimate the confidence score for each label $y_{trg}^{p,u} \in Y_{trg}^p$ and $y_{arg}^{p,v} \in Y_{arg}^p$, denoted as $C_{trg}^{p,u}$ and $C_{arg}^{p,v}$. When $C_{trg}^{p,u}$ or $C_{arg}^{p,v}$ is lower than the corresponding threshold S_{trg}^u or S_{arg}^v, the model correct the noisy label $y_{trg}^{p,u}$ or $y_{arg}^{p,v}$ into all labels with high confidence over the threshold:

$$y_{trg}^{p,u} \leftarrow \{l_{trg}^{p,i} | l_{trg}^{p,i} \in \textbf{logits}_{trg}^p, \text{sigmoid}(l_{trg}^{p,i}) > S_{trg}^i\} \tag{5}$$

$$y_{arg}^{p,v} \leftarrow \{l_{arg}^{p,j} | l_{arg}^{p,j} \in \textbf{logits}_{arg}^p, \text{sigmoid}(l_{arg}^{p,j}) > S_{arg}^j\} \tag{6}$$

Since there might be no $l_{trg}^{p,i}$ or $l_{arg}^{p,j}$ that has confidence score higher than the threshold in current p-th event sample. We temporarily remove the latent noisy label $y_{trg}^{p,u}$ or $y_{arg}^{p,v}$ and train a new event extraction model with corrected data in the next iteration.

2.5   Cooperative Global Pointer Network

When encountering extremely noisy labels, the confidence of clean labels could be affected, which brings accumulated error in each iteration due to the sample-selection bias. To alleviate this problem, we further propose the CoGPN. For each token $\mathbf{h}_i$, the span-specific GPN $g_s(\cdot)$ produces the *logits* to predict whether $\mathbf{h}_i$ corresponds to the start or end position of a trigger or argument. The cascade GPN $g_c(\cdot)$ generates the multi-head *logits* of the span and type concurrently, each head of cascade GPN focuses on the span representation of a certain event type or argument role. Specifically, we first use linear layers to obtain the head and tail position representation for the cascade GPN $g_c(\cdot)$ (Eq. 7, Eq. 8) and the span-specific GPN $g_s(\cdot)$ (Eq. 9, Eq. 10).

$$Q_{\alpha,h_\tau} = \mathbf{W}_{Q,h_\tau}\mathbf{H} + \mathbf{b}_{Q,h_\tau} \tag{7}$$

$$K_{\beta,h_\tau} = \mathbf{W}_{K,h_\tau}\mathbf{H} + \mathbf{b}_{K,h_\tau} \tag{8}$$

$$Q_\alpha = \mathbf{W}_Q\mathbf{H} + \mathbf{b}_Q \tag{9}$$

$$K_\beta = \mathbf{W}_K\mathbf{H} + \mathbf{b}_K \tag{10}$$

Here, $\mathbf{H} \in \mathbb{R}^{N \times H}$ denote the hidden embedding of whole event text T_p. $\mathbf{W}_{Q,h_\tau}$, $\mathbf{W}_{K,h_\tau}$, $\mathbf{W}_Q$, $\mathbf{W}_K \in \mathbb{R}^{1 \times H}$, $\mathbf{b}_{Q,h_\tau}$, $\mathbf{b}_{K,h_\tau}$, $\mathbf{b}_Q$ and $\mathbf{b}_K$ are trainable parameters. h_τ is one of the multi-head for particular type τ. Then, the *logits* for span $s[\alpha : \beta]$

corresponding to the position of a trigger or an argument is calculated as Eq. 11. While the *logits* for span $s[\alpha : \beta]$ of certain type τ is computed as Eq. 12:

$$\textbf{logits}_{span}^{\alpha,\beta} = (\mathcal{R}_\alpha Q_\alpha)^\top (\mathcal{R}_\beta K_\beta) = Q_\alpha^\top \mathcal{R}_{\beta-\alpha} K_\beta \tag{11}$$

$$\textbf{logits}_{h_\tau}^{\alpha,\beta} = (\mathcal{R}_\alpha Q_{\alpha,h_\tau})^\top (\mathcal{R}_\beta K_{\beta,h_\tau}) = Q_{\alpha,h_\tau}^\top \mathcal{R}_{\beta-\alpha} K_{\beta,h_\tau} \tag{12}$$

$\mathcal{R}_\alpha$ and $\mathcal{R}_\beta$ are rotary position embedding [10] that inject relative position information to the *logits*.

Different from named entity recognition, the labels of event extraction are relatively sparse. Therefore, we apply Circle loss [10] to alleviate label imbalance.

$$\mathcal{L}_{circle} = \log \left(1 + \sum_{(Q_\alpha, K_\beta) \in \mathcal{P}} e^{-\textbf{logits}^{\alpha,\beta}} \right) + \log \left(1 + \sum_{(Q_\alpha, K_\beta) \in \mathcal{N}} e^{\textbf{logits}^{\alpha,\beta}} \right) \tag{13}$$

Q_α, K_β represent the head and tail indexes of the span $s[\alpha : \beta]$. $\mathcal{P}$ denotes the set of spans with the ground-truth label. $\mathcal{N}$ denotes the set of spans that are not triggers or arguments. $\textbf{logits}^{\alpha,\beta}$ can be *logits* from the $g_s(\cdot)$ or $g_c(\cdot)$.

Algorithm 1. Cooperative Training Algorithm

Input: The training set $\mathcal{D}_{train}$, the span-specific GPN $g_s(\cdot)$, the cascade GPN $g_c(\cdot)$, noise rate γ, max training epoch $\mathcal{E}_{max}$.
Output: The final span-specific GPN $g_s(\cdot)$ and cascade GPN $g_c(\cdot)$.
1: **for** Each epoch $\mathcal{E} \leftarrow 1, 2, \cdots, \mathcal{E}_{max}$ **do**
2:      Shuffle the noisy training set $\mathcal{D}_{train}$ and generate $\mathcal{K}$ mini-batch $\mathcal{B}$
3:      Update the selection rate $\mathcal{R}$: $\mathcal{R} \leftarrow 1 - \frac{\mathcal{E}-1}{\mathcal{E}_{max}}\gamma$
4:      **for** Each mini-batch $\mathcal{B} \leftarrow \mathcal{B}_1, \mathcal{B}_2, \cdots, \mathcal{B}_\mathcal{K}$ **do**
5:          $\mathcal{D}_s \leftarrow \{\mathcal{I}_{sl}|\mathcal{I}_{sl} \in \mathcal{B}, \mathcal{L}_{circle}(g_s, \mathcal{I}_{sl}) \le \mathcal{L}_{circle}(g_s, \mathcal{B})[\mathcal{R}]\}$
6:          $\mathcal{D}_c \leftarrow \{\mathcal{I}_{sl}|\mathcal{I}_{sl} \in \mathcal{B}, \mathcal{L}_{circle}(g_c, \mathcal{I}_{sl}) \le \mathcal{L}_{circle}(g_c, \mathcal{B})[\mathcal{R}]\}$
7:          Update $g_s(\cdot)$ on $\mathcal{D}_c$ based on Eq. 13
8:          Update $g_c(\cdot)$ on $\mathcal{D}_s$ based on Eq. 13
9:      **end for**
10: **end for**
11: **return** $g_s(\cdot)$, $g_c(\cdot)$

During the training phase, the two GPNs mutually treat the small loss event samples as possible clean. Specifically, in each mini-batch $\mathcal{B}$, $g_s(\cdot)$ and $g_c(\cdot)$ select $\mathcal{R}$ percentage of training samples form $\mathcal{B}$. The value of selection rate $\mathcal{R} = 1 - \frac{\mathcal{E}-1}{\mathcal{E}_{max}}\gamma$ depends on the noisy rate γ (inferred via validation set), and linear decreasing along with current epoch $\mathcal{E}$, since deep neural networks initially fit clean samples, and then gradually fit noisy ones. As shown in Algorithm 1, the selection process is elaborated in lines 3 to 6. Then, the selected samples from $g_s(\cdot)$ are transferred to $g_c(\cdot)$ for parameter updates, and the selected samples from $g_c(\cdot)$ are transferred to $g_s(\cdot)$ simultaneously. The parameter updating process is described in lines 7 to 8 of Algorithm 1. In the inference phase, we add the $\textbf{logits}_{span}^{\alpha,\beta}$ to the $\textbf{logits}_{h_\tau}^{\alpha,\beta}$ as the fusion *logits*.

3 Experiments

3.1 Experimental Setup

Benchmark Setting. For evaluation, we apply Tolar to several strong EE methods and CoGPN on ACE 2005 [1] and MAVEN [2] with synthetic noise. In order to approximate real-world noise, we produce challenging synthetic noise instead of random noise. Specifically, for each sample in ACE 2005, we replace the span of triggers and arguments with other verbs or entities within the same sample. Meanwhile, we replace the types of triggers and arguments with other types in the identical dataset. Since MAVEN is an event detection dataset, only the labels of triggers will be replaced. The probability of this replacement is {20%, 40%, 60%, 80%} to simulate the increasing noise rate. Following prior work [11], we keep the splits of training, validation and test set for ACE 2005 and MAVEN. The detailed splits and the statistics of the two datasets are shown in Table 1. In order to optimize models with the validation set, we only generate noisy labels on the training set.

Table 1. Statistics of ACE 2005 and MAVEN.

Datasets	Documents	Sentences	Types	Roles	Samples	Train	Valid	Test
ACE 2005	599	15,789	33	35	5,349	529	30	40
MAVEN	4,480	49,873	168	-	118,732	2,913	710	857

Evaluation Metric. We adopt the criteria defined in previous work [6]: A trigger is correctly classified if its span and event type match those of a gold-standard trigger. An argument is correctly classified if its span and argument role match those of any of the reference argument mentions. The trigger classification task (T-C) and argument classification task (A-C) are evaluated via micro F1 scores.

Baselines. We compare our work to a number of competitive methods constituted by both event extraction models and noisy label learning methods. All event extraction methods use BERT [4] as encoder, except PAIE utilizes BART [5] as the backbone:

- **CRF**: A sequence labeling-based baseline that incorporates BERT and the additional Conditional Random Field (CRF) layer.
- **CasEE** [9]: A pointer network-based joint learning framework that solves overlapping event extraction with cascade decoding.

---

[1] https://catalog.ldc.upenn.edu/LDC2006T06.
[2] https://github.com/THU-KEG/MAVEN-dataset.

- **EEQA** [1]: A MRC-based method that extracts the event arguments in an end-to-end manner.
- **PAIE** [8]: A prompt-based model designed for event argument extraction task. We extend PAIE to event extraction including both trigger and argument classification via a pipeline-based paradigm.
- **S-model** [2]: This method estimates the noise transition matrix via an additional softmax layer that connects the correct labels to the noisy ones.
- **MentorNet** [3]: MentorNet provides a curriculum for student network to focus on the samples where labels are probably correct.
- **JoCoR** [12]: A joint training method that calculates a joint loss with co-regularization for each training example.

3.2   Main Results

Table 2 illustrates the main experimental results. Our Tolar significantly promotes various strong event extraction models. The F1 scores can be improved by 10.2% at most on ACE 2005 and 13.0% at most on MAVEN. Most noisy label learning methods encounter a sharp decline with the increase of noise rate, while Tolar maintains its performance until the noise rate reaches 80%. Meanwhile, Tolar enhances the F1 scores by 8.9% at most on ACE 2005 and 8.2% at most on MAVEN, compared with vanilla event extraction models. Though Tolar decreases a lot when the noise rate is 80%, CoGPN achieves 50.1% and 41.7% F1 scores on ACE 2005 and MAVEN respectively, obtaining further 15% performance gains compared with PAIE (the best performance among all baselines). The results indicate the superiority of CoGPN.

In addition, Table 2 shows that EEQA and PAIE are more stable than CRF and CasEE when dealing with increasing noise. We speculate that the well-designed templates of these methods fully utilize the priori knowledge of PLMs, which guarantees their stability in noisy settings. We also observe that S-model slightly enhances various event extraction models on ACE 2005, while playing a negative role on MAVEN. Considering MAVEN is a fine-grained event detection benchmark with 168 event types, it is quite difficult for S-model to acquire a noise transition matrix. MentorNet achieves competitive performance under low-level noisy circumstances while losing its advantages or even worse than vanilla event extraction models when the noise rate is over 50%. We suppose that sample-selection bias induces performance decline in high-level noisy scenarios.

3.3   Ablation Study

We verify the effectiveness of different components in Tolar + CoGPN by removing each module. (1) **event schema mapping**. We remove the event schema mapping in the first stage. (2) **self-adaptive iteration**. The data from the first stage will be used for training CoGPN directly without self-adaptive iteration. (3) **span-specific GPN**. We drop the span-specific GPN $g_s(\cdot)$ and only train the cascade GPN $g_c(\cdot)$. (4) **cooperative training**. We retain two GPNs without the cooperative training algorithm. To evaluate the performance of each

Table 2. Overall performance (F1%), where T-C and A-C denote trigger classification and argument classification. Results of all the methods are the average of random five times. We highlight the best result and underline the second best. *Only* denotes vanilla event extraction models without any noisy label learning methods.

Datasets		ACE 2005								MAVEN			
Noise Rate (%)		20		40		60		80		20	40	60	80
Task		T-C	A-C	T-C	A-C	T-C	A-C	T-C	A-C	T-C	T-C	T-C	T-C
Only	CRF	71.4	55.4	66.9	51.0	56.3	42.6	34.3	27.1	65.8	60.8	42.3	18.8
	CasEE	71.2	56.1	66.7	51.6	57.1	42.2	35.1	27.4	66.3	61.0	44.2	19.1
	EEQA	70.8	52.1	65.9	49.6	57.9	43.4	37.6	28.2	63.3	59.2	46.7	22.9
	PAIE	72.9	57.6	68.3	54.5	59.6	46.6	38.8	30.9	66.5	62.1	48.1	24.7
S-model +	CRF	71.8	55.2	64.8	48.6	53.2	40.6	33.7	26.3	61.0	57.6	37.5	14.7
	CasEE	71.3	56.3	66.9	50.4	56.8	43.0	32.1	27.0	61.3	58.9	41.8	16.2
	EEQA	71.0	52.4	67.4	51.8	58.5	43.9	39.8	27.6	61.8	57.8	44.2	20.5
	PAIE	73.6	58.3	69.0	55.4	60.1	47.7	40.4	30.1	63.4	60.2	46.6	23.4
MentorNet +	CRF	72.8	56.5	70.3	53.6	54.7	41.4	30.1	25.6	66.5	61.4	36.3	14.6
	CasEE	72.7	56.3	69.9	53.0	56.5	41.6	30.4	25.7	66.8	61.7	39.9	15.5
	EEQA	71.0	52.4	68.2	50.6	60.1	45.8	36.5	27.3	64.1	61.9	46.8	20.4
	PAIE	74.2	58.6	72.4	56.0	60.8	49.4	37.2	29.7	66.8	62.3	47.7	22.0
JoCoR +	CRF	71.5	55.6	67.4	51.5	54.4	40.7	35.8	27.2	65.3	60.9	44.4	19.1
	CasEE	71.4	56.1	66.8	52.2	58.3	40.5	36.6	28.2	66.3	61.4	45.9	19.6
	EEQA	70.9	52.2	65.3	49.8	56.7	45.4	36.9	27.6	64.0	59.4	47.0	21.4
	PAIE	73.4	57.9	72.3	57.4	61.2	47.9	40.6	31.6	65.8	61.9	48.8	24.2
Tolar +	CRF	73.6	56.7	71.4	55.5	63.7	50.8	38.4	30.6	66.5	62.5	49.3	20.4
	CasEE	73.4	56.9	71.3	56.1	65.0	50.6	38.9	30.7	**67.4**	<u>64.2</u>	52.0	22.3
	EEQA	72.3	52.7	71.0	52.4	<u>66.8</u>	52.9	42.6	34.0	64.8	62.1	54.9	25.8
	PAIE	**75.2**	**59.6**	<u>73.0</u>	<u>57.8</u>	66.5	<u>53.5</u>	<u>43.3</u>	<u>35.2</u>	67.1	63.8	<u>55.4</u>	<u>27.2</u>
	CoGPN	<u>74.0</u>	<u>58.8</u>	73.1	58.0	68.4	55.7	58.2	50.1	<u>67.2</u>	65.3	59.6	41.7

module, we divide ACE 2005 into three subsets, namely, I, II, III. I is relatively easy, where only the labels of triggers are replaced with noisy labels. II is a subset where both triggers and arguments can be noisy labels. However, We either replace the span of triggers and arguments with other verbs or entities within the same sample or replace their types with other types in ACE 2005. III is the most challenging, where all labels of triggers and arguments can be noisy.

The Table 3 reveals that: (1) Event schema mapping stage significantly improves the performance in I set and plays a positive role in II set and III set, thanks to the restricted logical rule in Eq. 2. (2) Self-adaptive iteration is vital, the F1 scores will drop by 2.3%–8.6% along with the increasing noise rate. (3) Removing the span-specific GPN will lead to an evident decline on II set and III set. It indicates the superiority of CoGPN that utilize both the span and type features. (4) When the noise rate comes to 60% or more, the F1 scores will decrease dramatically on both II set and III set. It demonstrates that the cooperative training algorithm ensures outstanding resistance to extreme noise.

Table 3. Ablation study on Tolar + CoGPN. We conduct trigger classification task on ACE 2005 (with synthetic noise).

Datasets	I				II				III			
Noise Rate (%)	20	40	60	80	20	40	60	80	20	40	60	80
Tolar + CoGPN	76.6	75.9	73.7	65.7	74.8	73.7	69.5	58.8	70.7	69.2	62.6	53.8
w/o event schema mapping	74.3	72.8	69.6	59.1	74.1	72.5	67.8	55.9	70.1	67.4	61.3	52.6
w/o self-adaptive iteration	74.8	71.3	64.5	59.7	71.5	69.0	60.9	51.3	68.4	65.3	54.5	47.2
w/o span-specific GPN	74.6	73.0	70.3	55.4	72.0	70.7	65.6	50.4	69.5	66.3	59.0	47.5
w/o cooperative training	77.2	74.8	70.4	52.2	75.5	73.2	65.3	45.5	71.2	67.8	58.4	40.9

4   Conclusion

This paper proposes a novel two-stage label rectification framework Tolar for noisy event extraction settings, which is an non-trivial yet practical scene. Tolar deals with explicit label inconsistency via event schema mapping and utilizes the self-adaptive iteration to cope with the latent semantic noise. Experimental results show the promising performance and efficiency of our methods to handle challenging synthetic noise.

References

1. Du, X., Cardie, C.: Event extraction by answering (almost) natural questions. In: EMNLP (2020)
2. Goldberger, J., Ben-Reuven, E.: Training deep neural-networks using a noise adaptation layer. In: ICLR (2017)
3. Jiang, L., Zhou, Z., Leung, T., Li, L.J., Fei-Fei, L.: Mentornet: learning data-driven curriculum for very deep neural networks on corrupted labels. In: ICML (2018)
4. Kenton, J.D.M.W.C., Toutanova, L.K.: BERT: pre-training of deep bidirectional transformers for language understanding. In: NAACL-HLT (2019)
5. Lewis, M., et al.: BART: denoising sequence-to-sequence pre-training for natural language generation, translation, and comprehension. In: ACL (2020)
6. Li, Q., Ji, H., Huang, L.: Joint event extraction via structured prediction with global features. In: ACL (2013)
7. Li, Q., et al.: A compact survey on event extraction: approaches and applications. arXiv preprint arXiv:2107.02126 (2021)
8. Ma, Y., et al.: Prompt for extraction? Paie: prompting argument interaction for event argument extraction. In: ACL (2022)
9. Sheng, J., et al.: CASEE: a joint learning framework with cascade decoding for overlapping event extraction. In: ACL-IJCNLP (2021)
10. Su, J., et al.: Global pointer: novel efficient span-based approach for named entity recognition. arXiv preprint arXiv:2208.03054 (2022)
11. Wang, X., et al.: MAVEN: a massive general domain event detection dataset. In: EMNLP (2020)
12. Wei, H., Feng, L., Chen, X., An, B.: Combating noisy labels by agreement: a joint training method with co-regularization. In: CVPR (2020)

A Unified Visual Prompt Tuning Framework with Mixture-of-Experts for Multimodal Information Extraction

Bo Xu[1], Shizhou Huang[1], Ming Du[1], Hongya Wang[1], Hui Song[1],
Yanghua Xiao[2,3], and Xin Lin[4(✉)]

[1] School of Computer Science and Technology, Donghua University, Shanghai, China
{xubo,duming,hywang,songhui}@dhu.edu.cn, 2202408@mail.dhu.edu.cn
[2] Shanghai Key Laboratory of Data Science, School of Computer Science,
Fudan University, Shanghai, China
shawyh@fudan.edu.cn
[3] Fudan-Aishu Cognitive Intelligence Joint Research Center, Shanghai, China
[4] East China Normal University, Shanghai, China
xlin@cs.ecnu.edu.cn

Abstract. Recently, multimodal information extraction has gained increasing attention in social media understanding, as it helps to accomplish the task of information extraction by adding images as auxiliary information to solve the ambiguity problem caused by insufficient semantic information in short texts. Despite their success, current methods do not take full advantage of the information provided by the diverse representations of images. To address this problem, we propose a novel unified visual prompt tuning framework with Mixture-of-Experts to fuse different types of image representations for multimodal information extraction. Extensive experiments conducted on two different multimodal information extraction tasks demonstrate the effectiveness of our method. The source code can be found at https://github.com/xubodhu/VisualPT-MoE.

Keywords: Multimodal information extraction · Mixture-of-Experts · Prompt learning · Social media

1 Introduction

Recently, multimodal information extraction (MIE) has gained increasing attention in social media understanding, as it helps to accomplish the task of information extraction by adding images as auxiliary information to solve the ambiguity problem caused by insufficient semantic information in short texts. Existing approaches focus on using multimodal interaction mechanisms to enhance the representation of text and images, which have achieved promising results in many multimodal information extraction tasks [16,20].

Despite their success, current approaches do not take full advantage of the information provided by the diverse representations of images. As shown in

© The Author(s), under exclusive license to Springer Nature Switzerland AG 2023
X. Wang et al. (Eds.): DASFAA 2023, LNCS 13945, pp. 544–554, 2023.
https://doi.org/10.1007/978-3-031-30675-4_40

Fig. 1, the current approaches of representing images can be broadly classified into two categories, namely pixel-level representations and semantic representations. The pixel-level representations can be obtained by using a convolutional neural network (CNN) to encode the entire image [7,14,16,19] or salient regions of the image [1,4,18], containing low-level visual information such as texture, shape and contour, presented as feature maps. The semantic representations contain high-level visual information related to the salient objects in the image and are represented as text embeddings, such as object labels, semantic structures, and image caption, which can be obtained using object detection [12], scene graph generation [20], and image caption generation [11], respectively. There is a significant semantic gap between the pixel-level representation and the semantic representation due to the different forms of the two, and the current approaches can only use one of them, which limits the image representation capability.

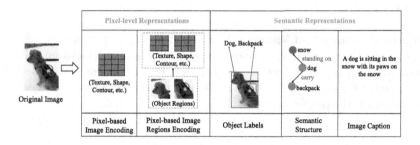

Fig. 1. Examples of different ways for representing an image.

In this paper, we argue that an image with diverse representations is worth a thousand words and that multimodal information extraction tasks can be accomplished more effectively using different types of image representations. However, there are two challenges in using these diverse types of image representations for multimodal information extraction. First, as mentioned above, there is a huge semantic gap between semantic and pixel-level representations. Second, their contribution to the accomplishment of each MIE task is different.

To address these issues, we propose a unified visual prompt tuning framework with mixture-of-experts (VisualPT-MoE) for multimodal information extraction tasks. Specifically, to eliminate the semantic gap between pixel-level representations and semantic representations, we project both types of representations as prompts. To fuse different types of image representations, we consider the weights of different image representations and combine the image representations with different weights by Mixture-of-Experts (MoE). In addition, considering that images are not always helpful for information extraction [13], we additionally introduce one prompt from the pseudo image representation to mitigate the interference caused by unhelpful images.

Our main contributions are summarized as follows:

- Firstly, to the best of our knowledge, we are the first to propose the use of both pixel-level and semantic representations of images with a huge semantic gap in multimodal information extraction tasks.
- Secondly, we propose a novel unified visual prompt tuning framework with mixture-of-experts for multimodal information extraction tasks, which can eliminate the semantic gap between pixel-level representations and semantic representations and consider the contributions of diverse image representations in accomplishing the specific multimodal information extraction task.
- Finally, we have conducted experiments on multimodal named entity recognition and multimodal relation extraction, and the experimental results show that our proposed method can significantly and consistently outperform the state-of-the-art on both tasks.

2   Overview

Our overall framework of VisualPT-MoE is shown in Fig. 2, which contains four main components: (1) diverse image encoders (i.e. image regions encoder, full image encoder, object detection encoder, image caption encoder and pseudo image encoder); (2) visual prompts fusion module; (3) visual-enhanced text encoder; (4) task-specific decoder.

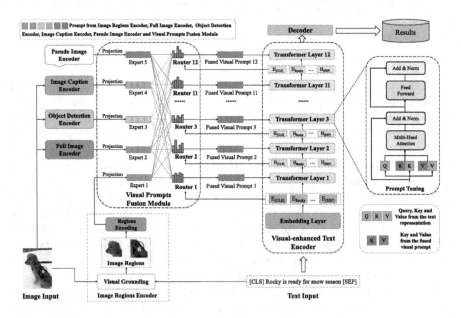

Fig. 2. Overall framework of VisualPT-MoE.

3   Method

3.1   Diverse Image Encoders

As shown in Fig. 2, we use five different image encoders to obtain diverse image representations, namely image regions encoder, full image encoder, object detection encoder, image caption encoder and pseudo image encoder.

For the image regions encoder, we first use a visual grounding model to obtain the top 3 significant regions and feed them into the regions encoding module, which consists of a convolutional neural network and a pooling layer, to obtain the pixel-level representation. Finally, we concatenate the representations of the 3 regions together to get the final pixel-level image regions representation $I_r \in \mathbb{R}^{2048 \times M_r}$, where $M_r = 3 \times 2 \times 2 = 12$.

For the full image encoder, we input the image into ResNet [3] to get the pixel-level full image representation $I_e \in \mathbb{R}^{2048 \times M_e}$, where $M_e = 49$ is the number of regions in the image and 2048 is the dimension of the representation of each region in the image.

For the object detection encoder, we first input the image into the Faster RCNN [9] and select the top 3 objects with the higher object classification scores as object labels. After that, we concatenate the names of the object labels as plain text and obtain the semantic representation of object labels $I_o \in \mathbb{R}^{768 \times M_o}$ by using the Embedding layer of BERT [2], where M_o is the length of the object labels text.

For the image caption encoder, we first input the image into VIT-GPT2[1] to obtain the image caption. Then we also treat the caption as plain text and obtain the semantic representation of the image caption $I_c \in \mathbb{R}^{768 \times M_c}$ by using the Embedding layer of BERT, where M_c is the length of the caption.

Considering the *text-image mismatch* problem [13,14], we propose a trainable pseudo image encoder to generate a pseudo image representation. Specifically, we initialize an Embedding layer of length M_p with 768 dimensions as the pseudo image encoder to obtain the pseudo image representation $I_p \in \mathbb{R}^{768 \times M_p}$, where M_p is the max length of the M_r, M_e, M_o and M_c.

3.2   Visual Prompts Fusion Module

The visual prompts fusion module is used to project the five image representations into different visual prompts (the image representations serve as prompts for the text, which is the same as the visual prefix in [1]) with the same dimensions and combine the different prompts to get the fused visual prompts.

Firstly, we unify the length of the different image representations as M. Then we use the trainable weight matrices $W_r \in \mathbb{R}^{d \times 2048}$, $W_e \in \mathbb{R}^{d \times 2048}$, $W_o \in \mathbb{R}^{d \times 768}$, $W_c \in \mathbb{R}^{d \times 768}$ and $W_p \in \mathbb{R}^{d \times 768}$ to project them as prompts:

$$e_1 = W_r I_r; e_2 = W_e I_e; e_3 = W_o I_o; e_4 = W_c I_c; e_5 = W_p I_p, \qquad (1)$$

---

[1] https://huggingface.co/nlpconnect/vit-gpt2-image-captioning.

where $\{e_1, e_2, e_3, e_4, e_5\} \in \mathbb{R}^{d \times M}$ are the weight matrices, $d = 768$.

Then we input the different prompts into the router module based on the mixture-of-experts (MoE) technique, which consists of a cross-modal interaction layer and a multilayer perceptron (MLP) with one hidden layer. The cross-modal interaction layer is used to obtain the relevance between the text representation and the image representations, which is defined as follows:

$$R^l = \left(\frac{1}{n+2} \sum_{i=1}^{n+2} H_i^{l-1}\right) \odot \left(\frac{1}{E} \sum_{j=1}^{E} e_i\right), \quad 1 \le l \le 12, \tag{2}$$

where $R^l \in \mathbb{R}^{M*d}$ is the l-th layer relevance between the text representation from the $l-1$ Transform layer and the image representations from different experts, E is the number of experts (in this paper, E is 5).

Next, we input R^l into an MLP with an activation function of Softamx to get the corresponding weights of each expert at each layer:

$$\alpha^l = Softmax(W_2^l \ (relu(W_1^l R^l)) + \epsilon), \tag{3}$$

where relu is an activation function, $W_1^l \in \mathbb{R}^{M*d \times M*d}$, $W_2^l \in \mathbb{R}^{E \times M*d}$, $\epsilon \in \mathbb{R}^E$ and $\epsilon \sim \mathcal{N}(0, \frac{1}{E^2})$, $\alpha^l \in \mathbb{R}^E$. Finally, we weight the representations of all experts to get the fused visual prompt of the l-th layer $P^l \in \mathbb{R}^{d \times M}$:

$$P^l = \sum_{i=1}^{E} \alpha_i e_i, \quad \alpha_i \in \alpha^l \tag{4}$$

We use the balance loss [15] to allow the router module to select each expert more equally during training, which regards the largest weight among the experts in every layer as the loss:

$$\mathcal{L}_{balance} = \sum_{l=1}^{12} Max(\alpha^l) \tag{5}$$

3.3 Visual-Enhanced Text Encoder

The visual-enhanced text encoder uses BERT [2] as the backbone. Specifically, we first add a [CLS] token and a [SEP] token at the beginning and end of the text input as $S \in \mathbb{R}^{768 \times (n+2)}$. Then we feed the S to the Embedding layer to get the text representation of the 0-th layer $H^0 \in \mathbb{R}^{768 \times (n+2)}$. Next, we input the text representation of $l-1$ layer H^{l-1} and the fused visual prompt of the l layer P^l into the l-th Transformer layer to obtain the visual-enhanced text representation of l-th layer H^l. The detailed process is as follows.

We first project the H^{l-1} as the "queries", "keys" and "values" of the l-th layer according to the attention mechanism [7]:

$$Q^l = W_Q^l H^{l-1}; K^l = W_K^l H^{l-1}; V^l = W_V^l H^{l-1}, \tag{6}$$

where $\{\boldsymbol{W}_Q^l, \boldsymbol{W}_K^l, \boldsymbol{W}_V^l\} \in \mathbb{R}^{d \times d}$ are the weight matrices. Then, we project $\boldsymbol{P}^l$ as additional "keys" $\boldsymbol{K}_p^l$ and "values" $\boldsymbol{V}_p^l$ to obtain $\boldsymbol{H}^l$:

$$\boldsymbol{K}_p^l = \boldsymbol{\phi}_k^l \boldsymbol{P}^l; \boldsymbol{V}_p^l = \boldsymbol{\phi}_v^l \boldsymbol{P}^l \tag{7}$$

$$\boldsymbol{H}^l = Softmax(\frac{(\boldsymbol{Q}^l)^T [\boldsymbol{K}_p^l; \boldsymbol{K}^l]}{\sqrt{d}}) [\boldsymbol{V}_p^l; \boldsymbol{V}^l]^T \tag{8}$$

where $\{\boldsymbol{\phi}_k^l, \boldsymbol{\phi}_v^l\} \in \mathbb{R}^{d \times d}$ are the weight matrices, $\boldsymbol{H}^l \in \mathbb{R}^{(n+2) \times d}$. After 12 layers of Transformer, we obtain the final representation $\boldsymbol{H}^{12} \in \mathbb{R}^{(n+2) \times d}$.

3.4   Task-Specific Decoder

Finally, depending on the specific multimodal information extraction task, the visual-enhanced text representation is fed to the task-specific decoder to obtain the final results.

For the multimodal named entity recognition (MNER), we regard MNER as a sequence labeling task following [1,7,14,16]. We use Softmax as a decoder to predict the MNER results $p(y_{ner}|\boldsymbol{H}^{12})$ and use cross entropy and balance loss (see Sect. 3.2) to obtain the final loss for this task:

$$\mathcal{L}_{ner} = \frac{1}{B} \sum_j^B \left(-\lambda \log p(y_{ner}^{(j)}|\boldsymbol{H}^{12}) + (1-\lambda)\mathcal{L}_{balance}^{(j)} \right), \tag{9}$$

where B is the batch size, λ is a hyper-parameter.

For the multimodal relation extraction (MRE), we take the representation of [CLS] in $\boldsymbol{H}^{12}$ as input and use the Softmax as the decoder to predict the MRE results $p(y_{re}|\boldsymbol{H}^{12})$ following [1] and use the same loss as MNER:

$$\mathcal{L}_{re} = \frac{1}{B} \sum_j^B \left(-\lambda \log p(y_{re}^{(j)}|\boldsymbol{H}^{12}) + (1-\lambda)\mathcal{L}_{balance}^{(j)} \right) \tag{10}$$

4   Experiment

4.1   Dataset

For the MNER task, we use `Twitter2015` [19] dataset and `Twitter2017` [7] dataset. There are four types of entities: Person (PER), Organization (ORG), Location (LOC) and others (MISC). In total, there are 4,000/1,000/3,357 and 3,373/723/723 sentences in train/development/test set contained in `Twitter2015` and `Twitter2017`, respectively.

For the MRE task, we use `MNRE` dataset [20]. It contains 9,201 sentences and 15,485 entity pairs with 23 types of relations. There are 12,247/1,624/1,614 entity pairs in train/development/test set, respectively.

4.2   Parameter Settings

We conduct all the experiments on NVIDIA GTX 2080 Ti GPUs with PyTorch 1.7.1. The parameters settings of our model are as follows:

- We use $BERT_{base}$ as the backbone of our visual-enhanced text encoder.
- We use ResNet50 for image regions encoding and image encoding. We use Faster-RCNN-ResNet50, VIT-GPT2 for object labels and image caption, respectively.
- We use the grid search in the development set to find the learning rate of visual-enhanced text encoder and visual prompts fusion module within $[1e^{-5}, 7e^{-5}]$, the learning rate of decoder within $[3e^{-4}, 3e^{-2}]$, the batch size within $[8,32]$, and the hyper-parameter λ within $[1e^{-4}, 1e^{-6}]$.

4.3   Baselines

To demonstrate the effectiveness of our model (VisualPT-MoE), we compare our model with several representative text-based models and multimodal models.

For the MNER, we compare four text-based models and ten multimodal models. The text-based NER models are: (1) CNN-BiLSTM-CRF [8], (2) HBiLSTM-CRF [5], (3) BERT-CRF and (4) MRC-MNER-Text [4]. The multimodal NER models are: (1) Ada-CNN-BiLSTM-CRF [19], (2) GVATT-HBiLSTM-CRF [7], (3) Ada-BERT-CRF [16], (4) MAF [14], (5) MRC-MNER [4], (6) UMT [16], (7) UMGF [18], (8) MEGA [20], (9) VisualBERT [6], and (10) HVPNet [1].

For the MRE, we compare two text-based RE models and five multimodal RE models. The text-based RE models are: (1) PCNN [17] and (2) MTB [10].

Table 1. Performance comparison on MNER and MRE. The results from [4] and [1].

Methods	Twitter2015			Twitter2017			MNRE		
	Precision	Recall	F1	Precision	Recall	F1	Precision	Recall	F1
CNN-BiLSTM-CRF	66.24	68.09	67.15	80.00	78.76	79.37	–	–	–
HBiLSTM-CRF	70.32	68.05	69.17	82.69	78.16	80.37	–	–	–
BERT-CRF	69.22	74.59	71.81	83.32	83.57	83.44	–	–	–
MRC-MNER-Text	76.35	69.46	72.24	87.12	84.03	85.55	–	–	–
PCNN	–	–	–	–	–	–	62.85	49.69	55.49
MTB	–	–	–	–	–	–	64.46	57.81	60.86
Ada-CNN-BiLSTM-CRF	72.75	68.74	70.69	84.16	80.24	82.15	–	–	–
GVATT-HBiLSTM-CRF	73.96	67.90	70.80	83.41	80.38	81.87	–	–	–
Ada-BERT-CRF	69.87	74.59	72.15	85.13	83.20	84.10	–	–	–
MAF	71.86	75.10	73.42	86.13	86.38	86.25	–	–	–
MRC-MNER	**78.10**	71.45	74.63	**88.78**	85.00	86.85	–	–	–
UMT	71.67	75.23	73.41	85.28	85.34	85.31	62.93	63.88	63.46
UMGF	74.49	75.21	74.85	86.54	84.50	85.51	64.38	66.23	65.29
MEGA	70.35	74.58	72.35	84.03	84.75	84.39	64.51	68.44	66.41
VisualBERT	68.84	71.39	70.09	84.06	85.39	84.72	57.15	59.48	58.30
HVPNet	73.87	**76.82**	75.32	85.84	87.93	86.87	83.64	80.78	81.85
VisualPT-MoE	76.11	75.16	**75.63**	86.89	**87.96**	**87.42**	**84.81**	**83.75**	**84.28**

The multimodal RE models are: (1) UMT [16], (2) UMGF [18], (3) MEGA [20], (4) VisualBERT [6], and (5) HVPNet [1].

4.4 Performance Comparison

Firstly, we compare VisualPT-MoE and HVPNet with other multimodal methods. From the Table 1, both methods perform better than other multimodal methods, which demonstrates the effectiveness of the prompt tuning framework.

Then, we compare VisualPT-MoE with HVPNet, our method performs better than HVPNet on three datasets. Especially on the MNRE dataset, our model significantly outperforms HVPNet by 2.43 points in the F1 score, which demonstrates that our method of fusing diverse image representations is effective.

4.5 Ablation Study

Table 2. Ablation study of our model. We use E1, E2, E3, E4 and E5 to represent the different experts from image regions encoder, full image encoder, object detection encoder, image caption encoder and pseudo image encoder, respectively.

Methods	Twitter2015			Twitter2017			MNRE		
	Precision	Recall	F1	Precision	Recall	F1	Precision	Recall	F1
References									
BERT-CRF	69.22	74.59	71.81	83.32	83.57	83.44	–	–	–
MTB	–	–	–	–	–	–	64.46	57.81	60.86
HVPNet	73.87	76.82	75.32	85.84	87.93	86.87	83.64	80.78	81.85
HVPNet w/o E1	73.15	75.23	74.18	85.11	87.56	86.32	64.67	60.63	62.58
HVPNet w/o E2	72.71	75.69	74.17	85.63	86.90	86.26	81.66	80.00	80.82
VisualPT-MoE	**76.11**	75.16	**75.63**	**86.89**	**87.96**	**87.42**	**84.81**	**83.75**	**84.28**
Using Single Image Representation									
E1 only	73.15	75.44	74.28	85.66	86.68	86.17	81.96	80.94	81.45
E2 only	72.13	76.42	74.21	85.16	87.49	86.31	61.80	62.19	62.00
E3 only	72.58	76.38	74.43	85.33	87.86	86.58	61.73	62.50	62.11
E4 only	73.10	75.65	74.35	85.45	86.97	86.21	63.71	65.00	64.35
E5 only	72.11	75.67	73.85	84.52	87.27	85.87	65.16	58.44	61.61
Fusing Multiple Image Representations by Averaging Strategy									
E3+E4+E5 w/ AVG	73.80	75.69	74.74	86.55	87.19	86.87	64.77	66.47	65.61
E1+E2+E3+E4 w/ AVG	74.01	75.86	74.93	85.67	88.08	86.86	83.39	82.34	82.86
E2+E3+E4+E5 w/ AVG	73.93	76.21	75.05	86.35	87.56	86.95	64.31	67.21	65.73
Fusing Multiple Image Representations by MoE Strategy									
E3+E4+E5 w/ MoE	74.29	76.04	75.15	86.32	87.79	87.05	66.36	65.11	65.73
E1+E2+E3+E4 w/ MoE	74.32	76.41	75.35	86.54	87.36	86.95	83.83	83.44	83.63
E2+E3+E4+E5 w/ MoE	74.58	**76.61**	75.58	86.56	87.65	87.10	67.60	68.34	67.97

Firstly, we investigate the effectiveness of single image representation on different tasks. As shown in "Using Single Image Representation" in Table 2, different representations have different effects on different tasks. Specifically, in the MNER datasets, the best results are achieved with the model using the representation of object labels (E3). In the MRE dataset, the best results are achieved with the model using the representation of pixel-level image regions (E1).

Secondly, we investigate the effectiveness of using diverse image representations on different tasks. As shown in both the "Fusing Multiple Image Representations by Averaging Strategy" and the "Fusing Multiple Image Representations by MoE Strategy" parts in Table 2, using multiple image representations does perform better than using a single image representation, which illustrates the effectiveness of using diverse image representations.

Finally, we investigate the effectiveness of the MoE fusion strategy in fusing different image representations on different tasks. From the table, we can find that using the same image representations, our MoE fusion strategy always performs better than the averaging fusion strategy, which illustrates the effectiveness of our MoE fusion strategy.

5   Conclusion

In this paper, we propose VisualPT-MoE, a unified visual prompt tuning framework with mixture-of-experts for multimodal information extraction. Specifically, we propose projecting different image representations into prompts to eliminate the semantic gap and introducing the prompt from the pseudo image representation to alleviate the interference caused by mismatched images. And, we use MoE to fuse all prompts to get the final prompt and establish the relationship between the image and the text. We conduct experiments and ablation studies to show that using diverse image representations can effectively improve model performance and MoE can effectively fuse different prompts.

Acknowledgement. This work is supported by the National Key Research and Development Program of China (No. 2021ZD0111004), the National Natural Science Foundation of China (No. 61906035), the Natural Science Foundation of Shanghai (No. 22ZR1402000) and the Science and Technology Commission of Shanghai Municipality Grant (No. 21511100101, 22511105901, 22511105902).

References

1. Chen, X., et al.: Good visual guidance make a better extractor: hierarchical visual prefix for multimodal entity and relation extraction. In: Findings of the Association for Computational Linguistics: NAACL, pp. 1607–1618 (2022)
2. Devlin, J., Chang, M.W., Lee, K., Toutanova, K.: BERT: pre-training of deep bidirectional transformers for language understanding. arXiv:1810.04805 (2018)
3. He, K., Zhang, X., Ren, S., Sun, J.: Deep residual learning for image recognition. In: Proceedings of the IEEE Conference on Computer Vision and Pattern Recognition, pp. 770–778 (2016)

4. Jia, M., et al.: Query prior matters: a MRC framework for multimodal named entity recognition. In: Proceedings of the 30th ACM International Conference on Multimedia (2022)
5. Lample, G., Ballesteros, M., Subramanian, S., Kawakami, K., Dyer, C.: Neural architectures for named entity recognition. In: Proceedings of NAACL-HLT, pp. 260–270 (2016)
6. Li, L.H., Yatskar, M., Yin, D., Hsieh, C.J., Chang, K.W.: VisualBERT: a simple and performant baseline for vision and language. arXiv:1908.03557 (2019)
7. Lu, D., Neves, L., Carvalho, V., Zhang, N., Ji, H.: Visual attention model for name tagging in multimodal social media. In: Proceedings of the 56th Annual Meeting of the Association for Computational Linguistics, pp. 1990–1999 (2018)
8. Ma, X., Hovy, E.: End-to-end sequence labeling via bi-directional LSTM-CNNs-CRF. In: Proceedings of the 54th Annual Meeting of the Association for Computational Linguistics (Volume 1: Long Papers), pp. 1064–1074 (2016)
9. Ren, S., He, K., Girshick, R., Sun, J.: Faster R-CNN: towards real-time object detection with region proposal networks. IEEE Trans. Pattern Anal. Mach. Intell. (2015)
10. Soares, L.B., Fitzgerald, N., Ling, J., Kwiatkowski, T.: Matching the blanks: distributional similarity for relation learning. In: Proceedings of the 57th Annual Meeting of the Association for Computational Linguistics, pp. 2895–2905 (2019)
11. Wang, X., et al.: Ita: image-text alignments for multi-modal named entity recognition. In: Proceedings of the 2022 Conference of the North American Chapter of the Association for Computational Linguistics: Human Language Technologies, pp. 3176–3189 (2022)
12. Wu, Z., Zheng, C., Cai, Y., Chen, J., Leung, H.F., Li, Q.: Multimodal representation with embedded visual guiding objects for named entity recognition in social media posts. In: Proceedings of the 28th ACM International Conference on Multimedia, pp. 1038–1046 (2020)
13. Xu, B., et al.: Different data, different modalities! reinforced data splitting for effective multimodal information extraction from social media posts. In: Proceedings of the 29th International Conference on Computational Linguistics, pp. 1855–1864 (2022)
14. Xu, B., Huang, S., Sha, C., Wang, H.: MAF: a general matching and alignment framework for multimodal named entity recognition. In: Proceedings of the Fifteenth ACM International Conference on Web Search and Data Mining, pp. 1215–1223 (2022)
15. Xue, F., Shi, Z., Wei, F., Lou, Y., Liu, Y., You, Y.: Go wider instead of deeper. In: Proceedings of the AAAI Conference on Artificial Intelligence, vol. 36, pp. 8779–8787 (2022)
16. Yu, J., Jiang, J., Yang, L., Xia, R.: Improving multimodal named entity recognition via entity span detection with unified multimodal transformer. In: Proceedings of the 58th Annual Meeting of the Association for Computational Linguistics, pp. 3342–3352 (2020)
17. Zeng, D., Liu, K., Chen, Y., Zhao, J.: Distant supervision for relation extraction via piecewise convolutional neural networks. In: Proceedings of the 2015 Conference on Empirical Methods in Natural Language Processing, pp. 1753–1762 (2015)
18. Zhang, D., Wei, S., Li, S., Wu, H., Zhu, Q., Zhou, G.: Multi-modal graph fusion for named entity recognition with targeted visual guidance. In: Proceedings of the AAAI Conference on Artificial Intelligence, vol. 35, pp. 14347–14355 (2021)

19. Zhang, Q., Fu, J., Liu, X., Huang, X.: Adaptive co-attention network for named entity recognition in tweets. In: Proceedings of the AAAI Conference on Artificial Intelligence, vol. 32 (2018)
20. Zheng, C., Feng, J., Fu, Z., Cai, Y., Li, Q., Wang, T.: Multimodal relation extraction with efficient graph alignment. In: Proceedings of the 29th ACM International Conference on Multimedia, pp. 5298–5306 (2021)

Is a Single Embedding Sufficient? Resolving Polysemy of Words from the Perspective of Markov Decision Process

Cheng Zhang, Zhi Chen, Dongmei Yan[✉], Jingxu Cao, and Quan Zhang

Tianjin University of Finance and Economics, Tianjin, China
{zhangcheng,ydongmei}@tjufe.edu.cn

Abstract. Polysemy is a widespread linguistic phenomenon. A word can carry a variety of semantic information and is recognized as a definite semantic in a specific context. However, in many models that are good at processing text sequences, words are usually treated as a single embedding, which ignores the polysemy characteristics of words and makes it difficult to model the process of semantic cognition. To fill this gap, this paper models a variety of RNNs from the perspective of the Markov decision process (MDP) and classifies them as Single-state RNN (SRNN), pointing out SRNN deficiencies in polysemy and semantic cognitive processes. A Polymorphic Recurrent Neural Network (PRNN) that can effectively simulate the process of human semantic cognition is proposed by improving the policy function. PRNN selects the specific semantics to be expressed according to the actual context in which the word is located. Extensive experimental results show that PRNNs are superior to RNNs in many natural language processing tasks. The analysis of specific cases shows how PRNNs simulate the process of human language cognition.

Keywords: Natural Language Processing · Recurrent Neural Network · Markov Decision Process · Word Embedding

1 Introduction

Natural Language Understanding (NLU) mainly contains two problems: polysemy and semantic cognition.

- **Polysemy:** This is an inherent property of natural language words. A word can contain multiple meanings at the same time.
- **Semantic Cognition:** This is a problem caused by polysemy. People need to determine the explicit semantics expressed by a word through a specific context.

Table 1 presents examples of polysemy and semantic cognition challenges faced in language modeling and machine translation. In the example of the

© The Author(s), under exclusive license to Springer Nature Switzerland AG 2023
X. Wang et al. (Eds.): DASFAA 2023, LNCS 13945, pp. 555–571, 2023.
https://doi.org/10.1007/978-3-031-30675-4_41

language model, the word apple can express both a kind of fruit and a kind of electronic product, and only under a particular context, can we tell the specific meaning. In the other example, the specific context again determines the word we refer determines what the word we refer to, such as readers or scientists. If polysemy is not considered, the word we may only be considered as similar to other personal pronouns such as they, from a grammatical perspective. In machine translation, a word can express multiple meanings, which means that the same word corresponds to several different translations during the machine translation process. For example, the word 'hot' has many meanings in English, it can describe temperature, it can be used to express taste, it can also describe a person's temper, etc. This leads to the need to combine the context in the translation process to accurately find the words corresponding to the hot meaning. In both cases, the problem of polysemy is obvious.

Table 1. Polysemy in NLP

Polysemy in language model		Polysemy in machine translation		
word	text	word	original-en	translation-cn
apple	I like to eat *apple*	hot	*hot* bath	热水浴
	I like to use *apple*.		Pepper is *hot*	胡椒是辣的
We	*We* are readers, they are riders		*hot* on pop music	热衷于流行音乐
	We are scientists		*hot* music apple	节奏强的音乐

From the above analysis, it is not difficult to see that the problems of polysemy and semantic cognition are widespread. However, in many current natural language understanding models, such as the Vanilla RNN (VRNN) [1], Long Short-Term Memory (LSTM) [2] and Gated Recurrent Unit (GRU) [3], words are usually mapped to a fixed embedding. As shown in the following formula, the word x_i is input into the embedded model g to generate the corresponding embedded e_i.

$$e_i = g(x_i) \tag{1}$$

There are two explanations for this mapping:

- Each embedding represents a single semantics. According to the transmission effect, it is easy to conclude that a word contains only a single semantic, which contradicts the inherent properties of natural language itself and cannot model the actual process of human semantic cognition.
- Each embedding is a fusion of multiple semantics. Although this method solves the problem of polysemy to a certain extent, it is elementary to cause a phenomenon of semantic chaos. For example, the word hot mentioned above has many completely different meanings. When its multiple semantics are combined and fixed embedded representation is used, it is difficult to explain the state of the embedding corresponding to the word. Therefore, it is difficult to determine whether a word accurately represents its meaning in a specific context.

In the cognitive process, humans must recognize a specific meaning based on the current environment or context in which the text is placed [4]. In the NLP task, the deterministic meaning of words in the sequence is determined as the sequence is learned. In the current study, Recurrent Neural Networks (RNN) and their variants are widely used in NLP tasks due to their advantages in long-sequence information processing. Different variants of RNN share a common mathematical formulation. Take RNNs that are good at processing natural language sequences as an example, in the iterative process of an RNN, at each time stamp i, the original input x_i is mapped to an embedding e_i via a lookup table, and then the new context vector h_i is generated based on e_i and the previous context information h_{i-1}. The process repeats until time t. RNN treats each input as a fixed embedding, in which all possible meanings are mixed up in an indistinguishable manner. From a human cognitive point of view, this does not accurately represent the multiple semantics of words. Ideally, a limited number of different meanings should be explicitly encoded so that the same word is represented by different embedding vectors in different contexts.

It can be seen that it is difficult for RNNs to meet the modeling of the two core NLU problems of polysemy and semantic cognition. To fill this gap, we propose to solve this problem from a new cognitive incentive perspective, model the RNN as a Markov decision process (MDP), and establish a mapping relationship between RNN and MDP, in the decision process from multiple different models trained on a large corpus generate embeddings that tend to have different semantics and select the embedding that best fits the context, so that a word can be mapped to multiple embeddings. As far as we know, we are the first to adopt this view. Our analysis shows that the current RNN can be classified as Single-state RNN (SRNN), which does not consider the ambiguity of input objects. From the MDP point of view, we find that RNNs with different types of recursive units only care about the state transition function and ignore the policy function in MDP. This insight provides a new way to build new models that overcome the limitations of current RNN structures. Specifically, we propose a Polymorphic Recurrent Neural Network (PRNN) and analyze it from the perspective of MDP. PRNN uses multiple pre-training models to build multiple embeddings (states) for the input object at each time and develops a strategy function in the process of analyzing from the MDP perspective, which can determine which embedding of the object should be used in the current state based on the object's environment. Then PRNN realizes the multi-semantic expression of words through multiple embedding. Finally, PRNNs are evaluated in text classification and machine translation tasks. The results show that PRNN outperforms existing RNNs in terms of effectiveness.

In summary, the main contributions of this paper are as follows:

1. A solution to explicitly modeling polysemy is given through the way that one word corresponds to multiple embeddings.
2. A solution for explicitly modeling semantic cognitive phenomena is given. Use the RNNsden layer vector of the recurrent neural network to select the semantics of the word in the current context.

3. A generalization framework for improving recurrent neural networks is given, different types of recurrent neural networks are modeled from the perspective of the Markov decision process, and their shortcomings in decision pre-trained found.

2   Related Work

Word embeddings are a by-product of modeling NLP tasks [5]. It is widely used because it can capture a certain degree of semantic and grammatical relationship between words. There has been much research on word embeddings to improve the accuracy of NLP tasks. In this chapter, based on the number of embedding used or generated by the model, a summary of related research will be made from two aspects: Single Embedding Model and Multiple Embedding Model.

- **Single Embedding Models:** Traditionally the word embedding-based models, such as Skip-Gram, CBOW model [5], Glove [6], and fastText [7], mostly assume that each word is represented by a unique vector. Using fixed embedding to represent words makes it difficult to accurately represent the multiple meanings of words in different contexts. Various recent models such as Bert [8]and ELMo [9] provided a context-sensitive way to build word vectors, but all possible meanings of a word are still mixed up within a fixed vector. Furthermore, various self-attention models such as Transformer [10], GPT [11] and Transformer-XL [12], are based on the multi-head mechanism, which can be roughly regarded as the treatment of polysemy. Although these attention models can better solve the long-distance dependency problem of sequences, they cannot handle continuous decision-making processes and ignore the information contained in some local features, which is crucial when dealing with a series of text sentences. Moreover, even though the attention-based model considers more factors than previous models, it still has a unique embedded representation of words. From the perspective of human cognition, it is still difficult to express the multiple semantics of words. Pittaras proposed a model of semantic frequency vector weighting for word embedding [13]. Although this way of enhancing semantics improves the accuracy of common semantic expressions of words, it reversely weakens the expression of other semantics of words, which limits the possibility of multiple semantic expressions of words.
- **Multiple Embedding Models:** Reisinger and Mooney introduced a multi-prototype vector space model (VSM), where word sense discrimination is applied by clustering context [14]. In 2014, Tian, Fei and others introduced a skip-gram model based on probability expansion, which can learn multiple prototype word vectors according to a priori probability and generate a fixed number of word meanings for each word using a clustering algorithm with parameters [15]. Neelakantan et al. Proposed a skip-gram polysemy expansion model in 2014, which is the first model that can automatically learn the number of word meanings [16]. In the training process, if the similarity between the current context and the existing word meaning vector is lower

than a certain threshold, the model will generate a new word meaning vector. However, these models lack cognitive motivation, thus difficult to fully utilize the context information to determine which state an object should exhibit in a particular environment in line with the human cognitive process. Subba and Kumari propose a heterogeneous stacking ensemble-based sentiment analysis framework using multiple word embeddings [17]. In the whole emotion analysis task, the process of text processing is completed by multiple models in parallel. On the whole, multiple embedded words represent the same word. Although there are multiple embeddings as a whole to represent the same word, in specific applications, each embedding corresponds to a separate classifier, which essentially still uses a fixed vector to represent the word.

The perception of words and the decision to choose which meaning to choose at each moment in a given context (action selection) will influence subsequent decisions, as well as the processing of subsequent textual tasks. Using multiple pre-trained models to generate embeddings, and combining the ability of RNN to process sequences with the ability of MDP to make continuous decisions, we propose a Polymorphic Recurrent Neural Network (PRNN) to deal with word polysemous phenomena.

3   Modeling RNN as Markov Decision Process

In this section, we give a brief introduction to MDP and formulate it with a definite policy function. We then show how a general structure of RNN is modeled as MDP. In particular, we analytically show that RNN using different types of recurrent units can be classified as Single-state Recurrent Neural Networks (SRNN) from the MDP perspective.

3.1   Markov Decision Process

A Markov decision process (MDP) [18] is formulated as a 5-tuple (S, A, P_a, R_a, γ) as shown in Table 2. When an action a_t is applied to the state s_{t-1}, only the next state s_t can be obtained. As a result, we can get $P_a(s_{t-1}, a_t, s_t) = 1$. Accordingly,

Table 2. Notations of Markov decision process.

S:	a set of states
A:	a set of actions
$P_a(s_{t-1}, a_t, s_t)$:	the probability that an action a_t in state s_{t-1} at time $t-1$ will lead to state s_t
$R_a(s_{t-1}, a_t, s_t)$:	a reward function, which is received after the transition from state s_{t-1} to state s_t, due to the action a_t.
γ:	a discount factor, which represents the difference of importance between the future and present rewards.

we can define a state transition function $T(s_{t-1}, a_t) = s_t$, applying a_t on s_{t-1} leads to the next state s_t through the function T.

The core problem of MDP is to find a policy function $\pi : s_{t-1} \Rightarrow a_t$, which specifies the action a_t that the decision maker will choose to apply on state s_{t-1}. The process of finding a policy, whether using policy iteration [19], value iteration [18] or reinforcement learning [20], depends on the reward function $R_a(s_{t-1}, a_t, s_t)$ and discount factor γ.

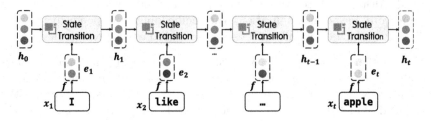

Fig. 1. A general architecture of RNN.

3.2   MDP with a Definite Policy Function

In the traditional definition of MDP, a reward function R_a is needed to obtain the reward after an action takes place in a certain state at a time, but the role of R_a varies in different situations.

- **MDP without a definite policy function.** The reward function is in place to help effectively find an optimal policy, e.g., for searching for the best path in the classic grid world [21]. The policy will eventually converge regardless of whether to use the policy iteration or value iteration to obtain the optimal policy.
- **MDP with a definite policy function.** The reward function will have no effect except for evaluating the quality of a policy function. With the given policy function, we choose a certain action at a state in a greedy way.

In other words, when the policy function is unknown, we rely on the reward function to solve the optimal policy. When a specific policy is found, a greedy search algorithm can be employed to select the best action, and the resulting state-action sequence in an episode is a fixed sequence. In addition, whether a decision process is MDP depends on whether the Markov property is satisfied in this process. That is, the action taken by the agent at a moment only depends on the current state, and does not depend on the previous states.

Given the above clarifications, MDP with a definite policy function can be formulated as a 4-tuple (S, A, T, π).

3.3   Single-State Recurrent Neural Network

A general network structure of RNN is shown in Fig. 1. The processing of a sequence $x_1, x_2, ..., x_t$ using RNN can be expressed by the following formula.

$$e_i = f(x_i)$$
$$h_i = T(h_{i-1}, e_i) \tag{2}$$

The process of mapping the original one-hot input x_i to an embedding e_i is represented by the function f. Subsequently, RNN process the sequence information and constructs new context information h_i according to the input e_i and previous context information h_{i-1}. At last, the context information sequence $h_1, h_2, ..., h_t$ or the latest hidden vector h_t is used for classification or regression.

According to the above description, RNN can be naturally modeled as MDP with a definite policy function, as shown in the following formula.

$$S = \{h_0, h_1, h_2, ..., h_{t-1}, h_t\}$$
$$A = \{e_1, e_2, e_3, ..., e_{t-1}, e_t\}$$
$$h_i = T(h_{i-1}, e_i^*) \tag{3}$$
$$e_i^* = \pi_{h_{i-1}}(e_i) = e_i$$

The state set S consists of each step of the hidden layer h_i of the sequence. The action set A consists of the embedding e_i that is generated by the input x_i at each step. At each time step the i, the input x_i only generates a single embedding (action) e_i due to the limitation of this structure, so only the action e_i can be selected as the best action e_i^*. The state transfer function T at time i takes the state h_{i-1} and action e_i^*, and results in the next state h_i. The core problem of MDP is to find a policy π specifying the action $\pi_{h_{i-1}}$ that the decision maker will choose to act on the state h_{i-1}. In RNN, when the input x_i is given, the corresponding optimal action e_i^* is fixed and determined according to f.

Whether SRNN or PRNN, the reward function of MDP is realized by the loss function of the whole network in the actual calculation process. The greater the calculated loss, the lower the reward of action a_i. In the training process, with the continuous optimization of the model, the loss of the loss function will be lower and lower, which also means that the total reward for all actions is increasing, thus making every step of the action get the best choice. Since the reward function is the same and replaced by the loss function of the neural network, it will not be repeated in the description of the PRNN model below.

Typical RNN architectures such as VRNN, LSTM and GRU can be viewed as different types of transfer functions (recurrent units) $h_t = T(h_{t-1}, e_t)$ to improve the performance of RNNs. They are summarized as follows:

VRNN implements $h_t = T(h_{t-1}, e_t)$ as:

$$h_t = sigmoid(Ue_t + Wh_{t-1}) \tag{4}$$

LSTM implements $h_t = T(h_{t-1}, e_t)$ by introducing 3 gates:

$$f_t = \sigma(W_f \cdot [h_{t-1}, e_t] + b_f)$$
$$i_t = \sigma(W_i \cdot [h_{t-1}, e_t] + b_i)$$
$$\widetilde{C}_t = tanh(W_c \cdot [h_{t-1}, e_t] + b_c)$$
$$C_t = f_t * C_{t-1} + i_t * \widetilde{C}_t \qquad (5)$$
$$o_t = \sigma(W_o[h_{t-1}, e_t] + b_o)$$
$$h_t = o_t * tanh(C_t)$$

GRU implements $h_t = T(h_{t-1}, e_t)$ by introducing 2 gates, namely reset gate and update gate:

$$z_t = \sigma(W_z \cdot [h_{t-1}, e_t])$$
$$r_t = \sigma(W_r \cdot [h_{t-1}, e_t])$$
$$\widetilde{h}_t = tanh(W \cdot [r_t * h_{t-1}, x_t]) \qquad (6)$$
$$h_t = (1 - z_t) * h_{t-1} + z_t)\widetilde{h}_t$$

The above analysis shows that RNN has been modeled as MDP. It can be seen that existing RNNs pay more attention to the transfer function T, and adopt an over-simplified policy function π. While well-designed gates help RNNs remember more information or forget certain things, the policy function is incapable of reflecting the polymorphism of information objects. Indeed, RNNs with different types of recurrent units can be uniformly classified as Single-state Recurrent Neural Networks (SRNN), in the sense that they treat an information object as having only a single fixed state. In reality, an object can have multiple meanings (states), and only in a certain context, the object shows a specific state. Therefore, we propose a Polymorphic Recurrent Neural Network (PRNN) to capture the polymorphism of information objects, with an effective policy function to determine which specific state an object is in based on the context.

4   Polymorphic Recurrent Neural Network

The proposed PRNN is a generalized form of SRNN. In order to better solve the ambiguity of words, it is necessary to express multiple meanings of words at the same time, and then select the most reasonable embedding as the final action for the current input according to the context information. We will first model PRNN as MDP and then explain its detailed structure.

4.1   Modeling PRNN as MDP

The modeling of PRNN as MDP is shown in the following formula.

$$
\begin{aligned}
S &= \{h_0, h_1, h_2, ..., h_{t-1}, h_t\} \\
A &= \{e_1^{d_1}, e_1^{d_2}, ..., e_1^{d_N}\}, \{e_2^{d_1}, e_2^{d_2}, ..., e_2^{d_N}\}, \\
&\quad ..., \{e_t^{d_1}, e_t^{d_2}, ..., e_t^{d_N}\} \\
h_i &= T(h_{i-1}, e_i^*) \\
e_i^* &= \pi_{h_{i-1}}(e_i^{d_1}, ..., e_i^{d_N})
\end{aligned}
\tag{7}
$$

The state set S consists of each step h_i of the hidden layer of the sequence. Each original input object x_i has N states (embeddings) $\{e_i^1, e_i^2, ..., e_i^N\}$ to represent the possible meanings of x_i, as compared to SRNN which uses a fixed e_i. Using a fixed e_i to represent the possible existence of a variety of meanings may cause several problems. For instance, the word apple can mean either an electronic product or a kind of fruit, but when mixing these two kinds of meanings in e_i, it is difficult to decide whether it should be closer to banana or dell. In PRNN, in order to further distinguish the different meanings of a limited number of input objects, the embeddings $\{e_i^1, e_i^2, ..., e_i^N\}$ are differentiated to generate new embeddings $\{e_i^{d_1}, e_i^{d_2}, ..., e_i^{d_N}\}$. The action set A consists of the differentiated embeddings of all input objects. This construction method of A is analogic to discretizing the different states (meanings) that an input x_i may exist in, hence avoiding the deficiencies of SRNN in the action set. Furthermore, after the construction of the action set is completed, it is crucial to select the action according to the current environment. That is, according to the context, the object state e_i^* that best conforms to the current context should be selected as the action. The policy function π of the PRNN takes the context h_{i-1} and the differentiated embeddings $\{e_i^{d_1}, e_i^{d_2}, ..., e_i^{d_N}\}$ of all possible meanings of the input object x_i

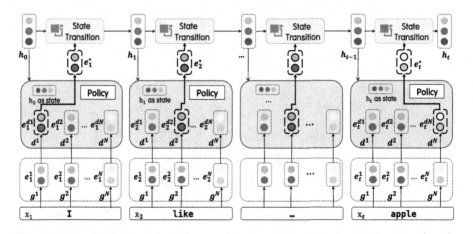

Fig. 2. Generalized form of PRNN when processing sequence information.

as parameters, and selects the optimal meaning e_i^* that most conforms to the current environment, as shown in the policy section of the Eq. 7. PRNN makes better use of context information than SRNN's policy function as shown in Eq. 3. The selected e_i^* is the optimal meaning that the current input object x_i best corresponds to the current environment h_{i-1}, and is also the action to be applied to h_{i-1}. Through the state transition function T, we can obtain the context of the next timestamp. The state transition function can be any SRNN's transfer function for a specific task.

This wraps up the modeling of PRNN as MDP. The advantages of PRNN over RNN can be seen by comparing Eq. 3 with Eq. 7. When PRNN does not entail different meanings of x_i(i.e. $N = 1$), it degenerates to an RNN with the same state transition function. Therefore, PRNN is a generalization of RNN from an MDP point of view.

4.2   Network Structure and Policy Function

Figure 2 shows the network structure of PRNN. Among them, $g^{1 \sim N}$ are different external pre-training models trained based on large corpora. External pre-trained models tend to represent different semantics due to their different architectures and their training corpora. Through multiple mutually independent functions $g^{1 \sim N}$, the corresponding multiple embeddings $\{e_i^1, e_i^2, ..., e_i^N\}$ are generated. These embeddings represent the various semantics that a word can appear in. Then, the embeddings generated by $g^{1 \sim N}$ are differentiated by multiplying different diagonal matrices. The independence of $g^{1 \sim N}$ and the differentiation process of the embeddings guarantee the discretization of the different meanings of the input x_i at each moment. The policy function π_{h_i} is then called for to select the most suitable meaning from the discrete meanings according to the context information h_i. Considering that the attention mechanism [22] has achieved great success, we adopt an attention mechanism in our policy function. Our policy function π_{h_i} is described as follows:

$$
e_i^{d_1}, e_i^{d_2}, ..., e_i^{d_N} = [d^1 e_i^1, d^2 e_i^2, ..., d^N e_i^N]
$$

$$
\pi_{h_i}(e_i^{d_1}, e_i^{d_2}, ..., e_i^{d_N}) = \sum_{j=1}^{N} \sigma(c_i)^j e_i^{d_j}
$$

$$
\sigma(x_i) = \frac{e^{x_i}}{\sum_k e^{x_k}}
$$

$$
c_i = [h_{i-1} e_i^{d_1}, h_{i-1} e_i^{d_2}, ..., h_{i-1} e_i^{d_N}]
$$

(8)

c_i consists of the dot product of each embedding e_i^j and the context h_{i-1}, indicating the importance of the different embeddings in the current context h_{i-1}. It is then converted to discrete probability values via the softmax function σ, and combined with the embedding e_i^j in the form of a weighted sum. The policy function π determines which embedding needs to select or pay more attention to. PRNN improves RNNs' structure by approving the policy function π, and establishing links between the context information and various states that may exist in the original input. This policy function allows PRNN to capture multiple states for an object and use context information to determine its current state.

5 Experiments

We evaluate our models on text classification tasks (AGNews[1] and IMDB [23]) and machine translation tasks (Multi30k [24]). For AGNews and IMDB, we inherited the initially provided training and validation sets and used accuracy as the performance measure. For Multi30k, we also use the default training and validation sets. Bilingual Evaluation Understudy (BLEU) [25] is selected as the evaluation metric to evaluate the effect of PRNN on Multi30k. BLEU is the most commonly used evaluation metric for machine translation tasks.

5.1 Experimental Parameters

To evaluate the PRNN and eliminate unnecessary interference, we simplify the entire network structure as much as possible when dealing with specific tasks, to better highlight the role of key elements of the PRNN. In the experiment, we chose the random embedding[2], GloVe and Gensim as three independent functions to realize the conversion of the original input to various states (embedding). Random embedding is generated from experimental datasets, while GloVe and Gensim models are externally trained pretrained functions, trained on different corpora, to provide embeddings with complementary semantics for words.

Algorithm 1. N-embeddings PRNN.

Input: Sequence $\{x_1, x_2, ..., x_t\}$, Hidden h_0
for $i = 1$ to t do
    for $j = 1$ to n do
        $e_{i}^{j} = g^{j}(x_i)$
        $e_{i}^{dj} = e_{i}^{j} d^{j}$
    end for
    $e_{i}^{*} = \pi_{h_{t-1}}(e_{i}^{d_1}, e_{i}^{d_2}, ..., e_{i}^{d_N})$
    $h_i = T(h_{i-1}, e_{i}^{*})$
    $o_i = \sigma(W_o h_i + b_o)$
end for
if dataset is Multi30k then
    Output: Sequence $\{o_1, o_2, ..., o_t\}$
else if dataset is AGNews or IMDB then
    Output: o_t
end if

---

[1] http://groups.di.unipi.it/~gulli/AG_corpus_of_news_articles.html.
[2] Implemented by PyTorch embedding.

At the same time, the embedding size and N embedding of different parameters are set for comparative experiments. All calculations are shown in Algorithm 1. The input sequence is defined as $x_1, x_2, ..., x_t$. Subscript t indicates the sequence length. We define N as the number of states (embeddings), and call it N-Embeddings. Embedding size indicates the dimension of each embedding. Note that when N-embedding is 1, the PRNN degenerates to an RNN using the same recursive unit. Therefore, we use the 1-Embedding as the baseline in each set of comparison experiments. For Multi30k, since the experimental results of the GloVe and Gensim models fluctuate greatly, we choose the random embedding as the baseline of 1-Embedding. LSTM and GRU are variants of the RNN model. We will extend the PRNN application to LSTM and GRU and conduct experiments to further verify our theoretical correctness.

5.2   Experimental Results and Analysis

The experimental results of AGNews and IMDB are shown in Table 3, when the N-Embedding increases, PRNN obtains better results than the baseline. In each set of comparative experiments, the PRNN accuracy with 3-Embeddings is higher than the baselines where the 3 generative embedding functions are used alone. The accuracy of most 2-Embedding is also higher than the baseline, and the rest is basically the same as the baseline. The experimental performance of

Table 3. Experimental results on AGNews and IMDB dataset. The bold part indicates that PRNN outperforms the baseline, and * indicates the best result in a set of runs.

Dataset	Model	Dimension	Accuracy for N-embedding					
			1	1	1	2	2	3
			Random-g^1	GloVe-g^2	Gensim-g^3	g^1&g^2	g^1&g^3	g^1&g^2&g^3
AGNews	RNN	50	89.13%	87.03%	89.82%	**90.3%**	89.74%	**90.9%***
		100	89.38%	89.28%	90.67%	**91.06%**	**90.75%**	**91.14%***
		200	90.23%	90.52%	90.56%	**90.72%**	90.49%	**90.73%***
	LSTM	50	89.9%	88.13%	90.77%	**90.69%**	**91.25%***	91.14%
		100	90.26%	90.76%	90.69%	**91.09%**	90.7%	**91.26%***
		200	90.48%	90.93%	90.85%	**90.9%**	90.73%	**91.22%***
	GRU	50	90.07%	91.34%	91.12%	90.81%	91.17%	**91.39%***
		100	90.34%	90.81%	90.69%	**91.11%**	**91.07%**	**91.15%***
		200	90.48%	90.92%	90.86%	90.77%	90.84%	**90.93%***
IMDB	RNN	50	77.49%	78.57%	81.98%	**82.96%**	81.19%	**83.05%***
		100	81.87%	82.3%	78.8%	81.36%	**82.34%**	**82.88%***
		200	82.3%	83.16%	77.07%	82.76%	82.63%	**83.27%***
	LSTM	50	80.8%	82.35%	82.15%	**83.86%**	**82.89%**	**84.24%***
		100	79.59%	81.1%	79.11%	**82.49%**	82.63%	**83.94%***
		200	83.3%	83.62%	84.42%	84.07%	83.63%	**84.43%***
	GRU	50	80.35%	82.95%	82.31%	**83.73%***	82.3%	82.53%
		100	78.96%	80.82%	80.67%	79.7%	80.15%	**82.84%***
		200	83.02%	83.36%	83.44%	**83.53%***	80.36%	83.44%

Table 4. Experimental results on Mult30k. The bold part indicates that PRNN outperforms the baseline, and * indicates the best result in a set of runs.

Dataset	model	Dimension	BLEU for N-embedding		
			1	2	3
Mult30k	RNN	100	27.69	**29.19***	**27.71**
		200	28.9	**29.33**	**30.48***
	LSTM	100	27.01	**27.64***	26.46
		200	27.54	**30***	**28.63**
	GRU	100	27.05	**29.28***	**27.87**
		200	27.22	**29.47**	**29.75***

Multi30k is the same as the previous datasets, and adding N-Embedding can improve the BLEU score, as shown in Table 4. Experimental results show that just one embedding is not enough, we do need more embeddings to express an object, and more contextual information to determine which embedding (state) to adopt.

In the calculation process, PRNN regards multiple embeddings as actions and selects the e_i^* that best matches the current semantics through the strategy function π combined with the context. We select a specific sentence from the ag dataset with 200-dimensional embeddings to observe the entire selection process, as shown in Fig. 3. As you can see, we provide embeddings generated by three different functions as actions. In the process of policy, due to the continuous change of the context, the scores of different embeddings in the current environment are also constantly changing. The score is actually a quantification of the performance of the embedding for the current environment, and the embedding with the highest score can better represent the current semantics. Compared to RNNs that use a single function to generate embeddings, PRNNs offer a better alternative. At the same time, different semantic environments make the embeddings of the same word selected by the policy function different. On the basis of the previous sentence, we choose the word 'it' to demonstrate, in the

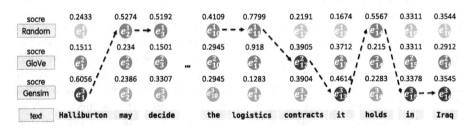

Fig. 3. In a complete sentence of AGNews, the PRNN strategy function continuously selects the optimal embedding process among the three embedding functions according to the current environment.

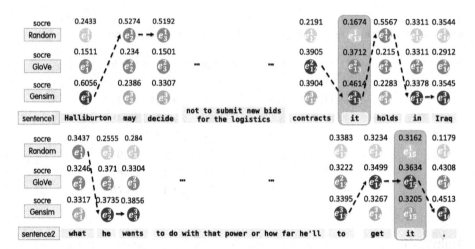

Fig. 4. The process of selecting the optimal embedding for the word 'it' in two sentences by the PRNN policy function among the three embedding functions.

first sentence 'it' refers to Halliburton, and in the second sentence 'it' refers to power. The process of policy selection in Sentence 1 and Sentence 2 in the case of 200-dimensional embeddings is shown in Fig. 4. For sentence 1, we tend to choose Gensim to generate the embedding to represent 'it', while for sentence 2, we tend to choose GloVe as the embedding to represent 'it'. Therefore, in the two sentences, 'it' has two different embedding representations. This is the specific embodiment of PRNN's advantages over RNN. From the perspective of MDP, we provide three kinds of embedding actions for the same word 'it' with different semantics, while RNN only has a single embedding with multiple semantics. Therefore, in the task, 'it' under the PRNN model can be represented by the corresponding embedding under different semantic environments, while RNN has no choice. This also fully proves that PRNN is the most appropriate embedding strategy based on the current context.

5.3 Stability of PRNN

In the process of model evaluation, the stability of the model is a very important evaluation index. The effect of the neural network model usually depends on the initialization parameters of the model. Some models can only obtain good experimental results under the carefully designed random number seed, which leads to the model being difficult to be widely used in more data sets and tasks. We choose 10 random seeds to experiment with RNN and 3-Embedded PRNN to observe the stability of the model in the case of using 100-dimensional embedding for IMDB.

As shown in Fig. 5, the accuracy of PRNN under all seeds is higher than the baseline, and it can be clearly seen that the fluctuation of the results is smaller. The mean and variance of the results of the 10 experiments were calculated, as

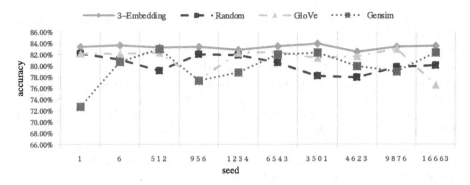

Fig. 5. With 100-dimensional embeddings for IMDB, ten experiments with different seeds are performed on a 3-Embedding PRNN and three separate embedding models RNN.

shown in Table 5. It can be clearly seen that the accuracy of the PRNN model is nearly 3% points higher than the baseline model, and the variance is only one of 10 of the limit model. This remarkable improvement in model stability and accuracy makes our model very reliable.

Table 5. Mean and variance of the 3-Embedding PRNN and three separate embedding model RNNs in ten experiments with different seeds, using 100-dimensional embeddings for IMDB.

	PRNN	Random	GloVe	Gensim
Mean	83.33%	80.28%	81.14%	79.83%
Variance	0.0016	0.0244	0.0499	0.097

6   Conclusions and Future Work

Through in-depth analysis of the objective difficulties existing in current natural language understanding, we model multiple variants of Recurrent Neural Network (RNN) as Single-state Recurrent Neural Network (SRNN) from the perspective of the Markov Decision Process and find that current SRNN play a deficiency role in polysemy and semantic cognition. The Polymorphic Recurrent Neural Network (PRNN) model in this paper is proposed by improving the number of word embeddings and the policy function. The PRNN model explicitly discretizes the various semantics contained in a word and can observe how the model performs the process of semantic cognitive decision-making in different contexts. This innovation provides a new perspective for further improvement of RNN and provides a progress ladder for revealing the black-box phenomenon of deep learning models in natural language processing. In addition, the PRNN model also provides a new way to fuse multiple pre-trained word vectors within the same model efficiently. PRNN has proved its effectiveness and stability through extensive experimental results.

In the future, we will apply PRNN to more datasets and tasks to verify its generality. Maintaining the number of embeddings for each word dynamically and expanding the policy function through reinforcement learning to obtain a more elegant network structure will improve PRNN in our following work. Then reduce the running time of the PRNN model and improve the effectiveness of PRNN, making its improvement on the basic model more obvious.

Acknowledgements. This work is funded in part by the Tianjin Municipal Education Commission of China (Grant No. 2021SK103).

References

1. Mikolov, T., Karafiát, M., Burget, L., Cernocký, J., Khudanpur, S.: Recurrent neural network based language model. ACM (2010)
2. Gers, A.F., Schmidhuber, J., Cummins, F.: Learning to forget: continual prediction with LSTM. Neural Comput. (2000)
3. Cho, K., et al.: Learning phrase representations using RNN encoder-decoder for statistical machine translation. Comput. Sci. (2014)
4. Dent, E.B.: The observation, inquiry, and measurement challenges surfaced by complexity theory. Social Science Electronic Publishing (2005)
5. Mikolov, T., Chen, K., Corrado, G., Dean, J.: Efficient estimation of word representations in vector space. Comput. Sci. (2013)
6. Pennington, J., Socher, R., Manning, C.: Glove: global vectors for word representation. In: Conference on Empirical Methods in Natural Language Processing (2014)
7. Bojanowski, P., Grave, E., Joulin, A., Mikolov, T.: Enriching word vectors with subword information. Trans. Assoc. Comput. Linguist. **5**, 135–146 (2017). https://doi.org/10.1162/tacl_a_00051
8. Devlin, J., Chang, M.W., Lee, K., Toutanova, K.: BERT: pre-training of deep bidirectional transformers for language understanding. CoRR (2018)
9. Peters, M., Neumann, M., Iyyer, M., Gardner, M., Zettlemoyer, L.: Deep contextualized word representations (2018)
10. Vaswani, A., et al.: Attention is all you need. arXiv (2017)
11. Brown, T.B., Mann, B., Ryder, N., Subbiah, M., Amodei, D.: Language models are few-shot learners. CoRR (2020)
12. Dai, Z., Yang, Z., Yang, Y., Carbonell, J., Salakhutdinov, R.: Transformer-XL: attentive language models beyond a fixed-length context (2019)
13. Pittaras, N., Giannakopoulos, G., Papadakis, G., Karkaletsis, V.: Text classification with semantically enriched word embeddings. Nat. Lang. Eng. **27**(4), 391–425 (2021)
14. Reisinger, J., Mooney, R.: A mixture model with sharing for lexical semantics. In: Conference on Empirical Methods in Natural Language Processing (2010)
15. Tian, F., et al.: A probabilistic model for learning multi-prototype word embeddings. Choose... (2014)
16. Neelakantan, A., Shankar, J., Passos, A., Mccallum, A.: Efficient non-parametric estimation of multiple embeddings per word in vector space. In: Empirical Methods in Natural Language Processing (2015)

17. Subba, B., Kumari, S.: A heterogeneous stacking ensemble based sentiment analysis framework using multiple word embeddings. Comput. Intell. **38**(2), 530–559 (2022). https://doi.org/10.1111/coin.12478
18. Bellman, R.: A Markovian decision process. Indiana Univ. Math. J. **6**(4), 15 (1957)
19. Kakade, S.M.: A natural policy gradient. In: Advances in Neural Information Processing Systems 14 [Neural Information Processing Systems: Natural and Synthetic, NIPS 2001, 3–8 December 2001, Vancouver, British Columbia, Canada] (2001)
20. Sutton, R.S., Barto, A.G.: Reinforcement learning: an introduction. IEEE Trans. Neural Netw. **9**(5), 1054–1054 (1998). https://doi.org/10.1109/TNN.1998.712192
21. Crook, P.A., Hayes, G.: Learning in a state of confusion: perceptual aliasing in grid world navigation (2003)
22. Bahdanau, D., Cho, K., Bengio, Y.: Neural machine translation by jointly learning to align and translate. Comput. Sci. (2014)
23. Pang, B., Lee, L.: Seeing stars: exploiting class relationships for sentiment categorization with respect to rating scales. arXiv (2005)
24. Elliott, D., Frank, S., Sima'An, K., Specia, L.: Multi30k: multilingual English-German image descriptions. In: Proceedings of the 5th Workshop on Vision and Language (2016)
25. Papineni, K., Roukos, S., Ward, T., Zhu, W.J.: Bleu: a method for automatic evaluation of machine translation (2002)

Unleashing Pre-trained Masked Language Model Knowledge for Label Signal Guided Event Detection

Mengnan Xiao, Ruifang He$^{(\boxtimes)}$, Junwei Zhang, Jinsong Ma, and Haodong Zhao

Tianjin Key Laboratory of Cognitive Computing and Application,
College of Intelligence and Computing, Tianjin University, Tianjin 300350, China
{mnxiao,rfhe,junwei,jsma,2021244138}@tju.edu.cn

Abstract. Event detection (ED) aims to recognize triggers and their types in sentences. Previous work employs distantly supervised methods or pre-trained language models to generate sentences containing events to alleviate data scarcity. Further, determining the spans and types of triggers is complex and may have deviations. In this paper, we propose to unleash Pre-trained Masked Language Model (PMLM) knowledge for label signal guided ED by a novel trigger augmentation. We directly generate triggers by leveraging the rich knowledge of PMLM through masking triggers. However, these newly replaced triggers may not correspond to the label of the masked trigger. To control such trigger augmentation noises, we design a label signal guided classification mechanism with event type-subtype guidance. To ensure the quality of generated triggers, a semantic consistency mechanism is introduced. Experimental results on the ACE2005 and FewEvent show the effectiveness of our proposed approach.

Keywords: Trigger augmentation · Label signal guided event classification · Sentence semantic consistency

1 Introduction

As a challenging subtask of event extraction, event detection (ED) aims to identify and classify triggers. As per the general ACE2005 annotation guideline: an event type contains one or more event subtypes. A sentence example is as follows: *"He lost an **election** to a dead man."* Here, *"**election**"* triggers a *"Personnel: Elect"* event where *"Personnel"* is the event type and *"Elect"* is the event subtype.

So far, many methods have been proposed, extending from feature-based approaches to advanced deep learning methods [8,11]. Although previous methods achieve success in many aspects, data scarcity is a growing challenge that can not be ignored as mainstream models become bigger and bigger. The lack of training data seriously hinders the performance of existing methods, which are under the supervised learning paradigm and eager for the large training dataset. To alleviate the problem, Liu et al. [6] propose a multilingual approach

© The Author(s), under exclusive license to Springer Nature Switzerland AG 2023
X. Wang et al. (Eds.): DASFAA 2023, LNCS 13945, pp. 572–581, 2023.
https://doi.org/10.1007/978-3-031-30675-4_42

by machine translation to bootstrap the source data. However, ensuring the mapping between tokens and labels across languages is complex and may have deviations. There also have been some efforts to enlarge training data for ED models by exploiting distantly supervised techniques [1,11,12]. Moreover, some work [8,13] leverages pre-trained language models to automatically generate training data for models. The common in these methods is to generate sentences containing events. However, there are two main weaknesses: 1) there are noises in the generated sentences and need extra mechanisms (such as knowledge distillation) to control; 2) ED is a token-level classification task, determining the spans and subtypes of triggers is difficult, and may have deviations.

To address the aforementioned problems, we explore directly generating proper triggers without changing the context, which can not only weaken noises but also reuse the labels of triggers in the original sentence. Inspired by Dai et al. [2], we propose a novel trigger augmentation approach by leveraging the existing pre-trained masked language model (PMLM) to automatically generate triggers. By replacing original triggers with generated ones, we can obtain candidate sentences with different triggers. Specially, we aim to fine-tune a PMLM on the existing training dataset by masking triggers so it can generate alternative triggers and corresponding scores. Yet trigger augmentation might still involve noises due to the complexity of natural language and the large vocabulary of PMLM. So we also design a **label signal guided classification mechanism** with **event type-subtype guidance**, including event type classification (ETC) and event subtype classification (ESC). The results of ETC serve as signals to guide ESC. Through the medium of ETC, we can calculate multiple times and finally select the maximum value of the product of ETC and ESC as the final result. In this manner, though the result of ETC is not correct, the final result may also be right. We also design a **sentence semantic consistency mechanism** that makes the semantics between the candidate and original sentence as similar as possible to ensure the quality of the generated triggers. With the right generated triggers, the semantics of sentences are naturally similar. Our contributions in this paper can be summarized as follows:

- Propose a novel trigger augmentation approach (called PMLMLS) for ED to directly generate alternative triggers by leveraging the knowledge of PMLM;
- Build a label signal guided classification mechanism with event type-subtype guidance for ED which helps control noises in trigger augmentation;
- Employ a sentence semantic consistency mechanism to ensure the quality of generated triggers;
- Experimental results on the ACE2005 and FewEvent demonstrate the effectiveness of our method and achieve state-of-the-art performance.

2   Methodology

Figure 1 shows the proposed PMLMLS model, which leverages the knowledge of the pre-trained masked language model (PMLM) to improve ED. The model consists of two stages: (1) **Trigger Augmentation**: to employ PMLM to generate

574   M. Xiao et al.

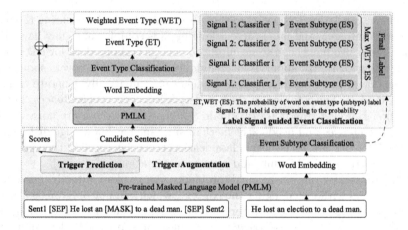

Fig. 1. The overview of our proposed PMLMLS.

alternative triggers and corresponding scores; (2) **Label Signal Guided Event Classification**: to utilize label signal to guide event type-subtype classification which helps control noises in (1).

2.1 Trigger Augmentation

As presented in Sec. 1, our motivation is to obtain proper candidate triggers without changing the context. The overall strategy is to mask the trigger with a special token and leverage PMLM to generate the candidates. Formally, assume that $x = [x_1, \ldots, x_i, \ldots, x_n]$ is a sentence of n tokens with only one trigger located at x_i, the masked sentence x' would have the form: $x' = [x_1, \ldots, [MASK], \ldots, x_n]$ where $[MASK]$ is the special token to symbolize the trigger. x' is then employed as the input of PMLM to obtain the representation h_{mask} of $[MASK]$:

$$h_{\text{mask}} = \text{PMLM}(x') \in R^d \tag{1}$$

where d denotes the dimension of the hidden layer in PMLM. Then we utilize PMLM head (i.e., LMhead) to obtain top k triggers $T = [t_1, \ldots, t_i, \ldots, t_k]$ and corresponding scores $s = [s_1, \ldots, s_i, \ldots, s_k]$:

$$(T, s) = \text{LMhead}(h_{\text{mask}}) \tag{2}$$

where LMhead is a pre-trained two-layer non-linear classifier with layer normalization and the output dimension is the size of the vocabulary of PMLM. The score s_i is the probability of LMhead on the corresponding candidate trigger t_i. Note that the sum of s is not equal to 1 and then we normalize s:

$$s_i = \frac{s_i}{\sum_{j=1}^{k} s_j} \in R \tag{3}$$

Before we fill T into $[MASK]$ and obtain k candidate sentences, we preliminarily judge the quality of T through $x_i \in T$ or not. If $x_i \notin T$, then the quality of T is unreliable and we will abandon it.

Considering that the trigger is usually the core word (verb or noun) of the sentence, there would be many choices in the scope of the vocabulary of PMLM. Sometimes it even generates candidates that are appropriate in the context but completely irrelevant to the original word with high scores (e.g. the example in the introduction). To help PMLM generate suitable candidates that are related to the original trigger, we add the previous and next sentences of x as a prompt to x'. The enriched x' would have the form: $x' = \left[\boldsymbol{Sent1}, [SEP], x_1, \ldots, [MASK], \ldots, x_n, [SEP], \boldsymbol{Sent2} \right]$ where $[SEP]$ is the special token to identify the span of sentences.

2.2   Label Signal Guided Event Classification

To control noises in trigger augmentation, we design a **label signal guided classification mechanism** with **event type-subtype guidance**.

Label Signal Guided Classification Mechanism: Considering that an event type consists of one or more event subtypes, we design a label signal guided classification mechanism, first event type classification (ETC) then event subtype classification (ESC). Formally, as per the pre-defined event schema, we have an event type set C and an event subtype set $\mathcal{Y}$. The overall goal is to predict all events in gold set $\mathcal{E}_x$ of the sentence x. We aim to maximize the joint likelihood of training data $\mathcal{D}$:

$$\prod_{x \in \mathcal{D}} \left[\prod_{(t,c,y) \in \mathcal{E}_x} p((t,c,y) \mid x) \right] = \prod_{x \in \mathcal{D}} \left[\prod_{t \in \mathcal{T}_x} \left[p(t \mid x) p(c \mid x, t) p(y \mid x, t, c) \right] \right]$$

(4)

where $\mathcal{T}_x$ denotes the triggers set occurring in x, t denotes the trigger in $\mathcal{T}_x$, c denotes the event type of t, and y denotes the event subtype of t. The result of ETC is leveraged as a signal to guide ESC. It is a tree with a layer height of 3, the root node is the trigger, and the second and third layers are event types and subtypes respectively. The children of the second layer node are the event subtypes contained in the event type, and the weights of edges are probabilities of ETC and ESC classifiers. When classification, the trigger selects a path to the leaf node in a depth-first search (DFS) based on the edge weight.

To control noises in the trigger augmentation, we do not only utilize the label corresponding to the maximum value of the ETC prediction result as a signal but the top m results as signals. When starting from each node, instead of choosing one path, we choose m paths as per the signals. Finally, the maximum value of the product of all edge weights on the search path is employed as the final result. In this manner, though the result of ETC is not correct, the final result may also be right. We can obtain the global optimal solution to a certain extent through multiple searches.

Event Type-Subtype Guidance Classification Network: As per the aforementioned mechanism, we build an event type-subtype guidance classification network containing ETC and ESC. The thought of ETC and ESC are similar while ETC is trained on candidate sentences and obtain event type results, ESC is trained on original sentences and obtain event subtype results as per the results of ETC. Assume that $\hat{X}$ is the candidate sentences obtained by the original sentence x after Sec. 2.1. Then we utilize the PMLM to obtain the hidden presentation of tokens in $\hat{X}$ and x:

$$\hat{H} = \text{PMLM}(\hat{X}) \qquad H = \text{PMLM}(x) \qquad (5)$$

where the PMLM is the one in Sec. 2.1, they share weights, $\hat{H}$ is the embedding of tokens in candidate sentences, and H is the embedding of tokens in the original sentence. Then $\hat{H}$ is used as the input of ETC to obtain the event type result $\hat{C}$:

$$\hat{C} = \text{ETC}(\hat{H}) \qquad (6)$$

where ETC is a two-layer non-linear classifier with dropout and layer normalization. In addition, we obtain the score of candidate sentence s by Eq. 2 and 3. Therefore we obtain the weighted probability over event type $\hat{p}$ by the product of $\hat{C}$ and s normalized by softmax($\cdot$):

$$\hat{p} = \text{softmax}\left(\sum_{i=1}^{z} s_i \hat{C}_i \right) \qquad (7)$$

Then the top m probability v and the corresponding event type label id ℓ of $\hat{p}$ consist of signals to guide ESC:

$$y = \max\left\{ v_i \cdot \text{softmax}(\text{ESC}_{\ell_i}(H)) | i = 1, \ldots, m \right\} \qquad (8)$$

where ESC contains L classifiers and each is a two-layer non-linear classifier with dropout and layer normalization. L denotes the number of event types, ESC_{ℓ_i} denotes choosing the ℓ_i-th classifier as per ℓ_i, $v_i \cdot \text{softmax}(\text{ESC}_{\ell_i}(H))$ denotes the product of probabilities, and y denotes the final event subtype result of tokens in x.

2.3   Training

This section describes the training of our model. In addition, to further make sure the quality of generated triggers, sentence semantic consistency is introduced.

Sentence Semantic Consistency: In Sec. 2.1, we preliminarily judge the quality of the candidate triggers by $x_i \in T$ or not. But for $x \in T \setminus \{x_i\}$, the quality can not be guaranteed. Considering the only difference between candidate and original sentences is triggers. Therefore, we try to make the semantics between the candidate and the original sentence as similar as possible. In this work, we utilize the mean squared error between $\hat{H}_{\text{cls}}$ and H_{cls} as a supervised target for the loss function:

$$\mathcal{L}_s = \frac{1}{|H_{\text{cls}}|} \sum_{i=1}^{|H_{\text{cls}}|} \left(H_{\text{cls},i} - \hat{H}_{\text{cls},i} \right)^2 \tag{9}$$

where $\hat{H}_{\text{cls}}$ and H_{cls} denote the semantics of candidate and original sentences respectively, $|H_{\text{cls}}|$ denotes the dimension of H_{cls}, and $H_{\text{cls},i}$ denotes the i-th element of H_{cls}.

Joint training: Finally, to train PMLMLS, the following combined loss function is employed:

$$\mathcal{L} = \mathcal{L}_{\text{ETC}} + \alpha \mathcal{L}_{\text{ESC}} + \beta \mathcal{L}_s \tag{10}$$

where $\mathcal{L}_{\text{ETC}}$ employs cross-entropy loss between the real and predicted event type labels, $\mathcal{L}_{\text{ESC}}$ employs the same loss on the real and predicted event subtype labels, α and β are the trade-off parameters.

3 Experiments

In this section, we explore the following questions:

Q1: Can PMLMLS better utilize the knowledge of PMLM to boost the performance of ED? Q2: Is every module essential? Q3: How do hyper-parameters affect the performance of PMLMLS?

3.1 Settings

Datasets: We conduct experiments on the event detection benchmark ACE2005, which has 599 English annotated documents and 8 event types total of 33 event subtypes. The same split as the previous work [8,11] is used.

In addition, we also conduct experiments on another benchmark FewEvent [3], which contains 70,852 instances for 19 event types graded into 100 event subtypes in total. To validate the performance of PMLMLS in the data scarcity scenario, we randomly select 30 instances for each event subtype in each trial. In a trial, the proportion of instances for each event subtype in the training, development, and test set are 70%, 10%, and 20% respectively.

For evaluation, we employ standard Precision (P), Recall (R), and the F_1 score following the previous work [8,11]. And we employ the average of 5 experimental results as the final result.

Baselines: To verify PMLMLS, we compare our method with models based on the aforementioned two strategies and other SOTA methods.

For ACE2005, we compare PMLMLS with several state-of-the-art models in three categories: (1) Multi-label classification model: **DMCNN** [1], **MLBiNet** [7], and **ED3C** [9]; (2) QA-based model: **RCEE_ER** [5]; (3) Data augmentation model: **GMLATT** [6], **DMBERT** [12], **DRMM** [10], **EKD** [11], and **GPTEDOT** [8]. For FewEvent, we compare PMLMLS with the following models: **PLMEE** [13], **DMBERT** [12], and **EEQA** [4].

Table 1. Overall performance (a) and ablation study (b) on the ACE2005 test set. In (a), * indicates models based on PLMs. In (b), all the models in this table utilize RoBERTa-base. (The same as below)

Model	P	R	F_1
DMCNN [1]	79.7	69.6	74.3
GMLATT [6]	78.9	66.9	72.4
DMBERT* [12]	77.6	71.8	74.6
RCEE_ER* [5]	75.6	74.2	74.9
DRMM* [10]	77.9	74.8	76.3
EKD* [11]	79.1	78.0	78.5
MLBiNet [7]	74.7	83.0	78.6
ED3C* [9]	75.1	83.5	79.1
GPTEDOT* [8]	82.3	76.3	79.2
PMLMLS (ours)*	76.6	82.8	**79.6**

(a) Overall performance

Model	P	R	F_1
ED	74.3	73.0	73.6
LSED	74.8	75.2	75.0
PMLMED^{-all}	73.4	79.0	76.1
PMLMED^{-cp}	76.2	78.0	77.1
PMLMED^{-ssc}	75.8	78.9	77.3
PMLMED	76.0	80.7	78.3
PMLMLS^{-all}	74.0	80.2	77.0
PMLMLS^{-cp}	76.8	80.5	78.6
PMLMLS^{-ssc}	76.6	80.5	78.5
PMLMLS	76.6	**82.8**	**79.6**

(b) Ablation study

Implementations: We choose RoBERTa-base as the pre-trained masked language model and experiment with MindSpore. The hidden state and dropout of ETC and ESC are set to 768 and 0.1 respectively. The trade-off parameters α and β are set to 0.6 and 0.2 respectively. The learning rate is set to 1e−5 for the Adam optimizer and the batch size of 4 is employed during training. k is set to 4 denotes trigger augmentation will generate 4 alternative triggers. m is set to 2 denotes ESC will compute 2 times as per the top 2 probability of ETC. The epoch is set to 50 and the early stop is set to 8.

3.2 Overall Performance

Table 1 (a) presents the performance of all baselines and PMLMLS on the ACE2005 test set. For *Q1*, we can observe that:

1) By fully leveraging the rich knowledge of the pre-trained masked language model and label signal guided classification, PMLMLS outperforms all baselines with simpler architecture. Our method, only using a shared PMLM, surpasses GPTEDOT [8] which utilizes two PLMs and achieves competitive performance with the new SOTA. Furthermore, compared with other models that need the extra complicated module to control noise (e.g. knowledge distillation), PMLMLS only utilizes a two-stage classification based on label signal.

2) By directly generating alternative triggers from the pre-trained masked language model, PMLMLS achieves better results compared to other data argumentation models. Our method improves F_1 by 1.0% and 0.4% over the SOTA EKD [11] based on distant supervision and GPTEDOT [8] based on GPT-2 respectively. Compared with generating sentences containing events, directly generating alternative triggers can weaken noise and reuse the label of the original sentence.

Table 2. Overall performance and ablation study on the FewEvent test set.

Model	P	R	F_1
PLMEE* [13]	60.1	58.2	59.1
DMBERT* [12]	60.3	58.4	59.3
EEQA* [4]	61.2	59.3	60.2
PMLMLS (ours)*	**62.0**	**60.3**	**61.1**

(a) Overall performance

Model	P	R	F_1
ED	60.2	53.3	56.5
LSED	60.7	54.1	57.2
PMLMED	57.4	59.6	58.5
PMLMLS	**62.0**	**60.3**	**61.1**

(b) Ablation study

Table 2 (a) presents the performance of PMLMLS on the FewEvent test set. We can see that: our proposed model has an improvement compared with all baselines, thus further confirming the advantages of PMLMLS for ED.

3.3 Ablation Study

To verify *Q2*, for ACE2005, first, for the importance of label signal, we take the following baselines: (1) ED: the base model based on the PMLM without trigger augmentation and label signal guided classification; (2) LSED: based on (1), LSED adds label signal guided classification. Second, based on the trigger augmentation, three components need to be evaluated, the previous and next sentences prompt (context prompt, cp), label signal guided classification (ls), and sentence semantic consistency (ssc) respectively. There are a total of 8 combinations, one of which is PMLMLS. Therefore, we choose the remaining 7 combinations as degradation experiments. They are (3) PMLMED^{-all}: the baseline model based on trigger augmentation, without cp, ls, and ssc; (4) PMLMED^{-cp}: based on (3), add ssc; (5) PMLMED^{-ssc}: based on (3), add cp; (6) PMLMED: based on (3), add cp and ssc; (7) PMLMLS^{-all}: the baseline model based on trigger augmentation and label signal guided classification, without cp and ssc; (8) PMLMLS^{-cp}: based on (7), add ssc; (9) PMLMLS^{-ssc}: based on (7), add cp.

For FewEvent, there is no concept of the document, and the training data is in the form of sentences, so there is no context prompt. Degradation experiments include: (1) ED: the baseline only utilizes RoBERTa-base; (2) LSED: based on (1), add label signal guided classification; (3) PMLMED, based on (1), add trigger augmentation. From Table 1 (b), we can observe that:

1) The trigger augmentation, cp, ssc, and ls are necessary for PMLMLS to achieve the highest performance. Remove any component, performance will decrease. In particular, the F_1 score decreases by 1.0%, 1.1%, 1.3%, and 4.6% when removing cp, ssc, ls, and trigger augmentation. Note that when removing trigger augmentation, cp and ssc will also remove.

2) Label signal guided classification is helpful at any time. There are 10 degradation experiments, and we can divide them into 5 groups: a) ED and LSED; b) PMLMED^{-all} and PMLMLS^{-all}; c) PMLMED^{-cp} and PMLMLS^{-cp}; d) PMLMED^{-ssc} and PMLMLS^{-ssc}; e) PMLMED and PMLMLS. The difference

Table 3. Performance of PMLMLS on the ACE2005 test set with different k and m.

	k						m		
	1	2	3	4	5	6	1	2	3
P	74.6	75.5	76.9	76.6	74.9	75.6	75.8	76.6	76.8
R	75.0	78.4	79.5	**82.7**	80.9	74.5	81.6	82.7	82.9
F_1	74.8	76.9	78.2	**79.6**	77.8	75.1	78.6	79.6	79.7

between the two experiments in each group is whether to perform label signal guided classification. We can see that the effect of using label signal guided classification in each set of experiments is better than not using and the average improvement is 1.3%.

3) Adding additional training data is an effective method for data scarcity. Yet it will inevitably introduce noises. The key is to control noises while increasing the training data. Compared with ED, PMLMED^{-all} adds additional training data without extra mechanisms to control noises, we can see that the F_1 score increases, but at the cost of a decrease in P. When additional mechanisms (cp, ssc, or both) are added to control noise, the scores of P, R, and F_1 increase over ED. In addition, from Table 2 (b), we can see that: Compared with ACE2005, the effect of each module is better in the scarcer FewEvent.

3.4   Parameter Analysis

To illustrate $Q3$, in addition to the hyperparameters of the neural network, two additional hyperparameters need to be set. They are the number of alternative triggers generated for the masked trigger k and the top m results of ETC consist of signals to guide ESC.

To study the importance of k, we experiment with different k on the ACE2005. From the left of Table 3, the highest performance of the proposed model is achieved when k is 4 which denotes trigger augmentation generates 4 alternative triggers for the masked trigger. More specially, when $k \leq 3$, as k increases, P, R, and F_1 increase. We can see the knowledge of the pre-trained masked language model can predict proper and various triggers, alleviate data scarcity and improve performance. When k equals 4, P drops slightly compared to k equals 3. Though achieving the highest, we can see it is a bit noisy but more profitable. When $k \geq 5$, noise dominates and affects the performance of the ED model.

To provide more insights into the influence of label signal guided classification, we conduct experiments with different m on the ACE2005. From the right of Table 3, we can see that with the increment of m, the performance of PMLMLS improves. That is because PMLMLS makes multiple judgments when making the final result, weakening the interference of noise. Note that using label signal guided classification will affect the parallelism and need more time since we need to select the corresponding classifier in ESC as per the results of ETC. Even though the F_1 score when $m = 3$ is higher than when $m = 2$, however, the improvement is slight. So we select $m = 2$ as the final result to balance F_1 and time costing.

4   Conclusions

In this paper, we propose a novel trigger augmentation method (called PMLMLS) for ED leveraging the rich knowledge of the pre-trained masked language model. Unlike other data augmentation methods that generate sentences containing events, PMLMLS directly generates alternative triggers by masking triggers to weaken noises from the source. We also design a label signal guided classification mechanism with event type-subtype guidance to alleviate the noises in trigger augmentation. Sentence semantic consistency is also introduced to ensure the quality of generated triggers. Comprehensive experimental results on the ACE2005 and FewEvent demonstrate the effectiveness of the proposed method.

Acknowledgments. Our work is supported by the National Natural Science Foundation of China (61976154) and CAAI-Huawei MindSpore Open Fund.

References

1. Chen, Y., Liu, S., Zhang, X., Liu, K., Zhao, J.: Automatically labeled data generation for large scale event extraction. In: ACL, pp. 409–419 (2017)
2. Dai, H., Song, Y., Wang, H.: Ultra-fine entity typing with weak supervision from a masked language model. In: ACL, pp. 1790–1799 (2021)
3. Deng, S., Zhang, N., Kang, J., Zhang, Y., Zhang, W., Chen, H.: Meta-learning with dynamic-memory-based prototypical network for few-shot event detection. In: WSDM, pp. 151–159 (2020)
4. Du, X., Cardie, C.: Event extraction by answering (almost) natural questions. In: EMNLP, pp. 671–683 (2020)
5. Liu, J., Chen, Y., Liu, K., Bi, W., Liu, X.: Event extraction as machine reading comprehension. In: EMNLP, pp. 1641–1651 (2020)
6. Liu, J., Chen, Y., Liu, K., Zhao, J.: Event detection via gated multilingual attention mechanism. In: AAAI, pp. 4865–4872 (2018)
7. Lou, D., Liao, Z., Deng, S., Zhang, N., Chen, H.: MLBiNet: A cross-sentence collective event detection network. In: ACL, pp. 4829–4839 (2021)
8. Pouran, Ben Veyseh, A., Lai, V., Dernoncourt, F., Nguyen, T.H.: unleash GPT-2 power for event detection. In: ACL, pp. 6271–6282 (2021)
9. Veyseh, P.B.A., Nguyen, M.V., Ngo Trung, N., Min, B., Nguyen, T.H.: Modeling document-level context for event detection via important context selection. In: EMNLP, pp. 5403–5413 (2021)
10. Tong, M., et al.: Image enhanced event detection in news articles. In: AAAI, pp. 9040–9047 (2020)
11. Tong, M., et al.: Improving event detection via open-domain trigger knowledge. In: ACL, pp. 5887–5897 (2020)
12. Wang, X., Han, X., Liu, Z., Sun, M., Li, P.: Adversarial training for weakly supervised event detection. In: NAACL:HLT, pp. 998–1008 (2019)
13. Yang, S., Feng, D., Qiao, L., Kan, Z., Li, D.: Exploring pre-trained language models for event extraction and generation. In: ACL, pp. 5284–5294 (2019)

Wukong-CMNER: A Large-Scale Chinese Multimodal NER Dataset with Images Modality

Xigang Bao, Shouhui Wang, Pengnian Qi, and Biao Qin[✉]

School of Information, Renmin University of China, Beijing, China
{baoxigang,wsh_inf,pengnianqi,qinbiao}@ruc.edu.cn

Abstract. So far, Multimodal Named Entity Recognition (MNER) has been performed almost exclusively on English corpora. Chinese phrases are not naturally segmented, making Chinese NER more challenging; nonetheless, Chinese MNER needs to be paid more attention. Thus, we first construct Wukong-CMNER, a multimodal NER dataset for the Chinese corpus that includes images and text. There are 55,423 annotated image-text pairs in our corpus. Based on this dataset, we propose a lexicon-based prompting visual clue extraction (LPE) module to capture certain entity-related visual clues from the image. We further introduce a novel cross-modal alignment (CA) module to make the representations of the two modalities more consistent through contrastive learning. Through extensive experiments, we observe that: (1) Discernible performance boosts as we move from unimodal to multimodal, verifying the necessity of integrating visual clues into Chinese NER. (2) Cross-modal alignment module further improves the performance of the model. (3) Our two modules decouple from the subsequent predicting process, which enables a plug-and-play framework to enhance Chinese NER models for Chinese MNER task. LPE and CA achieve state-of-the-art (SOTA) results on Wukong-CMNER when combined with W2NER [11], demonstrating its effectiveness.

Keywords: Multimodal Named Entity Recognition · Prompt Learning · Contrastive Learning

1 Introduction

Traditional Named Entity Recognition (NER) refers to the task of identifying and classifying the noun phrases that predefined semantic categories in unstructured texts, such as organizations, location and person names, etc. [20]. NER plays a fundamental role in many natural language processing (NLP) tasks including question answering, relation extraction and entity linking. Recently, billions of multimodal posts containing image-text pairs are shared in social media platforms. To extract relevant information from social media, multimodal named entity recognition (MNER) has attracted much attention. MNER extends

© The Author(s), under exclusive license to Springer Nature Switzerland AG 2023
X. Wang et al. (Eds.): DASFAA 2023, LNCS 13945, pp. 582–596, 2023.
https://doi.org/10.1007/978-3-031-30675-4_43

traditional text-based NER by incorporating visual clues from images as an aid, because visual context can help resolve ambiguous words in the texts.

So far, most of the research [18, 21, 22, 36] on multimodal named entity recognition is for the English corpus while Chinese multimodal named entity recognition lacks sufficient attention. There are billions of users on Chinese social media platforms and huge amounts of multimodal data are generated every day. Therefore, investigation on MNER for Chinese corpora is necessary. Compared with Chinese NER, Chinese MNER is aimed at social media data with short and rough context and colloquial language. So how to combine the attached multimodal data to enhance the text is very critical.

Sui [20] proposed a large-scale Chinese Multimodal NER Dataset with speech clues, which introduces acoustic modality as the supplement of the textual modality. In this paper, we also focus on Chinese multimodal NER. However, different from the above works, we pursue to couple Chinese textual modality with the visual modality. To promote the research of Chinese multimodal NER, we construct a Chinese multimodal NER dataset with images modality. To the best of our knowledge, this is the first Chinese multimodal named entity recognition dataset with image modality. Figure 1 shows some samples within our dataset.

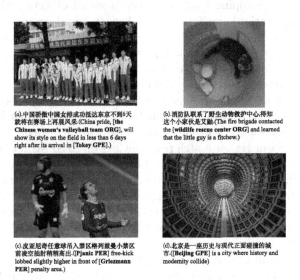

(a).中国骄傲中国女排成功抵达东京不到6天就将在赛场上再展风采.(China pride, [the **Chinese women's volleyball team ORG**], will show its style on the field in less than 6 days right after its arrival in [**Tokey GPE**].)

(b).消防队联系了野生动物救护中心,得知这个小家伙是艾鼬.(The fire brigade contacted the [**wildlife rescue center ORG**] and learned that the little guy is a fitchew.)

(c).皮亚尼奇任意球吊入禁区格列兹曼小禁区前凌空抽射稍稍高出.([**Pjanic PER**] free-kick lobbed slightly higher in front of [**Griezmann PER**] penalty area.)

(d).北京是一座历史与现代正面碰撞的城市.([**Beijing GPE**] is a city where history and modernity collide)

Fig. 1. Examples of image-text pairs in our Wukong-CMNER dataset.

Compared with other fields such as news articles, texts on social media have inherent problems such as colloquialism and short context. Existing works have achieved good performance, which mainly focus on incorporating visual representation into textual representation using cross-modal attention [15, 26, 33, 36]. Recently, some researchers [1, 2, 29] study the use of captions or a label set as a way to enrich the context for MNER. However, existing approaches still have the following two limitations:

One remarkable limitation is that they ignore the mapping relation between visual objects and named entities. In a sentence containing multiple entities and multiple types, there is more than one mapping between named entities and visual objects within an image. For example, in Fig. 1(c), there are two visual objects and named entities. Previous multimodal NER methods [17,36] representing the image with only one vector may mislead their models to extract different types of entities into the same type. Compared with these methods which combine image-level features into multimodal features, object-level features can reflect the mapping relation between visual objects and text words. However, if all visual objects are extracted from images as in these studies [3,38], many irrelevant vision clues may introduce noise to affect entity recognition.

The other limitation is the popular semantic gap problem. Since the representations of text and image come from different encoders, they are mapped to different semantic spaces. Accordingly, the inconsistent representations may prevent the model from establishing a good connection between the text and image. Thus, it is difficult to directly use these inconsistent representations for feature fusion and capture the correspondence between words in text and regions in image.

To capture the mapping relations between visual objects and textual entities (the first limitation), we introduce a lexicon-based prompting visual clue extraction (LPE) module. We first adopt the visual grounding toolkit [32] for extracting visual region objects using general words of pre-defined entity types. The extracted regions are more related to the named entities in the sentence. Unlike English, where there are spaces between characters, Chinese has a more complex composition and no explicit word boundary, which poses many difficulties for Chinese NER. To overcome this limitation, many works [12,13,16] pay attention to incorporating word information by utilizing lexicon features. In this work, we also use words that match from the lexicon to construct prompts. Then a gate mechanism utilizes the relevance between the lexicon-based prompts and region visual clues to refine the entity-related representation. To tackle the second issue, we propose a cross-model alignment (CA) module to make the presentations of text and visual more consistent. Specifically, we maximize the mutual information between the corresponding text-image pairs through a region-word contrastive loss.

This paper proposes two embedding enhanced modules (LPE and CA) for Chinese MNER. The two embedding enhanced modules decouple from the subsequent entity prediction. LPE and CA are devised as efficient embedding computing modules. Chinese NER model such as W2NER [11] and LEBERT [13] can combine with them to predict Chinese MNER. Such separable embedding and prediction enable our modules as a plug-and-play framework to enhance Chinese NER models.

The main contributions of this work can be summarized as follows:

- We construct Wukong-CMNER, the first human-annotated Chinese multimodal NER dataset with images modality, where each annotated sentence is paired with its corresponding image data. To the best of our knowledge, our

Wukong-CMNER is currently the largest multimodal NER dataset. Moreover, our Wukong-CMNER can foster research on Chinese MNER.

– We propose two general modules (LPE and CA) based on Wukong-CMNER, which can extract the proper visual clues and make the representations between the two modalities more consistent. And the two modules we proposed are based on self-supervised learning, without requiring any additional data annotations, and can be easily extended to other multimodal tasks. Moreover, our modules decouple from the subsequent predicting process, which can be combined with other unimodel Chinese NER models to tackle Chinese multimodel NER.

– We conduct extensive experiments on Wukong-CMNER, including traditional Chinese NER models and MNER models for English corpora. Experimental results demonstrate that our two modules can boost the performances of Chinese NER models when they are equipped with our modules to address Chinese multimodel NER task.

2   Related Work

2.1   Chinese NER

Compared with English, there is no explicit word boundaries in Chinese sentences, posing many difficulties to Chinese NER. Therefore, how to incorporate word information into a character-based models is the key challenge in Chinese NER. Zhang [37] first introduced a lattice LSTM to encode both characters and words for Chinese NER. Based on this method, many recent studies improve it by following efforts in terms of training efficiency [7,16], graph structure [4,8], model degradation [19], and Transformer-based lexical enhancement [12,17,28]. Liu et al. [13] integrated external lexicon knowledge into BERT layers directly by a Lexicon Adapter layer. Li et al. [11] presented a novel method by modeling NER task as word-word relation classification, pushing the state-of-the-art performances of Chinese NER. Different from previous methods, we explore the idea of using word information to extract entity-related visual clues from the image.

2.2   Multimodal NER

There has been vast prior research about MNER on social media, the critical challenge is how to combine text representation with image representation. Zhang et al. [36] first proposed a co-attention network to incorporate the visual information. Since text-irrelevant visual signals can bring noises, some works explore to extract text-related visual clues using the attention mechanism while restraining other visual features [15,18,33]. Besides, Sun et al. [21,22] introduced a text-image relation propagation method to filter text-related visual features. To address the semantic gap problem, Zheng et al. [38] leveraged adversarial learning to map two different representations into a shared space. Wu et al. [29]

leveraged object labels as visual features to bridge vision and language. Chen et al. [2,24] transformed images into captions. Wang et al. [20] utilized entity-related prompts for extracting proper visual clues. Since we study Chinese NER, we use lexicon information to design prompts and exploit them to capture visual clues from entity-related visual regions directly.

Table 1. The statistics of Wukong-CMNER dataset.

	Total	Train	Dev	Test
Image-text Pairs	53,554	35,233	9,231	9,090
Unique Tokens	5,781	5,369	4,240	4,241
Avg Sent Len	26.93	26.93	26.94	26.93
Max Sent Len	40	40	36	39
Entity	34,388	22,409	5,987	5,992
#ORG	9,742	6,381	1,677	1,684
#PER	12,013	7,781	2,144	2,088
#LOC	2,148	1,381	401	366
#GPE	10,485	6,866	1,765	1,854

3   Dataset Acquisition and Comparison

To study the task of Chinese multimodal NER containing image-text pairs, we annotate a new dataset called Wukong-CMNER, which is publicly available at https://github.com/10652835/Wukong_MNER. In this section, We first describe how we collect and select data (Sect. 3.1). We then present statistics and an analysis of the dataset (Table 1). Finally, we compare Wukong-CNNER with traditional Chinese NER datasets and multimodal NER datasets for the English corpus (Sect. 3.2).

3.1   Dataset Collection and Annotation

Wukong-CMNER is constructed in the following stages.

1) Obtaining Image-Text Pairs. We use Noah-Wukong dataset [6] as the source of image-text pairs. Noah-Wukong dataset is a large-scale Chinese cross-modal dataset, containing 100 million image-text pairs from the web, which is collected according to a high-frequency Chinese word list of 200K queries. We filter out sentences with special characters, such as (*person name*), and sentences less than 10 Chinese words. Since some image links in the Noah-Wukong dataset are not available for download, we discard them.

2) Tool Labeling. Hanlp [9] is a multilingual NLP library for researchers and companies, we use ERNIE [23] model trained on close-source Chinese corpus loaded with Hanlp to assign labels to entities in each sentence, including names of people (PER), location (LOC), organization (ORG), and geopolitical entities (GPE), which is consistent with Ontonotes [27].

3) Manual Inspection And Labeling. To ensure the quality of annotation, we employ 3 annotators to inspect and add complementary annotations to the results of the previous phase. For those sentences with different annotations, we discuss them one by one and finally reach the agreements for all cases.

4) Preparing Data Split. We split Wukong-CMNER into train, validation and test. To this end, we randomly divide 80%, 10% and 10% of image-text pairs for train, test, and validation, respectively. Table 1 shows the statistics of Wukong-CMNER dataset. We show the total number of image-text pairs. We also calculate the number of unique tokens and the statistics of tokens per caption.

3.2 Dataset Comparison

Since Wukong-CMNER is a Chinese NER dataset, we compare it with the four currently mainstream Chinese NER datasets, namely Ontonotes [27], MSRA [5], Weibo [10,19], Resume [37]. Meanwhile, since Wukong-CMNER dataset is also a multimodal dataset, we compare it with two English multimodal NER datasets Twitter-2015 [36] and Twitter-2017 [15], and the Chinese speech multimodal NER dataset CNERTA [20].

As shown in Table 2, we can find that Wukong-CMNER has the following advantages over other datasets: (1) Wukong-CMNER is the first Chinese multimodal MNER dataset that contains both images and text. (2) Wukong-CMNER is currently the largest public dataset for named entity recognition.

Table 2. A comparison between Wukong-CMNER and other widely-used NER datasets.

Dataset	Train	Dev	Test	Total	Language	Modality
MSRA	46,364	–	4,365	50,729	Chinese	Text
Ontonotes	15,724	4301	4,346	24,371	Chinese	Text
Weibo	1,350	271	270	1,891	Chinese	Text
Resume	3,821	463	477	4,761	Chinese	Text
Twitter-2015	4,000	1,000	3,257	8,257	English	Text+Image
Twitter-2017	3,373	723	723	4,819	English	Text+Image
CNERTA	34,102	4,440	4,445	42,987	Chinese	Text+Speech
Wukong-CMNER	35,233	9,231	9,090	53,554	Chinese	Text+Image

4   Methodology

4.1   Overview

The overall structure of our model is shown in Fig. 2, which consists of three main components: Firstly, a Lexicon-based Prompting Visual Clue Extraction

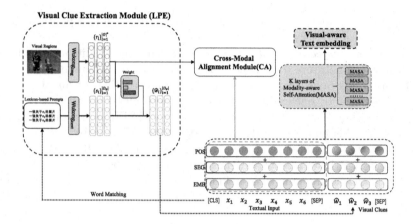

Fig. 2. Framework of our model.

Module (Sect. 4.2) is designed to extracted entity-related visual clues with a Vision-Language Pre-training (VLP) Model from the corresponding image. Secondly, a Cross-modal Alignment Module (Sect. 4.3) is proposed to make the representations of the two modalities more consistent through contrastive learning. Finally, the Cross-modal Feature Fusion Module (Sect. 4.4) fuses the input textual embeddings and the extracted entity-related visual clues to obtain the final visual-aware text representation for NER prediction or as input into other models for further processing.

Task Definition. Given a Chinese text $\mathbf{T} = \{x_1, x_2, ..., x_n\}$ and its associated image $\mathbf{I}$ as input, the goal of MNER is to extract a set of entities into one of the pre-defined types $\mathbf{y} = \{y_1, y_2, ..., y_n\}$. Following most existing work on MNER, we formulate the task as a sequence labeling problem.

Feature Extraction. For the Chinese sentence $\mathbf{T}$, we first convert each token x_i into word pieces and feed them into a pre-trained language model. After obtaining the embeddings, we employ max pooling to gain word representations based on the word piece embeddings. Finally, we obtain text representation $\mathbf{T_s} = \{w_1, w_2, ..., w_n\}$.

Since the global image features may express abstract concepts, they play the role of a weak learning signal. As for the image $\mathbf{I}$, we adopt the visual grounding toolkit for extracting local visual objects [34]. Furthermore, we introduce general words of pre-defined entity types (i.e., miscellaneous, person, location and organization) to encourage discovering more objects related to entities in the sentence. Then, we rescale extracted object images to 224×224 pixels as visual objects $\mathbf{O} = \{o_1, o_2, ..., o_m\}$. Finally, we utilize the pretrained Swin-Transformer [14] in Wukong [6] to extract regional features $\mathbf{V_r} = \{r_1, r_2, ..., r_m\}$ in an end-to-end manner.

4.2 Lexicon-Based Prompting Visual Clue Extraction Module

In Chinese NER, many recent studies [12,13,16] use word-matching methods to enhance character-based models, which has been proven to be effective. In this work, we exploit the lexicon information to design a set of entity-related prompts. We extract the entity-related visual clues with corresponding weights by comparing the relevance between the entity-related visual region and every prompt, as well as weaken the influence of task-irrelevant noises.

Prompt Design. Given a Chinese sentence with n characters $\mathbf{T} = \{x_1, x_2, \ldots, x_n\}$ and a Chinese Lexicon $\mathbf{D}$, we find out all the potential words inside the sentence by matching the character sequence with $\mathbf{D}$. Specifically, we first construct a Trie based on the $\mathbf{D}$, then traverse all character subsequences of the sentence and match them with Trie to obtain all potential words. Taking the truncated sentence " 南京市长 (Nanjing Mayor)" as an example, we can find out three different words inside it, " 南京 (Nanjing)", " 南京市 (Nanjing City)" and " 市长 (Mayor)". Subsequently, we obtain a potential entity dictionary $\mathbf{D_e}$ for each sentence.

We design the entity-related prompts for a sentence $\mathbf{T}$ as the form of " 一张 关于 $[w_i]$ 的图片 (an image of $[\bar{w}_i]$)", $\bar{w}_i \in \mathbf{D_e^T}$, where $\mathbf{D_e^T}$ represents the entity-related vocabulary obtained from the sentence $\mathbf{T}$ and w_i denotes a phrase from $\mathbf{D_e^T}$. For sentences that do not match any phrase, in other words, the dictionary $\mathbf{D_e}$ is empty, and hence we design the prompt as the form of " 一张关于空白的 图片 (an image of blank)".

Visual Clue Extraction. In this work, we adopt Wukong [6] as the VLP, which is pre-trained on the large-scale multi-modality Chinese dataset named Noah-Wukong Dataset. Given the input region image $\mathbf{r_i} \in \mathbb{R}^{C \times H \times W}$, we resort to the Swin-transformer image encoder of Wukong to obtain its embedding: $\mathbf{r_i} = \text{Wukong}_{\text{img}}(\mathbf{o_i})$; Meanwhile, we use Wukong's text encoder to obtain the embedding of prompt $\mathbf{s_i} = \text{Wukong}_{\text{text}}(\mathbf{P_i})$.

Wang [25] computed the relevance between global image and prompts. Since the global image features may express abstract concepts, we choose to obtain the relevance between entity-related region images $\mathbf{O}$ and every prompt $\mathbf{P_i}$ for the pairs$< \mathbf{T}, \mathbf{I} >$:

$$p(P_i|O) = \sum_{k=1}^{|O^*|} \frac{exp(< \mathbf{s_i}, \mathbf{r_k} > /\tau)}{\sum_{j=1}^{|\mathbf{D_e}|} exp(< \mathbf{s_j}, \mathbf{r_k} > /\tau)} \tag{1}$$

where O^* represents the region images set extracted from image $\mathbf{I}$, $< \cdot, \cdot >$ denotes the cosine similarity between two vectors and τ is a temperature parameter.

Finally, we consider the weighted embeddings of the prompt incorporating entity-related region image features as entity-related visual clues $\hat{w}_i = p(P_i|O) \times \mathbf{s_i}$. We proposed to fuse entity-related visual clues $\hat{w}$ with text representation $\mathbf{T_s}$ in Sect. 4.3.

4.3   Cross-Modal Alignment Module

Since the representations of text and image come from different encoders, the representations between them are inconsistent. The previous work cannot align the representations between the two modalities, our cross-modal alignment(CA) module try to bridge the semantic gap for cross-model consistent representations. Zhang et al. [35] proposed to capture inter-modality correspondences via multiple contrastive losses. Inspired by their work, we maximize the mutual information between the corresponding pairs through contrastive learning.

The inputs of the CA module are the text representations $\mathbf{T_s}$, the regional visual representations $\mathbf{V_r}$. We maximize the mutual information between image regions and words for constructing consistent representations. As we all know, it is difficult to directly maximize mutual information, and hence we choose to maximize the lower boundary the mutual information by optimizing contrastive losses. The updated text representation and image representation are used as the output of the CA module for subsequent cross-modal fusion.

Region-word Contrastive Loss. Intuitively, we believe that the entity-related regions in the individual image should be consistent with corresponding words in an input sentence. Thus, we propose the region-word contrastive loss to further make the text representation more consistent with the visual representation. Given a batch of $(\mathbf{V_r}, \mathbf{T_s})$ pairs, $\mathbf{T_s} = \{w_1, w_2, \ldots, w_n\}$ represents the text presentation containing n words and $\mathbf{V_r} = \{r_1, r_2, \ldots, r_m\}$ denotes the visual representation containing m entity-related regions. We use attention to learn connections between entity-related regions $\mathbf{V_r}$ and words in sentence $\mathbf{T_s}$. We first compute the cosine similarity between all words in the sentence and all entity-related regions in the corresponding image. We next compute the soft attention $\alpha_{i,j}$ for word representation w_i and region representation r_j as:

$$\alpha_{i,j} = \frac{exp(\rho_1 < w_i, r_j >)}{\sum_{k=1}^{|O^*|} exp(\rho_1 < w_i, r_k >)} \tag{2}$$

where ρ_1 is a sharpening hyper-parameter to reduce the entropy of the soft attention. Thus, the entity-related region feature for the ith word in the sentence is computed as $c_i = \sum_{j=1}^{|O^*|} \alpha_{i,j} r_j$. Then, the score function between all words in the sentence T and all entity-related regions in the corresponding image I is defined as:

$$S_{word}(I, T) = log(\sum_{h=1}^{|T^*|} exp(\rho_2 < w_h, c_h >))^{\frac{1}{\rho_2}} / \tau \tag{3}$$

where ρ_2 is a hyper-parameter that determines the weights of the most aligned word-region pair, $|T^*|$ represents the total number of words in the sentence $\mathbf{T}$. Finally, the region-word contrastive loss can be formulated as:

$$\mathcal{L}_{word}(I_i, T_i) = -log \frac{exp(S_{word}(I_i, T_i))}{\sum_{j=1}^{N} exp(S_{word}(I_i, T_j))} \tag{4}$$

4.4 Cross-Modal Fusion

In this section, we introduce how to fuse visual clues representation $\{\hat{w}_i\}_{i=1}^{|\mathbf{D}_e^T|}$ with text representation $\mathbf{T_s} = \{w_1, w_2, \ldots, w_n\}$ described in Sect. 4.1. Firstly, we concatenate the above two representations and add the representation of special tokens of [CLS] and [SEP]:

$$[\text{CLS}], w_1, w_2, \ldots, w_n, [\text{SEP}], \hat{w}_1, \ldots, \hat{w}_{|\mathbf{D}_e^T|}, [\text{SEP}]. \tag{5}$$

To capture the position information and the type information, we also add the position embedding and segmentation embedding into them. The concatenated representations are subsequently fed into multiple layers of modality-aware self-attention.

With the assumption that it is beneficial to use the information of target token types when computing the attention scores [25,31], we take the modality type into consideration when computing the attention query. Let x_i represent the input token that corresponds to w_i or $\hat{w}_i$. The attention weight corresponding to the jth token in the sequence is computed as:

$$\alpha_{ij} = softmax(\frac{(\mathbf{K}x_j)^T(\mathbf{Q}_{Q2V}(x_i))}{\sqrt{N + |\mathbf{D}_e^T|}}) \tag{6}$$

where $\mathbf{Q}_{Q2V}$ and $\mathbf{K}$ denote the query and key matrices respectively. The query matrix $\mathbf{Q}_{Q2V}$ contains four forms depending on the modality of the query and the value: $\mathbf{Q_{w2w}}, \mathbf{Q_{w2\hat{w}}}, \mathbf{Q_{\hat{w}2w}}$ and $\mathbf{Q_{\hat{w}2\hat{w}}}$. The final textual representation output is then calculated as:

$$x_i := \sum_{j=1}^{N+|\mathbf{D}_e^T|} \alpha_{ij}x_j \tag{7}$$

We apply the K layer of modality-aware self-attention to embed the concatenated embeddings, and obtain the visual-aware text embeddings from the last layer for prediction or as input for other models.

4.5 Model Training

In summary, we train the main task (MNER) and our self-supervised learning task (LPE and CA) jointly. The final loss function is computed as follows:

$$\mathcal{L}_{task} = \frac{1}{N}\sum_{i=1}^{N}(\lambda_c\mathcal{L}_{word} + (1 - \lambda_c)\mathcal{L}_{mner}) \tag{8}$$

where $\lambda_c \in [0, 1]$ is a hyperparameter and $\mathcal{L}_{mner}$ is the loss function used by the MNER task. The goal of model training is to minimize $\mathcal{L}_{task}$ loss.

5   Experiments

5.1   Experimental Setups

Datasets. We conduct extensive experiments on the proposed dataset, namely Wukong-CMNER.

Baselines. To demonstrate the necessity of studying Chinese multimodal named entity recognition and the effectiveness of two modules we proposed, we compare with extensive baselines including Chinese NER and multimodal NER baselines: LEBERT [13], LexiconAugment [16], Flat [12], MECT [28], W2NER [11], UMT [33], MAF [30]. Considering the structure of the model, we conduct ablation experiments on W2NER and LEBERT, that is, we combine our *LPE* and *CA* with W2NER and LEBERT, respectively.

Metrics. We use overall precision (**P**), recall (**R**) and F1 score (**F1**) to evaluate the performance of the models, which are widely used in many recent works.

5.2   Main Results

Table 3. Precision (%), Recall (%) and F1 score (%) of baselines and our method on Wukong-CMNER. ↑ means the points higher than the corresponding baselines without the proposed modules.

Model	Modality	Corpora	Wukong-CMNER		
			P	R	F1
LexiconAugment	Text	Chinese	74.39	71.83	73.09
LEBERT	Text	Chinese	78.25	80.93	79.57
Flat-Transformer	Text	Chinese	79.21	81.91	80.54
MECT	Text+Glyph	Chinese	72.86	76.41	74.59
W2NER	Text	Chinese	81.34	81.96	81.65
UMT	Text+Image	English	66.13	79.74	72.26
MAF	Text+Image	English	77.63	80.86	79.21
LEBERT+LPE+CA	Text+Image	Chinese	80.43	81.23	80.83(↑1.26)
W2NER+LPE+CA	Text+Image	Chinese	**81.40**	**83.95**	**82.66**(↑1.01)

Table 3 shows the results of baselines and some baselines enhanced by the proposed modules. From the table, we find:

(1)During the training, we replace pre-trained language model from bert-base-cased to bert-base-chinese. Comparing the performances of MNER model (UMT and MAF) designed for English corpus with that of Chinese NER model, even though visual modality has been introduced, they still perform worse than some Chinese named entity recognition models which only use textual modality. These results indicate that Chinese MNER datasets have their specific properties, which cannot be made full use by English MNER models and further illustrate the necessity of constructing a Chinese MNER dataset to investigate Chinese MNER.

(2)Introducing visual modality can boost the performance of the character-based models. After LEBERT and W2NER are equipped with our LPE and CA modules, their performances increase by 1.26% and 1.01% respectively. These experimental results demonstrate the effectiveness of introducing the visual modality in character-based NER models.

(3)W2NER equipped with LPE and CA can achieve the SOTA result on Wukong-CMNER. The result shows that our modules can provide character-based with proper visual information that does not contain in lexicon.

5.3   Ablation Study

To investigate the effectiveness of the LPE and CA modules, we perform comparisons between the two full models and its ablation methods.

As shown in Table 4, both the two full models benefit from the LPE module and CA module. Specifically, for the model LEBERT, without the LPE and CA module, **w/o LPE + CA** drops 1.26 F1 scores; without **w/o CA** drops 0.37 F1 scores. After the CA module and both two modules are removed from W2NER, its F1 scores drop by 0.38 and 1.01 respectively. These results indicate that both the LPE and CA module take effect. The LPE module captures proper visual clues to assist character-based NER models in identifying entities more accurately and the CA module makes the representation of the two modalities more consistent.

We also present the performance of each entity category in Table 4. In most entity categories, our methods show superior performance compared with baselines. All models perform poorly in the categories of ORG and LOC. We speculate that the reason may be that the models have difficulty in extracting visual clues related to these two types of entities.

Table 4. Ablation Study of our LPE and CA Modules. We turn off the CA module and both two modules on two full model respectively, which are represented as "w/o CA" and "w/o CA + LPE". We also present the model performance in terms of different entity types.

Methods	Wukong-CMNER				
	PER	GPE	ORG	LOC	F1
LEBERT+LPE+CA	86.76	83.63	59.28	73.91	80.83
w/o CA	86.67	83.31	58.93	72.02	80.46(↓0.37)
w/o LPE+CA	87.88	81.20	53.08	71.94	79.57(↓1.26)
W2NER+LPE+CA	**88.67**	**84.98**	**63.66**	**75.26**	**82.66**
w/o CA	88.38	84.93	63.52	75.79	82.28(↓0.38)
w/o LPE+CA	87.65	84.58	63.42	74.64	81.65(↓1.01)

6   Conclusion

In this paper, we explore Chinese multimodal NER with both textual and visual contents. To achieve this, we construct a large scale human-annotated Chinese multimodal NER dataset, named Wukong-CMNER. Based on the dataset, we propose a lexicon-based prompting visual clue extraction module to find entity-related visual clues. We further design a cross-modal alignment module to alleviate the semantic gap problem. Through extensive experiments, we find that the model designed for MNER on English corpora is not suitable for Wukong-CMNER. These results illustrate the necessity of our work. The experimental results on Wukong-CMNER show that LPE and CA take effect in improving the Chinese MNER performance. $LPE+CA$ combined with W2NER creates new SOTA on Wukong-CMNER.

Acknowledgements. This work is partially supported by the National Natural Science Foundation of China under Grant No. 61772534, partially supported by Public Computing Cloud, Renmin University of China.

References

1. Chen, D., Li, Z., Gu, B., Chen, Z.: Multimodal named entity recognition with image attributes and image knowledge. In: Jensen, C.S., et al. (eds.) DASFAA 2021. LNCS, vol. 12682, pp. 186–201. Springer, Cham (2021). https://doi.org/10.1007/978-3-030-73197-7_12
2. Chen, S., Aguilar, G., Neves, L., Solorio, T.: Can images help recognize entities? a study of the role of images for multimodal NER. arXiv:2010.12712 (2020)
3. Chen, X., et al.: Good visual guidance makes a better extractor: hierarchical visual prefix for multimodal entity and relation extraction. arXiv:2205.03521 (2022)
4. Ding, R., Xie, P., Zhang, X., Lu, W., Li, L., Si, L.: A neural multi-digraph model for chinese ner with gazetteers. In: ACL, pp. 1462–1467 (2019)
5. Gina-Anne, L.: The third international Chinese language processing bakeoff: word segmentation and named entity recognition. In: CLP, pp. 108–117 (2006)
6. Gu, J., et al.: Wukong: 100 million large-scale Chinese cross-modal pre-training dataset and a foundation framework. arXiv:2202.06767 (2022)
7. Gui, T., Ma, R., Zhang, Q., Zhao, L., Jiang, Y.G., Huang, X.: CNN-based Chinese NER with lexicon rethinking. In: IJCAI, pp. 4982–4988 (2019)
8. Gui, T., et al.: A lexicon-based graph neural network for Chinese NER. In: EMNLP, pp. 1040–1050 (2019)
9. He, H., Choi, J.D.: The stem cell hypothesis: dilemma behind multi-task learning with transformer encoders. arXiv preprint arXiv:2109.06939 (2021)
10. He, H., Sun, X.: F-score driven max margin neural network for named entity recognition in Chinese social media. arXiv:1611.04234 (2016)
11. Li, J., Fei, H., Liu, J., Wu, S., Zhang, M., Teng, C., Ji, D., Li, F.: Unified named entity recognition as word-word relation classification. In: AAAI. vol. 36, pp. 10965–10973 (2022)
12. Li, X., Yan, H., Qiu, X., Huang, X.: Flat: Chinese NER using flat-lattice transformer. In: ACL, pp. 6836–6842 (2020)

13. Liu, W., Fu, X., Zhang, Y., Xiao, W.: Lexicon enhanced Chinese sequence labeling using bert adapter. In: ACL, pp. 5847–5858 (2021)
14. Liu, Z., et al.: Swin transformer: hierarchical vision transformer using shifted windows. In: ICCV, pp. 10012–10022 (2021)
15. Lu, D., Neves, L., Carvalho, V., Zhang, N., Ji, H.: Visual attention model for name tagging in multimodal social media. In: ACL, pp. 1990–1999 (2018)
16. Ma, R., Peng, M., Zhang, Q., Huang, X.: Simplify the usage of lexicon in Chinese NER. In: ACL, pp. 5951–5960 (2020)
17. Mengge, X., Bowen, Y., Tingwen, L., Yue, Z., Erli, M., Bin, W.: Porous lattice-based transformer encoder for Chinese NER. In: COLING (2019)
18. Moon, S., Neves, L., Carvalho, V.: Multimodal named entity recognition for short social media posts. In: NAACL-HLT, pp. 852–860 (2018)
19. Peng, N., Dredze, M.: Named entity recognition for chinese social media with jointly trained embeddings. In: EMNLP, pp. 548–554 (2015)
20. Sui, D., Tian, Z., Chen, Y., Liu, K., Zhao, J.: A large-scale chinese multimodal ner dataset with speech clues. In: ACL, pp. 2807–2818 (2021)
21. Sun, L., et al.: RIVA: a pre-trained tweet multimodal model based on text-image relation for multimodal NER. In: COLING, pp. 1852–1862 (2020)
22. Sun, L., Wang, J., Zhang, K., Su, Y., Weng, F.: RpBERT: a text-image relation propagation-based BERT model for multimodal NER. In: AAAI, vol. 35, pp. 13860–13868 (2021)
23. Sun, Y., et al.: ERNIE: enhanced representation through knowledge integration. arXiv:1904.09223 (2019)
24. Wang, X., et al.: ITA: image-text alignments for multi-modal named entity recognition. arXiv:2112.06482 (2021)
25. Wang, X., et al.: Prompt-based entity-related visual clue extraction and integration for multimodal named entity recognition. In: Database Systems for Advanced Applications. DASFAA 2022. LNCS, vol. 13247, pp. 297–305. Springer, Cham (2022). https://doi.org/10.1007/978-3-031-00129-1_24
26. Wang, X., et al.: CAT-MNER: multimodal named entity recognition with knowledge-refined cross-modal attention. In: ICME, pp. 1–6. IEEE (2022)
27. Weischedel, R., et al.: OntoNotes release 5.0 ldc2013t19. web download. Philadelphia: Linguistic data consortium, 2013 (2013)
28. Wu, S., Song, X., Feng, Z.: MECT: multi-metadata embedding based cross-transformer for chinese named entity recognition. In: ACL, pp. 1529–1539 (2021)
29. Wu, Z., Zheng, C., Cai, Y., Chen, J., Leung, H., Li, Q.: Multimodal representation with embedded visual guiding objects for named entity recognition in social media posts. In: MM, pp. 1038–1046 (2020)
30. Xu, B., Huang, S., Sha, C., Wang, H.: MAF: a general matching and alignment framework for multimodal named entity recognition. In: WSDM, pp. 1215–1223 (2022)
31. Yamada, I., Asai, A., Shindo, H., Takeda, H., Matsumoto, Y.: Luke: deep contextualized entity representations with entity-aware self-attention. In: EMNLP (2020)
32. Yang, Z., Gong, B., Wang, L., Huang, W., Yu, D., Luo, J.: A fast and accurate one-stage approach to visual grounding. In: ICCV, pp. 4683–4693 (2019)
33. Yu, J., Jiang, J., Yang, L., Xia, R.: Improving multimodal named entity recognition via entity span detection with unified multimodal transformer. In: ACL (2020)
34. Zhang, D., Wei, S., Li, S., Wu, H., Zhu, Q., Zhou, G.: Multi-modal graph fusion for named entity recognition with targeted visual guidance. In: AAAI, vol. 35, pp. 14347–14355 (2021)

35. Zhang, H., Koh, J.Y., Baldridge, J., Lee, H., Yang, Y.: Cross-modal contrastive learning for text-to-image generation. In: CVPR, pp. 833–842 (2021)
36. Zhang, Q., Fu, J., Liu, X., Huang, X.: Adaptive co-attention network for named entity recognition in tweets. In: AAAI (2018)
37. Zhang, Y., Yang, J.: Chinese NER using lattice LSTM. In: ACL, pp. 1554–1564 (2018)
38. Zheng, C., Wu, Z., Wang, T., Cai, Y., Li, Q.: Object-aware multimodal named entity recognition in social media posts with adversarial learning. Multimedia **23**, 2520–2532 (2020)

CAB: Empathetic Dialogue Generation with Cognition, Affection and Behavior

Pan Gao[1], Donghong Han[1,2(✉)], Rui Zhou[3], Xuejiao Zhang[1],
and Zikun Wang[1]

[1] School of Computer Science and Engineering, Northeastern University,
Shenyang, China
handonghong@cse.neu.edu.cn
[2] Key Laboratory of Intelligent Computing in Medical Image of Ministry
of Education, Northeastern University, Shenyang, China
[3] Swinburne University of Technology, Hawthorn, Australia
rzhou@swin.edu.au

Abstract. Empathy is an important characteristic to be considered when building a more intelligent and humanized dialogue agent. However, existing methods did not fully comprehend empathy as a complex process involving three aspects: cognition, affection and behavior. In this paper, we propose CAB, a novel framework that takes a comprehensive perspective of cognition, affection and behavior to generate empathetic responses. For cognition, we build paths between critical keywords in the dialogue by leveraging external knowledge. This is because keywords in a dialogue are the core of sentences. Building the logic relationship between keywords, which is overlooked by the majority of existing works, can improve the understanding of keywords and contextual logic, thus enhance the cognitive ability. For affection, we capture the emotional dependencies with dual latent variables that contain both interlocutors' emotions. The reason is that considering both interlocutors' emotions simultaneously helps to learn the emotional dependencies. For behavior, we use appropriate dialogue acts to guide the dialogue generation to enhance the empathy expression. Extensive experiments demonstrate that our multi-perspective model outperforms the state-of-the-art models.

Keywords: Empathetic dialogue · Dialogue generation · Cognition affection and behavior

1 Introduction

Empathy is the ability to understand others' feelings, and respond appropriately to their situations . Previous studies have shown that empathetic dialogue models can improve user's satisfaction in several areas, such as customer service [14], healthcare community [26] and etc. Therefore, how to successfully implement empathy becomes one of the key issues to build an intelligent and considerate agent. In recent years, many studies have been conducted on the task of

© The Author(s), under exclusive license to Springer Nature Switzerland AG 2023
X. Wang et al. (Eds.): DASFAA 2023, LNCS 13945, pp. 597–606, 2023.
https://doi.org/10.1007/978-3-031-30675-4_44

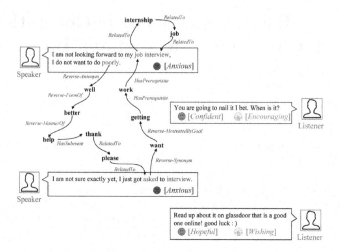

Fig. 1. A dialogue from the *EmpatheticDialogues* dataset. The cognitive ability is improved by retrieving entities (bold in black) and relationships (grey) from *ConceptNet* and building paths between critical keywords (red) to generate a high quality response under the influence of *anxious* and *confident* emotions and *wishing* dialogue act. (Color figure online)

empathetic dialogue generation, which are mainly divided into two categories: One is to enhance the understanding of a user's situation and emotion by leveraging knowledge from one or more external knowledge bases [11,15,21,25] or adding emotion causes as prior emotion knowledge [3,25]. This is to improve the cognitive ability. The issue of the existing work is that they overlook the importance of paths between users' critical keywords, which can actually reflect the contextual logic in the conversation. Although some studies [25] build paths between emotion concepts and cause concepts, they mainly focus on the causality aspect and ignore the fact that paths between any keywords can help. The second category is to design emotion strategies, such as mixture of experts [12], emotion mimicry [17] and multi-resolution emotions [10] to generate appropriate responses from the affection aspect. Unfortunately, these studies learn to respond properly mainly according to the speaker's emotion rather than both interlocutors' emotions. In this paper, we aim to improve the aforementioned weak aspects of the existing works to help advance the study of empathetic dialogue generation.

Psychological research shows that empathy is a complex mental process involving three aspects of interlocutors: cognition, affection and behavior [13]. Specifically, cognitive empathy refers to the ability to understand and interpret a user's situation [2]; affective empathy is an emotional reaction based on differentiating the emotions of oneself and others [13]; behavioral empathy means verbal or non-verbal forms of communication used in the empathetic dialogue [6]. Among the existing works, some only consider the aspects of congition and affection [21,28]; others mainly consider the aspect of behavior [1,27]. None of the existing works had comprehensively considered all the three aspects (cognition, affection, behavior), which we believe are all important. In the following, we elaborate in detail with the example in Fig. 1. The dialogue in Fig. 1 shows that (1) **Cognition**: The speaker is *anxious* about attending a job interview. In the first turn, there exists a path between <*job, interview*> with internship as a bridge to enhance the understanding of the keywords and the context. In the next turn, the paths between < *poorly, asked*> and < *asked, job*> are built to alleviate the problem that it is difficult to capture the contextual logic based on limited context. Thus, it can be seen that the paths , which establish the relationships between utterances, are critical to improve the cognitive ability. (2) **Affection**: In interpersonal conversations, responses are usually influenced by both interlocutors' emotions [5]. As shown in Fig. 1, in the second turn, instead of both sides falling into anxiety, the listener is able to perceive the speaker's emotion and accept the emotion difference between them, thus generating a response with more positive emotion (*hopeful*). Therefore, how to learn the emotional dependencies between the context and target response based on both interlocutors' emotions is critical for responding properly. (3) **Behavior**: Appropriate dialogue acts are used as communicative form to enhance empathy expression. For example, the listener inspires the speaker by *encouraging* and makes the speaker relaxed by *wishing*. Different from [27], we consider that all the responses (rather than some of them) are generated by the guiding of dialog act. In this way, we can guide dialogue generation better.

To this end, we propose a novel empathetic dialogue generation model including aspects of **C**ognition, **A**ffection and **B**ehavior (**CAB**) to achieve a comprehensive empathetic dialogue task. Specifically, since keywords are important to understand the contextual logic, our model builds paths between critical keywords through multi-hop commonsense reasoning to enhance the cognitive ability. Conditional Variational Auto Encoder (CVAE) model with dual latent variables is built based on both interlocutors' emotions, and then the dual latent variables are injected into the decoder together with the dialogue act features to produce empathetic responses from the perspective of affection and behavior. Our contributions are summarized as follows:

- To the best of our knowledge, we are the first to propose a novel framework for empathetic dialogue generation based on psychological theory from three perspectives: cognition, affection and behavior.
- We propose a context-based multi-hop reasoning method, in which paths are established between critical keywords to acquire implicit knowledge and learn contextual logic.
- We present a novel CVAE model, which introduces dual latent variables to learn the emotional dependencies between the context and target responses. After that, we incorporate the dialogue act features into the decoder to guide the generation.
- Experiments demonstrate that CAB generates more relevant and empathetic responses compared with the state-of-the-art methods.[1]

---

[1] Code and data are available at https://github.com/geri-emp/CAB.

2 Related Work

Recently, there has been numerous works in the task of empathetic dialogue generation proposed by Rashkin et al. [20]. Lin et al. [12] assign different decoders for various emotions, and fuse the output of each decoder with users' emotion weights. Majumder et al. [17] adopt emotion stochastic sampling and emotion mimicry to respond to positive or negative emotions for generating empathetic responses. Li et al. [10] construct an interactive adversarial learning network considering multi-resolution emotions and user feedback. Liu et al. [16] incorporate anticipated emotions into response generation via reinforcement learning. Gao et al. [3] adopt emotion cause to better understand the user's emotion. However, all of the above methods only consider the user's emotion and ignore the influence between both interlocutors' emotions in the dialogue.

Several studies have incorporated external knowledge into empathetic dialogue generation. Li et al. [11] employ multi-type knowledge to explore implicit information and construct an emotional context graph to improve emotional perception. Liu et al. [15] prepend the retrieved knowledge triples to the gold responses in order to get proper responses. However, these approaches retrieve knowledge triples without fully considering the contextual meaning of the words. Although Wang et al. [25] adopt *ConceptNet* to explore the emotional causality by commonsense reasoning between the emotion clause and the cause clause, the logical relationships between other utterances may be ignored. Sabour et al. [21] use *ATOMIC* for commonsense reasoning to better understand the user's situation and feeling, but reasoning on a whole dialogue history may neglect the important role of keywords in the context. To overcome the above proposed shortcomings, we propose a context-based multi-hop commonsense reasoning method to enrich contextual information and reason about the logical relationships between utterances.

3 Method

3.1 Task Formulation and Overview

In empathetic dialogue generation, each dialogue consists of a dialogue history $C = [S_1, L_1, S_2, L_2, \ldots, S_{N-1}, L_{N-1}, S_N]$ of 2N-1 utterances and a gold empathetic response $L_N = [w_N^1, w_N^2, \ldots, w_N^n]$ of n words, where S_i and L_i denote the i-th utterance of speaker and listener respectively. Our goal is to generate an empathetic response $R = [r_1, r_2, \ldots, r_m]$ based on the dialogue history C, the speaker's emotion e_s, the listener's emotion e_l, and the listener's dialogue act a_l.

We provide a overview of CAB in Fig. 2, which consists of five components: **(a) Emotional Context Representation.** The predicted emotions, e_s and e_l, are fed into context C by emotional context encoder to obtain the emotional context representation $\hat{H}_S$ and $\hat{H}_L$; **(b) Affection.** Then prior network and posterior network capture dual latent variables z_s and z_l, based on $\hat{H}_S$ and $\hat{H}_L$ in the test and training phase; **(c) Cognition.** To build paths P, we leverage *ConceptNet* to acquire external knowledge and incorporate it into C to obtain a

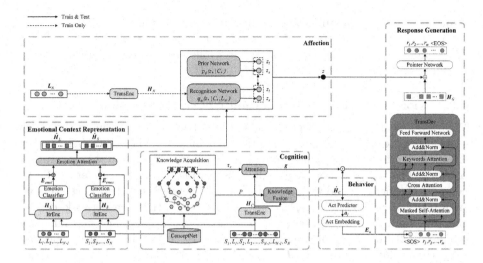

Fig. 2. The overall architecture of CAB.

knowledge-enhanced context representation $\hat{H}_C$; **(d) Behavior**. The dialogue act features E_a are distilled based on a predictor and the embedding layer; **(e) Response Generation**. The three-stage decoder generates an empathetic response R based on the aspects of affection, cognition and behavior.

We evaluate the model on *EmpatheticDialogues* [20], which is a publicly available benchmark dataset for empathetic dialogue generation. However, dialogues in this dataset do not contain labels of emotion and dialogue act for each listener's utterance, and we annotate emotion and dialogue act by Emoberta [7] and EmoBERT [27], respectively, to support the studies in this paper.

From Sect. 3.2 to Sect. 3.7, we introduce CAB briefly due to space limit. More model and experiment details are in the full version [4].

3.2 Emotional Context Encoder

Input Representation. We divide the dialogue history into two segments $C_S = [S_1, S_2, \ldots, S_N]$ and $C_L = [L_1, L_2, \ldots, L_{N-1}]$. Following the previous work [12], we first gain the embedding of speaker context, listener context, global context and gold response respectively. Then the embedding of speaker context and listener context are fed into the Transformer-based inter-encoder (**ItrEnc**) to obtain H_S and H_L, and the Transformer encoder (**TransEnc**) encodes the embedding of global context and gold response into H_C and H_N.

Emotion Classification. To understand the emotions of the speaker and the listener, we project the hidden representations of the first token from H_S and H_L into the emotion category distribution P_s and P_l to predict their emotions. Then we send the emotions to a trainable emotion embedding layer to obtain the emotion states embedding matrix E_{emos} and E_{emol}.

Emotion Self-attention. To make the latent variables in Sect. 3.3 incorporate both interlocutors' emotions, H_S and H_L are concatenated with E_{emos} and E_{emol} and then fed into a self-attention layer followed by a linear layer to obtain the emotional context representation $\hat{H}_S$ and $\hat{H}_L$.

3.3   Prior Network and Recognition Network (Affection)

We introduce dual latent variables $z_* \in \{z_s, z_l\}$ in CVAE, mapping the input sequences $C_* \in \{C_S, C_L\}$ into the output sequence L_N via z_*. Taking speaker as an example, we illustrate how to realize the prior network and the recognition network. The **prior network** $p_\theta(z_s|C_S)$ is parameterized by 3-layer MLPs to compute the mean μ'_s and variance σ'^2_s of z_s. The network structure of the **recognition network** $q_\varphi(z_s|C_S, L_N)$ is the same as that of the prior network, except that the input also includes H_N. In order to learn the emotional dependencies based on both interlocutors' emotions, we fuse z_s and z_l due to the emotional similarity coefficient β between E_{emos} and E_{emol} to obtain $z = \beta \cdot z_s + (1-\beta) \cdot z_l$.

3.4   Knowledge Acquisition and Fusion (Congnition)

Knowledge Acquisition. We first obtain the keyword set τ_{all} of size cw from C_S based on the TextRank algorithm [18]. Then we build paths as follows:

   a. Take one keyword in τ_{all} as the head entity $h_i \in \tau_{all}$, then feed the embedding of h_i and speaker context into ItrEnc to extract the semantic features of h_i. The Top-K knowledge triples in *ConceptNet* associated with h_i are retrieved based on a score and removed relation set [11].

   b. To ensure that the triples are logically related to other keywords τ_{other}, we first obtain the semantic features of $h_j \in \tau_{other}$ like step a. After ranking the triples by relevance between tail entity and h_j, we select Top-k triples. If the tail entity is same as h_j, which indicates there exists a one-hop path between h_i and h_j, we add them to the final keywords set τ_r (e.g. red circles in Fig. 2). If not, the tail entity is added to τ_{all} to continue finding the paths by repeating step a and b. Finally, we retain some paths P (e.g. the paths connected by grey arrows in Fig. 2) for futher fusion. The attention weight vector g is calculated to measure importance of each word in C with τ_r by the attention mechanism.

Knowledge Fusion. We first convert the paths into sequences. Then the sequences are fed into the two-layer Bi-GRU to obtain the knowledge representation H_k. Finally, following previous work [21], we concatenate H_k with context at token-level to learn the knowledge-enhanced context representation $\hat{H}_C$.

3.5   Dialogue Act Predictor and Representation (Behavior)

To guide the communicative form of empathetic dialogue generation, our model uses the first token of $\hat{H}_C$ to predict dialogue act a_l. Then, a_l is fed into the embedding layer to learn the dialogue act embedding representation E_a.

3.6   Response Generation

Finally, the aforementioned information E_a, g, z and $\hat{H}_C$ are applied at the Transformer-based decoder (**TransDec**) through the following three stages: **(1)** The embedding of the start-of-sequence token E_{SOS} and E_a are fed into a linear layer, then the high-level act features are adopted to guide the generation. **(2)** We design a multi-head keywords attention, which takes the output of the cross-attention layer as query, the dot-product over g and $\hat{H}_C$ as key and value. Then TransDec outputs the hidden state H_G. **(3)** To learn the emotional dependencies, we concatenate z and H_G at token-level and use pointer network [23] to output the probability distribution of each word in the vocabulary.

3.7   Training Objectives

We jointly optimiaze the emotion classification loss, dialogue act prediction loss, the loss of CVAE model and bag-of-word loss as:

$$\mathcal{L} = \gamma_1 \mathcal{L}_s + \gamma_2 \mathcal{L}_l + \gamma_3 \mathcal{L}_a + \gamma_4 \mathcal{L}(C_*, C_N; \theta, \varphi) + \gamma_5 \mathcal{L}_{bow} \tag{1}$$

where γ_1, γ_2, γ_3, γ_4 and γ_5 are hyper-parameters.

4   Experiments

4.1   Experimental Setup

Baselines. We compare our model with the state-of-the-art models as follows: (1) **Transformer** [22]: The vanilla Transformer with the pointer network trained by optimizing the generation loss. (2) **Multi-Trans** [20]: A variant of Transformer that includes emotion classification loss in addition to the generation loss to jointly optimize the model. (3) **MOEL** [12]: A model that includes several Transformer decoders, and the outputs are softly combined to generate responses. (4) **MIME** [17]: A model adopting emotion mimicry and emotion clusters to deal with positive or negative emotions. (5) **EmpDG** [10]: A generative adversarial network that considers multi-resolution emotion and introduces discriminators to supervise the training in semantics and emotion. (6) **KEMP** [11]: A model that uses two-type knowledge to help understand and express emotions. (7) **CEM** [21]: A method for generating empathetic responses by leveraging commonsense to improve the understanding of interlocutors' situations and feelings.

Implementation Details. We implement all models in PyTorch[2] with GeForce GTX 3090 GPU, and train models using Adam optimization [8] with a mini-batch size of 16. All common hyper-parameters are the same as the work in [12]. We adopt 300-dimensional pre-trained 840B GloVE vectors [19] to initialize the word embeddings, which are shared between the encoders and the decoder. The hidden size is 300 everywhere, and the size of latent variable is 200. We use the

---

[2] https://pytorch.org/.

Table 1. Results of the automatic evaluation, and w/o Cog/Aff/Beh indicate ablation experiments and the best results of all models are bold.

Models	PPL	DIST-1	DIST-2	EmoSA	EmoLA	ActA
Transformer	34.11	0.49	1.91	-	-	-
Multi-Trans	36.42	0.43	1.85	28.91	-	-
MOEL	36.59	0.60	3.12	32.33	-	-
MIME	37.52	0.32	1.22	34.88	-	-
EmpDG)	37.37	0.45	1.89	32.45	-	-
KEMP	36.39	0.66	3.08	36.57	-	-
CEM	36.11	0.66	2.99	39.07	-	-
CAB	34.36	**1.13**	**4.23**	**40.52**	**72.23**	41.72
W/o Cog	**33.88**	0.94	3.33	39.42	71.82	**43.09**
W/o Aff	34.98	1.12	3.97	34.25	-	37.25
W/o Beh	34.79	1.06	3.83	40.05	72.20	-

KL annealing of 15,000 batches to achieve the best performance. During test, the batch size is 1 and the maximum greedy decoding steps is 50.

Automatic Evaluation Metrics. We choose the widely used PPL [24], Distinct-1, Distinct-2 [9] as our main automatic metrics. **PPL** is used to estimate the generation quality of a model in general. **Distinct-1** and **Distinct-2** are used to measure the diversity of responses. Since emotion accuracy of speaker/listener (**EmoSA/EmoLA**) reflects the understanding of both interlocutors' emotions and dialogue act accuracy (**ActA**) can determine whether the proper dialogue acts are chosen to produce responses, we also report these metrics.

4.2   Results and Analysis

Automatic Evaluation Results. The overall automatic evaluation results are shown in the Table 1. Our model CAB outperforms the baselines on all metrics significantly. The lower PPL score implies that CAB has a higher quality of generation generally, reflecting the importance of considering empathy from multi-perspective. The remarkable improvements in Distinct-1 and Distinct-2 suggest that the introduction of external knowledge can be beneficial in improving the understanding of dialogue history and thus generating a wider variety of response. The higher accuracy of emotion classification verifies the validity of modelling both interlocutors' emotions separately.

Ablation Study. As shown in the bottom part of Table 1, we also conduct ablation experiments to explore the effect of each component. From the results, we can observe that all metrics decrease except for PPL, especially Distinct-1 and Distinct-2, when commonsense knowledge acquisition and fusion are removed (**w/o Cog**), suggesting that the paths capture additional information to enhance cognitive ability, thus improving the quality and diversity of responses. The increasing PPL score may be due to the introduction of knowledge, which may have an impact on the fluency of the generated responses. In addition, we find that only considering the speaker's emotion by removing the latent variable of listener (**w/o Aff**) yields lower emotion accuracy and higher PPL score, and thus it is difficult to generate appropriate responses without understanding both interlocutors' emotions exactly. All metrics decrease when we remove the classification of dialogue act and the dialogue act features fused at the decoder (**w/o Beh**), indicating the emphasis of the dialogue acts in improving empathy.

5   Conclusions

In this paper, we build paths by leveraging commonsense knowledge to enhance understanding of the user's situation, considering both interlocutors' emotions and guiding responses generation through dialogue act, namely by generating empathetic responses from three perspectives: cognition, affection and behavior. Extensive experiments based on benchmark metrics have shown that our method CAB outperforms the state-of-the-art methods, demonstrating the effectiveness of our method in improving empathy of the generated responses.

Acknowledgments. This work was supported by the National Natural Science Foundation of China (61672144, 61872072).

References

1. Chen, M.Y., Li, S., Yang, Y.: EmpHi: generating Empathetic Responses with Human-like Intents. In: NAACL-HLT 2022, pp. 1063–1074 (2022)
2. Elliott, R., Bohart, A.C., Watson, J.C., Murphy, D.: Therapist empathy and client outcome: An updated meta-analysis. Psychotherapy **55**(4), 399–410 (2018)
3. Gao, J., et al.: Improving Empathetic Response Generation by Recognizing Emotion Cause in Conversations. In: EMNLP 2021, pp. 807–819 (2021)
4. Gao, P., Han, D., Zhou, R., Zhang, X., Wang, Z.: CAB: empathetic dialogue generation with cognition, affection and behavior. arXiv preprint arXiv:2302.01935 (2023)
5. Ghosal, D., Majumder, N., Poria, S., Chhaya, N., Gelbukh, A.F.: DialogueGCN: a graph convolutional neural network for emotion recognition. In: EMNLP-IJCNLP, pp. 154–164 (2019)
6. Gladstein, G.A.: Empathy and counseling outcome: an empirical and conceptual review. Couns. Psychol. **6**(4), 70–79 (1977)
7. Kim, T., Vossen, P.: Emoberta: speaker-aware emotion recognition in conversation with roberta. (2021) arXiv preprint arXiv:2108.12009
8. Kingma, D.P., Ba, J.: Adam: a Method for Stochastic Optimization. In: 3rd International Conference on Learning Representations (2019)
9. Li, J., Galley, M., Brockett, C., Gao, J., Dolan, B.: A Diversity-Promoting Objective Function for Neural Conversation Models. In: NAACL HLT 2016, pp. 110–119 (2016)

10. Li, Q., Chen, H., Ren, Z., Ren, P., Tu, Z., Chen, Z.: EmpDG: multi-resolution Interactive Empathetic Dialogue Generation. In: ICCL-20, pp. 4454–4466 (2020)
11. Li, Q., Li, P., Ren, Z., Ren, P., Chen, Z.: Knowledge Bridging for Empathetic Dialogue Generation. In: AAAI-22, pp. 10993–11001 (2022)
12. Lin, Z., Madotto, A., Shin, J., Xu, P., Fung, P.: MoEL: Mixture of Empathetic Listeners. In: EMNLP-IJCNLP, pp. 121–132 (2021)
13. Liu, C., Wang, Y., Yu, G., Wang, Y.: A review of relevant theories of empathy and exploration of new dynamic models. Adv. Psychol. Sci. (5), 9 (2009)
14. Liu, S., Zheng, C., Demasi, O., Sabour, S., Huang, M.: Towards Emotional Support Dialog Systems. In: ACL/IJCNLP(1), pp. 3469–3483 (2021)
15. Liu, Y., Maier, W., Minker, W., Ultes, S.: Empathetic dialogue generation with pre-trained roberta-gpt2 and external knowledge. (2021) arXiv preprint arXiv:2109.03004
16. Liu, Y., Du, J., Li, X., Xu, R.: Generating Empathetic Responses by Injecting Anticipated Emotion. In: ICASSP, pp. 7403–7407 (2021)
17. Majumder, N., Hong, P., Peng, S., Lu, J., Poria, S.: MIME: MIMicking Emotions for Empathetic Response Generation. In: EMNLP(1), pp. 8968–8979 (2020)
18. Mihalcea, R., Tarau, P.: TextRank: bringing Order into Text. In: EMNLP, pp. 404–411 (2004)
19. Pennington, J., Socher, R., Manning, C.D.: Glove: global Vectors for Word Representation. In: EMNLP, pp. 1532–1543 (2014)
20. Rashkin, H., Smith, E.M., Li, M., Boureau, Y.: Towards Empathetic Open-domain Conversation Models: a New Benchmark and Dataset. In: ACL 2019, pp. 5370–5381 (2019)
21. Sabour, S., Zheng, C., Huang, M.: CEM: commonsense-Aware Empathetic Response Generation. In: AAAI-22, pp. 11229–11237 (2022)
22. Vaswani, A., et al.: Attention is All you Need. In: NIPS-17, pp. 5998–6008 (2017)
23. Vinyals, O., Fortunato, M., Jaitly, N.: Pointer Networks. In: NEURIPS 2015, pp. 2692–2700 (2015)
24. Vinyals, O., Le, Q.V.: A neural conversational model. (2015) arXiv preprint-arXiv:1506.05869
25. Wang, J., Li, W., Lin, P., Mu, F.: Empathetic response generation through graph-based multi-hop reasoning on emotional causality. Knowl. Based Syst. **233**, 107547 (2021)
26. Wang, L., et al.: Cass: Towards building a social-support chatbot for online health community. Proc. ACM Hum. Comput. Interact. **5**(CSCW1), 1–31 (2021)
27. Welivita, A., Pu, P.: A Taxonomy of Empathetic Response Intents in Human Social Conversations. In: COLING 2020, pp. 4886–4899 (2020)
28. Zheng, C., Liu, Y., Chen, W., Leng, Y., Huang, M.: CoMAE: a Multi-factor Hierarchical Framework for Empathetic Response Generation. In: Findings of the Association for Computational Linguistics: ACL/IJCNLP, pp. 813–824 (2021)

Multimodal Entity Linking with Mixed Fusion Mechanism

Gongrui Zhang⬤, Chenghuan Jiang⬤, Zhongheng Guan⬤,
and Peng Wang$^{(\boxtimes)}$⬤

Southeast University, Nanjing, China
{grzhang,chjiang,goufugui,pwang}@seu.edu.cn

Abstract. Many efficient multimodal entity linking (MEL) methods
have been developed in recent years. However, most MEL methods still
suffer from two drawbacks. On the one hand, the inconsistency of modal
encoding brings the semantic gap between modalities in the feature
space and blocks the multimodal fusion. On the other hand, previous
attention-based multimodal fusions cannot efficiently handle noise. To
address these issues, we propose a Multimodal Encoder Representation
from Transformers for Multimodal Entity Linking (Mert-MEL). Firstly,
we concatenate contexts of mentions and Wikidata abstracts of candidate entities as inputs. Then we utilize transformer encoders to extract
features of both textual and visual information, and employ contrastive
learning to better align feature spaces. We also incorporate phrase-level
text embeddings to get rich textual representations. Subsequently, we use
a combination of global fusion and bottleneck fusion to integrate multimodal information and extract key information instead of noise. Finally,
we send the fused embeddings to an MEL head to predict the matching
scores between the mention and the candidate entities, and then link the
mention to the candidate with the highest score. Experiments demonstrate that Mert-MEL prominently outperforms strong baselines on two
MEL datasets.

Keywords: Multimodal entity linking · Contrastive learning ·
Multimodal fusion

1 Introduction

Entity linking (EL) is a crucial downstream task of natural language processing
(NLP), which aims to link a given mention to the corresponding entity in a knowledge graph (KG) [25]. EL is widely applied in many areas, such as Web search
[2], information extraction [16], question answering [21] and so on. Most existing EL work only considers the text modality, which is able to achieve excellent
performance if the text contains enough information. However, in social media
platforms like forums and blogs, people tend to present multimodal information
such as a sentence with pictures or videos. The diversity of modalities guarantees the adequacy of information, while a single modality is possibly not able

© The Author(s), under exclusive license to Springer Nature Switzerland AG 2023
X. Wang et al. (Eds.): DASFAA 2023, LNCS 13945, pp. 607–622, 2023.
https://doi.org/10.1007/978-3-031-30675-4_45

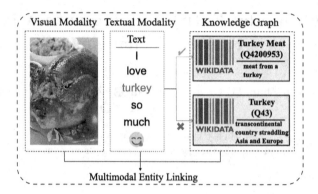

Fig. 1. An example of Multimodal Entity Linking. *turkey* links to the turkey meat.

to provide enough information. For example, as shown in Fig. 1, if a social post only contains the sentence *I love turkey so much*, it is hard to determine whether the *turkey* in the post is chicken or a country, as well as the EL models. But when the picture is given, we are confident to assert that *turkey* should be linked to *Turkey Meat (Q4200953)* instead of *Turkey (Q43)*. Thus, in order to develop EL models that have the ability to process multimodal information like human, multimodal entity linking (MEL) is proposed to improve the performance of raw EL tasks with the introduction of visual information [29].

The core of MEL models is fusing multimodal information. Most existing models apply convolution neural network (CNN) to encode images and apply bidirectional long short-term memory (BiLSTM) or pretrained language models (PLMs) to encode texts. And then they utilize co-attention mechanism and gated fusion layer to integrate the feature of two modalities into one embedding, which is used for downstream tasks. These models have achieved excellent results, but there are still two drawbacks:

– **Semantic gap between modalities.** Many models conventionally employ CNN and BiLSTM as encoders, while new architecture like Transformer [28] is proved to have better performance on encoding data [10,17]. Models applying the transformer-based BERT [8] to encode texts have gained improvement, while they still use CNN to encode images. As reported in [26], different architectures of two encoders could create a semantic gap between the feature space of the representations of two modalities, increasing the difficulty of multimodal fusion.
– **Noise during multimodal fusion.** Most MEL models use co-attention mechanism to integrate the two modalities in self-attention layers [29], where the query and key matrices are from different modalities. It is a fine-grained method to fuse their features, which also brings much noise from each modality into the final embedding. Additionally, the performance of self-attention is also affected by the semantic gap, because it is hard to determine the attention between embeddings in different feature space, and the false attention obviously brings noise.

To address the above issues, we propose a Multimodal Encoder Representation from Transformers for MEL (Mert-MEL). Specifically, we firstly use Wikidata to obtain abstracts of candidate entities, and construct textual inputs by concatenating the contexts of mentions and the abstracts. Then, in order to bridge the semantic gap, we utilize transformer encoders with the same architecture to extract features of visual information and token-level features of textual information. We also utilize one-dimensional convolution on the token-level features to generate phrase-level features. Then we use a mixed fusion mechanism to integrate the three features, which consists of two global fusion layers and a bottleneck fusion layer. In the global fusion layer, we concatenate the visual features and one of the two textual features respectively, and send them to a multi-layer transformer for fully information exchange. Then we reassemble the features and send them to the bottleneck fusion layer. The bottleneck fusion layer is also a multi-layer transformer but only allows them to exchange information through several units, which limits the flow of information and forces the model into collecting key information, instead of noise. Consequently, we obtain the embeddings with multimodal information from the bottleneck fusion layer. The above part of our model is called Mert. In order to further bridge the semantic gap and obtain better embeddings, we employ contrastive learning [4] to pretrain Mert. Finally, we send the embeddings generated by the pretrained Mert to an MEL head which consists of two fully connected layers and a softmax layer, to predict the matching score between the mention and the candidate entity. Then, the mention is linked to the candidate with the highest matching score.

The main contributions of our work can be summarized as follows:

- We propose a transformer-only MEL model Mert-MEL, unifying architectures of encoders for different modalities and bridging the semantic gap between them.
- We are among the first to apply bottleneck fusion to MEL, based on which we propose a mixed fusion mechanism for fusing information from different modalities and reducing noise during information exchange effectively.
- We conduct comprehensive experiments and analysis on real-world datasets, and results demonstrate that Mert-MEL outperforms strong baselines.

2   Related Work

Multimodal Fusion. Multimodal fusion is the process of fusing representations of different modalities into a single representation. Typically, it can be divided into three categories: early fusion, middle fusion and late fusion. Early fusion is simple but ignores the heterogeneity among different modalities [11]. Late fusion can capture heterogeneous information but has high computational cost [27]. Middle fusion mitigates the disadvantage of the above two strategies because it can capture the deep interactive information from different modalities [9,30]. The above fusion strategies indeed improve the model performance. However, they are unable to align the semantic space of the modalities well, which may bring much noise into the final representation. In our models, we employ the bottleneck fusion [23], which forces information of different modalities to pass through several bottleneck units with lower computational cost.

Contrastive Learning. Contrastive learning [4] is widely used to align the feature space of representations of different modalities. Alec et al. [24] proposes a simple but efficient contrastive pretraining method. It predicts the probability that a caption matches a given image. Singh et al. [26] proposes a language and vision alignment model which learns strong representations through joint pretraining on both unimodal and multimodal data, encompassing cross-modal alignment objectives and multimodal fusion objectives. Wang et al. [29] proposed a multimodal entity linking method with gated hierarchical multi-modal fusion and contrastive training, which is able to catch fine-grained inter-model correlations. In our work, we adapt the method in CLIP [24] to pretrain our model before input data into the MEL module. The method is efficient and help mitigate the semantic space between two modalities.

Multimodal Entity Linking. The MEL task maps mentions with multimodal information to entities in a structured knowledge base (KB) such as Wikidata [13]. Bunescu et al. [3] first proposed to link a named mention to an entity in a knowledge base. Existing EL methods can be divided into two categories: local approaches [6,7] and global approaches [14,15]. Local approaches disambiguate mentions individually using lexical mention-entity measures and contextual features. Global approaches make use of document-level features to disambiguate all the mentions in the documents. They can also model semantic relationships between entities in the KB. Moon et al. [22] pioneered the study of multi-modal entity linking. It leverages modal attention mechanism to fuse features of images, texts and characters. Zhang et al. [34] proposes a model which highlights the effect of removing the negative impact of noisy images and leverages multiple attention mechanisms to get richer information from the texts and images with mentions and their corresponding candidate entities.

3   Methodology

3.1   Preliminary

Let $\mathcal{D} = \{x_i\}_{i=1}^{N}$ denote the input multimodal samples, where $x_i = \{x_i^t, x_i^v\}$. The textual input $x^t = \{\mathbf{s}, (m_1, m_2, \cdots, m_n)\}$ contains a sentence $\mathbf{s} = \{s_1, s_2, \cdots, s_{l_s}\}$ and mentions $m_i \in \mathbf{s}$ in the sentence, where s_i means the i-th token of $\mathbf{s}$ and l_s is the length of the sentence. The visual input x^v is an image in RGB format.

The MEL task aims to link the mentions in a sentence to the corresponding entities in a KG. Let $\mathcal{K}$ denote the used KG, $e \in \mathcal{K}$ denote the ground truth entity of a mention m in the sentence s. θ denotes parameters of the model and $\hat{e}$ denotes the linked entity. Then the problem is to maximize the probability that the model links m to e:

$$\underset{\theta}{\text{argmax}}\, P(\hat{e} = e | x^t, x^v, \mathcal{K}, m, \theta) \tag{1}$$

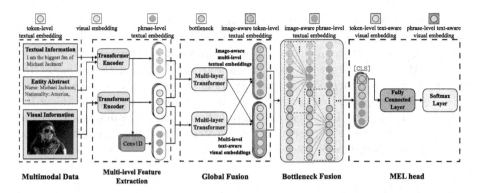

Fig. 2. The architecture of Mert-MEL

3.2 Overview of Mert-MEL

Figure 2 shows the architecture of Mert-MEL. The model consists of four components: (1) multi-level feature extraction, (2) global fusion, (3) bottleneck fusion, and (4) MEL head. Different from conventional model like GHMFC [29], our model treats MEL as a binary classification task. Therefore, we specifically design the input data at first. Then a sample goes through the multi-level feature extraction module. After that, the output embeddings are fused by the global fusion module. Subsequently, the fused multimodal representations go through the bottleneck fusion module for further refining. Finally, a MEL head is performed on the final image-aware textual embedding to predict the matching score between the mention and the candidate entity.

3.3 Input Design

In Mert-MEL, the semantic information of a mention m is represented by the whole sentence $\mathbf{s} = \{s_1, s_2, \cdots, s_{l^s}\}$ that m belongs to, where s_i is the i-th token and l^s is the sentence length. And an entity e is represented by its properties and the corresponding values in the KG, which is called the entity abstract and denoted by $\mathbf{a} = \{a_1, a_2, \cdots, a_{l^a}\}$, where a_i is the i-th token and l^a is the abstract length. For example, the entity *Michael Jackson* is represented as *Name: Michael Jackson, Gender: male, Occupation: singer, dancer* $\cdots$. Then we concatenate $\mathbf{s}$ and $\mathbf{a}$ as $x^c = (\texttt{[CLS]}, s_1, s_2, \cdots, s_{l^s}, \texttt{[SEP]}, a_1, a_2, \cdots, a_{l^a})$. Finally, x^c and the visual input x^v are fed into the model together.

Note that a sample is positive if the mention matches the entity, otherwise it is negative. During training, the number of positive samples and negative samples in a batch is equal, in order to avoid the imbalanced classification problem [19].

3.4 Preprocess

The Mert-MEL model takes a pair (or a batch of pairs) of image and text as input. The input image is in form of a matrix with RGB channels, and the text is a string. The processor contains two parts. The visual feature extractor do a series of transforming jobs. It resizes the image to a unified size of (C, H, W), where C, H and W mean channel, height and width respectively. Then it normalizes the image data with given mean and standard deviation. Finally, the image is divided into $P \times P$ patches. By default, $C = 3$, $W = H = 224$, and $P = 16$. The tokenizer for the input text generates the corresponding token IDs and attention masks. The sequence length is limited to a fix value (such as 64 or 128 in practice), while the rest part is truncated. The preprocessed images and sentences are directly sent to encoders.

3.5 Encoders

We use a transformer encoder to extract the contextual features of the text. The text encoder outputs the hidden state vectors $\mathbf{h}^t$ of the input texts, including the additional $\mathbf{h}^t_{\text{CLS}}$ for the [CLS] token.

The image encoder adopts the ViT architecture [10]. Similar to the text encoder, we flatten the $P \times P$ patches into a sequence, with an additional [CLS] token ahead, and feed it into the encoder. The image encoder outputs hidden state vectors $\mathbf{h}^v$ likewise.

3.6 Multi-level Feature Extraction

Generally, the semantic information of tokens close to an entity is beneficial to disambiguation, and the image is more likely to be related to a phrase in the sentence. Hence, inspired by the work in [29], we not only extract visual embeddings and token-level textual embeddings in this module, but also extract phrase-level textual embeddings.

The textual input x^c and visual input x^v are fed into the text encoder and the image encoder separately, then we get the token-level textual embeddings $\mathbf{h}^t \in \mathbb{R}^{l^c \times d}$ and visual embeddings $\mathbf{h}^v \in \mathbb{R}^{l^v \times d}$ respectively, where l^c and l^v are the length of x^c and the patch number of x^v respectively, d is the dimension of the feature space. Afterwards, k one-dimensional convolution kernels $[\mathbf{W}_1, \mathbf{W}_2, \cdots, \mathbf{W}_k]$ are performed on $\mathbf{h}^t$, whose kernel size are $[l_1, l_2, \cdots, l_k]$. Then the local feature of the j-th token extracted by the i-th kernel is

$$\mathbf{p}_{i,j} = \text{ReLU}\left(\mathbf{W}_i \cdot \mathbf{h}^c_{[j:j+l_i-1]} + \mathbf{b}_i\right). \tag{2}$$

After that, we perform max pooling on all local features extracted by the k kernels for each token, so as to retain the most prominent features.

$$\mathbf{p}_j = \max\{\mathbf{p}_{1,j}, \mathbf{p}_{2,j}, \cdots, \mathbf{p}_{k,j}\} \tag{3}$$

Now we get the phrase-level embeddings $\mathbf{h}^p = [\mathbf{p}_1, \mathbf{p}_2, \cdots, \mathbf{p}_{l^c}], \mathbf{p}_i \in \mathbb{R}^d$ of the textual input. The two levels of token embeddings constitute multi-level features of textual inputs, which enrich the semantic information of entities and contribute to disambiguation.

3.7 Global Fusion

We concatenate $\mathbf{h}^t$ and $\mathbf{h}^v$ as $[\mathbf{h}^t, \mathbf{h}^v]$, then put it into a multi-layer transformer. Finally, we get $[\mathbf{h}^t_v, \mathbf{h}^v_t]$, where $\mathbf{h}^t_v$ is called image-aware token-level textual embeddings and $\mathbf{h}^v_t$ is called token-level text-aware visual embeddings.

Similarly, we put $[\mathbf{h}^p, \mathbf{h}^v]$ into another multi-layer transformer and get image-aware phrase-level textual embeddings $\mathbf{h}^p_v$ and phrase-level text-aware visual embeddings $\mathbf{h}^v_p$.

$$[\mathbf{h}^t_v, \mathbf{h}^v_t] = \mathrm{Transformer}_t\left([\mathbf{h}^t, \mathbf{h}^v]\right) \tag{4}$$

$$[\mathbf{h}^p_v, \mathbf{h}^v_p] = \mathrm{Transformer}_p\left([\mathbf{h}^p, \mathbf{h}^v]\right) \tag{5}$$

In the end, we obtain the image-aware multi-level textual embeddings $\mathbf{T}^v$ by concatenating $\mathbf{h}^t_v$ and $\mathbf{h}^p_v$, and obtain the multi-level text-aware visual embeddings $\mathbf{V}^t$ by concatenating $\mathbf{h}^v_t$ and $\mathbf{h}^v_p$.

$$\mathbf{T}^v = [\mathbf{h}^t_v, \mathbf{h}^p_v] \tag{6}$$

$$\mathbf{V}^t = [\mathbf{h}^v_t, \mathbf{h}^v_p] \tag{7}$$

3.8 Bottleneck Fusion

After global fusion, we pair $\mathbf{T}^v$ and $\mathbf{V}^t$, and feed them into the bottleneck fusion module [23].

This module is intended for further information interchange between the two modalities, which forces the model into concentrating on the most relevant information in each modality. This module stacks several layers of transformer encoder. Specifically, we initialize bottlenecks $\mathbf{B} = [\mathbf{b}_1, \mathbf{b}_2, \cdots, \mathbf{b}_{|b|}]$, where $\mathbf{b}_i \in \mathbb{R}^d$ is a zero vector and $|b|$ is the number of bottlenecks. In each fusion layer, the concatenation of textual embeddings and bottlenecks goes through a transformer encoder, and then the resulted bottlenecks goes through another transformer encoder with visual embeddings. The update process can be written as follows:

$$\left[\mathbf{T}^v_{\ell+1} \| \hat{\mathbf{B}}_{\ell+1}\right] = \mathrm{Transformer}^v_{(\ell)}\left([\mathbf{T}^v_\ell \| \mathbf{B}_\ell]\right) \tag{8}$$

$$\left[\mathbf{V}^t_{\ell+1} \| \mathbf{B}_{\ell+1}\right] = \mathrm{Transformer}^t_{(\ell)}\left(\left[\mathbf{V}^t_\ell \| \hat{\mathbf{B}}_{\ell+1}\right]\right) \tag{9}$$

The final fused textual embeddings $\mathbf{T}^v_b$ are used in the later downstream task.

3.9 Contrastive Learning

So far, the above part of the whole model is called Mert, which is a multimodal encoder essentially. Before applying Mert to MEL, we firstly use contrastive learning to pretrain Mert, inspired by CLIP [24].

Given a batch containing m multimodal samples $\{x_1, \cdots, x_m\}$, let $\mathbf{E}^v_i, \mathbf{E}^t_i \in \mathbb{R}^d$ denote the L2-normalized embedding of the [CLS] token in $\mathbf{V}^t_b$ and $\mathbf{T}^v_b$ of the

i-th sample respectively. And $\mathbf{E}^v = [\mathbf{E}_1^v, \mathbf{E}_2^v, \cdots, \mathbf{E}_m^v]^T$, $\mathbf{E}^t = [\mathbf{E}_1^t, \mathbf{E}_2^t, \cdots, \mathbf{E}_m^t]^T$. $\mathbf{E}^v, \mathbf{E}^t \in \mathbb{R}^{m \times d}$. Then we calculate the scaled pairwise cosine similarities,

$$\mathbf{L} = \mathbf{E}^v \mathbf{E}^{t^T} e^t, \tag{10}$$

where t is a trainable temperature parameter, and $\mathbf{L} \in \mathbb{R}^{m \times m}$.

Let $\mathbf{y} = [1, 2, \cdots, m]$ be the labels, and then we obtain the total symmetric loss with the following formula:

$$\mathcal{L}_v = \text{CrossEntropyLoss}(\mathbf{L}, \mathbf{y}), \tag{11}$$

$$\mathcal{L}_t = \text{CrossEntropyLoss}(\mathbf{L}^T, \mathbf{y}), \tag{12}$$

$$\mathcal{L} = (\mathcal{L}_v + \mathcal{L}_t)/2. \tag{13}$$

The goal of pretraining is to minimize $\mathcal{L}$ so that the distance between a textual embedding and a visual embedding gets shortened if they are from the same sample, otherwise gets enlarged.

3.10   MEL Head

The MEL head utilizes the bottleneck-fused features $\mathbf{T}_b^v = [\mathbf{t}_{b1}^v, \mathbf{t}_{b2}^v, \cdots, \mathbf{t}_{b2l^c}^v]$, $\mathbf{t}_{bi}^v \in \mathbb{R}^d$, which is output by the pretrained Mert, to predict the matching score between the mention m and the candidate entity e.

We take out $\mathbf{h}_{\text{CLS}}$ from $\mathbf{T}_b^v$, which is the embedding of the [CLS] token and has fused the semantic information of m and e with the help of visual modality. In our model, $\mathbf{h}_{\text{CLS}}$ is exactly $\mathbf{t}_{b1}^v$, because we set [CLS] as the first token. Subsequently, we send $\mathbf{h}_{\text{CLS}}$ into two fully connected layers to obtain the logits. And then we get the probabilities that whether m matches e, after softmax.

$$\widetilde{\mathbf{h}}_{\text{CLS}} = \text{Tanh}(\mathbf{W_1}\mathbf{h}_{\text{CLS}} + \mathbf{b_1}) \tag{14}$$

$$\mathbf{l}_{m:e} = \mathbf{W_2}\widetilde{\mathbf{h}}_{\text{CLS}} + \mathbf{b_2} \tag{15}$$

$$\mathbf{p}_{m:e} = \text{Softmax}(\mathbf{l}_{m:e}) \tag{16}$$

where $\mathbf{W_1}, \mathbf{b_1}$ and $\mathbf{W_2}, \mathbf{b_2}$ are the weights and biases of the two fully connected layers respectively, and $\mathbf{l}_{m:e} = [l_{m:e}^t, l_{m:e}^f]$, where $l_{m:e}^t$ is the score that m matches e, and $l_{m:e}^f$ is the score that m does not match e. Here $l_{m:e}^t$ is called the matching score between m and e.

For training, we minimize the cross entropy loss:

$$\mathcal{L} = \text{CrossEntropyLoss}(\mathbf{p}_{m:e}, \mathbf{y}) \tag{17}$$

where $\mathbf{y}$ is the ground truth label.

For inference, we link m to the candidate $\widetilde{e}$ that has the maximum matching score:

$$\widetilde{e} = \text{argmax}_e \, l_{m:e}^t \tag{18}$$

4   Experiments

4.1   Datasets, Baselines and Settings

Datasets. The dataset we use to pretrain the Mert model is *flickr 30k* [33], which has been used for many multimodal tasks. It contains 31,783 images, which are photographs of daily activities, events and scenes. Each image accompanies 5 descriptive sentences from different perspectives.

We carry out experiments on two MEL datasets: Wiki-MEL and Richpedia-MEL [35]. In both datasets, the textual and visual descriptions of mentions are collected from Wikipedia, and each mention can be linked to an entity in Wikidata. The Wiki-MEL dataset contains over 22k samples, and the Richpedia-MEL dataset has about 17.8k samples. In the Richpedia-MEL dataset, most samples have multiple descriptive images, and we only take the first image of a sample as the visual input.

To obtain the candidate entities of a mention, we input the mention into the Wikidata look-up API, and then choose the first 100 entities of the returned results as the candidates, after removing the ground truth entity. And we use the first candidate to construct the negative sample of the mention. Finally, we collect essential information of all entities in the datasets to construct their abstracts, including basic properties of human, such as sex, occupation, languages, birth date, death date, and brief description. The statistics of datasets are summarized in Table 1.

Table 1. Summary statistics for the datasets.

Dataset	Samples	Mentions	Text length (avg.)	Mentions (avg.)	Neg. candidates (avg.)
Wiki-MEL	22,070	25,846	8.2	1.17	1.85
Richpedia-MEL	16,851	16,922	16.0	1.00	1.71

Baselines

- **ARNN** [12], which combines RNN implemented by GRU unit [5] with attention mechanism, and uses the encoder-decoder structure.
- **BERT** [8], which encodes tokenized text with multi-layer transformers, and has been pretrained with specific objectives.
- **BLINK** [32], which uses a bidirectional encoder to model entities and a cross encoder for EL, both based on BERT, to achieve zero-shot EL.
- **JMEL** [1], which mainly uses fully connected layers and simple concatenation, and encodes the contexts of mentions and entities separately.
- **DZMNED** [22], which employs modal attention mechanism to fuse embeddings from CNN and BiLSTM. It predicted entity in the KG on the basis of all extracted context information.
- **DZMNED-BERT**, a variant of DZMNED, which uses BERT to extract textual features.

- **HieCoATT-Alter** [21], which considers question hierarchy, and computes word-level, phrase-level and question-level features. It involves parallel co-attention and alternating co-attention mechanism.
- **MEL-HI** [34], which incorporates multiple attention mechanism in LSTM to improve visual representation with noise.
- **GHMFC** [29], which applies co-attention mechanism to extract multimodal features, and uses BERT to extract entity features. It performs entity linking according to cosine similarity between mentions and entities.
- **ViLT** [18], a simple multimodal model using transformer. It directly uses BERT and ViT to process textual and visual features, and sends the embeddings to a transformer encoder.
- **FLAVA** [26], which is a transformer-based multimodal model pretrained with many objectives, and is efficient on most vision recognition and language understanding tasks. This model serves as a basic component of Mert-MEL.
- **Mert-MEL**, the model we proposed in this paper.

Settings. Both image encoder and text encoder are transformer encoders with the ViT [10] architecture, and pretrained in FLAVA [26], which have 12 heads and 12 layers, and the intermediate size is 3072. The size of the input image is 224×224 and the patch size is 16×16. The dimensions of visual features and textual features are all set to 768. The phrase-level feature extractor consists of 3 Conv1D modules with kernel size 3 and padding 1, which ensures unchanged sequence length. The Multi-layer Transformer in Global Fusion has the same dimensions as the encoders, which is also pretrained in FLAVA but only has 6 layers. The bottleneck fusion layer has 4 bottlenecks, 8 heads and 6 layers with 0.1 dropout probability. At the end, the MEL head has 2 linear layers with size 768-768-2. For pretraining Mert and training Mert-MEL, we optimize all parameters by AdamW [20] with learning rate 5×10^{-5}, batch size 32, dropout 0.4, and epoch 5. All experiments are conducted on a NVIDIA® V100 GPU.

We use the Top-k accuracy as the evaluation metric, which can be calculated as follows:

$$\text{Accuracy}_{\text{top}-k} = \frac{1}{N} \sum_{i=1}^{N} \text{I}(e_i \in \widetilde{E}_k) \tag{19}$$

where N is the total number of samples, e_i is the ground truth entity of the i-th sample, and $\widetilde{E}_k$ is the set of candidates whose matching scores rank top k among all the candidates of the i-th sample. I is the indicator function, which is set to 1 if the receiving condition is satisfied, and 0 otherwise.

The source code is available at https://github.com/aanonymity/mert-mel.

4.2   Main Experimental Results

In the experiments we conduct, we evaluate on the performance of Mert-MEL, and compare it with other baselines. Table 2 shows the Top-1,5,10,20 accuracy results of all models on the two datasets.

Table 2. MEL results of the compared models at Top-1, 5, 10, 20 accuracy (%). (T: textual modal, V: visual modal, C: character). † stands for results produced by original implementations, and baselines with ‡ are implemented according to corresponding papers. Other baselines are implemented by the Transformers library [31].

Modalities	Models	Wiki-MEL				Richpedia-MEL			
		Top-1	Top-5	Top-10	Top-20	Top-1	Top-5	Top-10	Top-20
T	ARNN † [12]	32.0	45.8	56.6	65.0	31.2	39.3	45.9	54.5
T	BERT † [8]	31.7	48.8	57.8	70.3	31.6	42.0	47.6	57.3
T	BLINK † [32]	30.8	44.6	56.7	66.4	30.8	38.8	44.5	53.6
T+V	JMEL ‡ [1]	31.3	49.4	57.9	64.8	29.6	42.3	46.6	54.1
C+T+V	DZMNED ‡ [22]	30.9	50.7	56.9	65.1	29.5	41.6	45.8	55.2
C+T+V	DZMNED-BERT ‡	34.7	53.9	58.1	70.1	32.4	43.7	48.2	60.8
T+V	HieCoATT-Alter ‡ [21]	40.5	57.6	69.6	78.6	37.2	46.8	54.2	62.4
T+V	MEL-HI ‡ [34]	38.6	55.1	65.2	75.7	34.9	43.1	50.6	58.4
T+V	GHMFC † [29]	43.6	64.0	74.4	85.8	38.7	50.9	58.5	66.7
T+V	ViLT [18]	24.9	58.9	76.5	90.4	25.5	61.9	76.7	87.4
T+V	FLAVA [26]	37.0	73.4	85.5	95.2	56.9	81.5	87.7	**95.4**
T+ V	Mert-MEL	**43.9**	**80.1**	**92.8**	**98.0**	**59.1**	**82.3**	**88.7**	93.8

Firstly, we observe that models with visual modality generally outperforms text-only models. Multimodal models such as MEL-HI, GHMFC and our Mert-MEL have significantly higher Top-1 accuracy than text-only models such as ARNN, BERT and BLINK. Specifically, the Top-1 accuracy of Mert-MEL is 12.2% and 27.5% higher than that of BERT on Wiki-MEL and Richpedia-MEL respectively. This indicates the necessity to mine the underlying information in visual modality, which could largely improve the accuracy of MEL.

Secondly, models with advanced fusion method and stronger feature extractors perform better than others. While JMEL simply employs fully connected layers and embedding concatenation, GHMFC applies co-attention mechanism and gated fusion, whose Top-1 accuracy is 12.3% and 9.1% higher than JMEL on Wiki-MEL and Richpedia-MEL respectively. Our Mert-MEL applies a mixed fusion mechanism which can better deal with the noise than co-attention, and thus gains improvement in all metrics to varying degrees, compared with GHMFC. In addition, models employing the transformer-based encoders tend to have better performance. Specifically, DZMNED-BERT has 2–5% improvement in each metric compared with the raw DZMNED. FLAVA and Mert-MEL, which employ transformer encoder to extract textual and visual features, rank the top 2 in all metrics. This demonstrates that stronger feature extractors can prominently improve the performance of MEL models.

Thirdly, the method for entity linking is comparatively crucial to the results. Models such as BERT, GHMFC and HieCoATT-Alter extract multimodal mention features and entity features with different encoders respectively, and then perform entity linking based on the cosine similarities between the features. Obviously, this method hardly achieves high performance because the inconsistency of encoders brings a semantic gap between the feature space of mentions

618 G. Zhang et al.

and entities, which leads to the cosine similarity not working well. By contrast, Models with a different entity linking method, such as FLAVA and Mert-MEL, achieve much higher performance, where the Top-1 accuracy of FLAVA is 18.2% higher than that of GHMFC on Richpedia-MEL. These models concatenate the contexts of the mention and the entity abstract as the input text, and then use fully connected layers to calculate the matching score from the fused textual embeddings, which utilizes the ability of neural network to mine the underlying information from features.

Finally, the performance of Mert-MEL takes the lead in almost all metrics on the two datasets. Compared with the second best model FLAVA, Mert-MEL has 6.9% and 2.2% improvement in the Top-1 accuracy on the Wiki-MEL and Richpedia-MEL respectively. The image encoder and text encoder of Mert-MEL have the same architecture, which is helpful to align the feature space of the two modalities. The global fusion layer and the bottleneck fusion layer also make a big difference to the performance, which we will discuss in the next section.

4.3 Ablation Experiments

In this section, we study the effect of each module in Mert-MEL via ablation experiments on Wiki-MEL. Table 3 shows the performance of 5 incomplete models. Obviously, the removal of any module leads to lower performance.

Phrase-Level Feature Extraction. We extract phrase-level textual features in order to enrich the representation of texts and capture more accurate semantic information. Absence of phrase-level feature leads to all the Top-k accuracy dropping in varying degrees, among which the reduction of Top-1 accuracy reaches 4.5%. Phrase-level features not only enrich the textual information together with token-level feature, but also contribute to better attention weights in the cross-modal attention layer of global fusion and bottleneck fusion module, because the visual information may not only be related to a single word in the sentence, but more likely to be related to a phrase. The result proves that this multi-level feature extraction absolutely improves the performance of the whole model.

Global Fusion. The global fusion module stacks multiple layers of multi-head attention, which plays the role of co-attention between two modalities, and

Table 3. Results of Mert-MEL ablation experiments on Wiki-MEL (BF: Bottleneck Fusion, CL: Contrastive Training, PT: Phase-Level Encoder, GF: Global Fusion).

Models	Top-1	Top-5	Top-10	Top-20
Mert-MEL	**43.9**	**80.1**	**92.8**	**98.0**
- BF	36.1	72.8	86.9	94.7
- CL	38.8	78.0	88.5	96.2
- PT	39.4	79.3	89.8	96.6
- GF	36.5	76.8	90.3	97.2

ensures each embedding can absorb global information from another modality so as not to leave out key points. After removing this module, the destructive effect turns most notably in Top-1 accuracy, which decreases by 7.8%.

Bottleneck Fusion. The bottleneck fusion module is also crucial to our model. It sets several bottlenecks and limits features of the two modalities to communicate only through them. This forces the bottlenecks to collect the most important information and then refine the features. Consequently, the noise brought by global fusion is also eliminated. Without bottleneck fusion, the Top-1 and Top-5 accuracy of the model ranks last with a terrific reduction about 7%.

Contrastive Learning. The contrastive learning helps to align the feature space of the two modalities, making the model easier to be trained on MEL task. After removing this step, the Top-1 accuracy decreases by 5.1%. The reduction of performance proves that contrastive pretraining is an effective strategy.

Results of ablation experiments show that all the modules of Mert-MEL boost the performance of the whole model, especially the two fusion modules. The improvements brought by contrastive learning and phrase-level feature extraction are relatively small, but they still make a difference and cannot be ignored.

4.4 Parameter Sensitivity Analysis

In this section, we test the performance of our model with different parameters of bottleneck fusion module on Wiki-MEL, which is shown in Fig. 3. Firstly, we fix the number of Bottleneck Fusion layers to 6, epoch number to 5, and then train the model with the number of bottlenecks equal to 2, 4, 8, 16, 32, 64 and 128 respectively. Results indicate that setting 4 bottlenecks is the best choice. Fewer bottlenecks are not enough to contain key information, while too many bottlenecks means more noise comes in. This is consistent with the trend that the accuracy starts to fall when the number of bottlenecks grows over than 4.

After that, we fix the number of bottlenecks to 4, epoch number to 5, and then train the model with the number of bottleneck fusion layers equal to 2, 4, 6, 8, 10 and 12 respectively. Figure 3(b) shows that the performance of model reaches the peak when the number is 6. Fewer layers mean that the two modalities

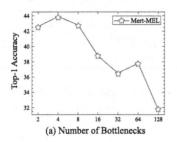

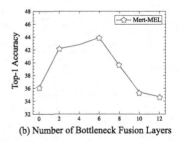

(a) Number of Bottlenecks   (b) Number of Bottleneck Fusion Layers

Fig. 3. Top-1 accuracy with different parameters of bottleneck fusion module.

have fewer opportunities to obtain information from each other. However, the Top-1 accuracy drops sharply when the number of layers increases to 8 from 6, proving that too many interactions between the two modalities bring in more noise, which is harmful for the performance. Both of the above two experiments demonstrate that it is necessary to restrict the communication between the two modalities.

4.5   Case Study

We further analyze the effectiveness of Mert-MEL by real-world cases in Fig. 4, which are chosen from the datasets and external sources of Wikipedia.

In case 1 and 2, our model correctly links the mentions to the corresponding entities. Specifically, in case 1, Mert-MEL mines the relationship between the camera in the image and the word *photojournalist* in the entity abstract, and gives the highest matching score to the ground truth entity. In case 2, the textual description of the mention leads to successful linking, which implicitly indicates that the mentioned person performed a part in a play and should be an actor.

In case 3, our model fails to predict the correct answer. The top two entities *Fan Cong* and *Zhu Mujie* have completely different labels, but have the same alias *Dashan*. While among the candidate entities whose labels are *Dashan*, the ground truth entity ranks first. In case 4, Mert-MEL assigns high matching scores to the entities whose names include *Philip VI* and who are noblemen. But it fails to distinguish the ground truth from these similar entities, partly owing to the failure to recognize *Philip VI of France* as a whole.

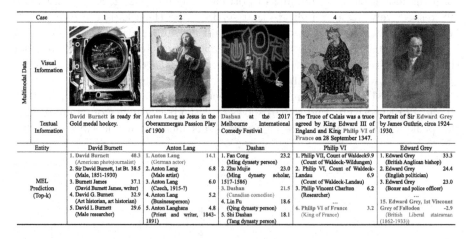

Fig. 4. Multimodal entity linking cases (Each case contains visual information, textual information, and an entity. The mention is in bold blue text. The MEL results are ranked, and the correct answer is in bold blue text. Some of the images have been cropped to fit the table). (Color figure online)

Lack of key information could lead to wrong predictions. In case 5, the ground truth entity gets an extremely low score. As a matter of fact, the textual information does not provide any clue that helps to make the correct prediction. It shows that Mert-MEL may not perform well when the given sentence has little correlation with abstracts of entities.

5   Conclusion and Future Work

In this paper, we propose Mert-MEL, an effective model for MEL. Mert-MEL is completely based on Transformer, which is used to extract and fuse features. In order to fully capture semantic information in the text, it extracts both token-level textual features and phrase-level textual features. Additionally, it employs a mixed fusion mechanism, which combines global fusion and bottleneck fusion. We also introduce contrastive learning to reduce the semantic gap between two modalities. Experiments show that our models have excellent performance, which are capable of mining and fusing the information of two modalities sufficiently, and reducing the noise in the extracted features effectively.

In the future, we will continue to find out a method to extract fine-grained mention-level features, in order to improve our models further. Besides, removing the MEL head, the remaining Mert can also be used for other multimodal downstream tasks by adding a task head theoretically, so we also hope to generalize our model to other tasks in the future.

References

1. Adjali, O., Besançon, R., Ferret, O., et al.: Multimodal entity linking for tweets. In: ECIR (2020)
2. Blanco, R., Ottaviano, G., Meij, E.: Fast and space-efficient entity linking for queries. In: WSDM (2015)
3. Bunescu, R., Pasca, M.: Using encyclopedic knowledge for named entity disambiguation. In: EACL (2006)
4. Chen, T., Kornblith, S., Norouzi, M., et al.: A simple framework for contrastive learning of visual representations. In: ICML (2020)
5. Cho, K., van Merriënboer, B., Gulcehre, C., et al.: Learning phrase representations using RNN encoder-decoder for statistical machine translation. In: EMNLP (2014)
6. Cucerzan, S.: Large-scale named entity disambiguation based on Wikipedia data. In: EMNLP-CoNLL (2007)
7. Daher, H., Besançon, R., Ferret, O., et al.: Supervised learning of entity disambiguation models by negative sample selection. In: CICLing (2017)
8. Devlin, J., Chang, M.W., Lee, K., et al.: BERT: pre-training of deep bidirectional transformers for language understanding. In: NAACL (2019)
9. Dolmans, T.C., Poel, M., van 't Klooster, J.W.J., et al.: Perceived mental workload classification using intermediate fusion multimodal deep learning. Front. Hum. Neurosci. **14** (2021)
10. Dosovitskiy, A., Beyer, L., Kolesnikov, A., et al.: An image is worth 16x16 words: transformers for image recognition at scale. In: ICLR (2021)

11. Dzogang, F., Lesot, M.J., Rifqi, M., et al.: Early fusion of low level features for emotion mining. Biomed. Inform. Insights **5**, BII-S8973 (2012)
12. Eshel, Y., Cohen, N., Radinsky, K., et al.: Named entity disambiguation for noisy text. In: CoNLL (2017)
13. Gan, J., Luo, J., Wang, H., et al.: Multimodal entity linking: a new dataset and a baseline. In: MM (2021)
14. Globerson, A., Lazic, N., Chakrabarti, S., et al.: Collective entity resolution with multi-focal attention. In: ACL (2016)
15. Guo, Z., Barbosa, D.: Entity linking with a unified semantic representation. In: WWW (2014)
16. Hoffart, J., Yosef, M.A., Bordino, I., et al.: Robust disambiguation of named entities in text. In: EMNLP (2011)
17. Khan, S., Naseer, M., Hayat, M., et al.: Transformers in vision: a survey. ACM Comput. Surv. **54**(10s) (2022)
18. Kim, W., Son, B., Kim, I.: Vilt: vision-and-language transformer without convolution or region supervision. In: ICML (2021)
19. Lemnaru, C., Potolea, R.: Imbalanced classification problems: systematic study, issues and best practices. In: ICEIS (2012)
20. Loshchilov, I., Hutter, F.: Decoupled weight decay regularization. In: ICLR (2019)
21. Lu, J., Yang, J., Batra, D., et al.: Hierarchical question-image co-attention for visual question answering. In: NIPS (2016)
22. Moon, S., Neves, L., Carvalho, V.: Multimodal named entity disambiguation for noisy social media posts. In: ACL (2018)
23. Nagrani, A., Yang, S., Arnab, A., et al.: Attention bottlenecks for multimodal fusion. In: NIPS (2021)
24. Radford, A., Kim, J.W., Hallacy, C., et al.: Learning transferable visual models from natural language supervision. In: ICML (2021)
25. Shen, W., Wang, J., Han, J.: Entity linking with a knowledge base: issues, techniques, and solutions. IEEE Trans. Knowl. Data Eng. **27**(2), 443–460 (2015)
26. Singh, A., Hu, R., Goswami, V., et al.: Flava: a foundational language and vision alignment model. In: CVPR (2022)
27. Snoek, C.G.M., Worring, M., Smeulders, A.W.M.: Early versus late fusion in semantic video analysis. In: MM (2005)
28. Vaswani, A., Shazeer, N., Parmar, N., et al.: Attention is all you need. In: NIPS (2017)
29. Wang, P., Wu, J., Chen, X.: Multimodal entity linking with gated hierarchical fusion and contrastive training. In: SIGIR (2022)
30. Wang, Y., Shen, Y., Liu, Z., et al.: Words can shift: dynamically adjusting word representations using nonverbal behaviors. In: AAAI (2019)
31. Wolf, T., Debut, L., Sanh, V., et al.: Transformers: state-of-the-art natural language processing. In: EMNLP (2020)
32. Wu, L., Petroni, F., Josifoski, M., et al.: Scalable zero-shot entity linking with dense entity retrieval. In: EMNLP (2020)
33. Young, P., Lai, A., Hodosh, M., et al.: From image descriptions to visual denotations: new similarity metrics for semantic inference over event descriptions. TACL **2**, 67–78 (2014)
34. Zhang, L., Li, Z., Yang, Q.: Attention-based multimodal entity linking with high-quality images. In: DASFAA (2021)
35. Zhou, X., Wang, P., Li, G., et al.: Weibo-mel, Wikidata-mel and Richpedia-mel: multimodal entity linking benchmark datasets. In: CCKS (2021)

Optimizing Empathetic Response by Generating and Integrating Emotion Feedback and Topic Discussion

Jing Li[1], Donghong Han[1,2]([✉]), Shi Feng[1], and Yifei Zhang[1]

[1] School of Computer Science and Engineering, Northeastern University, Shenyang, China
2110662@stu.neu.edu.cn, {handonghong,fengshi,zhangyifei}@cse.neu.edu.cn
[2] Key Laboratory of Intelligent Computing in Medical Image of Ministry of Education, Northeastern University, Shenyang, China

Abstract. Expressing empathy is a trait in human daily conversation, in which people are willing to give responses containing appropriate emotions and topics on the basis of understanding the interlocutor's situation. However, empathetic dialogue models trained by data-driven training method tend to generate general responses, which are usually monotonous and difficult to infuse emotions and topics concurrently. To solve this issue, we propose a novel model that generates two sub-responses, namely, emotion feedback and topic discussion, then integrates them to optimize empathetic responses. Specifically, in the sub-response generation stage, we introduce emotion lexicon and commonsense knowledge to make sub-responses focus on emotional words and topic-related words respectively, which drives the sub-responses to be contextually related from different perspectives. Afterward, we utilize cross attention to integrate the global information to optimize the final response. Our model is trained on the pre-trained language model BART. Experimental results show that our method can generate responses involving emotion and topic well, and compared with existing methods, empathy and relevance are improved. Our code is available at https://github.com/outsider-lj/edsgi_bart.

Keywords: empathetic dialogue generation · commonsense knowledge · pre-trained language model

1 Introduction

Enabling machines to communicate like humans is a long-term goal of open-domain dialogue generation. Empathy is an essential human trait, which reflects our ability to share the feelings of others [28]. The empathetic dialogue system aims to recognize user's emotion and situation, then generates responses accordingly. Such empathetic dialogue system can improve user's experience and establish long-term human-machine interaction, thereby it is widely used in social companion, psychological counseling and other fields. Recently, many

© The Author(s), under exclusive license to Springer Nature Switzerland AG 2023
X. Wang et al. (Eds.): DASFAA 2023, LNCS 13945, pp. 623–638, 2023.
https://doi.org/10.1007/978-3-031-30675-4_46

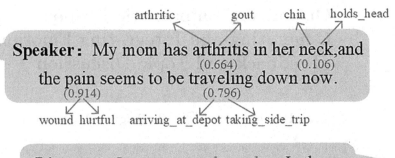

Fig. 1. An example of Empathetic Dialogue with commonsense knowledge. Emotion feedback is highlighted in green, topic discussion is highlighted in purple, and commonsense concepts are highlighted in blue. Numbers under words denote emotion intensity.

researches on empathetic dialogue generation have focused on user's situation cognition [25,31], user's emotion recognition [8,13,34] and empathetic response strategies [5,26,27].

To explore the specific characteristics of empathetic responses, we observe the real daily dialogue dataset Daily Dialogue [11] and the empathetic dialogue dataset Empathetic Dialogue [24]. We find that human's responses frequently include emotion feedback and topic discussion. Among them, the empathetic responses often start with emotion feedback and then have a topic discussion, as shown in Fig. 1. However, due to a lot of repetitive general sentences in the dataset and the data-driven training strategy of deep learning models, it's easy for the empathetic dialogue model to generate general responses but difficult to express both emotional and topic-related content. The responses such as "I'm sorry to hear that." which only express emotion and "What do you think?" which only discuss the topic, neither can well reflect the model's empathy, it's hard for users to feel well understood.

In this paper, we address the above problems by generating two sub-responses, namely emotion feedback and topic discussion, and then integrating them to optimize the final responses. First of all, we segment and categorize the responses in the dataset based on the rules in Sect. 4.1. In short, we treat the utterances with more emotionally expressive words (mostly adjectives or adverbs) as emotion feedback; the utterances with more topic-related words (mostly nouns or verbs) as topic discussions. Following the example in Fig. 1, the first reply contains a clear emotional word "sorry", but no topic-related word, so we treat it as a emotion feedback. There are no explicit emotional words in the second utterance, but "physical therapy" is related to the current topic "arthritis", so it is a topic discussion. Secondly, we introduce emotion lexicon and commonsense knowledge to help the model understand the contexts and generate sub-responses. In our opinion, the attention to "pain" can help the model respond with appropriate emotion, and the attention to "arthritis" and "traveling" can improve the semantic richness of topic discussion. At last, we think about that the separately generated sub-responses may have problems of semantic inconsistency and duplicate text. Therefore, inspired by the idea of deliberation [33], we integrate the sub-responses and further check to generate more smooth responses.

To sum up, we propose the Empathetic Dialogue Sub-response Generation and Integration (EDSGI) model. It includes one encoder and three decoders which are used to generate and integrate the two sub-responses. In the encoder, we utilize commonsense knowledge to expand the dialogue context and boost the understanding of the dialogue context. In the two sub-response decoders, the gated cross attention is used to pay attention to the emotional words and topic-related words of the context respectively. In the integration decoder, we adopt two cross attention to integrate sub-responses and adjust final responses respectively. As the pre-trained language model BART [6] has good language expression ability, and its structure is consistent with the Transformer [30], we train EDSGI based on BART.

Our contributions can be summarized as: (a) We propose a novel model EDSGI, which can generate and integrate emotion feedback and topic discussion. This model makes responses better incorporate both emotional expressions and topic discussions, thus reducing the general responses. (b) In the stage of sub-response generation, we introduce multi-type knowledge to improve the model's understanding of the dialogue context, and focus on emotional words and topic-related words respectively to promote diversity and contextual relevance of responses. (c) Automatic and human evaluation results show that our model has a better performance on empathy and relevance compared to previous methods.

2   Related Work

Rashkin et al. [24] firstly propose the empathetic dialogue generation task and open source the Empathetic Dialogue dataset. Since then, some empathetic dialogue datasets, such as EDOS [32] and PEC [37] have been conducted successively. The research on empathetic dialogue generation is mostly based on the Transformer [30]. Lin et al. [12] propose MoEL model which builds multiple decoders for different emotions, and integrates them into the final responses. Majumder et al. [18] propose to model response emotions by emotion clustering and emotion imitation. Gao et al. [2] utilize the gated attention to focus on words related to the emotion causes. Wang et al. [31] simultaneously model the emotion, topic and situation of the dialogue context. However, due to the small amount of data in the Empathetic Dialogue dataset, models based on simple transformer are difficult to fully learn the interaction patterns in empathetic dialogue, so the generated responses are often general and monotonous.

There are also many applications of pre-trained models in empathetic dialogue generation due to their strong language representation abilities. EmpTransfo [34] adds next-sentence emotion head and next-sentence language head to GPT2 [22]. Liu et al. [17] regard the responses generated by GPT as inputs for secondary training and inject predicted emotion during generation. Li et al. [10] firstly leverage causal emotion information to supplement input of GPT. Zaranis et al. [35] fine-tune T5 [23] by three objectives of response language modeling, sentiment understanding, and empathy forcing. The above studies take the pre-trained model as a whole, resulting in limited improvement. The model performance, that is empathy and topic relevance of generated responses, can be further improved.

External knowledge, such as emotion lexicon [19,20] and commonsense knowledge [3,29], can improve the model's ability of cognition, reasoning and generation, so it is often used in empathetic dialogue generation. Li et al. [8] exploit the words' emotion intensity knowledge to understand users' emotion states on the fine-grained level. Sobour et al. [25] use COMET [1] for commonsense reasoning to help the model understand the user's situation. Liu et al. [15] add knowledge to the beginning of the decoder input to guide empathetic responses generation. Li et al. [9] exploit multi-type knowledge to conduct an emotional context graph to perceive and express implicit emotions. Among them, few of the existing researches introduce multi-type knowledge. Meanwhile, the knowledge is often used to improve cognition, but it doesn't play a sufficient role on generation.

In this paper, we make use of the pre-trained model's language ability to alleviate the problem of small dataset. Meanwhile, we improve the internal structure of the pre-trained language model, which facilitates the introduction of knowledge and the integration of emotions and topics. In addition, we introduce emotion lexicon and commonsense knowledge, which are used to improve understanding of context and focus on emotional words and topic-related words during generation.

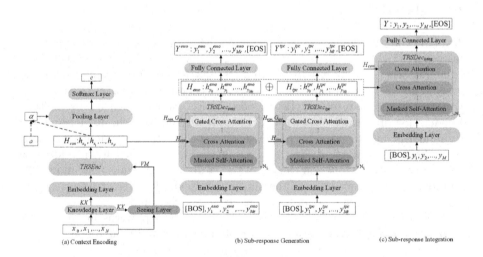

Fig. 2. An overall architecture of EDSGI.

3 Methology

The overview of EDSGI is shown in Fig. 2, it has three stages: (a) context encoding, (b) sub-response generation, and (c) sub-response integration. We formulate the task as follow: Given the dialogue context $D = \{U_1, U_2, \ldots, U_L\}$ containing L utterances and $U_i = \{w_1^i, w_2^i, \ldots, w_K^i\}$ with K tokens. Utterances with odd subscripts ($U_1, U_3 \ldots$) are user's utterances, even subscripts ($U_2, U_4 \ldots$) are empathetic responses. We concatenate the previous L utterances of D as input. Meanwhile, we insert [BOS] at the beginning of the sequence and [EOS] at the end of every utterance to form the input sequence $X = \{x_0, x_1 \ldots, x_N\}$. In stage (a), we utilize commonsense knowledge to extend the input sequence X, encode it and recognize the user's emotion e. In stage (b), we generate emotion feedback $Y^{emo} = \{y_1^{emo}, y_2^{emo}, \ldots, y_{Me}^{emo}\}$ and topic discussion $Y^{tpc} = \{y_1^{tpc}, y_2^{tpc}, \ldots, y_{Mt}^{tpc}\}$ respectively. In stage (c), we integrate Y^{emo} and Y^{tpc} to obtain the final empathetic response $Y = \{y_1, y_2, \ldots, y_M\}$.

3.1 Context Encoding

In stage (a), we firstly leverage ConceptNet [29] to extend dialogue context. ConceptNet is a common large-scale multilingual commonsense knowledge, each assertion of which is a form of (head concept, relation, tail concept, confidence score). We follow the rules in [15] to select up to three extremely relevant assertions for each no-stop word, every dialogue context includes up to 10 assertions. For encoding the knowledge without additional pre-training and involving context information, we follow K-BERT [14] to process the context sequence X through the knowledge layer and seeing layer to get KX and VM separately. The difference is, we treat all relations in selected assertions as semantically related, and do not insert specific relations to X. This reduces the length of KX. Therefore, KX is a sequence that concepts of selected assertions inject into their corresponding position in X. The visual matrix VM represents the attention relationship between words in KX, that is, the context words pay attention to all context words and their own related concepts, and the concepts pay attention to themselves and the context words linking them. In this way, the knowledge does not interfere with each other but can contain context information. The specific approach is intuitively shown in Fig. 3.

After getting KX and VM, we input KX to embedding layer which sums the word embedding (EW), soft position embedding (EP) [14] and dialogue state embedding (ES) of each token to get the embedding representation $EC(KX)$.

EDSGI uses the Transformer Encoder to learn the context semantic feature. Transformer Encoder is a stacked structure composed of multi-head self-attention and feed forward layers. For injecting the features of the knowledge into the corresponding words, we add VM to the multi-head self-attention of each layer as a mask matrix according to [14]. We obtain the encoder hidden state $H_{con} = \{h_{x_0}, h_{x_1}, \ldots, h_{x_N}\}$ which contains the information of commonsense knowledge and dialogue context:

$$H_{con} = TRSEnc(EC(KX), VM) \tag{1}$$

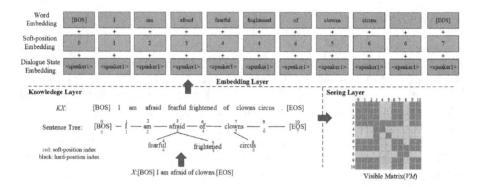

Fig. 3. The input and output form of knowledge layer, seeing layer and embedding layer [14].

In particular, after encoding, commonsense knowledge information is involved into the dialogue context words, so we only extract the features of N dialogue context words from the output of $TRSEnc$.

Emotion Classification: Recognizing the current users' emotion states is a indispensable step of empathetic dialogue generation. We utilize a simple attention which introduces a trainable vector ve to obtain the emotion attention weight α_i, and sum the hidden states h_{x_i} by weights α_i to get the user's emotion feature r:

$$\alpha_i = \frac{\exp(ve^T h_{x_i})}{\sum_{j=1}^{N} \exp(ve^T h_{x_j})} \tag{2}$$

$$r = \sum_{i=1}^{N} \alpha_i h_{x_i} \tag{3}$$

We use a linear layer with softmax operation to project r into an emotion category distribution. e denotes user's emotion label, the negative log likelihood loss is defined as:

$$\mathcal{L}_{emo} = -\log \mathcal{P}(e|X) \tag{4}$$

3.2 Sub-response Generation

After encoding dialogue context, we conduct two decoders $TRSDec_{emo}$ and $TRSDec_{tpc}$ to generate sub-responses. The normal Transformer Decoder layer includes multi-head masked self-attention, multi-head encoder-decoder cross attention and feed forward layers. In order to make the two sub-responses involve emotion and topic respectively, inspired by [2], we add gated cross attention above the original cross attention of the two decoders to enhance the attention of emotional words or topic-related words in the dialogue context. We utilize two different gate sequences $G_{emo} = \{g_0^{emo}, \ldots, g_i^{emo}, \ldots, g_N^{emo}\}$

and $G_{tpc} = \{g_0^{tpc}, \ldots, g_i^{tpc}, \ldots, g_N^{tpc}\}$ to control which word information that is selected from H_{con} by the two decoders, the gated cross attention weight of the i−th position on H_{con} can be calculated as:

$$z_i^{(l)} = \frac{g_i \odot \exp(q_l W_l^q (h_{x_i} W_l^h)^T)}{\sum_{j=1}^{N} g_j \odot \exp(q_l W_l^q (h_{x_j} W_l^h)^T)} \tag{5}$$

where g_i is the emotion gate or topic gate at the i−th position on H_{con}, q_l is the output of the cross attention in the l−th decoder layer, W_l^q and W_l^h are the trainable weight parameters of layer l.

We gain $G_{emo/tpc}$ by external knowledge. We take the emotion intensity calculated through the emotion lexicon NRC_VAD [19] as the emotion gate. NRC_VAD is an emotion lexicon containing Valence-Arousal-Dominance scores for more than 20,000 English words. We calculate the emotion intensity value of words x_i as follow:

$$\eta(x_i) = min\text{−}max(\|V(x_i) - \frac{1}{2}, \frac{A(x_i)}{2}\|_2) \tag{6}$$

where min-max() represents min-max normalization. $\| . \|_k$ denotes the l_k norm, $V(x_i)$ and $A(x_i)$ are the scores on the valence and arousal dimensions. If word x_i is not in NRC_VAD, we set $\eta(x_i)$ to 0. So the emotion gate $g_i^{emo} = \eta(x_i)$.

We set the topic gate as the binary gate, meanly $g_i^{tpc} \epsilon \{0, 1\}$. If we can search x_i in ConceptNet, $g_i^{tpc} = 1$, it means $TRSDec_{tpc}$ will focus on x_i during generating topic discussion. In this scheme, G_{tpc} is discrete, and gradient can't back-propagate during optimizing the model. Therefore, we use the Gumbel-softmax trick [4] to sample from the Gumbel distribution instead of directly sampling from the category.

We finally obtain the decoder hidden states $H_{emo} = \{h_{y_0}^{emo}, h_{y_1}^{emo}, \ldots, h_{y_{Me}}^{emo}\}$ and $H_{tpc} = \{h_{y_0}^{tpc}, h_{y_1}^{tpc}, \ldots, h_{y_{Mt}}^{tpc}\}$ from $TRSDec_{emo}$ and $TRSDec_{tpc}$, then utilize the fully collected network to predict the generation probability and get Y^{emo} and Y^{tpc}. The sub-response generation is based on the dialogue context X and previous generated words $y_{<i}^{emo/tpc}$, the loss functions are defined as:

$$\mathcal{L}_{emo\_re} = -\sum_{i=0}^{Me} \log \mathcal{P}(y_i^{emo} \mid y_{<i}^{emo}, X) \tag{7}$$

$$\mathcal{L}_{tpc\_re} = -\sum_{i=0}^{Mt} \log \mathcal{P}(y_i^{tpc} \mid y_{<i}^{tpc}, X) \tag{8}$$

3.3 Sub-response Integration

Since the sub-responses are generated by the two separate decoders, there may be some problems such as semantic repetition or inconsistency. According to [33], deliberation during decoding would involve global information and get better generation results, we design $TRSDec_{integ}$ to integrate Y^{emo} and Y^{tpc} obtained in Sect. 3.2. The sub-responses generated by EDSGI are different from the golden responses, so we take H_{emo} and H_{tpc} as input of $TRSDec_{integ}$. Based on the Transformer Decoder structure, we add the other cross attention between the original masked self-attention and cross attention, where the key and value are $[H_{emo} \bigoplus H_{tpc}]$, and the query is the output of masked self-attention layer. At this point, every decoder layer of $TRSDec_{integ}$ can obtain the information of the generated tokens through the masked self-attention, integrate the sub-responses through the first cross attention, and revise through the second cross attention. The final response Y is also obtained from fully collected network. The loss function is:

$$\mathcal{L}_{fin} = -\sum_{i=0}^{M} \log \mathcal{P}(y_i \mid y_{<i}, Y^{emo}, Y^{tpc}, X) \tag{9}$$

3.4 Model Training

EDSGI is actually a multi-task learning model, which includes the tasks of emotion recognition, sub-response generation and sub-response integration. The idea of EDSGI is to generate sub-responses first and then integrate them, so the training process is also divided into two stages. The loss function of the first stage is defined as:

$$\mathcal{L}_{sub} = \theta \mathcal{L}_{emo\_re} + \sigma \mathcal{L}_{tpc\_re} + \mu \mathcal{L}_{emo} \tag{10}$$

where θ, σ and μ are hyperparameters. In the case of obtaining the optimal sub-responses, we train the integration decoder, and the loss is L_{fin}. At this time, we do not perform gradient backpropagation on parts other than $TRSDec_{integ}$ and only update the parameters of $TRSDec_{integ}$.

4 Experimental Setup

4.1 Dataset

We apply Empathetic Dialogue dataset [24] for experiments. The dataset contains 24,850 open-domain multi-turn dialogues between two participants. And each group of data provides the speaker's current situation and fine-grained emotion labels of 32 categories. The train/valid/test subset is divided by 8:1:1 following the original dataset definition.

There is no distinction between emotion feedback and topic discussion in the Empathetic Dialogue dataset. Therefore, we define rules to divide each turn of the listener's responses into emotion feedback and topic discussions for training. We split responses by end punctuation at first, and then annotate the responses by the rule. We treat the non-question utterances that appear more than three times as emotion feedback. In addition, the sentences with more emotional words than concepts are classified as emotion feedback, where emotional words and concepts are filtered by NRC_VAD [19] and ConceptNet [29]. Correspondingly, the sentences in which the number of concepts exceeds the number of emotional words are regarded as topic discussions. We also regard the non-general responses ending with question marks as the exploration of topic content. If through the above judgment, the utterance does not belong to the two types, we divide it according to the length. Precisely, we regard utterances with more than 10 words as topic discussions, otherwise are emotion feedback. After data preprocessing, the ratio of emotion feedback to topic discussion in the train subset is 20,447 / 32,551.

To evaluate the accuracy of the above segmentation results, we randomly sample 100 dialogues from the preprocessed Empathetic Dialogue dataset. Then, we ask three annotators to annotate whether the categories of the sub-responses are correct. The evaluation criteria is based on the annotation rule in Sect. 4.1. If all categories of the response's sentences are correct, the score is 1 (else 0). The average score of three annotators is 78, which means the segmentation accuracy is 78%, and the Fleiss' kappa (κ) is 0.34, which reaches a fair agreement.

4.2 Comparison Models

We select advanced models based on the pre-trained language model in empathetic dialogue generation as comparison models: (1) BART [6]: a pre-trained model based on normal Transformer. (2) Multi-BART: it is trained on BART with multi-task strategy (emotion classification and text generation) like [24]. (3) EmpBot [35]: it is trained on T5 [23], with three training objectives: response language modeling, sentiment understanding and empathy forcing. (4) Roberta-GPT2 [15]: it uses Roberta [16] as encoder, GPT2 [22] as decoder, and injects commonsense knowledge to the beginning of the decoder inputs. (5) DialogGPT [36]: it is a common dialogue pre-trained model which is based on GPT structure and trained on large-scale dialogue datasets. (6) EmpGPT [17]: based on GPT2, it uses the idea of deliberation and injects expected emotion into response generation.

4.3 Implementation Details

Our training has two stages, the first is the training of the encoder and sub-response decoders, the second is the training of the integration decoder. In the second stage, we only optimize the integration decoder, the parameters of the rest of the model remain unchanged. We train on BART-base[1] model with 6 layers (N_L is 6.) and 12 heads. During training, we set the batch size to 8, and the learning rate to $2e^{-5}$. The hyperparameters θ, σ and μ are 0.5, 0.5 and 1 respectively. In addition, we reproduce some advanced empathetic dialogue generation models based on the hyperparameters mentioned in their papers. For fair comparison, we use top-k and top-p sampling methods for all generation, $k = 5$, $p = 0.9$.

---

[1] https://huggingface.co/facebook/bart-base.

4.4   Evaluation Metrics

Automatic Evaluation: BLEU [21] calculates n-gram overlap ratio of golden responses and generated responses. We use BLEU-avg (the average of BLEU-1,-2,-3,-4) to verify sentence similarity between generated and true responses in this paper. **Dist-1** and **Dist-2** [7] are the proportion of the distinct unigrams / bigrams in all the generated results which evaluate discourse diversity. **PPL** is short for perplexity which shows the uncertainty of generation.

Human Evaluation: Open-domain dialogue generation lacks reliable automatic evaluation metrics, so human evaluation plays an important role. Firstly, we use the common **human rating** method like [18]. Specifically, we randomly sample 100 dialogues from the test dataset. Three professional annotators use the Likert scale to rate the generated responses from different models on a scale of 1 to 5 for empathy, relevance, and fluency. The scores 1, 3, and 5 indicate unacceptable, moderate, and excellent performance. We calculate the average score of each aspect. In addition, we also perform **A/B test** for human evaluation. Three professional annotators choose the better response between two models for the same context. If both are good or bad to the same degree, they can choose a Tie. We resample 100 dialogues for every A/B test to ensure persuasiveness, it reflects the quality of responses intuitively.

5   Results and Discussions

5.1   Response-Generation Performance

Automatic Evaluation: The automatic evaluation results are shown in Table 1. We provide the PPL of generating the sub-responses and the final empathetic responses. Figure 4 shows the generation probability of the two types of responses. The words in the final response are mainly selected directly from sub-responses, so the probabilities are mostly 1. This leads EDSGI to own a low PPL in the integration process. Although the perplexity score of EDSGI is relatively worse due to the insufficient data of sub-responses, the other automatic scores of EDSGI are obviously better. The promotion on BLEU proves the rationality of our structure, which can better learn the reply pattern of true responses. Additionally, the higher values of Dist-1 and Dist-2 indicate our approach is effective for generating non-generic and diverse responses. Our models can generate responses containing both general emotional expressions and specific topic discussions. The results on BLEU and Dist-1/2 confirm our method finds a good balance between response diversity and similarity to real responses. It will be further verified in the case study. Moreover, EDSGI and EmpGPT both utilize the idea of deliberation. EmpGPT takes the responses generated by the GPT as input for secondary training. We only use the source data for training and the deliberation is mainly reflected in the integration of sub-responses, which makes better use of limited data.

Table 1. Results on automatic evaluation and human rating. All automatic evaluation results are the mean of 5 runs for a fair comparison. The results marked mean the best results in this table. The PPL of EDSGI contains the two-stage results. Emp., Rel. and Flu. are short for Empathy, Relevance, and Fluency. The Fleiss-Kappa of results in Emp., Rel., and Flu. are 0.21, 0.22, and 0.24, which all reach fair agreements.

Models	PPL	Dist-1	Dist-2	BLEU	Emp.	Rel.	Flu.
BART	16.25	2.36	13.11	6.38	3.65	3.64	4.50
MultiBART	14.75	2.34	12.52	6.00	3.37	3.35	4.42
EmpBot	12.02	2.50	13.09	6.17	3.16	3.15	4.47
Roberta-GPT2	16.89	2.11	12.32	6.98	3.21	3.14	4.32
DialogGPT	15.01	2.54	14.95	7.69	3.30	3.34	4.49
EmpGPT	**11.53**	2.59	14.84	8.28	3.69	3.76	**4.51**
EDSGI	14.62/2.51	**2.60**	**15.58**	**8.63**	**4.02**	**4.12**	4.42

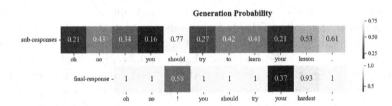

Fig. 4. The probability of every word in the generated responses.

Human Evaluation: From the human rating results shown in Table 1, our model is better than other models in empathy and relevance. It indicates that the targeted learning for dialogue's emotion and topic is effective in making responses take both the emotion (empathy) and topic (relevance) into account. The fluency of our model's responses is relatively low. The reason may be that the final responses of our model prefer to select words in sub-responses, inconsistency or logical problems are likely to occur. The specific analysis is given in Sect. 5.4. The A/B test results in Table 2 can demonstrate more directly that our responses are better than the other mentioned pre-trained models. In the comparison between our model and other models, the ratio of win is far more than the ratio of loss.

Table 2. Results on A/B test. The Fleiss-Kappa of results is 0.33, which reaches a fair agreement.

EDSGI vs	Win (%)	Loss (%)	Tie (%)
BART	44.33	27.00	28.67
MultiBART	53.00	29.67	17.33
EmpBot	53.33	28.67	18.00
Roberta-GPT2	51.00	27.00	22.00
DialogGPT	55.33	20.00	24.67
EmpGPT	53.67	29.33	17.00

Table 3. Results on ablation study. Here pip, kg, gcatt are short for pipeline, knowledge and gated cross attention. The best results are highlighted.

gcatt	kg	pip	PPL	Dist-1	Dist-2	BLEU
×	×	×	14.91 / 5.57	2.49	15.47	8.51
×	×	√	14.63 /**2.46**	2.46	15.03	8.54
×	√	√	**14.58** /2.54	2.53	15.47	8.41
√	√	√	14.62 / 2.51	**2.60**	**15.58**	**8.63**

5.2 Ablation Study

Effect of Training Strategy: Due to the two generation stages of EDSGI, the model can be trained by two strategies, pipeline training and joint training. The first strategy is to train the sub-response generation to the optimal, and then train the integration stage. The latter strategy optimizes all tasks together. The experimental results in Table 3 show that joint training has higher PPL. During joint training, we find that three tasks can't achieve the optimal at the same time, so we choose the model that has the minimum sum of all tasks' PPL. However, the integration is often underfitting and sub-response generation is often overfitting, so the final responses have many context-free words. Although the Dist-1/2 is higher than the pipeline, the final responses deviate greatly from the sub-responses and have the problem of irrelevant context. Besides, joint training also increases the influence between multiple tasks.

Effect of Knowledge: We introduce the knowledge layer and the seeing layer of KBERT to the empathetic dialogue generation for the first time. It can be seen in Table 3 that after adding knowledge, the values of Dist-1/2 are promoted, but the BLEU decreases. This confirms that the monotonous general utterances decrease, and the responses are more diverse.

Effect of Gated Cross Attention: In order to make the two sub-responses pay more attention to emotion and topic respectively, the gated cross attention is used in EDSGI. It can be seen from Table 3 that the similarity with golden responses and the diversity of responses both improve. It reveals that the use of gated cross attention makes knowledge play a better role.

5.3 Case Study

Table 4 shows some example responses generated by different models. Among them, the responses of EDSGI have more specific context-related words, such as "constant traffic trouble" and "group", which can reflect topic relevance. Meanwhile, responses have clear emotional expressions, such as "that is a good idea!". In Table 4, we can see that the final responses are the selection and adjustment of sub-responses. It reflects the function of the integration decoder, which replaces words with lower probability or repeated expressions to make the responses more fluent.

Table 4. Example responses generated by different models.

Emotion: Angry
Context: Speaker: Around town it seems like traffic has been horrible lately. Driving around town is just crazy.
Golden response: People just do not seem to pay attention when they are in a hurry
BART: I am sorry to hear that. I hope it goes better
MultiBART: What makes you so annoyed about driving
EmpBot: That is terrible. what do you do when you are out?
Roberta-GPT2: I hate that! I am so sorry you were in such pain!
DialogGPT: I know right? I am so sorry to hear that. I have to work late so i have to drive all night
EmpGPT: I am so sorry to hear that. what do you mean?
Emotion feedback: I have never been to a major city
Topic discussion: That is a nightmare. Are you in rush hour?
EDSGI: I have never been in a city like that. Are you in constant traffic trouble?
Emotion: Lonely
Context: Speaker: Hi, i feel so lonely sometimes because all my friends live in a different country
Listener: Oh, i am sure you are lonely. Maybe you can join some kind of club that lets you meet new friends?
Speaker: I was thinking about it! I wanted to join a group for local moms
Golden response: That is a good idea! This way you can also meet friends for yourself, but also maybe meet new friend's for your children to hang out with while you do with their moms!
BART: I hope you can find some good friends
MultiBART: I hope you can find someone soon!
EmpBot: I know what you mean. i hope you find a new place to live!
Roberta-GPT2: I bet you have a great time!
DialogGPT: I have friends who live in other parts of the world
EmpGPT: Yeah, that would be fun! It is a good idea!
Emotion feedback: That is a good idea
Topic discussion: I think it is a good idea to join some group
EDSGI: That is a good idea! I think it is a good group to join

5.4   Error Analysis and Outlook

Dataset Quality Problem: We segment the responses of the Empathetic Dialogue dataset based on rules. This method reduces the cost of manual annotation, but there is a problem of inaccurate segmentation. Some utterances that express both emotions and topics are difficult to define. In the future, we can annotate small amount of data manually, and utilize them to train the model to classify all sentences of responses.

Low Fluency: Low fluency mainly reflects on insufficient correlation between response sentences. In order to ensure that there is no content overlap between sub-responses, including emotions and topics respectively, there is no interaction between the sub-response decoders. But there will be a problem that the sub-response semantics are not consistent enough, which directly affects the fluency of the final responses. In the future, we plan to introduce a module which utilizes the gate mechanism to check whether the semantic of words in sub-responses is inconsistent and add a penalty to the loss.

6 Conclusion

In this paper, we propose a new model EDSGI, including context encoder, sub-response decoder and integration decoder. Among them, we introduce knowledge to improve the model's understanding of the dialogue context, use the gating strategy to make the sub-responses focus on emotional words and topic-related words respectively, and adopt the cross attention mechanism to integrate and deliberate sub-responses. EDSGI is trained on BART, experimental results demonstrate that our responses can incorporate both emotion and topic well, the semantic richness and empathy are improved.

Acknowledgements. This work is supported by the National Natural Science Foundation of China (61672144, 61872072).

References

1. Bosselut, A., Rashkin, H., Sap, M., Malaviya, C., Celikyilmaz, A., Choi, Y.: COMET: commonsense transformers for automatic knowledge graph construction. In: Proceedings of the 57th Conference of the Association for Computational Linguistics, pp. 4762–4779 (2019)
2. Gao, J., Liu, Y., Deng, H., Wang, W., Du, Y.C.J., Xu, R.: Improving empathetic response generation by recognizing emotion cause in conversations. In: Findings of the Association for Computational Linguistics: EMNLP2021, pp. 807–819 (2021)
3. Hwang, J.D., et al.: COMET-ATOMIC 2020: On symbolic and neural commonsense knowledge graphs. In: Proceedings of the 35th AAAI Conference on Artificial Intelligence, pp. 6384–6392 (2021)
4. Jang, E., Gu, S., Poole, B.: Categorical reparameterization with Gumbel-Softmax. In: Proceedings of the 5th International Conference on Learning Representations (2017)
5. Kim, H., Kim, B., Kim, G.: Perspective-taking and pragmatics for generating empathetic responses focused on emotion causes. In: Proceedings of the 2021 Conference on Empirical Methods in Natural Language Processing, pp. 2227–2240 (2021)
6. Lewis, M., et al.: BART: denoising sequence-to-sequence pre-training for natural language generation, translation, and comprehension. In: Proceedings of the 58th Annual Meeting of the Association for Computational Linguistics, pp. 7871–7880 (2019)
7. Li, J., Galley, M., Brockett, C., Gao, J., Dolan, B.: A diversity-promoting objective function for neural conversation models. In: Proceedings of the 2016 Conference of the North American Chapter of the Association for Computational Linguistics: Human Language Technologies, pp. 110–119 (2016)

8. Li, Q., Chen, H., Ren, Z., Ren, P., Tu, Z., Chen, Z.: EmpDG: multi-resolution inter-active empathetic dialogue generation. In: Proceedings of the 28th International Conference on Computational Linguistics, pp. 4454–4466 (2020)

9. Li, Q., Li, P., Ren, Z., Ren, P., Chen, Z.: Knowledge bridging for empathetic dialogue generation. In: Proceedings of the 36th AAAI Conference on Artificial Intelligence, AAAI 2022, pp. 10993–11001 (2022)

10. Li, Y., et al.: Towards an online empathetic chatbot with emotion causes. In: Proceedings of the 44th International ACM SIGIR Conference on Research and Development in Information Retrieval, pp. 2041–2045 (2021)

11. Li, Y., Su, H., Shen, X., Li, W., Cao, Z., Niu, S.: DailyDialog: a manually labelled multi-turn dialogue dataset. In: Proceedings of the 8th International Joint Conference on Natural Language Processing, pp. 986–995 (2017)

12. Lin, Z., Madotto, A., Shin, J., Xu, P., Fung, P.: MoEL: mixture of empathetic listeners. In: Proceedings of the 2019 Conference on Empirical Methods in Natural Language Processing and the 9th International Joint Conference on Natural Language Processing (EMNLPIJCNLP), pp. 121–132 (2019)

13. Lin, Z., et al.: CAiRE: an end-to-end empathetic chatbot. In: Proceedings of the 34th AAAI Conference on Artificial Intelligence, pp. 13622–13623 (2020)

14. Liu, W., et al.: K-BERT: enabling language representation with knowledge graph. In: Proceedings of the 34th AAAI Conference on Artificial Intelligence, pp. 2901–2908 (2020)

15. Liu, Y., Maier, W., Minker, W., Ultes, S.: Empathetic dialogue generation with pre-trained RoBERTa-GPT2 and external knowledge. arXiv preprint arXiv:2109.03004 (2021)

16. Liu, Y., et al.: RoBERTa: a robustly optimized BERT pretraining approach. arXiv preprint arXiv:1907.11692 (2019)

17. Liu, Y., Du, J., Li, X., Xu, R.: Generating empathetic responses by injecting antici-pated emotion. In: Proceedings of the IEEE International Conference on Acoustics, Speech and Signal Processing, ICASSP 2021, pp. 7403–7407 (2021)

18. Majumder, N., et al.: MIME: MIMicking emotions for empathetic response generation. In: Proceedings of the 2020 Conference on Empirical Methods in Natural Language Processing, pp. 8968–8979 (2020)

19. Mohammad, S.: Obtaining reliable human ratings of valence, arousal, and dominance for 20, 000 English words. In: Proceedings of the 56th Annual Meeting of the Association for Computational Linguistics, pp. 174–184 (2018)

20. Mohammad, S.: Word affect intensities. In: Proceedings of the 11th International Conference on Language Resources and Evaluation (2018)

21. Papineni, K., Roukos, S., Ward, T., Zhu, W.: BLEU: a method for automatic evaluation of machine translation. In: Proceedings of the 40th Annual Meeting of the Association for Computational Linguistics, pp. 311–318 (2002)

22. Radford, A., Wu, J., Child, R., Luan, D., Amodei, D., Sutskever, I.: Language models are unsupervised multitask learners. OpenAI blog 1(8), 9 (2019)

23. Raffel, C., et al.: Exploring the limits of transfer learning with a unified text-to-text transformer. Mach. Learn. Res. 21, 1–67 (2020)

24. Rashkin, H., Smith, E.M., Li, M., Boureau, Y.L.: Towards empathetic opendomain conversation models: a new benchmark and dataset. In: Proceedings of the 57th Annual Meeting of the Association for Computational Linguistics, pp. 5370–5381 (2019)

25. Sabour, S., Zheng, C., Huang, M.: CEM: commonsense-aware empathetic response generation. In: Proceedings of the 36th AAAI Conference on Artificial Intelligence, AAAI 2022, pp. 11229–11237. AAAI Press (2022)

26. Shen, L., Zhang, J., Ou, J., Zhao, X., Zhou, J.: Constructing emotional consensus and utilizing unpaired data for empathetic dialogue generation. In: Findings of the Association for Computational Linguistics: EMNLP 2021, pp. 3124–3134 (2021)
27. Shin, J., Xu, P., Madotto, A., Fung, P.: Generating empathetic responses by looking ahead the user's sentiment. In: Proceedings of the 2020 IEEE International Conference on Acoustics, Speech and Signal Processing, pp. 7989–7993 (2020)
28. Singer, T., Lamm, C.: The social neuroscience of empathy. Annal. New York Acad. Sci. **1156**, 81–96 (2010)
29. Speer, R., Chin, J., Havasi, C.: ConceptNet 5.5: an open multilingual graph of general knowledge. In: Proceedings of the 31th AAAI Conference on Artificial Intelligence, pp. 4444–4451 (2017)
30. Vaswani, A., et al.: Attention is all you need. In: Advances in Neural Information Processing Systems, pp. 5998–6008 (2017)
31. Wang, Y.H., Hsu, J.H., Wu, C.H., Yang, T.H.: Transformer-based empathetic response generation using dialogue situation and advanced-level definition of empathy. In: Proceedings of the 12th International Symposium on Chinese Spoken Language Processing, pp. 1–5 (2021)
32. Welivita, A., Xie, Y., Pu, P.: A large-scale dataset for empathetic response generation. In: Proceedings of the 2021 Conference on Empirical Methods in Natural Language Processing, pp. 1251–1264 (2021)
33. Xia, Y., et al.: Deliberation networks: sequence generation beyond one-pass decoding. In: Proceedings of Advances in Neural Information Processing Systems 30, pp. 1784–1794 (2017)
34. Zandie, R., Mahoor, M.H.: Emptransfo: A multi-head transformer architecture for creating empathetic dialog systems. In: Proceedings of the 33th International Florida Artificial Intelligence Research Society Conference, pp. 276–281 (2020)
35. Zaranis, E., Paraskevopoulos, G., Katsamanis, A., Potamianos, A.: EmpBot: A t5-based empathetic chatbot focusing on sentiments. arXiv preprint arXiv:2111.00310 (2021)
36. Zhang, Y., et al.: DIALOGPT : large-scale generative pre-training for conversational response generation. In: Proceedings of the 58th Annual Meeting of the Association for Computational Linguistics: System Demonstrations, ACL 2020, pp. 270–278 (2020)
37. Zhong, P., Zhang, C., Wang, H., Liu, Y., Miao, C.: Towards persona-based empathetic conversational models. In: Proceedings of the 2020 Conference on Empirical Methods in Natural Language Processing, pp. 6556–6566 (2020)

Improving Event Representation with Supervision from Available Semantic Resources

Shuchong Wei[1,2], Liangjun Zang[2](✉), Xiaobin Zhang[1,2],
Xiaohui Song[1,2], and Songlin Hu[2]

[1] School of Cyber Security, University of Chinese Academy of Sciences,
Beijing, China
[2] Institute of Information Engineering, Chinese Academy of Sciences, Beijing, China
{weishuchong,zangliangjun,zhangxiaobin,songxiaohui,husonglin}@iie.ac.cn

Abstract. Learning distributed representations of events is an indispensable but challenging task for event understanding. Existing studies address this problem by either composing the embeddings of event arguments as well as their attributes, or exploiting various relations between events like co-occurrence and discourse relations. In this paper we argue that the knowledge learned from sentence embeddings and word semantic meanings could be leveraged to produce superior event embeddings. Specifically, we utilize both natural language inference datasets for learning sentence embeddings and the knowledge base WordNet for word semantics. We propose a **M**ulti-**L**evel **S**upervised **C**ontrastive **L**earning model (**MLSCL**) for learning event representations. Our model fuses diverse semantic resources at the levels of sentences, events, and words in an end-to-end way. We conduct comprehensive experiments on three similarity tasks and one script prediction task. Experimental results show that MLSCL achieves new state-of-the-art performances on all tasks consistently and higher training efficiency than prior competitive model SWCC.

Keywords: event representation · event similarity · ontology knowledge

1 Introduction

Learning distributed representations of events in the form of (subject, predicate, object) is a challenging but valuable task that supports various applications for event comprehension, such as event detection, script prediction, and event induction. Event representations (i.e., event embeddings) could be derived by modeling the interactions of event arguments [1,5], leveraging related event knowledge [4], and modeling discourse relations among events [8]. It is common practice to exploit document-level co-occurrence information of events in corpus as supervision signal [9]. While recent developments have enabled impressive improvements on this task, diverse types of available semantic resources including datasets for

© The Author(s), under exclusive license to Springer Nature Switzerland AG 2023
X. Wang et al. (Eds.): DASFAA 2023, LNCS 13945, pp. 639–648, 2023.
https://doi.org/10.1007/978-3-031-30675-4_47

sentence representation learning task and ontology knowledge base have not been fully leveraged by existing works.

The intuition of our method can be described as follows: From one point of view, the task of learning sentence representation enjoys rich and carefully annotated resources, such as the datasets of natural language inference (NLI). We argue that leveraging these datasets contributes to learning event representations since events consisting of (subject, predicate, object) could be viewed as some kinds of short sentences. Therefore, it is natural to employ SimCSE [10], which demonstrates that a contrastive objective can be extremely effective when coupled with pre-trained language model to produce superior sentence embeddings, to learn event representations based on labeled NLI datasets.

From another point of view, ontology knowledge plays an important role in understanding events, especially new events that people have no idea about. For example, "police department catch robber" could be regarded as an instance of "administrative unit get hold of scoundrel" through conceptualizing "police department" to "administrative unit", "catch" to "get hold of", and "robber" to "scoundrel" respectively. Both events are analogous to each other, and thus should share closer event embeddings. Intuitively, introducing the semantic ontology knowledge base *WordNet* [20] could facilitate event representation learning.

To make full use of existing semantic resources for more supervision, we propose a simple but effective method by combining sentence-level, event-level and word-level embedding enhancement strategies in a single contrastive learning framework. Specifically, our model consists of three strategies at different granularity: (1) At sentence-level, we follow SimCSE training method on several NLI datasets which cover numerous sentence pairs in high-quality. (2) At event-level, we extract events from NewYork corpus and consider co-occurring events as positive pairs as well as introduce co-occurrence frequencies as the strength of the connection between two events. (3) At word-level, we leverage ontology knowledge in WordNet to make event semantic abstraction. The conceptualized events are viewed as positive samples for anchor event. Our model aims to pull the embedding of a given anchor event closer to its positive events, and push away negative events.

The contributions of this paper could be summarized in three folds.

1. To the best of our knowledge, we are the first to fuse sentence-level, event-level, and word-level embedding enhancement strategies with kinds of datasets or knowledge bases in a single model to learn event embeddings.
2. To alleviate event sparsity in text corpus, we devise a novel augmentation method to conceptualize events, which generates positive and negative samples for a target event based on the ontology knowledge base WordNet.
3. We evaluate our model on two event similarity tasks and the script prediction task. The model achieves state-of-the-art performances with substantial improvement compared to prior methods. Furthermore, we verify the training efficiency improvement after performing sentence-level strategy compared to SWCC [9] indeed.

2   Related Work

Event representation learning aims to learn distributed embeddings for structured events represented as a (subject, predicate, object) triple. Training neural networks for robust event embeddings is effective to capture event-level semantics under the assumption of the embeddings of similar events are close to each other while those of dissimilar events are far away from each other in the same vector space.

Some previous studies learn event embeddings based on the interactions of event arguments and introduce related knowledge of event semantic property. [1] proposed to use a novel neural tensor network which can learn the semantic compositionality over event arguments to train event embeddings. [2] fed the concatenation of word embeddings for each event argument to a neural network to acquire event embeddings. [5] designed tensor-based composition models, which combine the subject, predicate, and object to produce the final event representation. There are also a number of studies utilizing external knowledge to enhance event embeddings. [3] proposed to leverage attributes and properties of entities involved in events to improve event embeddings. [6] captured fine-grained event properties using both event-level features (e.g., sentiment polarity of a given event) and entity-level features (e.g., animacy of participants). [4] had the capability of distinguishing events with a subtle difference with intent and sentiment knowledge. [11] incorporated an event knowledge base, which uses triple facts to describe various relations between events. [12] proposed to learn similar inference patterns instead of exact matching to exploit knowledge in event knowledge base.

Event representations could also be generated by modeling event-event relations, such as easily accessible co-occurrences of events as well as discourse relations among events. Different applications deal with different structures of events, including event pairs, event chains, and event graphs. [7] integrated temporal information over an event chain and utilized LSTM to generate event representations. [13] was the first to construct an event evolutionary graph based on narrative event chains, and then use a graph neural network to learn event embeddings. [17] used a self-attention mechanism to recognize and model event segments besides event pairs. [8] proposed a multi-relational event embedding approach by investigating 11 fine-grained syntactic relations. [16] enhanced the event embeddings by mining their connections at multiple granularity levels following [13]. [18] constructed a heterogeneous graph network to model event chains with three relation types. [15] employed a variational auto-encoder to model the scenario-level knowledge (also called event contexts) to obtain event representations. [14] traced the events back to their original texts and exploited the texts' informative constituents describing the events to obtain more comprehensive event semantic embeddings. Contrastive learning has recently been applied to the task of event representation learning. [19] performed data augmentation on two types of structures, i.e., event chains and event graphs, which alleviates the issue of data sparsity and insufficient labeled data. [9] extended

Event Augmentation at Word-Level

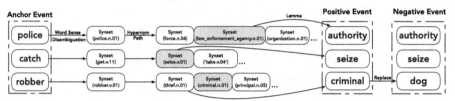

Fig. 1. Visualize the procedure of word-level event augmentation through WordNet by taking a specific example.

the InfoNCE loss with multiple weighted positive samples, and used a prototype-based clustering method to gather events with similar topics.

3   Methodology

To make full use of existing semantic resources, we propose a contrastive learning framework to learn better event embeddings with sentence-level, event-level and word-level strategies. We will elaborate on them in the following descriptions successively.

At sentence-level, we utilize sentence pairs in the datasets of NLI to train the pre-trained encoder for powerful representation ability. Thanks to the release of model checkpoint by [10], we could adopt the off-the-shelf model with no extra time cost to serve as our pre-trained encoder. The following mention of pre-trained encoder is denoted as "Event-RoBERTa". Then we use "Event-RoBERTa" as our backbone and carry out event-level and word-level fine-tuning.

At event-level, we consider event pairs that co-exist in the text corpus are related semantically and may share information such as similar topics or scenarios. Hence, we introduce a co-occurrence frequency matrix to enhance event embeddings. The co-occurrence loss is defined based on InfoNCE [10] as Eq. 1:

$$\mathcal{L}_1 = -\log \frac{freq \cdot g(z_a, z_c)}{g(z_a, z_c) + \sum_{neg \in \mathcal{N}(a)} g(z_a, z_{neg})}$$

$$g(z_a, z_c) = e^{h(z_a, z_c)/\tau_1}$$
(1)

where $freq$ is a normalized frequency and a larger value means two events co-occur more frequently. $g(z_a, z_c)$ is a similarity function (e.g., cosine similarity) between anchor event embedding z_a and co-occurring event embedding z_c both encoded by Event-RoBERTa. $\mathcal{N}(a)$ is the set of in-batch negative event samples. τ_1 is the temperature coefficient. Additionally, We feed the anchor event to the encoder twice using different dropout masks and compute standard InfoNCE loss.

At word-level, we introduce ontology knowledge to provide guidance for learning event representations. As well known, WordNet includes various ontology knowledge organized into sets of word sense, each of which groups several synonym words called "synset". For the purpose of introducing this type of knowledge to enlarge the positive set of the anchor event, it's desirable to develop an ontology-based augmentation strategy. We decompose the strategy into the following steps and present a concrete example in Fig. 1:

(1) For an anchor event (subject, predicate, object), we employ word sense disambiguation on the event by [21] to mount its arguments to w^s, w^p, w^o respectively, where w represents a synset.

(2) We compute d-depth transitive closures of each synset to collect the hypernyms as related synsets.

$$\mathrm{hyper}(w^{role}) = [w^{role}, h_1^{role}, h_2^{role}, \cdots, h_l^{role}] \tag{2}$$

where $\mathrm{hyper}(\cdot)$ serves as the closure collecting function and h_l is more general than specific synset h_{l-1}. The superscript $role$ denotes s, p, or o. The length of each composed hypernyms is l.

(3) From each hypernym closure, we sample one synset randomly and obtain its lemma (a lemma is a canonical form or morphological form of a word in linguistics), which represents a specific word included in the synset. The composition of lemmas generates a new event that could be regarded as a positive event performing semantic abstraction for the anchor event.

(4) Calculate the WordNet-based semantic similarity between anchor event and positive event. We decompose WordNet-based similarity to the sum of path similarity between event arguments, which could be formulated as:

$$\mathrm{wn\_sim}(e_1, e_2) = \sum_{t \in \{s,p,o\}} \frac{1}{\mathrm{md}(w_1^t, w_{share}) + \mathrm{md}(w_2^t, w_{share}) + 1} \tag{3}$$

where w_{share} is the first synset in the intersection of $\mathrm{hyper}(w_1^t)$ and $\mathrm{hyper}(w_2^t)$ (the nearest synset to arrive by taking w_1^t and w_2^t as source synset respectively). And $\mathrm{md}(\cdot)$ computes the minimum distance from w^t to w_{share}.

(5) We corrupt the positive event by replacing a lemma with a random lemma to be a hard negative event and also calculate the WordNet-based similarity.

Given the positive event and hard negative event augmented with ontology knowledge, we modify similarity function in Eq. 1 with the injection of WordNet-based similarity as Eq. 4:

$$f(z_a, z_o) = e^{h(z_a, z_o) \cdot \mathrm{wn\_sim}(e_a, e_o)/\tau_2} \tag{4}$$

where $\mathrm{wn\_sim}(\cdot)$ computes the WordNet-based similarity for anchor event and augmented event, and a softmax layer is applied to normalize with others among the batch. It is noted that $\mathcal{N}_2(a)$ not only includes the in-batch negatives but also contains a hard negative event.

To sum up, we jointly fine-tune the model with the sum of loss:

$$\mathcal{L} = \mathcal{L}_1 + \mathcal{L}_2 \tag{5}$$

Table 1. Experimental results on event similarity tasks. The best results are in bold.

Model	Hard Similarity (Accuracy %)		Transitive Similarity (ρ)
	Original	Extended	
Role-Factor Tensor	43.5	20.7	0.64
KGEB	52.6	49.8	0.61
FEEL	58.7	50.7	0.67
NTN-IntSent	77.4	62.8	0.74
UniFAS	78.3	64.1	0.75
MulCL	78.3	64.3	0.76
SWCC	80.9	72.1	0.82
MLSCL	**83.5**	**75.1**	**0.82**

4   Experiments

Following the common practice in event representation learning [4,15], we conduct evaluations on two event similarity tasks and a script prediction task to compare the performance of our approach against a variety of event embedding models developed in recent years.

4.1   Dataset and Implementation Details

At event-level, we collect statistics on the co-occurrence of two events in the same document as the same as [9]. At word-level, we perform word sense disambiguation by [21] to link words (or spans) involved in event arguments to ontology senses in WordNet. It ended up with 129,807 distinct events with their corresponding synsets. We follow the training method of SimCSE with loss adjustment well-suited for event-level and word-level fine-tuning. We conduct experiments on Nvidia Tesla V100 GPU. We train our model with a batch size of 64 using an Adam optimizer. The learning rate is set to 2e−6 for the event representation model. The temperature coefficient is set to 1.0, 1.0 at event-level and word-level respectively. The depth of the transitive closure is set to 5.

4.2   Performance on Event Similarity Tasks

We evaluate the representation ability of our proposed approach, which determines whether it improves the classification accuracy for similar and dissimilar event pairs. Two related tasks are available to assess that are consistent with [5]: (1) Hard Similarity Task and (2) Transitive Sentence Similarity.

Hard Similarity Task reports $Accuracy \in [0,1]$ metric, which indicates the fraction of cases where the similar pair receives a higher cosine score than the dissimilar pair.

Transitive Sentence Similarity uses the metric of Spearman's correlation ($\rho \in [-1,1]$) to measure the relatedness between cosine similarity of event pair calculated by model and the annotated similarity score.

Table 2. Evaluation performance on the MCNC task. The best result is in bold.

Model	Accuracy (%)
PPMI	30.52
BiGram	29.67
Word2Vec	37.39
SWCC	44.50
MLSCL	**47.2**

Comparison Methods. According to the introduction of related work in Sect. 2, we compare the performance of our approach against the following competitive baselines: **Role-Factor Tensor** [5], **KGEB** [3], **FEEL** [6], **NTN-IntSent** [4], **UniFAS** [15], **MulCL** [19], **SWCC** [9].

Overall Performance. Table 1 reports the overall performances of baseline models on both the hard similarity task and the transitive sentence similarity task. In a word, MLSCL achieves state-of-the-art performances and substantially improves accuracy on two datasets of hard similarity task.

It is worth noting that our model outperforms knowledge-based models [3, 4,6] by a large margin. Since these baseline models only use knowledge about properties of event instances, it implies that semantic resources we utilized are effective at improving event representation.

Let us zoom in on the comparison of our model MLSCL against the most competitive baseline SWCC. From the perspective of implementation, they both use co-occurrence-based contrastive learning. MLSCL adopts additional sentence-level and word-level strategies for powerful pre-trained encoder by contrastive learning method, while SWCC uses prototype-based clustering to put events sharing similar topics together. For the two datasets of hard similarity task, MLSCL improves the accuracy from 80.9% to 83.5% and from 72.1% to 75.1%. It means that our learning strategies, including the absorption of supervision signals from NLI datasets and the introduction of ontology knowledge in WordNet are more effective than prototype-based clustering in terms of gathering semantically similar or relevant events. For the transitive similarity task, our model has a similar performance with SWCC in terms of Spearman's correlation. On one hand, this task relies more on event co-occurrence information which is shared for both MLSCL and SWCC. On the other hand, the human-crafted annotation scores may not be absolutely consistent with the actual situation.

4.3 Performance on Script Prediction Task

We also conduct inferring experiments on the Multiple Choice Narrative Cloze (MCNC) task [8] to evaluate the generalization of the event representations for script knowledge. To be specific, we adopt the summation of the individual event embeddings as the sequence representation and calculate its similarity with candidate event embeddings to evaluate their correlation. The candidate event

Table 3. Ablation study for MLSCL on the three datasets.

Model	Hard Similarity (Accuracy %)		Transitive Similarity (ρ)
	Original	Extended	
RoBERTa	31.3	22.7	0.41
w/ sentence-level	54.8	53	0.77
w/ sentence-level + word-level	74.8	68.8	0.71
w/ sentence-level + word-level + event-level	**83.5**	**75.1**	**0.82**

corresponding to the highest similarity score will be selected as the predicted next event. We employ several other unsupervised learning models to serve as baselines: **PPMI** [22], **BiGram** [23], **Word2Vec** [24], **SWCC** [9].

The prediction accuracy of each model is shown in Table 2. MLSCL outperforms other baselines under the zero-shot transfer setting by a large margin, suggesting that it improves the quality of the event representations and has much more generalization to the downstream task.

4.4 Ablation Study

As aforementioned, our model MLSCL is built on multi-level contrast learning method. To verify the effectiveness of each level, we perform an ablation study by adding a certain level one by one and reporting the corresponding performance for all event similarity tasks. From Table 3, we could observe that adding each level would lead to a significant improvement in performance. It demonstrates that all three levels are indispensable in our framework. They are complementary to each other in improving the quality of event embeddings. The lack of sentence-level transfer fine-tuning results in a performance drop, which indicates that it also helps the model generalize better.

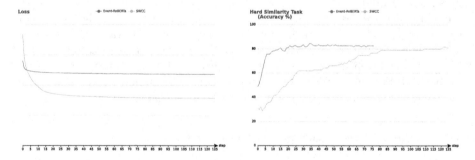

Fig. 2. The loss change while the training steps increase.

Fig. 3. The accuracy change of hard similarity task while the training steps increase.

4.5 Training Efficiency

We have demonstrated above that our model outperforms all baseline models consistently on three benchmark datasets and all components of it are indispensable. To investigate its efficiency of training, we compare Event-RoBERTa with SWCC by observing how its loss and evaluating metric changes as training steps increase. As shown in Fig. 2, the loss of Event-RoBERTa decreases sharply with increasing training steps and converges faster than SWCC. Figure 3 shows that Event-RoBERTa achieves much better performance than SWCC at the same training step. The reason may be that Event-RoBERTa has fewer parameters than SWCC with the removal of the prototype clustering module, and it has been trained on NLI datasets for sentence-level fine-tuning in advance.

5 Conclusion

In this work, we propose multi-level supervised contrastive learning framework MLSCL to generate embeddings that are more informative for events. Furthermore, we evaluate the performance of our model on two tasks about event similarity and script prediction. We found that great improvement could be achieved by utilizing semantic resources. Additionally, Event-RoBERTa has much higher training efficiency than its most competitive baseline SWCC by performing event-level and word-level strategies. In future work, we will explore other knowledge bases to verify the necessity and effectiveness of introducing ontology knowledge.

References

1. Ding, X., Zhang, Y., Liu, T., Duan, J.: Deep learning for event-driven stock prediction. In: Twenty-Fourth International Joint Conference on Artificial Intelligence (2015)
2. Modi, A.: Event embeddings for semantic script modeling. In: Proceedings of the 20th SIGNLL Conference on Computational Natural Language Learning, pp. 75–83 (2016)
3. Ding, X., Zhang, Y., Liu, T., Duan, J.: Knowledge-driven event embedding for stock prediction. In: Proceedings of COLING 2016, the 26th International Conference on Computational Linguistics: Technical Papers, pp. 2133–2142 (2016)
4. Ding, X., Liao, K., Liu, T., Li, Z., Duan, J.: Event representation learning enhanced with external commonsense knowledge. In: Proceedings of the 2019 Conference on Empirical Methods in Natural Language Processing and the 9th International Joint Conference on Natural Language Processing (EMNLP-IJCNLP), pp. 4894–4903 (2019)
5. Weber, N., Balasubramanian, N., Chambers, N.: Event representations with tensor-based compositions. In: Proceedings of the AAAI Conference on Artificial Intelligence, vol. 32 (2018)
6. Lee, I.-T., Goldwasser, D.: Feel: featured event embedding learning. In: Proceedings of the AAAI Conference on Artificial Intelligence, vol. 32 (2018)
7. Wang, Z., Zhang, Y., Chang, C.Y.: Integrating order information and event relation for script event prediction. In: Proceedings of the 2017 Conference on Empirical Methods in Natural Language Processing, pp. 57–67 (2017)

8. Lee, I.-T., Goldwasser, D.: Multi-relational script learning for discourse relations. In: Proceedings of the 57th Annual Meeting of the Association for Computational Linguistics, pp. 4214–4226 (2019)
9. Gao, J., Wang, W., Yu, C., Zhao, H., Ng, W., Xu, R.: Improving event representation via simultaneous weakly supervised contrastive learning and clustering. arXiv preprint arXiv:2203.07633 (2022)
10. Gao, T., Yao, X., Chen, D.: SimCSE: simple contrastive learning of sentence embeddings. In: Proceedings of the 2021 Conference on Empirical Methods in Natural Language Processing, pp. 6894–6910 (2021)
11. Lv, S., Zhu, F., Hu, S.: Integrating external event knowledge for script learning. In: Proceedings of the 28th International Conference on Computational Linguistics, pp. 306–315 (2020)
12. Zhou, Y., Geng, X., Shen, T., Pei, J., Zhang, W., Jiang, D.: Modeling event-pair relations in external knowledge graphs for script reasoning. In: Findings of the Association for Computational Linguistics: ACL-IJCNLP 2021, pp. 4586–4596 (2021)
13. Li, Z., Ding, X., Liu, T.: Constructing narrative event evolutionary graph for script event prediction. arXiv preprint arXiv:1805.05081 (2018)
14. Bai, L., Guan, S., Guo, J., Li, Z., Jin, X., Cheng, X.: Integrating deep event-level and script-level information for script event prediction. arXiv preprint arXiv:2110.15706 (2021)
15. Zheng, J., Cai, F., Chen, H.: Incorporating scenario knowledge into a unified fine-tuning architecture for event representation. In: Proceedings of the 43rd International ACM SIGIR Conference on Research and Development in Information Retrieval, pp. 249–258 (2020)
16. Wang, L., et al.: Multi-level connection enhanced representation learning for script event prediction. In: Proceedings of the Web Conference 2021, pp. 3524–3533 (2021)
17. Lv, S., Qian, W., Huang, L., Han, J., Songlin, H.: Sam-net: integrating event-level and chain-level attentions to predict what happens next. In: Proceedings of the AAAI Conference on Artificial Intelligence, vol. 33, pp. 6802–6809 (2019)
18. Zheng, J., Cai, F., Ling, Y., Chen, H.: Heterogeneous graph neural networks to predict what happen next. In: Proceedings of the 28th International Conference on Computational Linguistics, pp. 328–338 (2020)
19. Zheng, J., Cai, F., Liu, J., Ling, Y., Chen, H.: Multistructure contrastive learning for pretraining event representation. IEEE Trans. Neural Netw. Learn. Syst. (2022)
20. Miller, G.A.: WordNet: An Electronic Lexical Database. MIT Press, Cambridge (1998)
21. Loureiro, D., Jorge, A.: Language modelling makes sense: propagating representations through wordnet for full-coverage word sense disambiguation. arXiv preprint arXiv:1906.10007 (2019)
22. Chambers, N., Jurafsky, D.: Unsupervised learning of narrative event chains. In: Proceedings of ACL-08: HLT, pp. 789–797 (2008)
23. Jans, B., Bethard, S., Vulic, I., Moens, M.-F: Skip n-grams and ranking functions for predicting script events. In: Proceedings of the 13th Conference of the European Chapter of the Association for Computational Linguistics (EACL 2012), pp. 336–344. ACL, East Stroudsburg, PA (2012)
24. Mikolov, T., Chen, K., Corrado, G., Dean, J.: Efficient estimation of word representations in vector space. arXiv preprint arXiv:1301.3781 (2013)

Select, Extend, and Generate: Generative Knowledge Selection for Open-Domain Dialogue Response Generation

Sixing Wu[1], Ping Xue[2], Ye Tao[2], Ying Li[2,3(✉)], and Zhonghai Wu[2,3]

[1] National Pilot School of Software, Yunnan University, Kunming, China
wusixing@ynu.edu.cn
[2] School of Software and Microelectronics, Peking University, Beijing, China
[3] National Research Center of Software Engineering, Peking University, Beijing, China
li.ying@pku.edu.cn

Abstract. Incorporating external commonsense knowledge can enhance machines' cognition and facilitate informative dialogues. However, current commonsense knowledge-grounded dialogue generation works can only select knowledge from a finite set of candidates retrieved by information retrieval (IR) tools. This paradigm suffers from: 1) The knowledge candidate space is limited because IR tools can only retrieve existing knowledge from the given knowledge base, and the model can only use the retrieved knowledge; 2) The knowledge selection procedure lacks enough interpretability to explain the selected result. Moreover, with the increasing popularity of pre-trained language models (PLMs), many knowledge selection methods of non-PLM models have become incapable because of the input/structure restrictions of PLMs. To this end, we propose a simple but elegant *SEG-CKRG*, and introduce a novel PLM-friendly *Generative Knowledge Selection (GenSel)* to select knowledge via a generative procedure. Besides selecting the knowledge facts from the retrieved candidate set, *GenSel* can also generate newly extended knowledge. *GenSel* also improves interpretability because the output of the knowledge selection is a natural language text. Finally, *SEG-CKRG* uses *GPT-2* as the backbone language model. Extensive experiments and analyses on a Chinese dataset have verified the superior performance of *SEG-CKRG*.

Keywords: dialogue generation · knowledge-grounded

1 Introduction

Open-domain dialogue response generation (RG) models enable machines to converse with humans using natural language and play an important role in human-computer interaction [43]. However, machines lack enough real-world knowledge cognition because they can only access the parametric knowledge of a model besides the dialogue history [45]. Thus, machines struggle to thoroughly understand the semantics of dialogue histories and generate informative responses.

© The Author(s), under exclusive license to Springer Nature Switzerland AG 2023
X. Wang et al. (Eds.): DASFAA 2023, LNCS 13945, pp. 649–664, 2023.
https://doi.org/10.1007/978-3-031-30675-4_48

Seeking information from external knowledge sources is an effective solution [50], i.e., knowledge-grounded dialogue response generation (KRG) [4,17].

Compared to RG models, the superiority of KRG models derives from the ability to use external knowledge [42]. The general paradigm of KRG can be summarized as three stages [7,39]: 1) *Knowledge-Retrieval stage:* it first employs an efficient Information Retrieval (IR) tool to retrieve a set of knowledge candidates in a coarse-grained way. The retrieved knowledge candidates contain much irrelevant information because IR tools only consider the literal feature; 2) *Knowledge-Selection stage:* To filter out irrelevant information and select contextually-relevant knowledge, KRG also has a knowledge selection stage using more fine-grained methods; 3) *Response Generation stage:* it finally generates the target response by accessing the dialogue history and selected knowledge. Among such three stages, the second knowledge selection stage plays the most crucial role in the research of KRG and has received much attention [10,23,28].

This paper focuses on commonsense knowledge-grounded dialogue response generation (CKRG). Despite many successes [42,46], CKRG still suffers from several challenges, especially in the era of pre-trained language models (PLMs) [14,19]. First, the knowledge candidate space (i.e., the knowledge can be selected and used when generating the response) is fixed and limited. On the one hand, IR tools can only retrieve knowledge candidates already existing in the knowledge base. On the other hand, the model can only use the knowledge candidate already retrieved by IR tools. This may lead to insufficient knowledge coverage [42]. Second, in the knowledge selection stage, previous CKRG works [46,50] often use deep but complex networks, which lack enough interpretability to explain the knowledge selection procedure. For example, it is hard to determine which knowledge facts have been selected. Finally, although PLMs are powerful, they also bring many thorny restrictions to the downstream applications [15], such as the length (most PLMs can only operate at most 512/1024 tokens), the input format (must be plain text), the network structure, and so on. Consequently, many knowledge selection methods originally proposed for non-PLM-based models have become incapable in the era of PLMs; then, knowledge selection can only rely on the external network or the implicit self-attention mechanism [23,49].

Considering these challenges, we propose *SEG-CKRG*, a simple but elegant CKRG model. As shown in Fig. 1, *SEG-CKRG* introduces a novel *Generative Knowledge Selection (GenSel)* mechanism, which regards knowledge selection as a generative problem. *GenSel* uses a PLM to explicitly generate contextually-relevant knowledge based on the dialogue history and the knowledge candidate set retrieved by IR tools. By regarding this task as a generative problem, *GenSel* can not only select knowledge from the candidate set retrieved by IR tools, but can also extend the knowledge by externalizing the inherent knowledge of PLMs. Then, *SEG-CKRG* generates the target response conditioned on both the generated knowledge and the retrieved knowledge. Considering both the generative knowledge selection procedure and the dialogue generation procedure are generative problems, we can train/infer *SEG-CKRG* in an end-to-end fashion. We pre-

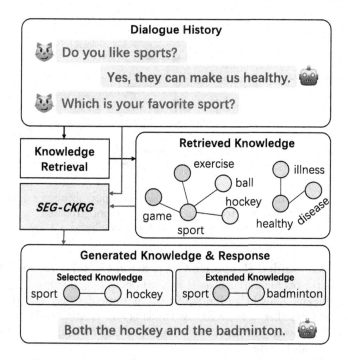

Fig. 1. An example. *SEG-CKRG* can use *Generative Knowledge Selection* to select the existing knowledge and extend the new knowledge, then generates the response.

train two GPT-2 models [27] as the backbone PLMs. To boost the knowledge representation density and the infusing of two generative procedures, we propose an *Efficient Input Representation* technique and a *Dual-Head Generator* technique, respectively.

We conduct extensive experiments on a Chinese conversational dataset *Weibo-ConceptNet* [41], whose dialogues have been aligned to a commonsense knowledge base, ConceptNet [32]. Experimental results have verified that *SEG-CKRG* has significantly outperformed previous state-of-the-art models, and *GenSel* can not only accurately select the knowledge but also generate new contextually-relevant knowledge. We also bring extensive analyses to investigate our approach further.

2 Methodology

2.1 Preliminary

Response Generation (RG). Suppose $\mathcal{D} = \{(H_i, R_i)\}^N$ is a conversational corpus, where $H_i = (h_1, \cdots, h_{|H_i|})$ is the dialogue history, $R_i = (r_1, \cdots, r_{|R_i|})$ is the response. Then, RG learns a conditional language model $P_{RG}(R_i|H_i)$ to generate R_i conditioned on H_i: $P_{RG}(R_i|H_i) = \prod P_{RG}(r_t|r_{<t}, H_i)$.

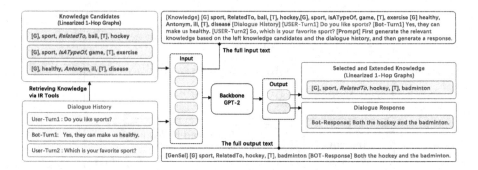

Fig. 2. An overview of *SEG-CKRG*. We show the input/output examples. In this example, *SEG-CKRG* has selected a knowledge fact '(sport, RelatedTo, Hockey)' and extended a '(sport, RelatedTo, badminton)' in the example.

Knowledge-Grounded Response Generation (KRG). RG models tend to generate generic responses such as 'I don't know.' [13] because P_{RG} can only use the insufficient knowledge hidden in the parameters θ_{RG} and the dialogue history. To address this issue, KRG methods try to seek more knowledge from the external knowledge base, such as encyclopedic knowledge [7], commonsense knowledge [32], and so on [26].

More specifically, in the commonsense knowledge-grounded dialogue response generation (CKRG) scenario, there is a knowledge base $\mathcal{K} = \{k_i = (e_i^h, e_i^r, e_i^t)\}^M$, where k_i is a commonsense fact triplet, e_i^h, e_i^r, and e_i^t are the corresponding head entity, relation, and tail entity, respectively. Then, for each dialogue history H_i, we need to employ an IR tool to retrieve a set of commonsense facts $K_i = \{k_{i,j}\}^L$, $L << M$ form $\mathcal{K}$. Finally, the problem of CKRG is given by:

$$P_{CKRG}(R_i|H_i) = \prod P_{CKRG}(r_t|r_{<t}, H_i, K_i) \qquad (1)$$

where P_{CKRG} is a conditional language model with the ability to access the knowledge K_i. Although K_i is the filtered results via IR tools, IR tools can only consider the token-level literal feature. Thus, a more fine-grained context-aware knowledge selection procedure is needed in P_{CKRG}. In non-PLM CKRG works, this procedure can be explicitly modeled and then integrated into P_{CKRG}. For example, [50] employs graph attention network [34]. In the era of PLM, limited by the input format and network structure, this procedure can only be implicitly performed by the integrated self-attention mechanism or external tools, bringing less interpretability but more limitations to the knowledge selection procedure.

2.2   Problem Definition and Overview

As shown in Fig. 2, unlike previous CKRG works, *SEG-CKRG* introduces a novel *Generative Knowledge Selection (GenSel)* mechanism, which regards knowledge selection as a generative problem. The objective of *SEG-CKRG* is:

$$P_{GenSel}(K_i^G|H_i, K_i) \cdot P_{ResGen}(R_i|H_i, K_i, K_i^G) \qquad (2)$$

where $P_{GenSel}(K_i^G|H_i, K_i)$ first generates the contextually-relevant knowledge K_i^G conditioned both the dialogue history H_i and the retrieved knowledge K_i; subsequently, $P_{ResGen}(R_i|H_i, K_i, K_i^G)$ generates the target response R_i.

2.3   Generative Knowledge Selection

SEG-CKRG uses a generative method to explicitly select and extend knowledge. Similar to other generation tasks, it is a conditional language modeling problem:

$$P_{GenSel}(K_i^G|H_i, K_i) = \prod P_{GenSel}(k_t^G|k_{<t}^G, H_i, K_i) \qquad (3)$$

Efficient Input Representation. Most PLMs can only accept plain texts as input, which means the structural commonsense knowledge must be linearized to plain text. Thus, the input of P_{GenSel} is given by:

$$S_i = [\omega_K(K_i), \omega_H(H_i), Prompting] \qquad (4)$$

where $\omega_H(H_i)$ linearizes the dialogue history with role (human/bot) labels and turn identifiers; *Prompting* is a prompting text[1] [52] to hint the PLM about the following generation action; $\omega_K(K_i)$ linearizes the structural $K_i = \{k_{i,j} = (e_{i,j}^h, e_{i,i}^r, e_{i,i}^t)\}^M$ to a sequence. To reduce the loss of structural information and improve the representation density, $\omega(K_i)$ uses a graph-level pattern:

$$\omega(K_i) = (\omega_G(g_{i,1}); \omega_G(g_{i,2}); \cdots ; \omega_G(g_{i,j}); \cdots)$$

$$\omega_G(g_{i,j}) = ([G], e_{i,j}^{hg}, e_{i,j}^{rg}, e_{i,j,1}^{tg}, [T], e_{i,j,2}^{tg}, \cdots) \qquad (5)$$

where K_i is first compressed as a set of 1-hop graphs $G_i = \{g_{i,j} = e_{i,j}^{hg}, e_{i,j}^{rg}, \{e_{i,j}^{tg}\}\}$; namely, $\forall k \in K_i$ that have the same head entity $e_{i,j}^{hg}$ and the same relation $e_{i,j}^{rg}$ are placed to the corresponding 1-hop graph $g_{i,j}$; then, $g_{i,j}$ is sequentially linearized and concatenated with a graph separator $[G]$ and a tail entity separator $[T]$. Compared to previous triplet-level patterns [42,52], our graph-level pattern can reduce the length of the linearized knowledge and achieve higher representation density. Higher representation density means more knowledge facts can be included under the same length limitation.

Generation. The goal is to generate the linearized contextually-relevant knowledge sequence $\omega_K(K_i^G)$. We adopt a widely-used auto-regressive GPT-2 [27] to implement $P_{GenSel}(K_i^G|H_i, K_i)$ and generate the $\omega_K(K_i^G)$:

$$\omega_K(K_i^G) = GPT2(S_i) = GPT2([\omega_K(K_i), \omega_H(H_i), Prompting]) \qquad (6)$$

In the training stage, we use a weakly-supervised way [51,52] to construct the generation goal K_i^G. Given a knowledge candidate set K_i retrieved by IR

---

[1] The translated text is 'First generate the relevant knowledge based on the left knowledge candidates and the dialogue history, and then generate a response.'.

tools, if there is a knowledge candidate $k \in K_i$ whose head entity and tail entity appear in the dialogue history H_i and the dialogue response R_i, respectively; then this k is added to the target K_i^G.

During the generation, K_i^G is fully generated based on the K_i and H_i. Intuitively, the generated K_i^G can select the relevant knowledge from K_i. Besides, as a generative model, GPT2 can also extend to generate the relevant knowledge that is not included in the K_i^G, which is an inherent feature of generative language models [9]. Meanwhile, the generated $\omega_K(K_i^G)$ is a natural language text, which can explicitly explain the results of knowledge selection and extension.

2.4   Dialogue Response Generation

Finally, we use the same GPT2 to generate the dialogue response R_i based on the dialogue history H_i, the retrieved knowledge K_i, and the generated contextually-relevant knowledge K_i^G. We feed the input S_i^{DG} to the GPT2, estimate $P_{ResGen}(r_t | R_{<t}, H_i, K_i, K_i^G)$, and then generate the R_i:

$$S_i^{DG} = [\omega_K(K_i), \omega_H(H_i), Prompting, \omega_K(K_i^G)]$$

$$R_i = GPT2(S_i^{DG}) = GPT2([\omega_K(K_i), \omega_H(H_i), Prompting, \omega_K(K_i^G)]) \tag{7}$$

where the generation head $\mathbf{W^R}$ is newly introduced compared to Eq. 6. This is because two generative procedures have different generation spaces, two separate generation heads help avoid confusion. Such a two-head generation mechanism is called *Dual-Head Generator*.

2.5   Training

Two generative procedures can be jointly trained in an end-to-end fashion by sharing the same GPT-2. We have pre-trained two different GPT-2 models and our *SEG-CKRG* on two Nvidia RTX-3090 GPUs:

General GPT2: The general-purpose or dialogue-oriented *base size*[2] Chinese GPT-2 resources are not very abundant [31]. Consequently, we first pre-train a Chinese GPT2 for our experiments. We implement a *base size* GPT2 language model network using the Huggingface transformer library[3] and PyTorch. There are 12 layers of 768-dimensional (for both the hidden states and embeddings) and 12-head Transformer layers. The vocabulary includes 30,000 subwords and 200 special symbols (placeholders). For efficiency, the maximum input length is limited to 512 tokens. This GPT-2 is first pre-trained on massive Chinese unsupervised data, including massive open-released news, movie/product comments, and Wikipedia data. In total, there are 18.4M sessions and 5.22B tokens. During the training, the batch size is 512, the number of total training steps is 80,000, and the optimizer is AdamW. After 4,000 warm-up steps, the learning rate will reach 2e−4; then, the learning rate will linearly decay to 0.

---

[2] a *base size* PLM models always has about 100M parameters.

[3] https://huggingface.co/.

Dialogue-Oriented GPT2: We also fine-tune a dialogue-oriented GPT-2. We use the Chinese conversational pre-training corpus *LCCC-large* released by [37], which includes 7.2M/4.7M sessions of single/multi-turn dialogues and 380M tokens in total. This GPT-2 is initialized from our general GPT-2, the batch size is 512, the number of total training steps is 180,000, and the optimizer is AdamW. It has the same learning rate strategy as GPT-2, except for the highest learning rate is decreased to $1.5e-4$.

SEG-CKRG: Finally, *SEG-CKRG* is fine-tuned on the general GPT2 (by default) or the dialogue-oriented GPT2 (in ablation study). The batch size is set to 32, the maximum training epoch is set to 15. The best epoch on the validation set is adopted in the following test stage.

3 Experiment

3.1 Settings

Dataset. We test models on a Chinese dataset *Weibo-ConceptNet* [41], which has been aligned to a well-known commonsense knowledge base ConceptNet [32]. The training/validation/test set includes 102K/5.6K/5.6K single-turn dialogues. Each utterance has 10.3 words on average. The commonsense graph has 696,466 facts, 27,189 entities, and 26 relations. On average, each dialogue has 77.7 candidate facts that are retrieved from ConceptNet.

Comparison Models. We first selected several non-PLM baselines: *1) Seq2Seq*: an attentive Seq2Seq RG Model [3,24]; *2) PGN*: Seq2Seq + Pointer-Genetor copy network [29]; *3) ConKADI*: a KRG model with the felicitous knowledge selection mechanism [41]; *4) GOKC*: a KRG model with a novel knowledge copy mechanism [1]. We also selected several fine-tuned *base*-size PLM methods: *5) BERT2Seq, 6) BERT-PGN*: We changed the encoder of *Seq2Seq* and *PGN* to the *'hfl/chinese-bert-wwm-ext'* [5] BERT encoder [6]. *7) CDial-GPT2*: An open-released conversational GPT-2 RG models [37]. We select the GPT-2 configuration *'GPT2LCCC-base'*. *8) MHKD-GPT2*: A PLM-based KRG models [42], which is based on *CDial-GPT2*.

Implementation. We use the official codes for *ConKADI, GOKC, CDial-GPT2*, and *MHKD-GPT2*, and we re-implement the remaining models using PyTorch. For non-PLM models, we use a 2-layer 768d bi-GRU/LSTM[4] encoder, 2-layer 768d GRU/LSTM decoder, Adam optimizer, 1e-4 learning rate. For all baselines, we use 32 batch size, up to 20 epochs, and finally select the best model on the validation set. Due to the different requirements, *BERT2Seq, BERT-PGN* use the corresponding BERT tokenizer and vocab, *Seq2Seq, PGN, GOKC*, and

---

[4] our codes use GRU, the others keep the original setting.

ConKADI use the original tokenizer and vocab, *CDial-GPT2* and *MHKD-GPT2* use *CDial-GPT2*'s tokenizer and vocab. The implementation of our SEG-CKRG will be released at https://github.com/pku-sixing/DASFAA23_GenSel.

Automatic Evaluation Metrics. Different models use different tokenizers; thus, we conduct character-level evaluations to avoid such differences. We use the following automatic metrics: *1) F1*: it is the F-measure of character-overlapping relevance [1]; *2) BLEU-4*: it is the 4-gram BLEU to evaluate the precision-oriented relevance [25]; *3) ROUGE*: we use ROUGE-L to evaluate the recall-oriented relevance [18]; *4) EM-A/G/X*: we use embedding evaluate the semantic relevance, the embedding is computed using Average/Greedy/Extrema [21]; *5) DI-1/2*: we use Distinct-1/2 to evaluate the diversity [12]; *6) Ent*: we use 4-gram entropy to evaluate the informativeness [30]; *7) Mean*: following [42], we compute the geometric mean of all previous scores to evaluate the overall performance.

3.2 Automatic Evaluation

Table 1. Automatic Evaluation Results. **First**/Second denotes the first/second best.

Model	F1	ROUGE	BLEU-4	EM-A	EM-G	EM-X	DI-1	DI-2	Ent	Mean
Seq2Seq	16.20	12.40	1.09	0.869	0.677	0.649	0.32	3.22	8.61	2.09
PGN	16.56	12.65	1.23	0.872	0.676	0.651	**0.58**	8.13	9.55	2.54
GOKC	18.13	14.95	1.47	0.881	0.684	**0.695**	0.35	7.95	10.35	2.56
ConKADI	<u>19.20</u>	14.60	1.94	0.885	0.679	0.664	0.38	**11.22**	**12.04**	<u>2.81</u>
BERT2Seq	17.49	13.21	1.93	0.877	0.670	0.658	0.26	2.77	8.83	2.18
BERT-PGN	18.76	13.72	<u>2.52</u>	<u>0.892</u>	0.674	0.664	0.36	6.91	9.69	2.64
CDial-GPT2	14.79	12.31	1.61	0.866	0.675	0.653	0.26	3.69	8.47	2.13
MHKD-GPT2	18.77	<u>16.60</u>	2.45	0.874	<u>0.690</u>	0.667	0.28	4.13	9.43	2.46
SEG-CKRG	**21.02**	**17.15**	**3.11**	**0.896**	**0.708**	<u>0.689</u>	0.48	9.94	<u>11.13</u>	**3.09**

As reported in Table 1, *SEG-CKRG* has achieved tier-1 results (the first and the second best) in all metrics and significantly outperformed previous methods in the *Mean* score, demonstrating the best overall performance and effectiveness. In addition, rather than pursuing the best score on a single-dimensional metric or only using some handpicked metrics, the philosophy of *SEG-CKRG* is multi-dimensional because a single automatic metric is not reliable [21].

Relevance: In the three overlapping-based metrics (i.e., F1, BLEU-4, and ROUGE), *SEG-CKRG* has the best results because our approach can simultaneously seek information from both the pre-trained language model and the external knowledge source to help the dialogue generation. In another three embedding-based relevance metrics (i.e., EMB-A/G/X), *SEG-CKRG* also has the best overall performance, showing the dialogue responses generated by our approach are more semantically relevant to the ground truth. Besides, we can also find that PLM-based models have better relevance performance than non-PLM-based models in the mass. Indicating the necessity of using PLMs in CKRG.

Table 2. Human Annotation Results. **Scores** denotes *SEG-CKRG* is significantly better (sign-test, p-value < 0.005). The 2/3 agreement ratio (at least 2 judges gave the same) is 95.4%, the 3/3 ratio is 54.8.2%.

%	Fluency			Rationality			Informativeness		
Compare to	*Lose*	Tie	*Win*	*Lose*	Tie	*Win*	*Lose*	Tie	*Win*
Seq2Seq	36.7	22.3	**41.0**	31.0	11.7	**57.3**	31.3	7.0	**61.7**
GOKC	8.3	6.0	**85.7**	9.0	7.0	**84.0**	15.0	4.0	**81.0**
ConKADI	11.0	5.3	**83.7**	25.0	3.0	**72.0**	38.6	1.4	**60.0**
BERT-PGN	36.7	6.6	**56.7**	42.3	2.3	**55.4**	45.6	2.4	52.0
CDial-GPT	20.6	9.4	**60.0**	38.6	5.7	**55.7**	41.0	3.0	**56.0**
MHKD-GPT	27.6	14.7	**57.7**	38.3	2.7	**59.0**	40.6	3.4	**56.0**
Human	33.6	31.4	**35.0**	46.3	14.0	39.7	62.3	8.7	29.0

Diversity and Informativeness: The situation is different in this part. *ConKADI* and our *SEG-CKRG* notably surpass other models. Between such two models, *SEG-CKRG* is slightly lower than *ConKADI*, and the reason can be summarized as 1) *SEG-CKRG* does not sacrifice the relevance to improving diversity and informativeness; 2) *SEG-CKRG* does not use any copy mechanism. Copy mechanism can copy words from the dialogue history or the external knowledge directly, which can significantly boost diversity and informativeness in the automatic evaluation. For example, compared with *Seq2Seq/BERT2Seq*, the copy variant *PGN/BERT-PGN* has more notable improvements in such metrics. However, we find previous copy works tend to repeat the given query rather than extend the new information, and then we decide not to equip this mechanism.

3.3 Human Evaluation

We employed three well-educated native-speaker to evaluate the practical generation quality of *SEG-CKRG*. The criteria include three dimensions: *1) Fluency*: is this response grammatically correct and fluent? *2) Rationality:* does this response logically conform to the current dialogue context? *3) Informativeness:* can this response provide enough meaningful information?

As reported in Table 2, we sampled 100 comparison cases[5] and compared *SEG-CKRG* with the three best baselines in the automatic evaluation (ConKADI, BERT-PGN, and GOKC) and the naive Seq2Seq. We have several findings: 1) Although Seq2Seq is the naive baseline, the comparison result is not the worst, especially in terms of fluency. This is because the task and the network of Seq2Seq are simple but stable; 2) Compared to GOKC and ConKADI, *SEG-CKRG* has notable advantages, indicating the importance of introducing the PLMs to CKRG; 3) Compared to BERT-PGN, *SEG-CKRG* is still better, demonstrating the effectiveness of using external knowledge. Finally, we also

---

[5] 5*100 pair-wise comparisons in total.

Table 3. Generated knowledge types. # is the average counting per response.

# Original	#Actual	#Generated	#Selected	#Extend
77.7	52.5	1.282	1.157	0.124

Table 4. Ablation Study.

#	Setting	ROUGE	EMBED-X	DIST2	Mean
0	*Full*	17.15	0.689	9.94	3.09
Different Backbones					
1	*DialogueGPT2*	16.72	0.686	9.69	3.05
2	*FromScratch*	15.64	0.682	5.12	2.56
Different Knowledge Accessing					
3	*w/o GenSelKnow*	16.52	0.690	8.11	2.95
4	*w/o SelKonw*	12.24	0.659	9.70	2.66
5	*w/o All (general GPT2)*	12.44	0.665	8.03	2.56
6	*w/o All (dialogue GPT2)*	12.62	0.665	7.76	2.59

compare *SEG-CKRG* with the human-generated ground-truth response. *SEG-CKRG* is comparable to the human in terms of fluency. However, *SEG-CKRG* is still behind the human in terms of rationality and informativeness. This shows we still have a large room to improve CKRG in future works.

3.4   More Analyses

Generated Knowledge. As reported in Table 3, we counted the types among the knowledge generated by our *Generative Knowledge Selection (GenSel)*. The original dataset has 77.7 facts per dialogue, but our model could only use 52.5 randomly sampled facts because of the restriction of the backbone GPT-2. We can find *GenSel* generated 1.282 facts per response on average, and about 9.7% of them are extended knowledge. It indicates 1) *GenSel* can indeed extend the knowledge and 2) using IR tools to explicitly retrieve knowledge candidates is still necessary.

Backbone GPT2. We have pre-trained two GPT-2 models; the first general GPT-2 is pre-trained on the general unsupervised data, and the second dialogue-oriented GPT-2 is further fine-tuned on a large-scale conversational dataset [37]. As shown in Table 4, if we train a *SEG-CKRG* based on the second dialogue-oriented GPT-2 (#1), although the perplexity is lower (9.93 vs 11.04), the real performance is even worse than the *SEG-CKRG* trained on the general GPT-2 (#0). We think our GenSel prefers the general backbone. Meanwhile, we also train a *SEG-CKRG* without initializing from a pre-trained GPT-2 (#2). The results are worse in every dimension, indicating the necessity of pre-training.

Table 5. Case Study. For *SEG-CKRG*, we report the generated knowledge (*K:*) and response (*R:*). We use red/blue color to highlight the selected/extended knowledge.

Query #1	水果茶好喝吗? Does fruit tea taste good?
Seq2Seq	还可以，我觉得挺好喝的。Not bad, I think it's good.
ConKADI	还行吧我觉得喝茶水果的最好吃的。 Not bad, I think drinking tea and fruit the best foods.
BERT-PGN	我觉得挺好喝的就是水果茶比较好喝。 I think it's very good, that is the fruit tea is good.
SEG-CKRG	*K:* [G]水果 HasProperty 酸 [G] 樱桃 InvIsA 水果 [G] fruit, HasProperty,sour [G] fruit, InvIsA, cherry *R:* 还不错,酸酸甜甜的,酸樱桃味的很好喝。 Not bad. It's sour and sweet. The sour cherry taste is good.
Query #2	为啥我不喜欢用拍照? Why don't I like taking pictures?
Seq2Seq	因为你好看啊。Because you look beautiful.
ConKADI	因为你不喜欢拍照啊，我喜欢手机。 Because you don't like taking pictures. I like mobile phones.
BERT-PGN	因为手机拍的太好看了。 Because the pictures taken by the phone are so beautiful.
SEG-CKRG	*K:*[G]拍照 InvCapableOf 手机 [G]taking pictures, InvCapableOf, phone *R:* 我也不喜欢,手机拍出来的效果也不好。 I don't like it either. The pictures taken by the phone are bad.

Knowledge Accessing. As shown in Table 4, we first remove the usage of knowledge generated by *Generative Knowledge Selection (GenSel)* (#3), and the performance is worse. It verifies *GenSel* can indeed help *SEG-CKRG* to select the existing knowledge and to extend to generate new relevant knowledge. Subsequently, similar to [52], we remove the usage of retrieved knowledge and try to externalize the implicit knowledge inherent in the backbone PLM (#4). It can be seen that the performance of (#4) is significantly worse than the original model. Thus, we think it is still necessary to explicitly access external knowledge even in the era of PLMs. Finally, we discard all knowledge usages and purely fine-tune two RG models based on our general GPT-2 (#5) and dialogue-oriented GPT-2 (#6). Both (#5) and (#6) are worse than previous model variants, showing the value of modeling dialogue response with external knowledge. We also find if we do not use any external knowledge, the dialogue-oriented GPT-2 is better than the general GPT-2, which differs from the previous (#0 vs #1). It verifies *SEG-CKRG* prefers to general GPT-2 again. In addition, (#2) and (#5) have similar overall performance, showing that explicit commonsense knowledge and implicit PLM knowledge play similar importance in our *SEG-CKRG*.

3.5 Case Study

We report two cases in Table 5 In the first case, *SEG-CKRG* generated two commonsense knowledge facts. We can find one knowledge fact is selected from the retrieved knowledge candidate set, but the other does not exist in the retrieved

Table 6. Error Case. Same notations as the Case Study.

Query #1	一句话证明你是搞游戏的。 One sentence proves that you are a game enthusiast.
Seq2Seq	对啊，我也是这么想的? Yes, I think so too.
ConKADI	没有啊，电脑的时候电脑剧，你怎么看的。 No, it's a computer drama. What do you think of it.
BERT-PGN	哈哈哈，我就是这么想的，不知道为什么。 Hahaha, that's what I think. I don't know why.
SEG-CKRG	K:[G]游戏Synonym玩 [G]game, Synonym, play R: 我玩的不是游戏 是情怀。 What I play is not a game, but faith!
Query #2	哈哈哈哈你家的男朋友呢? Ha ha ha, where's your boyfriend?
Seq2Seq	你男朋友是谁? Who is your boyfriend?
ConKADI	我家的男朋友在家里。 My boyfriend stays at home.
BERT-PGN	我家男朋友都不知道怎么回事了，我还是个孩子。 My boyfriend doesn't know what's going on. I'm still a child.
SEG-CKRG	K:[G]男朋友 Causes 拍照 [G]boyfriend, Causes, taking pictures R: 哈哈哈哈 拍照技术好。 Haha, haha, he is good at taking pictures

knowledge candidate set. This case verifies that our *Generative Knowledge Selection* mechanism can not only select knowledge but also extend the new knowledge. Subsequently, we can find the response generated by *SEG-CKRG* is highly relevant to the dialogue history and such generated knowledge facts, demonstrating the important role of the generated knowledge. Besides, compared to baselines, *SEG-CKRG* is the only one that can keep fluency, rationality, and informativeness at the same time. In the second case, we can find although ConKADI, BERT-PGN, and *SEG-CKRG* tried to introduce the new knowledge concept *phone*, only *SEG-CKRG* answered a proper response.

Limitations. Although *SEG-CKRG* has surpassed all baselines, we also find a limitation in the current work, i.e., *Error Propagation*. *SEG-CKRG* sequentially generates the selected/extended knowledge and the dialogue response. Thus, if irrelevant knowledge has been generated in the first knowledge generation procedure, the next response generation procedure will be impacted. We report two typical error cases in Table 6. In the first case, *SEG-CKRG* generated a new but incorrect knowledge fact '(game, Synonym, play)' in the knowledge generation procedure. *SEG-CKRG* wrongly predicted the relation between 'game' and 'play', where the correct relation should be 'CapableOf'. But fortunately, this level of error has little impact on the following dialogue response generation. *SEG-CKRG* still generated a better response than other baselines. In the next case, *SEG-CKRG* has generated an existing but contextually-irrelevant knowledge fact. This error has significantly impacted the relevance of the generated response. Without considering the dialogue query, the response generated by *SEG-CKRG* is still fluent.

4   Related Work

Knowledge-Grounded Response Generation (KRG): Due to the inability to access enough knowledge, traditional RG models [33,35] always generate safe but boring responses in spite of the given query [12,13]. Consequently, KRG models try to solve this issue by accessing the external knowledge bases [20,36, 48]. Commonsense knowledge is a popular knowledge type in the current research [32,51], which helps a model to understand the dialogue, extend the topic, and then generate informative responses [38,40,41,44,46,50].

Pretrained Language Models (PLMs): PLMs such as BERT [6], RoBERTa [22], GPTs [2,27], and BARTs [11] have shown the dominate advantages in many NLP tasks [16]. PLMs can transfer the knowledge learned from massive unsupervised corpus to the open-domain RG models and bring significant improvements [8,31,37,47]. As for KRG models, previous works have shown PLMs can further prompt the text knowledge-grounded dialogue response generation [4,49] and the commonsense knowledge-grounded dialogue response generation [52].

Knowledge Selection: It is a research focus in KRG [7]. Non-PLM KRG models often adopt specific modules to conduct this job. [50], and [46] adopt graph neural networks, [41] uses the posterior response to help the learning of knowledge selection, [20], and [1] introduces copy networks to select the knowledge, [10] proposes a sequential knowledge selection paradigm, [28] proposes a global-to-local paradigm. In the era of PLMs, most knowledge selection methods that are originally designed for non-PLM KRG models become incompatible due to the restrictions of PLMs. Thus, the knowledge selection can only rely on the self-attention implemented by the Transformers of PLMs [23] or use the external module [49]. Meanwhile, such works can only select knowledge from a fixed and limited knowledge space, and the selection procedure is not very transparent. Different from such works, *SEG-CKRG* proposes a PLM-friendly *Generative Knowledge Selection* mechanism, which regards knowledge selection as a generative problem. Thus, our method can not only select the existing knowledge but also extend the new knowledge. Another difference is our work can explain the selection result using the human understandable natural language. In addition, although *TBS* [52] uses a PLM to generate knowledge, it does not include any knowledge selection procedure. The methodology of *TBS* is similar to our model (#4) in Table 4. Please refer to the corresponding results.

5   Conclusion

We propose an end-to-end CKRG model *SEG-CKRG*. Unlike previous works that can only use the limited and fixed knowledge retrieved by IR tools, *SEG-CKRG* introduces a novel *Generative Knowledge Selection (GenSel)* mechanism to select existing knowledge and extend new knowledge in a generative way.

More importantly, the knowledge selection/extension procedure has higher interpretability than previous works because the output is a natural language text. *SEG-CKRG* is implemented based on two Chinese GPT-2s pre-trained by ourselves. Finally, experimental results have shown the very competitive performance of *SEG-CKRG*.

Our future work includes three directions. First, we will continue to address the mentioned limitation; Second, we will explore and verify the effectiveness of *GenSel* in more different types of knowledge, such as text-based and table-based knowledge; Third, we are considering jointly modeling the CKRG task and the conversational relation extraction simultaneously by extending the potential of *GenSel*.

References

1. Bai, J., Yang, Z., Liang, X., Wang, W., Li, Z.: Learning to copy coherent knowledge for response generation. In: AAAI 2021 (2021)
2. Brown, T.B., et al.: Language models are few-shot learners. CoRR abs/2005.14165 (2020). https://arxiv.org/abs/2005.14165
3. Cho, K., van Merrienboer, B., Bahdanau, D., Bengio, Y.: On the properties of neural machine translation: encoder-decoder approaches. In: Wu, D., Carpuat, M., Carreras, X., Vecchi, E.M. (eds.) SSST@EMNLP 2014 (2014)
4. Cui, L., Wu, Y., Liu, S., Zhang, Y.: Knowledge enhanced fine-tuning for better handling unseen entities in dialogue generation. In: EMNLP 2021, November 2021
5. Cui, Y., Che, W., Liu, T., Qin, B., Yang, Z.: Pre-training with whole word masking for Chinese bert. IEEE/ACM TASLP (2021)
6. Devlin, J., Chang, M., Lee, K., Toutanova, K.: BERT: pre-training of deep bidirectional transformers for language understanding. In: NAACL-HLT 2019 (2019)
7. Dinan, E., Roller, S., Shuster, K., Fan, A., Auli, M., Weston, J.: Wizard of wikipedia: Knowledge-powered conversational agents. In: ICLR 2019 (2019)
8. Gu, X., Yoo, K.M., Ha, J.: Dialogbert: Discourse-aware response generation via learning to recover and rank utterances. In: AAAI2021 (2021)
9. Ippolito, D., Kriz, R., Sedoc, J., Kustikova, M., Callison-Burch, C.: Comparison of diverse decoding methods from conditional language models. In: ACL 2019, July 2019
10. Kim, B., Ahn, J., Kim, G.: Sequential latent knowledge selection for knowledge-grounded dialogue. In: ICLR 2020 (2020)
11. Lewis, M., et al.: BART: denoising sequence-to-sequence pre-training for natural language generation, translation, and comprehension. In: ACL 2020 (2020)
12. Li, J., Galley, M., Brockett, C., Gao, J., Dolan, B.: A diversity-promoting objective function for neural conversation models. In: NAACL 2016, June 2016
13. Li, J., Monroe, W., Jurafsky, D.: A simple, fast diverse decoding algorithm for neural generation. CoRR abs/1611.08562 (2016). http://arxiv.org/abs/1611.08562
14. Li, J., Tang, T., Zhao, W.X., Nie, J., Wen, J.: A survey of pretrained language models based text generation. CoRR abs/2201.05273 (2022). https://arxiv.org/abs/2201.05273
15. Li, J., Tang, T., Zhao, W.X., Wei, Z., Yuan, N.J., Wen, J.R.: Few-shot knowledge graph-to-text generation with pretrained language models. In: Findings of ACL-IJCNLP 2021 (Aug 2021)

16. Li, J., Tang, T., Zhao, W.X., Wen, J.: Pretrained language models for text generation: A survey. CoRR abs/2105.10311 (2021). https://arxiv.org/abs/2105.10311
17. Liang, Y., Meng, F., Zhang, Y., Chen, Y., Xu, J., Zhou, J.: Infusing multi-source knowledge with heterogeneous graph neural network for emotional conversation generation. In: AAAI 2021 (2021)
18. Lin, C.Y.: ROUGE: a package for automatic evaluation of summaries. In: Text Summarization Branches Out, pp. 74–81. Association for Computational Linguistics, Barcelona, Spain, July 2004
19. Lin, T., Wang, Y., Liu, X., Qiu, X.: A survey of transformers. CoRR abs/2106.04554 (2021). https://arxiv.org/abs/2106.04554
20. Lin, X., Jian, W., He, J., Wang, T., Chu, W.: Generating informative conversational response using recurrent knowledge-interaction and knowledge-copy. In: ACL 2020 (2020)
21. Liu, C.W., Lowe, R., Serban, I., Noseworthy, M., Charlin, L., Pineau, J.: How NOT to evaluate your dialogue system: An empirical study of unsupervised evaluation metrics for dialogue response generation. In: EMNLP 2016, November 2016
22. Liu, Y., et al.: Roberta: a robustly optimized BERT pretraining approach. CoRR abs/1907.11692 (2019). http://arxiv.org/abs/1907.11692
23. Lotfi, E., Bruyn, M.D., Buhmann, J., Daelemans, W.: Teach me what to say and I will learn what to pick: Unsupervised knowledge selection through response generation with pretrained generative models. CoRR abs/2110.02067 (2021). https://arxiv.org/abs/2110.02067
24. Luong, T., Pham, H., Manning, C.D.: Effective approaches to attention-based neural machine translation. In: EMNLP 2015 (2015)
25. Papineni, K., Roukos, S., Ward, T., Zhu, W.: Bleu: a method for automatic evaluation of machine translation. In: ACL, pp. 311–318. ACL (2002)
26. Qin, L., Liu, Y., Che, W., Wen, H., Li, Y., Liu, T.: Entity-consistent end-to-end task-oriented dialogue system with KB retriever. In: Inui, K., Jiang, J., Ng, V., Wan, X. (eds.) EMNLP-IJCNLP 2019 (2019)
27. Radford, A., Wu, J., Child, R., Luan, D., Amodei, D., Sutskever, I.: Language models are unsupervised multitask learners (2019)
28. Ren, P., Chen, Z., Monz, C., Ma, J., de Rijke, M.: Thinking globally, acting locally: Distantly supervised global-to-local knowledge selection for background based conversation. In: AAAI 2020, pp. 8697–8704 (2020)
29. See, A., Liu, P.J., Manning, C.D.: Get to the point: Summarization with pointer-generator networks. In: Barzilay, R., Kan, M. (eds.) ACL 2017 (2017). 10.18653/v1/P17-1099
30. Serban, I.V., et al.: A hierarchical latent variable encoder-decoder model for generating dialogues. In: AAAI 2017 (2017)
31. Shao, Y., et al.: CPT: a pre-trained unbalanced transformer for both Chinese language understanding and generation. CoRR abs/2109.05729 (2021). https://arxiv.org/abs/2109.05729
32. Speer, R., Havasi, C.: Conceptnet 5: a large semantic network for relational knowledge. In: The People's Web Meets NLP, Collaboratively Constructed Language Resources (2013)
33. Sutskever, I., Vinyals, O., Le, Q.V.: Sequence to sequence learning with neural networks. In: Advances in Neural Information Processing Systems 27 (2014)
34. Velickovic, P., Cucurull, G., Casanova, A., Romero, A., Liò, P., Bengio, Y.: Graph attention networks. In: ICLR 2018 (2018)
35. Vinyals, O., Le, Q.V.: A neural conversational model. CoRR abs/1506.05869 (2015). http://arxiv.org/abs/1506.05869

36. Wang, S., et al.: Modeling text-visual mutual dependency for multi-modal dialog generation. CoRR abs/2105.14445 (2021). https://arxiv.org/abs/2105.14445

37. Wang, Y., et al.: A large-scale chinese short-text conversation dataset. In: Zhu, X., Zhang, M., Hong, Yu., He, R. (eds.) NLPCC 2020. LNCS (LNAI), vol. 12430, pp. 91–103. Springer, Cham (2020). https://doi.org/10.1007/978-3-030-60450-9_8

38. Wu, S., Li, Y., Xue, P., Zhang, D., Wu, Z.: Section-aware commonsense knowledge-grounded dialogue generation with pre-trained language model. In: COLING 2022, pp. 521–531. International Committee on Computational Linguistics (2022). https://aclanthology.org/2022.coling-1.43

39. Wu, S., Li, Y., Zhang, D., Wu, Z.: Improving knowledge-aware dialogue response generation by using human-written prototype dialogues. In: Cohn, T., He, Y., Liu, Y. (eds.) Findings of EMNLP 2020 (2020)

40. Wu, S., Li, Y., Zhang, D., Wu, Z.: Generating rational commonsense knowledge-aware dialogue responses with channel-aware knowledge fusing network. IEEE ACM Trans. Audio Speech Lang. Process. **30**, 3230–3239 (2022). https://doi.org/10.1109/TASLP.2022.3199649

41. Wu, S., Li, Y., Zhang, D., Zhou, Y., Wu, Z.: Diverse and informative dialogue generation with context-specific commonsense knowledge awareness. In: ACL 202 (2020)

42. Wu, S., Wang, M., Li, Y., Zhang, D., Wu, Z.: Improving the applicability of knowledge-enhanced dialogue generation systems by using heterogeneous knowledge from multiple sources. In: WSDM 22 (2022)

43. Yan, R.: "Chitty-chitty-chat bot": deep learning for conversational AI. In: IJCAI 2018 (2018)

44. Young, T., Cambria, E., Chaturvedi, I., Zhou, H., Biswas, S., Huang, M.: Augmenting end-to-end dialogue systems with commonsense knowledge. In: AAAI 2018 (2018)

45. Yu, W., et al.: A survey of knowledge-enhanced text generation. CoRR abs/2010.04389 (2020). https://arxiv.org/abs/2010.04389

46. Zhang, H., Liu, Z., Xiong, C., Liu, Z.: Grounded conversation generation as guided traverses in commonsense knowledge graphs. In: ACL 2020 (2020)

47. Zhang, Y., et al.: DIALOGPT: large-scale generative pre-training for conversational response generation. In: Proceedings of the 58th Annual Meeting of the Association for Computational Linguistics: System Demonstrations, July 2020

48. Zhao, X., Wu, W., Tao, C., Xu, C., Zhao, D., Yan, R.: Low-resource knowledge-grounded dialogue generation. In: ICLR 2020 (2020)

49. Zhao, X., Wu, W., Xu, C., Tao, C., Zhao, D., Yan, R.: Knowledge-grounded dialogue generation with pre-trained language models. In: EMNLP 2020 (2020)

50. Zhou, H., Young, T., Huang, M., Zhao, H., Xu, J., Zhu, X.: Commonsense knowledge aware conversation generation with graph attention. In: IJCAI 2018 (2018)

51. Zhou, P., et al.: Commonsense-focused dialogues for response generation: an empirical study. In: SIGdial 2021 (2021)

52. Zhou, P., et al.: Think before you speak: explicitly generating implicit commonsense knowledge for response generation. In: ACL 2022, May 2022

Unify the Usage of Lexicon in Chinese Named Entity Recognition

Wenjia Wu[1,2], Changyou Zhang[1(✉)], Shuzi Niu[1], and Lin Shi[1,2]

[1] Institute of Software, Chinese Academy of Sciences, Beijing, China
changyou@iscas.ac.cn
[2] University of Chinese Academy of Sciences, Beijing, China

Abstract. Lexicon plays a critical role in Chinese Named Entity Recognition (CNER). The major reason lies in that words in the lexicon, lexicon words for short, are highly related to entity mention boundaries. Most lexicon enhanced CNER approaches focus on introducing lexicon words to the input and hidden layers. However, existing lexicon enhanced methods make the method hard to be adaptable and put weak lexicon constraints on architectures. To tackle these challenge, we propose a unified lexicon enhanced CNER framework. Specifically, lexicon word identification (LWI) task is proposed to locate and classify textual references to lexicon words. Similar to CNER task, this task is formalized either as sequence labeling or character relation classification, adopts CRF or Co-Predictor (Character relation classification) as the task specific layer, and is optimized with log-likelihood function of sequence probability or cross entropy of character pair relation type distribution. LWI task shares the input and hidden layers with CNER task. The whole framework is pretrained with LWI task and fine-tuned with CNER task. Experimental results on two benchmark CNER datasets show the better effectiveness and flexibility than state-of-the-art baselines.

Keywords: Named Entity Recognition · Lexicon augmentation · Lexicon based Word Identification · Lexicon based Pre-training

1 Introduction

Named Entity Recognition (NER) is fundamental to many downstream tasks, such as question answering and knowledge graph construction. As information extraction subtask, NER locates textual references to named entities, i.e. mentions, in unstructured text and classify them into predefined categories for named entities. How to locate these mentions is key to NER. Mention boundaries are also word boundaries [19]. In other words, word boundaries provide prior knowledge for mention locations of named entities. Different from English NER, Chinese NER (CNER) task deals with sequences of Chinese characters without explicit word boundaries. Thus Chinese NER task is more challenging.

Early Chinese Word Segmentation (CWS) based CNER methods often perform word segmentation first to derive the word boundaries and then sequence

© The Author(s), under exclusive license to Springer Nature Switzerland AG 2023
X. Wang et al. (Eds.): DASFAA 2023, LNCS 13945, pp. 665–681, 2023.
https://doi.org/10.1007/978-3-031-30675-4_49

labeling to word sequence like in English. Although derived word boundaries provide knowledge for CNER, obviously the derived knowledge is unreliable and contains errors which will be cascaded to the following NER step. Thus current mainstream CNER methods are based on character. Recent attempt to incorporate word information into CNER is Lexicon enhanced CNER methods. Lexicon knowledge is often utilized to improve the performance of CNER.

How to incorporate lexicon knowledge into current methods is often highly dependent on model architectures. Without loss of generality, a typical model architecture is usually divided into three parts: input layer, hidden layers, minimal parameterized task specific layer (task specific layer). Here the task specific layer is to predict the final score with model parameters as few as possible, such as softmax function and Linear layer for classification task. Lexicon words are often introduced to either the input or hidden layer. Lexicon word features are used in the embedding layer of lattice-lstm [12] or as word cells in the hidden layer of lattice-lstm [19]. Lexicon words are encoded in either nodes or edges of graph neural networks [5,16]. Other methods takes lexicon words as weak supervision for character representation learning in the attention mechanisms of Convolution Networks or Transformers [4,10]. Generally, lexicon enhanced networks designed for one kind are not suitable for other kinds.

However, current lexicon enhanced CNER methods are often related to a certain model, which has two drawbacks. On one hand, these specific model lexicon enhanced CNER methods are not applicable directly and adaptable to new model architectures. Almost no model architectures always keep their state-of-the-art performances for a certain CNER task. Model specific lexicon enhanced CNER methods provide little guidance for future work. On the other hand, whether lexicon word features in the input or hidden layer are taken as implicitly soft constraints on models. Therefore, how to make lexicon enhanced CNER methods more flexible and take full advantage of lexicon knowledge is challenging.

To tackle these challenges, we propose a universal lexicon enhanced CNER framework to adapt to various model architectures. The proposed framework, namely **LWICNER**, includes a **L**exicon **W**ord **I**dentification task parallel to CNER task, referred to **LWI**, an additional LWI task specific layer to the network architecture, and rich lexicon word supervision with corresponding loss functions for pre-training.

Specifically, in light of word segmentation task [1,14,17] in multitask learning scenario, the proposed LWI task is to locate and classify the textual references to lexicon words in unstructured text. Due to highly related to CNER task, the LWI task is formalized either as sequence label or as character relation classification. Then LWI task specific layer is designed as CRF or character relation classification layer like co-predictor [9] correspondingly. Both LWI and CNER task share the input and hidden layers. Next, supervision information from named entities and lexicon words are transformed into either a target tag sequence in sequence label or a character relation matrix in character relation classification. Correspondingly, the loss function of the LWI task is defined as the log-likelihood function of sequence probability or cross entropy between predicted and target character pair relation type distribution. Finally, the whole framework is learned with LWI task for pre-training and trained with CNER task for fine-tuning.

To verify its effectiveness, we conduct comprehensive experiments under sequence labeling and character relation classification implementation of our proposed framework with different model architectures. For sequence labeling, we adopts lattice-lstm [19] network architecture while we utilizes W2NER as the backbone network for character relation classification. Experimental results on public benchmark datasets, Weibo and Resume, indicate that the proposed framework outperforms state-of-the-art CNER approaches on both datasets. Compared with other lexicon enhanced methods, its performance improvement is significantly higher. Our code and data are released[1].

The main contributions of the proposed framework can be summarized as follows:

- A lexicon word identification task is proposed, which is highly related to CNER task, and can be formalized as either sequence labeling or character relation classification.
- LWI specific layer is implemented as CRF or character relation classifier. The corresponding loss function is log-likelihood function for tag sequence or cross entropy for character pair relation.
- Model parameters are first pre-trained with the LWI task and then fine-tuned with the CNER task, which is better than jointly training on both CNER and LWI task.
- Comprehensive experiments on benchmark CNER datasets are conducted to show its better effectiveness and flexibility than state-of-the-art baselines.

2   Related Work

According to the basic unit of input sequences fed into CNER architecture, existing CNER approaches fall into two categories: word based and character based CNER methods. As mentioned in the previous section, word based CNER method involves performing word segmentation first and then NER from the derived word sequence similar to English NER, which will lead to cascaded errors. Here we focus on word enhanced methods including word segmentation and lexicon enhanced methods.

Character based approaches use the character sequences as input. [11] learns multi-prototype Chinese character embeddings and applies these features to CNER task. [3] introduces the character-level and radical-level representations into the BiLSTM-CRF network. [8] regards each Chinese character as an image, extracts the features through Convolution neural network, and combines extracted features to character-level features with the help of attention mechanism. W2NER [9] is a unified NER architecture to retreat the NER task as word-word relation classification to recognize three major types named entities, i.e. flat, overlapped and discontinuous named entities.

---

[1] https://github.com/morediligent/LWICNER.

Word Segmentation enhanced CNER methods improve the performance of CNER task by introducing word segmentation task to the model in multitask learning scenario. [1] take the word segmentation and adversarial transfer learning based model into CNER to eliminate the noise interference by using three LSTM models in private CNER task space, CWS mission task and shared space. [17] proposes a CNN-LSTM-CRF architecture, learns the model parameters with CNER and CWS task jointly to improve the prediction accuracy of entity boundaries. [14] introduces CWS task as a supplement to the CNER task for Chinese social media and jointly optimizes two task loss functions.

Lexicon enhanced CNER approaches predict the entity boundaries and categories of entity mentions with the help of lexicon knowledge. Lattice-LSTM [19] is the first to utilize the lexicon and generate a classical lattice structure. Motivated by this idea of Lattice-LSTM [19], many related studies appear to apply the lexicon into CNER task. [10] proposes a flat structure which consists of spans instead of lattice structure, and is composed of transformer layers with special position encoding. [4] solves the conflicting candidate lexicon word problem through CNN, which runs easily in parallel. [5] adds a global node to capture the global sentence semantic information, and extends the dependency distance based on the graph structure. [12] connects the lexicon information with the character embedding to solve the complex sequence modeling problem. Along this line, [18] makes Transformer Encoder accommodate both character-level and word-level features with distance-aware and un-scaled attention mechanism. In all, existing approaches adopt the dynamic architecture, which is not easily adaptable.

3   Formalization of CNER Task

As mentioned before, how to locate these textual references (mentions) is key to NER task. According to mention location methods, existing approaches often reduce CNER task to sequence labeling and character-character relation classification method. The former is dominating especially for flat NER and the latter is universal for nested, overlapping and flat NER methods.

3.1   Sequence Labeling

Given an input character sequence $s = c_1, c_2, \ldots, c_n$, where c_j means the jth character, sequence labeling method [7,19] is to obtain a tag sequence $\hat{y} = \hat{y}_1, \hat{y}_2, \ldots, \hat{y}_n$ to locate mentions with a certain tagging scheme, such as BIOES and BMES tagging scheme. The tag sequence $\hat{y}$ is either produced with strictly each label per character from conditional random fields [7,19] or generated loosely by sequence-to-sequence methods [15].

Take the latter tagging scheme for instance. Given the character sequence s ="龙门石窟潜溪寺", which means Qianxi Temple near Longman Grottoes, its target tag sequence under the BMES scheme is y =B-Location, M-Location, M-Location, E-Location, B-Location, M-Location, E-Location. For each character c_j, its target (ground truth) and predicted tag is denoted as y_j and $\hat{y}_j$.

3.2   Character Relation Classification

Character relation classification method, originated from word-word relation classification method [9], is to classify the character pair into pre-defined relation types by constructing a character graph to locate mentions in each input sequence.

Given a character sequence $s = c_1, c_2, \ldots, c_n$, it is to predict the relation type between each character pair $\{(c_i, c_j)\}_{i,j=1}^{n}$, and derive a character relation matrix $\hat{r} = (\hat{r}_{ij})_{n \times n}$ to describe the character relation graph, where $\hat{r}_{ij}$ means the relation type between character c_i and c_j. The predefined character relation type set is denoted as $R = \{0, \ldots, 19\}$. Each $r_{ij} \in R$ means tail character c_i lies r_{ij} characters behind head character c_j of the same entity mention. $\hat{r}_{ij} = 0$ means no relations.

The input character sequence $s =$"龙门石窟潜溪寺 " is the same as above, its target relation matrix is as Eq. (1). For each sequence s, its target (ground truth) and predicted relation matrix is denoted as r and $\hat{r}$.

$$r = \begin{pmatrix} 0 & 1 & 0 & 0 & 0 & 0 & 0 \\ 0 & 0 & 1 & 0 & 0 & 0 & 0 \\ 0 & 0 & 0 & 1 & 0 & 0 & 0 \\ 3 & 0 & 0 & 0 & 0 & 0 & 0 \\ 0 & 0 & 0 & 0 & 0 & 1 & 0 \\ 0 & 0 & 0 & 0 & 0 & 0 & 1 \\ 0 & 0 & 0 & 0 & 6 & 0 & 0 \end{pmatrix} \tag{1}$$

4   Proposed Method

Based on the observation of high correlation between lexicon word and named entity mention boundaries, we propose a novel Lexicon Word Identification task, incorporate it to the CNER backbone.

4.1   Observation

Lexicon words provides reliable boundary and semantics knowledge to CNER task who has no explicit word boundaries. Here we utilize the reliable knowledge as supervision information. As mentioned before, the supervision information for CNER task is either a tag sequence y or a character relation matrix r for each sequence s.

We locate the word mentions in the input sequence s through exacting string matching method. The lexicon is denoted as $\mathcal{D}$, which is a word set $\mathcal{D} = \{w\}$ with each word $w = c_1, c_2, \ldots, c_m$ as a character sequence. For a word $w \in \mathcal{D}$, we determine whether it appears in sequence s as whether w is a sub-string of s. In this way, we derive the word mentions in sequence s denoted as $\mathcal{D}_s$. For instance, $s =$"龙门石窟潜溪寺 " and the derived word mentions $\mathcal{D}_s = \{$"龙门 ", "石窟 ", "龙门石 "$\}$.

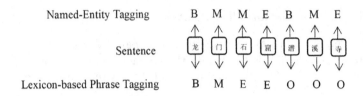

Fig. 1. The labels using the BMES tagging pattern. The upper labels are generated from the named entity information and the nether labels are generated from the lexicon based words information.

	龙	门	石	窟	潜	溪	寺
龙		NNC	NNC	NNC			
门							
石							
窟	THC-*						
潜						NNC	NNC
溪							
寺				THC-*			

	龙	门	石	窟	潜	溪	寺
龙		NNC	NNC	NNC			
门	THC						
石	THC						
窟			THC				
潜							
溪							
寺							

Fig. 2. Illustration of Character relation matrix for CNER(right) and LWI(right) task. NNC (Next-Neighbor-Character) and THC (Tail-Head-Character) are relation types.

With sequence labeling method for entity mentions, word mentions $\mathcal{D}_s$ are transformed to a tag sequence γ under a certain tagging scheme for each sequence s. The target tag sequence is $\gamma = B, M, E, E, O, O, O$ to identify lexicon word in sequence $s =$"龙门石窟潜溪寺" and y to locate entity mentions in s as shown in Fig. 1. The correlation $\rho_s(\gamma, y)$ between γ and y is computed as Eq. (2), where $I(\cdot)$ is the indicator function that equals to 1 when the condition holds and 0 otherwise. Statistical information over benchmark datasets are shown in Fig. 3.

$$\rho_s(\gamma, y) = \frac{\sum_{i=1}^{n} I(\gamma_i == y_i)}{n} \tag{2}$$

With character relation classification method for entity mentions, word mentions $\mathcal{D}_s$ are formatted as a relation matrix τ for each sequence s. For sequence $s =$"龙门石窟潜溪寺", the relation matrix τ and r are to locate word and entity mentions in s respectively as shown in Fig. 2. The correlation $\rho_r(\tau, r)$ between τ and r is computed as Eq. (3). Statistical correlation information over benchmark datasets are shown in Fig. 4.

$$\rho_r(\tau, r) = \frac{\sum_{i=1}^{n} \sum_{j=1}^{n} I(\tau_{ij} == r_{ij})}{n^2} \tag{3}$$

Statistical results in Figs. 3 and 4 over these correlation values for both benchmark datasets suggests that lexicon word boundaries are highly correlated with

Fig. 3. The correlation of the target tag sequence between LWI and CNER task using BMES tagging scheme.

Fig. 4. The correlation of the target character relation matrix between LWI and CNER task.

named mention boundaries. Therefore we propose a novel Lexicon Word Identification task (LWI), which is highly related to CNER task.

4.2 Formalization of LWI Task

Lexicon word Identification task is to locate and classify textual references to words in the lexicon $\mathcal{D}$. For each sequence s, we construct its word mention set $\mathcal{D}_s$ through exact match between lexicon words and s efficiently with the algorithm [19].

Formalized as sequence labeling, LWI task aims to decode a tag sequence $\hat{\gamma} = \hat{\gamma}_1, \hat{\gamma}_2, \ldots, \hat{\gamma}_n$ with each tag per character based on the learned model for s. Here we define the tag set as $G = \{B, M, E, S, O\}$ to indicate whether the current character is a beginning, middle, end, single, stop character of a lexicon word. We use $\hat{\gamma}$ and γ to distinguish the predicted and target (ground truth) tag sequence. We train the model under the supervision of γ which is derived from $\mathcal{D}_s$ for sequence s.

Formalized as character relation classification, LWI task is to classify each character pair (c_i, c_j) into pre-defined relation types $\hat{\tau}_{ij} \in A$, obtain a relation matrix $\hat{\tau} = (\hat{\tau})_{n \times n}$ like Eq. (1) as a character graph. Here we define $A = \{0, 1, \ldots, 19\}$. Each relation type $a \in A$ means tail character c_i lies a characters behind head character c_j of the same word mention. Specially $a = 1$ means c_i is next to c_j. To differentiate the predicted and ground truth relation matrix, we utilize $\hat{\tau}$ and τ. LWI models are trained under the supervision of τ derived from the word mention set $\mathcal{D}_s$.

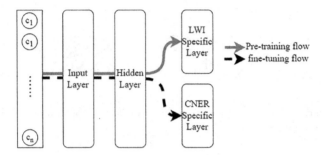

Fig. 5. The Proposed LWICNER framework with (1) red curve as the data flow in pre-training stage; (2) black curve as the data flow in fine-tuning stage. (Color figure online) .

4.3 Network Architecture

Traditionally the proposed model architecture is mainly composed of input, hidden, LWI and CNER task specific layers as Fig. 5.

Input Layer. Each character sequence s is fed into the input layer, such as BERT or word2vec lookup table. The character embedding in this sequence is obtained as $X = (\mathbf{x}_i)_{n \times 1}$ with $\mathbf{x}_i \in \mathbf{R}^d$ as the d-dim embedding vector of character c_i.

Hidden Layers. The major kinds of hidden layers include Graph Neural Networks [5,16], Lattice-LSTM [19], Transformer [9,10] and Convolution Networks [2,9]. For some methods, hidden layers deal with only one kind like Lattice-LSTM [19]. For other methods, hidden layers are often made up of several kinds like W2NER [9], which includes Transformer, bi-LSTM and Dilated Convolution layers. Character representations or character pair representations are updated through these layers as

$$\mathbf{H} = \mathrm{HL}(\mathbf{X}) \in \mathbf{R}^{n \times d'},$$

$$\mathbf{P} = \mathrm{HP}(\mathbf{X}) \in \mathbf{R}^{n \times n \times d'}$$

respectively. HL and HP are the function of hidden layers for character and character pair respectively.

CNER Specific Layer. Either character representations $\mathbf{H}$ or pair representations $\mathbf{P}$ is fed into the entity recognition task specific layer with two typical implementation methods denoted as CRF and Co-Predictor respectively.

(1) *CRF* is the dominating implementation method. With character representations $\mathbf{H}$ as input, it calculates the probability that the decoded tag sequence is y for each input sequence s as Eq. (4). y' means one possible decoded sequence. $\{\mathbf{W}_{\mathrm{CRF}}^t\}_{t \in T} \cup \{b_{\mathrm{CRF}}^{(t_i,t_j)})\}_{t_i,t_j \in T}$ are all parameters of the CNER task specific layer. T is the set of all possible categories pre-defined based

on named entities under a certain scheme. For BMES scheme, $T = \{$B-Location, $\ldots$, E-Person$\}$.

$$P(y|s) = \frac{\exp(\sum_j \mathbf{W}_{\text{CRF}}^{y_j} \mathbf{h}_j + b_{\text{CRF}}^{(y_{j-1}, y_j)})}{\sum_{y'} \exp(\sum_j \mathbf{W}_{\text{CRF}}^{y'_j} \mathbf{h}_j + b_{\text{CRF}}^{(y'_{j-1}, y'_j)})} \qquad (4)$$

(2) *Co-Predictor* [9] is a universal method to accommodate all kinds of entities. It takes both character representations $\mathbf{H}$ and pair representations $\mathbf{P}$ as input. Character pair scores are obtained from directly linear transformation of pair representations P denoted as MLP($\mathbf{P}$) and indirectly biaffine transformation of character representations $\mathbf{H}$ denoted as $B(\mathbf{h}_i, \mathbf{h}_j)$. For each sequence s, these predicted character pair scores over different relation types are normalized as a probability distribution $P(\hat{\mathbf{r}}_{ij})$ through softmax in Eq. (5).

$$P(\hat{\mathbf{r}}_{ij}) = \text{softmax}(B(\mathbf{h}_i, \mathbf{h}_j) + \text{MLP}(\mathbf{P}_{ij})) \qquad (5)$$

LWI Specific Layer. Either character representations $\mathbf{H}$ or character pair representations $\mathbf{P}$ is fed into the following lexicon word identification task specific layer with two similar implementation methods to CNER specific layer, i.e. CRF_w and Co-Word-Predictor.

(1) *CRF* takes character representations $\mathbf{H}$ as input, and calculates the probability that the decoded tag sequence is γ for each input sequence s as Eq. (6). γ' means one possible decoded sequence. $\{\mathbf{W}_{\text{CRF}_w}^g\}_{g \in G} \cup \{b_{\text{CRF}_w}^{(g_i, g_j)})\}_{g_i, g_j \in G}$ are all parameters of the LWI task specific layer denoted as CRF_w. G is the set of all possible categories pre-defined under a certain scheme.

$$P(\gamma|s) = \frac{\exp(\sum_j \mathbf{W}_{\text{CRF}_w}^{\gamma_j} \mathbf{h}_j + b_{\text{CRF}_w}^{(\gamma_{j-1}, \gamma_j)})}{\sum_{\gamma'} \exp(\sum_j \mathbf{W}_{\text{CRF}_w}^{\gamma'_j} \mathbf{h}_j + b_{\text{CRF}_w}^{(\gamma'_{j-1}, \gamma'_j)})} \qquad (6)$$

(2) *Co-Predictor* [9] takes both character representations $\mathbf{H}$ and pair representations $\mathbf{P}$ as input for LWI task denoted as Co-Word-Predictor. Character pair scores are obtained from directly linear transformation of pair representations P denoted as $\text{MLP}_w(\mathbf{P})$ and indirectly biaffine transformation of character representations $\mathbf{H}$ denoted as $B_w(\mathbf{h}_i, \mathbf{h}_j)$. For each sequence s, these predicted character pair scores over different relation types are normalized as a probability distribution $P(\hat{\tau}_{ij})$ through softmax in Eq. (7).

$$P(\hat{\tau}_{ij}) = \text{softmax}(B_w(\mathbf{h}_i, \mathbf{h}_j) + \text{MLP}_w(\mathbf{P}_{ij})) \qquad (7)$$

4.4 Training Paradigm

Considering the homogeneity of both LWI and CNER tasks, we adopt the pre-training and fine-tuning paradigm. Specifically, we learn these parameters from the input, hidden and LWI specific layers with the optimization of L_{*w}, where $* = \{s, r\}$ is the implementation method of task specific layer, in the pre-training stage. Here data flows according to the red curve in Fig. 5. Then we train these parameters from the input, hidden and CNER specific layers with the optimization of L_* in the fine-tuning stage, where data flow is transformed according to the dotted black curve in Fig. 5.

Both task specific layers are implemented as CRF and CRF_w separately for CNER and LWI task. Given the training set $\{(s, y)\}$ for CNER task, its loss function is defined as the log-likelihood function of $P(y|s)$ with L_2 regularization as Eq. (8). All the parameters in the model architecture are denoted as Θ. Given the training set $\{(s, \gamma)\}$ for LWI task, its loss function is defined as the log-likelihood function of $P(\gamma|s)$ with L_2 regularization as Eq. (9). All the parameters in the model architecture are denoted as Θ_w. The only difference between Θ and Θ_w lies in the parameter difference between CRF and CRF_w.

$$L_s = \sum_{(s,y)} \log P(y|s) + \frac{\lambda}{2}\|\Theta\|^2 \tag{8}$$

$$L_{sw} = \sum_{(s,\gamma)} \log P(\gamma|s) + \frac{\lambda}{2}\|\Theta_w\|^2 \tag{9}$$

Both task specific layers are implemented as Co-Predictor and Co-Word-Predictor respectively for CNER and LWI task. For the training set $\{(s, r)\}$ of CNER task, its loss function is defined as the cross entropy between the ground truth and predicted relation label distribution for each character pair (c_i, c_j) in Eq. (10), i.e. $\mathbf{r}_{ij}$ and $P(\hat{\mathbf{r}}_{ij})$ respectively. $\mathbf{r}_{ij}$ is a one-hot vector to denote which relation type the character pair belongs to. For the training set $\{(s, \tau)\}$ of LWI task, its loss function is defined as the cross entropy between the ground truth one-hot vector τ_{ij} and predicted relation label distribution $P(\hat{\tau}_{ij})$ for each character pair (c_i, c_j) in Eq. (11).

$$L_r = -\sum_{(s,r)} \frac{1}{n^2} \sum_{i=1}^{n} \sum_{j=1}^{n} \mathbf{r}_{ij} \log P(\hat{\mathbf{r}}_{ij}) \tag{10}$$

$$L_{rw} = -\sum_{(s,\tau)} \frac{1}{n^2} \sum_{i=1}^{n} \sum_{j=1}^{n} \tau_{ij} \log P(\hat{\tau}_{ij}) \tag{11}$$

5   Experiments

To demonstrate the effectiveness and flexibility of our proposed framework, we conduct comprehensive experiments on benchmark datasets. Experimental setting and results are shown in the following subsections.

5.1   Experimental Setup

Datasets. Weibo [6,13] and Resume [19] are two public benchmark datasets for CNER task. Corpus in Weibo [6,13] is from social media, which contains

1.4(73.8), 0.27(14.5), 0.27(14.8) thousands of sentences (characters) for training, validation and testing separately. Sentences in Resume [19] is from personal resume data. It is divided into three parts with 3.8(124.1)k training sentences (characters), 0.46(13.9)k validation sentences (characters) and 0.48(15.1)k testing sentences (characters).

Lexicon. We use the Chinese Giga-Word[2] as the lexicon $\mathcal{D}$. It is composed of 708k words including 5.7k one-character words, 291.5k two-character words, 278.1k three-character words, 129.1k other words [12]. The Giga-Word is a general and professional lexicon.

Table 1. Hyper-parameter Setting for LWICNER-L

Parameter	Value	Parameter	Value
char emb size	30	bigram emb size	50
lattice emb size	50	LSTM hidden size	200
char dropout	0.5	lattice dropout	0.5
LSTM layer	1	learning rate lr	0.015

Model Architecture. To investigate the flexibility of our proposed framework, we consider two kinds of models architectures with dominating task specific layers: CRF and co-predictor. Specifically, we adapt Lattice-LSTM [19] and W2NER [9] to our proposed framework LWICNER, and obtain two implementation methods denoted as LWICNER-L and LWICNER-W respectively.

Table 2. Hyper-parameter setting for LWICNER-W

Parameter	Value	Parameter	Value
clip grad norm	1e–3	emb dropout	0.5
weight decay	0	bert learning rate	5e–6
con dropout	0.5	learning rate	1e–3
dist emb size	10	batch size	4
type emb size	20	warm factor	0.1

Baseline Methods. To show the effectiveness of our proposed methods including LWICNER-L and LWICNER-W, we compare them with two kinds of baselines. One is the character based CNER methods without word information, such as W2NER [9]. The other is the lexicon enhanced methods, including Lattice-LSTM [19], Tender [18], LGN [5], FLAT [10], softLexicon [12]. Some performances of baselines are reported in the original paper while others are reproduced according to the hyper-parameter setting in the original paper. All the performances are derived on one GPU 3080 with 10GB memory.

---

[2] https://catalog.ldc.upenn.edu/LDC2011T13.

Implementation Details. We choose the hyper-parameter setting of our proposed LWICNER-L and LWICNER-W with the best performance on validation set shown in Tables 1 and 2. The performance is measured by P(precision), R(recall) and F1(F1-score).

5.2 Effectiveness Study

All the performances listed in Table 3 are the same as that reported in their original papers. Performances of our proposed LWICNER-W in terms of three evaluation metrics are reported on test sets of both datasets as Table 3. LWICNER-W is better than LWICNER-L in most cases, so here we only show the performances of LWICNER-W.

Table 3. Results of State-of-the-art Baselines and LWICNER-W on Benchmark datasets.

	Resume			Weibo		
	P	R	F1	P	R	F1
Lattice LSTM	94.81	94.11	94.46	53.04	62.25	58.79
Tender	–	–	95.00	–	–	58.17
LGN	95.37	94.84	95.11	57.14	66.67	59.92
FLAT	–	–	95.86	–	–	68.55
Soft-Lexicon	96.08	96.13	96.13	70.94	67.02	70.05
W2NER	96.01	96.01	96.01	71.65	67.70	69.62
Ours	**96.08**	**96.26**	**96.17**	**73.91**	**69.14**	**71.45**

Our proposed framework LWICNER, implemented as LWICNER-W, outperforms all the baseline methods listed in Table 3. Among all the baselines, SoftLexicon and W2NER are two representative methods of CNER methods with and without lexicon knowledge, and perform better than other baselines.

Compared with the best baseline method without lexicon knowledge W2NER, the performance improvement is 0.0729%, 0.260%, 0.167%, 3.15%, 2.13%, 2.63% in terms of Precision, recall and F1-score on Resume and Weibo. The performance gain of LWICNER-W is owing to the addition of lexicon word identification task because it shares the same network architecture with W2NER except the LWI specific layer. Thus it is effective to utilize the lexicon knowledge.

Compared with the best baseline with lexicon knowledge SoftLexicon, the performance improvement is 0%, 0.135%, 0.0416%, 4.19%, 3.16%, 2.00% in terms of Precision, recall and F1-score on Resume and Weibo. SoftLexicon injects the lexicon word features into the input layer, and the lexicon constraints become weaker as layers go deeper. LWICNER-W incorporates lexicon knowledge directly to task specific layer. Performance comparison results indicate that LWICNER makes a fuller use of lexicon knowledge than SoftLexicon.

5.3 Flexibility Study

With the original codes released by the authors, we reproduce the performances of Lattice-LSTM and W2NER with the hyper-parameter setting in Tables 1 and 2 respectively. Our proposed framework LWICNER is implemented as two kinds of network architectures, namely Lattice-LSTM and W2NER, and obtains corresponding methods LWICNER-L and LWICNER-W. We compare the LWICNER-L with its backbone network Lattice-LSTM under the same hyper-parameter setting and results are shown in Table 4. We compare LWICNER-W with its backbone network W2NER under the same hyper-parameter setting and results are shown in Table 5.

Table 4. Performance comparison between LWICNER-L and its backbone

	Resume			Weibo		
	P	R	F1	P	R	F1
Lattice LSTM	96.51	94.42	94.71	63.64	**47.34**	54.50
LWICNER-L	96.50	**95.06**	**94.77**	**67.62**	45.89	**54.68**
Improvement(%)	−0.01036	0.6778	0.06335	6.254	−3.063	0.3303

For most metrics on two datasets, LWICNER-L performs better than Lattice-LSTM. F1-score of LWICNER-L is consistently higher than that of Lattice-LSTM on both datasets. However, comparison results are inconsistent under precision and recall metrics. For Resume, precision of Lattice-LSTM is slightly better while recall of Lattice-LSTM is 3.063% higher on Weibo. LWICNER-L provides harder lexicon constraints on the network architectures than Lattice-LSTM, which is more likely to prompt the precision than recall on Weibo.

Table 5. Performance comparison between LWICNER-W and its backbone

	Resume			Weibo		
	P	R	F1	P	R	F1
W2NER	96.01	96.01	96.01	71.65	67.70	69.62
LWICNER-W	**96.08**	**96.26**	**96.17**	**73.91**	**69.14**	**71.45**
Improvement(%)	0.07291	0.2604	0.1666	3.154	2.127	2.629

LWICNER-W is always better than its corresponding backbone model W2NER under all the three metrics on both datasets due to the usage of lexicon word knowledge. Especially the precision improvement on Weibo is slightly higher than recall improvement. This coincides with the result from the above table. In other words, our proposed framework is more likely to improve the precision.

Through the above analysis, the performance gain of the proposed framework is significantly high for both network architectures. This suggests our proposed

LWICNER is flexible enough for these kinds of network architectures. Besides, LWICNER-W is better than LWICNER-L due to the limitation of sequence labeling formulation. Sequence labeling implemented as Lattice-LSTM is friendly to flat NER while character relation classification implemented as W2NER is suitable for all kinds of NER, such as discontinuous, overlapping and flat NER. LWICNER implemented as Lattice-LSTM cannot take full advantage of lexicon words with many overlapped lexicon words.

Compared with overlapped entities mentions in CNER task, overlapped words are common in the lexicon. For example, "居留" and "居留权" are nested, i.e.totally overlapped, while "龙门石", "石窟" are partly overlapped. Thus the improvement space of LWICNER-L are left for future.

5.4   Influence of Sparsity Phenomena

Some sentences contain only one entity mention, while others contain a few. The former is referred to as sparsity phenomena. Take sequence labeling for example, we compute the ratio of the number of named entity labels ("B", "M", "E", "S") to the total number of all tagged labels ("B", "M", "E", "S", "O") on Resume and Weibo, as shown in Table 6. Here we referred to the ratio as named entity ratio.

Table 6. The named entity ratio on the Resume and Weibo

	Named entity labels			Total labels			Ratio of named entity labels to total labels(percent)		
	Train	Dev	Test	Train	Dev	Test	Train	Dev	Test
Resume	79012	8444	9910	127916	14353	15577	61.76	58.83	63.61
Weibo	4951	971	1078	75128	14779	15112	6.59	6.57	7.13

We find that the named entity ratio on Weibo is much lower than on Resume, which means Weibo is more sparse than Resume. That's why the performances on Resume are higher than these on Weibo. Named entity supervision information provides enough knowledge for CNER on Resume while little knowledge for CNER on Weibo. That's the reason of the higher performance gain of lexicon-enhanced methods on Weibo.

The sparsity phenomena on Weibo also explains why precision improvement is larger than recall improvement. Recall improvement depends on the relationship between input sequences in the sparse scenario. However, the proposed framework only focuses on augmenting the knowledge of each input sequences. Thus the framework works better under the measure of precision in the sparse scenario. Incorporation of lexicon words into the network architectures will help make these training sequences related implicitly like Lattice-LSTM.

5.5   Ablation Study

We conduct an ablation study for training paradigm to explore why pretrain-finetune paradigm is used here. We compare the proposed framework LWICNER with jointly training methods, which learns model parameters to optimize the loss sum of LWI and CNER tasks denoted as $L_* + L_{*w}$. Specifically, there are two implementations under different formalization methods. Performances of LWICNER and $L_s + L_{sw}$ are shown in Table 7 while these of LWICNER-W and $L_r + L_{rw}$ are demonstrated in Table 8.

Table 7. Different Training paradigms with Lattice-LSTM backbone

	Resume			Weibo		
	P	R	F1	P	R	F1
LWICNER-L	96.50	95.06	94.77	67.62	45.89	54.68
$L_s + L_{sw}$	94.77	94.48	94.62	65.41	42.03	51.18

It is obvious that the proposed framework LWICNER performs consistently better than jointly training paradigm $L_* + L_{*w}$ on both datasets under all the metrics in Tables 7 and 8. The underlying reason is the highly correlation between LWI and CNER task in terms of supervision format, network architecture and loss function.

Table 8. Different Training paradigms with W2NER backbone

	Resume			Weibo		
	P	R	F1	P	R	F1
LWICNER-W	96.07	95.89	95.98	73.91	69.14	71.45
$L_r + L_{rw}$	95.63	95.21	95.42	72.02	66.51	69.91

6   Conclusion

To accommodate different model architectures of CNER task, a unified lexicon enhanced CNER framework is proposed. The framework includes the Lexicon word identification task, and its homogeneous implementation to CNER task. The formalization, task specific layer and loss function are designed for this novel task. Different from existing lexicon based methods, the proposed framework is more flexible that it work better for two major kinds of architectures on benchmark datasets. More importantly, it achieves better performances than state-of-the-art CNER methods on benchmark datasets. The remaining question is how to improve the framework in the entity-sparse scenario.

Acknowledgements. This research work was funded by the National Natural Science Foundation of China under Grant No. 62072447.

References

1. Cao, P., Chen, Y., Liu, K., Zhao, J., Liu, S.: Adversarial transfer learning for Chinese named entity recognition with self-attention mechanism. In: Proceedings of the 2018 Conference on Empirical Methods in Natural Language Processing, pp. 182–192 (2018)
2. Chen, Y.: Convolutional neural network for sentence classification. Master's thesis, University of Waterloo (2015)
3. Dong, C., Zhang, J., Zong, C., Hattori, M., Di, H.: Character-based LSTM-CRF with radical-level features for Chinese named entity recognition. In: Lin, C.-Y., Xue, N., Zhao, D., Huang, X., Feng, Y. (eds.) ICCPOL/NLPCC -2016. LNCS (LNAI), vol. 10102, pp. 239–250. Springer, Cham (2016). https://doi.org/10.1007/978-3-319-50496-4_20
4. Gui, T., Ma, R., Zhang, Q., Zhao, L., Jiang, Y.G., Huang, X.: CNN-based Chinese NER with lexicon rethinking. In: IJCAI, pp. 4982–4988 (2019)
5. Gui, T., et al.: A lexicon-based graph neural network for Chinese NER. In: Proceedings of the 2019 Conference on Empirical Methods in Natural Language Processing and the 9th International Joint Conference on Natural Language Processing (EMNLP-IJCNLP), pp. 1040–1050 (2019)
6. He, H., Sun, X.: F-score driven max margin neural network for named entity recognition in Chinese social media. arXiv preprint arXiv:1611.04234 (2016)
7. Huang, Z., Xu, W., Yu, K.: Bidirectional LSTM-CRF models for sequence tagging. arXiv preprint arXiv:1508.01991 (2015)
8. Jia, Y., Ma, X.: Attention in character-based BiLSTM-CRF for Chinese named entity recognition. In: Proceedings of the 2019 4th International Conference on Mathematics and Artificial Intelligence, pp. 1–4 (2019)
9. Li, J., et al.: Unified named entity recognition as word-word relation classification. In: Proceedings of the AAAI Conference on Artificial Intelligence, vol. 36, pp. 10965–10973 (2022)
10. Li, X., Yan, H., Qiu, X., Huang, X.: Flat: Chinese NER using flat-lattice transformer. arXiv preprint arXiv:2004.11795 (2020)
11. Lu, Y., Zhang, Y., Ji, D.: Multi-prototype Chinese character embedding. In: Proceedings of the Tenth International Conference on Language Resources and Evaluation (LREC2016), pp. 855–859 (2016)
12. Ma, R., Peng, M., Zhang, Q., Huang, X.: Simplify the usage of lexicon in Chinese NER. arXiv preprint arXiv:1908.05969 (2019)
13. Peng, N., Dredze, M.: Named entity recognition for Chinese social media with jointly trained embeddings. In: Proceedings of the 2015 Conference on Empirical Methods in Natural Language Processing, pp. 548–554 (2015)
14. Peng, N., Dredze, M.: Improving named entity recognition for Chinese social media with word segmentation representation learning. arXiv preprint arXiv:1603.00786 (2016)
15. Shen, Y., Yun, H., Lipton, Z.C., Kronrod, Y., Anandkumar, A.: Deep active learning for named entity recognition. arXiv preprint arXiv:1707.05928 (2017)
16. Sui, D., Chen, Y., Liu, K., Zhao, J., Liu, S.: Leverage lexical knowledge for Chinese named entity recognition via collaborative graph network. In: Proceedings of the 2019 Conference on Empirical Methods in Natural Language Processing and the 9th International Joint Conference on Natural Language Processing (EMNLP-IJCNLP), pp. 3830–3840 (2019)

17. Wu, F., Liu, J., Wu, C., Huang, Y., Xie, X.: Neural Chinese named entity recognition via CNN-LSTM-CRF and joint training with word segmentation. In: The World Wide Web Conference, pp. 3342–3348 (2019)
18. Yan, H., Deng, B., Li, X., Qiu, X.: TENER: adapting transformer encoder for named entity recognition. arXiv preprint arXiv:1911.04474 (2019)
19. Zhang, Y., Yang, J.: Chinese NER using lattice LSTM. arXiv preprint arXiv:1805.02023 (2018)

A Prompt-Based Representation Individual Enhancement Method for Chinese Idiom Reading Comprehension

Ying Sha[1,2,3,4], Mingmin Wu[1,2,3,4], Zhi Zeng[1,2,3,4], Xing Ge[1,2,3,4],
Zhongqiang Huang[1,2,3,4], and Huan Wang[1(✉)]

[1] College of Informatics, Huazhong Agricultural University, Wuhan, China
{shaying,hwang}@mail.hzau.edu.cn,
{wmm_nlp,zengmouren,2020317010018,sta}@webmail.hzau.edu.cn
[2] Key Laboratory of Smart Farming for Agricultural Animals, Wuhan, China
[3] Hubei Engineering Technology Research Center of Agricultural Big Data,
Wuhan, China
[4] Engineering Research Center of Intelligent Technology for Agriculture,
Ministry of Education, Wuhan, China

Abstract. Chinese idiom is a distinctive language phenomenon, which usually consists of four Chinese characters and expresses a non-compositional and metaphorical meaning. Therefore, Chinese idioms pose unique challenges for Chinese machine reading comprehension. To address this issue, researchers proposed a Chinese idiom cloze task and a large-scale Chinese idioms dataset ChID. Existing methods have proposed a number of models and achieved reasonable performance on ChID. However, they fall short of fully exploring the precise representations and distinctions of the meanings of idioms, especially idioms with similar meanings. In this paper, we propose a prompt-based representation individual enhancement method (PRIEM). This method fuses the context-specific representation and the generic definition representation of the idioms, and uses the prompt method to guide the model in learning the metaphorical meanings of idioms purposefully. To further improve the distinction representations of idioms with similar meanings, PRIEM adopts a method of idiom representation mapping and decomposing based on orthogonal projection to obtain the common and individual representations of idioms respectively. Experimental results on ChID show that our model outperforms state-of-the-art models.

Keywords: Chinese idiom · reading comprehension · prompt method · orthogonal projection · individual representation

1 Introduction

As a special linguistic phenomenon, Chinese idioms pose unique challenges for Chinese machine reading comprehension [11]. They mainly derive from ancient

© The Author(s), under exclusive license to Springer Nature Switzerland AG 2023
X. Wang et al. (Eds.): DASFAA 2023, LNCS 13945, pp. 682–698, 2023.
https://doi.org/10.1007/978-3-031-30675-4_50

literature, and usually consist of four characters [15]. Now new idioms are still emerging, mainly based on Internet slang or hot events, but the number is small. The generation of idioms is a slow process, and with the passage of time, the meaning and scenarios of idioms gradually evolve. The metaphorical meanings of some Chinese idioms are consistent with their literal meanings, such as "衣食住行" (literal meaning: "clothing, food, housing and transportation"). But more idioms have metaphorical meanings that are far from their literal meanings, such as "盲人摸象" (literal meaning: "a blind man touches an elephant";metaphorical meaning: "inadequate understanding of things and reckless speculation"). Due to the common non-compositionality and metaphorical meaning of Chinese idioms [25], the key to Chinese idiom reading comprehension is not only to understand the context but also to get prior knowledge of the idioms. And we also should pay more attention to the big difference between literal and metaphorical meanings [9, 22].

To address this issue, Zheng et al. [25] created the Chinese idiom cloze task and the dataset ChID to evaluate the ability of machines to comprehend Chinese idioms. As shown in Table 1, this task is to choose one idiom that best matches the context from seven idiom options given a sentence with blanks, and the candidate set of 7 idioms is specified by the ChID dataset. Currently, researchers mainly use the deep learning model to obtain the contextual representation of the blanks. Zheng et al. [25] used Bi-LSTM [26] to get the hidden state which is used to deduce the probability of selecting one candidate idiom. The pre-trained BERT model adopts a self-attention mechanism [21], and performs better than Bi-LSTM. Then Wang et al. [23] proposed a method of integrating the idiom's definition, its character representation and its context based on the BERT model. Long et al. [12] proposed a method of using near-synonym idioms to alleviate the inconsistency between literal and metaphorical meanings, which constructed a synonym graph based on idioms with high semantic similarity.

For the Chinese idiom cloze task, the key is to get the idiom representation that can comprehensively consider various factors that form idioms, and make a distinction between idioms with similar semantics. Currently, deep learning models are still insufficient in mining the precise representations and accurate meanings of idioms. Introducing the definitions of idioms can help the model learn the semantics of idioms, but simply concatenating the original passages and idioms' definitions cannot ensure the model learns the correspondence between the candidate idiom and the passage and the grammatical function that the idiom plays in the passage. This leads to the poor generalization representations of idioms. Using synonym graphs based on idioms with high semantic similarity is beneficial for understanding the semantics of idioms but increases the difficulty of distinguishing idioms with similar semantics.

To address the limitations above, we propose a prompt-based representation individual enhancement method (PRIEM) for the Chinese idiom cloze task. We first introduce the definitions of idioms using a prompt template "成语 [MASK] 的定义是 D " (meaning: "The definition of idiom [MASK] is D"). By using this method, the model learns to take into account the generic meaning and the grammatical function of an idiom when filling in the blank. Then,

Table 1. An example of Chinese idiom cloze.

Passage & Blanks: 突然从身后看台传来一小球迷＿＿＿＿＿＿的加油声，惹得国乒男 队主教练也不禁回头张望。 Suddenly from the stands behind came a small fan's ＿＿＿＿＿ cheering. The head coach of the national table tennis men's team couldn't help but look back.

No.	idioms	literal meaning	metaphorical meaning (general definition)
1	"闻风破胆"	hear the wind,break the gall	extreme fear of a certain force
2	"声嘶力竭"	shout oneself hoarse	shout with all your might
3	"咬牙切齿"	gnash one's teeth	hate to the extreme
4	"凶神恶煞"	fiends and demons	very vicious person
5	"怒气冲天"	rage to the sky	very furious
6	"甜言蜜语"	sweet words honeyed phrases	nice words to please or to coax
7	"大声疾呼"	urgent shout	appeal to society

Ground Truth: "No.2"

we fuse the context-specific representation of the idiom and the generic definition representation of the idiom by using a mapping-based method. As a result, we obtain an extended representation of the idiom across multiple contexts. Next, we decompose the original representation of idioms in the candidate idioms Set into common representation and individual representation through an orthogonal projection. Finally, we use the joint projection method to adjust the weights of the original representation and the individual representation, so that we can get a more comprehensive and precise representation of the idiom. Experiments on the ChID dataset show that our model outperforms the state-of-the-art models, especially when the candidate idioms are near-synonyms.

Our main contributions are summarized as follows:

- We propose a prompt method to introduce the definition of idiom, and fuse the contextual representations of idiom across multiple contexts, which enhances the generalization of representation.
- We propose a method of idiom representation mapping and decomposing based on orthogonal projection to enhance the individual representation and the distinctions between idioms.
- Experimental results on ChID show that our model outperforms state-of-the-art models. The accuracy on dataset **Test** reaches 95.7%, and the accuracy on dataset **Sim**, which has more synonyms or near-synonyms idioms, is significantly improved to 97.9%.

2 Related Work

Cloze test is a typical reading comprehension task that is essential to assessing machine reading ability [2,4]. Researchers have created a number of cloze-style reading comprehension datasets [6,8] to facilitate cloze research. Chinese idioms and English slang are special forms of language with unique expressions and a long history [16]. The non-literal deep metaphor of Chinese idioms has brought great challenges to the research on chinese idiom cloze task [18]. ChID has settings similar to CLOTH [24], where the answers are selected from the given options. But unlike most existing cloze test corpora, the answers from ChID usually do not appear in the context.

Zheng et al. [25] proposed the Chinese idiom cloze task and used Bi-LSTM, Attentive Reader, and Standford Attentive Reader as benchmark models. Attentive Reader [6] added attention mechanism [14] to Bi-LSTM, and Stanford Attentive Reader [2] adopted bilinear function as a matching function to calculate attention weight. With the emergence of the BERT model [5], researchers have gradually adopted the BERT model to solve the Chinese idiom Cloze task. The first BERT-based dual-embedded idiom cloze model [19] learned the dual-embedding of idioms to predict the idioms in the blanks. However, the basic BERT models still face challenges in dealing with long sequences and understanding metaphorical meanings of idioms. Subsequently, by introducing external knowledge, idiom definitions and idioms' characters can be used to correct the misuse of idioms [23]. Long et al. [12] found that the literal meaning of many idioms was significantly different from their metaphorical meaning, so they constructed a synonym graph according to the meanings of idioms and encoded the idiom into a new representation. However, it is more challenging for this method to distinguish idioms with similar meanings.

Idioms also play a certain grammatical role in sentences, such as subject, object, and attribute. Although the above methods introduce the definitions of idioms, they do not consider the grammatical functions of idioms in sentences. Therefore, these models cannot obtain a more complete and comprehensive contextual representation of the idiom in different contexts. In this paper, we use the prompt [10] method to transfer the knowledge gained from pre-training to the idiom cloze task. The generalization ability of idiom representation can be enhanced by fusing the generic definition representation of idioms based on the prompt template. We also propose a method of representation mapping and decomposing based on orthogonal projection to eliminate the commonality between idioms and enhance the individual representation.

3 PRIEM Model

In order to introduce external knowledge and guide the model to learn the metaphorical meanings of idioms purposefully, we construct the prompt template for the idiom definition to obtain the generic representation of idioms. Thus, we can effectively solve the problem of deep metaphors and obtain the good generalization of idiom representation in different contexts.

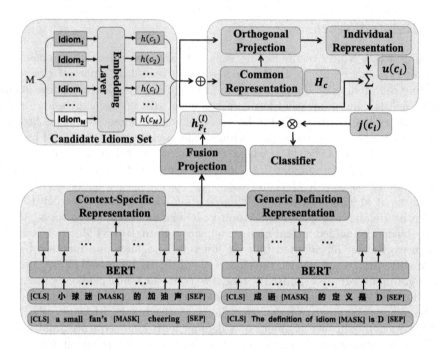

Fig. 1. Architecture of our model, M is the number of idioms in the candidate set.

Next, although many idioms are composed of different characters, they have similar meanings. But there are subtle semantic differences between these idioms with similar meanings. So we decompose the representation of idioms to extract their common features representation and individual features representation. And in order to alleviate the problem of data imbalance, we propose a joint projection method to adjust the weights of individual representations of idioms.

3.1   Model Frame

The framework of the PRIEM model is shown in Fig. 1. The input is a passage with a blank: "[CLS] 小球迷 [MASK] 的加油声 [SEP] , " and the candidate idioms set C : "1.闻风破胆, " "2.声嘶力竭 " , etc. The output is the best matching idiom selected from the candidate idioms set C to fill up the blank. The model consists of two parts: Idioms Fusion Representation Learning (the bottom of Fig. 1) is to fuse the idiom representation across multiple contexts, and Idioms Individual Representation Learning (the top of Fig. 1) is to decompose the idiom representation based on orthogonal projection.

The first part is based on the BERT model. The input on the left side is the original passage, which is used to obtain the context-specific representation in a specific context. The input on the right side is the prompt template of the generic definition, which is used to obtain the generic representation in a static context. Two special tokens [CLS] and [SEP] are used as the boundary markers leading and ending the input passage sequence, and the blank in the input passage is replaced by [MASK] token. After obtaining the representations of two inputs through the BERT model, we perform the representation fusion under different semantic dimensions. In the second part, We adopt the representation mapping and decomposing method based on orthogonal projection to extract the individual representations of idioms by eliminating the commonalities among the candidate idioms.

3.2   Learning Idioms Fusion Representation

The purpose of this part is to obtain the generalized representation of idioms across multiple contexts, which is mainly composed of the following three steps: obtaining the context-specific representation of idiom in a specific context, obtaining the generic definition representation of idiom in a static context, and fusing the above two representations. These steps are described in detail below.

Idioms Context-Specific Representation. For the passage set T, the t-th $(t \in T)$ sequence can be expressed as $O_t = \{[CLS], w_1, w_2, \cdots, [MASK], \cdots, w_n, [SEP]\}$. We use the BERT to process this sequence and obtain the hidden representation for [MASK] in the last (l-th) encoding hidden layer as the context-specific representation $h_{O_t}^{(l)} \in \mathbb{R}^{N \times d}$, where N is the maximum length of the sequence and d is the dimension of the hidden layer.

$$h_{o_t}^{(0)} = \text{Embedding}\,(O_t) \tag{1}$$

$$h_{O_t}^{(l)} = \text{Transformer}\left(h_{O_t}^{(i-1)}\right), \quad i \in \{1, \ldots, l\} \tag{2}$$

The context-specific representation $h_{O_t}^{(l)}$ of the idiom is based on the original passage, so the representation of the same idiom in different contexts should be different. Since idioms are a condensed special language phenomenon, understanding the accurate representation of idioms requires additional background knowledge rather than simple context. This makes the representation $h_{O_t}^{(l)}$ is highly context-specific and does not sufficiently exploit the precise representation of the idiom. Therefore, we need to fuse the representations of the same idiom in different contexts to enhance the generalizability of the idiom representation. Next, we need to learn the generic definition representation of idioms in the static context by introducing the definition of idioms.

Idioms Generic Definition Representation. By using prompt method, we construct the prompt template of idiom definition to get generic definition representation of idioms in the static context. The structure of the prompt template is similar to that of the original passage sequence, and also has a [MASK] token, which guides the model to learn how to choose an appropriate idiom to fill in the blank and also let the model feel the grammatical functions of idioms in sentences.

For the t-th ($t \in T$) sequence in the passage set T, we create the corresponding prompt template $P_t = \{[\text{CLS}] \text{成语} [\text{MASK}] \text{的定义是 D[SEP]} \}$ (meaning: [CLS] The definition of idiom [MASK]is D[SEP]]), where D represents the definition of the corresponding idiom. For example, the prompt template of the sequence in Fig. 1 can be "[CLS] 成语 [MASK] 的定义是嗓子喊哑 ,气力用尽,形容竭力呼喊 [SEP] "(meaning:[CLS]The definition of idiom[MASK]is hoarse voice, exhaustion of strength, to describe trying to shout[SEP]). We use the BERT to process this prompt template, and obtain the hidden representation for [MASK] in the last (l-th) encoding hidden layer as the generic definition representation $h_{P_t}^{(l)} \in \mathbb{R}^{N \times d}$, where N is the maximum length of the sequence and d is the dimension of the hidden layer.

$$h_{P_t}^{(0)} = \text{Embedding}\,(P_t) \tag{3}$$

$$h_{P_t}^{(l)} = \text{Transformer}\left(h_{P_t}^{(i-1)}\right), \quad i \in \{1, \ldots, l\} \tag{4}$$

Idioms Representation Fusion. Finally, we fuse the context-specific representation and the generic definition representation using a representation mapping method. In other words, we map the generic definition representation $h_{P_t}^{(l)}$ of the idiom in the static context into the space of the context-specific representation $h_{O_t}^{(l)}$ of the idiom, and get the fusion representation $h_{F_t}^{(l)}$ as follows:

$$h_{F_t}^{(l)} = \text{Map}\left(h_{P_t}^{(l)}, h_{O_t}^{(l)}\right) \tag{5}$$

where Map is a projection function that projects vector a to vector b :

$$\text{Map}\,(a,b) = \frac{a \cdot b}{|b|} \frac{b}{|b|} \tag{6}$$

3.3   Learning Idioms Individual Representation

Even idioms with similar semantics often have some subtle differences in meaning in actual scenarios. This is mainly because the idioms' meanings originated from different old stories, and the process of generating metaphorical meanings is also different. As a result, although these idioms have similar semantics, they may express different emotional positions or apply to different scenarios.

As a special linguistic phenomenon, the Chinese idioms should be represented as a subspace of Chinese semantic space. So, we map the representation of idioms in the candidate Set C into this subspace by using the orthogonal projection [17]. In this way, the representation of the idiom is decomposed into the common representation and the individual representation, which are orthogonal to each other. The common representation represents the baseline of this subspace, and the individual representation represents the individualized representation of the idiom relative to the baseline.

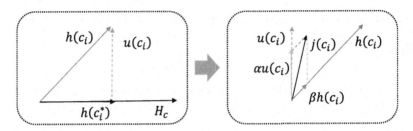

Fig. 2. The orthogonal projection and the joint projection. On the left is the orthogonal projection obtaining the individual representation, and on the right is the joint projection obtaining the joint representation.

The process of orthogonal projection is shown in the left diagram of Fig. 2 (Take 2-dimensional space as an example). $h(c_i)$ is the original representation vector of an idiom; H_c is the common representation vector of idioms in the candidate idioms set, which is the average of the weighted sum of the representation vectors of all idioms in the candidate set; $h(c_i^*)$ is the projection of the original representation vector onto the common representation vector; $u(c_i)$ is the individual representation vector of the idiom after the orthogonal projection, which is orthogonal to the common representation vector.

Firstly, the candidate idioms are encoded into a low-dimensional vector by an embedding layer as follows:

$$h(c_i) = \text{Embedding}(c_i), \quad i \in \{1, \dots, M\} \tag{7}$$

where M is the number of idioms in the candidate idioms set.

Secondly, the common representation H_c of the idioms is obtained as follows:

$$H_c = \frac{1}{M} \sum_{i=1}^{M} w_{c_i} h(c_i) \tag{8}$$

where w_{c_i} represents the learned weight of the i-th idiom representation.

We use the representation mapping method mentioned above (see equation (6)) to obtain the projection $h(c_i^*)$.

$$h(c_i^*) = \text{Map}(h(c_i), H_c) \tag{9}$$

Thirdly, the individual representation $u(c_i)$ of each idiom in the candidate set C is obtained by using orthogonal projection as follows:

$$u(c_i) = \text{Map}(h(c_i), (h(c_i) - h(c_i^*))), \quad i \in \{1, \dots, M\} \tag{10}$$

Finally, in order to further improve the generalization ability of the model and take into account the individual and the common representations among idioms thoroughly. We take the weighted sum of the original representation and individual representation as the final representation. The process of the joint projection is to obtain the joint representation $j(c_i)$ as shown in the right diagram of Fig. 2.

$$j(c_i) = \alpha u(c_i) + \beta h(c_i) \tag{11}$$

where α is the weight of the individual representation $u(c_i)$ of the idiom, β is the weight of the original representation $h(c_i)$ of the idiom, and $\alpha + \beta = 1$.

3.4 Selecting the Correct Candidate Idiom

Selecting the correct candidate idiom can be regarded as a classification problem. After obtaining the fusion representations $h_{F_t}^{(l)}$ and the joint representation $j(c_i)$, we then calculate the probability of selecting candidate idiom c_i ($c_i \in C$) among all candidates given the context O_t as follows:

$$P(c_i \mid O_t) = \frac{\exp\left(w \cdot \left(j(c_i) \otimes h_{F_t}^{(l)}\right) + b\right)}{\sum_{i'=1}^{M} \exp\left(w \cdot \left(j(c_{i'}) \otimes h_{F_t}^{(l)}\right) + b\right)} \tag{12}$$

where $w \in \mathbb{R}^d$ is the model parameter, $b \in \mathbb{R}$ is the bias parameter, and $\otimes$ is the element-wise multiplication. The training goal is to minimize the cross-entropy loss between the ground truth and the prediction as follows:

$$\text{loss} = -\sum_{i=1}^{M} c_g \log P(c_i \mid O_t) \tag{13}$$

M is the number of idioms in the candidate set, and c_g is the one-hot vector of the ground truth.

3.5 Method Integration

First, for the passage set T, the t-th ($t \in T$) original passage O_t and the corresponding prompt template passage P_t are selected as the input of the model. After getting the idiom's context-specific representation $h_{O_t}^{(l)}$ and the idiom's generic definition representation $h_{P_t}^{(l)}$, we can obtain the fusion of these two representations $h_{F_t}^{(l)}$, which is used as an input of the classifier. Second, after getting the common representation H_c of idioms in the candidate idioms set C and the individual representation $u(c_i)$ by using orthogonal projection, we can obtain the joint representation $j(c_i)$ by using joint projection, which is used as another input of the classifier. Finally, according to the similarity between the fusion representation $h_{F_t}^{(l)}$ and the joint representation $j(c_i)$, the candidate idiom with the highest probability is selected as the prediction result of the model. Pseudocode is shown as Algorithm 1.

Algorithm 1. PRIEM

Input: O_t — Original context passage.
      P_t — Definition-based passage.
      C — Candidate set of idioms.
Output: S — A set of matching idioms.
1: $S = \varnothing$.
2: **for** each iteration t **do**
3:     $h_{O_t}^{(l)} = \text{Transformer}(O_t)$;
4:     $h_{P_t}^{(l)} = \text{Transformer}(P_t)$;
5:     Calculate $h_{F_t}^{(l)}$ by Equation (5);
6:     **for** each c_i in C **do**
7:         $h(c_i) = \text{Embedding}(c_i), (c_i \in C)$;
8:         Calculate H_c by Equation (8);
9:         $h(c_i^*) \leftarrow h(c_i)$ mapping to H_c
10:       Calculate $u(c_i)$ by Equation (10);
11:       $j(c_i) = \alpha u(c_i) + \beta h(c_i)$;
12:     **end for**
13:     Select c_i by Equation (12);
14:     $S = S + c_i$
15: **end for**
16: **return** S

Table 2. ChID dataset statistics.

	In-domain				Out-of-domain	Total
	Train	Dev	Test/Ran/Sim	Total	Out	
Passages	520,711	20,000	20,000	560711	20,096	580,807
Tokens per passage	99	99	99	99	127	100
Distinct idioms	3848	3458	3502	3848	3626	3848

4 Experiments

4.1 Experimental Setup

We use the large-scale idiom reading comprehension dataset called ChID [25] which is divided into in-domain data and out-of-domain data. The in-domain data contains a training set **Train**, a validation set **Dev** and a test set **Test**. The out-of-domain data is test set **Out**. In addition, we also evaluated the generalization ability of the model on **Ran** and **Sim**. These two datasets have the same passages as **Test**, but the method of constructing the candidate idioms set is different. In **Ran**, there is no similarity between the candidate idioms and the ground truth. On the contrary, in **Sim**, the metaphorical meanings between the candidate idioms and the ground truth are very close. The idioms with similar meanings in the candidate idioms set are more likely to be the interference option, so **Sim** is more challenging than **Ran**. The detailed statistics of ChID are shown in Table 2.

We use a pre-trained model with whole word masking BERT-wwm [3], with 128 as the maximum length of the input passage. The initial learning rate is 5e-5, and the warm-up steps are 1000. The optimizer AdamW [13] is used based on a Warm-up Linear Schedule. The model takes an average of 5 epochs to accomplish the training. Our code is available online.

Table 3. Performance comparison in terms of accuracy on ChID. (PRIEM-p-o-j : $\alpha = 0.15$)

Model	Dev	Test	Ran	Sim	Out
Human	–	87.1	97.6	82.2	86.2
LM	71.8	71.5	80.7	65.6	61.5
AR	72.7	72.4	82.0	66.2	62.9
SAR	71.7	71.5	80.0	64.9	61.7
BERT-wwm	75.4	75.7	83.7	70.2	66.1
SKER	76.0	76.3	87.0	68.8	68.3
BTSM	81.9	81.8	92.9	74.1	72.0
CM	83.0	83.1	92.3	76.1	77.6
PRIEM-prompt	84.4	84.5	89.6	78.8	78.2
PRIEM-op	94.2	94.5	76.5	95.8	90.9
PRIEM-p-o	**95.8**	**95.7**	74.4	**97.9**	**92.6**
PRIEM-p-o-j	84.3	84.7	**93.1**	79.2	78.4

4.2 Experimental Results and Discussion

Methods Comparison. We compared our model with the following methods. The first three baselines all use Bi-LSTM as their backbones. The next four methods are all based on the BERT model. SKER, BTSM and CM are the latest methods.

Language Model (LM): It uses Bi-LSTM [7] to obtain the hidden states of the blanks, and uses them to select candidate Chinese idioms.

Attentive Reader (AR): This method uses Bi-LSTM with the attention mechanism [6].

Standard Attentive Reader (SAR): An improvement on AR [2]. It uses a bilinear function as a matching function to get the attention weights.

BERT-wwm: An upgrade on the BERT model using Whole Word Masking (WWM) [3], masking the whole word instead of masking Chinese characters.

Synonym Knowledge Enhanced Reader (SKER): A synonym graph is constructed using the idiom representation [12], and then the graph is encoded to replace the original representation of the idiom.

BERT-based Two-stage Model (BTSM): A two-stage model is to fine-tune the idiom-oriented BERT [20] on a specific idioms dataset.

Correcting the Misuse (CM): The definitions of idioms are introduced, and the attribute attention mechanism [1,23] is used to balance the different representations of idioms.

PRIEM-prompt: Our model only uses the prompt method.

PRIEM-op: Our model only uses the orthogonal projection.

PRIEM-p-o: Our model uses both the prompt method and the orthogonal projection.

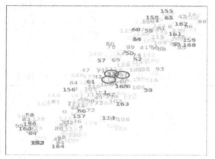 

(a) Representation distribution of 200 idioms before orthogonal projection.

(b) Representation distribution of 200 idioms after orthogonal projection.

Fig. 3. Representation distribution of idioms before and after orthogonal projection. Each number represents an idiom. The three idioms circled are "风平浪静" (No.53, general definition: "Life and situation are stable."), "天下太平" (No.107, general definition: "The whole society is peaceful and tranquil.") and "一帆风顺" (No.132, general definition: "It went very smoothly without any hindrance."), and their meanings are very close. Obviously, the representations of these three idioms are very close before the orthogonal project. But after the orthogonal, their representations are far away from each other. Therefore, the idioms with similar meanings are easier to distinguish.

PRIEM-p-o-j: Our model uses the prompt method, the orthogonal projection, and joint projection.

Evaluation Metric: The evaluation metric is accuracy, which is the percentage of predicted idioms in the test set that are identical to the ground truth.

Experimental Results. The experimental results are shown in Table 3. The results of Human, LM, AR, and SAR are taken directly from [25]. The results of our PRIEM-p-o on **Dev**, **Test**, **Sim**, and **Out** are significantly better than those of other models. Especially on **Test**, **Sim**, and **Out**, the accuracy is improved by 20.0%, 27.7%, and 26.5% compared with the BERT-wwm baseline method. Instead of constructing a synonym graph to represent idioms as in SKER model, we use the orthogonal projection to obtain distinguishable idiom representations. Instead of simply concatenating the definitions of the idioms and the original passages as in CM model, we construct the prompt template to introduce the definitions of the idioms. It can guide the model to learn the metaphorical meanings of idioms purposefully and also consider the grammatical functions of idioms.

As shown in Table 3, our model (PRIEM-p-o) achieves the best performance on **Sim**, which shows that PRIEM-p-o is good at distinguishing idioms with similar meanings. However, PRIEM-p-o does not perform well on **Ran**. Because our model pays too much attention to the individual representation of idioms through orthogonal projection, and does not reserve enough common representation of idioms, thus resulting in insufficient semantic representation of idioms as a whole. Therefore, we propose PRIEM-p-o-j which uses the joint projection to take into account both the original representation and the individual representation of idioms, achieving the best performance on **Ran**.

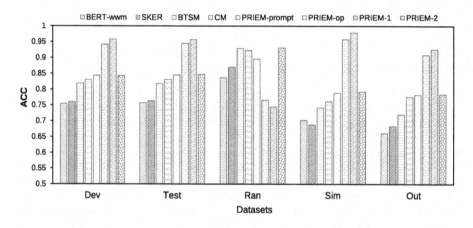

Fig. 4. Performance comparison of ablation experiments.

4.3 Ablation Study

In order to further evaluate the contribution of each part in our method, four variants of PRIEM are provided in Table 3.

The results of the ablation studies in Table 3 reveal that these three parts are all helpful for this task. To show the effect of our PRIEM-op method on distinguishing idioms with similar meanings, we selected 200 idiom samples. We used the t-SNE algorithm to draw the distributions of 200 idiom representations before and after the orthogonal projection, as shown in Fig. 3. It can be seen that the distributions of idiom representations after orthogonal projection are more dispersed, so the idioms with similar meanings are easier to distinguish.

As shown in Fig. 4, PRIEM-p-o on **Dev**, **Test**, **Sim** and **Out** achieve the best performance, which verifies that the prompt method and the orthogonal projection are helpful to the performance of the model. And the orthogonal projection has more contribution to the performance improvement of the model. It shows that the orthogonal projection can obtain the individual representation of idioms, so as to effectively distinguish the idioms with similar meanings. However, these two methods are less helpful for **Ran**, mainly because the candidate idioms in **Ran** are not similar to the ground truth idiom.

PRIEM-p-o-j achieves an accuracy of 93.1% on **Ran**, which is the best performance among all methods. That is because PRIEM-p-o-j uses the joint projection to take into account both the original representation of idioms and the individual representation of idioms. The fundamental reason that no method performs well in all datasets is that there are huge differences in the selection strategies of candidate idioms in the datasets. Thus, PRIEM-p-o-j is actually a dynamic compromise of the original representation and the individual representation.

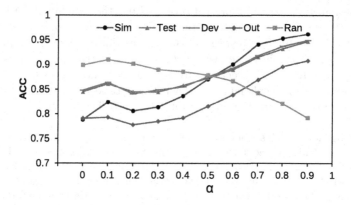

Fig. 5. Impact of the weight α of the individual representation.

4.4 Joint Projection Parameter Adjustment

Our model PRIEM-p-o performs significantly better on **Dev**, **Test**, **Sim** and **Out** than the previous methods, but does not perform well on **Ran**. Thus, We propose a joint projection method to make the model capable of taking into account both the original representation and the individual representation of idioms, while alleviating the imbalance of candidate idioms. The parameter $\alpha \in [0, 1]$ is the weight of the individual representation. When α is adjusted, the accuracy on each dataset is shown in Fig. 5.

When $\alpha \in (0, 0.2]$, the performance on the five datasets is consistent, decreases slightly. Because the individual representation of idiom accounts for a small proportion, the introduction of individual representation is not enough, which will interfere with the model to make the right choice.

When α exceeds 0.2, it is clear that the performance on **Ran** changes in the opposite direction to the other four datasets. The performance on **Dev**, **Test**, **Sim**, and **Out** is getting better, and the performance on **Ran** is getting worse. This is due to the difference between **Ran** and other datasets. The larger α is, the further the model learns the distinction between idioms with similar meanings. But the idioms in **Ran** are not similar. Therefore, the performance on **Ran** is greatly affected by the proportion of reserving their original representation, and is the best without orthogonal projection.

5   Conclusion

In this paper, we propose a prompt-based representation individual enhancement method for the Chinese idiom cloze task. We introduce external knowledge to enhance the generalization of the idiom representation by fusing the context-specific representation and the generic definition representation of the idioms. We also use orthogonal projection to increase the distinction of the candidate idiom representation, and use joint projection to adjust the weight of the individual representation of idiom. Experimental results show that our model performs state-of-the-art on different test sets of the Chinese idiom reading comprehension dataset ChID. The ablation experiments also verify the effectiveness of three parts of our method. The precise representation of idioms can be obtained by our method, which is beneficial to the downstream tasks of NLP. And our method can also be applied to the reading comprehension of English slang.

Acknowledgments. This work was supported by the Fundamental Research Funds for the Central Universities 2662021JC008, the National Natural Science Foundation of China (No. 62272188), and the National Social Science Foundation of China under Grant 19BSH022. Thank the experimental teaching center of College of informatics in Huazhong Agricultural University for providing the experimental environment and computing resources.

References

1. Bahdanau, D., Cho, K., Bengio, Y.: Neural machine translation by jointly learning to align and translate. In: Bengio, Y., LeCun, Y. (eds.) 3rd International Conference on Learning Representations, ICLR 2015, San Diego, CA, USA, 7–9 May 2015, Conference Track Proceedings (2015). http://arxiv.org/abs/1409.0473
2. Chen, D., Bolton, J., Manning, C.D.: A thorough examination of the CNN/daily mail reading comprehension task. arXiv preprint arXiv:1606.02858 (2016)
3. Cui, Y., Che, W., Liu, T., Qin, B., Yang, Z.: Pre-training with whole word masking for Chinese BERT. IEEE/ACM Trans. Audio Speech Lang. Process. **29**, 3504–3514 (2021)
4. Cui, Y., et al.: A span-extraction dataset for Chinese machine reading comprehension. arXiv preprint arXiv:1810.07366 (2018)
5. Devlin, J., Chang, M.W., Lee, K., Toutanova, K.: BERT: pre-training of deep bidirectional transformers for language understanding. arXiv preprint arXiv:1810.04805 (2018)

6. Hermann, K.M., et al.: Teaching machines to read and comprehend. In: Advances in Neural Information Processing Systems 28 (2015)
7. Hochreiter, S., Schmidhuber, J.: Long short-term memory. Neural Comput. **9**(8), 1735–1780 (1997)
8. Lai, G., Xie, Q., Liu, H., Yang, Y., Hovy, E.: Race: large-scale reading comprehension dataset from examinations. arXiv preprint arXiv:1704.04683 (2017)
9. Li, X.: Conceptual metaphor theory and teaching of English and Chinese idioms. J. Lang. Teach. Res. **1**(3), 206–210 (2010)
10. Liu, P., Yuan, W., Fu, J., Jiang, Z., Hayashi, H., Neubig, G.: Pre-train, prompt, and predict: a systematic survey of prompting methods in natural language processing. CoRR abs/2107.13586 (2021). https://arxiv.org/abs/2107.13586
11. Liu, S., Zhang, X., Zhang, S., Wang, H., Zhang, W.: Neural machine reading comprehension: methods and trends. Appl. Sci. **9**(18), 3698 (2019)
12. Long, S., Wang, R., Tao, K., Zeng, J., Dai, X.Y.: Synonym knowledge enhanced reader for chinese idiom reading comprehension. arXiv preprint arXiv:2011.04499 (2020)
13. Loshchilov, I., Hutter, F.: Decoupled weight decay regularization. In: 7th International Conference on Learning Representations, ICLR 2019, New Orleans, LA, USA, 6–9 May 2019. OpenReview.net (2019). https://openreview.net/forum?id=Bkg6RiCqY7
14. Luong, M.T., Pham, H., Manning, C.D.: Effective approaches to attention-based neural machine translation. arXiv preprint arXiv:1508.04025 (2015)
15. Min, F.: Cultural issues in Chinese idioms translation. Perspect. Stud. Transl. **15**(4), 215–229 (2007)
16. Moore, R.L., Bindler, E., Pandich, D.: Research note: language with attitude: American slang and Chinese lǐyu1. J. Socioling. **14**(4), 524–538 (2010)
17. Qin, Q., Hu, W., Liu, B.: Feature projection for improved text classification. In: Proceedings of the 58th Annual Meeting of the Association for Computational Linguistics, pp. 8161–8171 (2020)
18. Sag, I.A., Baldwin, T., Bond, F., Copestake, A., Flickinger, D.: Multiword expressions: a pain in the neck for NLP. In: Gelbukh, A. (ed.) CICLing 2002. LNCS, vol. 2276, pp. 1–15. Springer, Heidelberg (2002). https://doi.org/10.1007/3-540-45715-1_1
19. Tan, M., Jiang, J.: A BERT-based dual embedding model for Chinese idiom prediction. arXiv preprint arXiv:2011.02378 (2020)
20. Tan, M., Jiang, J., Dai, B.T.: A BERT-based two-stage model for Chinese chengyu recommendation. Trans. Asian Low-Res. Lang. Inf. Process. **20**(6), 1–18 (2021)
21. Vaswani, A., et al.: Attention is all you need. In: Advances in Neural Information Processing Systems 30 (2017)
22. Wang, L., Yu, S.: Construction of Chinese idiom knowledge-base and its applications. In: Proceedings of the 2010 Workshop on Multiword Expressions: from Theory to Applications, pp. 11–18 (2010)
23. Wang, X., Zhao, H., Yang, T., Wang, H.: Correcting the misuse: a method for the Chinese idiom cloze test. In: Proceedings of Deep Learning Inside Out (DeeLIO): the First Workshop on Knowledge Extraction and Integration for Deep Learning Architectures, pp. 1–10 (2020)
24. Xie, Q., Lai, G., Dai, Z., Hovy, E.: Large-scale cloze test dataset created by teachers. arXiv preprint arXiv:1711.03225 (2017)

25. Zheng, C., Huang, M., Sun, A.: ChID: a large-scale Chinese idiom dataset for cloze test. arXiv preprint arXiv:1906.01265 (2019)
26. Zhou, P., et al.: Attention-based bidirectional long short-term memory networks for relation classification. In: Proceedings of the 54th Annual Meeting of the Association for Computational Linguistics (Volume 2: Short papers), pp. 207–212 (2016)

Cross-Modal Contrastive Learning for Event Extraction

Shuo Wang[1,2], Meizhi Ju[4], Yunyan Zhang[4], Yefeng Zheng[4], Meng Wang[1,3(✉)], and Guilin Qi[1,3]

[1] School of Computer Science and Engineering, Southeast University, Nanjing, China
`{wangs,meng.wang,gqi}@seu.edu.cn`
[2] Southeast University-Monash University Joint Research Institute, Suzhou, China
[3] Key Laboratory of Computer Network and Information Integration (Southeast University),
Ministry of Education, Nanjing, China
[4] Tencent Jarvis Lab, Shenzhen, China
`{yunyanzhang,yefengzheng}@tencent.com`

Abstract. Event extraction aims to extract information of triggers and arguments from texts. Recent advanced methods leverage information from other modalities (e.g., images and videos) besides the texts to enhance event extraction. However, the different modalities are often misaligned at the event level, negatively impacting model performance. To address this issue, we firstly constructed a new multi-modal event extraction benchmark Text Video Event Extraction (TVEE) dataset, containing 7,598 text-video pairs. The texts are automatically extracted from video captions, which are perfectly aligned to the video content in most cases. Secondly, we present a Cross-modal Contrastive Learning for Event Extraction (CoCoEE) model to extract events from multi-modal data by contrasting text-video and event-video representations. We conduct extensive experiments on our TVEE dataset and the current benchmark VM2E2 dataset. The results show that our proposed model outperforms baseline methods in terms of F-score. Furthermore, the proposed cross-modal contrastive learning method improves event extraction in each single modality. The dataset and code will be released once upon acceptance.

Keywords: Cross-modal · Event Extraction · Contrastive Learning

1 Introduction

Event Extraction (EE) aims to identify triggers and associated arguments, playing a crucial role in downstream tasks such as timeline summarization [10, 15] and text summarization [2, 4]. Most research focuses on the text modality of EE [6, 16], and some extract events from image and video modalities, neglecting that event extraction can be complemented and enhanced by considering cross-modal information. Initial efforts towards the multi-modal EE mainly considered image due to its simplicity in collecting text-image pairs with related contents. Beyond the image, the video modality covers more EE-related information. Therefore, recent research moved from the image to video modality coupled with the text to combine multi-modal information to enhance EE.

© The Author(s), under exclusive license to Springer Nature Switzerland AG 2023
X. Wang et al. (Eds.): DASFAA 2023, LNCS 13945, pp. 699–715, 2023.
https://doi.org/10.1007/978-3-031-30675-4_51

In multi-modal event information, different words and phrases in sentences can trigger the same event type, whereas the events of the same type mostly occur in a similar scene with similar entities in different videos. For example, in Fig. 1 (a), the *Arrest* event is triggered by the word "arrested" in the sentence, which is different from the word "detained" triggering the same *Arrest* event type. But in videos, the *Arrest* event content is almost the same between Fig. 1 (a) and (b). Similarly, event types that are hard to detect from videos may have similar sentence descriptions. Learning the event distributions from the text modality can help extract these challenging video events. Hence, cross-modal learning can enhance EE by learning the event distribution incorporated in different modalities.

With the success of contrastive learning in cross-modal research [8, 28, 30], [1] proposed to pre-train on videos with their auto-generated Automatic Speech Recognition (ASR) transactions in a contrastive learning manner to pair texts and videos for event extraction. However, current cross-modal contrastive learning methods cannot ensure the alignment of the event contents across modalities. This inevitably introduces mis-alignments of events for paired instances, negatively impacting the EE models. Meanwhile, they constructed contrastive samples equally without considering the difference in event contents with anchor samples. This limits the learning ability of the contrastive methods on event extraction since events composed in different contrastive samples may have diverse levels of distinction with anchor samples. For example, in Fig. 1, the difference in event contents between (a) and (c) is more significant compared with (d) and (c). It is because (a) thoroughly describes an *Arrest* event, which is totally different from the instance (c) that describes the *Demonstrate* event. While (d) contains the *Attack* event, which is not the same as sample (c), they are similar in the *Demonstrate* event contents.

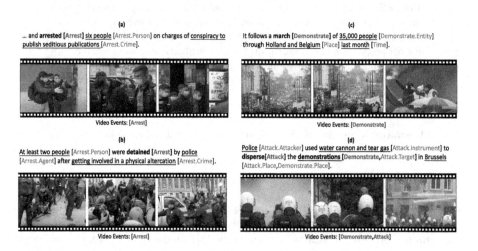

Fig. 1. Four samples from the proposed TVEE dataset. Every sample has a sentence labelled with text events and a video segment with video events. Triggers are marked in bold with event types in red colour, arguments are marked with underlines with roles in green colour, and video event type sets are listed below videos. (Color figure online)

To address these issues, we firstly construct a novel dataset named **Text Video Event Extraction (TVEE)**, which is composed of pairs of sentences and videos with aligned events, i.e., sentence and video in a pair describe the same event contents. To encode the task-specific (i.e., EE) multi-modal representation, we present a **Cross-modal Contrastive Learning Event Extraction (CoCoEE)** model with three modules: text event extractor, video event extractor and **Cross-modal contrastive Learner** (CoLearner). The two event extractors deal with the extraction of event triggers and arguments from sentences and the extraction of events from videos, respectively. The CoLearner contrasts text-video and event-video representations with different sample weights by comparing their event contents to enhance both single- and multi-modal event extraction.

In a nutshell, we summarize our contributions as follows:

- We build a benchmark dataset named TVEE. To the best of our knowledge, it is the first dataset that pairs texts and videos using the same event descriptions to guarantee the event content alignment. The dataset consists of 7,598 pairs annotated with 36 event types and 36 argument roles.
- We present a contrastive learning model with three modules that assigns different weights to contrastive samples based on the occurrences of events to extract events by contrasting cross-modal representations.
- We conduct experiments on two benchmark datasets TVEE and VM2E2 [1]. The proposed model outperforms the state-of-the-art (SOTA) on both single- and multi-modal EE in terms of F-score, showing the effectiveness of the proposed contrastive method.

2   Model

We present the proposed model in Fig. 2, which contains three modules: (1) The text event extractor is a stack of the BERT model and two Conditional Random Field (CRF) layers for labeling the input text sequence with event types and argument roles (Sect. 2.2). (2) The video event extractor is a Transformer-based model to extract events from videos (Sect. 2.3). (3) The CoLearner contrasts text-video and event-video representations by comparing event contents and assigns weights to samples condition on the occurrences of event types (Sect. 2.4).

2.1   Task Definition

Given a pair of sentence and video (x, v), the task aims to extract events from each modality, where events are defined as:

Text Event(s). A text event e constitutes one trigger r associated with arguments (i.e., entities). Every argument has an argument role for an event type, which reveals the relation between the argument entity and the event.

Video Event(s). Considering the temporal change of entity positions, extracting arguments from videos is a hard sub-task of video event extraction, so we do not extract video arguments. Hence we denote event with event type for video modality.

2.2 Text Event Extractor

The module takes in sentences and outputs triggers and arguments using the corresponding trigger and argument extractors.

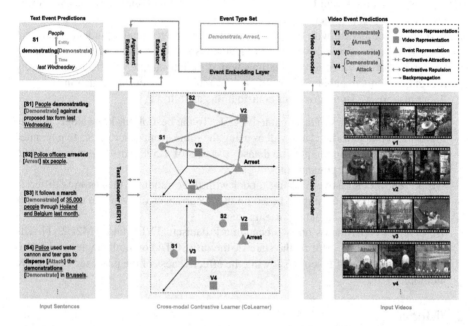

Fig. 2. An overview of the proposed CoCoEE model consisting of the text event extractor, the video event extractor and the CoLearner. The text event extractor is composed of a text encoder (BERT), a trigger extractor and an argument extractor, which is shown in brown. The video event extractor performs in a sequence-to-sequence manner with an encoder and a decoder, which is shown in purple. The CoLearner contrasts text-video and event-video representations based on their event contents. We only show positive pairs $s_2 - v_2$, $s_1 - v_3$, $Arrest - v_2$ (the same event), and negative pairs $s_1 - v_2$, $s_1 - v_4$, $Arrest - v_3$, $Arrest - v_4$ (different events). Please note, s_1 also contains the *Demonstrate* event as v_4. However, the dominant event of v_4 is *Attack*, therefore, it forms a negative pair to s_1 with a small weight. (Color figure online)

Trigger Extractor. Given an input sentence x, we firstly feed the sentence to the BERT model (i.e., text encoder) to produce the contextualized representation $\mathbf{x} \in \mathbb{R}^{n \times d_t}$, where d_t is the embedding dimension of a token and n is the number of input tokens. Then a CRF layer is stacked on top of the text encoder to label triggers with

$$\mathcal{L}_t = -\frac{1}{N} \sum_{i=1}^{N} \log P(y_i | \mathbf{x_i}),$$

where N is the size of the training set and y_i is the ground truth label sequence of x_i.
Argument Extractor. Given a trigger r and its event type e, we obtain the trigger vector representation $\mathbf{r}$ using the span vector in $\mathbf{x}$ and get the representation $\mathbf{e}$ of event

type e with an embedding layer. Then $\mathbf{r}$ and $\mathbf{e}$ are concatenated with the sequence representation $\mathbf{x}$. The argument entities are labeled by another CRF layer:

$$\mathcal{L}_{\mathrm{a}} = -\frac{1}{N} \sum_{i=1}^{N} \sum_{j=1}^{n_E} \log P(y_i | \mathbf{x}_i; \mathbf{r}_i^j; \mathbf{e}_i^j),$$

where n_E is the size of the ground truth event set of s. The overall loss of the text event extraction is formulated as:

$$\mathcal{L}_{\mathrm{text}} = \mathcal{L}_{\mathrm{t}} + \mathcal{L}_{\mathrm{a}}.$$

2.3   Video Event Extractor

Given a video v, the module calculates its representation and generates the video events. Specifically, we follow [1] to extract features in terms of video-level, frame-level object label, frame-label region and frame-level object coordinates. A concatenation of the feature embeddings are fed to the video encoder. A Transformer encoder layer is used as the video encoder to calculate the video representation $\mathbf{v}_{\mathrm{out}}$. Then a Transformer-based decoder is introduced to generate the event type sequences. The loss of the video event extractor, which is represented as $\mathcal{L}_{\mathrm{video}}$, is calculated following [19] with a standard teacher-forcing strategy with cross-entropy loss.

2.4   Cross-Modal Contrastive Learner

The CoLearner aims to enhance event extraction on the text and video modalities by contrasting their event contents. Specifically, we design two loss functions to contrast text-video and event-video representations and incorporate event contents to assign weights to contrastive samples. To contrast representations between texts and videos, for a sentence x, the representation of the first output token of the BERT model is used as the global representation $\mathbf{s}_{\mathrm{g}} \in \mathbb{R}^{d_t}$, for a video v, we use the average pooling output of the representation output by the video encoder $\mathbf{v}_{\mathrm{out}}$ as its global representation $\mathbf{v}_{\mathrm{g}} \in \mathbb{R}^{d_v}$. d_t and d_v are the dimensions of the global text representation and global video representation, respectively. Then two learnable networks f_t and f_v are used to project the two representations into a shared embedding space with dimension d to get $\mathbf{s} = f_t(\mathbf{s}_{\mathrm{g}}) \in \mathbb{R}^d$ and $\mathbf{v} = f_v(\mathbf{v}_{\mathrm{g}}) \in \mathbb{R}^d$, respectively.

Contrastive Losses. Intuitively, the distance between $\mathbf{s}$ and $\mathbf{v}$ describing similar events should be closer in the shared embedding space than the distance between $\mathbf{s}$ and $\mathbf{v}'$ with unrelated events. Based on this intuition, a text-video contrastive loss is defined, leveraging each other modality to enhance text and video representations by matching texts and videos conditioned on their event contents.

To compare cross-modality event contents, considering that text triggers and video event types are not directly comparable, we contrast their event type sets. For a sample, its event type set is the union set of the two modalities. Specifically, for a sample, if the similarity of its event content with the anchor sample is higher than θ, then it is positive, otherwise negative. The method of calculating event content similarity between samples and the sample weight μ will be introduced in the following content. In this way, vectors of text-video pairs describing similar event contents are pulled together, and pairs with different event contents are pushed apart. The text-video contrastive learning loss is defined as:

$$\mathcal{L}_g(\mathbf{s}, \mathbf{v}) = \mathbb{E}_{\mathbf{s}'}[\mu_t(k, l) \cdot S(\mathbf{s}', \mathbf{v}) + \mu_t(i, l) \cdot S(\mathbf{s}, \mathbf{v}) + \epsilon]_+$$
$$+\mathbb{E}_{\mathbf{v}'}[\mu_t(i, j) \cdot S(\mathbf{s}, \mathbf{v}') + \mu_t(i, l) \cdot S(\mathbf{s}, \mathbf{v}) + \epsilon]_+,$$

where i, j, k, l are the indexes of samples $\mathbf{s}$, $\mathbf{v}'$, $\mathbf{s}'$ and $\mathbf{v}$, respectively, $S(\cdot, \cdot)$ is the cosine distance function. ϵ is the margin parameter that requires the distances between anchor-positive and anchor-negative to be larger than ϵ.

For text event extraction, argument extraction relies on both global text representation and event representation, where global text representation is refined by $\mathcal{L}_g$. Like text-video contrastive learning, representations of an event and the video depicting it tend to be closer than irrelevant videos. We employ event contrastive learning by matching event-video pairs rather than event-text pairs because that event representation is trained along with text event extraction, which does not need to be contrasted with text. Specifically, for an event type e, we push apart its representation from the negative sample video representation $\mathbf{v}'$ and reduce the distance from the positive sample video $\mathbf{v}$. To contrast events and videos, the method of selecting negative samples is similar to $\mathcal{L}_g$: if the similarity of the event and the event content of a video is bigger than θ, then they are positive samples for each other, otherwise negative.

The event-video contrastive learning loss is defined as:

$$\mathcal{L}_e(\mathbf{e}, \mathbf{v}) = \mathbb{E}_{\mathbf{e}'}[\mu_e(\mathbf{e}', i) \cdot S(\mathbf{e}', \mathbf{v}) + \mu_e(\mathbf{e}, i) \cdot S(\mathbf{e}, \mathbf{v}) + \epsilon]_+$$
$$+\mathbb{E}_{\mathbf{v}'}[\mu_e(\mathbf{e}, j) \cdot S(\mathbf{e}, \mathbf{v}') + \mu_e(\mathbf{e}, i) \cdot S(\mathbf{e}, \mathbf{v}) + \epsilon]_+,$$

where i, j are the indexes of the samples v and v', respectively.

The overall loss of CoLearner is defined as:

$$\mathcal{L}_{\text{CoLearner}} = \sum_{(s, v) \in D} \lambda_1 \mathcal{L}_g(\mathbf{s}, \mathbf{v}) + \sum_{v \in D} \sum_{e \in E_{all}} \lambda_2 \mathcal{L}_e(\mathbf{e}, \mathbf{v}),$$

where λ_1 and λ_2 are hyperparameters to balance weights of $\mathcal{L}_g$ and $\mathcal{L}_e$, respectively, and D is the training set.

Sample Weight. As mentioned above, treating contrastive samples chosen based on their event contents equally is not reasonable because different pairs of samples have various degree of difference. To address this problem, we assign weight score to the contrastive sample with index j by measuring the similarity between its event content and the anchor sample with index i. The similarity of event contents between two samples is defined as:

$$Sim(i, j) = \frac{|E_i \cap E_j|}{|E_i \cup E_j| + \delta},$$

where δ is a small constant used to avoid the denominator to be 0 and E_i is the event type set of the i-th sample. For $\mathcal{L}_g$, the sample weight can be presented as:

$$\mu_t(i,j) = Sim(i,j) - \theta.$$

Similarly, the similarity between an event type and the event content of the i-th sample is defined as:

$$Sim(e,i) = 1 - \frac{|E_i - \{e\}|}{|E_i| + \delta},$$

where $E_i - \{e\}$ means E_i except event type e.

For $\mathcal{L}_e$, the sample weight is calculated by:

$$\mu_e(e,i) = Sim(e,i) - \theta.$$

2.5 Training and Inference

For the text event extractor, during the training phase, the trigger extractor and argument extractor both use ground truth annotations to train the model. While in the inference phase, the argument extractor uses the extracted event triggers and types from the trigger extractor to predict arguments.

The two event extractor losses and CoLearner loss are jointly optimized:

$$\mathcal{L} = \lambda_3 \mathcal{L}_{\text{text}} + \lambda_4 \mathcal{L}_{\text{video}} + \lambda_5 \mathcal{L}_{\text{CoLearner}},$$

where λ_3, λ_4 and λ_5 are hyperparameters to balance the losses.

To reduce the gap between the training phase and inference phase, we freeze the video encoder when contrasting video representations with event type embeddings.

3 TVEE Dataset

3.1 Data Collection

Event Schema. We follow the event schema from the ACE2005 benchmark [23], which contains 8 superior event types and 33 event types. However, there are many videos that cannot be covered by the current event types, so we add three more event types, which are *Contact.Speech*, *Disaster.Disaster* and *Accident.Accident*.

Data Source. We collect data from the On Demand News[1] channel that contains international news videos with broad coverage of event types. In addition, news from this channel generally has multiple sentences describing events embedded as captions in the videos. As a result, we collected 24,129 news videos and split them into frames. As sentences are embedded in pictures as captions, we then employ an Optical Character Recognition (OCR) tool[2] to extract sentences from frames. The sentences are located

---

[1] https://www.youtube.com/c/ondemandnews.

[2] https://cloud.tencent.com/product/ocr-catalog.

at the bottom of the frames. Hence we cut the frames and extracted sentences from the bottom portion. To further alleviate the OCR mistakes, we choose the longest continuous frames corresponding to the same sentence as a single video segment. Then we drop the frames without captions and video segments less than three frames. The remaining 7,598 instances are kept as our sentence-video pairs.

3.2 Data Annotation

To annotate events from texts, we follow the ACE2005 [23] to annotate triggers, event types, entities and argument roles in a two-stage iterative manner. To save the annotation cost, we adopt the state-of-the-art information extraction model ONEIE [13] to obtain pseudo-event annotations from raw sentences. We annotated all event types in a specific video segment for video annotation.

In the first annotation stage, we employed ten expert annotators to correct the pseudo labels, supplement event annotations missed by the ONEIE model and another ten annotators to annotate videos. In the second annotation stage, six experienced annotators were invited to double-check the annotations to guarantee the annotation quality. Then we asked four annotators to evaluate 100 samples selected at random, resulting in inter-annotator agreements of 83.4% and 85.6% for the text and video modalities, respectively. The statistics of TVEE are listed in Table 1.

Table 1. Statistics of TVEE. The sentence length is measured in tokens, while the video length is measured in seconds.

Statistics	Item						
	# Instances	# Events	#Average Events/Instance	Average Length	Max Length	Min Length
Sentence	7,598	6,584	0.87	17.0	43	12
Video	7,598	5,487	0.72	6.7	7	4

4 Experiments

In this section, we describe the experimental setup in Sects. 4.1–4.4. Then we show and analyse the main results in Sect. 4.5. To demonstrate the efficiency of the proposed methods, we conduct ablation studies on the TVEE dataset in Sect. 4.6, which details the contribution of the CoLearner and event type-wise benefits.

4.1 Datasets

We conduct experiments on the TVEE and VM2E2 datasets [1]:

- **TVEE.** The TVEE is randomly split into training, development and test sets with a ratio of 8:1:1.

– **VM2E2.** VM2E2 is a text-video dataset with 13,239 sentences and 860 videos for multi-modal event coreference resolution and event extraction. Among them, a total of 562 sentence-video pairs are constructed conditioning on the same event type that shared between modalities. Each pair contains one event only. In addition, VM2E2 contains 16 multi-modal event types, which are defined from the LDC ontology. We follow [1] to split the sentence-video pairs into training and test sets, which contain 411 and 151 samples, respectively.

4.2 Evaluation Metrics

We evaluate the model with *Precision (P)*, *Recall (R)* and *F-score (F1)* for both single- and multi-modal settings. For text event extraction, a trigger prediction is considered correctly extracted if the text span and event type both match one of the ground-truth text triggers; an argument is considered correctly extracted when the text span, argument role and event type all match one of the ground-truth text arguments [12]. For video event extraction, an event prediction is considered correct if its event type matches one of the ground-truth video events. For multi-modal setting, an event prediction is considered correct if it matches one of the golden text events or ground-truth video events (only event type).

Table 2. The results of the proposed CoCoEE on the TVEE and VM2E2 [1] test sets in comparison with three SOTA methods, EEQA [5], JSL [18] and JMMT [1]. Best results are highlighted in bold. All metrics are presented in percentages (%).

Dataset	Training	Model	Text Evaluation						Video Evaluation			Multi-modal Evaluation		
			Trigger			Argument								
			P	R	F1	P	R	F1	P	R	F1	P	R	F1
TVEE	Text	EEQA [5]	**81.5**	70.6	75.7	47.7	**56.8**	51.9	–	–	–	81.5	70.6	75.7
		CoCoEE$_T$	76.0	76.6	76.3	62.9	44.2	51.9	–	–	–	76.0	76.6	76.3
	Video	JSL [18]	–	–	–	–	–	–	48.2	51.6	49.8	48.2	51.6	49.8
		CoCoEE$_V$	–	–	–	–	–	–	49.1	**60.7**	54.3	49.1	60.7	54.3
	Multi-modal	JMMT [1]	74.3	**80.2**	77.1	50.1	54.9	52.3	55.4	57.0	56.2	87.2	88.6	87.9
		CoCoEE	80.7	76.4	**78.5**	65.6	45.4	**53.6**	56.4	57.4	**56.9**	**92.9**	**92.9**	**92.9**
VM2E2	Text	EEQA [5]	44.3	40.1	42.1	15.2	18.6	16.7	–	–	–	44.3	40.1	42.1
		CoCoEE$_T$	41.5	45.6	43.5	20.5	15.3	17.5	–	–	–	41.5	45.6	43.5
	Video	JSL [18]	–	–	–	–	–	–	21.2	18.6	19.8	21.2	18.6	19.8
		CoCoEE$_V$	–	–	–	–	–	–	27.3	31.2	29.1	27.3	31.2	29.1
	Multi-modal	JMMT [1]	39.7	**56.3**	46.6	17.9	**24.3**	20.6	32.4	**37.5**	34.8	76.1	69.5	72.7
		CoCoEE	**47.3**	47.7	**47.5**	**26.7**	18.5	**21.8**	**33.2**	37.2	**35.1**	**78.2**	**75.6**	**76.9**

4.3 Compared Methods

Text Event Extraction. To compare the event extraction in the text modality, we adopt the following two baselines, which only consider sentences in TVEE:

- **EEQA.** We use the SOTA model EEQA from [5] as our baseline for text EE. To rebuild EEQA on new event types, we follow [5] to define new event type question templates.
- **CoCoEE$_T$.** This baseline is the text event extraction module of CoCoEE without the CoLearner, i.e., it consists of a BERT encoder and two CRF decoders.

Video Event Extraction. To compare the event extraction in the video modality, we use the following two baselines, which only consider videos in TVEE:

- **JSL.** We follow [1] to use a SOTA model of grounded image extraction JSL [18] as the video event extraction baseline to extract events from key frames. For a fair comparison, we only keep the event detection part from the JSL model.
- **CoCoEE$_V$.** This baseline is the video event extraction module of CoCoEE without the CoLearner, i.e., it generates event type sequences with a video encoder and decoder.

Multi-modal Event Extraction. We compare the SOTA model of text-video event extraction JMMT [1] with CoCoEE on both TVEE and VM2E2. When comparing the event types from videos, we modify the JMMT model to generate sequences of event types while argument sequences are ignored for direct comparison.

4.4   Implementation Details

We use the *BERT_base* model[3] to produce contextualized representations. We extract five video features following [1], including (1) 2D Video Features, (2) 3D Video Features, (3) Object Region Features, (4) Object Label Features, (5) Object Coordinate Features.

In our experiments, we use the encoder and decoder of the *T5_base*[4] as the video encoder and decoder, respectively. Event extraction losses are much higher than contrastive learning losses when training together. To balance losses of different modules, we set $\lambda_1, \lambda_2, \lambda_3, \lambda_4$ as 1.0 and λ_5 as 1000.0 to make sure that they are the same in magnitude. All losses are jointly optimized using the Adam optimizer with a learning rate of 0.00001. We trained models for 60 epochs on a Tesla V100 GPU about 10 h with a batch size of 16. The parameters are searched on the validation set by selecting the highest sum value of text trigger classification F1, text argument classification F1 and video event classification F1.

4.5   Main Results

Table 2 presents the overall results of our model in comparison with related work on both TVEE and VM2E2 test sets. CoCoEE outperforms related SOTA models in extracting events in terms of F1, thus achieving the best results in both single- and multi-modal EE.

---

[3] https://huggingface.co/bert-base-uncased.
[4] https://huggingface.co/t5-base.

Text Evaluation. In text evaluation, although the two baselines EEQA and CoCoEE$_T$ have different performances on their precision and recall, they are almost the same in terms of F1, where CoCoEE$_T$ is a bit higher than EEQA. Compared with EEQA and CoCoEE$_T$, our CoCoEE gains consistent improvements on F1 in terms of trigger extraction and argument extraction on both test sets, indicating the effectiveness of the CoLearner on text EE. Furthermore, CoCoEE outperforms JMMT on text evaluation on both test sets for its higher precision than JMMT, which shows the superiority of our proposed model. As JMMT integrates inputs of the two modalities that complement each other on event information and jointly predicts events, it can produce a higher recall than CoCoEE.

Video Evaluation. CoCoEE outperforms JSL and CoCoEE$_V$ on both TVEE and VM2E2 datasets, showing that the CoLearner can also improve video EE. As improvements, CoCoEE gains better precision and recall compared with JMMT, leading to SOTA F1 in the video evaluation.

On VM2E2, it is similar to the results on TVEE that CoCoEE has an incremental increase of both precision and F1 from JST, CoCoEE$_V$ and JMMT. Unlike on TVEE, JMMT performs the best recall, which is also because of the integration of inputs of the two modalities and the joint event prediction. Since most videos have only one event due to their short time, the video event extraction is similar to a classification task. Hence the recall values of CoCoEE$_T$ and JMMT are almost the same.

Multi-modal Evaluation. CoCoEE outperforms all the baselines when used in the multi-modal evaluation, illustrating the effectiveness of CoCoEE on multi-modal EE. Performances of models, which are trained with either texts or videos, are shared in the multi-modal setting. As our video EE focuses on event types only, the multi-modal evaluation also only predicts event types without considering textual triggers or arguments, which will cause a much higher value of precision, recall and F1 than uni-modal results.

Fig. 3. Four examples from the TVEE dataset. The main region of events in videos are labeled by red boxes. (Color figure online)

4.6 Ablation Study

To better illustrate the contribution of all modules of the CoCoEE, we conduct ablation studies with the following five settings on text evaluation and video evaluation[5]:

- **Pair-CL.** This setting selects the original pairs from TVEE as positive samples, and the others as negative, i.e., selecting contrastive samples based on their corresponding relations rather than event contents.
- **-CoLeaner.** This setting extracts events using single-modal modules without CoLearner, i.e., for text evaluation, using text event extractor to label triggers and arguments; for video evaluation, using video event extractor to generate event type sequences.
- **-$\mathcal{L}_g + \mu$.** The text-video contrastive learning loss $\mathcal{L}_g$ and sample weight μ are removed.
- **-$\mathcal{L}_e + \mu$.** The event-video contrastive learning loss $\mathcal{L}_e$ and sample weight μ are removed.
- **-μ.** The sample weight μ is removed.

Table 3. The results of ablation studies on the TVEE test set. Best results are highlighted in bold. All metrics are presented in percentages (%).

Setting	Text Evaluation						Video Evaluation		
	Trigger			Argument					
	P	R	F1	P	R	F1	P	R	F1
-CoLearner	76.0	76.6	76.3	62.9	44.2	51.9	49.1	60.7	54.3
Pair-CL	75.2	78.3	76.7	61.5	45.3	52.2	48.2	60.3	53.6
-$\mathcal{L}_e + \mu$	76.3	**78.5**	77.5	62.3	**46.1**	53.0	54.3	57.8	56.0
-$\mathcal{L}_g + \mu$	76.5	77.1	76.8	64.1	45.6	53.3	49.1	**60.7**	54.3
-μ	78.2	77.7	78.0	**66.4**	44.7	53.4	54.3	57.8	56.0
CoCoEE	**80.7**	76.4	**78.5**	65.6	45.4	**53.6**	**56.4**	57.4	**56.9**

The Pair-CL outperforms -CoLearner setting on text evaluation but underperforms on video evaluation. The reason may be that the text semantic understanding can be enhanced by contrasting the corresponding relations with videos [28]. However, the performance of the video evaluation with lower inherent complexity [14] may be damaged by the greater proportion of negative samples when using the pair relation to select contrastive samples. The -μ setting outperforms -$\mathcal{L}_e + \mu$ and -$\mathcal{L}_g + \mu$ settings on the overall EE results, which shows that the combination of text-video and event-video contrastive learning can benefit both trigger extraction and argument extraction for text EE and video EE. The performance of video EE in the -$\mathcal{L}_e + \mu$ setting is same to -μ, because when we optimize the $\mathcal{L}_e$ loss, parameters of the video encoder are frozen.

---

[5] Because the multi-modal evaluation only focuses on event type extraction, it can't show the performance of every module, we perform ablation study on text evaluation and video evaluation.

Effects of Text-video Contrastive Learning. Compared with Pair-CL and -CoLearner settings, by contrasting samples based on event contents, the $-\mathcal{L}_e + \mu$ setting, which exploits global representations, obtains improvement on both trigger extraction and argument extraction on the text and video evaluation.

Table 4. F-scores of text trigger extraction on different superior event types with -CoLearner setting and CoCoEE. Values in bold means that it is bigger than the other setting. All metrics are presented in percentages (%).

Event Type	-CoLearner	CoCoEE	Event Type	-CoLearner	CoCoEE
Business	**4.0**	0.0	Justice	**85.4**	83.3
Conflict	73.2	**78.5**	Contact	**84.5**	82.8
Personnel	**66.1**	60.2	Transaction	13.1	**18.6**
Life	92.1	**95.1**	Movement	74.1	**80.6**

Effects of Event-video Contrastive Learning. Compared with Pair-CL and -CoLearner settings, the $-\mathcal{L}_g + \mu$ improves text argument extraction, which shows the effectiveness of learning event representations by contrasting with videos on text argument extraction. Meanwhile, the performance of video EE in the $-\mathcal{L}_g + \mu$ setting is the same to -CoLearner due to the frozen video encoder parameters.

Effects of Sample Weight. When introducing the contrastive sample weight μ, CoCoEE increases the performances on F-scores of text EE and video EE compared with $-\mu$, which shows the necessity of weighing samples.

Effects of Cross-modal Contrastive Learning. As shown in Table 3, most cross-modal contrastive learning settings outperform the single-modal setting, demonstrating that contrasting EE across modalities has better performance than extracting events that only considers a single modality.

Effects of CoLearner on Different Text Event Types. We compare the performance of -CoLearner and CoCoEE settings on text event types for text evaluation, which is shown in Table 4. The F-scores are improved with CoCoEE on four text event types, where *Movement* obtains the most improvement and *Personnel* declines the most. By observing videos of these event types, it turns out that it is easier to identify events by a human from the videos corresponding to the improved event types than the declined ones. We list two examples from TVEE in Fig. 3 (a) and (b). The crowd gathered in (a) is the main content in the video, which indicates a *Conflict.Demonstrate* event directly; however, in (b) the *Business.Start-Org* event can only be identified by the red rope (in the red box) from the third picture. A intuitive description of visual events can help improve the text representation. Otherwise, the videos will introduce noise to the event extraction model. Therefore, we can conclude that the performance of cross-modal enhancement on text modality largely depends on the clarity of events in videos.

Table 5. F-scores of event extraction on different superior event types with -CoLearner setting and CoCoEE. Values in bold means that it is bigger than the other setting. All metrics are presented in percentages (%).

Event Type	-CoLearner	CoCoEE	Event Type	-CoLearner	CoCoEE
Business	0.0	0.0	Justice	21.3	**25.5**
Conflict	25.2	**28.2**	Contact	**70.6**	65.4
Personnel	3.4	**5.1**	Transaction	8.3	**11.2**
Life	67.5	**70.2**	Movement	**33.1**	26.5
Disaster	67.3	**71.3**	Accident	**54.1**	43.2

Effects of CoLearner on Different Video Event Types. We compare the performance of -CoLearner and CoCoEE settings on video event types for video evaluation, which is shown in Table 5. We can see that six event types gain improvements with *Justice* event type improving the most; *Contact*, *Movement* and *Accident* decline compared with -CoLearner results. Similar to text event extraction, by observing data in TVEE, we found that the improved event types always have intuitive text event descriptions in sentences, but the declines do not. For example, in Fig. 3, videos in example (c) and (d) describe *Justice.Convict* and *Contact.Meet* events obviously. The *Justice.Convict* events are mostly mentioned in sentences with direct triggers. However, because *Contact.Meet* event is not so important for text as other event types, many sentences do not mention it directly. As a result, event types with specific triggers and intuitive event descriptions in corresponding sentences are possible to gain improvement in video event extraction. In contrast, contrastive learning on other event types may introduce noise to the model.

5   Related Work

5.1   Event Extraction

Most event extraction research focuses on the sentence level. Early efforts on event extraction mainly used common convolutional neural network, recurrent neural network and their variants [16, 17] to tackle the extraction of triggers and arguments. With the success of Pretrained Language Models (PLMs), Transformers-based models such as BERT have been employed to improve the task [9, 22]. In computer vision field, event extraction is performed as *Situation Recognition* [18], which classifies images with actions (visual events) and extracts frames (arguments) consisting of entities and roles. [20] classified events and extracted frames from videos. Unlike [20] that only extracted one event from a video, we extract all events from a specific video.

To learn better representation, [24] leveraged contrastive learning for the Automatic Speech Recognition (ASR) of massive unlabeled data. To utilize knowledge from other modalities, some studies introduced multi-modal data to perform multi-modal event extraction. [29] demonstrated the effectiveness of extracting events with visually based entity data. [21] proposed a dual recurrent multi-modal model to improve text event detection with external news images. [11] extracted events from both text and image data jointly by projecting them into a common embedding space in an unsupervised way. They also constructed a text-image multi-modal event extraction dataset called M2E2. Most similar work to ours is [1], which proposed a Transformer-based model to extract events from text and video data jointly. It uses a pre-trained text-video retrieval model to retrieve the most relevant text-video pairs. Meanwhile, they released a text-video event extraction dataset namely VM2E2 that pairs sentences and videos based on the event type. Our work is different from [1] in mainly two aspects. Firstly, sentences from our proposed TVEE dataset are embedded as cations in videos, which are perfectly aligned to the video event contents in most cases, thus saving the sentence-video alignment cost. Secondly, our TVEE dataset is ten times larger than the VM2E2 dataset. Thirdly, we target to enhance representations of the two modalities with contrastive learning for further event extraction, rather than integrating them and predicting events jointly, hence we have a better performance on single-modality too.

5.2 Contrastive Learning

Contrastive learning methods have shown the effectiveness in representation learning via pulling together positive samples to anchor samples and pushing apart negative samples in the representation space [3,7]. Contrastive learning has achieved impressive performance in many natural language processing tasks, such as question answering [26] and information extraction [24,25]. [24] pre-trained for EE in semantic structures by contrastive learning to learn event structures. [25] pre-trained a model which leveraged contrastive learning to learn triggers and arguments based on their events. However, they only focused on learning EE in the limited text modality, neglecting the rich event information in other modalities.

Constrastive learning has also been demonstrated to perform greatly in multi-modal tasks. [27] introduced a contrastive learning based model to not only learn inter-modal similarities but also take intra-modal representation into account. [28] proposed a text-video match model exploiting rich information in videos to learn better text constituents representation for unsupervised grammar induction. However, it only focused on using videos to enhance text representations. Meanwhile, they treated contrastive samples equally, which dose not take the distinction of different contrastive samples pairs into account. Different from their work, in this paper, we conduct cross-modal contrastive learning and assign weights to contrastive sample pairs by measuring the difference between their event contents. Moreover, event representations are also learnt by contrasting with videos to improve argument extraction.

6 Conclusion

In this work, we contrasted text video pairs to assist event extraction by considering their event information. We introduced a new dataset called TVEE, which consists of sentence-video pairs describing the same events. We will publicly release the dataset to stimulate further research on multi-modal event extraction and other tasks. Meanwhile, we proposed a contrastive learning based model composed of two contrastive losses and a negative sample weighting function. Experiments on two multi-modal event extraction datasets showed that our model could improve event extraction and outperformed the baselines on this task.

Acknowledgments. Supported by the National Key Research and Development Program of China (No. 2022YFF0712400), the National Natural Science Foundation of China (No. 62276063), and the Natural Science Foundation of Jiangsu Province under Grants No. BK20221457.

References

1. Chen, B., et al.: Joint multimedia event extraction from video and article. In: Proceedings of the Conference on Empirical Methods in Natural Language Processing, pp. 74–88 (2021)
2. Chen, H., Shu, R., Takamura, H., Nakayama, H.: GraphPlan: story generation by planning with event graph. In: Proceedings of the 14th International Conference on Natural Language Generation, pp. 377–386 (2021)
3. Chen, T., Kornblith, S., Norouzi, M., Hinton, G.: A simple framework for contrastive learning of visual representations. In: International Conference on Machine learning, pp. 1597–1607 (2020)
4. Daiya, D.: Combining temporal event relations and pre-trained language models for text summarization. In: IEEE International Conference on Machine Learning and Applications, pp. 641–646 (2020)
5. Du, X., Cardie, C.: Event extraction by answering (almost) natural questions. In: Proceedings of the Conference on Empirical Methods in Natural Language Processing, pp. 671–683 (2020)
6. Du, X., Rush, A.M., Cardie, C.: GRiT: generative role-filler transformers for document-level event entity extraction. In: Proceedings of the 16th Conference of the European Chapter of the Association for Computational Linguistics: Main Volume, pp. 634–644 (2021)
7. He, K., Fan, H., Wu, Y., Xie, S., Girshick, R.: Momentum contrast for unsupervised visual representation learning. In: Proceedings of the IEEE/CVF Conference on Computer Vision and Pattern Recognition, pp. 9729–9738 (2020)
8. Huang, P.Y., Patrick, M., Hu, J., Neubig, G., Metze, F., Hauptmann, A.: Multilingual multimodal pre-training for zero-shot cross-lingual transfer of vision-language models. arXiv preprint arXiv:2103.08849 (2021)
9. Kenton, J.D.M.W.C., Toutanova, L.K.: BERT: pre-training of deep bidirectional transformers for language understanding. In: Proceedings of the Conference of the North American Chapter of the Association for Computational Linguistics, pp. 4171–4186 (2019)
10. Li, M., et al.: Timeline summarization based on event graph compression via time-aware optimal transport. In: Proceedings of the Conference on Empirical Methods in Natural Language Processing, pp. 6443–6456 (2021)
11. Li, M., et al.: Cross-media structured common space for multimedia event extraction. In: Proceedings of Annual Meeting of the Association for Computational Linguistics, pp. 2557–2568 (2020)
12. Li, Q., Ji, H., Huang, L.: Joint event extraction via structured prediction with global features. In: Proceedings of Annual Meeting of the Association for Computational Linguistics, vol. 1, pp. 73–82 (2013)
13. Lin, Y., Ji, H., Huang, F., Wu, L.: A joint neural model for information extraction with global features. In: Proceedings of Annual Meeting of the Association for Computational Linguistics, pp. 7999–8009. Association for Computational Linguistics, (2020)

14. Lu, J., Batra, D., Parikh, D., Lee, S.: ViLBERT: pretraining task-agnostic visiolinguistic representations for vision-and-language tasks. In: Advances in Neural Information Processing Systems, pp. 13–23 (2019)
15. Martschat, S., Markert, K.: A temporally sensitive submodularity framework for timeline summarization. In: Proceedings of the Conference on Computational Natural Language Learning, pp. 230–240 (2018)
16. Nguyen, T.H., Cho, K., Grishman, R.: Joint event extraction via recurrent neural networks. In: Proceedings of Annual Meeting of the Association for Computational Linguistics, pp. 300–309 (2016)
17. Nguyen, T.H., Grishman, R.: Event detection and domain adaptation with convolutional neural networks. In: Proceedings of the Annual Meeting of the Association for Computational Linguistics and the International Joint Conference on Natural Language Processing, pp. 365–371 (2015)
18. Pratt, S., Yatskar, M., Weihs, L., Farhadi, A., Kembhavi, A.: Grounded situation recognition. In: European Conference on Computer Vision, pp. 314–332 (2020)
19. Raffel, C., et al.: Exploring the limits of transfer learning with a unified text-to-text transformer. J. Mach. Learn. Res. **21**(140), 1–67 (2020)
20. Sadhu, A., Gupta, T., Yatskar, M., Nevatia, R., Kembhavi, A.: Visual semantic role labeling for video understanding. In: Proceedings of the IEEE/CVF Conference on Computer Vision and Pattern Recognition, pp. 5589–5600 (2021)
21. Tong, M., et al.: Image enhanced event detection in news articles. Proceed. AAAI Conf. Artif. Intell. **34**(5), 9040–9047 (2020)
22. Wadden, D., Wennberg, U., Luan, Y., Hajishirzi, H.: Entity, relation, and event extraction with contextualized span representations. arXiv preprint arXiv:1909.03546 (2019)
23. Walker, C., Strassel, S., Medero, J., Maeda, K.: ACE 2005 multilingual training corpus. Linguist. Data Consort. Philadelp. **57**, 45 (2006)
24. Wang, Z., et al.: CLEVE: contrastive pre-training for event extraction. In: Proceedings of Conference on Empirical Methods in Natural Language Processing and International Joint Conference on Natural Language Processing. vol. 1, pp. 6283–6297 (2021)
25. Yao, S., Yang, J., Lu, X., Shuang, K.: Contrastive learning for event extraction. In: International Conference on Machine Learning and Soft Computing, pp. 167–172 (2022)
26. Yeh, Y.T., Chen, Y.N.: QAInfomax: learning robust question answering system by mutual information maximization. In: Proceedings of Conference on Empirical Methods in Natural Language Processing and International Joint Conference on Natural Language Processing, pp. 3370–3375 (2019)
27. Zhang, H., Koh, J.Y., Baldridge, J., Lee, H., Yang, Y.: Cross-modal contrastive learning for text-to-image generation. In: Proceedings of the IEEE Conference on Computer Vision and Pattern Recognition, pp. 833–842 (2021)
28. Zhang, S., Song, L., Jin, L., Xu, K., Yu, D., Luo, J.: Video-aided unsupervised grammar induction. In: Proceedings of Annual Meeting of the Association for Computational Linguistics, pp. 1513–1524 (2021)
29. Zhang, T., et al.: Improving event extraction via multimodal integration. In: Proceedings of ACM International Conference on Multimedia, pp. 270–278 (2017)
30. Zolfaghari, M., Zhu, Y., Gehler, P., Brox, T.: CrossCLR: cross-modal contrastive learning for multi-modal video representations. In: Proceedings of the IEEE/CVF International Conference on Computer Vision, pp. 1450–1459 (2021)

Distinguishing Sensitive and Insensitive Options for the Winograd Schema Challenge

Dong Li[1], Pancheng Wang[1], Liangliang He[1], Kunyuan Pang[1], Shasha Li[1,2], Jintao Tang[1,2(✉)], and Ting Wang[1(✉)]

[1] College of Computer Science and Technology,
National University of Defense Technology, Changsha, China
{lidong1,wangpancheng13,heliangliang19,pangkunyuan10,shashali,
tangjintao,tingwang}@nudt.edu.cn
[2] Key Laboratory of Software Engineering for Complex Systems, Changsha, China

Abstract. The Winograd Schema Challenge (WSC) is a popular benchmark for commonsense reasoning. Each WSC instance has a component that corresponds to the mention of the correct answer option of the two options in the context. We observe that the answers of many instances are insensitive to the options. In this paper, based on this observation, we propose an approach based on fine-tuning the pre-trained language model for WSC by distinguishing sensitive and insensitive options. First, we split WSC instances into option-sensitive and insensitive categories, and use option expanding and option masking strategies to weaken the options so that the model does not pay attention to options when they are insensitive during fine-tuning. Second, we treat the two categories as intermediate-task of each other, and use transfer learning to improve the performance. We fine-tune BERT-Large and T5-XXL with our approach on WINOGRANDE, a new dataset of WSC, and the experiment shows our method outperforms baselines by a large margin, achieving state-of-the-art, which indicates the effectiveness of our instance-distinguishing strategy.

Keywords: WSC · Transfer Learning · Option Weakening

1 Introduction

The Winograd Schema Challenge (WSC) is a pronoun resolution problem, introduced for testing AI agents for commonsense knowledge [6]. The latest WSC dataset WINOGRANDE (WG) [14] is a larger and more difficult variant than the original WSC. WG is formatted as a fill-in-the-blank problem where the blank corresponds to the mention of the correct answer option of the two options.

We observe that the answers of many WSC instances are insensitive to the options. The insensitivity is reflected in two aspects: (1) The answer is irrelevant to the content of the options, and changing the options to other contents, such

© The Author(s), under exclusive license to Springer Nature Switzerland AG 2023
X. Wang et al. (Eds.): DASFAA 2023, LNCS 13945, pp. 716–726, 2023.
https://doi.org/10.1007/978-3-031-30675-4_52

as swapping each other, remains the same. (2) The answer is irrelevant to the mention of options but depends on the location and the order in where the options are mentioned in the sentence. Take Table 1 as an example. The first block of Table 1 shows two option-insensitive instances and the second block shows an option-sensitive instance.

Table 1. Examples of option-sensitive and option-insensitive instances. Option 1 is red and option 2 is blue. The first line of each division is the original instance from WINOGRANDE, and the second line is the new instance after swapping options.

Sentence	option1/option2	answer
Option-Insensitive		
Emily never had as much money to spend as *Carrie*, because _ had a good job.	Emily/Carrie	2
Carrie never had as much money to spend as *Emily*, because _ had a good job.	Carrie/Emily	2
The commodities trader decided to buy *wool* and sell *cotton* because the _ was priced low.	wool/cotton	1
The commodities trader decided to buy *cotton* and sell *wool* because the _ was priced low.	cotton/wool	1
Option-Sensitive		
I picked up a bag of *peanuts* and *raisins* for a snack. I wanted a sweeter snack out so I stored the _ for now.	peanuts/raisins	2
I picked up a bag of *raisins* and *peanuts* for a snack. I wanted a sweeter snack out so I stored the _ for now.	raisins/peanuts	1

In this paper, based on the above observation, we divide a WSC dataset into the option-sensitive subset and option-insensitive subset. Then we propose two methods on the two subsets:

For the option-insensitive subset, we design two **Option Weakening** strategies to reduce the effect of options during the answer inference process, so that the model does not pay attention to these insensitive options when fine-tuning the pre-trained language model. Furthermore, we regard the option-sensitive subset and option-insensitive subset as each other's intermediate-task and take advantage of **Intermediate-Task Transfer Learning** [2,8,11] to improve the fine-tuning effect on the pre-trained language model. See Sect. 3 for details.

We fine-tune BERT [3] and T5 [12,13] model on WG, a new dataset of WSC, with our approach, and the experiment shows our method outperforms baselines by a large margin, achieving state-of-the-art.

In a nutshell, our main contributions in this work are threefold:

- We observe the difference between option-sensitive and insensitive instances of WSC and leverage intermediate-task transfer learning to joint train both option-sensitive and option-insensitive instances.

- We propose a variety of option weakening strategies when training option-insensitive instances, so the model is not or rarely pays attention to options.
- Experimental results on WG show that our approach substantially outperforms baselines and achieves state-of-the-art performance.

2   Related Work

There are 4 group methods that have been used to solve the WSC.

The first group approach is a rule and feature-based approach [10], the performances of these approaches heavily rely on the coverage and quality of the manually defined rules and features.

The second group approach is a knowledge-based approach. These approaches leverage different commonsense knowledge resources such as search engines [4] or knowledge base [16] to solve WSC questions in an explainable way.

The third group approach is a neural approach [7,9]. These approaches rely on neural networks and deep learning, but they do not use any human-defined features.

The fourth group approach is a language model approach. [15] are the first to use pre-trained language models. They compute the probabilities of each candidate for each pronoun using language models, by predicting the resolution of the ambiguity with higher probability.

[8] propose a new multitask benchmark, RAINBOW with the six datasets, NLI, COSMOSQA, HELLASWAG, PIQA, and SOCIALIQA, and WINO-GRANDE. [8] summarize three Intermediate-task Transfer Learning strategies: multitask training, sequential training, multitask fine-tuning and analyzing the characteristics of the three methods.

[5] use the latest advances in language modeling to build a single pre-trained QA model, UNIFIEDQA. UNIFIEDQA advocates for a unifying view of QA formats by building a format-agnostic QA system. [5] treat WINOGRANDE as Multiple-choice QA (MC), and use the format containing questions that come with candidate options. We follow the same format.

3   Our Approach

We approach WSC by fine-tuning the pre-trained language model on the training set. We observe the difference between option-sensitive and insensitive instances, and our motivation is to distinguish them. How to distinguish whether an instance is option-sensitive or insensitive is our first work.

We note that if both options are person names (except for names that contain specific meanings, such as Snow White and Pinocchio), the instance is option-insensitive, and we divide it into INS (option-**ins**ensitive), otherwise it may be option-insensitive or may be option-sensitive. We tried some methods to distinguish them, but the results were not good. It is a difficult task, so we divide it into SEN (option-**sen**sitive) without any distinction. In the following, we give an empirical analysis of such treatment.

3.1 Option Weakening

For the INS subset, we design Option Weakening strategy to reduce the effect of options during the answer inference process, so that the model does not pay attention to these insensitive options when fine-tuning the pre-trained language model. Option Weakening consists of two strategies: Option Masking and Option Expanding.

Option Expanding (OE). Option Expanding constructs some pseudo instances for an instance. The options of these instances are replaced with other words, and the other content of the sentence and the answer remain the same. When we fine-tune the pre-trained language model on these instances with different options but the same context and answer, the model may be informed that the options are unimportant and redundant. Exchange options is a simple method, which we call Option Expanding by Exchanging.

Option Masking (OM). Option Masking constructs a new instance by masking options with $symbol_1$ and $symbol_2$ for an instance. We consider three different types of $symbol_1$ and $symbol_2$, which are *indefinite reference symbols, anonymous symbol* and *unused symbol*.

Indefinite reference symbols refer to indefinite names that can refer to anyone, such as Tom, Dick and Harry, Joe Blow and Joe Shmoe in the English tradition.

Anonymous symbols refer to those symbols that contain unknown and anonymous semantics, such as X and Y. X represents mystery, unknown and unsolved(e.g., Mr. X, Dr. X). Y has a similar meaning when it appears with X. In our work, we mainly use this OM method for ablation experiments.

Unused symbols refer to unused symbols reserved in the symbol table. The pre-trained language model has not processed these symbols in the pretraining stage, and the semantics of these symbols are blank.

3.2 Intermediate-Task Transfer Learning

Intermediate-task transfer learning is a widely applied technique, which first fine-tunes the pre-trained language models on an intermediate-task before the target task of interest. After dividing a WSC dataset into the INS subset and SEN subset, we regard them as each other's intermediate task and improve the fine-tuning effect on the pre-trained language model by transfer learning (Fig. 1).

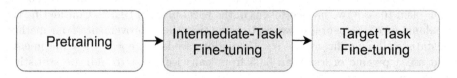

Fig. 1. Overview of Intermedia-Task Transfer Learning.

We design four training strategies, which are Single-Task and three intermediate-task transfer learning strategies, following [8]. These strategies are shown below:

(1) **Single-Task:** independent training on INS or SEN, respectively
(2) **Multitask Training:** training on INS and SEN all at once. In this case, Multitask Training is equivalent to not distinguishing between INS and SEN, which is the baseline of our experiment below.
(3) **Sequential Training:** first training on INS, and then continuing to train on the SEN alone and vice versa.
(4) **Multitask Fine-tuning:** first training on INS and SEN all at once through multitask training, and then continuing to fine-tune on the INS or SEN alone.

4   Experiments

In this work, We fine-tune BERT and T5 with our approach on WG, a new dataset of WSC. WG dataset is split into training, development, and test sets. The training set is available in five different sizes, XS, S, M, L, and XL. The primary measure performance metric is the Area Under the (learning) Curve, which is computed by accuracies for models trained on all different sizes.[1] The label of the test set is unpublished, and researchers can submit data to the Allen AI leaderboard once a week, and the website will test and publish the accuracy and AUC score.

4.1   Implementation Details

We use the PyTorch 1.6.0 implementation of pre-trained language model, BERT-large and T5-XXL. To identify INS from the WG, we use the NER flair tool [1]. We set Adam optimizer with an initial learning rate of $1e^{-5}$.

Considering the training cost of BERT and T5 models, the purpose of our experiments on BERT is to verify the effects of various strategies of our approach, while the purpose of experiments on T5 is to further verify the conclusions on BERT and make some comparisons and additions.

BERT. We fine-tune BERT on the training set with XL size, using the batch size of 32, early stopping with wait step of 10, the primary measure performance metric is accuracy. We experiment with four Option Weankening strategies. They are Option Expanding, Option Masking with X and Y, Option Masking with Tom and Dick, and Option Masking with token 104 and token 105. Token 104 and token 105 are two unused tokens in the vocabulary of BERT. Considering the randomness of the neural network, to show that the improvement of our method is statistically significant, we set 15 random seeds to repeat the experiments, and use a p-value of less than 0.05 from pairwise t-test to indicate statistical significance.

---

[1] https://github.com/allenai/mosaic-leaderboard/tree/master/winogrande/evaluator.

Table 2. Meaning of various symbols

Symbol	Mean
WG	WINOGRANDE set
WG_{dev}	WG development set
WG_{test}	WG test set
WG_{xl}	WG training set with XL size
WG_{train}	Combination of $WG_{xs,s,m,l,xl}$
$WG_{dev \cdot ins}$	the INS division of WG_{dev}
OM	Option Masking
$WG_{xl \cdot ins}^{m(X,Y)}$	OM with X and Y on $WG_{xl \cdot ins}$
Model-A	fine-tune model on A
Model-A-B	fine-tune model on A and B sequentially

Table 3. Statistic of the WINOGRANDE dataset.

	#INS	#SEN	#All	#INS / #All
train_xs	100	60	160	0.63
train_s	378	262	640	0.59
train_m	1440	1118	2558	0.56
train_l	5710	4524	10234	0.56
train_xl	23424	16974	40398	0.58
dev	734	533	1267	0.58
test	1068	699	1767	0.60

T5. We fine-tune T5 on all different sizes (XS, S, M, L, XL) training set, the primary measure performance metric is the AUC, using the batch size 24, early stopping with wait step of 10. We fine-tune T5 on WG and experiment with one Option Weankening strategy, Option Masking with X and Y. Considering the training cost of T5-XXL and using the AUC as the primary measure metric, some randomness has been eliminated, we set a random seed.

Symbols Table. Table 2 describes the meanings of the various symbols that appear in this work.

4.2 Analysis of the Dataset

Table 3 shows the statistical data of various divisions of the WG dataset. It is clear that in the WG training, development, and test set, the proportions of the instances of INS are approximately 60%. It will not lead to the problem of the different distribution of training set and test set in our subsequent work.

722   D. Li et al.

Table 4. The performance on WG$_{dev}$ and WG$_{test}$ on BERT. The first two columns represent models trained on INS and SEN, respectively. The data in column 3 is the accuracy on the development set. The first line of data is H0 of t-test, and the following lines are H1. The data in column 4 is the accuracy of H1 minus H0. The * in the upper right corner of the number indicates that there is a statistically significant difference between H0 and H1(p-value <= 0.05).

Model		Dev		Test
INS	SEN	ACC	H1-H0	Acc
Multitask Training (baseline)		66.59(0.65)	H0	65.20
Single-Task + OM	Multitask Training	67.27(0.55)	0.68*	
Single-Task + OM	Sequential Training	67.90(0.72)	1.31*	**66.89**
Single-Task + OM	Multitask Fine-tune	**67.91**(0.59)	1.32*	

Table 5. The performance on WG$_{dev}$ and WG$_{test}$ on T5. The measure metric is the AUC score in %.

Model	Dev	Test
T5-WG$_{train}$ (baseline)	87.91	87.63
Sequential Training + OM	**89.34**	**88.29**

4.3 Overall Results

Tables 4 and 5 show the performance of our approach on WG.

The two parts of our work, option weakening and intermediate task transfer learning, are independent of each other. The overall effect is determined by the effect of the two methods.

Table 4 shows some combinations of methods with a statistically significant improvement on the baseline. Single-Task training and Option Masking on INS and sequential training on SEN improve the accuracy by 1.31% on the development set and 1.69% on the test set.

Table 5 shows sequential training and Option Masking improve the performance. We raise the AUC score of baseline from 87.91% to 89.43% on the development set, and from 87.63% to 88.29% on the test set, achieving SOTA.

4.4 Ablation Study

Option Weakening. Table 6 and Table 7 show the performance on Option Weakening for BERT-Large and T5, respectively.

From the first block of Table 6, we can see that all four Option Weakening methods are effective, and the Option Masking with Tom and Dick improves accuracy the most, but the standard deviation of this method is indeed the highest. From the second block, We can find no statistically significant difference between the four methods. Considering the mean, standard deviation, and

Table 6. The performance on Option Weakening for BERT-Large. $m(X,Y)$ stands for Option Masking with X and Y, $m(Tom, Dick)$ stands for Option Masking with Tom and Dick, $m(unused1, unused2)$ stands for Option Masking with token 104 and 105, ee stands for Option Expanding by Exchanging. The data in column 2 are the accuracy on development set in %.

Model	Acc	H1-H0
t-test to None with OW on $WG_{dev \cdot ins}$		
BERT-WG$_{xl \cdot ins}$	65.91(1.28)	H0
BERT-WG$_{xl \cdot ins}^{m(X,Y)}$	66.79(0.78)	0.88*
BERT-WG$_{xl \cdot ins}^{m(unused1, unused2)}$	66.87(0.86)	0.92*
BERT-WG$_{xl \cdot ins}^{m(Tom, Dick)}$	**67.18**(1.04)	1.27*
BERT-WG$_{xl \cdot ins}^{ee}$	66.62(0.85)	0.71*
t-test to m(X,Y) with other OW on $WG_{dev \cdot ins}$		
BERT-WG$_{xl \cdot ins}^{m(X,Y)}$	66.79(0.78)	H0
BERT-WG$_{xl \cdot ins}^{m(unused1, unused2)}$	66.87(0.86)	0.08
BERT-WG$_{xl \cdot ins}^{m(Tom, Dick)}$	**67.18**(1.04)	0.39
BERT-WG$_{xl \cdot ins}^{ee}$	66.62(0.85)	−0.171
t-test to None with m(X,Y) on $WG_{dev \cdot sen}$		
BERT-WG$_{xl \cdot sen}$	**66.09**(1.80)	H0
BERT-WG$_{xl \cdot sen}^{m(X,Y)}$	64.29(1.46)	−1.80*

training cost, we think Option Masking with X and Y is the most appropriate Option Weakening method. So we do not use the other three Option Weakening methods to continue our experiments.

From the last block of Table 6 and Table 7, we can see that Option Masking method harms the accuracy on WG_{sen}, but the accuracy on WG_{sen} is still much higher than the random accuracy of 50%. It enlightens us that the proportion of option-insensitive instances in WG_{sen} is not low. Using empirical estimates, the accuracy of option-sensitive instances on Model T5-WG$_{xl \cdot sen}^{m(X,Y)}$ is approximately equal to 50%. Considering the extreme case, the accuracy of

Table 7. The performance on Option Masking with X and Y for T5. The data in column 2 are the AUC of accuracy fine-tuned on $WG_{xs,s,m,l,xl}$. The data in column 3 are the accuracy fine-tuned on WG_{xl}.

Model	AUC	ACC
T5-WG$_{train \cdot ins}$	88.34	93.46
T5-WG$_{train \cdot ins}^{m(X,Y)}$	**89.56**	**93.74**
T5-WG$_{train \cdot sen}$	**87.05**	**93.06**
T5-WG$_{train \cdot sen}^{m(X,Y)}$	78.63	82.36

option-insensitive instances on Model T5-WG$_{xl \cdot sen}^{m(X,Y)}$ is 100%, then the proportion of option-insensitive instances shall not be less than 64.72%. If the accuracy of option-insensitive instances on Model T5-WG$_{xl \cdot sen}^{m(X,Y)}$ is 93.06%, the proportion shall be 75.15%.

Table 8. The performance on WG$_{dev \cdot ins}$ of Intermediate-task Transfer Learning for BERT-Large.

Transfer Learning	Acc	H1-H0
Multitask Training(baseline)	66.59	H0
Single-Task Training	65.99	−0.60
Sequential Training	66.13	−0.46
Multitask Fine-tuning	**67.83**	1.24∗

Table 9. The performance on WG$_{dev \cdot ins}$ of Intermediate-task Transfer Learning for T5-XXl.

Transfer Learning	AUC
Multitask Training (baseline)	87.90
Single-Task Training	87.80
Sequential Training	**88.99**

Intermediate-Task Transfer Learning. Table 8 and Table 9 show the performance of Intermediate-task Transfer Learning for BERT-Large and T5-XXL. From Table 8, we can see that multitask fine-tuning is significantly better than the baseline(multitask training). From Table 9, we can see that sequential training is effective for T5.

5 Conclusion

In this paper, we distinguish sensitive and insensitive options for the Winograd Schema Challenge (WSC). We divide WG, a WSC dataset, into the option-sensitive and option-insensitive subsets and propose Option Weakening and Intermediate-task Transfer Learning strategies on the two subsets. We conduct extensive experiments and the results illustrate the effectiveness of our proposed strategies.

In our work, splitting SEN into Option-Insensitive and Option-Sensitive subset is a difficult task. Through our experiments, we find that even conducting Option Weakening on SEN, the accuracy is still much higher than the random accuracy of 50%. Combined with the data set analysis, we believe that Option-Insensitive subset accounts for a large proportion in SEN. According to the empirical estimation method, the proportion is more significant than 64.72%, and with a high probability of being less than 75.15%. If we can split SEN with high accuracy, we will get a better result. From another perspective, distinguishing Option-Insensitive and Option-Sensitive in SEN is also a task that requires common sense for reasoning and is worth investigating. In essence, our method exploits useful features and weakens useless features. It is a method of applying feature engineering to the large-scale pre-trained language model, which is a problem worth studying.

Acknowledgements. We would like to thank the anonymous reviewers for their helpful comments. This work was supported by the National Key Research and Development Project of China (No. 2021ZD0110700).

References

1. Akbik, A., Bergmann, T., Blythe, D., Rasul, K., Schweter, S., Vollgraf, R.: Flair: an easy-to-use framework for state-of-the-art NLP. In: Proceedings of the 2019 Conference of the North American Chapter of the Association for Computational Linguistics (Demonstrations), pp. 54–59 (2019)
2. Chang, T.-Y., Chi-Jen, L.: Rethinking why intermediate-task fine-tuning works. In: Findings of the Association for Computational Linguistics: EMNLP 2021, pp. 706–713 (2021)
3. Devlin, J., Chang, M.-W., Lee, K., Toutanova, K.: BERT: pre-training of deep bidirectional transformers for language understanding. arXiv preprint arXiv:1810.04805 (2018)
4. Emami, A., De La Cruz, N., Trischler, A., Suleman, K., Cheung, J.C.K.: A knowledge hunting framework for common sense reasoning. In: EMNLP (2018)
5. Khashabi, D., et al.: UnifiedQA: crossing format boundaries with a single QA system. In: Findings of the Association for Computational Linguistics: EMNLP 2020, pp. 1896–1907 (2020)
6. Levesque, H., Davis, E., Morgenstern, L.: The Winograd schema challenge. In: Thirteenth International Conference on the Principles of Knowledge Representation and Reasoning (2012)
7. Liu, Q., Jiang, H., Ling, Z.-H., Zhu, X., Wei, S., Hu, Y.: Combing context and commonsense knowledge through neural networks for solving Winograd schema problems. In: 2017 AAAI Spring Symposium Series (2017)
8. Lourie, N., Le Bras, R., Bhagavatula, C., Choi, Y.: Unicorn on rainbow: a universal commonsense reasoning model on a new multitask benchmark. In: Proceedings of the AAAI Conference on Artificial Intelligence, 35, pp. 13480–13488 (2021)
9. Opitz, J., Frank, A.: Addressing the Winograd schema challenge as a sequence ranking task. In: Proceedings of the First International Workshop on Language Cognition and Computational Models, pp. 41–52 (2018)
10. Peng, H., Khashabi, D., Roth, D.: Solving hard coreference problems. In: Proceedings of the 2015 Conference of the North American Chapter of the Association for Computational Linguistics: Human Language Technologies, pp. 809–819 (2015)

11. Pruksachatkun, Y., et al.: Intermediate-task transfer learning with pretrained language models: when and why does it work? In: Proceedings of the 58th Annual Meeting of the Association for Computational Linguistics, pp. 5231–5247 (2020)
12. Raffel, C., et al.: Exploring the limits of transfer learning with a unified text-to-text transformer. arXiv preprint arXiv:1910.10683 (2019)
13. Roberts, A., Raffel, C., Shazeer, N.: How much knowledge can you pack into the parameters of a language model? In: Proceedings of the 2020 Conference on Empirical Methods in Natural Language Processing (EMNLP), pp. 5418–5426 (2020)
14. Sakaguchi, K., Le Bras, R., Bhagavatula, C., Choi, Y.: Winogrande: an adversarial Winograd schema challenge at scale. In: Proceedings of the AAAI Conference on Artificial Intelligence, vol. 34, pp. 8732–8740 (2020)
15. Trinh, T.H., Le, Q.V.: A simple method for commonsense reasoning. arXiv preprint arXiv:1806.02847 (2018)
16. Zhang, H., Ding, H., Song, Y.: Sp-10k: a large-scale evaluation set for selectional preference acquisition. In: Proceedings of the 57th Annual Meeting of the Association for Computational Linguistics, pp. 722–731 (2019)

Semi-supervised Learning for Fine-Grained Entity Typing with Mixed Label Smoothing and Pseudo Labeling

Bo Xu[1], Zhengqi Zhang[1], Ming Du[1(✉)], Hongya Wang[1], Hui Song[1], and Yanghua Xiao[2,3]

[1] School of Computer Science and Technology, Donghua University, Shanghai, China
{xubo,duming,hywang,songhui}@dhu.edu.cn, 2202405@mail.dhu.edu.cn
[2] Shanghai Key Laboratory of Data Science, School of Computer Science, Fudan University, Shanghai, China
shawyh@fudan.edu.cn
[3] Fudan-Aishu Cognitive Intelligence Joint Research Center, Shanghai, China

Abstract. Distant supervision (DS) has been proposed to automatically annotate data and achieved significant success in fine-grained entity typing(FET). Despite its efficiency, distant supervision often suffers from the *noisy labeling* problem. To solve the *noisy labeling* problem, existing approaches assume the existence of "clean" and "noisy" sets in the training data and use different types of methods to utilize them. However, they still suffer from the *confirmation bias* problem in the "noisy" set and the *false positive* problem in the "clean" set. To address these issues, we propose a novel semi-supervised learning method with mixed label smoothing and pseudo labeling for distantly supervised fine-grained entity typing. Specifically, to solve the *false positive* problem on the "clean" set, we propose a mixed label smoothing method to smooth the labels of the "clean" set to train the FET model. To solve the *confirmation bias* problem on the "noisy" set, we do not consider the labels in the "noisy" set and use a pseudo labeling technique to deal with the "noisy" set. Extensive experiments conducted on three widely used FET datasets show the effectiveness of our proposed approach. The source code is publicly available at https://github.com/xubodhu/NFETC-SSL.

1 Introduction

Fine-grained entity typing (FET) is an essential task in natural language processing that aims to classify an entity mentioned in a sentence into a predefined set of fine-grained types. The extracted entity type information can be used for many downstream applications, such as entity linking, relation extraction and question answering. To reduce manual efforts in labeling training data, distant supervision has been adopted to automatically annotate a large number of unlabeled mentions in the training corpus. Specifically, an unlabeled entity mention will be linked to an existing entity in the knowledge base, and then all possible types of the entity will be assigned to the entity mention.

© The Author(s), under exclusive license to Springer Nature Switzerland AG 2023
X. Wang et al. (Eds.): DASFAA 2023, LNCS 13945, pp. 727–736, 2023.
https://doi.org/10.1007/978-3-031-30675-4_53

Despite its efficiency, distant supervision often suffers from the *noisy labeling* problem, as it assigns labels in a context-agnostic manner. To address the *noisy labeling* problem on FET, from a data perspective, existing approaches assume the existence of "clean" and "noisy" sets in the training data and use three different types of methods to utilize them. The first is a class of supervised methods that try to select the clean set of them to train the FET model [4,8,16]. However, these methods greatly reduce the amount of training data and do not take full advantage of the annotated data. Some of these methods even require additional manual annotation. The second is a class of supervised methods that use both "clean" and "noisy" data to train the FET model. They model the losses for "clean" and "noisy" sets separately and use partial label loss to train the "noisy" set [9,17]. Despite making full use of the annotated data, they suffer from the *confirmation bias* problem on the "noisy" set. Since they assume that the type with the highest probability among the candidate types in the "noisy" set is the correct type and use it as the optimization target during the training process. The third is a class of semi-supervised methods that treat the "clean" set as labeled data and the "noisy" set as unlabeled data to train the FET model [2]. Despite their success, the semi-supervised methods suffer from the *false positive* problem on the "clean" set. Since they assume that samples whose candidate types can form a single path in the type hierarchy are "clean". However, the "clean" samples are not always true, they also contain noise [16].

To address these issues, we propose a novel semi-supervised learning method with mixed label smoothing and pseudo labeling for distantly supervised fine-grained entity typing. We first divide the training data into "clean" and "noisy" sets according to the previous strategy [2,9,16,17], and then propose two novel strategies to deal with both sets, respectively. Specifically, to solve the *false positive* problem on the "clean" set, we propose a mixed label smoothing method, including hierarchical label smoothing and online label smoothing methods, to generate smoothed labels to train the FET model to mitigate the overfitting of the model to noisy labels, while considering the hierarchical and correlation relationships between the labels. To solve the *confirmation bias* problem on the "noisy" set, we do not consider the labels in the "noisy" set and treat them as unlabeled data. Then we a pseudo labeling technique to deal with the unlabeled data to regularize the FET model. We demonstrate the effectiveness of our method through the experiments. Experiment results on three public benchmarks show that our framework has achieved state-of-the-art results.

2   Overview

2.1   Problem Definition

We follow the same setting adopted by [2,9,10,17,19]. The input is a knowledge base Ψ with type hierarchy $\mathcal{Y}$, and an automatically labeled training corpus (samples) $\mathcal{D}$ obtained by distant supervision with $\mathcal{Y}$. The output is a *type-path* in $\mathcal{Y}$ for each mention in a test sentence from a corpus $\mathcal{D}_t$.

Specifically, we define $\Gamma = \{t_1, t_2, \cdots, t_K\}$ as all candidate *type-path* labels, where K is the total number of paths. Each label is a *type-path* from the root node to the terminal node, and a terminal node could be either a leaf node or a non-leaf node (e.g. /person/artist/actor and /person/artist). The training corpus $\mathcal{D}$ consists of triples with form $\{(m_i, c_i, \boldsymbol{y}_i)\}_{i=1}^N$. For each training sample, we denote the context sentence as a word sequence $c_i = \{w_1, w_2, \cdots, w_n\}$, and entity mention $m_i = \{w_j, \cdots, w_k\}$ as a continuous sub-sequence from the context sentence. We treat samples labeled with single type path (i.e. triples $(m_i, c_i, \boldsymbol{y}_i)$ in $\mathcal{D}$ whose corresponding $\|\boldsymbol{y}_i\|_1 = 1$) as "clean" set and others as "noisy" set, where the "clean" set can be considered labeled data D_L and the "noisy" set can be considered unlabeled data D_U. The main challenge of distantly supervised entity typing systems is to exploit both the "clean" set and "noisy" set to obtain a high-performance classifier.

2.2 Framework

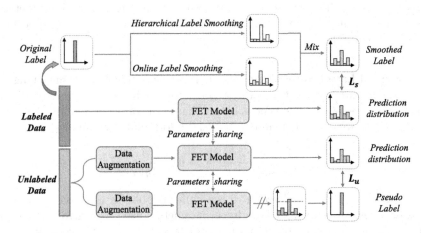

Fig. 1. The overall framework of our semi-supervised learning method. L_s and L_u are the cross-entropy loss functions on the labeled data and unlabeled data, respectively.

The framework of our semi-supervised learning method is shown in Fig. 1. We first divide the training data into "clean" and "noisy" sets according to the previous strategy [2,9,16,17], and treat the "clean" set as labeled data and the "noisy" set as unlabeled data. Then we train the FET model using the labeled data D_L, while regularizing the model using the unlabeled data D_U.

For training on labeled data, we propose a mixed label smoothing method to smooth the original labels and use the smoothed labels to train the FET model to alleviate model overfitting to noisy labels. Specifically, the mixed label smoothing method consists of a hierarchical label smoothing(HLS) method and an online label smoothing(OLS) method. The hierarchical label smoothing method considers the prior knowledge of the type hierarchy, while the online label smoothing method considers the relevance of the types. For training on unlabeled data, we first perform two data augmentation methods to the same unlabeled sample to generate two augmented samples and then use one of them to generate hard pseudo labels. The model is trained on the other one to align the predictions with the pseudo labels.

In this paper, for a fair comparison with previous works [2,10,17,19], we adopt NFETC [17] as the FET model.

3   Method

3.1   Mixed Label Smoothing for Labeled Data

According to the observation, each sample may suffer from the noisy labeling problem in labeled data. To solve the problem, label smoothing has been proposed to mitigate the label noise. It can prevent the model from outputting overly confident predictions and becoming less confident in both the noisy and correct labels [6]. In this paper, we propose a mixed label smoothing method, which consists of a hierarchical label smoothing method and an online label smoothing method.

Hierarchical Label Smoothing. We first use hierarchical label smoothing [16] (HLS) to generate static soft labels to smooth the original labels, since predicting an ancestor type of the true type is better than some other unrelated types. The process of hierarchical label smoothing can be formulated as follows:

$$\check{y} = \frac{1-\alpha-\beta}{||\boldsymbol{y}||_1} \cdot \boldsymbol{y} + \alpha \cdot \boldsymbol{u} + \frac{\beta}{||\widehat{\boldsymbol{y}}||_1} \cdot \widehat{\boldsymbol{y}}, \tag{1}$$

where $\boldsymbol{y}$ is the original label vector, $\boldsymbol{u}$ is a uniform label distribution and $\widehat{\boldsymbol{y}}$ is the ancestor label vector, α and β are hyperparameters used to control the weights of uniform label distribution $\boldsymbol{u}$ and ancestor label vector $\widehat{\boldsymbol{y}}$, respectively.

Online Label Smoothing. We also consider the correlation between types and use online label smoothing [18] (OLS) to smooth the original labels. Since predicting a related type of the true type is also better than some other unrelated types. In this paper, we propose a variant of online label smoothing that filters out low-confidence predictions to generate dynamic soft labels during the training process.

Formally, for the t-th training epoch, we initialize an all-zero asymmetric correlation matrix $\boldsymbol{S}^t \in \mathbb{R}^{K \times K}$, where K is the total number of types, and each row $\boldsymbol{S}_k^t$ represents the correlation between all types and the k-th type. For each labeled sample (m_i, c_i, y_i^*), where y_i^* is the distantly supervised type, $\boldsymbol{p}_i$ is the predicted label distribution obtained by the FET model. When the sample is correctly classified by the model and its prediction probability $p(y_i^*|m_i, c_i)$ is greater than a predefined threshold ϕ, we update the row of y_i^* in $S_{y_i^*}^t$, which can be formulated as:

$$S_{y_i^*}^t = S_{y_i^*}^t + \mathbb{I}(p(y_i^*|m_i, c_i) > \phi)p_i \qquad (2)$$

At the end of t-th training epoch, we normalize the cumulative matrix S^t row by row as represented by:

$$S_{k,k'}^t \leftarrow \frac{S_{k,k'}^t}{\sum_{l=1}^{K} S_{k,l}^t} \qquad (3)$$

Finally, we smooth the original label vector y at the $t + 1$ epoch as follows:

$$\breve{y}^{t+1} = (S^t)^T y \qquad (4)$$

Mixed Label Smoothing. Finally, we combine both the static smoothed labels and the dynamic smoothed labels as the final smoothed labels to train the model. The mixed label smoothing at the t-th training epoch can be represented by:

$$\widetilde{y}^t \leftarrow \gamma \tilde{y} + (1 - \gamma)\breve{y}^t, \qquad (5)$$

where γ is used to balance HLS and OLS. The training loss on labeled data at the t-th training epoch can be represented by:

$$L_s^t = H(\widetilde{y}^t, p), \qquad (6)$$

where H is the cross-entropy loss function. It constrains the model not to be overconfident in labels of the labeled data and avoids overfitting to noisy labels.

3.2 Pseudo Labeling for Unlabeled Data

For the "noisy" set, existing methods relabel them by selecting the type with the highest probability from the available candidate types, which leads to the *confirmation bias* problem. Therefore, in this paper, we do not consider the labels of the "noisy" set and treat them as unlabeled data. Inspired by FixMatch [12], we use the pseudo labeling technique to generate reliable labels for unlabeled data to regularize the FET model.

Specifically, for an unlabeled sample $x = (m, c)$, we first apply two data augmentation methods to the same unlabeled sample to generate a pair of augmented samples. In this paper, we use *dropout* [3] as the data augmentation method, which can generate a pair of latent vectors with similar semantics but different representations. After that, we feed the pair of augmented samples to the FET model and obtain the prediction distributions $p_1(y|\omega(x))$ and $p_2(y|\omega(x))$, respectively. Next, we use one of them to generate a hard (one-hot distribution) pseudo labels based on *argmax* operation. The FET model is trained on the other one to align the prediction distribution with the hard pseudo labels. Therefore, the training loss on unlabeled data can be expressed as:

$$L_u = \mathbb{I}(max(p_1(y|\omega(x)))) > \tau)H(\widehat{p}_1(y|\omega(x)), p_2(y|\omega(x))), \tag{7}$$

where H is the cross-entropy loss function and $\widehat{p}_1(y|\omega(x))$ is the hard pseudo label of the unlabeled sample x. We use the threshold τ to control the model to train only hard pseudo labels with high confidence for filtering out noisy labels.

3.3   Training Process

The overall training process of our semi-supervised learning method is shown in Algorithm 1.

Algorithm 1. The Training Process Algorithm

Inputs: The labeled data D_L and unlabeled data D_U, hyperparameters of HLS and OLS α, β, γ and ϕ, the hyperparameters of SSL τ and λ, the number of epochs E.
Output: the fine-grained entity typing model f_θ.

1: **Initialize** parameters θ for the model f_θ, training epoch $t \leftarrow 1$;
2: **for** $t \leftarrow 1$ *to* E **do**
3:     **Initialize** $S^t = 0$
4:     **for** *iteration* $\leftarrow 1$ *to* $\lceil N/B \rceil$ **do**
5:         Sample a batch of $B_L \subset D_L$ and $B_U \subset D_U$;
6:         Generate smoothed label $\widetilde{y}^t$ of B_L according equation 5;
7:         Output predicted probability distribution p through f_θ;
8:         Calculate the overall loss by equation 8 and update the parameters θ;
9:         **for** $i \leftarrow 1$ *to* $|B_L|$ **do**
10:             Update S^t through equation 2;
11:         **end for**
12:     **end for**
13:     Normalize S^t according to equation 3;
14: **end for**

We first initialize parameters θ for the fine-grained entity typing model f_θ and the training step t (Step 1). Then we iterate for E epochs to update the parameter θ of the model. Specifically, for t-th epoch, we first initialize the all-zero matrix S^t (Step 3). In each iteration, we take a batch of data from the labeled data and the unlabeled data respectively (Step 5). Then we generate smoothed labels for the batch B_L of the labeled data by mixed label smoothing (Step 6). The predicted probability distribution p can be output by feeding B_L and B_U together into the fine-grained entity typing model f_θ (Step 7). Our training loss consists of L_s and L_u. We train these jointly and update the parameters θ (Step 8). The final loss function is defined as follows:

$$L = L_s + \lambda L_u, \tag{8}$$

where λ is a hyperparameter used to balance the two losses. Additionally, we update S^t with the predicted probability distribution with high confidence according to Eq. 2 (Step 9–11) and apply row-wise normalize the S^t by Eq. 3 at the end of t-th training epoch (Step 13).

4 Experiment

4.1 Datasets and Metrics

We conduct experiments on three widely used distantly supervised fine-grained entity typing datasets, namely Wiki [5], OntoNotes [14] and BBN [13]. We use the strict accuracy, loose macro-averaged F1 score and loose micro-averaged F1 score to evaluate the performance of the FET model, which are widely used in many recent works [2,9,17,19].

4.2 Baselines

We compare our method *NFETC-SSL* with several state-of-the-art FET methods, including *AFET* [9], *Attentive* [11], *AAA* [1], *NDP* [15], *Box* [7], *NFETC* [17], *NFETC-CLSC* [2], *NFETC-VAT* [10] and *NFETC-AR* [19].

4.3 Performance Comparison

Table 1. Performance Comparison on The Three Benchmarks.

Methods	Wiki			OntoNotes			BBN		
	ACC	Ma-F1	Mi-F1	ACC	Ma-F1	Mi-F1	ACC	Ma-F1	Mi-F1
Attentive [11]	59.7	80.0	75.4	51.7	71.0	64.9	48.4	73.2	72.4
AFET [9]	53.3	69.3	66.4	55.3	71.2	64.6	68.3	74.4	74.7
AAA [1]	65.8	81.2	77.4	52.2	68.5	63.3	65.5	73.6	75.2
NDP [15]	67.7	81.8	78.0	58.0	71.2	64.8	72.7	76.4	77.7
Box [7]	–	81.6	77.0	–	77.3	70.9	–	78.7	78.0
NFETC [17]	68.9	81.9	79.0	60.2	76.4	70.2	73.9	78.8	79.4
NFETC-CLSC [2]	–	–	–	62.8	77.8	72.0	74.7	80.7	80.5
NFETC-AR [19]	70.1	83.2	80.1	64.0	78.8	73.0	76.7	81.4	81.5
NFETC-VAT [10]	-	-	-	63.8	78.7	73.0	76.7	80.7	80.9
NFETC-SSL	**71.1**	**84.4**	**80.7**	**64.4**	**79.7**	**74.3**	**77.8**	**82.1**	**82.4**

We report the metrics of strict accuracy (ACC), loose macro-averaged F1 score (Ma-F1) and loose micro-averaged F1 score (Mi-F1) on three benchmarks. Table 1 shows the overall performance. We highlight the statistically significant best scores of each metric in bold. We find that our method achieves the new state-of-the-art performance on three benchmarks, demonstrating that our method can indeed effectively train the FET model. The detailed analysis is as follows.

Firstly, compared with the basic NFETC method, our method has a significant improvement (improving the strict accuracy on the three benchmarks from 68.9 to 71.1 (Wiki), 60.2 to 64.4 (OntoNotes) and 73.9 to 77.8 (BBN), respectively). That indicates the necessity of noise reduction in the distantly supervised FET task and the effectiveness of our proposed method.

Secondly, compared with other NFETC-based methods, our method *NFETC-SSL* performs better than *NFETC-CLSC* and *NFETC-VAT* under most metrics when using the same backbone and similar hyperparameters settings, indicating that our semi-supervised learning method is better than theirs. Our method *NFETC-SSL* performs better than the state-of-the-art method *NFETC-AR*, indicating that the pseudo-labels predicted by the model are not all reliable due to they are trained with pseudo-labels of all the data and we set a threshold τ.

4.4 Ablation Study

To investigate the effectiveness of each component proposed in our framework, we perform comparisons between the full model and its ablation methods on OntoNotes and BBN.

Table 2. Ablation Study of Our *NFETC-SSL* Method on OntoNotes and BBN.

Method	OntoNotes			BBN		
	Stric-Acc	Macro-F1	Micro-F1	Strict-Acc	Macro-F1	Micro-F1
NFETC-SSL	**64.4**	**79.7**	**74.3**	**77.8**	**82.1**	**82.4**
w/o L_u	63.0	78.3	72.8	76.7	81.1	81.5
-w/o OLS	62.2	77.8	72.1	76.1	80.7	81.0
-w/o HLS	60.5	75.9	69.6	75.9	80.2	81.0
-w/o MLS	57.8	74.4	67.7	75.5	80.3	80.1

As shown in Table 2, *NFETC-SSL* benefits from the pseudo labeling module and mixed label smoothing(MLS) module. Specifically, without the pseudo labeling module (**w/o** L_u), which means removing the effect of consistency regularization, the Strict-Acc score, Macro-F1 score and Micro-F1 score drop 1.4, 1.4 and 1.5 on OntoNotes, respectively. A similar situation can also be found on BBN. That shows that the pseudo labeling module is beneficial to the training of our FET model.

Based on **w/o** L_u, we investigate the effectiveness of our mixed label smoothing(MLS) module, which consists of hierarchical label smoothing and online label smoothing. Without considering the online label smoothing (**w/o OLS**), the Strict-Acc score, Macro-F1 score and Micro-F1 score drop 0.8, 0.5 and 0.7 on OntoNotes, respectively. That shows that it is reasonable to introduce OLS to consider the correlations between types on different paths. Without considering hierarchical label smoothing (**w/o HLS**), the Strict-Acc score, Macro-F1 score and Micro-F1 score drop 2.5, 2.4 and 3.2 on OntoNotes, respectively.

This shows that it makes sense for HLS to introduce prior knowledge of the type hierarchy. And without using the mixed label smoothing module (**w/o MLS**), the Strict-Acc score, Macro-F1 score and Micro-F1 score drop 5.2, 3.9 and 5.1 on `OntoNotes`, respectively. That demonstrates the effectiveness of our MLS module.

5 Conclusion

In this paper, we study the *noisy labeling* problem on the fine-grained entity typing (FET) task and propose a novel semi-supervised learning method with mixed label smoothing and pseudo labeling for distantly supervised fine-grained entity typing. Specifically, we consider the "clean" set as labeled data and the "noisy" set as unlabeled data. For the labeled data, we propose to use mixed label smoothing to generate smooth labels for model training due to the *false positive* problem in the labeled data. For unlabeled data, we do not consider its original labels to avoid the *confirmation bias* problem. We use the model's predicted labels as pseudo labels and encourage the model to have low entropy on unlabeled data, enhancing confidence in the correct label. We conduct extensive experiments and ablation studies to demonstrate the effectiveness and robustness of our method.

Acknowledgement. This work is supported by the National Natural Science Foundation of China (No. 61906035), the Natural Science Foundation of Shanghai (No. 22ZR1402000) and the Science and Technology Commission of Shanghai Municipality Grant (No. 22511105902).

References

1. Abhishek, A., Anand, A., Awekar, A.: Fine-grained entity type classification by jointly learning representations and label embeddings. In: Proceedings of the 15th Conference of the European Chapter of the Association for Computational Linguistics: Volume 1, Long Papers. pp. 797–807 (2017)
2. Chen, B., Gu, X., Hu, Y., Tang, S., Hu, G., Zhuang, Y., Ren, X.: Improving distantly-supervised entity typing with compact latent space clustering. In: Proceedings of the 2019 Conference of the North American Chapter of the Association for Computational Linguistics: Human Language Technologies, Volume 1 (Long and Short Papers). pp. 2862–2872 (2019)
3. Gao, T., Yao, X., Chen, D.: SimCSE: Simple contrastive learning of sentence embeddings. In: Empirical Methods in Natural Language Processing (EMNLP) (2021)
4. Gillick, D., Lazic, N., Ganchev, K., Kirchner, J., Huynh, D.: Context-dependent fine-grained entity type tagging. arXiv preprint arXiv:1412.1820 (2014)
5. Ling, X., Weld, D.S.: Fine-grained entity recognition. In: Twenty-Sixth AAAI Conference on Artificial Intelligence. pp. 94–100 (2012)
6. Lukasik, M., Bhojanapalli, S., Menon, A., Kumar, S.: Does label smoothing mitigate label noise? In: International Conference on Machine Learning. pp. 6448–6458. PMLR (2020)

 7. Onoe, Y., Boratko, M., McCallum, A., Durrett, G.: Modeling fine-grained entity types with box embeddings. In: Proceedings of the 59th Annual Meeting of the Association for Computational Linguistics and the 11th International Joint Conference on Natural Language Processing (Volume 1: Long Papers). pp. 2051–2064 (2021)
 8. Onoe, Y., Durrett, G.: Learning to denoise distantly-labeled data for entity typing. In: Proceedings of the 2019 Conference of the North American Chapter of the Association for Computational Linguistics: Human Language Technologies, Volume 1 (Long and Short Papers). pp. 2407–2417 (2019)
 9. Ren, X., He, W., Qu, M., Huang, L., Ji, H., Han, J.: Afet: Automatic fine-grained entity typing by hierarchical partial-label embedding. In: Proceedings of the 2016 Conference on Empirical Methods in Natural Language Processing. pp. 1369–1378 (2016)
10. Shi, H., Tang, S., Gu, X., Chen, B., Chen, Z., Shao, J., Ren, X.: Alleviate dataset shift problem in fine-grained entity typing with virtual adversarial training. In: Proceedings of the Twenty-Ninth International Conference on International Joint Conferences on Artificial Intelligence. pp. 3898–3904 (2021)
11. Shimaoka, S., Stenetorp, P., Inui, K., Riedel, S.: An attentive neural architecture for fine-grained entity type classification. In: Proceedings of the 5th Workshop on Automated Knowledge Base Construction. pp. 69–74 (2016)
12. Sohn, K., Berthelot, D., Carlini, N., Zhang, Z., Zhang, H., Raffel, C.A., Cubuk, E.D., Kurakin, A., Li, C.L.: Fixmatch: Simplifying semi-supervised learning with consistency and confidence. Advances in neural information processing systems **33**, 596–608 (2020)
13. Weischedel, R., Brunstein, A.: Bbn pronoun coreference and entity type corpus. Linguistic Data Consortium, Philadelphia 112 (2005)
14. Weischedel, R., Palmer, M., Marcus, M., Hovy, E., Pradhan, S., Ramshaw, L., Xue, N., Taylor, A., Kaufman, J., Franchini, M., et al.: Ontonotes release 5.0 ldc2013t19. Linguistic Data Consortium, Philadelphia, PA 23 (2013)
15. Wu, J., Zhang, R., Mao, Y., Guo, H., Huai, J.: Modeling noisy hierarchical types in fine-grained entity typing: A content-based weighting approach. In: IJCAI. pp. 5264–5270 (2019)
16. Xu, B., Zhang, Z., Sha, C., Du, M., Song, H., Wang, H.: A three-stage curriculum learning framework with hierarchical label smoothing for fine-grained entity typing. In: International Conference on Database Systems for Advanced Applications. pp. 289–296. Springer (2022)
17. Xu, P., Barbosa, D.: Neural fine-grained entity type classification with hierarchy-aware loss. In: Proceedings of the 2018 Conference of the North American Chapter of the Association for Computational Linguistics: Human Language Technologies, Volume 1 (Long Papers). pp. 16–25 (2018)
18. Zhang, C.B., Jiang, P.T., Hou, Q., Wei, Y., Han, Q., Li, Z., Cheng, M.M.: Delving deep into label smoothing. IEEE Transactions on Image Processing **30**, 5984–5996 (2021)
19. Zhang, H., Long, D., Xu, G., Zhu, M., Xie, P., Huang, F., Wang, J.: Learning with noise: Improving distantly-supervised fine-grained entity typing via automatic relabeling. In: IJCAI. pp. 3808–3815 (2020)

Meta-learning Siamese Network
for Few-Shot Text Classification

Chengcheng Han[1], Yuhe Wang[1], Yingnan Fu[1], Xiang Li[1(✉)], Minghui Qiu[2],
Ming Gao[1,3], and Aoying Zhou[1]

[1] School of Data Science and Engineering, East China Normal University, Shanghai,
China
{52215903007,51205903068,52175100004}@stu.ecnu.edu.cn,
{xiangli,mgao,ayzhou}@dase.ecnu.edu.cn
[2] Alibaba Group, Hangzhou, China
minghui.qmh@alibaba-inc.com
[3] KLATASDS-MOE, School of Statistics, East China Normal University, Shanghai,
China

Abstract. Few-shot learning has been used to tackle the problem of
label scarcity in text classification, of which meta-learning based methods
have shown to be effective, such as the prototypical networks (PROTO).
Despite the success of PROTO, there still exist three main problems:
(1) ignore the randomness of the sampled support sets when computing prototype vectors; (2) disregard the importance of labeled samples;
(3) construct meta-tasks in a purely random manner. In this paper, we
propose a Meta-Learning Siamese Network, namely, *Meta-SN*, to address
these issues. Specifically, instead of computing prototype vectors from the
sampled support sets, Meta-SN utilizes external knowledge (e.g. class
names and descriptive texts) for class labels, which is encoded as the
low-dimensional embeddings of prototype vectors. In addition, Meta-SN
presents a novel sampling strategy for constructing meta-tasks, which
gives higher sampling probabilities to hard-to-classify samples. Extensive
experiments are conducted on six benchmark datasets to show the clear
superiority of Meta-SN over other state-of-the-art models. For reproducibility, all the datasets and codes are provided at https://github.
com/hccngu/Meta-SN.

Keywords: text classification · few-shot learning · meta-learning

1 Introduction

Text classification is a pivotal task in natural language processing, which aims to
predict labels or tags for textual units (e.g., sentences, queries, paragraphs and
documents). It has been widely used in various downstream applications, such as
Relation Extraction [40] and Information Retrieval [28]. With the rapid development of deep learning, these approaches generally require massive labeled data
as training set, which is manually expensive to derive. To address the problem,
few-shot learning [39] has been proposed, which aims to train classifiers with
scarce labeled data. Previous studies [5,37,38] have shown that meta-learning

© The Author(s), under exclusive license to Springer Nature Switzerland AG 2023
X. Wang et al. (Eds.): DASFAA 2023, LNCS 13945, pp. 737–752, 2023.
https://doi.org/10.1007/978-3-031-30675-4_54

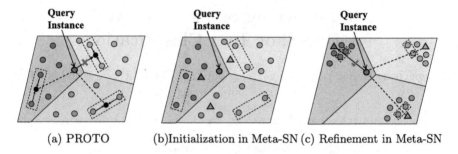

(a) PROTO   (b)Initialization in Meta-SN (c) Refinement in Meta-SN

Fig. 1. A comparison between PROTO and our method (Meta-SN) in a 3-way 2-shot classification task. The dashed box represents the support set randomly sampled from the training data. **Left**: The black dot represents the prototype vector calculated from the support set. Since the given query instance whose true label is purple is closest to the estimated prototype vector of the green class, PROTO misclassifies it to green. This essentially attributes to the randomness of the sampled support sets. **Middle**: The triangles represent the initialized prototype vectors computed from the external descriptive texts of classes, which are independent to the sampled support sets. **Right**: After refinement with the Siamese network, prototype vectors and samples are mapped into a low-dimensional space, where the inter-class distance between different prototype vectors are enlarged and the intra-class distance between samples and the corresponding prototype vector is shortened. (Color figure online)

techniques can be effectively used in few-shot learning. In meta-learning, the goal is to train a model based on different meta-tasks constructed from the training set and generalize the model to classify samples in unseen classes from the test set. Each meta-task contains a *support set* and a *query set*. Specifically, the support set is similar to the training set in traditional supervised learning but it only contains a few samples (instances)[1]; the query set acts as the test set but it can be used to compute gradients for updating model parameters in the training stage. As a representative meta-learning method, the prototypical network (PROTO) [35] first generates a *prototype vector* for each class by averaging the embeddings of samples in the support set of the class. Then it computes the distance between a query instance in the query set and these prototype vectors. Finally, it predicts the query instance to the class with the smallest distance.

Despite the success, there are three main problems in PROTO. First, the true prototype vector of each class should be intuitively fixed. However, the computation of prototype vectors could be adversely affected by the randomness of the sampled support sets, which could lead to the incorrect prediction of queries' labels (see Fig. 1a). Second, when calculating prototype vectors, all the samples in the support set are given the same weight. This fails to distinguish the importance of samples when predicting query instances' labels. Third, meta-tasks are randomly constructed from the training data. This could lead to simple

---

[1] We interchangeably use sample and instance in this paper.

meta-tasks composed of samples that are easy to be classified, which are thus insufficient to generalize the model.

In this paper, to address the issues, we propose a **Meta**-Learning **S**iamese Network, namely, **Meta-SN**. Instead of estimating prototype vectors from the sampled support sets, we compute these vectors by utilizing external knowledge on the class labels (see Fig. 1b), which includes class names and related descriptive texts (e.g., *Wiki titles* and *Wiki texts*) as shown in Table 1. This eliminates the dependence of prototype vector estimation on the sampled support sets and also the adverse impact of randomness. After that, we further refine these prototype vectors with a Siamese Network. In particular, we map both samples and prototype vectors into a low-dimensional space, where the interclass distance between different prototype vectors is enlarged and the intra-class distance between samples and their corresponding prototype vectors is shortened (see Fig. 1c). Further, we learn the importance of a sample in the support set based on its average distance to the query set. The closer a sample is to the query set, the more important the sample is for label prediction, and the larger the weight should be assigned. Finally, we adopt a novel sampling strategy to construct meta-tasks which assigns higher sampling probability to the hard-to-classify samples. On the one hand, the closer the distance between different prototype vectors, the more difficult the corresponding classes can be separated. On the other hand, the more distant a sample in the support set is to the prototype vector, the more difficult the classification task will be. Therefore, we give higher sampling probabilities to hard-to-classify tasks to help generalize our model. The main contributions of the paper are summarized as follows:

- We propose a novel **Meta**-learning **S**iamese Network (**Meta-SN**) for few-shot text classification. Instead of estimating prototype vectors from the sampled support sets, Meta-SN constructs the prototype vectors with the external descriptive information of class labels and further refines these vectors with a Siamese network. This alleviates the adverse impact of sampling randomness.
- We present an effective sampling strategy to construct meta-tasks, which assigns higher sampling probability to the hard-to-classify samples. This boosts the model's generalization ability. We further learn the importance of a labeled sample by considering its average distance to the query set.
- We evaluate the performance of our model on six benchmark datasets, including five text classification datasets and one relation classification dataset. Experimental results demonstrate that Meta-SN can achieve significant performance gains over other state-of-the-art methods.

Table 1. An example for 3-way 2-shot text classification on the Huffpost dataset, where only two support instances are given in each of the three classes. The ground-truth label of the query instance is Class B. External knowledge on class labels includes class names and related descriptive texts from Wikipedia, which are used to generate prototype vectors in Meta-SN.

Support set	
(A) *Politics*	(1) Trump's Crackdown On Immigrant Parents Puts More Kids In An Already Strained System. (2) Ireland Votes To Repeal Abortion Amendment In Landslide Referendum.
(B) *Entertainment*	(1) Hugh Grant Marries For The First Time At Age 57. (2) Mike Myers Reveals He'd 'Like To' Do A Fourth Austin Powers Film.
(C) *Sports*	(1) U.S. Olympic Committee Ignored Sexual Abuse Complaints Against Taekwondo Stars: Lawsuit. (2) MLB Pitcher Punches Himself In Face Really Hard After Blowing Game.
Query instance	
Which class?	'Crazy Rich Asians' Trailer Is Already A Magnificent Masterpiece
External knowledge (class name and related descriptive texts)	
(A)	**Politics** is the set of activities that are associated with making decisions in groups
(B)	**Entertainment** is a form of activity that holds the attention and interest of an audience
(C)	**Sports** pertain to any form of competitive physical activity or game

2 Related Work

The mainstream approaches for few-shot text classification are based on meta-learning. In this section, we first introduce the background of meta-learning and then review how to apply meta-learning in few-shot text classification.

2.1 Meta-learning

Meta-learning, also known as "learning to learn", refers to improving the learning ability of a model through multiple meta-tasks so that it can easily adapt to new tasks. Existing approaches can be grouped into three main categories:

Metric-based Methods. These kinds of methods aim to learn an appropriate distance metric to measure the distance between query samples and training samples. The label of a query sample is then predicted as that of the training sample with the smallest distance. The representative methods include Siamese Network [19], Matching Network [38], PROTO [35] and Relation Network [37]. Among these models, PROTO is simple-to-implement, fast-to-train and can achieve state-of-the-art results on several FSL tasks. Based on PROTO, our proposed method is also a metric-based method.

Optimization-based Methods. These kinds of methods learn how to optimize. Instead of simply using a traditional optimizer, such as stochastic gradient descent (SGD), they train a meta-learner as an optimizer or adjust the optimization process. A representative method is MAML [5], which emulates the quick adaptation to unseen classes during the optimization process. Other optimization-based models include Reptile [30], iMAML [31] and MetaOpt-Net [21].

Model-based Methods. Model-based methods learn a hidden feature space and predict the label of a query instance in an end-to-end manner, which lacks interpretability. Compared with optimization-based methods, model-based methods could be easier to optimize but less generalizable to out-of-distribution tasks [15]. The representative model-based methods include MANNs [33], Meta networks [29], SNAIL [26] and CPN [7].

2.2 Few-Shot Text Classification

Few-shot text classification has received great attention recently [9,34]. In particular, meta-learning has been applied to solve the problem [1,8]. For example, DS-FSL [1] adds distributional signatures (e.g. word frequency and information entropy) to the model within a meta-learning framework. MEDA [36] jointly optimizes the ball generator and the meta-learner, such that the ball generator can learn to produce augmented samples that best fit the meta-learner. MLADA [10] integrates an adversarial domain adaptation network with a meta-learning framework to improve the model's adaptive ability for new tasks and achieves the superior performance. There also exist methods that further extend PROTO to the problem. For example, HATT-Proto [6] learns weights of instances and features by introducing the instance-level and feature-level attention mechanism, respectively. LM-ProtoNet [4] adds a triplet loss to PROTO to improve the model generalization ability. IncreProtoNet [32] combines the deep neural network with PROTO to better utilize the training data. LaSAML [24] improves performance of PROTO by incorporating label information into feature extractors. LEA [14] derives meta-level attention aspects using a new meta-learning framework. ContrastNet [2] introduces a contrastive learning framework to learn discriminative representations for texts from different classes. Despite the success, all these methods disregard the randomness of the sampled support sets when computing prototype vectors and employ a completely random construction of meta-tasks. This adversely affects their wide applicability in various real-world tasks.

3   Background

In this section, we give a formal problem definition and summarize the standard meta-learning framework for few-shot classification [38].

Problem Definition. Given a set of labeled samples from a set of classes $\mathcal{Y}_{train}$, our goal is to develop a model that learns from these training data, so that we can make predictions over new (but related) classes, for which we only have a few annotations. These new classes are denoted as $\mathcal{Y}_{test}$, which satisfies $\mathcal{Y}_{train} \cap \mathcal{Y}_{test} = \emptyset$.

Meta-training. In meta-learning, we emulate the real testing scenario with meta-tasks during meta-training, so our model can learn to quickly adapt to new classes. To create a training meta-task, we first sample N classes from $\mathcal{Y}_{train}$.

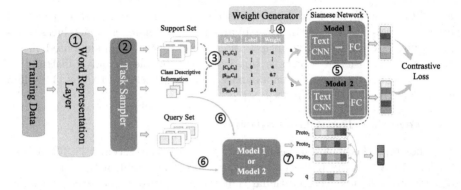

Fig. 2. The overall architecture of Meta-SN for a 3-way 2-shot problem. For details of each step, see Sect. 4.

After that, for each of these N classes, we sample K instances as the support set $\mathcal{S}$ and L instances as the query set $\mathcal{Q}$. The support set is used as training data while the query set is considered as testing data. Our model is updated based on the loss over these testing data. Given the support set, we refer to the task of making predictions over the query set as *N-way K-shot classification*.

Meta-testing. In the testing stage, we also use meta-tasks to test whether our model can adapt quickly to new classes. To create a testing meta-task, we first sample N new classes from $\mathcal{Y}_{test}$. Similar as in meta-training, we then sample the support set and the query set from the N classes, respectively. Finally, we evaluate the average performance on the query set across all testing meta-tasks.

4   Algorithm

In this section, we describe our Meta-SN algorithm. We first give an overview of Meta-SN, which is illustrated in Fig. 2. Meta-SN first represents each word with a d-dimensional embedding vector (Step ①), based on which the embeddings of samples and prototype vectors are initialized. Then it constructs meta-tasks by giving higher sampling probability to tasks that are hard to classify (Step ②). After that, it generates sample pairs based on the support sets and prototype vectors (Step ③), and also assigns weights to these sample pairs (Step ④). Meta-SN further employs a Siamese Network to map samples and prototype vectors into a space that is much easier to be classified (Step ⑤). Finally, Meta-SN feeds a query and all the prototype vectors into the Siamese Network (Step ⑥) to derive their embeddings, based on which the probability of the query in each class is calculated (Step ⑦). The overall procedure of Meta-SN is summarized in Algorithm 2. Next, we describe each component in detail.

4.1   Word Representation Layer

In this layer, we use pre-training models, such as fastText [16] and BERT [3], to represent each word with a d-dimensional embedding vector. Then for each class c_j, we can construct its initial embedding vector $f_0(c_j)$ by averaging the embeddings of words contained in its descriptive texts. Similarly, we can also derive the initial embedding vector $f_0(s_i)$ for a sample s_i.

4.2   Task Sampler

This module is used to construct meta-tasks from training data. Different from previous works that construct meta-tasks in a completely random manner, we assign higher sampling probability to tasks that are hard to classify. Intuitively, the closer the distance between prototype vectors, the more difficult the corresponding classes can be separated; the more distant a sample in the support set is to the prototype vector, the more difficult the task will be. Therefore, our goal is to construct more meta-tasks, where prototype vectors are close to each other and samples in a class are far away from the corresponding prototype vector.

To achieve the goal, we first define a probability score $p_{i,j}^c$ to capture the correlation between classes c_i and c_j:

$$p_{i,j}^c = \frac{e^{-dis(f_0(c_i),f_0(c_j))}}{\sum_{k=1}^{|\mathcal{C}|} e^{-dis(f_0(c_i),f_0(c_k))}}, \tag{1}$$

where $|\mathcal{C}|$ is the number of classes in the training set and $f_0(\cdot)$ represents the initial embedding vector.

After that, given a class c_i, we define a probability score p_{ij}^s to sample the j-th instance s_j in the class:

$$p_{i,j}^s = \frac{e^{dis(f_0(c_i),f_0(s_j))}}{\sum_{k=1}^{m_i} e^{dis(f_0(c_i),f_0(s_k))}}, \tag{2}$$

where m_i is the number of instances in class c_i.

Based on Eqs. 1 and 2, we utilize a greedy algorithm to construct meta-tasks. We first randomly sample a class c_1 in $\mathcal{Y}_{train}$. Then we sample the second class c_2 with the probability scores $\{p_{1,j}^c\}_{j=2}^{|\mathcal{C}|}$. From Eq. 1, a class that is closer to c_1 has a higher probability to be sampled. Next, we sample the third class c_3 based on the mean probability distribution $\{\frac{p_{1,j}^c + p_{2,j}^c}{2}\}_{j=3}^{|\mathcal{C}|}$, which indicates that a class with a small distance to both c_1 and c_2 will be more likely to be sampled. We repeat the process to sample N classes in total. After N classes are derived, for each class c_i, we constitute the support set $\mathcal{S}_i$ and the query set $\mathcal{Q}_i$ by sampling K and L instances via the probability distribution $\{p_{i,j}^s\}_{j=1}^{m_i}$, respectively. From Eq. 2, a sample that is more distant from c_i has a higher sampling probability. The pseudocode of meta-task sampling is summarized in Algorithm 1.

Algorithm 1. Task_Sampler

Input: Training data $\{\mathcal{X}_{train}, \mathcal{Y}_{train}\}$; N classes in a meta-task; K samples in each class in the support set and L samples in each class in the query set.

Output: a meta-task including a support set $\mathcal{S}$, a query set $\mathcal{Q}$ and a set $\mathcal{C}_{target}$ of their label information.

1: $\mathcal{S}, \mathcal{Q}, \mathcal{C}_{target} \leftarrow \emptyset, \emptyset, \emptyset$;

2: $c_1 = \text{Sample}(Random, \mathcal{Y}_{train}, 1)$;   ▷ $\text{Sample}(P, \mathcal{Y}, N)$ denotes selecting N elements from $\mathcal{Y}$ with the probability P.

3: $\mathcal{C}_{target} \leftarrow \mathcal{C}_{target} \cup c_1$;

4: **for** $i \in [2, N]$ **do**

5:    Calculate $P = \{\frac{\sum_{k=1}^{i-1} p_{k,j}^c}{i-1}\}_{j=i}^{|\mathcal{C}|}$ by Equation 1.

6:    $c_i = \text{Sample}(P, \mathcal{Y}_{train} \backslash \mathcal{C}_{target}, 1)$;

7:    $\mathcal{C}_{target} \leftarrow \mathcal{C}_{target} \cup c_i$;

8: **end for**

9: **for** $c_i \in \mathcal{C}_{target}$ **do**

10:    Calculate $P = \{p_{i,j}^s\}_{j=1}^{m_i}$ by Equation 2.

11:    $\mathcal{S} \leftarrow \mathcal{S} \cup \text{Sample}(P, \mathcal{X}_{train}^{c_i}, K)$;   ▷ $\mathcal{X}_{train}^{c_i}$ denotes samples with label c_i in $\mathcal{X}_{train}$.

12:    $\mathcal{Q} \leftarrow \mathcal{Q} \cup \text{Sample}(P, \mathcal{X}_{train}^{c_i} \backslash \mathcal{S}, L)$;

13: **end for**

14: **return** $\mathcal{S}, \mathcal{Q}, \mathcal{C}_{target}$

4.3   Constructing Sample Pairs

After meta-tasks are sampled, we next generate sample pairs. For each meta-task, we first construct sample pairs from the prototype vector set Φ. Specifically, we pair prototype vectors ϕ_i with ϕ_j, and denote the pair as $\langle \phi_i, \phi_j \rangle$. Further, we pair each sample in the support set $\mathcal{S}$ with all the prototype vectors in Φ. Specifically, we pair sample s_i with prototype vector ϕ_j, and denote the pair as $\langle s_i, \phi_j \rangle$. If the two items in a pair are in the same class, we denote the label of the pair as 1; otherwise, 0. For each pair, we generate its weight by the *weight generator* (see Sect. 4.4).

4.4   Weight Generator

This module is used to learn weights for sample pairs. For a sample pair $\langle s_i, \phi_j \rangle$, we define the weight of the sample pair to be inversely proportional to the average distance between s_i and the query set $\mathcal{Q} = \{q_l\}_{l=1}^L$:

$$w_{\langle s_i, \phi_j \rangle} = \text{softmax}\left[-\frac{1}{L} \sum_{l=1}^{L} dis(f_\theta(s_i), f_\theta(q_l)) \right], \quad (3)$$

Algorithm 2. Meta-SN Training procedure

Input: Training data $\{\mathcal{X}_{train}, \mathcal{Y}_{train}\}$; T meta-tasks and ep epochs; N classes in the support set or the query set; K samples in each class in the support set and L samples in each class in the query set; the model parameter θ.

Output: The model parameter θ after training.

1: Randomly initialize the model parameters θ;
2: **for** each $i \in [1, ep]$ **do**
3:      **for** each $j \in [1, T]$ **do**
4:          $\mathcal{S}, \mathcal{Q}, \mathcal{C}_{target} \leftarrow$ `Task_Sampler`$(\mathcal{X}_{train}, \mathcal{Y}_{train}, N, K, L)$;
5:          Construct sample pairs by $\mathcal{S}, \mathcal{C}_{target}$;
6:          Calculate w by Equation 3;
7:          Input sample pairs to the Siamese Network;
8:          Calculate $\mathcal{L}_c$ by Equation 4;
9:          Update θ to θ' by Equation 5;
10:         Input $\mathcal{Q}, \mathcal{C}_{target}$ to the model with parameter θ';
11:         Calculate $\mathcal{L}_{ce}$ by Equation 6;
12:      **end for**
13:      Update θ by Equation 7;
14: **end for**
15: **return** θ

---

where $f_\theta(\cdot)$ is the embedding vector of a sample generated from the Siamese network (see Sect. 4.5). We use the `Softmax` function to normalize the weight over all the samples in the support set. Further, for a sample pair $\langle \phi_i, \phi_j \rangle$, we manually set its weight to a hyper-parameter α, which can be used to control the distance between two different class prototype vectors.

4.5 Siamese Network

Siamese network contains two identical sub-networks that have the same network architecture with shared parameters to be learned. Each sub-network consists of a TextCNN [17] and a fully connected (FC) layer. In practice, sample pairs are taken as the input of the Siamese network, where each sample is fed into a sub-network.

To optimize the Siamese network, given a set of sample pairs $\{\langle x_{il}, x_{ir} \rangle\}_{i=1}^n$ with a label set $\{y_i\}_{i=1}^n$, we utilize the contrastive loss function defined in Eq. 4, which aims to enlarge the distance between two samples in zero-labeled pairs and shortens that between two samples in one-labeled pairs.

$$\mathcal{L}_c(\theta) = \sum_{i=1}^n w_{\langle x_{il}, x_{ir} \rangle} [y_i dis(f_\theta(x_{il}), f_\theta(x_{ir}))$$
$$+ (1 - y_i) max(0, \delta - dis(f_\theta(x_{il}), f_\theta(x_{ir})))], \tag{4}$$

Table 2. Statistics of datasets.

Dataset	# tokens/example	# samples	# train cls	# val cls	# test cls
HuffPost	11	36900	20	5	16
Amazon	140	24000	10	5	9
Reuters	168	620	15	5	11
20 News	340	18820	8	5	7
RCV1	372	1420	37	10	24
FewRel	24	56000	65	5	10

Here, θ denotes the trainable parameters of the Siamese network. We also introduce a margin δ. For zero-labeled pairs, they can only contribute to the loss function if their distance is smaller than the margin. We update model parameters θ with SGD:

$$\theta' = \theta - \alpha\nabla_\theta\mathcal{L}_c(\theta), \tag{5}$$

where α is the learning rate. With one-step update, θ becomes θ'. Based on θ', the Siamese network can map query instances and prototype vectors into low-dimensional embedding vectors. After that, we calculate the probability logits of a query q_i in class c_j and feed the results into a `cross-entropy` function:

$$\mathcal{L}_{ce}(\theta') = \sum_{i=1}^{L} -log(\frac{e^{-dis(f_{\theta'}(q_i),f_{\theta'}(c_j))}}{\sum_{k=1}^{N} e^{-dis(f_{\theta'}(q_i),f_{\theta'}(c_k))}}). \tag{6}$$

Following the optimization strategy in MAML [5], we update θ by:

$$\theta \leftarrow \theta - \beta\nabla_\theta\frac{1}{T}\sum_{t=1}^{T}\mathcal{L}_{ce}(\theta'_t), \tag{7}$$

where β is the meta learning rate and T represents the number of meta-tasks in each epoch. This boosts the generalization ability of the model to unseen classes with only one-step update. The overall procedure of Meta-SN is summarized in Algorithm 2.

5   Experiments

In this section, we comprehensively evaluate the performance of Meta-SN. In particular, we compare the classification accuracy of Meta-SN with seven other methods on six benchmark datasets to show the effectiveness of our model.

5.1   Datasets

We use six benchmark datasets: HuffPost [27], Amazon [13], Reuters-21578 [22], 20 Newsgroups [20], RCV1 [23] and FewRel [11]. In particular, the first five are for text classification while the last one is for few-shot relation classification. Statistics of these datasets are summarized in Table 2. All processed datasets and their splits are publicly available.

5.2   Experiment Setup

Baselines. We compare Meta-SN with seven state-of-the-art methods, which can be grouped into three categories: (1) *metric-based methods*: PROTO [35], HATT-Proto [6] and ContrastNet [2]; (2) *optimization-based methods*: MAML [5]; and (3) *model-based methods*: Induction Networks [8], DS-FSL [1] and MLADA [10]. Specifically, HATT-Proto extends PROTO by adding instance-level and feature-level attention to the prototypical network. Contrast-Net introduces a contrastive learning framework to learn discriminative representations for texts from different classes. Induction Networks learns a class-wise representation by leveraging the dynamic routing algorithm in the meta-learning training procedure. DS-FSL is a model that utilizes distributional signatures of words to extract meta-knowledge. MLADA integrates an adversarial domain adaptation network with a meta-learning framework to improve the model's adaptability to new tasks.

Implementation Details. We implemented Meta-SN by PyTorch. The model is initialized by He initialization [12] and trained by Adam [18]. We run the model with the learning rates 0.2 for contrastive loss and 0.00002 for cross-entropy loss on all the datasets. We apply early stopping when the validation loss fails to improve for 20 epochs. Since ContrastNet adopts BERT as the pre-training model for word embeddings while most other competitors like HATT-Proto, DS-FSL and MLADA use fastText in their original papers, we implemented both fastText-based and BERT-based[2] Meta-SN for fair comparison. In Siamese network, we follow [17] to use the 1-dimensional filter of sizes [1, 3, 5], each with 16 feature maps in CNN. We set the dimensionality of the fully connected layer to 64 and the number of meta-tasks T in each epoch to 3. We also fine-tune α (weight of sample pairs composed of two prototype vectors) by grid search over $\{1, 3, 5, 7, 9\}$ and set it to 5 on all the datasets. For Induction Network and DS-FSL, we report their results from [1]. For other competitors, part of their results are derived from the original papers; for the datasets where results are absent, we use the original codes released by their authors and fine-tune the parameters of the models. We run all the experiments on a single NVIDIA v100 GPU. In our experiments, we set K to 1 in 1-shot task, 5 in 5-shot task and L to 25. We evaluate the model performance based on 1,000 meta-tasks in meta-testing and report the average accuracy over 5 runs.

5.3   Classification Results

We report the results of 5-way 1-shot classification and 5-way 5-shot classification in Table 3. From the table, Meta-SN achieves the best results across all the datasets. For example, in the fastText-based comparison, Meta-SN achieves an average accuracy of 69.1% in 1-shot classification and 85.2% in 5-shot classification, respectively. In particular, it outperforms the runner-up model MLADA by

---

[2] We use the pretrained `bert-base-uncased` model for all datasets.

Table 3. Mean accuracy (%) of 5-way 1-shot classification and 5-way 5-shot classification over all the datasets. We highlight the best results in bold.

	Methods	HuffPost		Amazon		Reuters		20News		RCV1		FewRel		Average	
		1 shot	5 shot	1 shot	5 shot	1 shot	5 shot	1 shot	5 shot	1 shot	5 shot	1 shot	5 shot	1 shot	5 shot
fastText-	MAML [5]	35.9	49.3	39.6	47.1	54.6	62.9	33.8	43.7	39.0	51.1	51.7	66.9	42.4	53.5
	PROTO [35]	35.7	41.3	37.6	52.1	59.6	66.9	37.8	45.3	32.1	35.6	49.7	65.1	42.1	51.1
	Induct [8]	38.7	49.1	34.9	41.3	59.4	67.9	28.7	33.3	33.4	38.3	50.4	56.1	40.9	47.6
	Hatt-Proto [6]	41.1	56.3	59.1	76.0	73.2	86.2	44.2	55.0	43.2	64.3	77.6	90.1	56.4	71.3
	DS-FSL [1]	43.0	63.5	62.6	81.1	81.8	96.0	52.1	68.3	54.1	75.3	67.1	83.5	60.1	78.0
	MLADA [10]	45.0	64.9	68.4	86.0	82.3	96.7	59.6	77.8	55.3	80.7	81.1	90.8	65.3	82.8
	Meta-SN	**54.7**	**68.5**	**70.2**	**87.7**	**84.0**	**97.1**	**60.7**	**78.9**	60.0	86.1	**84.8**	**93.1**	69.1	85.2
BERT-	ContrastNet [2]	52.7	64.4	75.4	85.2	86.2	95.3	71.0	81.3	65.7	87.4	85.3	92.7	72.7	84.3
	Meta-SN	**63.1**	**71.3**	**77.5**	**89.1**	**87.9**	**96.7**	**72.1**	**83.2**	**67.3**	**88.9**	**86.8**	**94.6**	**73.6**	**87.3**

Table 4. Ablation study: mean accuracy (%) of 5-way 1-shot classification and 5-way 5-shot classification over all the datasets. We highlight the best results in bold. All the results are based on fastText.

Models	HuffPost		Amazon		Reuters		20News		RCV1		FewRel		Average	
	1 shot	5 shot	1 shot	5 shot	1 shot	5 shot	1 shot	5 shot	1 shot	5 shot	1 shot	5 shot	1 shot	5 shot
Meta-SN-rpv	46.0	61.5	62.9	80.1	75.0	89.9	52.2	70.0	52.3	79.1	76.1	85.9	60.8	77.9
Meta-SN-ew	51.4	64.9	68.4	83.3	81.1	93.4	57.9	74.4	57.7	82.5	81.7	89.3	66.4	81.3
Meta-SN-rts	52.1	66.1	69.3	84.5	81.6	95.1	60.0	76.8	58.7	84.2	79.7	88.8	66.9	82.6
Meta-SN-ln	53.8	68.0	69.6	87.1	83.3	96.0	59.8	78.0	59.5	85.4	84.3	92.6	68.4	84.5
Meta-SN	**54.7**	**68.5**	**70.2**	**87.7**	**84.0**	**97.1**	**60.7**	**78.9**	**60.0**	**86.1**	**84.8**	**93.1**	**69.1**	**85.2**

a notable 3.8% and 2.4% improvement in both cases. When compared against the PROTO model, Meta-SN leads by 27.0% and 34.1% on average in 1-shot and 5-shot classification, respectively. These results clearly demonstrate that our model is very effective in improving PROTO. While Hatt-Proto upgrades PROTO by learning the importance of labeled samples, it disregards the randomness of the sampled support sets when computing prototype vectors and constructs meta-tasks randomly, which degrades its performance. Further, in the BERT-based comparison, Meta-SN also outperforms ContrastNet over all the datasets. All these results show that Meta-SN, which generates prototype vectors from external knowledge, learns sample weights and constructs hard-to-classify meta-tasks, can perform reasonably well.

5.4   Ablation Study

We conduct an ablation study to understand the characteristics of the main components of Meta-SN. One variant ignores the randomness of the sampled support sets and directly uses the mean embedding vectors of samples in the support sets as the prototype vectors. We call this variant **Meta-SN-rpv** (random prototype vectors). To show the importance of weight learning for samples, we set equal weights for all sample pairs and call this variant **Meta-SN-ew** (equal weights). Another variant removes the task sampler and constructs meta-tasks in a completely random way. We call this variant **Meta-SN-rts** (random task sampler). Moreover, to study how external knowledge affects the classification results, we

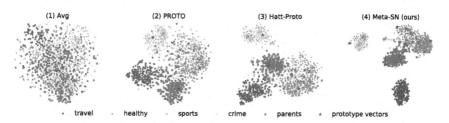

Fig. 3. The t-SNE visualization comparison of sentence embeddings in meta-testing on the HuffPost dataset. Note that these classes are unseen in meta-training. All the results are based on fastText.

only extract knowledge from label names to construct prototype vectors and call this variant **Meta-SN-ln** (label names). This helps to study the model robustness towards the richness of external descriptive information. The results of the ablation study are reported in Table 4. From the table, we observe: (1) Meta-SN significantly outperforms Meta-SN-rpv in all the comparisons across datasets. This shows the importance of eliminating the adverse impact induced by the randomness of sampled support sets when calculating prototype vectors. (2) Meta-SN also beats Meta-SN-ew clearly. For example, in 5 shot classification task on Amazon, the accuracy of Meta-SN is 87.7% while that of Meta-SN-ew is only 83.3%. This shows the importance of weight learning for sample pairs. (3) Meta-SN leads Meta-SN-rts in all the 1-shot classification and 5-shot classification tasks. This is because Meta-SN-rts constructs meta-tasks randomly while Meta-SN focuses more on the hard-to-classify meta-tasks to boost the model training. (4) The average performance gaps between Meta-SN-ln and Meta-SN on both 1-shot and 5-shot classification tasks are 0.7%. This shows that the inclusion of external descriptive texts for class labels can bring only marginal improvement on the classification results. Although we use only class names to derive the initial embeddings of class prototype vectors, Meta-SN can progressively refine these embeddings and thus lead to superior performance.

5.5 Visualization

We next evaluate the quality of generated embeddings of Meta-SN. Specifically, Fig. 3 uses t-SNE [25] to visualize the sentence embeddings of the query set generated from different methods on the HuffPost dataset. For other datasets, we observe similar results that are omitted due to the space limitation.

Figure 3(1) shows the results of *AVG*, which generates the sentence embedding by directly averaging the embeddings of the words contained in the sentence. From the figure, embeddings of sentences with different class labels are entangled with each other. While PROTO (Fig. 3(2)) and Hat-Proto (Fig. 3(3)) can produce higher quality of sentence embeddings, it still fails to distinguish some classes. Further, our method Meta-SN can generate embeddings that are clearly separated on all the datasets, as shown in Fig. 3(4). These results show the superiority of Meta-SN in generating high-quality embeddings.

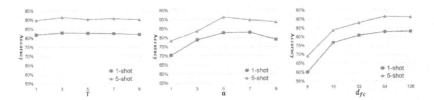

Fig. 4. Hyperparameter sensitivity study on the HuffPost dataset. Here, T is the number of meta-tasks in each training epoch, α is the weight controlling the distance between prototype vectors, and d_{fc} is the output embedding dimensionality of query samples. All the results are based on fastText.

5.6 Hyper-parameter Sensitivity Analysis

We end this section with a sensitivity analysis on the hyper-parameters. In particular, we study three main hyper-parameters: the number of meta-tasks T used in each training epoch, the weight α controlling the distance between prototype vectors, and the output embedding dimensionality d_{fc} of query samples. In our experiments, we vary one hyper-parameter with others fixed. The results on the HuffPost dataset are shown in Fig. 4. For other datasets, we observe similar results, which are omitted due to the limited space. From the figure, we see:

(1) Meta-SN gives very stable performance over a wide range of T values. This shows that Meta-SN is insensitive to the number of meta-tasks in meta-training.

(2) With the increase of α, the distance between prototype vectors of different classes is zoomed in and the performance of Meta-SN generally becomes better. We also notice a mild dip when α is set too large. This is because a large value of α enforces the model to focus more on enlarging the distance between different class prototype vectors, but disregards the importance of shortening the distance between a sample and its corresponding class prototype vector.

(3) As the output embedding dimensionality d_{fc} increases, Meta-SN achieves better performance. This is because when the dimensionality is small, the embedding vectors cannot capture enough information for classification.

6   Conclusion

In this paper, we studied the few-shot text classification problem and proposed a meta-learning Siamese network Meta-SN. Based on PROTO, Meta-SN maps samples and prototype vectors of different classes into a low-dimensional space, where the inter-class distance between different prototype vectors is enlarged and the intra-class distance between samples and their corresponding prototype vectors is shortened. We generated prototype vectors based on the external descriptive texts of class labels instead of from the sampled support sets.

We learned the importance of samples in the support set based on their distances to the query set. We also put forward a novel meta-task construction method, which samples more hard-to-classify meta-tasks to boost training. We conducted extensive experiments to show that Meta-SN can significantly outperform other competitors on six benchmark datasets w.r.t. both text classification and relation classification tasks. Future work includes applying Meta-SN to other fields, such as computer vision, and exploring other representative meta-learning methods to improve their performance in few-shot text classification.

Acknowledgments. This work has been supported by the National Natural Science Foundation of China under Grant No. U1911203, 61977025 and 62202172.

References

1. Bao, Y., Wu, M., Chang, S., Barzilay, R.: Few-shot text classification with distributional signatures. In: ICLR (2020)
2. Chen, J., Zhang, R., Mao, Y., Xue, J.: ContrastNet: a contrastive learning framework for few-shot text classification. In: AAAI (2022)
3. Devlin, J., Chang, M.W., Lee, K., Toutanova, K.: Bert: Pre-training of deep bidirectional transformers for language understanding. arXiv:1810.04805 (2018)
4. Fan, M., Bai, Y., Sun, M., Li, P.: Large margin prototypical network for few-shot relation classification with fine-grained features. In: CIKM, pp. 2353–2356 (2019)
5. Finn, C., Abbeel, P., Levine, S.: Model-agnostic meta-learning for fast adaptation of deep networks. In: ICML, pp. 1126–1135 (2017)
6. Gao, T., Han, X., Liu, Z., Sun, M.: Hybrid attention-based prototypical networks for noisy few-shot relation classification. In: AAAI, pp. 6407–6414 (2019)
7. Garnelo, M., et al.: Conditional neural processes. In: ICML, pp. 1704–1713 (2018)
8. Geng, R., Li, B., Li, Y., Zhu, X., Jian, P., Sun, J.: Induction networks for few-shot text classification. In: EMNLP-IJCNLP, pp. 3902–3911 (2019)
9. Geng, X., Chen, X., Zhu, K.Q.: MICK: a meta-learning framework for few-shot relation classification with small training data. In: CIKM, pp. 415–424 (2020)
10. Han, C., Fan, Z., Zhang, D., Qiu, M., Gao, M., Zhou, A.: Meta-learning adversarial domain adaptation network for few-shot text classification. In: Findings of ACL, pp. 1664–1673 (2021)
11. Han, X., et al.: FewRel: a large-scale supervised few-shot relation classification dataset with state-of-the-art evaluation. In: EMNLP, pp. 4803–4809 (2018)
12. He, K., Zhang, X., Ren, S., Sun, J.: Delving deep into rectifiers: surpassing human-level performance on ImageNet classification. In: ICCV, pp. 1026–1034 (2015)
13. He, R., McAuley, J.J.: Ups and downs: modeling the visual evolution of fashion trends with one-class collaborative filtering. In: WWW, pp. 507–517 (2016)
14. Hong, S., Jang, T.Y.: LEA: meta knowledge-driven self-attentive document embedding for few-shot text classification. In: NAACL, pp. 99–106 (2022)
15. Hospedales, T.M., Antoniou, A., Micaelli, P., Storkey, A.J.: Meta-learning in neural networks: a survey. CoRR abs/2004.05439 (2020)
16. Joulin, A., Grave, E., Bojanowski, P., Douze, M., Jégou, H., Mikolov, T.: FastText.zip: compressing text classification models. CoRR abs/1612.03651 (2016)
17. Kim, Y.: Convolutional neural networks for sentence classification. In: EMNLP, pp. 1746–1751 (2014)

18. Kingma, D.P., Ba, J.: Adam: a method for stochastic optimization. In: Bengio, Y., LeCun, Y. (eds.) ICLR (2015)

19. Koch, G., Zemel, R., Salakhutdinov, R.: Siamese neural networks for one-shot image recognition. In: ICML Deep Learning Workshop (2015)

20. Lang, K.: NewsWeeder: learning to filter Netnews. In: ICML, pp. 331–339 (1995)

21. Lee, K., Maji, S., Ravichandran, A., Soatto, S.: Meta-learning with differentiable convex optimization. In: CVPR, pp. 10657–10665 (2019)

22. Lewis, D.: Reuters-21578 text categorization test collection, distribution 1.0. http://www.research/.att.com (1997)

23. Lewis, D.D., Yang, Y., Rose, T.G., Li, F.: RCV1: a new benchmark collection for text categorization research. JMLR **5**, 361–397 (2004)

24. Luo, Q., Liu, L., Lin, Y., Zhang, W.: Don't miss the labels: label-semantic augmented meta-learner for few-shot text classification. In: ACL, pp. 2773–2782 (2021)

25. Van der Maaten, L., Hinton, G.: Visualizing data using t-SNE. JMLR **9**(11), 2579–2605 (2008)

26. Mishra, N., Rohaninejad, M., Chen, X., Abbeel, P.: A simple neural attentive meta-learner. arXiv:1707.03141 (2017)

27. Misra, R.: News category dataset (2018). https://doi.org/10.13140/RG.2.2.20331.18729

28. Mitra, B.: Neural models for information retrieval. arXiv:1705.01509 (2017)

29. Munkhdalai, T., Yu, H.: Meta networks. In: ICML, pp. 2554–2563 (2017)

30. Nichol, A., Achiam, J., Schulman, J.: On first-order meta-learning algorithms. arXiv:1803.02999 (2018)

31. Rajeswaran, A., Finn, C., Kakade, S., Levine, S.: Meta-learning with implicit gradients. arXiv:1909.04630 (2019)

32. Ren, H., Cai, Y., Chen, X., Wang, G., Li, Q.: A two-phase prototypical network model for incremental few-shot relation classification. In: COLING, p. 1618 (2020)

33. Santoro, A., Bartunov, S., Botvinick, M., Wierstra, D., Lillicrap, T.: Meta-learning with memory-augmented neural networks. In: ICML, pp. 1842–1850 (2016)

34. Schick, T., Schütze, H.: Exploiting cloze-questions for few-shot text classification and natural language inference. In: EACL, pp. 255–269 (2021)

35. Snell, J., Swersky, K., Zemel, R.S.: Prototypical networks for few-shot learning. arXiv:1703.05175 (2017)

36. Sun, P., Ouyang, Y., Zhang, W., Dai, X.: MEDA: meta-learning with data augmentation for few-shot text classification. In: IJCAI, pp. 3929–3935 (2021)

37. Sung, F., Yang, Y., Zhang, L., Xiang, T., Torr, P.H., Hospedales, T.M.: Learning to compare: Relation network for few-shot learning. In: CVPR, pp. 1199–1208 (2018)

38. Vinyals, O., Blundell, C., Lillicrap, T., Kavukcuoglu, K., Wierstra, D.: Matching networks for one shot learning. arXiv:1606.04080 (2016)

39. Wang, Y., Yao, Q., Kwok, J.T., Ni, L.M.: Generalizing from a few examples: a survey on few-shot learning. CSUR **53**, 1–34 (2020)

40. Wu, Y., Bamman, D., Russell, S.: Adversarial training for relation extraction. In: EMNLP, pp. 1778–1783 (2017)

Author Index

© The Editor(s) (if applicable) and The Author(s), under exclusive license
to Springer Nature Switzerland AG 2023
X. Wang et al. (Eds.): DASFAA 2023, LNCS 13945, pp. 753–756, 2023.
https://doi.org/10.1007/978-3-031-30675-4

Printed in the United States
by Baker & Taylor Publisher Services